2nd Annual International Conference on Information System and Artificial Intelligence (ISAI 2017)

Journal of Physics: Conference Series Volume 887

Tianjin, China
14 - 16 July 2017

Editors:

Yanghua Xiao
Shaoqing Hu
Yuhan Dong

ISBN: 978-1-5108-4791-0
ISSN: 1742-6588

Printed from e-media with permission by:

Curran Associates, Inc.
57 Morehouse Lane
Red Hook, NY 12571

proceedings.com

Some format issues inherent in the e-media version may also appear in this print version.

Printed by Curran Associates, Inc. (2017)

For permission requests, please contact the Institute of Physics at the address below.

Institute of Physics
Dirac House, Temple Back
Bristol BS1 6BE UK

Phone: 44 1 17 929 7481
Fax: 44 1 17 920 0979

techtracking@iop.org

Additional copies of this publication are available from:

Curran Associates, Inc.
57 Morehouse Lane
Red Hook, NY 12571 USA
Phone: 845-758-0400
Fax: 845-758-2633
Email: curran@proceedings.com
Web: www.proceedings.com

Table of contents

Volume 887

The 2nd Annual International Conference on Information System and Artificial Intelligence (ISAI2017)

14–16 July 2017, China

Accepted papers received: 17 August 2017
Published online: 8 September 2017

Preface

011001
OPEN ACCESS
The 2nd Annual International Conference on Information System and Artificial Intelligence (ISAI2017)

011002
OPEN ACCESS
Committee

011003
OPEN ACCESS
Keynote speakers

011004
OPEN ACCESS
Peer review statement

Artificial Application

012001
OPEN ACCESS
Profitability analysis of KINGLONG nearly 5 years

Mei Zhang and Jinghua Wen.....1

012002
OPEN ACCESS
Component reliability parameters estimation considering weather factors

Yuan Wang.....7

012003
OPEN ACCESS
Features research of ideal ahead-detecting method used in mining roadway and exploitation of a new technology --DFIP

Weijie Zhang and Zhimin Liu.....14

012004
OPEN ACCESS
Requirement analysis for the one-stop logistics management of fresh agricultural products

Jun Li, Hongmei Gao and Yuchuan Liu.....23

012005
OPEN ACCESS
An aggregation-decomposition bayesian stochastic optimization model for cascade hydropower reservoirs using medium-range precipitation forecasts

Yong Peng, Wei Xu and Xiaoli Zhang.....31

012006
OPEN ACCESS
Development of remote data acquisition system based on OPC for brake test bench

Yiwei Wang, Mengling Wu, Chun Tian and Tianhe Ma.....39

012007
OPEN ACCESS
Research and application of mobile teaching platform

Ping Yang and Hongjiao Xue.....46

012008
OPEN ACCESS
Combing VFH with bezier for motion planning of an autonomous vehicle

Feng Ye, Jing Yang, Chao Ma and Haijun Rong.....53

012009
OPEN ACCESS
Agricultural SWOT analysis and wisdom agriculture design of chengdu

Qian Zhang, Xiangyu Chen, Shaoming Du, Guowei Yin, Feng Yu, Guicai Liu, Jin Gong and Fujun Han.....63

012010
OPEN ACCESS
Geomagnetic matching navigation algorithm based on robust estimation

Weinan Xie, Liping Huang, Zhenshen Qu and Zhenhuan Wang.....69

012011
OPEN ACCESS
Design and realization of high voltage disconnector condition monitoring system

Jinrui Shi, Tianyang Xu, Shuixian Yang and Buoyang Li.....77

012012
OPEN ACCESS
Study to the simulation for pedestrian-vehicle collision accident of the second crushing by PC-CRASH

Jiangyang Gu, Xiaoli Chao and Xianguo Hu.....83

012013
OPEN ACCESS
Word2vec and dictionary based approach for uyghur text filtering

Turdi Tohti, Yunxing Zhao and Winira Musajan.....90

012014
OPEN ACCESS
SNMG: a social-level norm-based methodology for macro-governing service collaboration processes

Ji Gao, Hexin Lv, Zhiyong Jin and Ping Xu.....96

012015
OPEN ACCESS
Determination method for aero-engine optimal operation performance interval

Xuyun Fu, Guangyao Zhang, Yapeng Tian and Shisheng Zhong.....102

012016
OPEN ACCESS
Improving prediction accuracy of cooling load using EMD, PSR and RBFNN

Limin Shen, Yuanmei Wen and Xiaohong Li.....110

012017
OPEN ACCESS
A social-level macro-governance mode for collaborative manufacturing processes

Ji Gao, Hexin Lv, Zhiyong Jin and Ping Xu.....121

012018
OPEN ACCESS
Design of agricultural product quality safety retrospective supervision system of Jiangsu province

Kun Wang.....129

012019
OPEN ACCESS
A novel approach to describing and detecting performance anti-patterns

Jinfang Sheng, Yihan Wang, Peipei Hu and Bin Wang.....135

012020
OPEN ACCESS
Application of harmonic detection technology in methane telemetry

Yuehua Huo and Weiqiang Fan.....141

012021
OPEN ACCESS
Design of vehicle overload detection system based on geophone

Siquan Hu, Min Kong and Chundong She.....147

012022
OPEN ACCESS
Research on SEU hardening of heterogeneous Dual-Core SoC

Kun Huang, Keliu Hu, Jun Deng and Tao Zhang.....153

012023
OPEN ACCESS
The prediction of the impact of climatic factors on short-term electric power load based on the big data of smart city

Yunfei Qiu, Xizhong Li, Wei Zheng, Qinghe Hu, Zhanmeng Wei and Yaqin Yue.....159

012024
OPEN ACCESS
Research on optimal driver steering model based on Multi-Point preview

Jun Gu and Aijing Ma.....165

012025
OPEN ACCESS
Medium- and long-term electric power demand forecasting based on the big data of smart city

Zhanmeng Wei, Xiyuan Li, Xizhong Li, Qinghe Hu, Haiyang Zhang and Pengjie Cui.....173

012026
OPEN ACCESS
Research progress of control of condensate depression for condenser

Ying Liu, Run Liang and Fengyu Li.....180

012027
OPEN ACCESS
Analysis and improvement of face detection based on surf cascade

Siquan Hu, Caihong Zhang and Lei Liu.....186

012028
OPEN ACCESS
The enhancement of TextRank algorithm by using word2vec and its application on topic extraction

Xiaolei Zuo, Silan Zhang and Jingbo Xia.....192

012029
OPEN ACCESS
A monocular vision system based on cooperative targets detection for aircraft pose measurement

Zhenyu Wang, Yanyun Wang, Wei Cheng, Tao Chen and Hui Zhou.....199

012030
OPEN ACCESS
Maturity stage distinction of pear based on visible/near infrared spectroscopy technology

Weiwei Xu, Min Lin, Yongmei Huang, Huijun Liu and Xinqi Zhou.....205

012031
OPEN ACCESS
Hybrid feature selection for supporting lightweight intrusion detection systems

Jianglong Song, Wentao Zhao, Qiang Liu and Xin Wang.....212

012032
OPEN ACCESS
The research of emergency managers' risk attitude based on genetic neural network

Chen Pan, Chuanliang Jia, Jie Su and Junlin Chen.....219

012033
OPEN ACCESS
An underwater acoustic beacons positioning method using single hydrophone

Jia Liu, Feng Xu, Qian Zhang and Lijun Jiang.....226

012034
OPEN ACCESS
Analysis and optimization of calibration method of digital energy meter

Songxue Hou, Yuyou Liu, Yunying Xu, Shunchao Wang and Dan Xu.....231

Network Communication

012035
OPEN ACCESS
Research on backbone node deployment for Wireless Mesh Networks in dynamic environments

Meiyi Li and Shengling Cao.....238

012036
OPEN ACCESS
Design and implement of mobile equipment management system based on QRcode

Runze Yu, Xiaohui Duan and Bingli Jiao.....246

012037
OPEN ACCESS
A novel low-scattering and wideband monopole antenna based on artificial magnetic conductor

Pei Yao, Binzhen Zhang, Junping Duan and Qiong Bai.....254

012038
OPEN ACCESS
The design and implementation of network attack and defense platform based on cloud desktop

Zining Yan, Yunhan Wang, Shijiao Shao and Boquan Li.....261

012039
OPEN ACCESS
A channel estimation scheme for MIMO-OFDM systems

Chunlong He, Chu Tian, Xingquan Li, Ce Zhang, Shiqi Zhang and Chaowen Liu.....272

012040
OPEN ACCESS
Statistics-based email communication security behavior recognition

Junkai Yi, Yueyang Su and Xianghui Zhao.....280

012041
OPEN ACCESS
The Intelligent Technologies of Electronic Information System

Xianyu Li.....287

012042
OPEN ACCESS
On construction method of shipborne and airborne radar intelligence and related equipment knowledge graph

Ruizhe Hao and Jian Huang.....292

012043
OPEN ACCESS
Link prediction in social network based on local information and attributes of nodes

Yingying Liang, Lan Huang and Zhe Wang.....299

012044
OPEN ACCESS
Review on open source operating systems for internet of things

Zhengmin Wang, Wei Li and Huiliang Dong.....305

012045
OPEN ACCESS
Hybrid employment recommendation algorithm based on Spark

Zuoquan Li, Yubei Lin and Xingming Zhang.....311

012046
OPEN ACCESS
Near infrared spectroscopy detection of the content of wheat based on improved deep belief network

Wenwen Li, Min Lin, Yongmei Huang, Huijun Liu and Xinqi Zhou.....319

012047
OPEN ACCESS
Research on key technologies of data processing in internet of things

Yangqing Zhu and Peiying Liang.....325

012048
OPEN ACCESS
Human resource recommendation algorithm based on ensemble learning and Spark

Zihan Cong, Xingming Zhang, Haoxiang Wang and Hongjie Xu.....331

012049
OPEN ACCESS
Deep neural nets based power amplifier non-linear pre-distortion

Zhenyu Wang, Yanyun Wang, Chunfeng Song, Tao Chen and Wei Cheng.....338

012050
OPEN ACCESS
Research on technology environment improvement of related industries based on internet governance

Jing Zhang and Zhongliang Guan.....344

012051
OPEN ACCESS
Petri net modeling of encrypted information flow in federated cloud

Abdul Rauf Khushk and Xiaozhong Li.....352

012052
OPEN ACCESS
Predicting protein amidation sites by orchestrating amino acid sequence features

Shuqiu Zhao, Hua Yu and Xiujun Gong.....356

012053
OPEN ACCESS
Efficient detection of dangling pointer error for C/C++ programs

Wenzhe Zhang.....363

012054
OPEN ACCESS
Research and implementation of role-playing teaching mode supported by gamification

Xu Cui, Zhenglei Zhang and Lei Sun.....368

012055
OPEN ACCESS
Study on an effective algorithm of acquiring and tracking BD2 medium-frequency signals

Yanfei Liu, Yu Yang, Shiqiang Jiang, Ke Jiang and Jieling Wang.....374

012056
OPEN ACCESS
A new algorithm based on bipartite graph networks for improving aggregate recommendation diversity

Lulu Ma and Jun Zhang.....381

012057
OPEN ACCESS
An ontology based trust verification of software license agreement

Wenhuan Lu, Xiaoqing Li, Zengqin Gan and Jianguo Wei.....388

012058
OPEN ACCESS
A pruning algorithm for Meta-blocking based on cumulative weight

Fulin Zhang, Zhipeng Gao and Kun Niu.....399

Machine Learning

012059
OPEN ACCESS
Generation of Weibull distribution clutter based on correlated Gaussian sequence

Bin Wang and Fengming Xin.....405

012060
OPEN ACCESS
A multi-group firefly algorithm for numerical optimization

Nan Tong, Qiang Fu, Caiming Zhong and Pengjun Wang.....413

012061
OPEN ACCESS
A novel user-interest model based on mixed measure

Junkai Yi, Yacong Zhang, Mingyong Yin and Xianghui Zhao.....421

012062
OPEN ACCESS
A risk evaluation model and its application in online retailing trustfulness

Ruyi Ye and Yingcheng Xu.....428

012063
OPEN ACCESS
An ICP algorithm based on block path closest point search

Kuisheng Wang, Xing Li, Hongwei Lei and Xiaorui Zhang.....432

012064
OPEN ACCESS
A clustering algorithm based on maximum entropy principle

Yang Zhao and Fangai Liu.....438

012065
OPEN ACCESS
A hybrid genetic algorithm for solving bi-objective traveling salesman problems

Mei Ma and Hecheng Li.....446

012066
OPEN ACCESS
Heterogeneous massive feature fusion on grassmannian manifold

Haichao Huang, Hongning Liu, Xiaoyun Kong, Xingdan Lou and Zepeng Wang.....453

012067
OPEN ACCESS
A method for turbine blade temperature data segmentation

Chi Feng, Li Wang and Shan Gao.....459

012068
OPEN ACCESS
Real-time vehicle detection and tracking in video based on faster R-CNN

Yongjie Zhang, Jian Wang and Xin Yang.....465

012069
OPEN ACCESS
Vehicle positioning based on UWB technology

Siquan Hu, Min Kang and Chundong She.....470

012070
OPEN ACCESS
Fault diagnosis of wind bearing based on multi-scale wavelet kernel extreme learning machine

Siwen Zhu and Bin Jiao.....476

012071
OPEN ACCESS
Software detection method based on running state

XiaoLin Zhao, Quanbao Chen, Chun Shan, Ting Wang and Yiman Zhang.....483

012072
OPEN ACCESS
Aided diagnosis methods of breast cancer based on machine learning

Yue Zhao, Nian Wang and Xiaoyu Cui.....494

012073
OPEN ACCESS
Compact dual-band bandpass filter based on signal-interference techniques

Xingbing Ma and Ting Jiang.....500

012074
OPEN ACCESS
Adaptive control for time-delay teleoperation systems with uncertain dynamics

Shan Liu, Xia Zhang, Wenfeng Zheng and Bo Yang.....506

012075
OPEN ACCESS
Prediction of a service demand using combined forecasting approach

Ling Zhou.....513

012076
OPEN ACCESS
Research on regional intrusion prevention and control system based on target tracking

Yanfei Liu, Jieling Wang, Ke Jiang, Yanhui He and Zhilin Wu.....519

Image, Robotics and Voice Recognition

012077
OPEN ACCESS
Tweaked residual convolutional network for face alignment

Wenchao Du, Ke Li, Qijun Zhao, Yi Zhang and Hu Chen.....526

012078
OPEN ACCESS
Intelligent detection and removing of straight periodic streaks on digital images

Zehao Zhao, Hui Sheng and Xianfeng Xu.....537

012079
OPEN ACCESS
Multi-task convolution network for face alignment

Yang Sun, Xuan Zhang and Chongrong Li.....541

012080
OPEN ACCESS
Research status of operational environment partitioning and path planning for multi - robot systems

Xuefeng Dai, Qi Fan and Dahui Li.....548

012081
OPEN ACCESS
Research status of multi - robot systems task allocation and uncertainty treatment

Dahui Li, Qi Fan and Xuefeng Dai.....554

012082
OPEN ACCESS
The analysis and detection of hypernasality based on a formant extraction algorithm

Jiahui Qian, Fanglin Fu, Xinyi Liu, Ling He, Heng Yin and Han Zhang.....559

012083
OPEN ACCESS
Halftone information hiding technology based on phase feature of space filling curves

Jianhua Hu, Peng Cao, Zhihong Dong and Xiaohe Cao.....567

012084
OPEN ACCESS
Design of dual DC motor control system based on DSP

Peicheng Shi, Suo Wang, Zengwei Xu and Ping Xiao.....574

012085
OPEN ACCESS
Dynamic formant extraction of wa language based on adaptive variational mode decomposition

Meijun Fu, Huazhen Dong and Wenlin Pan.....581

012086
OPEN ACCESS
Vibration factors impact analysis on aerial film camera imaging quality

Jun Xie, Wei Han, Zhonglin Xu, Haifeng Tan and Mingquan Yang.....593

012087
OPEN ACCESS
Design of verification platform for wireless vision sensor networks

Juanjuan Ye, Fei Shang and Chuang Yu.....599

012088
OPEN ACCESS
The modeling and solution of course of action generation based on IN-DEA

Lujun Wan, Wen Li, Jiangbin Dai and Aqian Huang.....605

012089
OPEN ACCESS
Rock images classification by using deep convolution neural network

Guojian Cheng and Wenhui Guo.....614

012090
OPEN ACCESS
Intelligent acoustic data fusion technique for information security analysis

Ying Jiang, Yize Tang, Wenda Lu, Zhongfeng Wang, Zepeng Wang and Luming Zhang.....620

012091
OPEN ACCESS
Efficient large-scale graph data optimization for intelligent video surveillance

Quanhong Shang, Shujun Zhang, Yanbo Wang, Chen Sun, Zepeng Wang and Luming Zhang.....626

012092
OPEN ACCESS
Research on control law accelerator of digital signal process chip TMS320F28035 for real-time data acquisition and processing

Shuangle Zhao, Xueyi Zhang, Shengli Sun and Xudong Wang.....631

012093
OPEN ACCESS
The vision guidance and image processing of AGV

Tongqing Feng and Bin Jiao.....636

012094
OPEN ACCESS
Research on quasi-dynamic calibration model of plastic sensitive element based on neural networks

Fang Wang, Deren Kong, Lixia Yang and Zouzou Zhang.....643

012095
OPEN ACCESS
A image watermark resisting geometrical attacks based on quaternion moments

Ge Wu.....651

012096
OPEN ACCESS
Identification of maturity of cuiguan pear based on RobustICA and PSO-ELM

Yu Zhang, Yongmei Huang, Min Lin, Huijun Liu and Xinqi Zhou.....656

The International Conference on Information Systems and Artificial Intelligence is an academic conference on the advancement of artificial intelligence and information systems. The conference includes the fields of machine vision, natural language processing, speech recognition and the industry application of AI, and is of great significance to the development of intelligent machines, intelligent human-computer interaction, and effective use of information, human health and social security.

The primary goal of the conference is to promote research and developmental activities in information system and artificial intelligence and another goal is to promote scientific information interchange between researchers, developers, engineers, students, and practitioners working all around the world. The conference will be held every year to make it an ideal platform for people to share views and experiences in information system and artificial intelligence and related areas. In order to further promote the development of artificial intelligence and information systems disciplines and strengthen domestic and international academic exchanges and cooperation.

The 2017 International Conference on Artificial Intelligence and Information Systems was held in Tianjin from 14 to 16th July, 2017. I would like to thank the members of the Committee on Artificial Intelligence and Information Systems in 2017 and the reviewers for their careful review of the paper and the suggestions for improvement.

The conference received 280 full-paper and abstract submissions from China and many other countries or regions over the world. Through a rigorous peer-review process, all submissions were refereed based on their quality of content, level of innovation, significance, originality and legibility. 96 papers were accepted and their authors were invited to present their papers at the conference. We invited five keynote speakers to deliver their recent research progress, who are internationally recognized leading experts in their own research areas. The proceedings are published by IOP Conference Series: Journal of Physics: Conference Series (JPCS)

Let us hope that the participants of ISAI2017 will have fruitful interactions among each other and I look forward to meeting you all in the forthcoming years.

Prof.Yanghua Xiao
Fudan University

IOP Conf. Series: Journal of Physics: Conf. Series **887** (2017) 011002 doi:10.1088/1742-6596/887/1/011002

General Chairs
Prof. Junming Shao, University of Electronic Science and Technology of China, China
Dr. Liang Deng, Shanghai University of Electric Power, China
Prof. Xunwei Zhou, Beijing Union University, China

Editor
Dr. Yanghua Xiao, Fudan University, China

Technical Program Committee Chair
Prof. Andrew Kusiak, the University of Iowa, U. S. A

Co-Chair
Dr. Liang Deng, Shanghai University of Electric Power, China
Dr. Qiang Cheng, Southern Illinois University, U.S.A

Co-Editor
Dr. Shaoqing Hu, Queen Mary University of London, UK
Dr. Yuhan Dong, Graduate School at Shenzhen, Tsinghua University, China

Technical Program Committee
Prof. Junming Shao, University of Electronic Science and Technology of China, China
Prof. Andrew Kusiak, the University of Iowa, U. S. A
Dr. Shaoqing Hu, Queen Mary University of London UK
Dr. Lisu Yu, Southwest Jiaotong University, China
Dr. Zhenyu Wang, North China Electric Power University, Beijing, China
Dr. Zhixiao Wang, Xi'an University of Technology, China
Dr. Xuanxuan Tang, PLA University of Science and Technology, Nanjing, China
Dr. Hongxing Peng, South China Agricultural University, China
Dr. Lixin Gao, Wenzhou University, China
Dr. Yinghua Zhou, Chongqing University of Posts and Telecommunications, China
Dr. Liang Deng, Shanghai University of Electric Power, China
Dr. Xiaofeng Zhang,, Harbin Institute of Technology Shenzhen Graduate School, China
Dr. Yuhan Dong, Graduate School at Shenzhen, Tsinghua University, China
Dr. Yongjun Li, Northwestern Polytechnical University, China
Dr. Jun Ye Sichuan, University of Science & Engineering, China
Dr. Imran Memon, Zhejiang University, China
Dr. Xuanxuan Tang, PLA University of Science and Technology, China
Dr. Ding Wang, PLA Information Engineering University, China
Dr. Fengxiang Xu, Wuhan University of Technology, China
Dr. Lu Han, Central University of Finance & Economics, China
Dr. Zhaoguang Peng， College of Engineering Wayne State University, China
Dr. Zhe Chen, Department of Electronic Engineer, DUT, China
Dr. Jingyu Yang,S henyang Aerospace University, China
Dr. Xizhe Zhang, Northeastern University, China

Dr. Zhiwei Zhang, Purdue University, China
Dr. Murat Dener, Gazi University, Turkey
Dr. Ong Pauline, Universiti Tun Hussein Onn Malaysia, Johor, Malaysia
Dr. Cheng Li, Centre of Pattern Recognition and Data Analytics, Deakin University, China
Dr. Abiel Aguilar-González, National Institute of Astrophysics, Optics and Electronics, Mexico
Dr. Ştefan Ţălu, The Technical University of Cluj-Napoca, Romania
Dr. Marlen Promann, Purdue University, U. S. A
Dr. Linzhi Jiang,West An Hui University, China
Dr. Ahmed Nidhal Khdiar, College of Engineering Kufa University, Iraq
Prof. Wen-Tsai Sung, Electrical Engineering, National Chin-Yi University of Technology, Taiwan
Dr. Chien-Hung Yeh, Department of Photonics Feng Chia University, Taiwan
Dr. Lingwei Xu, Qingdao University of Science & Technology, China
Prof. Chungneng Huang, National University of Tainan, China
Dr. Weimo Liu,George Washington University, U. S. A
Dr. Qiang Cheng, Southern Illinois University, U. S. A
Dr. Saeed Tavakoli, University of Sistan and Baluchestan, Iran
Prof. Chingliang Su, Da Yeh University, Taiwan
Dr. Klimis Ntalianis, Athens University of Applied Sciences (TEI of Athens), Greece
Dr. Saman Shojae Chaeikar, Khajeh Nasir Toosi University of Technology, Iran
Dr. Umashanger Thayasivam, Rowan University, U. S. A
Dr. Zain Anwar Ali, Nanjing University of Aeronautics & Astronautics, Pakistan
Prof. Hui-Mi Hsu, National Ilan University, Taiwan
Dr. Lihui Shi, University of Washington, U. S. A
Prof. Sanjeevikumar Padmanaban, University of Johannesburg, Indian
Dr. Umashanger Thayasivam, Rowan University, U. S. A
Prof. Da Yan, the University of Alabama at Birmingham, China
Dr. Jan Kubicek, VSB – Technical University of Ostrava, Czech Republic
Dr. Chi-Seng Lam, University of Macau, Macao
Dr. Bin Liu, Dalian University of Technology, China
Dr. Husam Shaheen, Tishreen University, Syrian Arab Republic
Prof. Badre Bossouufi, Mohamed I University, Morocco
Prof. Minshiang Hwang, Asia University, Taiwan
Prof. Soundararajan Ezekiel, Indiana University of Pennsylvania, U. S. A
Dr. Wenjie Zhang, Minnan Normal University, China
Dr. Luis Gomez Deniz, University of Las Palmas Gran Canaria, Spain
Dr. Mawkae Hor, Kainan University, Taiwan
Dr. Aidil Azwin Bin Zainul Abidin, University Tenaga Nasional (UNITEN), Malaysia
Dr. Xiangdong Chen, Nanjing University of Posts and communications, China
Dr. P. Ganesh Kumar, Linyi Top Network Co Ltd, India
Dr. Ray-Hwa Wong, Hwa-Hsia University of Technology, Taiwan
Prof. Cheng-Yuan Tang, Huafan University, Taiwan
Dr. Tomasz Andrysiak, UTP University of Science and Technology, Poland
Prof. Min-Shiang Hwang, Asia University, Taiwan
Prof. Xunfei Jiang, Earlham College, U. S. A

IOP Conf. Series: Journal of Physics: Conf. Series **887** (2017) 011003 doi:10.1088/1742-6596/887/1/011003

Keynote Speaker 1:Prof. Xunwei Zhou

Beijing Union University, Beijing, China

Speech Title:

Inverse Hypothetical Inference Based Mutually-Inversistic Machine Learning

Abstract:

Mutually-inversistic machine learning is based on inverse hypothetical inference of mutually-inversistic logic constructed by the author. In mutually-inversistic logic, there are 8 kind of hypothetical inferences, all of which are from the major premise and the minor premise to infer the conclusion. For each hypothetical inference, there are 2 inverse hypothetical inferences, one is from the minor premise and the conclusion to the major premise, the other is from the major premise and the conclusion to the minor premise. Therefore, there are 16 kind of mutually-inversistic machine learning models altogether. FOIL, decision tree learning in inductive learning, rough set learning, genetic algorithm, inverse resolution in inductive logic programming are special cases of mutually-inversistic machine learning.

Biographical sketch:

Professor Zhou, professor in Institute of Information Technology, College of Information, Beijing Union University from 2005 to 2012, professor emeritus in Institute of Information Technology, College of Information, Beijing Union University since 2012. He construct mutually-inversistic logic and apply it to the fields of computer science, artificial intelligence, and automation, such as automatic theorem proving, logic programming, planning and scheduling, expert systems, semantic network, program verification, natural language understanding, hardware verification, machine learning, data mining, uncertainty reasoning, program refinement, many-valued computer, modern control theory, etc. He published about 80 papers, authored 4 monographs on them. He is a senior member of China Computer Federation, a member of Technical Committer of Artificial Intelligence Theory of Chinese Association for Artificial Intelligence, a director of Beijing Logical Association.

IOP Conf. Series: Journal of Physics: Conf. Series **887** (2017) 011003 doi:10.1088/1742-6596/887/1/011003

Keynote Speaker 2:Prof. Junming Shao
University of Electronic Science and Technology of China, China

Speech Title:
Synchronization-based Data Mining

Abstract:
Advances of modern technologies produce huge amounts of data in various fields, increasing the need for efficient and effective data mining tools to uncover the information contained implicitly in the data. This talk mainly aims to introduce innovative and solid algorithms for data mining from a novel perspective: synchronization. Synchronization is a prevalent phenomenon in nature that a group of events spontaneously come into co-occurrence with a common rhythm through mutual interactions. The mechanism of synchronization allows controlling of complex processes by simple operations based on interactions between objects. Inspired by the concept of synchronization, this talk presents Sync (Clustering by Synchronization), a novel approach to clustering. In combination with the Minimum Description Length principle (MDL), it allows discovering the intrinsic clusters without any data distribution assumptions and parameters setting. In addition, relying on the different dynamic behaviors of objects during the process towards synchronization, the algorithm SOD (Synchronization-based Outlier Detection) is further proposed. The outlier objects can be naturally flagged by the definition of Local Synchronization Factor (LSF). To cure the curse of dimensionality in clustering, a subspace clustering algorithm ORSC is introduced which automatically detects clusters in subspaces of the original feature space. Beyond, in order to reveal the underlying patterns in graphs, a graph partitioning approach Attractor is presented.

Biographical sketch:
His research interests include data mining and machine learning, especially for clustering in high-dimensional data, graph mining and brain structural network analysis. He has received his PhD degree with highest honer ("Summa Cum Laude") at the University of Munich, Germany. He became the Alexander von Humboldt Fellow in 2012. Currently, he is professor of Computer Science at the University of Electronic Science and Technology of China. His research interests include data mining, machine learning and neuroimaging. He not only published papers on top-level data mining conferences like KDD, ICDM, SDM, IEEE/TKDE, but also published data mining-related interdisciplinary work in leading journals including Brain, Neurobiology of Aging, and Water Research.

IOP Conf. Series: Journal of Physics: Conf. Series **887** (2017) 011003 doi:10.1088/1742-6596/887/1/011003

Keynote Speaker 3:Prof. Yanghua Xiao
Fudan University, Shanghai, China

Speech Title:
Enabling Machines to Understand Human Language by Knowledge Graphs

Abstract:
One of the bottlenecks in machine intelligence is that machines have limited cognitive capability to understand data or text in the form of human language. One of the prerequisites to power machines with natural language understanding capability is the availability of the large scale knowledge bases of rich semantic information. Now, with more and more online knowledge bases (also known as knowledge graphs) being published, we have the opportunity to build smart machines that can understand human language. In this talk, I will review our recent progress to enable machines with the cognitive ability to understand natural language and discuss some open problems.

Biographical sketch:
Prof. Yanghua Xiao got his PHD degree in software theory from Fudan University, Shanghai, China, in 2009. He now is an associate professor of computer science at Fudan University. He is one of young scientists of Chinese 973 program. Hisresearch interest includes big data management and mining, graph database, knowledge graph. He was a visiting professor of Human Genome Sequencing Centerat Baylor College Medicine, and visiting researcher of Microsoft Research Asia. Hewon the Best Phd Thesis Nomination of CCF (Chinese Computer Federation) inyear 2010, CCF2014 Natural Science Award (second level), ACM (CCF) Shanghai distinguished young scientists nomination award. Recently, he has published 70+papers in international leading journals and top conferences.

IOP Conf. Series: Journal of Physics: Conf. Series **887** (2017) 011003 doi:10.1088/1742-6596/887/1/011003

Keynote Speaker 4: Prof. Xiong Li
Academy of Armored Force Engineering, Beijing, China

Speech Title:
Agent-based Modeling and Simulation forNetworked Equipment Support

Abstract:
Networked equipment support is a new support mode that has the typical feature of complex adaptive system with various intelligent support entities in networked warfare space. Based on analysis on the process of networked equipment support, agent and multi-agent system technology is used, and agent-based networked equipment support models including interaction protocol model and behavior simulation model are designed. Thus, agent-based networked equipment support simulation is implemented by using EINSTein simulation experiment platform. The feasibility and efficiency of the proposed approach is validated by case study.

Biographical sketch:
Xiong Li now is a professor of Academy of Armored Force Engineering. His Research Areas including Agent-based Modeling, Multi-agent Systems, Information System Analysis, Software Modeling and Design, Distributed Systems. One of his academic awards is PhD thesis entitled "Study on Warfare System Modeling Based on Meta-agents Interaction Chain" was nominated as one of the 30 Chinese PLA most excellent doctor theses in 2010 and the 100 national most excellent doctor theses in 2011, respectively.

IOP Conf. Series: Journal of Physics: Conf. Series **887** (2017) 011004 doi:10.1088/1742-6596/887/1/011004

Peer review statement

All papers published in this volume of *Journal of Physics: Conference Series* have been peer reviewed through processes administered by the proceedings Editors. Reviews were conducted by expert referees to the professional and scientific standards expected of a proceedings journal published by IOP Publishing.

IOP Conf. Series: Journal of Physics: Conf. Series **887** (2017) 012001 doi:10.1088/1742-6596/887/1/012001

Profitability analysis of KINGLONG nearly 5 years

Mei Zhang [1,a] and Jinghua Wen[2,b]

[1]Information Institute Gui Zhou University of Finance and Economics Guiyang, China,
[2]Information Institute Gui Zhou University of Finance and Economics Guiyang, China,
[a]zm_gy@sina.com
[b]75301022@qq.com

Abstract. Profitability analysis for measuring business performance and forecast its prospects play an important role. In this paper, the research instance King Long Motor in understanding the basic theory on the basis of financial management, to take a combination of theory and data analysis methods, combined with a measure of profitability related indicators of King Long Motor company's profitability do a specific analysis to identify factors constraining the profitability of Kinglong company exists and the motivation to improve profitability, which made recommendations to improve the profitability of Kinglong car company to promote the company's future can be better and faster development.*)*

1. Introduction

Profitability is usually refers to the ability of business to make a profit [1], but it reflects the strength of the business performance is good or bad, the central issue is also of concern to all parties concerned companies.

Operator profitability through analysis, we can see that the quality of the company's development, identify problems and take effective measures to solve the problem, and lessons learned, to improve the overall efficiency. Investors on the strength of the company's profitability is particularly important to see that they get investment income usually depends on how much the level of corporate profits, their investment income and profitability are inextricably linked. The more robust corporate earnings, they get more benefits [2]. Thus, relative to other indicators, they are more concerned about profitability. Company's ability to timely repayment of debt interest, and he has a close relationship profitability. Therefore, when creditors lend money to understand the strength of the company's profitability, to ensure timely recovery of funds lent to avoid losses. Meanwhile, for the government, profit is the direct source of national revenue, profitability directly affect state revenue, help countries to better exercise social management functions, but also the employee collective welfare facilities have improved the important guarantee not only help safeguard the interests of enterprise employees, but also to attract more people, thus improving the overall efficiency of the company.

Development of the automotive industry, both for developed and developing economies have a relationship and responsibility cannot be ignored, the development of the automotive industry must promote the development of many related industries [3]. Throughout recent years, China's economic development, market demand for the car or larger, or the promising development. We want to understand the development of the automotive industry, profitability analysis is one of the most important part of its development by profitability analysis can be the most intuitive and effective understanding of the quality of the ability of the industry to make a profit and business performance. On the basis of analysis to identify potential problems in the development, and to be addressed and improved, so that the automotive industry can better development, in order to stimulate the development of related industry sectors to promote economic development.

However, the profitability of the automotive industry in recent years, in-depth analysis of the literature is very limited. Wu Ling [4] Comparative analysis of the use of financial theory and charts, the first comparative analysis of the two companies in terms of profitability, cash generation ability, value creation, growth capacity and risk control ability of the financial indicators. Then using DuPont analysis system, the ROE of both companies in the comparison year deep-seated reasons (ROE) change decomposed and analysis; and finally, the use of financial strategy matrix theory, the two companies in recent years used Specific financial strategy has also been analyzed. Wang Shan [5] First defines the basic meaning of the capital structure and operating performance of the previous studies carried out abroad preliminary carding theoretically analyzed the relationship between capital structure and business performance among; at the same time, for Kinglong car conducted a comprehensive SWOT analysis capabilities. Then, analyzed the current situation Kinglong car capital structure and operating performance, select the Shanghai and Shenzhen-listed 19 auto companies in 2010 - annual report 2012 financial data analysis using multiple regression analysis based on Logistic Model and Cluster Analysis the automotive industry capital structure and operating performance relationship between empirical research. Zhang Tian Ming [6] to small businesses A company for the study, based on a detailed analysis of business background, development survey, based on research A company operating around the implementation of the target customer management strategy case study change target customers, customer profitability and changes in corporate profitability, which draw some preliminary conclusions, the implementation of corporate strategies affect the target customer profitability, customer profitability generated. Sun makes Peng [7] by choosing Nanjing C auto 4S shop as an example, the profitability of the business as a research direction, financial management theory as the research base, combined with the theory of business management, to take the theoretical and data analysis method of combining, internal and external factors affecting C Nanjing Automobile 4S shop profitability carried out a detailed analysis, and then from the enterprise business and management point of view put forward countermeasures and suggestions to enhance the C Nanjing Automobile 4S shop profitability.

2. Basic Theory

Profitability refers to the ability to obtain corporate profits, it is a comprehensive index, reflecting the quality of different aspects of the development of enterprises [8]. It is mainly used to analyze profitability through investment and operating profitability of the two financial indicators.

2.1. Investment Profitability Indicators

Investment profitability indicators including return on equity, return on total assets, total assets net profit margin ratio analysis three indicators [9].

ROE, is the ratio of net profit to the average owner's equity, represents the invested capital gains obtained, the index level is directly related to the level of income. Investors will also look at whether to invest or continue to invest according to the index.

The total return on assets is defined as the ratio of total income received average total assets, reflects the ability of enterprises to use the asset re-creation of value, the asset management is an important indicator to assess the profitability of the strength of the enterprise. The level of the index, indicating that the quality of enterprise asset management, as well as the strength of profitability.

Total assets net profit margin is the ratio of the average total net assets, the use of all assets that the company obtained profit level. The level of the index, reflecting the level of asset management efficiency, obtained the level of profits, the level of costs, the level of operating efficiency.

Capital Maintenance and Appreciation refers to the owner's equity ratio at the end with the beginning of the owner's equity, indicating increase or decrease of capital under management in the business range. It's low, reflecting the level of debt security, the greater its higher corporate borrowing space for the future development of enterprises to provide a guarantee of funding.

2.2. Operating Profitability Indicators

Operating profitability indicators including operating margin, operating margin, operating margin, cost margins four ratios analysis indicators [10]. Operating margin is defined as operating gross profit and net operating income ratio, also known as gross margin. Its level, indicating the strength of the ability of enterprises to obtain gross profit, as well as the level of costs. The operating margin refers to the ratio of operating profit and operating income. This indicator shows after removal of operating costs, profits acquired business efficiency is an important indicator of good or bad. Operating margin is the ratio of net profit and revenue. It represents the ability to create business revenue net of its height, determine the business to obtain the ability to create profit strength. Cost rate margin is defined as the ratio of total corporate profit and the cost of the total cost, per unit cost represents income obtained. It is higher, then the total profit is greater than the total cost, the greater the greater the difference between profits, business profitability is stronger.

3. Profitability Analysis of King Long

3.1. Profitability Data Analysis

Between 2009 and 2013, king Long for five consecutive years of key financial data are shown in Table 1, there are the interests of the relevant data is shown in table 2.

Table 1. King long profitability data sheet (million)

Project	*2009*	*2010*	*2011*	*2012*	*2013*
Operating income	117.7107	161.5614	189.2491	191.6870	208.1230
Operating cost	102.7757	140.7802	166.7163	167.9785	182.1423
Sales expense	5.8633	8.0668	9.4051	10.6807	11.5259
Management costs	3.9952	5.3115	6.5663	7.8340	7.6974
Financial expenses	0.0474	-0.1130	-0.1645	-0.7420	-0.1074
Business tax and surcharges	0.6602	0.9296	1.0303	1.1695	1.5653
Business profit	3.3942	5.4037	5.1146	3.4629	3.8360
The total profit	3.6532	5.6983	5.7904	5.2160	5.3435
Net profit	3.0556	5.0049	4.9656	4.5134	4.4918
Interest expense	0.0038	0.0015	0.0029	0.0023	0.0037
Total assets	94.447	124.2903	132.8088	146.5987	160.8870

As can be seen from Table 1, King Long in the past five years, remain profitable, a loss does not occur. Operating profit, total profit and net profit 2009 - 2010 growth trend were tested, but starting in 2011 - 2013 in general are showing a downward trend. Net profit declined, indicating King Long profit growth is poor. Operating profit in the increase or decrease in the case of this five-year period is not ideal, we can see that revenue is a major part of the profits, sales revenue decrease resulting decrease in operating profit. Reduce sales revenue was mainly due to the excitation current competitive environment, as well as the factors themselves KingLong joint action, other income such as investment income also increased the company's profitability.

Table 2. Kinglong owner's equity related data table (billion)

	2009	**2010**	**2011**	**2012**	**2013**
The total owners' equity	24.9954	29.9130	33.7275	37.4432	39.8525

3.2. Profitability Index Analysis

3.2.1. Index of Investment Profitability: 2009 - 2013 Results in Table 3 and Figure 1.

Table 3. Kinglong automobile investment profitability analysis table (%)

	2009	**2010**	**2011**	**2012**	**2013**
ROE	12.6857	16.7316	15.6054	12.0540	11.6225
ROA	4.1747	5.2118	4.5067	3.7352	3.4780
ROTA	3.4881	4.5763	3.8628	3.2307	2.9216
VCPA	113.1831	120.0234	116.6004	113.3820	111.9965

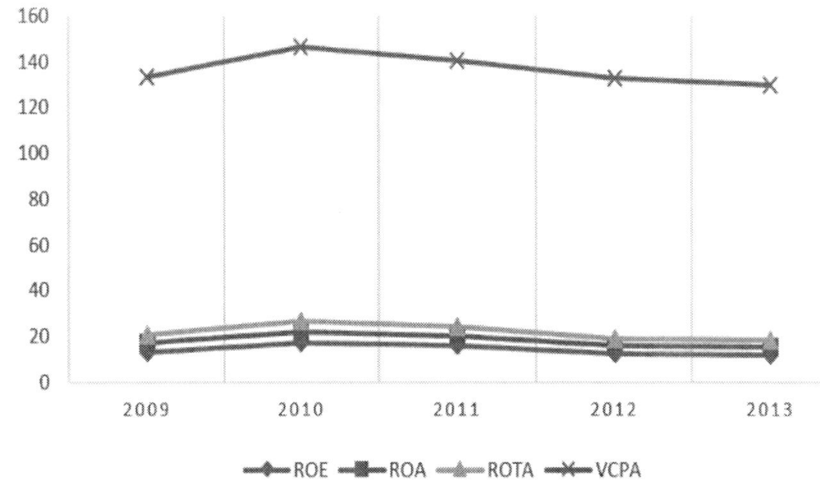

Figure 1. Kinglong capacity chart analysis of investment income (%)

From Figure 1 and Table 3 above can know: From 2009-2010 Rate of Return on Common Stockholders' Equity (ROE),Return On Total Assets(ROA),Return on Total Assets(ROTA)An upward trend, while in the 2010--2013 years with a downward trend, on the whole show a downward trend. ROE is a dynamic and comprehensive index, the index's decline indicates an ability to recreate KingLong net assets profit is not high, but on the whole is still relatively stable, and indicating KingLong profitability in the next period of time is quite stable. Decreased rate of return on total assets show the use of all of the assets of the ability to obtain economic benefits KingLong began to decline, more and more low-level input and output. Total assets decreased net interest margin of profit that the company is unstable, problems controlling costs and the level of corporate management level, the company is no good profitability. Kinglong car companies operate in five-year period, have achieved a capital preservation, but also make a capital gain. But Value of capital preservation and appreciation (VCPA) In addition to 2010, but have shown a downward trend, indicating that the capital invested enterprise security and value-added instability, unstable corporate profitability, the development is not stable.

3.2.2. Profitability Index of Business: 2009-2013 Calculation Results are shown in Table 4 and Figure 2.

Table 4. KingLong business vitality capacity analysis (%)

	2009	**2010**	**2011**	**2012**	**2013**
OGM	12.6878	12.8626	11.9064	12.3683	12.4833
OPR	2.8835	3.3446	2.7025	1.8065	1.8431
NIR	2.5958	3.0978	2.6238	2.3545	2.1582
RPC	3.2232	3.6769	3.1546	2.7904	2.6345

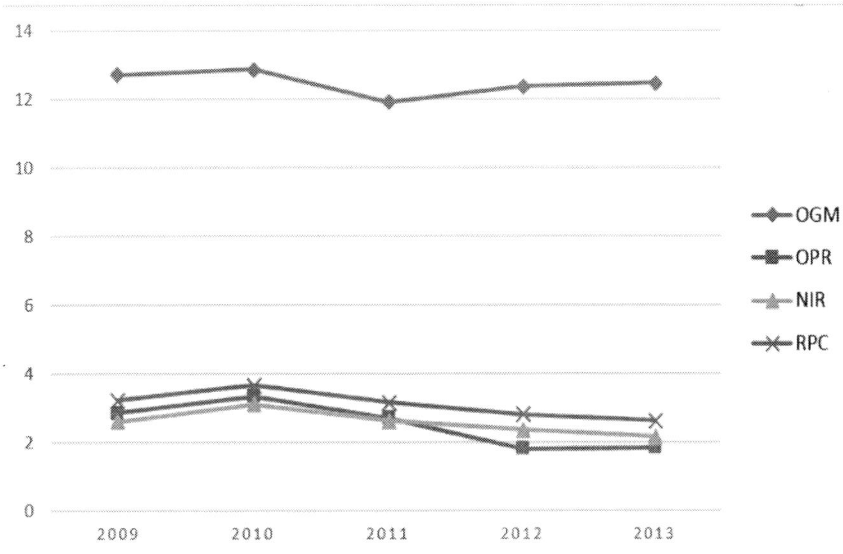

Figure 2. KingLong business activity analysis (%)

Operating margin (OGM) of KingLong showing the first decline and then rise again downward trend from above in Figure 2 and Table. Explain the company's profitability is unstable, the company did not lead to a good ability to control operating costs are too high, less competitive, and therefore unstable protect profits. Operating profit rate (OPR), In addition to 2010, the overall operating profit margin tended to decrease. Able to see KingLong managers to profit from the business is getting worse. The operating margin as a corporate profitability and the most important part, directly affect the stability of the profitability of the enterprise, and therefore profitability KingLong from the overall point of view is unstable. Net interest rate (NIR), In addition to 2010, the overall downward trend, there is no better description of the company's overall profitability. The decline was mainly due to product mix KingLong unreasonable, nor increase the number of high-end products according to market demand, adjust the overall structure of the product. Resulting product does not meet the market demand, which led to sales decrease, and finally to reduce the margin to reduce sales revenue. Ratio of Profits to Cos (RPC). On the whole show a downward trend, showing a small peak in 2010 but in 2011 dropped significantly, indicating that the total cost of enterprises are increasing, improper cost management.

4. Conclusion

Through the profitability indicators Jinlong car analysis, we can clearly see the operations and profitability of the company's situation and problems. For the company's current development situation, to improve the company's profitability can be improved from the following aspects.

First, improve the main business, optimize resource allocation. In the limited resources in how to better and more efficient allocation of resources is every business should pay attention to. The main business income is the most important part of the profits, especially manufacturing enterprises, the main business income stability, and long-term, predictable resistance, much of the strength of the decision of the profitability of enterprises. Therefore, Jinlong car in order to improve profitability and better development can continue to improve the proportion of main business income is an important part of it.

Secondly, reduce costs, improve asset turnover rate. On the basis of the team Jinlong car profitability analysis, can be seen to improve the return on total assets will improve the company's profitability. Want to improve the overall return on assets, asset turnover should increase speed and reduce operating costs. You can establish a sound product sales strategy, Jinlong Company adjusted according to market demand, market demand-oriented discover new developments, the biggest meet market demand, create more revenue.

Finally, change management model Jinlong car. Enterprise management mode directly affects the ability of the strength of corporate profits, companies have chosen what management ways and means for the future development of the company plays a vital role. Excellent management model can bring good economic benefits to the enterprise, to get more profits, while enhancing the competitiveness of companies and determine the direction of future development of enterprises. Jinlong car should be based on its own development, to create a business model in line with its development, increase sales capacity and cost management and control products, reduce the cost ratio, and promote the improvement of operating profit.

5. Acknowledgment

Thank Project Supported by Regional Science Fund of National Natural Science Foundation of China (Project approval number: 41261094. Thank Project Supported by Science and Technology Fund of Guizhou Provincial Science and Technology Agency (No:QianKeHeBasics (2016)1020). Thank Project Supported by Science and technology talents foundation of Natural Science of Guizhou Provincial Department of Education (QianJiaoHeKYZi (2016)069).

6. References

[1] Zhou shuo. "Profitability analysis" [J], Business Economics, 2013 (4).
[2] Section Trillian. "Profitability analysis to explore" [J], Economic Forum, 2012 (7).
[3] Pan Heng listed companies profitability analysis - Compare Dongfeng Motor and Changan Automobile profitability [J]. Jiangsu Teachers College (Natural Science), 2008, 14 (3): 75-81.
[4] Wu neighbors. Yutong Bus and Jinlong car finance comparison analysis [D] .2012, Southwest Jiaotong University.
[5] Shan Wang Jinlong Automobile Capital Structure and Performance of [D] .2014, Central South University.
[6] Zhang Tian Ming car service a company profitability analysis [D] .2014, East China University of Technology.
[7] Sun Peng make Nanjing C car 4S store profitability analysis [D] .2013, Jilin University.
[8] Zhang pear. "Profitability analysis" [J], economic and technological cooperation, 2014 (02).
[9] Zhang Dongmei."A discussion of corporate profitability evaluation of" [J], test weekly, 2011.
[10] Repair Ze Rui, paragraph Hongrui. "Profitability analysis" [J], modern economic information, 2014.
[11] Wang Qingxia. "Profitability analysis" [J], the financial sector (Academic Edition), 2014.
[12] Shinong Wu, Wu Yuhui. "CEO of financial analysis and decision-making" [M], Beijing University Press, 2013.
[13] Qin Xiaohong, Zhang Xiao Sheng, Zhang Qiang. "Excel 2007 financial management and analysis of typical examples" [M], Electronic Industry Press, 2009.
[14] Zhou Jinlan, Hubei loyalty. "Financial Analysis Experimental Course" [M], Dongbei University of Finance and Economics Press, 2011.
[15] Zhang Jinchang. "Financial Analysis and Decision" [M], Economic Science Press, 2002

IOP Conf. Series: Journal of Physics: Conf. Series **887** (2017) 012002 doi:10.1088/1742-6596/887/1/012002

Component reliability parameters estimation considering weather factors

Yuan Wang

School of electronic engineering, Xi'an Aeronautical University Xi'an, China
E-mail:578101618@qq.com

Abstract. In order to study influence of power system reliability parameters under the complex weather, this paper presents a component reliability parameter estimation method based on the one directional S-rough law and F-decomposition law. The S-rough rule handles the failure rate and the uncertain weather conditions. According to this, it establishes reliable parameter estimation model. Then the power system reliability evaluation adds the F-decomposition law analysis to form the interfere metrics and analyzes influence of component reliability parameters under different weather combinations. In the numerical example verification, the results show that the proposed model can predict the transmission line failure rate, and reveals the influence regularity of component reliability parameter. Based on the objective data (climate in many years), the proposed method can eliminate the influence of subjective factor. Evaluation results are more objective, and provide important information for the power system reliability evaluation.

1. Introduction

In recent years, there are blackout accidents which caused huge economic loss and social influence at home and abroad, so it is very important to enhance the power system reliability evaluation. It is the most basic work to get original reliability parameters of reliability research. Usually, the original reliability parameters adopt average values. However, the parameters are dynamic randomness with the greatly influence of many factors: climate factors, human factors, geographical conditions, the quality of the product itself, load level, operating conditions, etc. [1-2].

Until now, many researches have studied the selection of original reliability parameters prediction. Literature [3] focused on prediction and correction of original reliability parameters, rarely on how to make use of the reliability of the existing original parameters to evaluate the launch of the new project. Literature [4] get the failure rate of the whole line by segment length weighted. Literature [5] analyzed impact factors of the transmission line by the empirical formula. It only mathematical pulsed the impact factors, because of complex reality, it can can't reflect the comprehensive effect of meteorological factors.

Z. Pawlak professor, a polish mathematician professor, put forward rough sets theory in 1982. It is a mathematical tool to deal with fuzzy and uncertain information, to discover the implicit knowledge and reveal the potential regularity. The theory has been applied in power system: power system load forecasting [6], fault line selection [7], etc. According to the theory, Kaiquan Shi professor presented the one directional singular rough set (S-rough set), the bidirectional singular rough set, function S-rough set and the F-decomposition law theory(F is a given system)[8], with dynamic collection description. Component reliability parameters are uncertain. The one directional singular rough set theory can well solve such uncertain law problem with dynamic properties. So this paper adopts this theory.

Aiming at the shortcomings of the existing reliability parameter estimation methods, this paper adopts the following method to solve: Based on the historical statistical data, the article analyzes the characteristics of random change in the weather; Using the one-way S-rough law to analyze method, considering the weather factor of component reliability parameters estimation model is set up; Using F-decomposition rule to analyze the influence of climate on reliability parameters, this paper predicts transmission line failure rate. The model proposed in this paper provides a new way for the transmission line reliability parameters of reasonable evaluation of new production.

2. The One Directional Singular Rough Set Theory [9]

In the power system component reliability parameter estimation, when external factors that affect the reliability parameters don't change, the impact factor is described by a set of attributes A, and the corresponding factor in attribute values is described by X. The one directional S- rough rule obtains component failure rate model, namely the relationship between the impact factor and the component failure rate. When the dynamic external factors change, the attribute set A becomes A^f, and the attribute value X becomes X^f. The F-decomposition rough law analyzes the relationship between the impact factor and the reliability parameters.

2.1. The One Directional S-Rough Law

It assumes that A is the attribute set of law $p(t)$, and X is the attribute value vector of law $p(t)$. In another word, $X=\{x_1,x_2,\ldots,x_m\}$ is value of the attribute $A=\{a_1,a_2,\ldots,a_m\}$, where m is the number of attributes. It gets the the failure rate set $[\lambda]=\{\lambda_1,\lambda_2,\ldots\lambda_n\}$, where n is the number of lines, based on the component failure model. By rule generation model, it gets rough law $p(t)$ of element failure rate.

2.2. The F-Decomposition Law

In a dynamic system, a is the attribute set with dynamic characteristics, and f is the family of dynamic mobility. If $A^f = A \cup \{a'\} = \{a_1, a_2, \ldots, a_r, a'\}$, where a' is the new increase (decrease) the properties, A^f is attribute complement set of A. Similarly, $X^f=\{x_1,x_2,\ldots x_r,x'\}$ is vector f - decomposition of the attribute value vector X, and then $p(t)^f$ is the f - decomposition law of $p(t)$. If $\overline{p(t)}$ is the F-decomposition up law of $p(t)$, and $\underline{p(t)}$ is the F - decomposition under law of $p(t)$, $(\underline{p(t)}, \overline{p(t)})$ is the F-decomposition rough law.

3. Component Reliability Parameter Model

Equipment working of the transmission system outdoor, especially the overhead line, is greatly affected by the weather such as thunderstorms, hail, and fog, therefore it is necessary to take into account element reliability parameters in polymorphic weather.

According to the characteristics of the reliability impact factor, the impact factors of overhead transmission line failure rate are due to the following categories: a_1 shows storm weather; a_2 shows heavy rain the weather; a_3 shows ice and snow; a_4 shows shows high temperatures; a_5 shows foundation reason and others; a_6 shows launch time; a_7 shows voltage grade, where $a_1 \sim a_4$ is the weather data and $a_5 \sim a_7$ is the non-weather data. It assumes that the attribute set of the impact factors is $A=\{a_1,a_2,a_3,a_4,a_5,a_6,a_7\}$, the attribute value vector of the impact factors is $X=\{x_1,x_2,x_3,x_4,x_5,x_6,x_7\}$. It assumes that the weight of each attribute is $H=(h_1,h_2,h_3,h_4,h_5,h_6,h_7)^T$, and $\sum_{i=1}^{7} h_i = 1$.

3.1. Component Failure Rate Model

If the power system transmission line number is n, the matrix $[X]$ of attribute characteristic value are obtained by the attribute values vector X of every line.

$$[X]=\{X_1^T,X_2^T,\ldots,X_n^T\}=\begin{bmatrix} x_{11} & x_{12} & \cdots & x_{1n} \\ x_{21} & x_{22} & \cdots & x_{2n} \\ \vdots & \vdots & \ddots & \vdots \\ x_{m1} & x_{m2} & \cdots & x_{mn} \end{bmatrix}$$

The efficiency target optimization γ_{ij} of each eigenvalue x_{ij} in the matrix of attribute characteristic value is then calculated by:

$$\gamma_{ij} = x_{ij} \Big/ \left(\underset{j}{\vee} x_{ij} + \underset{j}{\wedge} x_{ij} \right) \quad (1)$$

In Equation 1, $\vee$ is taking big operation, and $\wedge$ is taking small operation.

Target optimization matrix Γ is obtained as follows:

$$\Gamma = \begin{bmatrix} \gamma_{11} & \gamma_{12} & \cdots & \gamma_{1n} \\ \gamma_{21} & \gamma_{22} & \cdots & \gamma_{1n} \\ \vdots & \vdots & \ddots & \vdots \\ \gamma_{m1} & \gamma_{m2} & \cdots & \gamma_{mn} \end{bmatrix} \quad (2)$$

In Equation (2), γ_{ij} is target optimization.

According to the ordering method approaching the ideal solution (TOPSIS), it constructs the ideal optimization (Excellent decision of target optimization) and the negative ideal optimization.

The ideal optimization is

$$G=(g_1,g_2,\ldots,g_m)^T=(\gamma_{11}\vee\gamma_{12}\vee\ldots\vee\gamma_{1n},\\ \gamma_{21}\vee\gamma_{22}\vee\ldots\vee\gamma_{2n},\ldots,\gamma_{m1}\vee\gamma_{m2}\vee\ldots\vee\gamma_{mn})^T$$

the negative ideal optimization is

$$B=(b_1,b_2,\ldots,b_m)^T=(\gamma_{11}\wedge\gamma_{12}\wedge\ldots\wedge\gamma_{1n},\\ \gamma_{21}\wedge\gamma_{22}\wedge\ldots\wedge\gamma_{2n},\ldots,\gamma_{m1}\wedge\gamma_{m2}\wedge\ldots\wedge\gamma_{mn})^T$$

The weight matrix gets $[H]$ from the attribute weights.

$$[H]=\{H_1^T,H_2^T,\ldots,H_n^T\}=\begin{bmatrix} h_{11} & h_{12} & \cdots & h_{1n} \\ h_{21} & h_{22} & \cdots & h_{2n} \\ \vdots & \vdots & \ddots & \vdots \\ h_{m1} & h_{m2} & \cdots & h_{mn} \end{bmatrix}$$

Each weight value in the weight matrix is put into the equation (3), and it can obtain the failure rate λ_i.

$$\lambda_i = \frac{1}{1+\left\{ \dfrac{\sum_{i=1}^{m} \left(h_i\left(g_i - r_{ij}\right)\right)^p}{\sum_{i=1}^{m} \left(h_i\left(r_{ij} - b_i\right)\right)^p} \right\}^{\frac{2}{p}}} \begin{pmatrix} i=1,2,\ldots,m \\ j=1,2,\ldots,n \end{pmatrix} \quad (3)$$

In Equation (3), the distance parameter is 2. Then component failure rate set $[\lambda]=\{\lambda_1,\lambda_2,\ldots,\lambda_n\}$ is obtained.

When the impact factor set $[A]$ is expanding dynamically, The impact factor value sequence set is formed, namely, $[X]=[X]_0 \supseteq [X]_1 \supseteq [X]_2 \supseteq \ldots \supseteq [X]_{k-1} \supseteq [X]_k$. It gets a set of component failure rate sequence $\{\{\lambda_1, \lambda_2, \ldots, \lambda_n\}_0, \ldots, \{\lambda_1, \lambda_2, \ldots, \lambda_n\}_k\}$.

3.2. The Rough Law Generation Model of Component Failure Rate

In the sample of the components failure rate, the weather data is the changing factor and the non-weather data is the relatively stable factor. It forms a new sample set (X_i^f, λ_i^f) because of the weather data as unilateral dynamic data.

The data $[\lambda]$ is put into the Lagrange interpolation formula, it gets a law generation $p(t)$ of $[\lambda]$ as follow:

$$p(t)=\sum_{j=1}^{n} \lambda_j \prod_{\substack{i,j=1 \\ i \neq j}}^{n} \left(\frac{t-i}{j-i} \right) \tag{4}$$

In Equation (4), t is line number.

If considered the weather data and the non-weather data, it can obtain λ_i from Eq. (3). Then it can obtain law $p(t)$ from the rough law generation model Eq. (4). It is called the upper law $\overline{p(t)}$. Similarly, if considered the non-weather data, it is gotten law $p(t)$ and then the lower law $\overline{p(t)}$ by employing Eq. (3) and Eq. (4). Thus, the one directional S-rough law $(\underline{p(t)}, \overline{p(t)})$ can be formed.

3.3. Component Failure Rate Prediction

The weather data can be forecast next year. With weather sample values which are known over the years and normal sampling method, x_{n+1} can be obtained, because the relatively stable non-weather data is set the average value. It can predict component failure rate λ_{n+1} the next year, according to the component failure rate model which is proposed from 3.1.

4. Rough Law Index

The law energy is a reflection of the device performance, and can be used as the basis for choosing components. The law interference degree reflects the influence of different weather conditions on the transmission system.

The law energy w: supposing that $p(t)$ is the law on the interval [a, b], w is the law energy of $p(t)$ as follow

$$w=\int_a^b p(t)dx \tag{5}$$

The interference degree ρ: supposing that $p(t)^f$ is f-decomposition law generated by $p(t)$, ρ is the interference degree $p(t)^f$ of $p(t)$ as follow:

$$\rho=\frac{\|X^f\|_2}{\|X\|_2}-1 \tag{6}$$

In Equation (6), X is the corresponding attribute value vector $p(t)$, and $\|X\|_2$ is the norm of the vector X.

5. Illustrative Example

The line failure sample data's are listed in Table I. The weather data are showed in Figure 2 from 2009 to 2014.

Table 1. The failure rate of each line and its influence factors

n	The annual average storm days/d	The annual average rainstorm days/d	The annual average days of snow and ice/d	The high temperature days(≥35°C) /d	The foundation and other reasons/(km·a)	The average production time/a	Voltage grade/ kV	Line failure rate
1	24	22	0	27	8	14	220	0.083
2	21	23	2	20	5	10	35	0.078
3	7	10	9	10	4	6	115	0.063
4	13	2	28	2	2	6	35	0.059
5	22	21	1	24	6	12	220	0.081
6	14	6	18	7	2	2	115	0.073

5.1. Test of the Rationality of the Prediction of Unidirectional S- Rough Sets

First of all, by using the rough law generation model of component failure rate, the failure rate of fluctuation range value, as shown in Figure 1, can be forecasted by the failure rate law $p(t)$ per line in Table I, which is compared with the actual values in Table 1 of transmission line.

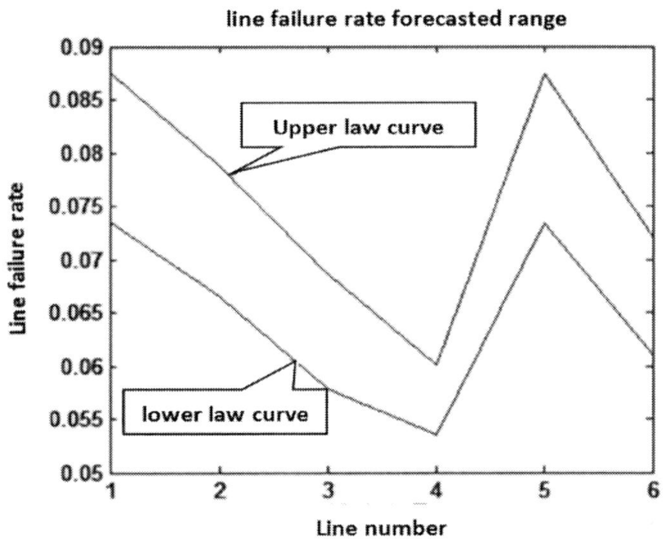

Figure 1. Line failure rate forecast by the S- rough law range

As can be seen from Figure 1 and Table 1, the actual values of transmission line failure rate are within the range of the upper and lower failure rate, which is calculated by the one directional S-rough law. It confirms the rationality of transmission line failure rate.

5.2. Prediction of Line Failure Rate in 2015

The weather data is forecast in 2015, according to the normal sampling method and the historical weather samples as shown in figure 2. Thus, it can get the attribute values $X=(3.8762, 2.7338, 10.5329, 28.386, 5, 10, 220)$, where the non-weather data sets the average value.

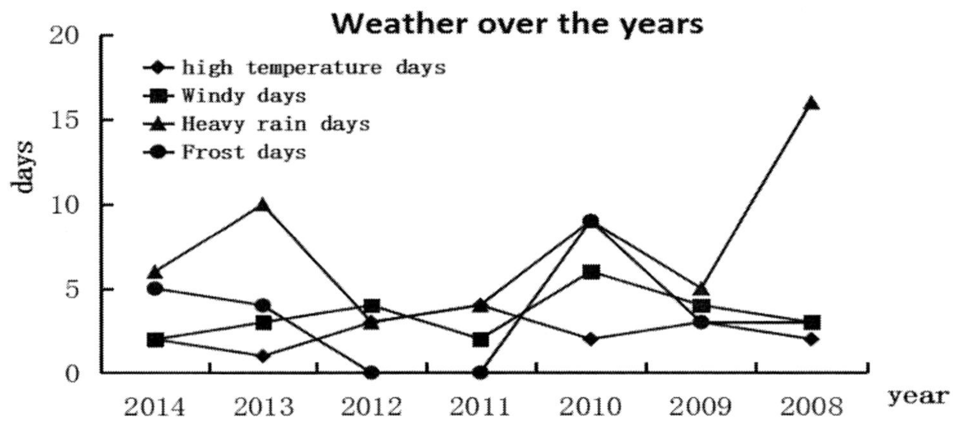

Figure 2. Weather in many years

The failure rate fluctuation range value of line in 2015 is [0.061, 0.073], by employing the component failure rate model and the law generation model.

5.3. Impact Analyzation of the Reliability Parameters under Different Weather

It can get the energy law w and interference value ρ, because of influence of dynamic climate factors, considering the role of a combination of factors, selecting four kinds climatic factors tested, and using F-decomposition law to analyze transmission line failure rate under different weather. Analysis of the impact of weather on the transmission line failure rate is listed in Table II, according to the w and ρ. In order to analyze the influence degree of various factors, it sets the following 6 kinds of situation reliability assessment.

Where:

$$a = \{a_1 \text{The storm weather}, a_2 \text{The heavy rain weather}, a_3 \text{ The ice and snow}, a_4 \text{The high temperature weather}, a_5 \text{The foundation}, a_6 \text{ The production time}, a_7 \text{The voltage grade}\}$$

$$\text{case1}: a_{case1} = \{a_1, a_2, a_3, a_4, a_5, a_6, a_7\}$$

$$\text{case2}: a_{case2} = \{a_4, a_5, a_6, a_7\}$$

$$\text{case3}: a_{case3} = \{a_3, a_5, a_6, a_7\}$$

$$\text{case4}: a_{case4} = \{a_2, a_5, a_6, a_7\}$$

$$\text{case5}: a_{case5} = \{a_1, a_5, a_6, a_7\}$$

$$\text{case6}: a_{case6} = \{a_5, a_6, a_7\}$$

Table 2. Data analysis of transmission line under different weather conditions

	Line failure rate	The law energy	The interference degree
Case1	[0.0612, 0.0731]	0.15913	0.1953
Case2	[0.0515,0.0568]	0.03733	0.1857
Case3	[0.0624, 0.0674]	0.03356	0.1922
Case4	[0.0598, 0.0546]	0.03754	0.1874
Case5	[0.0576, 0.0524]	0.03657	0.1859
Case6	[0.0456, 0.0507]	0.03549	--

As shown in Table 2, the influence of Frost factor is the most important for reliability parameters. The second is Heavy rains factor, and the third is High temperature factor and the Storm factor. As shown from the interference degree, it is a result of the combined action of many kinds of weather factors, not the sum total of the single factors, when four kinds of climate factors are working together. As shown from the law energy, the less important the influence factors and the smaller the energy laws.

Then performance of transmission line is better. The smaller reliability parameter estimation range, the more accurate reliability parameters.

6. Conclusion

The proposed method is applied to analysis example to do reliability parameter estimation. The following conclusions are obtained: (1) the one directional singular rough set can reasonably predict transmission line failure rate based on weather uncertainty. (2) The law energy and the interference degree reflect the influence of the component reliability parameters under the weather fators. So taking corresponding measures, it improve system reliability and the ability to cope with adverse weather.

This paper presents reliability parameter estimation method based on the one directional singular rough set theory in different weather. The one directional S-rough law analyzes line failure rate and weather uncertainty. The F- decomposition law analyzes impact of weaher on reliability parameter. So reliability parameter uncertainty has the certainty measurement. When the one directional singular rough set theory dealing with the uncertainty of transmission system reliability parameters, it just needs simple weather days. Thus, it obtains reliability parameter based on objective data, eradicating the effect of subjective factor. The evaluation results are more objective.

Acknowledgment

This research thanks the support of the Foundation of Xi'an Aeronautical University (2017KY1222).

7. References

[1] TANG Siqing, ZHANG Mi, LI Jianshe, et al. Review of blackout in Hainan on September 26th-cause and recommendations [J]. Automation of Electric Power Systems, 2006, 30(1): 1-7, 16.

[2] PANG Zhun, LI Bangfeng, YU Yue, et al. Study on operating modes of Hainan power grid during typhoon periods [J]. Power System Technology, 2007, 31(7): 46-50.

[3] ZHANG Yong-jun, CHEN Chao, XU Liang. Prediction of original reliability parameters of power system based on fuzzy clustering and similarity [J]. Power System Protection and Control, 2011, 39(8): 1-5.

[4] HAN Weiheng,LIU Junyong,ZHANG Jianming,et al.Power system time-section reliability assessment analysis considering topography and icing under freezing disaster weather[J.]Power System Protection and Control, 2010, 38(15):81-86...

[5] XIE Yunyun, XUE Yusheng, WEN Fushuan, et al. Space-time evaluation for impact of ice disaster on transmission line fault probability [J]. Automation of Electric Power Systems, 2013, 37(18): 32-41, 98.

[6] NIU Dongxiao, WANG Jianjun, LI li, et al. Short-term load forecasting using adaptive ANN based on rough set and decision tree [J]. Electric Power Automation Equipment, 2009, 29(10): 30-34.

[7] Pang Qingle,Sun Tongjing,Zhong Maiying,et al.Fault Line detection based on rough set theory in indirectly grounding power system [J]. Proceedings of CSEE. 2007, 27(4): 60-64.

[8] HUANG Shun-liang, SHI Kai-quan. Rough Law Energy and F-decomposing Rough Law Measurement [J]. Computer Science, 2009, 36(1): 177-180.

[9] HUANG Shun-liang, SHI Kai-quan. One direction S-rough decision law and decision law mining [J]. Systems Engineering and Electronics, 2008, 30(5):858-862.

IOP Conf. Series: Journal of Physics: Conf. Series **887** (2017) 012003 doi:10.1088/1742-6596/887/1/012003

Features research of ideal ahead-detecting method used in mining roadway and exploitation of a new technology --DFIP

Weijie Zhang[1] and Zhimin Liu[2]

[1]School of Mechanical-Electrical Engineering, North China Institute of Science and Technology, Yanjiao Beijing East 101601, China, wzhangweijie@163.com
[2]College of Mechanical and Equipment Engineering, Hebei University of Engineering, Handan 056038, China, liuzhiminhd@126.com

Abstract. In order to study the new high efficiency advanced detection technology of coal roadway, the characteristics of ideal advance detection are analyzed. Dual- frequency IP method advanced detection technology is based on the Induced polarization effect of coal (rock). The detection range of the angle and length scanning are achieved by simulation comply with the character of electric field superposition. So, the dual frequency IP method can meet the requirement of high efficiency advance detection of coal roadway. The research results provide theoretical support for the development of new advanced detection equipment.

1. Introduction

According to Coal Mine Safety Regulation, hydrologic conditions of coal mine must be surveyed and analyzed comprehensively before mining production [1]. How to effectively predict geologic anomalies, especially hydrologic conditions such as karst caves and karst collapses, of areas which must be excavated has become a hot issue in the mining technology. If the hidden geologic anomalies are failed to be detected or predicted, the security of workers' lives and colliery property will face the geologic dangers, such as gas outburst, roof falling and water inrush. The situation is even worse in the procedure of roadway excavation, because these hidden geologic anomalies are hardly to be found or located in macroscopic geological survey of mines. The existing geophysical prospecting methods include drilling method, direct current method, transient electromagnetic method, etc. Each geophysical prospecting method has its specific conditions of use and detection characteristics. However, there is still a large gap between the exploration and excavation of the heading face. In order to improve the situation and increase the excavation efficiency, many researchers focus on the study of effective ahead-detection technology and the development of suitable detector used in roadway excavation at present.

2. Characteristics of Ahead-Detection Used in Mechanized Excavation Face and Ideal Features of Ahead-Detection

2.1. Characteristics of Ahead-Detection Used in Mechanized Excavation Face

The working environment of mechanized excavation face of coal mine roadway has special characteristics [2]:

1) High quantity of mechanical equipment

There are many mechanical equipment working at the mechanized excavation face, such as transportation machinery, roadway support equipment, drilling machines and dust settling devices.

2) Confined space for working

Generally, the dimension of coal mine roadway's excavation face is about 4 meters in width and 3~3.5 meters in height, which is a confined space for many machines and workers. However, many working procedures, such as the excavation, the drilling, the roof supporting, the dust settling and the ahead-detection, must be conducted in the confined space.

3) Efficiency imbalance among working procedures

At present, it is hardly to combine all the working procedures together, which leads to the separation of the ahead-detection, the excavation and the roof supporting. As a result, the excavation efficiency is extremely restricted by stopping for the ahead-detection and the roof supporting.

4) High accident rate

Geologic anomalies hidden underground make the roadway's excavation faces harmful dangers, like water inrush, gas outburst and roof falling.

5) High dust content

Huge amount of dust could be produced during the excavation and the drilling process. And the dust settling device can only prevent the dust in a very limited extent.

At present, there are several ahead-detection methods used in the mechanized excavation face, for example, drilling method, direct current prediction method and transient electromagnetic method. However, restricted by the special working circumstance of coal mine roadway, almost all of these methods hardly get high prediction accuracy.

1) Separation of ahead-detection and the excavation

Constrained by the characteristics of presently used ahead-detection methods, the ahead-detection and the excavation are conducted separately in different working times. Meanwhile, complicated working process and long detection period of these ahead-detection methods extremely restricts the excavation efficiency. Even worse, in order to raise efficiency, the excavation was conducted without the ahead-detection in some coal mines, which has resulted in grave working accidents and serious casualties.

2) Low prediction accuracy

Affected easily by the geological conditions of roadway's roof and floor, hidden geologic anomalies' prediction behind cutting face through present ahead-detection methods are usually inaccurate. On the other hand, full-directional detection of excavation face is hardly realized through present ahead-detection methods. For example, Transient Electromagnetic Method can only detect the geologic conditions right behind cutting face. But it's hard to find anomalies hidden in the roadway's roof or floor.

2.2. Ideal Features of Ahead-Detection

In the area faced possible threat of water inrush, if the water is in no condition to be drained, the waterproof pillar must be preserved properly during the roadway's excavation. The size of the waterproof pillar is mainly determined by geologic structures and hydrologic conditions of coal mine [3]. And the width of the waterproof pillar of faults should be no less than 20 meters according to the regulation [4]

Through the analysis of special features of coal mine and the characteristics of present ahead-detection methods used in mechanized excavation face, several essential characteristics of the ideal ahead-detection method are concluded as follows:

1) Ahead Prediction

First of all, long distance prediction of geologic structures should be realized in the ideal ahead-detection technology. Considering the working arrangement of coal mine, the effective detection distance must be longer than the sum of maximum excavation footage per day and minimum horizontal safety distance.

The excavation velocity is determined by many factors, such as coaly features, excavation machine status and worker states. Presently, the working system called "three shifts" is conducted in the coal mining

IOP Conf. Series: Journal of Physics: Conf. Series **887** (2017) 012003 doi:10.1088/1742-6596/887/1/012003

enterprises. The workers work in three shifts of eight hours per day, two shifts for the excavation and one for equipment maintenance. Generally, the average excavating distance for one shift is 10 meters and 30 meters at most. Then the maximum daily excavation footage is less than 60 meters. Meanwhile, the minimum horizontal safety distance is mainly determined by water contents of hydrologic structure and hydraulic press. According to research data, the value of the safety distance is about 20~30 meters. In conclusion, the effective detection distance should be longer than 80~90 meters (Fig. 1).

2) Real-time Detection

There are two kinds of stop time in the process of excavation. One is equipment maintenance shift mentioned above, the other one is the roof support procedure which must be conducted after a few meters of excavation. The real-time detection means to fully use these stoppage times to conduct ahead-detection, not only in the longtime stoppage for equipment maintenance, but also in the short excavation intervals for roof supporting. So the time spent on detection should be limited in an hour and the results of prediction should be figured out and presented in ten minutes. In this way, ahead detection could be conducted many times in the whole excavation procedure and real-time ahead-detection can be realized in a certain extent. And the accuracy of ahead-detection will increase and the detection distance of every single prediction can reduce to 30~50 meters, which can effectively decrease the working power of the detector.

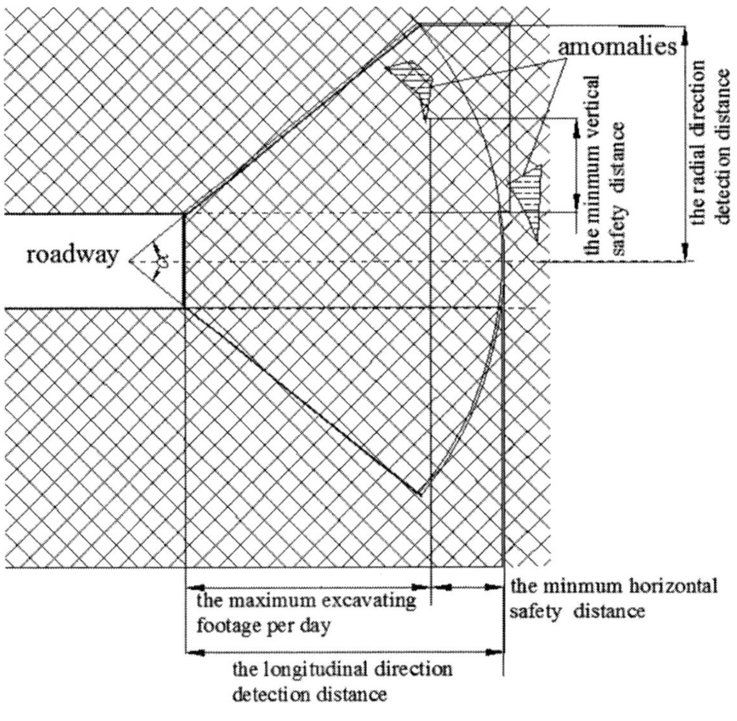

Figure 1. The range of detection

3) Directional Detection

The geologic anomalies can be located in direction and in distance by ideal ahead detection. The direction of ahead detection can be effectively controlled to scan the conical area behind the cutting face longitudinally and laterally, which can accurately predict the direction of the hidden geologic anomalies. On the other hand, the distance between the geologic anomaly and excavation face could be predicted by scanning the conical detection area contained horizontal and vertical safety distance (see Fig. 1).

4) High Predicting Accuracy

The ideal ahead detection predicts not only the types of geologic anomalies, such as accumulated water, faults, collapse and fragmentation, but also the location of the geologic dangers and water contents. Meanwhile, the false and miss alarm can be eliminated as possible to increase the efficiency of detection;

5) Intelligent Operation:

The results of a head-detection can be graphically displayed and easily understood. Foolproof operation interface, automatic judgment of dangerous level and speech alarm are also applied in the ideal ahead detection facility.

6) Integration of Ahead-detection and the Excavation

The ideal ahead-detection facility can be integrated with road header, which reduces the number of the equipment working at the mechanized excavation face. The integration of ahead-detection and the excavation will extremely increase the excavation efficiency.

7) Convenient Usability

The ahead-detection facility has features of light weight, compact size, low cost and convenient operation. And it's easy to be installed on the road header.

3. Detection Principle of DFIP Used in Mechanized Excavation of Coal Mine

Dual-Frequency Induced Polarization Method (DFIP) is based on the Induced Polarization Effect of rocks and minerals [5]. The induced polarization information contained geologic features of detection area could be measured in the form of dual-frequency current signal. From the information, useful current parameters could be extracted through amplification, frequency selection and detection. And the value of PFE (Percent Frequency Effect), which can characterize the geologic features of detection area, can be figured out properly [6].

3.1. Arrangement Of Detection Probes

As shown in the Fig. 2(a) is the arrangement method of detection probes on the excavation face. A0(+)stands for emitting probe (measuring probe), A1(+)stands for shielding probe series and B(-)is grounding electrode which is set hundreds of meters away from the excavation face. The shielding probe series contain eight symmetrically distributed shielding probes, as shown in Fig2 (b).

During detection, induced currents with identical frequency and polarity are separately emitted to detection area simultaneously from A0 and A1. As a result, the electric field is formed in the rocks and minerals of detection area. According to the principle of like poles repelling, the currents from A0 and A1 will interact with each other, resulting in the focusing of emitting current and the formation of a conical detection area (as shown in Fig. 2(a)). In this way, the direction of induced current can be controlled properly and "full space effect" [2] of electric field could be limited in a certain extent.

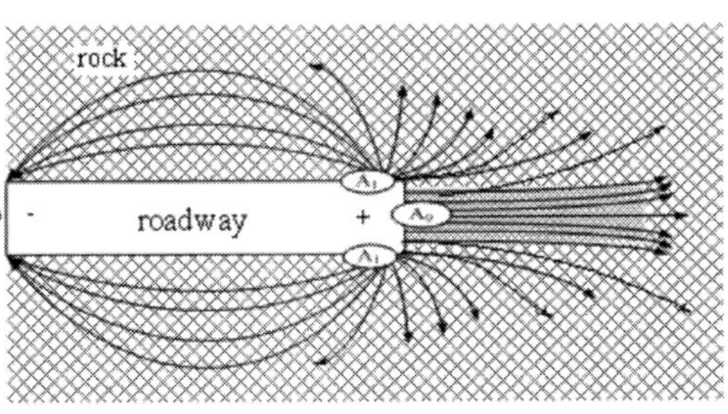

(a)The probes

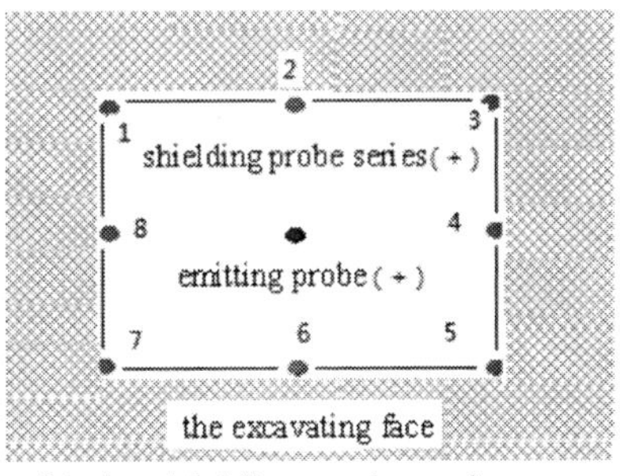

(b) the shielding probe series

Figure 2. The arrangement of the probes

As shown in Fig. 3, under the conjunct effect of emitting current and shielding currents, the trend of electric field lines reflect the focusing effect of electric field and the controlling effect of detection direction.

In addition to emitting induced current, A0 is still in charge of receiving the polarization information of rocks and minerals.

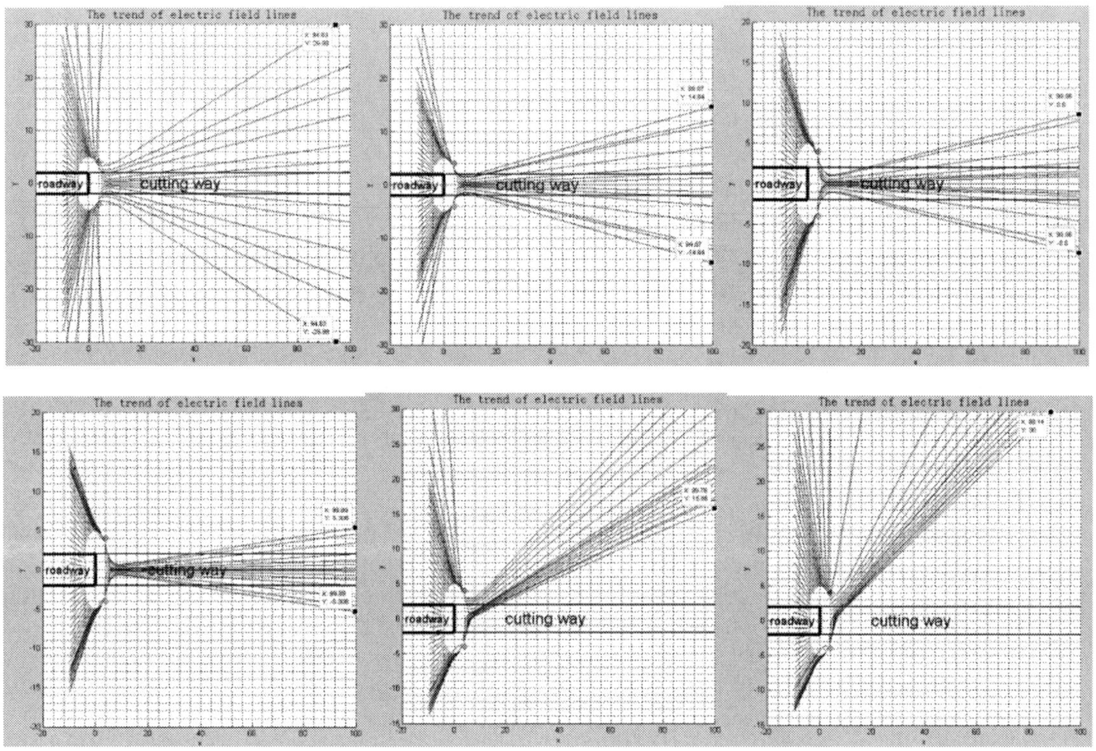

Figure 3. Focus and deflection of electric filed

3.2. Maintaining the Integrity of the Specifications

Considering the complexity of geological condition and diversity of affecting factors in the mechanized excavation face, several conditions are presumed in the research:

(1) Effected by induced current with ordinary frequency, the geologic mass (coal and rocks) behind the excavating face is regarded as Linear Time-Invariant system which has features of isotropy and homogeneous linearity; (2) The resistivity (ρ) of the geologic mass is considered to be constant when the geologic anomalies non-existed; (3) The influences of equipment, temperature, humidity and free electronic could be neglected.

The model of detection current field emitted from emitting probe and shielding probes in excavation face is illustrated in Fig. 4. In order to analyze the electric field property of point M seated at the right of emitting probe, the position of emitting probe is chose as the origin O of the model's coordinate system.

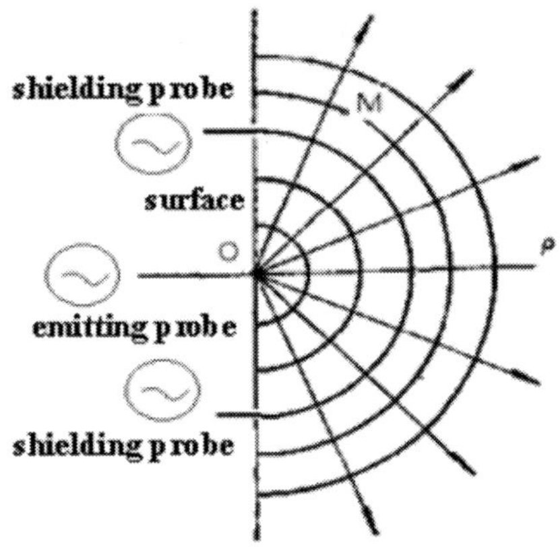

Figure 4. The modeling in excavating face

Based on the half-space property of electric field generated by point power on earth surface, the electric field emitted by emitting probe could be figured out because emitting probe is placed on the excavation surface. In the case of only electric field of emitting probe being considered, the electric potential of point M, which is away from origin O in distance R, can be figured out in formula (1).

$$\varphi_0 = \frac{\rho I_0}{2\pi}\frac{1}{R} \tag{1}$$

Because the shielding probes will be inserted in the coal bed, their electric fields are approximately considered as a full-spaced field. Thus, the electric potential of M generated by shielding currents can be calculated in formula (2). In the formula, stands for the current intensity emitted by shielding probe *ith*, and stands for the distance between M and shielding probe *ith*:

$$\varphi_p = \frac{\rho}{4\pi}\sum_{i=1}^{8}\frac{I_i}{R_i} \tag{2}$$

In (2) is the electric current strength of the *ith* shielding probe.

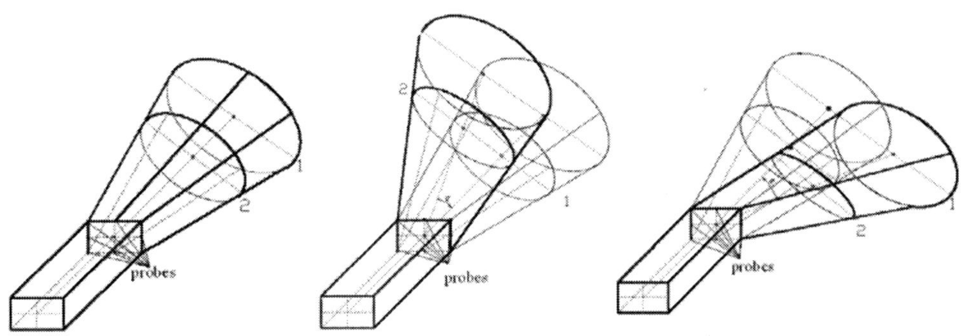

(a) (b) (c)

Figure 5. The angle and length scanning

IOP Conf. Series: Journal of Physics: Conf. Series **887** (2017) 012003 doi:10.1088/1742-6596/887/1/012003

Based on an overall consideration of formulas (1) and (2), we could get the electric potential of point M under the interaction of emitting current and shielding currents, as illustrated in formula (3). And formula (4) expresses the current intensity of point M.

$$\varphi = \varphi_0 + \varphi_p = \frac{\rho I_0}{2\pi} \frac{1}{R} + \frac{\rho}{4\pi} \sum_{i=1}^{8} \frac{I_i}{R_i} \tag{3}$$

$$E = -\nabla \varphi \tag{4}$$

Formulas (3) and (4) express the mathematical model of electric field in the mechanized excavation face. The model is founded on the basis of superposition of half-spaced electric field generated by emitting current and full-spaced field generated by shielding currents.

4. Control Strategy of Detecting Location

The direction and the distance of detection are two aspects of detection location. The location of geologic anomalies can be fixed when the two aspects are confirmed.

4.1. Angle Scanning

Angle scanning means through the control of current intensities of shielding currents, the direction of the focus electric field's axis can be turned in angles to realize the full-directional scanning of the excavation face.

When the current intensities of all shielding currents are equal, the focus axis is forward and the direction of ahead-detection will be right ahead, as shown in Fig 5(a).

If the current intensities of shielding currents emitted from shielding probes of one side increase, the detection direction will turn to the other side.

As shown in the Fig. 5 (b), when the current intensities of right-sided shielding currents increase (or left-sided shielding current intensities decrease), the axis of the detection conical area will turn to left in a deflection angle α.

When the left-sided shielding current intensities increase (or right-sided current intensities decrease), the axis will turn to right in a deflection angle α, as shown in the Fig. 5 (c).

The more relative difference between the current intensities of right-sided and left-sided shielding currents, the angle α will be greater.

Above and under-sided detection can also be realized in the same way. Therefore, the angle scanning is realized through changing intensities of shielding currents.

4.2. Length Scanning

Because the attenuation of electric field depends on the features of geologic mass and hidden geologic structures of detection area, the increase of detection distance is implemented by increasing emitting current intensity. Therefore, the length scanning is realized through the control of current intensity.

When the direction of ahead-detection is fixed, the detection area will expand with the increase of detection current intensity. As shown in the Fig. 5 (a) and(c), the detection boundary will expand from section 2 to section 1 as the increase of current intensity.

When the detection current intensity is constant, the detection direction can be controlled to realize scanning in a conical area with certain radius, as shown in Fig. 5 (b).

In conclusion, through the control of the intensities of detection currents and the close sequence of detection probes, a conical area scanning of detection can be realized in different depths and angles. Thus, the direction and distance of geologic anomalies can be judged through the combination of scanning in depths and in angles.

5. The Constituents of DFIP Facility Used in Mechanized Excavation of Coal Mine Roadway and the Realization of Integration with Road Header

5.1. The Constituents of DFIP Facility

The main constituents of DFIP facility used in mechanized excavation face of coal mine roadway are as follows:

1) Emitting Facility

Emitting facility, which is consisted of induced current generator, emitting probe, shielding probe series and earth electrode, is in charge of the emission of induced current and shielding currents. It's also in charge of the control of the frequency and the intensity of the induced current, as well as the close sequence of shielding probes.

2) Receiving Facility

The polarization information of geologic structures in the detection area can be received by receiving facility in form of current signals. Receiving facility will deal with these signals and extract useful information to figure out and display prediction results.

3) Data Interpretation Software

Data interpretation software is used for further processing the gathered data to get more adequate and detailed interpretation. The result will guide the excavation process by providing the prediction of geologic anomalies' location and dangerous level.

5.2. The Realization of Integration with Roadheader

In order to realize the integration with road header, the structural features and the working mechanism of boom-type road header is considered in the arrangement method of detection probes. Meanwhile, the special characteristics of roadway is also utilized.

For example, a special hydraulic cylinder is installed on the road header. The emitting probe is set on the head of the piston rod which could be pushed on the excavation face when detection conducting. Meanwhile, the anchors set around the roadway are taken advantage as shielding probes.

6. Conclusion

This paper emphasizes on the research of ideal features of ahead-detection used in mechanized excavation face of coal mine and putts forward essential features of the ideal ahead-detection method. The geologic anomalies hidden behind the excavation face can be located both in direction and in distance through the detection. The new ahead-detection technology can provide reliable geologic data and security assurance for coal mine roadway excavation.

7. Acknowledgment

The authors would like to thank the support from the Youth Fund of North China Institute of Science and Technology (No.3142015098) mind Hebei Natural Science Foundation of China (D2017402158).

8. References

[1] State Administration of Work Safety. *The Mine Safety Rules* [M].Beijing: China Legal Publishing House,2016.2

[2] Zhang Wei-jie, Hao Ming-rui Discussion on New Pilot Detection Technology of Mine Roadway Based on Double Frequency Induced Method[J].Coal science and technology 2010.03.73-79

[3] Personnel training department of State Administration of Work Safety . The flood prevention of mine[M]. Xuzhou: China University of Mining and Technology Press,2002.07

[4] Coal research institute in xi'an . Coal mine safety brochures (the fifth) The flood prevention[M].Beijing:China Coal Industry Publishing House,1992

[5] He Ji-shan. Double Frequency Induced Method[M].Beijing:High Education Publishing House,2006.01

[6] Gao Jian-ping. The Experimental Study on Applying Double Frequency Induced Polarization Method Detecting Coal Mine Gob[D].Taiyuan. Taiyuan University of Technology ,2007

[7] Liu Qing-xia Underground Electrical Lead Survey Method And Its Application[J] Coal Geology & Exploration 2001m29(5):60-62.

[8] Yang Er-bin, Yang Huan-hong ,Liu Rong-hui. The basic and application of the electromagnetic field(the second) [M].Beijing: China power Publishing Housem2009.09

[9] Fu Liang-kui. Electrical Prospecting tutorial[M].Beijing: Geological Publishing House,1983

[10] K. Titov ,V. Komarov,V. Tarasov,A. Levitski Theoretical and experimental study of time domain-induced polarization in water-saturated sands[J] Journal of Applied 50 (2002) 417-433

[11] Slater,L.,and D.P.Lesmes, IP interpretation in environmental investigations[J]:Geophysics, 2002a, 67,77-88.

[12] Weijie Zhang, Mingrui Hao, Weizhen Zhang, Miao Wu .Study on Realtime Ahead Detection Technology for Mining Roadway Based on Double Frequency Induced Method[J] The 2nd International Conference on Mechanic Automation and Control Engineering

[13] Mangi, F. A., Xiao, S., Jamro, D. A., Khan, S. A., Imran, M., & Kakepoto, G. F. (2016, July). Manipulating Electromagnetic Wave Linear-to-Circular Polarization Conversion Transmitter Based on Periodic Strips Array. In Information Science and Control Engineering (ICISCE), 2016 3rd International Conference on (pp. 1342-1345). IEEE.

IOP Conf. Series: Journal of Physics: Conf. Series **887** (2017) 012004 doi:10.1088/1742-6596/887/1/012004

Requirement analysis for the one-stop logistics management of fresh agricultural products

Jun Li[1], Hongmei Gao[2]* and Yuchuan Liu[3]

[1]Department of Economic Management, Tianjin Agricultural University, Tianjin, China, e-mail: ycliu_icloud@me.com
[2]Department of Economic Management, Tianjin Agricultural University, Tianjin, China, e-mail: gaohongmei@126.com*, Corresponding author
[3]Department of Information Management, Tainan University of Technology, Tainan, Taiwan, e-mail: t00258@mail.tut.edu.tw

Abstract. Issues and concerns for food safety, agro-processing, and the environmental and ecological impact of food production have been attracted many research interests. Traceability and logistics management of fresh agricultural products is faced with the technological challenges including food product label and identification, activity/process characterization, information systems for the supply chain, i.e., from farm to table. Application of one-stop logistics service focuses on the whole supply chain process integration for fresh agricultural products is studied. A collaborative research project for the supply and logistics of fresh agricultural products in Tianjin was performed. Requirement analysis for the one-stop logistics management information system is studied. The model-driven business transformation, an approach uses formal models to explicitly define the structure and behavior of a business, is applied for the review and analysis process. Specific requirements for the logistic management solutions are proposed. Development of this research is crucial for the solution of one-stop logistics management information system integration platform for fresh agricultural products.

1. Introduction

The heightened awareness of food-related safety issues among food consumers drives the demand for more information about the vertical food supply chain about the origin and handling of the basic commodities and food products generated and consumed throughout the world. Developments of agro-food industries are facing global challenges that can only be supported by information technologies. The major IT development lines, the support potential of their integration, organizational requirements for the utilization, and possible consequences for the future organization of the agro-food sector were reviewed [1]. Traceability is an essential subsystem of quality management, and must be managed by setting up a traceability system, which keeps the data tracking of product routes and of selected attributes. A traceability system can consist two elements, the routes of the product, path along which products can be identified throughout the manufacturing, distribution and retail procedures, and the extent of traceability wanted [2]. Food traceability requires that all stakeholders within the food supply chain, including agriculture and feed producers, food manufacturers, retailers, etc., must be able to identify the source of all raw materials and ingredients to whom the products have been sold. The food companies must apply identification systems and data handling procedures and these must be integrated into their quality management system. The sector encompassing information technology (IT) centers ought to find a reasonable compromise between the simple, step by step passing of traceable unit IDs for the neighboring actors, and the accumulated enormously huge databases of the actors. The traceability system is to provide services for the supply chain actors on cooperative basis of the mutual interests [3]. In addition, the IT centers must support the supply chain and value chain management, as well as the work of the authorities that are responsible for the human health. Opara [4] reviewed the concepts of supply chain management and traceability in agriculture and highlighted the technological challenges, including food product label and identification, activity/process characterization, information systems for data capture, analysis, storage and communication, and the integration of the overall traceable supply chain in implementing traceable agricultural supply chains. Wang et al. [5] addressed that the values on traceability can be integrated with the supply chain management processes to manage the business process and improve its performance. Bosona and Gebresenbet [6] summarized the literature review on the food traceability issues. The definition, driving forces, barriers, benefits, traceability technologies, improvements, and performances of food traceability system had been discussed. It was pointed out that the development of full chain food traceability system is quite complex in nature, and a deeper understanding of real processes from different perspectives such as economic, legal, technological, and social issues are essential. Consequently, studies on the integration of traceability activities with food logistics activities, the linkage between traceability system and food manufacturer, standardization of data capturing and communication protocol for different drivers, and performance evaluation frameworks for food traceability system need to be focused.

A new model and prototype of a new Farm Information Management System, which meets the changing requirements for advising managers with formal instructions, recommended guidelines and documentation requirements for various decision making processes, was developed [7, 8]. As achieving end-to-end traceability across the supply chain is quite a challenge from a technical, a co-ordination and a cost perspective, Kelepouris et al. [9] suggested a radio frequency identification (RFID) technology and outlined both information data model and system architecture that made traceability feasible and easily deployable across a supply chain. Based on an integration of alphanumerical codes and RFID technology, the traceability system for Parmigiano Reggiano (the famous Italian cheese) was developed [10]. Manthou et al. [11] provided empirical insights regarding the use of Internet-based applications in the agri-food supply chain by focusing on the Greek fruit canning sector. The companies' perceptions regarding perceived benefits, constrained factors and motivation factors towards the use of Internet-based applications were studied. A PDA-based Record-keeping and Decision-support System for traceability in cucumber production was developed on Windows Mobile platform invoking a Geographic Information System (GIS) control [12]. Two agricultural production enterprises were chosen as case study to evaluate the system and the results show that the efficiency of production record-keeping and decision-support is improved by the simple and friendly system. The state-of-the-art review in the recent advancements of food processing and packaging industry in the fields of smart packaging and materials, automation and control technology, standards, and their application scenarios, and production management principles and their improvements was proposed [13].

The logistics and information flow play an important role in the fresh agricultural products supply chain. The purpose of this research is to further provide a complete study on the issues and solutions for the one-stop logistics service management of fresh agricultural products. A full life-cycle business-to-technology method, model-driven business transformation MDBT, is both a business transformation methodology and a set of innovative technologies that allow business strategies to be realized by choreographing workflow tools and human activities. MDBT uses formal models to explicitly define the structure and behavior of a business component. Kumaran et al [14] presented a new approach to IT service workflow automation and a new generation of service-delivery management systems based on the model-driven transformational approach and service-oriented architecture. The MDBT approach is applied to propose an integrated solution. Development of this research is crucial for the solution of information and logistics management in fresh agricultural products supply chain business.

2. Model-Driven Business Transformation Approach

Model-driven business transformation, MDBT developed by IBM research, uses formal models to explicitly define the structure and behavior of a business component. These models can be employed to monitor, analyze, and improve its performance, and leverages these models in the construction of its IT systems. Each layer constitutes a different level of abstraction, performs a well-defined function, and has a different audience. The strategy layer defines the goals and objectives of the business system. The operation layer describes the operations performed by the business system to achieve the goals. The composition layer is an abstraction of the computational elements for the business operations. The implementation layer specifies how the computational elements are implemented on IT platform. Figure 1 shows the MDBT framework, including the separation of concerns, connections between model layers, and the closed-loop architecture using the business process modeling component.

In the MDBT approach, the transformation process begins with the identification of the strategic goals and objectives of the business component. This leads to a set of initiatives that support these goals. These initiatives determine the definition, analysis, optimization, and implementation of the business operations of the organization such that the strategic goals can be achieved. Formal definition of the business operations and the operational KPIs (key performance indicators) is the next step of transformation process which was referred as the business operation model. A business operation model is different from the more familiar workflow models. A business operation model, on the other hand, defines the key business artifacts and the operations performed on these artifacts. The third step of solution composition in MDBT is the judicious use of technology to support the execution of business operations. This involves the generation of a platform-independent solution composition model and the realization of this model on a specific software platform. The final step in MDBT is to create an implementation of the IT solution on a specific IT platform. Once the solution is deployed, business owners can monitor and analyze business performance using KPIs and continuously improve the models, both at the business and IT levels, based on this performance analysis.

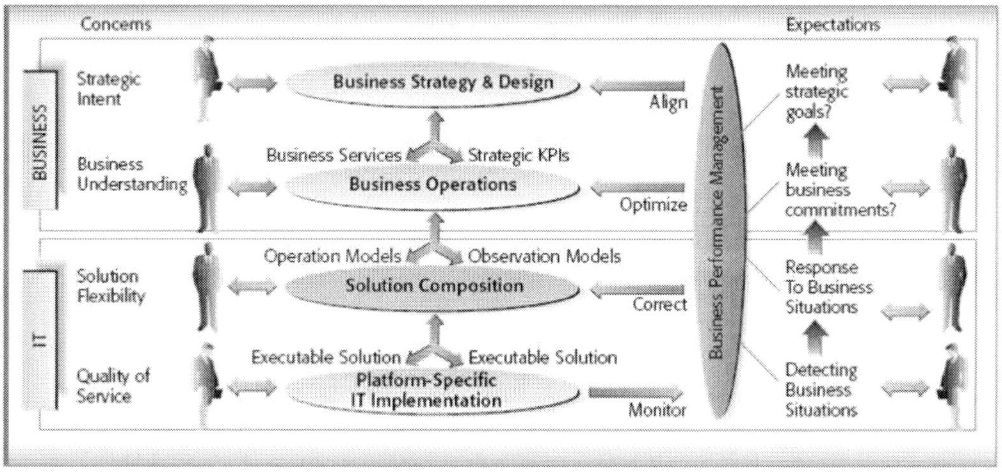

Figure 1. Model-driven business transformation framework, from [14]

3. Solution Analysis for One-Stop Logistics Service

The infrastructure of fresh agricultural products logistics supply chain management services can be summarized as shown in Fig. 2. The solid lines stand for the logistic flow of the products, and the dashed lines are the required information flow to assure the food safety. As indicated in Fig. 2, the traceability information is required during the farming process of the agricultural products. The process and ingredient information is necessary if the products are sent to the food industry for more process. The logistics information during the transportation and storage for the supply chain stages must ensure the completeness of the traceability management. By applying MDBT approach, the definition of goals and objectives should be firstly analyzed in strategy layer. As the project team reviewed the current fresh agricultural products supply chain in Tianjin, the first two issues are the product loss during the logistics procedure and cost (and/or effectiveness) of the cold chain logistics. The objectives are consequently to improve the effectiveness and efficiency for fresh agricultural products logistics management. One-stop supply chain management service is an integrated model of the logistics services. The fresh agricultural products are characterized by perishability, timeliness and regional nature, with intensive information requirements during the whole supply chain procedures. One-stop logistics service system is studied to reduce logistics cost, facilitate logistics efficiency, and promote the logistics development of fresh agricultural products.

The second step of business operation analysis for MDBT is to perform the business operation. Considering the supply chain process shown in Fig. 2, some observations can be made:

• End customers can receive the fresh agricultural products via an O2O scenario directly from the e-commerce provided by the agricultural corporations. The transportation procedures can be minimized so that the product loss caused can also be reduced.

• The O2O scenario also provides an optimized product combination both for the customer and supplier. The products delivered can be optimized according to the farming production condition and the customers' order requirement. The balance of the supply and demand can be optimized and the effectiveness of farming production can be improved.

• The HMR, home meal replacement, is more and more popular as the necessity of ready-to-eat convenience increased significantly. The traceability information during the farming and food process are becoming a competitive issue for HMR brand marketing.

Detail solution requirements are further studied through the supply chain processes. Figure 2 shows that the logistics information flow between the supply chain enterprises can be shared and integrated for further enhancement of the supply chain management efficiency. The information system must be designed to cope with the intricate farming data, food processing information, and the rigmarole product transportation and storage information. The platform independent solution requirements for one-stop logistics service of fresh agricultural products are proposed in this paper. The system platform should cover information requirements of the farming, food processing, and logistics management of the supply chain processes. Schematic diagram of the one-stop logistics service management for fresh agricultural products supply chain is suggested as shown in Fig. 3. The specific functional requirements from configuration management analysis include:

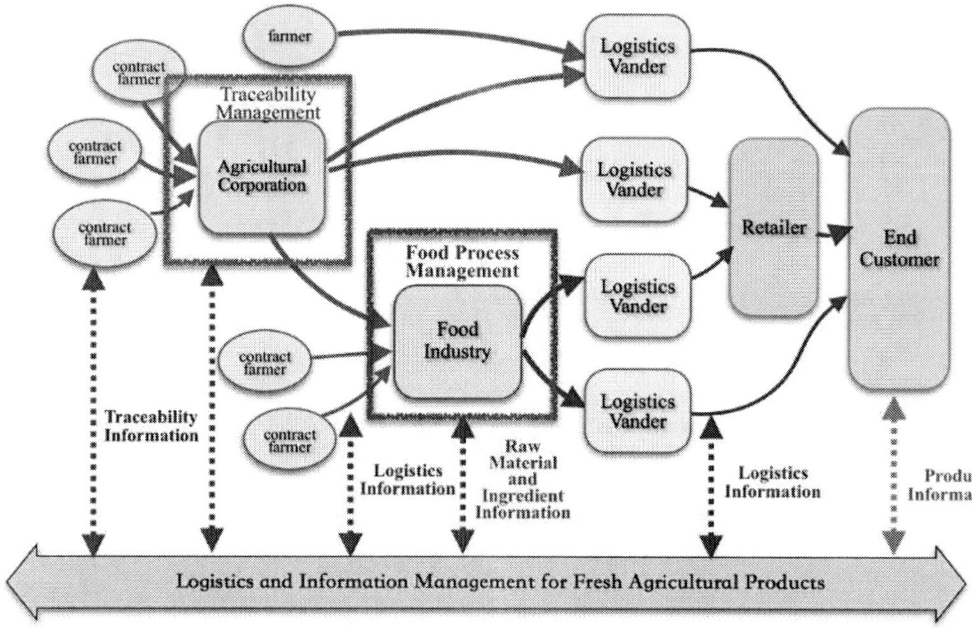

Figure 2. Infrastructure of the one-stop logistics management for the fresh agricultural products logistics management.

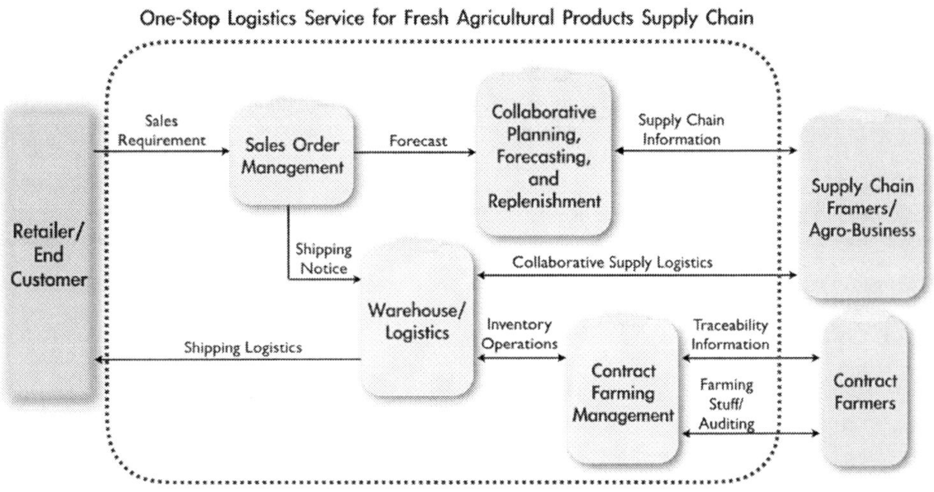

Figure 3. System integration architecture for the one-stop information management platform for traceability and logistics management of the fresh agricultural products supply chain.

3.1. Functional Analysis for the One-Stop Logistics Service

The one-stop logistics service system not only coordinate the links between the various nodes, but also help to make accurate estimation on the customers' needs through effective information analysis. The cold chain logistics companies should provide procurement, distribution, logistics and other value-added services to meet customers' demand. The requirement of the one-stop logistics service can be summarized as:

• One-stop procurement service can achieve the purpose of door-to-door procurement. It is much more complex for fresh agricultural products because of the storage conditions and variety in the combinations and substitute of product types. One-stop procurement service provides the opportunity for industries to obtain all procurement from their supplier to utilize the operation effectiveness.

• Constructing a one-stop distribution channel is to integrate different supply chain services platforms based on the satisfaction of customer needs. A win-win situation can be reached that the brand owners can reach more customers, and the customers can have more choices of product resources.

• The logistics management under the One-stop supply chain service platform cannot become possible without the collection, analysis, processing and timely update of the supply chain data. Information of the products, financial, and logistics can be collected through proper information system design. One should note that the main challenge is the information integration of the all venders in the supply chain.

• Food traceability can usually be found by attaching a 2-dimensional label, QR (quick response) code is the usual case, on the package of the food or agro-products. A mobile farming information system to collect the farming data and directly transmitted to the traceability system by cell phone and two dimensional codes was proposed in [15]. The basic traceability information is consisted of the farmer, cropland, and crop planted. All the farming activities are performed on the cropland. All the required farming operations are encoded into QR codes and every farming operation is transformed into distinct QR code label. As the usage of pesticide is critical for the pesticide residual can be harmful to human, the relationship of allowable pesticide for crop is constructed to guarantee that safety of pesticide usage. The main information entities and their relationships of the traceability information system are shown in Fig. 4. The basic traceability information is consisted of the farmer, cropland, and crop planted. All the farming activities are performed on the cropland. The farming activities can be divided into three categories: the farming operations for all kind of crops like seeding and pruning without the need of further attribute records; the fertilizing operations which the fertilizer and amount used need to be recorded; and the disease prevention operations with the use and records of pesticides. All the required operations are encoded into QR codes and every farming operation is transformed into distinct QR code label. Different fertilizers and pesticides are all encoded into distinct QR code. By scanning the proper QR code, the farmer can easily upload the operation messages into the data collection system.

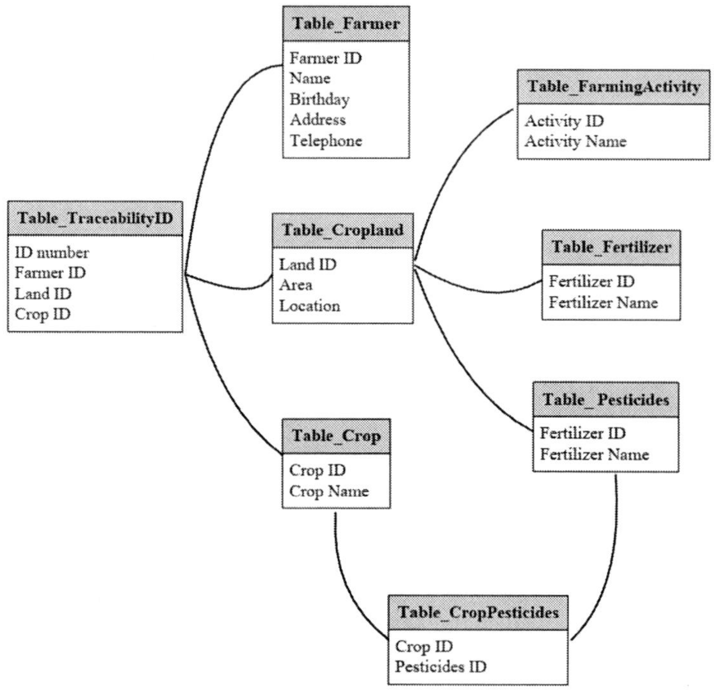

Figure 4. Main database entities for the mobile traceability data collection system, from [15]

3.2. Supporting System for The One-Stop Logistics Service

The support systems of one-stop supply chain logistics service platform for fresh agricultural products include cold chain logistics enterprise, information technology, logistics network, corporate reputation and professional team. The logistics service network provides an important foundation, such as the layout of distribution network, cold storage facilities, and information network, for one-stop supply chain services. Considering the complexity of the fresh agricultural products supply chain, high-quality and professional staff team is vital for the one-stop supply chain services. The operation team that is adaptive to the changing market and customer requirements is crucial to develop the logistics service platform. Effective project management is indispensable for the system design and implementation of the project consists of design, analysis, simulation, and verification of information systems. Industries must keep abreast on controlling logistics risk and improving the effectiveness and efficiency of logistics enterprises and customer service. The replenishment (CPFR) is the major solution to overcome the highly uncertain property of the supply and demand for fresh agricultural products. The key requirements for the system shown in Fig. 3 can be summarized as followed:

- CPFR for the fresh agricultural products supply chain with the considerations of the farming production and the market requirements for proper farming control and products dispatching is necessary to minimize the unbalance of the supply-and-demand of fresh agricultural products.
- Detail analysis for the traceability information of contract and supply farmers is necessary to assure the software systems can achieve properly the control and traceability purpose.
- The warehouse management of fresh agricultural products is quite different from the general grocery items. The storage condition, labeling, and packing need to be carefully reviewed.
- The integrated platform consists of different venders of the supply chain, the effective interface definition and communication protocol design are crucial for the success of the information systems.

3.3. Strengthen the One-Stop Logistics Information Service

Considering the specific characteristics of the fresh agricultural products, effective professional staff team, that is adaptive to the market changing, customer requirements, and the operational process to the last mile of customer, is vital to support the one-stop supply chain service. An efficient team assessment management to ensure the successful implementation of one-stop service system is necessary. Venders in the fresh agricultural products supply chain are usually located far away from food industries, retailer, and customers. Transportation of fresh agricultural products can be summarized in two manners. The first type is transported directly by the food industries (or agricultural corporations). The other is to be delivered by the third-party services such as DHL. The quality of transportation and fresh agricultural products must be controlled and guaranteed during transportation. The cold chain logistics is hence become inevitable. As to provide and integrate the logistics information, the cold chain logistics vehicle need to be upgraded with information accession and transmission. Some suggestions to strengthen the one-stop service information platform are listed:

- It is a great challenge to share information and reduce fluctuations among the different corporations in the fresh agriculture product supply chain. The fresh cold chain related enterprises should re-coordinate to achieve risk-sharing and benefit-sharing. Detailed contract should be compromised to encourage the supply chain members to work with the occurrence of unexpected events.
- Solution enhancement of the one-stop logistics service can be divided into two aspects: the hardware and software solutions. Hardware solutions should be designed to cope with that integrate various types of equipment involved in material flow including transportation, handling, and storage.
- Software solution is to provide an efficient one-stop logistics management platform system for the integration of complex logistics information including the products receiving and recording, warehouse entrance check and record, transportation management, and storage management of automatic storage location assignment, load/unload to storage location, lot split, and inventory check.

4. Summary and Conclusion

The fresh agricultural products supply chain is reviewed and the one-stop logistics management requirements are studied based on the project experience in Tianjin. Because of the importance food safety and the complex business characteristics through the food supply chain, the MDBT approach to provide a service-oriented-architecture solution is proposed in this paper. Three major considerations for the one-stop logistics management platform, including the functional analysis, supporting system, and strengthen for the one-stop logistics service, are suggested. With the proposed functional requirements, the results of this paper can be fruitful for the further design of the fresh agricultural products logistics information management systems.

5. Acknowledgment

This work was financially supported by the NPIE 2015 (National Training Program of Innovation and Entrepreneurship) for Undergraduates of the Ministry of Education of PRC (project no. 201510061019) and Tianjin Administration of Foreign Affairs (project no. 20171200248).

6. References

[1] G. Schiefer, "New technologies and their impact on the agri-food sector: an economists view," Comput. and Electron. in Agr., Vol. 43, no. 2, pp. 163-172, 2004.
[2] T. Moe, "Perspectives on traceability in food manufacture," Trends in Food Science & Technology, Vol. 9, pp. 211-214, 1998.
[3] M. Varga and B. Csukas, "On the way toward the sector spanning agrifood process traceability," Agricultural Informatics, vol. 1, no. 1, pp. 8-18, 2010.
[4] L. U. Opara, "Traceability in agriculture and food supply chain: a review of basic concepts, technological implications, and future prospects," J. of Food Agric. Environ., vol. 1, no. 1, pp.101-106, 2003.
[5] X. Wang, D. Li, and L. Li, "Adding value of food traceability to the business: a supply chain management approach," Int. J. of Serv. Oper. and Informat., vol. 4, no. 3, pp. 232-257, 2009.
[6] T. Bosona and G. Gebresenbet, "Food traceability as an integral part of logistics management in food and agricultural supply chain," Food Control, vol. 33, pp. 32-48, 2013.

IOP Conf. Series: Journal of Physics: Conf. Series **887** (2017) 012004 doi:10.1088/1742-6596/887/1/012004

[7] C. G. Sørensena, S. Fountasb, E. Nashf, L. Pesonend, D. Bochtisa, S. M. Pedersene, B. Bassoc, S. B. Blackmoreg, "Conceptual model of a future farm management information system," Comput. and Electron. in Agr., vol. 72, no. 1, pp. 37-47, 2010.
[8] Q. A. Arain, I. Memon, Z. Deng, M. H. Memon, F. A. Mangi, and A. Zubedi, "Location monitoring approach: multiple mix-zones with location privacy protection based on traffic flow over road networks," *Multimedia Tools and Applications*, pp. 1-45, 2017.
[9] T. Kelepouris, K. Pramatari, and G. Doukidis, "RFID-enabled traceability in the food supply chain," Ind. Mana. Data Syst., vol., 107, no. 2, pp. 183-200, 2007.
[10] A. Regattieri, M. Gamberi, and R. Manzini, "Traceability of food products: General framework and experimental evidence," J. of Food Eng., vol. 81, no. 2, pp. 347-356, 2007.
[11] V. Manthou, A. Matopoulos, and M., "Vlachopoulou, Internet-based applications in the agri-food supply chain: a survey on the Greek canning sector," J. of Food Eng., vol. 70, no. 3, pp. 447-454, 2005.
[12] M. Li, J. P. Qian, X. T. Yang, C. H. Sun, and Z. T. Ji, "A PDA-based record-keeping and decision-support system for traceability in cucumber production," Comput. and Electron. in Agr., vol. 70, no. 1, pp.69-77, 2010.
[13] N. P. Mahalik and A. N. Nambiar, "Trends in food packaging and manufacturing systems and technology," Trends Food Sci. Tech., vol. 21, no. 3, pp. 117-128, 2010.
[14] S. Kumaran, P. Bishop, T. Chao, P. Dhoolia, P. Jain, R. Jaluka, H. Ludwig, A. Moyer, and A. Nigam, "Using a Model-Driven Transformational Approach and Service-Oriented Architecture for Service Delivery Management", IBM System Journal, vol. 46, no. 3, pp. 513-529, July 2007.
[15] Y. C. Liu and H. M. Gao, "Development and Applications of Mobile Farming Information System for Food Traceability in Health Management," in Applied Computing in Medicine and Health, edited by D. Al-Jumeily, A. Hussain, C. Mallucci, and C. Oliver, Elsevier, Aug. 2015, pp. 244-268.

IOP Conf. Series: Journal of Physics: Conf. Series **887** (2017) 012005 doi:10.1088/1742-6596/887/1/012005

An aggregation-decomposition bayesian stochastic optimization model for cascade hydropower reservoirs using medium-range precipitation forecasts

Yong Peng[1], Wei Xu[2] and Xiaoli Zhang[3]

[1]School of Hydraulic Engineering, Dalian University of Technology, Dalian, China, pengyong@dlut.edu.cn

[2]College of River and Ocean Engineering, Chongqing Jiaotong University, Chongqing, China xuwei19850711@163.com

[3]School of Water Conservancy, North China University of Water Resources and Electric, Zhengzhou, China, zxl436@163.com

Abstract. The forecast information is essential to improve the utilization efficiency of hydropower resources. To address the uncertainties of forecasting inflow, the Aggregation-Decomposition Bayesian Stochastic Dynamic Programming (AD-BSDP) model is presented in the present paper by using the 10-days precipitation value of the Quantitative Precipitation Forecasts from Global Forecast System (GFS-QPFs). The application in China's Hun River cascade hydropower reservoirs shows that the GFS-QPFs are beneficial for hydropower generation and the performance of AD-BSDP is more efficiency and reliability than the others models.

1. Introduction

With the development of the social and economic, now the demand for electricity is increasing. The hydroelectric power, clean energy, has become more and more popular. In the hydropower generation operation, the Traditional Rule Curve uses water level as decision variable, the amount of information used in the decision diagram is limited. While considering rainfall and runoff forecasting information, in reservoir operation, is therefore an effective measure to improve the utilization of hydropower resources.

However, Quantitative Precipitation Forecasts (QPF) and runoff forecasts are not the accurate information. The accuracy of rainfall-runoff forecast model is affected by numbers of factors. Thus in the hydropower optimal operation, the uncertainty of forecasting inflow needs to be represented in the optimization model. Karamouz [1, 2] suggested the use of Bayesian Decision Theory incorporating new information in the interpretation of the flow probabilities. Karamouz and Vasiliadis [3] proposed a Bayesian Stochastic Dynamic Programming (BSDP) model, which embeds Bayesian theory in the SDP formulation. Kim and Palmer [4] constructed a BSDP model to investigate the value of seasonal flow forecasts in hydropower generation. Mujumdar and Nirmala [5] constructed a BSDP in cascade hydropower reservoirs by aggregated individual inflows to reduce the complexity of the uncertainties addressing.

In SDP model, especially in multi-reservoir cases, the complexity of the model and high computational costs due to curse of dimensionality are the important limiting factors for the application. The approach to overcome this problem is aggregate the storage and inflow of each reservoir to yield an equivalent reservoir, according to the hydrological or geographical features. While the way of aggregation inflow and storage is an important factor to affect the efficiency of the equivalent reservoir. Tejada-Guibert et al [6] and Mujumdar and Nirmala [5] taken aggregate inflow as a state variable to reduce the complexity of the model and solve the computationally tractable, considering the storage of individual reservoirs separately. This approach spatially accounts the storage of each reservoir in the system, and the operation policies are derived for each individual reservoir. Comparing the aggregate inflow, the aggregate method of storage is more difficult for the bias of operation policies in each reservoir. Therefore the method, convert the storage and inflow of each reservoir into potential electrical energy, is presented to help multi-reservoir to derive operation policies [7-8]. While this approach lack of consideration the individual characteristics of the reservoirs and the amount of water to release from the aggregate reservoir. In order to solve this problem, another method is presented by using historical date or simulation date to build relationship among the individual reservoirs. The main purpose of this paper is to investigate the medium-range QPFs application in cascade hydropower reservoirs operation. There are two-fold to attain the goal: ① the aggregation-decomposition method is implemented to overcome the complexity of the SDP model and high computational costs by aggregation cascade reservoirs as an equivalent reservoir. ② the Bayesian Decision Theory is used to address the uncertainties of flow and forecast inflow. Based on the equivalent reservoir, the Aggregation-Decomposition Bayesian Stochastic Dynamic Programming (AD-BSDP) model is constructed. In order to verify the efficiency and reliability of the equivalent reservoir and Bayesian Decision Theory, the Aggregating Flow Stochastic Dynamic Programming (AF-SDP) and Aggregation-Decomposition Stochastic Dynamic Programming (AD-SDP) are constructed. This study takes China's Hun River cascaded hydropower reservoirs as a case example. The real time QPFs (10 days lead time) of the Global Forecast System (QPFs-GFS) are utilized to forecast the 10 days inflow. Based on the performance of Traditional Rule Curve, the performances of the AF-SDP, AD-SDP and AD-BSDP by using medium-range QPFs information are evaluated and compared.

2. Case Study

2.1. Hun River Cascade Hydropower Reservoirs System

In this study, Hun River cascade hydropower reservoirs are taken as an application example. It is located in the lower reaches of the Hun river basin in the northeast of China. Fig. 1 shows the location of the cascade hydropower reservoirs and gauging stations. The basin covers an area of approximately 15000 km^2. The main features of this reservoirs system are shown in Fig.1 and Table 1.

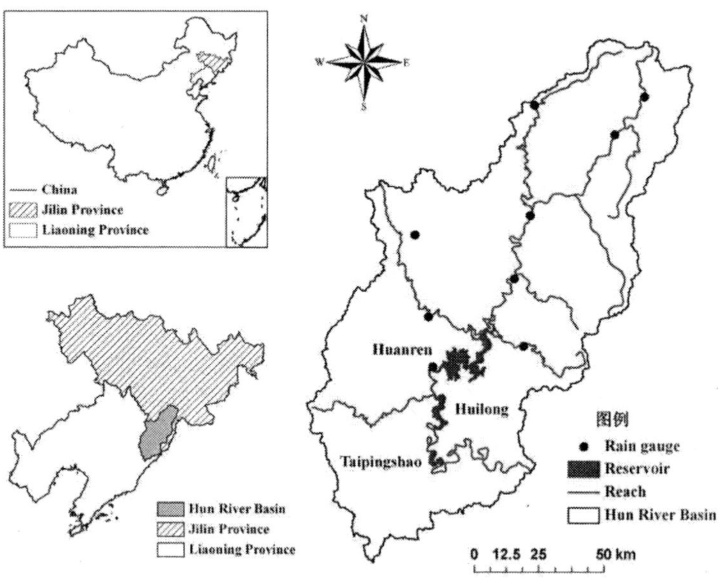

Figure 1. Location and main features of Hun River cascade reservoirs system

2.2. Rainfall-Runoff Models

According to the hydrology and meteorology of the Hun River basin, the inflow is roughly divided into dry season and wet season. During dry season (from October to next April), the inflow, low and stable, is mainly supplied by underground runoff. During wet season (from May to September), the inflow, high and instable, is mainly affected by the rainfall. Therefore, the multiple linear regression model and Xinanjiang model [7] are applied to establish the inflow forecasting model during dry season and wet season respectively.

Table 1. Basic paramters of Hun River cascade hydropower reservoirs

Characteristic	Huanren	Huilong	Taipingshao
Basin area (km^2)	10400	2100	528
Total storage (Mm^3)	3460	123	182
Usable storage (Mm^3)	2199	90	164
Dead storage (Mm^3)	1380	72	145
Normal pool level (m)	300	221	191.5
Dead pool level (m)	290	219	190
Installed capacity (MW)	222	72	161
Firm output (MW)	33	18	25
Turbine capacity (m^3/s)	416	330	534

2.3. Datas

The Global Forecast System (GFS) developed by the U.S. National Centers for Environmental Prediction, which is a global Numerical Weather Prediction (NWP) computer model. The data of GFS are available on NOAA FTP SERVERS for free, and such information forms the basis for non-state weather companies, e.g., Wunderground.com, Weatheronline.co.uk, and t7online.com. The 10 days QPFs-GFS data from 2001 to 2010 are obtained from t7online.com.

The observed precipitation and inflow originates from the Hun River cascade hydropower development authority. The observed precipitation data from 1968 to 2010 are obtained in the upstream of Huanren. The observed inflow data of each reservoir are also obtained from 1968 to 2010. The inflow of intermediate catchment from Huanren to Huilong (IC-HH) and Huilong to Taipingshao (IC-HT) are obtained by back stepping.

3. Reservoir Operation Optimization Model Object Function

3.1. Object Function

The performance measure $B(\cdot)$ of cascade hydropower reservoirs should be a penalty function for a given combination of available state variables. If the calculated power generation $b(\cdot)$ (MW) is less than the system firm output of e (i.e., 76 MW), the value, $B(\cdot)$ (MWH), will be penalized. The objective function of the Stochastic Dynamic Programming can be written as Eq. 2, which is the maximum expected generation of hydropower system return from the start time step to the end of the planning horizon [8].

$$B(k_t,q_t,l_t)=[b(k_t,q_t,l_t)-\alpha\cdot\{Max(e-b(k_t,q_t,l_t),0)\}^{\beta}]\cdot\Delta t \quad (1)$$

$$Max\left[\sum_{t=1}^{T}E[B(k_t,q_t,l_t)]\right] \quad (2)$$

Where, k_t and l_t are the storage combination of individual reservoirs at the beginning and the end of the time step t, respectively; q_t is the inflow combination; α and β are the penalty factors; Δt is the time in hours for decision interval t.

3.2. AF-SDP Model

In multi-reservoir system, application of SDP models bears high complexity and computational costs. Mujumdar and Nirmala [5] takes the aggregate inflow and storage of individual reservoirs as state variables to reduce the complexity of the application. It takes the aggregate inflow as state variable instead of each corresponding to the inflow to an individual reservoir. Based on this method, the Aggregate Flow Stochastic Dynamic Programming (AF-SDP) model is constructed in this study.

In the AF-SDP model, the inflows are aggregated by Eq.3 and the aggregate inflow is discretized into a number of class intervals. Each class interval is represented by a representative value and the vector of the aggregate inflow class interval $\hat{Q}_t$ is shown by Eq.4. The storages of individual reservoirs are discretized into a number of class intervals by Eq.5. The matrix, combination of the storage class intervals among the reservoirs, is shown in Eq.6.

$$Q_t = \sum_{z=1}^{Z} q_t^z \tag{3}$$

$$\hat{Q}_t = [Q_t^1, Q_t^2, \cdots\cdots, Q_t^{\mu}] \tag{4}$$

$$\begin{cases} B_{t-1}^1 = A_t^1 = \{v_1^1, v_2^1, \cdots\cdots, v_{\beta^1}^1\} \\ B_{t-1}^2 = A_t^2 = \{v_1^2, v_2^2, \cdots\cdots, v_{\beta^2}^2\} \\ \vdots \\ B_{t-1}^Z = A_t^Z = \{v_1^Z, v_2^Z, \cdots\cdots, v_{\beta^Z}^Z\} \end{cases} \tag{5}$$

$$\begin{bmatrix} l_{t-1}^1 \\ l_{t-1}^2 \\ \vdots \\ l_{t-1}^n \end{bmatrix} = \begin{bmatrix} k_t^1 \\ k_t^2 \\ \vdots \\ k_t^n \end{bmatrix} = \begin{bmatrix} v_1^1 & v_1^2 & \cdots & v_1^z \\ v_1^1 & v_1^2 & \cdots & v_2^z \\ \vdots & \vdots & \vdots & \vdots \\ v_{\beta^1}^1 & v_{\beta^2}^2 & \cdots & v_{\beta^z}^z \end{bmatrix} \tag{6}$$

Where superscripts 1, 2 and z refer to reservoir 1, reservoir 2 and reservoir z. q_t^z is the inflow of reservoir z at time step t. Q_t is the aggregate inflow of reservoirs system at time step t. $\hat{Q}_t$ is the vector of representative value of aggregate inflow discretization class intervals. μ is the number of class intervals. B_{t-1}^z and A_t^z are the vector of representative value of the storage class intervals for reservoir z at the end of time step t-1 and the beginning of time step t separately. β^z is the number of storage class intervals for reservoir z. l_{t-1}^n and k_t^n represent the nth scenario of storage combination of the individual reservoir at the end time step t-1 and the beginning time step t separately.

AF-SDP model recursive equation is defined in Eq.8.

$$\begin{cases} m = \mu \\ n = \prod_{z=1}^{Z} \beta^z \end{cases} \tag{7}$$

$$\begin{aligned} f_{opt}\left(k_t^i, F_t^j\right) = \underset{l_t}{Max}\Big\{ & B_t\left(k_t^i, F_t^j, l_t^w\right) \\ & + \sum_{f}^{m} P_{t+1}^f [Q_{t+1}^f / F_t^j] \cdot f_{opt}\left(l_t^w, Q_{t+1}^f\right) \Big\} \end{aligned} \tag{8}$$

Where, the m and n represent the computational costs of this model. The m is determined by the number of aggregate inflow discretization class intervals, and the n is determined by number of reservoir in system and the number of individual reservoir storage discretization class intervals. The

$P_{t+1}^{f}(Q_{t+1}^{f} / F_{t}^{j})$ suggests that the probability of the inflow (Q_{t+1}) at time step t+1 belongs to class interval f, when the forecasting inflow (F_t) at time step t belongs to class interval j.

3.3. AD-SDP model

Based on the AF-SDP model, the Aggregation-Disaggregation Stochastic Dynamic Programming (denoted by AD-SDP) aggregates the storage and inflow of the cascade reservoirs to yield an equivalent reservoir. The inflow of the equivalent reservoir is presented by aggregate inflow, which is mentioned above. The minimum and maximum reservoir storages of the equivalent reservoir are represented by Eq.9. The vectors of equivalent reservoir storage class interval are shown in Eq.10.

$$\begin{cases} V_{\max} = \sum_{z=1}^{Z} v_{\max}^{z} \\ V_{\min} = \sum_{z=1}^{Z} v_{\min}^{z} \end{cases} \tag{9}$$

$$\begin{bmatrix} L_{t-1}^{1} \\ L_{t-1}^{2} \\ \vdots \\ L_{t-1}^{\phi} \end{bmatrix} = \begin{bmatrix} K_{t}^{1} \\ K_{t}^{2} \\ \vdots \\ K_{t}^{\phi} \end{bmatrix} = \begin{bmatrix} V_{t}^{1} \\ V_{t}^{2} \\ \vdots \\ V_{t}^{\phi} \end{bmatrix} \quad V_{\min} \le V_{t}^{1}, V_{t}^{2}, \cdots V_{t}^{\phi} \le V_{\max} \tag{10}$$

$$\begin{cases} m = \mu \\ n = \phi \end{cases} \tag{11}$$

Where the $v_{\min}^{z}$ (Dead storage) and $v_{\max}^{z}$ (Usable storage) represent the minimum storage and maximum storage of reservoir z. The superscripts ϕ is the number of the storage class intervals. L_{t-1}^{ϕ} and K_{t}^{ϕ} represent the representative value of the storage in class interval ϕ for equivalent reservoir at the end of time step t-1 and the beginning of time step t separately.

The AD-SDP model is constructed by taking the aggregate storage and aggregate inflow as state variables. The K_{t}^{i} and L_{t}^{j} are the aggregate storages at the beginning and end of time step t. In order to measure the system performance in model iteration, the aggregate variables are disaggregated by the method of conditional expectation, which is introduced as below. The recursive equation of AD-SDP is defined in Eq.12.

$$\begin{aligned} f_{opt}\left(K_{t}^{i}, F_{t}^{j}\right) = \underset{L_t}{Max}\Big\{ & B_{t}\left(K_{t}^{i}, F_{t}^{j}, L_{t}^{w}\right) \\ & + \sum_{f}^{m} P_{t+1}^{f}[Q_{t+1}^{f} / F_{t}^{j}] \cdot f_{opt}\left(L_{t}^{w}, Q_{t+1}^{f}\right) \Big\} \end{aligned} \tag{12}$$

3.4. AD-BSDP model

In AF-SDP and AD-SDP model, these assume that the aggregate inflow constitutes a simple Markov Process and the uncertainty of forecasting inflow is not considered. In AD-BSDP model, the Bayesian Decision Theory (BDT) was used to address the uncertainty of forecasting inflow and the random in nature through both the posterior flow transition probability ($P(Q_t \mid F_t, Q_{t-1})$) and the predictive probability of forecasts ($P(Q_{t+1} \mid Q_t)$). The recursive equation of AD-BSDP model is defined in Eq.13.

$$\begin{aligned} f_{opt}\left(K_{t}^{i}, Q_{t-1}^{j}, F_{t}^{d}\right) = \underset{L_t}{Max}\Big\{ & \sum_{c}^{m} P\left[Q_{t}^{c} / F_{t}^{d}, Q_{t-1}^{j}\right] \cdot \\ & \left(B_{t}\left(K_{t}^{i}, Q_{t}^{c}, L_{t}^{w}\right) + \sum_{f}^{m} P_{t+1}^{f}\left[Q_{t+1}^{f} / Q_{t}^{c}\right] \cdot f_{opt}\left(L_{t}^{w}, Q_{t}^{c}, Q_{t+1}^{f}\right) \right) \Big\} \end{aligned} \tag{13}$$

4. Results

4.1. Quantitative Inflow Forecasting

The inflow forecasting models is applied to estimate the 10 days average inflow of Huanren. The observed rainfall and runoff data from 1968 to 2000 are used to calibrate the parameters of hydrological models. And the observed data from 2001 to 2010 are used to verify. Based on the work of calibration and verification, the forecasting data (10 days QPFs of GFS collected from 2001 to 2010) are applied to forecast the 10 days inflow of Huanren reservoir. The inflow of IC-HH and IC-HT are evaluated by the forecasting inflow of Huanren with the ratio of 10% and 5% respectively.

The Nash-Sutcliffe efficiency (NSE) is used to measure the performance of the hydrologic models during calibration, verification and forecasting. Results show that during dry season (from October to next April), the NSE value of calibration, verification and forecasting are only about 0.7. The main reason is hard to forecast the inflow during snowmelt for lack of meteorological data. Such as the data of temperature, the main factor affects snowmelt. During wet season (from May to September), the hydrological models perform well and the NSE value of calibration and verification exceed 0.9. The NSE value of forecasting reduces from 0.92 to 0.74 by convert the 10 days rainfall input data from observed to QPFs of GFS. Errors of the QPFs are the main source which influences accuracy of forecasting.

4.2. Simulation Results

The operating policies for Hun River cascade hydropower reservoirs are derived with the objects of maximize the total power supply and guarantee system stable running. Annual Total Hydropower Generation (ATHG), reliability and Annual Total Deficit Hydropower Generation (ATDHG), the important indicators for the managers in this study case, are chosen to study the performance of the system. Reliability of the system under a given policy is defined as the probability that the system output is satisfactory. The ATDHG is defined as the annual total deficit to the total firm power commitment during operation.

In order to comparing the performance of the operation policies (i.e., AF-SDP, AD-SDP, AD-BSDP), the existing Traditional Rule Curve (TRC), only taken the control water level as state variable, is selected as a model for comparison. The system ratio is the ratio by comparing the system indicator of operation policies with TRC. The operation policies are used to simulate the system operation by observed and forecasting inflow from 2001 to 2010 respectively. The ATHG of individual reservoirs and the system are listed in Table 2. The reliability and ATDHG of individual reservoirs and the system are listed in Table 3.

Results show that the performances of operation policies, deriving by AF-SDP, AD-SDP and AD-BSDP, are super than that of TRC. The state variable of TRC only is water level at current time step. The inflow information in future is not considered in strategy decision. So the decision-strategy is prone to spill, especially in wet season. The performance indicators of system ratio are demonstrated that the forecasting inflow information improved the utilization of hydropower resources in the study case.

The performance of AD-SDP is better than that of AF-SDP. The reasons are the AD-SDP model aggregates the reservoirs system as an equivalent reservoir and addresses the uncertainty of flow by Markov chain and forecasting inflow is treated as accurate information. In real-time operation, the obtained aggregate output needs to disaggregate among the reservoirs with the objectives of minimizing the spillages and maximizing the storage of the system. Thus, the disaggregate policies can control the release of Huanren to avoid or reduce spillages in Huilong and Taipingshao. Hence it produces higher ATHG and Reliability at Huilong and Taipingshao than AF-SDP.

The performance of the AD-BSDP is obviously better than that of AD-SDP. Because Quantitative Precipitation Forecasts (QPF) and runoff forecasts are not the accurate information. In the AD-SDP model, uncertainty of the flow was addressed by Markov chain and the forecasting inflow was treated as accurate information. While the AD-BSDP model used the BDT to address the uncertainties based on the equivalent reservoir.

Table 2. Simulation results of the Athg for observed and forecasting inflow from 2001 to 2010.

Scenarios	Models	Inflow information	ATHG (MWH) Huanren	Huilong	Taipingshao	System	System ratio (%)
1	TRC	—	372.58	227.13	386.68	986.39	—
2	AF-SDP	observed	428.93	231.78	382.85	1043.57	5.80
3	AD-SDP	observed	423.03	247.19	408.19	1078.42	9.33
4	AD-BSDP	observed	429.12	252.04	418.60	1099.76	11.49
5	AF-SDP	forecasted	411.78	229.13	372.74	1013.65	2.76
6	AD-SDP	forecasted	404.07	244.33	401.69	1050.09	6.46
7	AD-BSDP	forecasted	419.58	248.87	417.79	1086.24	10.12

Table 3. Simulation Results of the Reliability and Atdhg for Observed and Forecasting Inflow from 2001 To 2010

Scenarios	Reliability (%) Huanren	Huilong	Taipingshao	System	System ratio (%)	ATDHG (KWH) Huanren	Huilong	Taipingshao	System	System ratio (%)
1	83.05	81.38	76.66	81.11	—	192	103	191	477	—
2	84.44	83.83	86.39	88.88	9.58	136	54	84	252	-47.17
3	85.00	87.44	89.17	93.61	15.41	128	37	69	164	-65.62
4	85.56	87.17	88.61	92.77	14.38	120	41	75	184	-61.43
5	81.94	82.17	81.94	86.66	6.84	163	70	112	323	-32.29
6	79.17	86.61	84.21	92.50	14.04	175	43	75	191	-59.96
7	83.33	87.17	87.78	92.77	14.38	136	41	76	185	-61.22

5. Summary and Conclusions

This study has focused on the deriving operation policies for cascade hydropower reservoirs by using medium-range precipitation forecasts. In view of the curse of dimensionality, the aggregation-decomposition method is used to overcome this problem by aggregation cascade reservoirs as an equivalent reservoir. Firstly the AF-SDP has been constructed by aggregating inflow, which is used to verify the efficiency and reliability of AD-SDP based on equivalent reservoir. In SDP model, the uncertainty of forecasting inflow is not considered. Then the AD-BSDP constructed by embedding the Bayesian Decision Theory in SDP to address the uncertainty of forecast inflow. Finally, the Hun River cascade hydropower reservoirs are taken as an example to evaluate the performance of the models. The analytical results indicate that the aggregation-disaggregation method is useful to overcome the curse of dimensionality in cascade hydropower reservoirs, and the approach, Bayesian Decision Theory, is critical for medium-range QPFs application in reservoir operation.

6. Acknowledgment

This work is supported by the National Natural Science Foundation of China (Grant No. 91547111, 51379027, 51609025) and Scientific Research Foundation for High-level Talents of North China University of Water Resources and Electric Power (Grant No. 201702013).

7. References

[1] Karamouz M, "Forecast uncertainty in reservoir operation," In: Proceedings of the 15th Annual Water Resources Conference, Critical Water Issues and Computer Applications, 1988, ASCE, New York, pp: 265-268.

[2] Karamouz M, "Bayesian Decision Theory and Fuzzy Sets Theory in Systems Operation," Proceedings of the 17th Annual Water Resources Conference, Optimizing the Resources for Water Management, 1990, ASCE, New York.

[3] Karamouz M and Vasiliadis HV, "Bayesian stochastic optimization of reservoir operation using uncertain forecasts," Water Resources Research, vol. 28, May 1992, pp: 1221-1232.
[4] Kim YO and Palmer RN, "Value of Seasonal Flow Forecasts in Bayesian Stochastic Programming," Journal of Water Resources Planning and Management, vol. 123, May 1997,pp: 327-335.
[5] Mujumdar, P.P. and Nirmala, B., 2007. "A Bayesian Stochastic Optimization Model for a Multi-Reservoir Hydropower System," Water Resources Management, vol. 21, Dec. 2007, pp: 1465-1485, doi: 10.1007/s11269-006-9094-3 .
[6] Tejada-Guibert J, Johnson S, Stedinger J. "The value of hydrologic information in stochastic dynamic programming models of a multireservoir system," Water Resour Res, Vol. 31, Oct. 1995, pp: 2571-2579.
[7] Zhao, R. "The Xinanjiang model applied in China," Journal of Hydrology, Vol, 135, Sep. 1992, pp: 371-381.
[8] Tang, G. et al.. "Value of Medium-range Precipitation Forecasts in Inflow Prediction and Hydropower Optimization," Water Resources Management, Vol. 24, Jan. 2010, pp: 2721-2742.

IOP Conf. Series: Journal of Physics: Conf. Series **887** (2017) 012006 doi:10.1088/1742-6596/887/1/012006

Development of remote data acquisition system based on OPC for brake test bench

Yiwei Wang[1], Mengling Wu[2], Chun Tian[3] and Tianhe Ma[4]

[1]Institute of Railway and Urban Mass Transit, Tongji University, Shanghai, China, 102237@tongji.edu.cn

[2]Institute of Railway and Urban Mass Transit, Tongji University, Shanghai, China, wuml_sh@163.com

[3]Institute of Railway and Urban Mass Transit, Tongji University, Shanghai, China, chtian77@163.com

[4]Institute of Railway and Urban Mass Transit, Tongji University, Shanghai, China, 15900756470@163.com

Abstract. The 1:1 train brake system test bench can be used to carry out brake-related adhesion-slid control, stability test, noise test and dynamic test. To collect data of the test bench, a data acquisition method is needed. In this paper, the remote data acquisition system of test bench is built by LabVIEW based on OPC technology. Unlike the traditional hardwire way connecting PLC acquisition module with sensors, the novel method is used to collect data and share them through the internal LAN built by Ethernet switches, which avoids the complex wiring interference in an easy, efficient and flexible way. The system has been successfully applied to the data acquisition activities of the comprehensive brake system test bench of CRRC Nanjing Puzhen Haitai Brake Equipment Co., Ltd., and the relationship test between the adhesion coefficient and the slip-ratio is realized. The speed signal, torque signal and brake disc temperature can be collected and displayed. The results show that the system is reliable, convenient, and efficient, and can meet the requirements of data acquisition.

1. Introduction

At present, the methods of studying the characteristics of train wheel and rail adhesion mainly consists of two kinds, namely, theoretical research and experimental research. However, the theoretical study is constrained by the real wheel rail contact conditions. In order to make up for the lack of theoretical research, CRRC Nanjing Puzhen Haitai Brake Equipment Co., Ltd. set up a comprehensive test platform for the braking system. The railway vehicle was used as the experimental object to simulate the contact condition of the rail wheel and vehicle wheel. In the course of the experiment, many physical parameters of the test bench need to be collected, processed and displayed for operator reference. Therefore an effective data acquisition system is necessary for the experiment.

With the development of computer communication and control technology, data acquisition system in the industrial field has attained mature development. OPC (Object Linking and Embedding (OLE) for Process Control) provides a software interface standard for data source and data user communications, which establishes a communication bridge for process control data sources and Windows-based applications.

This paper is based on the LabVIEW platform for software design and the OPC protocol for data transmission. The software with good human-computer interaction interface was developed to achieve real-time communications. The physical parameters were shared by Ethernet, reducing the hard wire transmission wiring complexity and low stability, to realize a multi-point, real-time, efficient remote data collection. As the OPC structure is the client / server model, each OPC client connects to the OPC server management device through the OPC standard interface, so the system needed to establish the OPC interface, connections between the OPC server and OPC client applications, as shown in Figure 1.

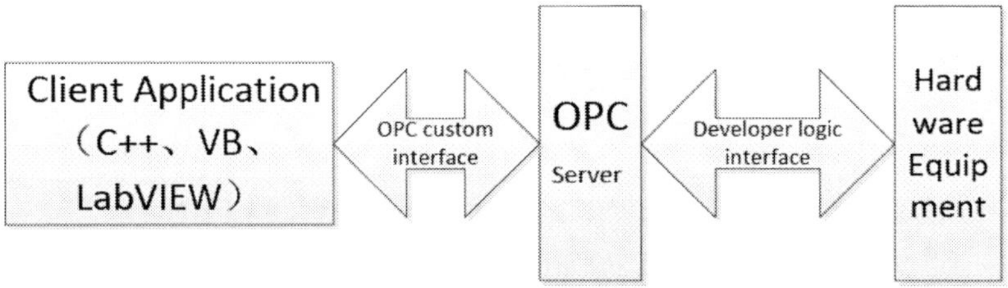

Figure 1. OPC clients and OPC servers

2. System Architecture and Design

2.1. System Architecture

The data acquisition system of brake test bench is composed of sensors, data acquisition card, serial interfaces, other hardwares and host computer configuration software. Among them, the hardware part of the data collection consists of computers, high-speed acquisition modules and signal isolation modules, which make the full use of high-speed, high-capacity storage and powerful computing analysis functions of computers and increase efficiency. During the experiment, the rail wheel speed signal, the torque signal, the brake disc temperature, the brake cylinder pressure and other signals are collected. The collected data are uploaded to the detection computer at the same time for data analysis, calculation and display. The control computer can save the test data automatically according to the requirement.

2.2. Scheme Design

The data flow of the collection way is displayed in Figure 2. In the figure, the force sensors, displacement sensors, signal conditioning modules and the acquisition cards transfer the acceleration, pressure and other physical parameters through the EPM shielded cable in the form of 4-20mA current signal; The camera and the temperature sensors collect the related data and transfer them to the high-speed data acquisition card through the RS485 serial interfaces; PCI high-speed data acquisition cards communicate with control computer by switch in the Ethernet.

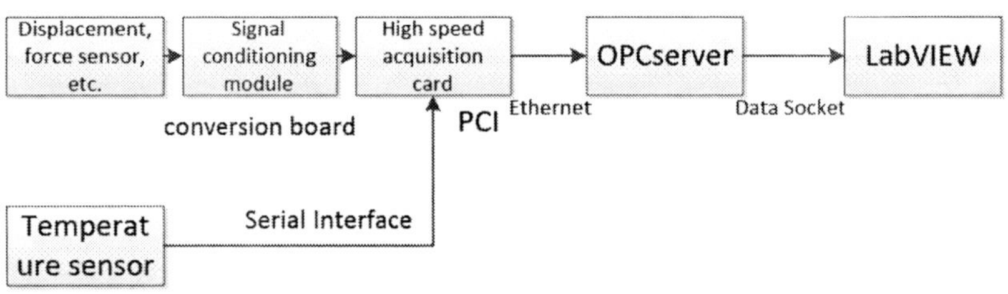

Figure 2. Flow chart of data acquisition

In order to solve the data transmission between the host computer and the lower computer, this paper uses NI Data Socket function to realize the data communication between LabVIEW and OPC Server. Data Socket is a network communication technology for the field of measurement and control. Using it can provide a platform-independent solution transferring data over the network, solve the problem of data acquisition, processing and display of remote instruments and equipment at the speed of 640kbps in 10M networks, which can achieve a very good transmission effect and avoid the failure of hardware facility.

3. System Hardware Components

3.1. Sensors

The data acquisition system needs to detect the physical variables and convert them into usable signals, thus requires a large number of sensors to meet the requirements of information transmission, processing, storage, display and control. The usage and selection of sensors are the primary part of the data acquisition system.

The system needs to collect signals such as speed, torque, pressure, temperature, etc., the key physical variables and the sensors used are shown in Table 1. The rotational torque measuring position is located between the motor end and the rail wheel, between the rail wheel and the gear box and between the brake discs and the flywheels group. Temperature detection uses infrared temperature detectors, which are non-contact sensors.

Table1. Key physical variables and sensors

Key parameters	Sensors	Range
Wheel speed [km/h]	Speed sensor	3000r/min
Rail Wheel speed [km/h]	Speed sensor	3000r/min
Torque meter torque [kN*m]	Torque meter	20 kN•m
Brake Cylinder Pressure[kPa]	Pressure sensor	1MPa
Brake disc temperature [°C]	Infrared temperature sensor	-40-1030°C

3.2. High-Speed Data Acquisition Module

The lower computer uses PCI6225 data acquisition card using the PCI bus (Peripheral Component Interconnect). It can be directly inserted in any compatible computer within any PCI slot. The NI PCI-6225 data acquisition card is a low-cost multi-function M-series data acquisition card with 16-bit, 80-channel analog and 24 Road digital, which is optimized for applications requiring cost control. In order to use the NI PCI-6255 for optimum performance, the system uses an EPM shielded cable 50CC01 for connector connection, as recommended by NI. The lower computer collects the real-time data from the PCI high-speed acquisition card through the DAQ assistant in LabVIEW, setting the sampling number, channel and other parameters, and then connects with the host computer through Ethernet.

3.3. Serial Interface

Serial interface transmits the data one by one. It is characterized by a simple communication line, which achieves two-way communication.

The system uses RS-485 bus interface, it is a balanced transmitting and differential receiving serial bus. The transmission speed, distance, anti-common mode interference capacity are suitable for the test bench. The support for serial communication of LabVIEW is realized by VISA (virtual instrument software architecture), the functions include: VISA configuration serial interface, VISA write, VISA read and VISA off. The key is to read the buffer and convert the data format, in order to access the data in LabVIEW, which is shown in Figure 3.

IOP Conf. Series: Journal of Physics: Conf. Series **887** (2017) 012006 doi:10.1088/1742-6596/887/1/012006

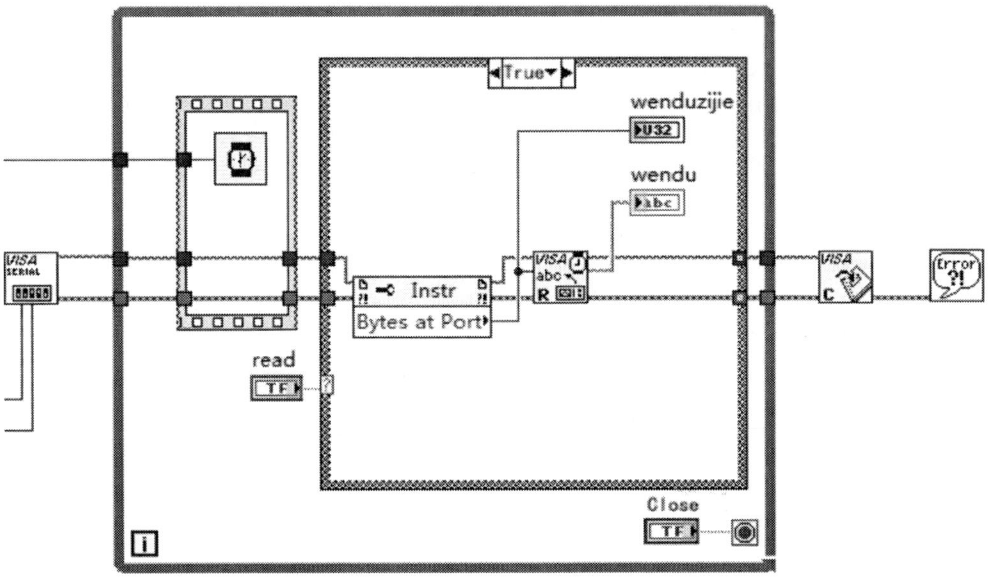

Figure 3. Serial read program block diagram

4. System Software Design

4.1. Data Acquisition

NI OPC Sever provides many manufacturer drivers, the Data Socket function in LabVIEW also uses OPC server for communication. Data Socket can effectively support different applications on local computers and exchange data between multiple applications on different computers in the same LAN, to achieve data exchange across process, across computers.

The specific software design method is: to define transmission protocol DSTP for the transmission of data across the network, the transmission protocol is in the form of URL type, which is similar to http, ftp.

The system collects the card corresponding to the URL dstp: //192.168.0.121/wavecard (responsible for acceleration, pressure) and dstp: //192.168.0.121/wendu (responsible for the temperature and other physical variables). LabVIEW obtains data items named "wavecard" and "wendu" from the OPC server via the above URL. Then the Data Socket function is used for the software, including open Data Socket, read Data Socket and close Data Socket.

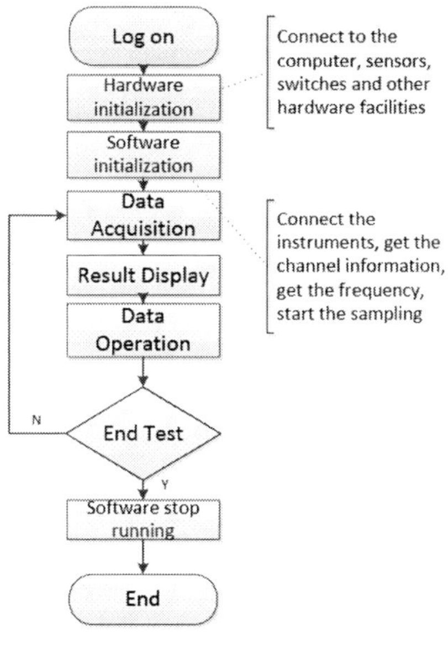

Figure 4. Software flow chart

It is noteworthy that the collected data by the acquisition card of the system are saved as the global variables during the software design process. That is because the global variables can be accessed anywhere in the brake test bench data acquisition system, and the addresses of global variables are fixed, thus improving the read and write efficiency. However the disadvantage is the occupied memory of global variables affecting the memory property. Considering the number of global variables is small, this memory problem can be avoided.

Figure 5 shows the remote data access programmatic implementation (the lower machine IP address is 192.168.0.121); Figure 7 is the software front panel of test bench data acquisition system. As the physical variables can be grouped in accordance with the wheel number, the tab navigation bar with 6 associated views can be set in the front panel to display different corresponding parameters, namely the rail wheel (including two tab bars), flywheel (including two tab bars), brake performance, temperature. Operators can not only check the corresponding physical variables in the form of real-time data, historic data, waveform but also take values based on the result and enlarge or shrink the waveforms.

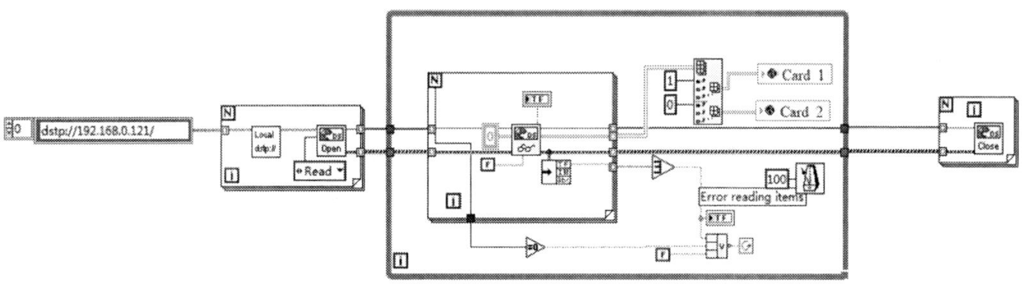

Figure 5. Remote data interactive diagram

4.2. Data Operation
In order to obtain the relationship between wheel rail adhesion coefficient and slip rate, the data acquisition system needs to calculate the relevant data. According to the calculation method of wheel rail adhesion coefficient of railway vehicles, the braking adhesion coefficient is defined as the ratio of maximum braking force to axial load. The slip rate can be calculated according to the formula

$\varsigma = \frac{v - w \cdot r}{v}$, where v is the vehicle speed, ω is the wheel speed, r is the wheel rolling radius, shown in Figure 6, the sub VI calculates the adhesion coefficient and the slip rate.

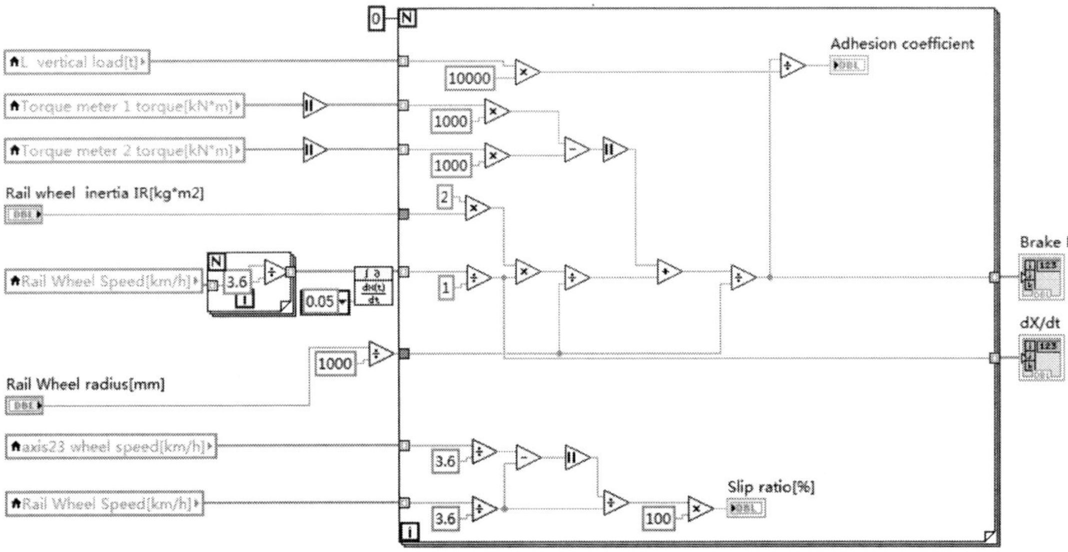

Figure 6. Data operation program diagram

5. Acquisition System Test

To verify the performance of the data acquisition system, the SWTB-400 bogie was selected as the test bogie, the vertical load using the adjusting weight up to 9 ton, and a number of adhesive tests were operated. During the test process, the traction motor was controlled by direct torque control method using frequency converter to maintain a constant speed, the brake cylinder pressure rises simultaneously, so that the wheels were decelerated and the velocity differences between vehicle wheels and rail wheels were made to simulate the "roll and slip" state of motion. At the same time, the system calculated adhesion, slip rate, vertical load and other physical variables.

Taking the test data as an example, the measurement and control performance of the test bench data acquisition system is shown in Figure 7, where the left ordinate is the brake cylinder pressure [kPa], and the right ordinate is the velocity [km / h]. As can be seen from the curve, the initial speed is 300km / h, the brake cylinder pressure remains stable at the range of 230kPa to 300kPa in the braking process. The relationship between the adhesion coefficient and the slip ratio calculated by the data operation module is shown in Figure 8, it can be seen that the adhesion coefficient is maintained between 0 and 0.3 throughout the test. During the test, the data acquisition system has strong anti-interference ability and high synchronization performance; there is no crash or stagnation phenomenon. The different physical variables can be displayed in real time. From the braking process it can be concluded that the system for the related data testing and acquisition is stable and reliable.

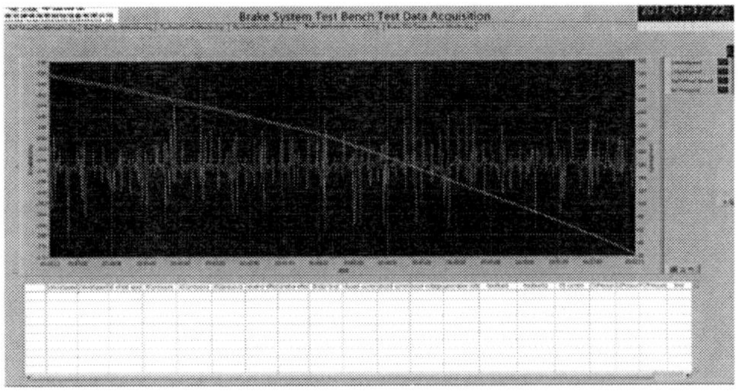

Figure 7. Test data

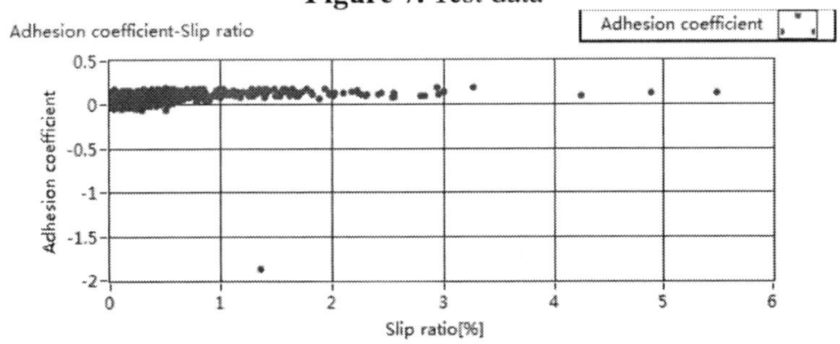

Figure 8. Diagram of adhesion coefficient-slip ratio

6. Conclusion

In this paper, the data acquisition system of the brake test bench was designed, which realized the functions of remote data acquisition, multi-point monitoring, LAN sharing and so on, using LabVIEW as development platform. It provides ideas for remote data acquisition. Tests show that the data acquisition system based on OPC technology is feasible. It realized real-time field data acquisition, operation and data exchange on different platform, reducing the cost and improving the efficiency of the test. It has been proved that the system has the characteristics of friendly man-machine interaction and high efficiency of data communication. The system also provides convenience for industrial production and provides solutions for professional research.

7. Acknowledgement

Thanks for the support of National Key Technology Research and Development Program (Grand No. 2015BAG12B01-20).

8. References

[1] Beyon J Y. LabVIEW programming, data acquisition and analysis [M]. Prentice Hall PTR, 2000.
[2] Lieping Z, Aiqun Z, Yunsheng Z. On remote real-time communication between MATLAB and PLC based on OPC technology[C]//Control Conference, 2007. CCC 2007. Chinese. IEEE, 2007: 545-548.
[3] Qi-wen H E. Usage of OPC technology in LabVIEW 8.0 DSC module [J] [J]. Computer Engineering and Design, 2006, 22: 066.
[4] Zhang W, Chen J, Wu X, et al. Wheel/rail adhesion and analysis by using full scale roller rig [J]. Wear, 2002, 253(1): 82-88.
[5] Zheng L, Nakagawa H. OPC (OLE for process control) specification and its developments[C]// Sice 2002. Proceedings of the, Sice Conference. IEEE Xplore, 2002:917-920 vol.2.

IOP Conf. Series: Journal of Physics: Conf. Series **887** (2017) 012007 doi:10.1088/1742-6596/887/1/012007

Research and application of mobile teaching platform

Ping Yang[1] and Hongjiao Xue[2]

[1]Beijing Polytechnic, Beijing 100176, China, E-mail: yangpingeda@126.com
[2]Beijing Polytechnic, Beijing 100176, China, E-mail: kisssun2016@126.com

Abstract. The application of mobile technology in university digital campus is ripe. This article mainly introduced the necessity of teaching platform based on mobile Internet in the teaching of higher vocational education, and the key to the construction of the feasibility of mobile learning platform, which is a feasible and effective teaching model under the new situation, worthy of promotion. The design and application of teaching platform based on mobile Internet is the change of educational ideas and working methods, and is the new starting point of Higher Vocational education.

1. Background

As mobile applications flourish, mobile devices such as smartphones and tablet PC are gaining popularity, and the wireless bandwidth increases lead to the costs continue to decline. In this case, the application of mobile technology in university digital campus is ripe [1].

Mobile digital campus based on the application of digital campus, the school teachers and students daily work, study and life, using the latest achievements in the development of mobile applications, resource sharing, information exchange, remote education, information technology, mobile terminal show in the unified guidance of the school teaching idea and information development thought. Further promote the teaching, research, management and other work of the school, and form a diversified, humanistic and intelligent information service environment. Effectively support the cause of the school, develop well and quickly, and improve the quality of teaching and work efficiency.

Mobile teaching platform is a new model of mobile learning experience, which can improve the quality and efficiency of school teaching. It is a useful and active attempt to develop high-quality resources, develop teaching methods and open learning environment. The latest achievements in the development of mobile applications using mobile teaching platform, resource sharing, information exchange, remote education, information technology, mobile terminal display in the unified guidance of development of teaching concepts and information school idea, promote the school teaching, scientific research, management and other work further, the formation of diversity, culture and intelligent teaching environment. The application of mobile teaching platform is not only the new direction of technological research and development, but also the change of educational idea and working mode.

2. Design of Teaching Platform Based on Mobile Internet

With the acceleration of the urbanization process, the speed of urban rail transit construction in our country is swift and violent. According to the relevant planning, the urban rail transit operation mileage exceeds 3100 km at the end of 2014. Beijing, Shanghai, Guangzhou, Tianjin, Suzhou, Nanjing and other 40 large and medium cities have or are building subways, and more and more cities are joining the ranks of urban rail transit construction. The urban rail transit industry has entered a stage of all-round high-speed development, and the shortage of related professional and technical personnel of rail transportation requires the support of a large number of high-quality professionals.

In order to meet the increasingly strong demand of rail transit talent, all colleges and universities have opened the related professionals of city rail transportation, and built the teaching laboratory and practice base, through systematic training, complete function, can increase the intensity of teaching practice. However, the urban rail transit training equipment is large, and the training conditions are limited, which limits the students' hands-on learning ability.

At present, the application of modern teaching technology at home and abroad is developing towards multimedia, networking, informatization, diversification of educational technology application mode and popularization of distance education mode. In twenty-first Century, China made great efforts to promote education informatization, and research the educational teaching model to train high-quality professionals. Under the background of educational reform, mobile education as a new education method will surely become a supplement of the traditional education, and that as a new form of distance education will let the learners to get rid of the restriction of space and time to study at any time and in any place according to the need of autonomous learning. Mobile learning represents the trend of learning in the future, and it will be a new and individualized learning style in distance learning.

Design of mobile teaching platform is a new remote teaching model based on the mobile Internet to provide a good reference for the curriculum construction, to make the communication between teachers and students more effective, to reform the classroom teaching and education, to promote the innovation of learning ability in the new information environment and teachers' teaching methods, in order to promote good the establishment of efficient and feasible new teaching mode[2].

The mobile internet teaching platform based on mobile digital campus server operating environment, establish the platform and the mobile terminal connection and communication mechanism, provide the user authentication, mobile application management, application management, release and upgrade data transfer session, auditing and other functions, the platform should have the ability to expand.

2.1. Design Idea

I)Design from the following three aspects:

a) Research on the multimedia mobile training mode of urban rail transit based on the mobile operating system;

b)Structure multimedia mobile training platform of cross space urban rail transit;

c) Application of multimedia interactive training course for urban rail transit on Intelligent equipment.

II)Five functions of mobile teaching platform:

a) Teachers and students identity authentication;

b) Interaction between teachers and students;

c) The assignment of exercises

d) Examination questions are handed down and handed in

e) Achievement push.

2.2. Mobile Instructional Platform Design Framework

The platform design framework is listed in figure 1.

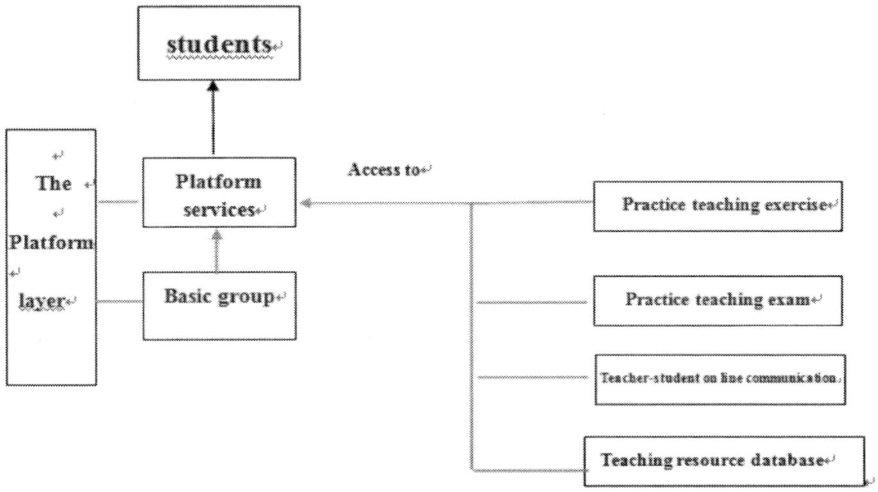

Figure 1. Mobile instructional platform design framework

The object of mobile teaching platform is the components of services content and the basic composition.

The following specific design:

a) investigation and study

The main task is to analyze and study the real practice teaching environment, extract the elements needed for the implementation of the system and collect system application requirements.

b) platform design

The main task is to analyze the various resources and requirements collected at the previous stage, abstract and extract the object model, and build the platform model based on it.

c) system implementation and verification

Through technical means, the design requirements of the platform are achieved and compared to the real city rail transit multimedia interactive environment inspection and correction system.

In the real environment to test the network and hardware equipment, the software system for performance capacity, pressure test and trial operation to ensure that the system functions can be used normally, and meet the needs of professional practice teaching, and find and correct the potential problems as much as possible before the formal run[3].

Physical design framework based on mobile internet teaching platform, showing in figure 2.

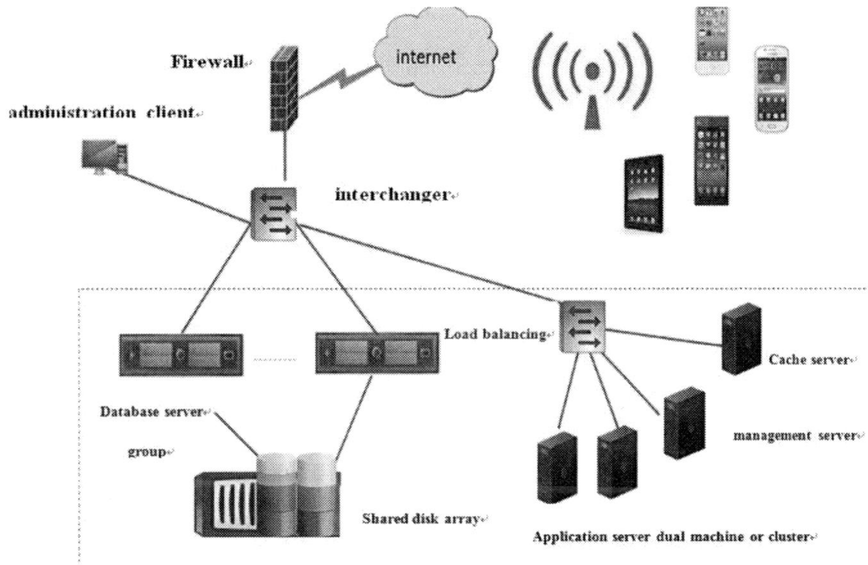

Figure 2. Physical design framework based on mobile internet teaching platform

3. Construction Of Mobile Teaching Platform

3.1. Key Technology of Platform Design
Component diagram of technology development.

Platform design using Java technology with the advanced development and management tools to complete the implementation of the software development life cycle, the establishment of data interface, powerful system management, data access, data transmission application management platform provides data exchange tasks for the client. J2EE platform is a distributed, high reliability and advanced solution for enterprise applications. The main code of J2EE platform adopts Java programming language and server Java technology[4].

Native App is better at accessing and being compatible because it is located above the platform layer. It can support online or offline, message push or local resource access, and the capture of camera dialing function.

Web App doesn't need to be installed, and its adaptability to device fragmentation is better than Native App. It only needs to execute WAP technology in any mobile browser by means of XHTML, CSS and JavaScript.

The cell phone used by the user is not within the scope of the client's development. The corresponding client program cannot be installed, and limited data access is accessed through the WAP browser access mode. WAP format is mainly developed by using WAP2.0 or HTML5 technology.

Hybrid App is between web-app and native-app, although it looks like a Native App, only one UI WebView, which access is a Web App. Hybrid App (hybrid mobile application mode) with Native App good user experience advantage "and" cross platform Web App the advantage of development"[5].

3.2. Plan of platform construction
The terminal application platform to achieve docking and session management, user authentication, business components, business logic and application management, application release and upgrade management functions, and realize data sharing with the docking data center and application systems, and data preprocessing.

a) protocol conversion

The platform provides a set of perfect data communication interface mechanism, which can easily handle any form of agreement, protocol, and also provide a realtime database as the core of the distributed client server access and service mechanism for third party companies such as: DDE, OPC, TCP/IP, ODBC, DbCom and API access control access function data.

b) data processing

The real-time database based on network distributed structure provides open access to the outside world and supports the application of C/S and B/S. The real-time data of each system can be conveniently gathered together to provide a reliable universal real-time data platform for the further development and application of data resources.

The platform is designed to handle constantly updated, rapidly changing data and event processing. Various operations can be performed in a variety of ways, including data processing, historical data storage, statistical processing, alarm processing, and service requests.

Provide time for real-time database driven scheduling and resource allocation algorithm using real-time technology, according to different application requirements and characteristics, design the real-time data model, real-time transaction scheduling and resource allocation strategy, real-time data communication using the theory research and systematic, structured in-depth.

The introduction of virtual memory technology (reduced disk I/O), memory allocation and management method, data dictionary and special structural design, and the use of multithreading and parallel processing way to further strengthen the efficiency of data processing.

c) Data conversion

The data converter provides methods for converting data from the source to the target system. This function has traditionally been done by programmers, which is extremely inefficient, because the need to write different data extraction and loading programs for each data source. The platform provides a general solution for data conversion, to generate the program code that is loaded in a graphic way, which visually and efficiently generate a special data converter, thus reducing the 70% to 80% of the work.

The specific process can be divided into three steps: the first is from the data source (business system underlying), extracted data will be needed; and then converted to a target data storage format of the data; finally transformed data will be loaded into the data warehouse, in order to solve the problem of the dirty data. After data conversion step, it also adds data cleaning function

D) Business system data access

Service system data access consists of real-time data and non real time data access.

- Non real time data, the use of shared data center system for data synchronization input docking mode. The shared data center through the data routing path, establishing Middle Library, data structure and business system comparison needed, finally perform synchronization data, the business system data set according to the rules of the data center onto the digital campus platform.
- The real-time data is provided by the service system interface, and the platform server implements the interface, and provides the call processing method for the digital campus client. Of course, real time data docking uses the WAP model, and the business functions are directly handled by the WAP program of the business system.

e) User information management

The module provides actions to add and delete functions to users.

Manual management: you can manually add, modify, and delete, query user information;

Import and export: you can import user information files written in the specified format into the system, generate user information, or export user information to the specified format file;

f) Role and authority management

The module provides role and authority management.The system administrator can set different roles according to the operational needs of the system, and give the role permissions.

When adding a user, the corresponding role name can be selected, and the user can enjoy the permission of the corresponding role.

Roles and permissions management include but not limited: roles definition (role name, role description, creator, creation time, etc), role query, menu permissions, personnel allocation, menu authority assignment, etc.

3.3. Test of Platform Feasibility

The platform feasibility test is divided into four phases:

- Unit testing: test each unit to make sure that each module works properly

- Integration testing: assembly test the modules that have been tested to ensure that the functional logic process is running properly
- Regression testing: perform regression testing on the tested version to ensure the stability of the version
- Acceptance testing: the final test of the deployed software to ensure the normal flow of the on line version

Results: compatibility supports Android4.0 and above, more than ios8.0 system.

Front and back test results can be executed.

4. Application Prospect

After the successful development of the platform, the professional practice teaching mode is also a change and promotion, and it is the change of teacher education idea and working mode. It can improve the level of the application of information technology teachers, update teaching concepts, improve teaching methods, improve teaching effectiveness, and improve the level of the application of information technology teachers, update teaching concepts, improve teaching methods, improve teaching effectiveness using in teaching and learning interaction between teachers and students in the limited space. It also encourage students to active learning, autonomous learning by means of information, enhance the ability to analyze and solve the problems by using information technology, which let students make full use of the mobile platform for teacher-student interaction, necessary training exercises and examination in the limited space. Students can take advantage of the limited time and strengthen their study.

The four design goals of mobile teaching practice platform is "interface unification, resource sharing, auxiliary teaching, mobile service".

A. Interface unification: digital campus platform is a public platform for information gathering and interaction, including the formulation, publication and management of business and technical standards. To protect the business system of different development companies and different development technologies, it is convenient, timely and efficient to connect with the digital campus platform, and to promote the continuous enrichment and perfection of mobile applications.

Resource sharing: on the basis of resource data electronization, the campus network and mobile learning platform are used to let the resources flow. Not only the transmission and reception, but also the establishment of "one to many" or "many to many" business information flow model according to the business requirements, forming a mobile information layer of cross system knowledge sharing.

C. Auxiliary teaching: using the latest information technology means, to strengthen the teaching process of new, intuitive and vivid way of information interaction, so that the entire teaching process has become vivid and three-dimensional. Students use debris time in a relatively relaxed state, combining with individual actual learning situation, targeted access to knowledge and information, auxiliary teaching purposes further achieved.

D. Mobile services: provide one-stop mobile services for campus services, teaching staff and students and parents. Improve work efficiency, improve management efficiency, improve decision-making efficiency, and improve information utilization. Actively display the school resources and research and technological achievements, to provide the community with more abundant and useful knowledge achievements, and enhance the service ability of the society.

Mobile teaching platform design and application of the new times, the distance teaching model based on the mobile Internet, to provide a good reference for the course construction, the communication between teachers and students more effective, conducive to classroom teaching and education reform, promote the innovation of learning ability in the new information environment and teachers' teaching methods, to promote in order to establish efficient and feasible new teaching mode.

Mobile digital platform is not only a new direction in technology research and development, but also a change in educational philosophy and working style. With the development of mobile and communication technology, the trend from digital learning to mobile learning is becoming more and more obvious. The growing popularity of mobile phones, the increasingly powerful function, to break the time and space constraints of mobile learning provides good hardware support. Therefore, the development of mobile learning platform based on WAP technology has a good application prospect. It is expected that the mobile internet teaching platform will have a broader application prospect.

IOP Conf. Series: Journal of Physics: Conf. Series **887** (2017) 012007 doi:10.1088/1742-6596/887/1/012007

5. Reference

[1] Xiang Wenjie: Review and prospect of mobile Internet development, telecommunications technology, 2009 (1):66-69
[2] Xiao Zhihui: Mobile Internet research introduction, telecommunication science, 2009 (10):30 -36
[3] Chen Haiming, Cui li, Xie Haibin, A comparative study of the architecture and implementation methods of Internet of things, Journal of computer science, 2013(1)
[4] He Shuiyan: Java programming, Mechanical Industry Press (2016)
[5] Mark Summerfield：Qt advanced programming, Electronic Industry Press (2011)

IOP Conf. Series: Journal of Physics: Conf. Series **887** (2017) 012008 doi:10.1088/1742-6596/887/1/012008

Combing VFH with bezier for motion planning of an autonomous vehicle

Feng Ye[1], Jing Yang[2], Chao Ma[3] and Haijun Rong[4]

[1]School of Electronic Information Engineering, Xi'an Jiaotong University, Xi'an,China,yf87953246@stu.xjtu.edu.cn
[2]School of Electronic Information Engineering, Xi'an Jiaotong University, Xi'an, China, jasmine_yang@126.com
[3]School of Electronic Information Engineering, Xi'an Jiaotong University, Xi'an, China, Machao0919@stu.xjtu.edu.cn
[4]School of Aerospace, Xi'an Jiaotong University, Xi'an, China, hjrong@xjtu.edu.cn

Abstract. Vector Field Histogram (VFH) is a method for mobile robot obstacle avoidance. However, due to the nonholonomic constraints of the vehicle, the algorithm is seldom applied to autonomous vehicles. Especially when we expect the vehicle to reach target location in a certain direction, the algorithm is often unsatisfactory. Fortunately, the Bezier Curve is defined by the states of the starting point and the target point. We can use this feature to make the vehicle in the expected direction. Therefore, we propose an algorithm to combine the Bezier Curve with the VFH algorithm, to search for the collision-free states with the VFH search method, and to select the optimal trajectory point with the Bezier Curve as the reference line. This means that we will improve the cost function in the VFH algorithm by comparing the distance between candidate directions and reference line. Finally, select the closest direction to the reference line to be the optimal motion direction.

1. Introduction

Motion planning is an important part of autonomous vehicle research. It is problem is to generate continuous motion that connects start state and target state, while avoiding collision with known obstacles [1]. The research in robot motion planning can be traced back to the late 60's.Over so many years, many motion planning algorithms were proposed, such as A*,D*[2] and RRT[3]. However, it is quite difficult to take the vehicle's non-holonomic constraints into considerations.

In contrast, the parameter curve is simpler and it can satisfy the dynamic constraints well. Bezier Curve is a typical example. Cornell team proposed the two-dimensional cubic Bezier curve to represent trajectories. Long Han present a trajectory planning algorithm based on Bezier curve [4]. For intelligent vehicles, the ability of avoiding obstacles becomes very important, so more and more researcher focus on the study of obstacle avoidance algorithm. For example, the MIT's intelligent vehicle uses RRT as trajectory planning to avoid obstacle. However, it is still difficult for intelligent vehicles to reach an expected state in complex environment.

The Bezier Curve earliest applied to automotive body design by Pierre Bezier in 1962.It is a typical method of generating trajectories by analytic method. The advantage of Bezier Curve is that it can be generated fast and it is smooth enough which is very important for intelligent vehicle when there is no obstacle or the obstacles are sparse. However, when the curve is generated, only the pose and curvature constrains of the endpoints are taken into account, the obstacles information are not taken into account. Therefore, the method itself does not have the obstacle avoidance function. What's more, the Bezier curve does not have the ability of continuous turning, so this method cannot solve motion planning problem in the complex environment.

VFH algorithm was proposed by Johann Borenstein for the mobile robot's motion planning [5].It is a method based on obstacle grid which used to represent environment.VFH algorithm has strong ability of avoiding obstacles and its planning space is continuous. After that, many improved versions are emerging. Iwan Ulrich et al. proposed VFH+ algorithm and VFH* algorithm which have large influence. Its main contribution is to add the cost function to the original VFH algorithm [6]. Dong Jie et al. proposed IVFH* for dynamic obstacles avoidance [7]. However, the vehicle is an non-holonomic constrains system, some states of active region in the original algorithm are unreachable for the vehicle. What's more, we often hope that the vehicle reaches the target point with the heading we want, but the VFH algorithm performs not very well.

Through the description of VFH algorithm and Bezier Curve. We can see that the Bezier Curve is an analytic method, VFH algorithm is based on the sampling pattern and generates a trajectory consisting of discrete points. So in this paper, we propose an algorithm that combines VFH algorithm with Bezier Curve. It is also a combination of discrete method and analytic method. We combine the advantages of both, so that it can be applied to the intelligent vehicle's motion planning to adapt to complex traffic scenes.

2. Problem Formulation

This section we describe the definition of motion planning. As shown in Fig .1, we present the state space of system as compact sets $X \subset R^n$ and the control space as $U \subset R^m$ .Then we have [8]

$$\dot{x}(t) = f(x(t), u(t)), x(0) = X_{init}, x(t_f) = X_{target}$$

Where $x(t) \in X, u(t) \in U$ and X_{init} is the initial state, X_{target} is the target state. $X_{obs} \subset X$ denotes that these states collide with obstacles, so collision-free state can be defined as $X_{free} \subset X \setminus X_{obs}$ .The motion planning task is to find an state sequence that satisfies the following condition:

$$\begin{aligned} & x(t) \in X_{free}, \forall t \in [0, t_f] \\ & x(0) = X_{init} \\ & x(t_f) = X_{target} \\ & u(t) \in U \end{aligned}$$

Where x is called the trajectory and $x(t)$ is the feasible state for our intelligent vehicle and u is called the control input sequence and $u(t)$ is also feasible for the vehicle.

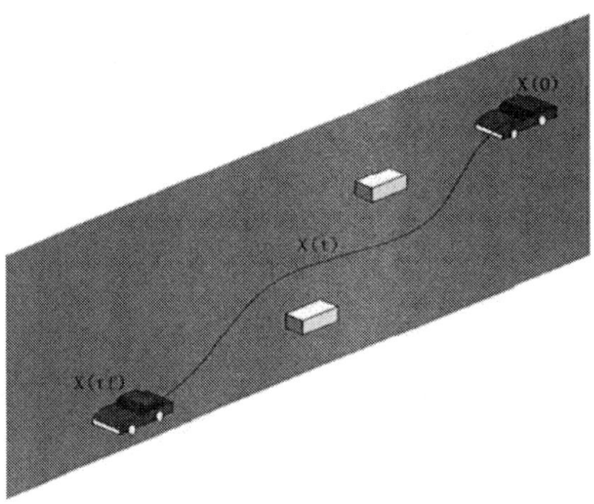

Figure1. Motion planning diagram

3. Algorithm Description

In this section, the algorithm based on Bezier curve and VFH algorithm is presented. We first make a brief introduction about Bezier Curve and VFH algorithm, and then describe our algorithm in details. Our algorithm is implemented in the following steps: Firstly, the Bezier curve clusters are generated according to the start point and target point, and then select the smoothest Bezier curve as the reference line. Secondly, we use VFH algorithm to find all the candidate directions. Thirdly, we select the direction closest to the reference line as the new motion direction and determine whether the vehicle to reach the end. If not, we will return to step two until the vehicle reaches the end point.

3.1. Third-Order Bezier Curve

Low-order Bezier Curve is simple, but the planning ability is bad. High-order Bezier Curve has strong planning ability, but the control point selecting is lack of basis and difficulty to control. To sum up, we choose the four control points of the third-order Bezier Curve planning.

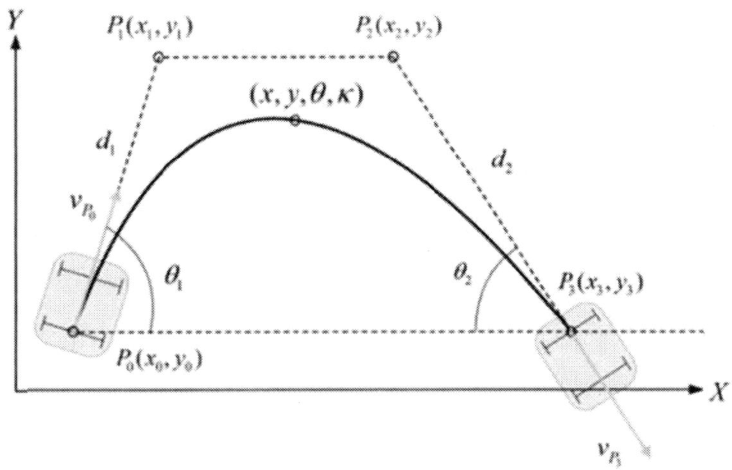

Figure 2. Third-order Bezier Curve

Given four control points $P_0(x_0, y_0, \theta_0)$, $P_1(x_1, y_1)$, $P_2(x_2, y_2)$ and $P_3(x_3, y_3, \theta_3)$ as shown in Fig .2, we can get the X and Y coordinates of third-order Bezier Curve:

$$\begin{aligned} x(t) &= \sum_{i=0}^{3} b_{i,n}(t)x_i \\ &= x_0(1-t)^3 + 3x_1 t(1-t)^2 + 3x_2 t^2(1-t) + x_3 t^3, t \in [0,1] \end{aligned} \tag{1}$$

$$\begin{aligned} y(t) &= \sum_{i=0}^{3} b_{i,n}(t)y_i \\ &= y_0(1-t)^3 + 3y_1 t(1-t)^2 + 3y_2 t^2(1-t) + y_3 t^3, t \in [0,1] \end{aligned} \tag{2}$$

(1), (2) Can be organized into third-order polynomials with respect to the parameter t:

$$\begin{aligned} x(t) = {} & [(x_3 - x_0) + 3(x_1 - x_2)]t^3 + 3(x_0 - 2x_1 + x_2)t^2 \\ & + 3(x_1 - x_0)t + x_0, t \in [0,1] \end{aligned} \tag{3}$$

$$\begin{aligned} y(t) = {} & [(y_3 - y_0) + 3(y_1 - y_2)]t^3 + 3(y_0 - 2y_1 + y_2)t^2 \\ & + 3(y_1 - y_0)t + y_0, t \in [0,1] \end{aligned} \tag{4}$$

For convenience of expression, (3), and (4) is reduced to the following form:

$$x(t) = a_3t^3 + a_2t^2 + a_1t + a_0, t \in [0,1] \tag{5}$$

$$y(t) = b_3t^3 + b_2t^2 + b_1t + b_0, t \in [0,1] \tag{6}$$

In (5), (6):

$$\begin{aligned} a_3 &= (x_3 - x_0) + 3(x_1 - x_2), & a_2 &= 3(x_0 - 2x_1 + x_2) \\ a_1 &= 3(x_1 - x_0), & a_0 &= x_0 \\ b_3 &= (y_3 - y_0) + 3(y_1 - y_2), & b_2 &= 3(y_0 - 2y_1 + y_2) \\ b_1 &= 3(y_1 - y_0), & b_0 &= y_0 \end{aligned} \tag{7}$$

The third-order Bezier Curve is shown in Fig.2. Where P_0, P_1, P_2 and P_3 correspond to the curve four control point. θ_0 is the inclination angle of the straight line of the line segment P_0P_1, θ_2 is the inclination angle of the straight line of the line segment P_2P_3 .In the following, we will briefly introduce some important features of the Bezier Curve:

First, Bezier Curves are guaranteed to pass through the first and last control points. If the first control point corresponds to the starting position of the vehicle, the last control point corresponds to the vehicle's target position, then the Bezier Curve can ensure that the trajectory over the initial position and ultimately reach the target position. Second, Bezier Curves endpoints P_0 and P_3 are tangent to edges P_0P_1 and P_2P_3 of control polygon $P_0P_1P_2P_3$ respectively. If the tangent of the Bezier Curve and the heading of the motion trajectory are linked, then the Bezier Curve can satisfy the heading of the motion trajectory at the starting state P_0 and target state P_3. Third, the third-order Bezier Curves are not only self-continuous, but also their first derivative and second derivative are continuous, which fully meet the requirements of trajectory continuity in restricted vehicle systems. Fourth, it is convenient to convert the kinematic constraints into curve constraints, thus adding the vehicle kinematic constraints to the motion planning process.

From the Fig.2, we can see that control point P_1 must be tangent at point P_0, control point P_2 must be tangent at point P_3.But the distance between P_0 and P_1 can be changed, and the distance between P_2 and P_3 can also be changed. So by changing the location of the control point P_1 and P_2, we can get the curves cluster connecting P_0 and P_3.

After obtaining the cluster of curves, we need to further select the smoothest curve .The smoothest curves can be abstracted into optimization problems:

$$tra_{optimal} = \arg\min_{tra_i} \sum \kappa^2 \tag{8}$$

Where $tra_{optimal}$ denotes the smoothest curve , $\sum_{tra_i} \kappa^2$ denotes the sum of square curvature at all points on the i-th curve if we use point sets to represent a curve.

Next we derive the curvature calculation formula. According to the definition of curvature:

$$\kappa(t) = \frac{|x'(t)y''(t) - x''(t)y'(t)|}{[x'(t)^2 + y'(t)^2]^{3/2}} \tag{9}$$

so the curvature of start point P_0 is:

$$
\begin{aligned}
\kappa(0) &= \frac{|x'(0)y''(0) - x''(0)y'(0)|}{[x'(0)^2 + y'(0)^2]^{3/2}} \\
&= \frac{2(x_0y_1 + x_1y_2 + x_2y_0 - x_0y_2 - x_1y_0 - x_2y_1)}{3[(x_1 - x_0)^2 + (y_1 - y_0)^2]^{3/2}} \\
&= \frac{2}{3} \cdot \frac{(y_3 - y_0)\cos\theta_1 - (x_3 - x_0)\sin\theta_1 + d_2\sin(\theta_1 - \theta_2)}{d_1^2}
\end{aligned}
\tag{10}
$$

the curvature of target point P_3 is:

$$
\begin{aligned}
\kappa(1) &= \frac{|x'(1)y''(1) - x''(1)y'(1)|}{[x'(1)^2 + y'(1)^2]^{3/2}} \\
&= \frac{2(x_1y_2 + x_2y_3 + x_3y_1 - x_1y_3 - x_2y_1 - x_3y_2)}{3[(x_3 - x_2)^2 + (y_3 - y_2)^2]^{3/2}} \\
&= \frac{2}{3} \cdot \frac{(x_3 - x_0)\sin\theta_2 - (y_3 - y_0)\cos\theta_2 + d_1\sin(\theta_1 - \theta_2)}{d_2^2}
\end{aligned}
\tag{11}
$$

so the relationship between d_1 and d_2 is:

$$
d_2 = \frac{(x_3 - x_0)\sin\theta_1 - (y_3 - y_0)\cos\theta_1 + \frac{3}{2}\kappa(0)d_1^2}{\sin(\theta_1 - \theta_2)} \tag{12}
$$

At this point, we can calculate the sum of square curvature of each curve and select the smoothest curve as the reference line.

3.2. Combining VFH Algorithm with Bezier Curve

We use VFH algorithm to search collision-free region and reference line to select the optimal motion direction. So we will introduce something about VFH to show how we find the collision-free region. This section includes the following contents: First, the definition of active region. Second, find collision-free region. Third, select the optimal motion direction. Fourth, smooth the trajectory.

First, we will introduce the active region. Whether the rectangular active region of the VFH algorithm or the circular active region of the VFH+ algorithm cannot satisfy the vehicle kinematics constraints. Because some of the status points in the active region are unreachable for the vehicle.

Taking into account the kinematics constraints of the vehicle, we use a new active region which suitable for the vehicle. As shown in Fig. 3, the new active region is defined by a sector window in [9].

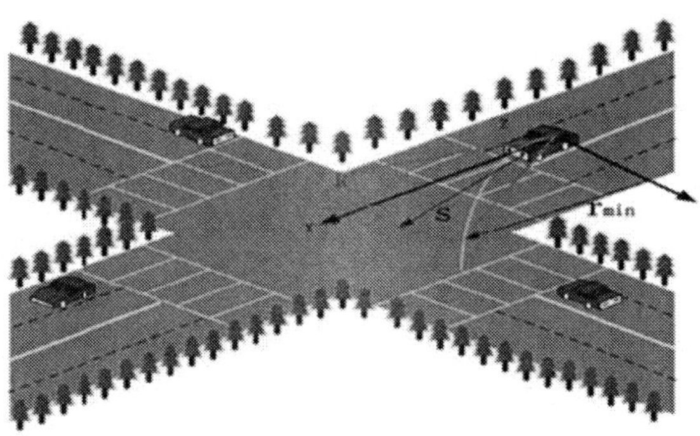

Figure 3. New active region

The sector active region Z can be expressed as formulas:

$$Z = P(\rho, \varphi) \tag{13}$$

P represents a polar coordinate system in which the center of the rear axle of the vehicle is the origin of the coordinates and the heading angle direction is 0°, ρ denotes the distance range of the active area, and $\rho \in [0, \rho_{\max}]$, $\rho_{\max}$ denotes the maximum active distance, φ denotes the angular range of the active region, and has

$$\varphi \in [-\arcsin\frac{s}{2r_{\min}}, \arcsin\frac{s}{2r_{\min}}] \tag{14}$$

In (14), s represents the search step, that is, the distance traveled by the vehicle every time. $r_{\min}$ represents the inherent minimum turning radius of the vehicle, which can be approximated as a constant in the case where the vehicle speed is not particularly high. The trajectory point must be on the arc $R = P(s, \varphi)$, obviously, all the points on the arc R are reachable to the vehicle.

By comparison we can see that the sector active region is smaller than the original active region. So it considers fewer obstacle grids, making the algorithm more efficient.

Second, we need to find the collision-free region. In the VFH algorithm, a two-dimensional grid is used to represent the environment [10]. Each grid contains a probability value. This probability value reflects the presence of obstacles in the grid credibility, the higher the probability value, indicating the presence of obstacles here the greater the possibility.

In order to search for the collision-free region, first, we need to represent the grid occupied by obstacles inside the sector active region as a vector whose direction depends on the relative position of the grid and the vehicle center point (VCP) [5]:

$$\beta_{i,j} = \arctan(\frac{y_i - y_0}{x_i - x_0}) \tag{15}$$

where (x_0, y_0) represents the current position of the VCP, and (x_i, y_i) represents the position of the obstacle grid $C_{i,j}$. The magnitude $m_{i,j}$ of the obstacle grid $C_{i,j}$ is defined by:

$$m_{i,j} = (c_{i,j})^2 (a - bd_{i,j}) \tag{16}$$

In (16), a, b are positive constants and chosen such that $a - bd_{\max} = 0$ where $d_{\max} = \rho_{\max}$, $c_{i,j}$ denotes the certainty value of the obstacle grid $C_{i,j}$, $d_{i,j}$ represents the distance between obstacle grid $C_{i,j}$ and VCP.

Figure 4. Dividing the region into several sectors

Next, we divide the active region into several sectors and count the obstacle density for each sector, we can get the initial polar histogram. As shown in Fig. 4, selecting an angular resolution α divides the vehicle's active region into $n = \frac{360}{\alpha}$ sectors. In order to ensure that n is a positive integer, angular resolution α is often taken as a divisor of 360.Corresponding to each sector, there is an index k and an obstacle density h_k, where k is related to the discrete direction $\rho = k \cdot \alpha$ of the sector and defined as (17):

$$k = INT(\frac{\beta_{i,j}}{\alpha}) \tag{17}$$

INT stands for rounding. The obstacle density h_k of the sector k is defined as (18):

$$h_k = \sum_{C_{i,j} \in k} m_{i,j} \tag{18}$$

By calculating h_k, we can get the initial polar histogram. However, due to the discrete nature of the obstacle grid, the initial histogram is uneven, which may affect the determination of the final motion direction. So we introduce a low-pass filter to smooth the initial polar histogram.

Finally, we use a threshold to determine candidate directions based on the smoothed polar histogram [11]. If the obstacle density of sector lowers than the threshold, we regard the sector as candidate directions. The sector is also the collision-free region.

Third, we use the cost function to find the optimal motion direction. Through the threshold we can see from the polar histogram which direction is collision-free? This section we will describe how to use the reference line (Bezier Curve) to select the best direction.

When we select the best direction, we should take the width of vehicle into account [9].Assuming that the search step of the vehicle is 8 meters and the angle of the active region is 60 degrees. If the active region is divided equally into 20 sectors, the arc length corresponding to each sector is about 0.42 meters. The width of vehicle is about 2 meters, so the arc length of five sectors to meet the vehicle's width. This require that the two sectors of the selected sector, which are adjacent to the left, and two sectors(five sectors in total) that are adjacent to the right are also collision-free. We use $C_n \in C_{free}$ to represents that the n-th sector is collision-free. If the n-th sector satisfies the requirement of vehicle's width, we can express it as

$$\{C_{n-2}, C_{n-1}, C_n, C_{n+1}, C_{n+2}\} \in C_{free} \tag{19}$$

For each candidate direction that satisfies (17), we can calculate the shortest distance between this direction and reference line through (18):

$$d_{i,\min} = \min\{d_{i,1}, d_{i,2}, ..., d_{i,n}\}, i = 1,2,...,m \tag{20}$$

This is also our cost function. We assume that the reference line is represented by n points, and the number of sectors satisfying (19) is m . i denotes the i-th sector satisfying (19), $d_{i,n}$ denotes the distance between the i-th sector and the n-th point of the reference line, $d_{i,\min}$ denotes the shortest distance between the i-th sector and the reference line.

Finally, we can calculate which sector is closest to the reference line by (21):

$$k = \arg\min_i(d_{i,\min}), k = 1,2,3,...., m \tag{21}$$

And the k-th sector is our optimal motion direction.

Fourth, we need to smooth the trajectory. In general, the trajectory of the vehicle is represented by a set of points. The trajectory points we obtain from procedure three is sparse and discrete. So we use the B-spline curve based on the trajectory points to smooth the trajectory [12].

The basis function of third-order B-spline curve is:
sequenceWe use the trajectory point obtained from procedure 3 as the control point of B-spline curve. Then we can obtain a dense smooth trajectory. This trajectory is feasible for vehicle.

4. Experiment Result

In this section, we will show the performance of proposed algorithm in different traffic scenarios. There are two kinds of environment, structured environment that the road shape is regular and unstructured environment which has irregular shape of the road and unclear edge.

The pictures in Fig. 5 show the performance of our algorithm in structured environment. Fig. 5(a) shows that when the traffic scenario is very easy without any obstacles, the vehicle always runs straight in a lane and always remains in the middle of the lane. The position of the yellow vehicle denotes the initial state and the position of the blue vehicle denotes the target state. In Fig. 5(b), there is an obstacle in the lane, so the vehicle change the lane to right then change to the original lane when exceed the obstacle .We can see that when the vehicle reach the target state, the vehicle's heading satisfies our requirement which we expect to be parallel to the lane line. This avoids causing shock. Fig. 5(c) shows a complex scenario with multiple obstacles. Our algorithm performs well in this scene and the heading of the target state is what we want.

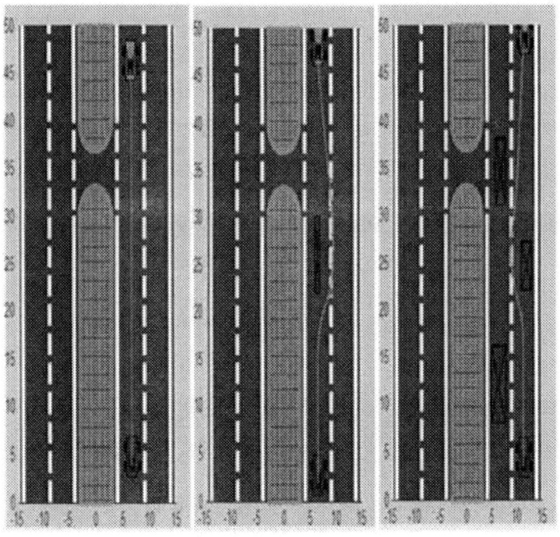

(a) (b) (c)

Figure 5. (a)Lane-keeping. (b)One obstacle avoidance. (c)Multi-obstacles avoidance.

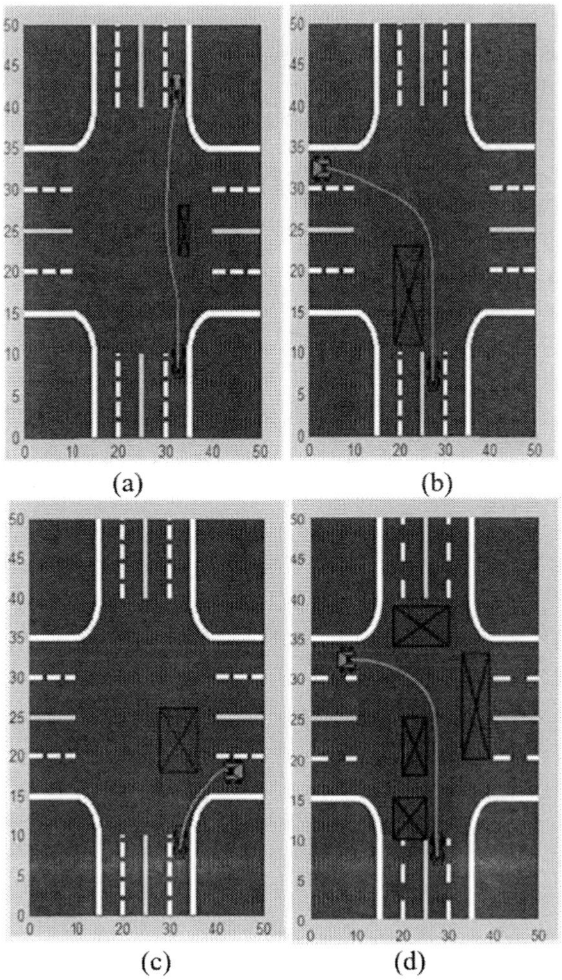

Figure 6. (a) Going-straight. (b)Turning-left. (c)Turning-right. (d)Multi-obstacles avoidance.

Fig. 6 shows the performance of our algorithm in intersection which is unstructured environment. There are four intersection scenes in Fig .6, including going-straight, turning-left, turning-right and multiple obstacles avoiding. Fig. 6(a) presents the result of our algorithm in the going-straight scene, and the vehicle performs well because the reference line is a straight line passing through the start point and the target point under this circumstance. Fig .6(b) shows the result of our algorithm in the turning-left scene and Fig .6(c) shows the result of our algorithm in the turning-right scene. Because there is no obstacle in the two scenes, the constraint effect of the reference line becomes very obvious. We can see that our trajectory is almost coincident with the reference line from Fig .6(b) and Fig .6(c).Fig .6(d) presents the result of our algorithm in the multiple obstacles avoiding. We can see from the picture that when the vehicle does not search for an obstacle, it goes almost along the reference line and it will avoid the obstacle when search for an obstacle due to the VFH algorithm. This is why we combine the VFH algorithm with Bezier Curve.

5. Conclusion

This paper presents a algorithm combined VFH algorithm with Bezier Curve to enhance the ability of intelligent vehicles to adapt to complex environments. Our algorithm has advantages from VFH algorithm and Bezier Curve. VFH algorithm has a strong ability to avoid obstacles which can find collision-free trajectories in complex multi-obstacles environment. Bezier Curves are not only self-continuous, but also their first derivative and second derivative are continuous, which fully meet the requirements of trajectory continuity of the constrained vehicle system. So the trajectories generated by our algorithm are continuous and have strong ability to avoid multi-obstacles.

6. Acknowledgment
This work is funded in part by National Nature Science Foundation of China (Grant No. 61403300), 973 Program (No. 2015CB351703), National Science Council of ShaanXi Province (Grant No. 2014JM8337) and the Fundamental Research Funds for the Central Universities.

7. References

[1] Howard T M, Pivtoraiko M, Knepper R A, et al. Model-Predictive Motion Planning[J], *IEEE ROBOTICS & AUTOMATION MAGAZINE*, 2014.
[2] Ferguson D, Stentz A. Field D*: An Interpolation-based Path Planner and Replanner[C]//International symposium on Robotics Research (ISRR), 2005.
[3] L. Ma, J. Xue, K. Kawabata, J. Zhu, C. Ma, N. Zheng, "Efficient Sampling-based Motion Planning for On-road Autonomous Driving," IEEE Transactions on Intelligent Transportation Systems, 2014 .
[4] Han L, Yashiro H, Nejad H T N, et al. Bezier curve based path planning for autonomous vehicle in urban environment[C], *Intelligent Vehicles Symposium (IV), 2010 IEEE. IEEE,* 2010: 1036-1042.
[5] Borenstein J, Koren Y. The vector field histogram-fast obstacle avoidance for mobile robots [J], *Robotics and Automation, IEEE Transactions on*, 1991, 7(3): 278-288.
[6] Ulrich I, Borenstein J. VFH+: Reliable obstacle avoidance for fast mobile robots[C], *Robotics and Automation, 1998. Proceedings. 1998 IEEE International Conference on. IEEE*, 1998, 2: 1572-1577.
[7] Jiea D, Xueming M, Kaixiang P. IVFH*: Real-time dynamic obstacle avoidance for mobile robots[C], *Control Automation Robotics & Vision (ICARCV), 2010 11th International Conference on. IEEE*, 2010: 844-847.
[8] Karaman S, Frazzoli E. Optimal kinodynamic motion planning using incremental sampling-based methods[C], *Decision and Control (CDC), 2010 49th IEEE Conference on. IEEE*, 2010: 7681-7687.
[9] P. Qu, J. Xue, L. Ma, C. Ma, "A Constrained VFH Algorithm for Motion Planning of Intelligent Vehicles" IEEE Transactions on Intelligent Transportation Systems, 2014.
[10] Elfes A. Using occupancy grids for mobile robot perception and navigation [J], *Computer*, 1989, 22(6): 46-57.
[11] Theodor Nitulescu, Stefan Talu, Applications of descriptive geometry and computer aided design in engineering graphics, Cluj-Napoca, Romania, Risoprint Publishing house, 2001, ISBN 973-656-102-X.
[12] Stefan Talu, Cristina Racocea, Axonometric representations with applications in technique, Cluj-Napoca, Romania, MEGA Publishing house, 2007. ISBN 978-973-7867-98-8.

IOP Conf. Series: Journal of Physics: Conf. Series **887** (2017) 012009 doi:10.1088/1742-6596/887/1/012009

Agricultural SWOT analysis and wisdom agriculture design of chengdu

Qian Zhang[1], Xiangyu Chen[2], Shaoming Du[3], Guowei Yin[4], Feng Yu[5], Guicai Liu[6], Jin Gong[7] and Fujun Han[8]

[1]Institute of Agricultural Information and Economics, Beijing Academy of Agriculture and Forestry Sciences, Beijing, China, zhangqianshuxue@126.com
[2]Institute of Agricultural Information and Economics, Beijing Academy of Agriculture and Forestry Sciences, Beijing, China, 31899100@qq.com
[3]Department of market and economic information, Ministry of Agriculture, Beijing, China, dushm@agri.gov.cn
[4]Department of market and economic information, Ministry of Agriculture, Beijing, China, guoweiyin@qq.com
[5]Institute of Agricultural Information and Economics, Beijing Academy of Agriculture and Forestry Sciences, Beijing, China, yuf@agri.ac.cn
[6]Department of market and economic information, Ministry of Agriculture, Beijing, China, liugc@agri.gov.cn
[7]Institute of Agricultural Information and Economics, Beijing Academy of Agriculture and Forestry Sciences, Beijing, China, gongj@agri.ac.cn
[8]Department of market and economic information, Ministry of Agriculture, Beijing, China, hanfj@agri.gov.cn

Abstract. According to the status of agricultural information, this paper analyzed the advantages, opportunities and challenges of developing wisdom agriculture in Chengdu. By analyzed the local characteristics of Chengdu agriculture, the construction program of Chengdu wisdom agriculture was designed, which was based on the existing agricultural informatization. The positioning and development theme of Chengdu agriculture is leisure agriculture, urban agriculture and quality agriculture.

1. Introduction

Wisdom Agriculture is a new direction for agricultural development, and it has not yet formed a unified definition at home and abroad [1-2]. Integrated the related literatures description [3-5], wisdom agriculture can be described as: the comprehensive application of internet of things, cloud computing, large data, mobile internet and other modern information technology. Combined with expert intellectual resources, it achieved intelligent production, network operation, digital management and precision service. Wisdom agriculture will lead the development of modern agriculture, and to be the inevitable trend of future agricultural development. At present, the focus of wisdom agriculture research only in the technical level [6-7], in the application, it was still in the experimental demonstration stage [8], and there were few mature regional applications.

Wisdom agriculture's regional application should base on the existing information facilities. How to develop wisdom agriculture in modern cities, which maximize used existing information technology and highlighted the advantages and own characteristics of different cities, was needed to think about before overall design.

2. Agricultural Informatization in Chengdu

Chengdu is located in the central of Sichuan province, belonging to the subtropical humid monsoon climate zone. It is rich in calories, abundant rainfall and four seasons distinct, and mainly by the plains, hills and mountains. In 2015, the city has a population of 14.482 million, land area of 18.185 million hectares and the total crop area of the crop is 683,900 hectares. Per capita disposable income of rural residents was 17747 yuan, and with an increase of 10.2%.

Chengdu is currently focusing on the development of modern agriculture, and accelerates the construction of national urban modern agricultural demonstration city. Agricultural informatization of

Chengdu has been greatly improved; however, it still has a big gap with the 2020 development goals. Therefore, Chengdu must seize the historical development opportunities, vigorously develop wisdom agriculture and improve the level of agricultural information, so as to speed up the realization of Chengdu urban agriculture modernization.

3. Agricultural SWOT Analysis of Chengdu

The SWOT analysis was proposed by the professor of management at the University of San Francisco in the early 1980s. "SWOT" is shorthand of Strengths, Weakness, Opportunity and Threat. SWOT analysis is a kind of situation analysis, that is, listed a variety of major internal advantages, disadvantages and threats witch closely related with the research object by enumeration. And then, use the idea of system analyzed the various factors to match each other, from which a series of corresponding conclusions. The conclusion is usually with a certain degree of decision-making [9].

3.1. Strengths Analysis

First, Chengdu has superior natural geographical conditions. It is located in the central region of Sichuan, with good terrain, diverse terrain and fertile soil, where the winter climate is warm, summer without heat, perennial air moist, and very suitable for crop growth. As a key city for regional development, Chengdu has obvious geographical advantages.

Second, Chengdu has broad prospects for agriculture. Chengdu insisted on taking agriculture as basic industries and strategic industries, and focused on changing the mode of agricultural development. Chengdu has formed a series of characteristic agriculture, such as vegetables, seasonal fruits, tea, kiwi, edible fungus, flower seedlings and aquatic products. Many agricultural and sideline products were exported to the country, renowned overseas.

Third, Chengdu has a huge market potential. As a large central city in the west China, Chengdu wisdom agriculture will drive more than 10 billion yuan market demand in future. Among them, farming, aquaculture and other production links will grow more than 3 billion yuan; Processing, marketing, logistics, distribution and other circulation will grow more than 6.5 billion yuan.

Fourth, Chengdu has plenty of intellectual resources. There were a lot of research institutes, such as Sichuan University, University of Electronic Science and Technology, and so on. It has a large number of local information technology enterprises, and also attracted many foreign information technology enterprises. In addition, there were high-quality enterprise resources, such as Intel, dell, Huawei, China Unicom and so on. All of these provided an intellectual resource reserves for the development of Chengdu wisdom agriculture.

3.2. Weakness Analysis

First, the level of Chengdu's agricultural informatization is not high. The large data integration applications were low and had not yet established the innovative research and development application system of agricultural networking. Agricultural integrated information service should be further improved. Small-scale agricultural management subjects were lack of consciousness for information investment and their ability were not strong.

Second, the maturity of agricultural informatization technology was low. At present, Chengdu agricultural information infrastructure construction process was relatively slow, and cannot meet the needs of wisdom agriculture construction, especially in the information entry rate and computer penetration rate.

Third, Chengdu was lack of agricultural information technology complex talents. Wisdom agriculture is a new thing with high technological content, and there was an urgent need for a large number of complex talents, which understand modern information technology and agricultural technology, and good at operating and management.

3.3. Opportunity Analysis

First, the government attached importance to agricultural informatization. Vigorously develop agricultural and rural information was the urgent need of government to promote agricultural modernization and build a well-off society. In current and future period of time, promote the construction of wisdom agriculture was the important measures to speed up the information and agricultural modernization.

Second, Chengdu has strong support of municipal government. The government put agricultural information work in an important prominent position and substantial support for it. The government has issued a number of documents to orderly advancement related work, which provided a strong policy support and financial protection.

Third, Chengdu has a complete agricultural information conditions. Since 2014, Chengdu has more than 30 cable connecting other provinces, and metropolitan area network export total bandwidth of more than 1.5T. The communication network transmission capacity ranked the forefront of the Midwest China. The 4.4 million city residents access to fiber, and wireless communication network coverage 100% administrative village.

3.4. Threat Analysis

First, the key technology of information technology and equipment depended on imports. In general, China was more or less depended on the import of information technology core equipment and software, in the internet of things, large data, cloud computing, 4G communications and other fields. Intelligent application technology research was needed to carry out in field planting, facilities cultivation and agricultural operations.

Second, the level of digitization of agricultural management needs to be improved. Big data is a necessary condition for the construction of wisdom agriculture. Although the level of informatization in Chengdu has been significantly improved in agricultural administrative examination and approval and management, the production data and business information of the enterprises and bases didn't communicate with the outside and the government. Therefore cannot achieve data sharing and centralized management.

4. Positioning Analysis of Chengdu Wisdom Agriculture

Design of Chengdu wisdom agriculture should closely integrated with its agricultural production, highlight local features, so as to build the leading cities in China of wisdom agriculture. Chengdu wisdom agriculture will highlight three positioning of leisure agriculture, urban modern agriculture and quality agriculture.

4.1. Leisure Agriculture

The geographical and climate environment have provided a unique condition for the development of leisure and tourism in Chengdu. In recent years, rural tourism in Chengdu has also developed rapidly: in 2015, Chengdu total received 191 million domestic and foreign tourists, and achieved total tourism income 2040.19 billion yuan. Among them, 95,191,800 rural tourists were received, which accounting for 49.75% of the total number and income of 20.06 billion yuan. Chengdu held 117 various rural tourism events, built 1489 new villages [10-11], operated 93 star country hotels and 10 agricultural theme parks[12].

Table1. Chengdu rural tourism and income

	Chengdu Tourism Scale and Income			
Year	***Rural visitors (million)***	***City visitors (million)***	***Rural tourism revenue (million yuan)***	***City tourism revenue (million yuan)***
2012	8427.2	12200	12849	1050.78
2013	8559.5	15500	14294	1330.66
2014	8896.43	18600	16097	1663.37
2015	9519.18	19100	20006	2040.19

Leisure agriculture is a new type of agricultural production and management, which used of agricultural landscape resources and agricultural production conditions to develop agriculture. Develop leisure agriculture in Chengdu can full use its human history, natural resources and other prominent advantages. The leisure agriculture can also effectively promote the rural industry interaction, the agricultural production efficiency, farmer employment income and new rural scientific development. Therefore, leisure agriculture is one of the positioning of wisdom agriculture in Chengdu.

4.2. Urban Modern Agriculture

The urbanization rate of Chengdu was 71.5% in 2015, which was higher than the province 23.8% and higher than the national average 15.4%. The rural labor output was 200.2 million, and non-agricultural employment increased by 87,000. According to the relevant evaluation system [9], Chengdu has a certain potential to develop urban agriculture.

Urban modern agriculture a kind of modern agriculture, which closely rely on the city's various advantages, and provide agricultural products for the market and good ecological environment for the public. Chengdu citizens have been pursuit the health and leisure lifestyle, which is very fit for the development of modern urban agriculture. Chengdu wisdom agriculture, on the one hand, should protect the quality and supply of agricultural products, on the other hand, should be fully satisfied citizens' spiritual needs and ecological demands.

4.3. Quality Agriculture

Chengdu agricultural products have always focused on quality and brand building. By the end of 2015, the certification of pollution-free agricultural products, green food and organic agricultural products has totaled 1213. Among them, there were 378 pollution-free agricultural products, 246 Green food and 589 organic agricultural products. The green ecological agriculture was based 50 million mu, with output value of 5 billion yuan. Chengdu has a total of 450 agricultural products brand.

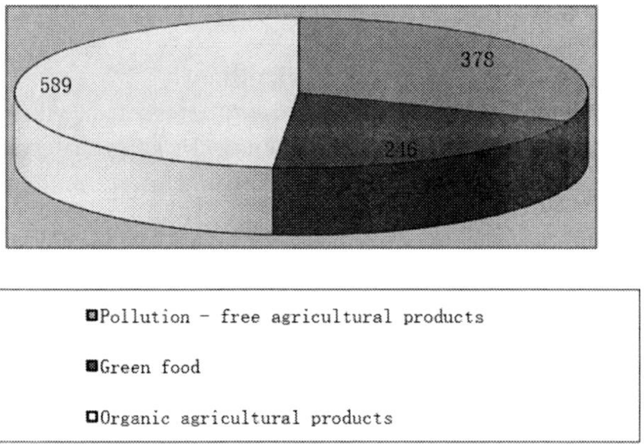

Figure1. Chengdu quality certification distribution in 2015.

Quality agriculture is a kind of high yield, high quality and efficient modern agriculture, which based on the agriculture scale and regionalization. It focused on famous brand and product quality, and took agricultural products quality and safety as the core. The key tasks of Chengdu wisdom agriculture one was to strengthen the quality and safety of agricultural products, and strengthen the source management with information technology, so as to ensure the quality of agricultural products and seize the future of agricultural development.

5. Chengdu Wisdom Agriculture Design

The overall design of Chengdu wisdom agriculture highlighted the regional agricultural characteristics, and focused on the key areas of intelligent agriculture. Through the further application of information technology, it achieved intelligent management and services which covering the whole industry chain. The chain included breeding, planting, purchasing, storage, processing, transportation, sales and so on. The overall structure of Chengdu wisdom agriculture was summed up as three "one": one data center, one integrated application and one supervision map.

5.1. One Data Center

Build Chengdu wisdom agriculture cloud data center. Through data fusion and data self-built, build a cloud data center that brings together all the agricultural data in Chengdu, so as to achieve unified management of data resources and open sharing. These data included production data, environmental data, circulation data, technology data and service data.

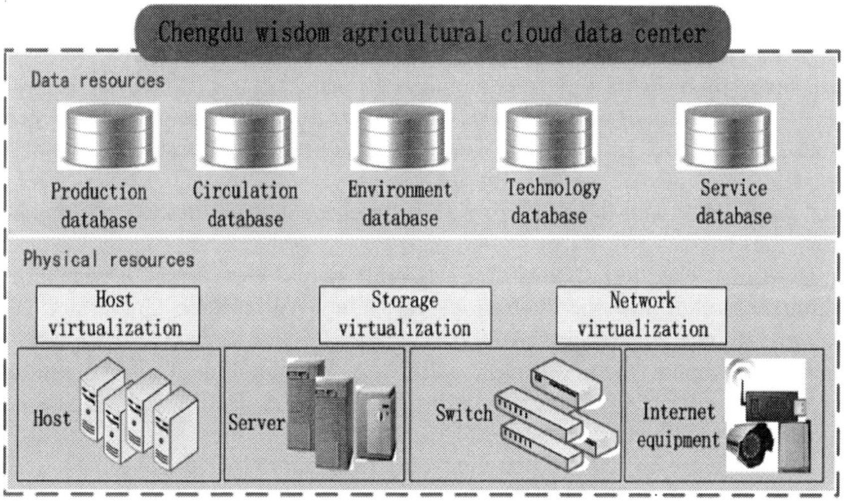

Figure 2. Architecture diagram of cloud data center.

5.2. One Integrated Application

On the basis of the cloud data center Chengdu wisdom agriculture achieved the integration application of all related agriculture activities, which with comprehensive application of a variety of information technology, such as mobile information service, 3S technology, wireless awareness, internet of things, location based service and so on. These applications were covered agricultural industry chain, the whole industry type and all agriculture-related services. The integration applications were divided into three categories: agricultural production information application, agricultural products circulation and transaction information application, agriculture-related services information application.

5.3. One Supervision Map

In order to improve supervision, command and dispatch management level of Chengdu agriculture, real-time master the city's agricultural production, management and service dynamics, Chengdu wisdom agriculture was needed to achieve a map of supervision. The map real-time and intuitive showed the entire Chengdu agricultural production situation, which provided users with agricultural resources display, statistical data analysis, agricultural disaster warning, production commanding, agricultural scheduling and other services. The map improved the agricultural early warning response and emergency command capability, and provided scientific basis for leadership decision-making and command.

6. Conclusion and Suggestions

The construction of Chengdu wisdom agriculture needed to be based on the local location advantages, urban characteristics and information technology foundation. To promote the adjustment of agricultural structure with information technology, and further optimize the mode of agricultural development, so as to constantly promote the new form of agricultural industry. Through strengthening the overall planning, do top-level design, develop relevant favorable policies, while encouraging multi-capital access, All-round promoted Chengdu wisdom agriculture development.

7. References

[1] Qin Huaibin, Li Daoliang, Guo Li. "Recent Advances in Development and Key Technologies of Internet of Things in Agriculture", Research on Agricultural Mechanization, vol. 4, 2014, pp. 246-248.

IOP Conf. Series: Journal of Physics: Conf. Series **887** (2017) 012009 doi:10.1088/1742-6596/887/1/012009

[2] Wang Yujie, "Internet of Things and Wisdom Agriculture". Beijing: China Agricultural Publishing House, 2014, pp. 61-68.
[3] Li Daoliang, "Internet of Things and Wisdom Agriculture", Agricultural Engineering, vol. 2, no. 1, 2012, pp. 1-6.
[4] Sun Zhongfu, Du Keming, Zheng Feixiang, Yin Shouyi, "Perspectives of Research and Application of Big Data on Smart Agriculture", Journal of Agricultural Science and Technology, vol. 15, no. 6, 2013, pp. 63-71.
[5] Zhang Songmao, "Research on Construction of Modern Intelligent Agricultural System". Nanjing: Nanjing University of Posts and Telecommunications, 2013.
[6] Zhang Xin, Chen Lansheng, Zhao Jun, "Design and Application of Intelligent Agricultural Greenhouse Based on Internet of Things Technology", Journal of Chinese Agricultural Mechanization, vol. 36, no. 5, 2015, pp. 90-95.
[7] Peng Gaili, "Application and Research of IOT in Intelligent Agricultural". Zhengzhou: Zhengzhou University, 2012.
[8] Ding Jing, Bai Ling, Zhu Jing, "Development Strategy and Development Mechanism of Wisdom Agriculture in Anhui Province", Journal of Anhui Agriculture University (Social Sciences Edition), vol. 25, no. 2, 2016, pp. 14-20.
[9] Wen Hua, Jiang Cuihong, Wang Ailing, Jia Jinsong, "Establishment of Evaluation Index System of Metropolis Modern Agriculture and Its Diagnosis in Beijing", Research of Agricultural Modernization, vol. 29, no. 2, 2008, pp. 156-158.
[10] Chengdu Tourism Bureau, "2015 Chengdu Tourism Economy Analysis Report", March 2016.
[11] Chengdu Tourism Bureau, "2014 Chengdu Tourism Economy Analysis Report", March 2015.
[12] Wu ronghui, Yang jianbo, Wang yao, "Thoughts on the Development of Leisure Agriculture of Chengdu Under the New Situation", Sichuan Agriculture Science and Technology, vol. 2, 2017, pp. 71-74.

Geomagnetic matching navigation algorithm based on robust estimation

Weinan Xie, Liping Huang, Zhenshen Qu and Zhenhuan Wang

Control Science and Engineering Department, Harbin Institute of Technology, Harbin, China.

Email: xieweinan@hit.edu.cn

Abstract. The outliers in the geomagnetic survey data seriously affect the precision of the geomagnetic matching navigation and badly disrupt its reliability. A novel algorithm which can eliminate the outliers influence is investigated in this paper. First, the weight function is designed and its principle of the robust estimation is introduced. By combining the relation equation between the matching trajectory and the reference trajectory with the Taylor series expansion for geomagnetic information, a mathematical expression of the longitude, latitude and heading errors is acquired. The robust target function is obtained by the weight function and the mathematical expression. Then the geomagnetic matching problem is converted to the solutions of nonlinear equations. Finally, Newton iteration is applied to implement the novel algorithm. Simulation results show that the matching error of the novel algorithm is decreased to 7.75% compared to the conventional mean square difference (MSD) algorithm, and is decreased to 18.39% to the conventional iterative contour matching algorithm when the outlier is 40nT. Meanwhile, the position error of the novel algorithm is 0.017° while the other two algorithms fail to match when the outlier is 400nT.

1. Introduction

With the rapid development of science and technology, navigation technology has been widely applied in military and civil field in recent decades. Nowadays there are many kinds of navigation technology such as Inertial Navigation System (INS), Global Position System (GPS) [1], Celestial Navigation System[2], and Terrain Contour Matching (TERCOM)[3] and so on. But these systems all have their respective shortcomings. As we all know, INS is a kind of autonomous navigation method. However, its location error accumulates along with the elapse of time. GPS technology is with high positioning accuracy, and it can output all-weather navigation information. Yet it is unable to locate when the GPS receivers are in the remote areas where the GPS signals are weak, and it is vulnerable to the interference of the enemy. Celestial Navigation System can't be used when the weather is terrible. TERCOM can't work well when the vehicle moves on the water or the flat terrain.

The geomagnetic field has certain magnitude and direction, which is consistent to all kinds of vector field. It is a kind of solid source of the navigation information. More importantly, it nearly exists on the entire earth's surface[4]. Therefore the geomagnetic navigation can be used in the places where the other navigation methods can't work. Moreover, the geomagnetic navigation has many other advantages such as strong independence, low cost and good concealment. Thus it is favored by more and more scholars in recent decades. Reference [5] presents the process of geomagnetic matching navigation. MSD algorithm is used in [6] to implement the geomagnetic matching. In [7], a numerical iterative method takes the place of the traversal method in MSD algorithm. Reference [8] considers the initial longitude, latitude and heading errors, and it applies the iterative method to realize the fast geomagnetic matching. Reference [9] utilizes affine model transposition to realize geomagnetic matching with higher precision.

However, some false data (namely the outliers) are usually measured because of the external disturbance influence and the performance limitations of the geomagnetic sensor when geomagnetic field is surveyed. It can seriously affect the geomagnetic matching navigation precision and badly disrupt its reliability. Based on [8], a fast geomagnetic matching algorithm that can remove the influence of the outliers is realized in this paper.

This paper will be structured as follows. Section II introduces the geomagnetic matching navigation principle briefly. Section III introduces the robust estimation theory, and in this section, how to design the weight function is also showed. Section IV describes the mathematical principle of the new algorithm in detail. Section V summarizes the processes of the novel navigation algorithm based on robust estimation. Section VI shows the simulation results. The last section is conclusions.

2. The Principle of Geomagnetic Matching Navigation

Figure 1 shows the geomagnetic matching navigation system. This system consists of three modules: INS module, geomagnetic map module and geomagnetic sensor module. Then the information of the three modules is fused together by matching algorithm, and the result is acquired. The concrete processes are as follows [5].

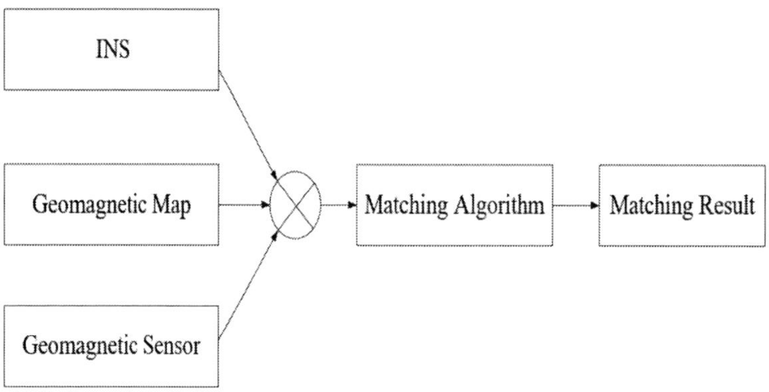

Figure 1. The geomagnetic matching navigation system.

(1) Obtain the geomagnetic map of the matching area before the vehicle moves, and save them in the vehicle computer.

(2) As the vehicle passes through the matching area, the carried geomagnetic sensor gathers a series of real-time geomagnetic data.

(3) Output the INS information, the geomagnetic map and the real-time geomagnetic data to the matching algorithm and calculate matching result.

3. Robust Estimation

As mentioned earlier, there are some outliers in the surveyed geomagnetic data sometimes. But traditional algorithms such as MSD algorithm are all based on the principle of the mean square error. These algorithms can work effectively only when the measured data follow normal distribution, and they can't remove the affection of the outliers. Therefore this paper employs the robust estimation theory in statistics to design new algorithm which can eliminate the influence of the outliers.

The main task of the robust estimation [10] is to design the weight function. The measured data [11] can be usually divided into the normal data, the available data and the outliers in the robust estimation theory. The independent variable of the weight function is correspondingly divided into the protected area, the reducing area and the refusal area. The basic efficiency of the proposed algorithm is guaranteed by the measured data in the protected area, which are in the majority in the survey data. The robustness of the proposed algorithm is guaranteed by the measured data in the reducing area. And the measured data in the refusal area can enhance the algorithm robustness. In this paper, the weight function is selected as follows:

$$\psi(x^2)=\frac{x^2}{\sigma+x^2} \tag{1}$$

where σ is an adjustable parameter, and it is determined by the value of the outliers.

Then the derivative of equation (1) is as follows:

$$\frac{d\psi(x^2)}{dx}=\frac{2x\sigma}{(\sigma+x^2)^2} \tag{2}$$

Figure 2 presents the weight function curve and Figure 3 shows its derivative curve. It can be seen that both the weight function and its derivative provide a gradual transition between the normal data and the outliers.

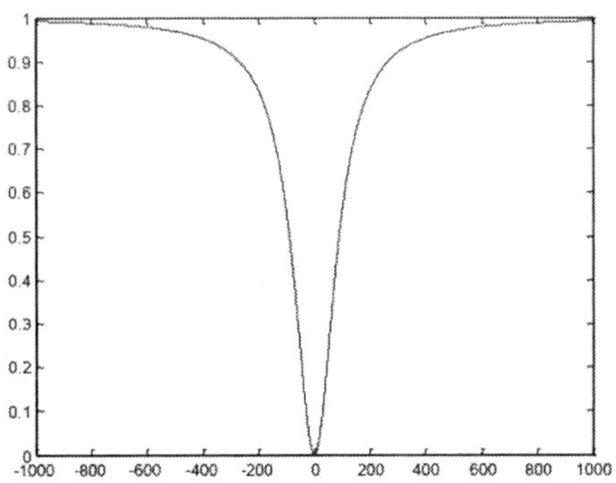

Figure 2. The weight function.

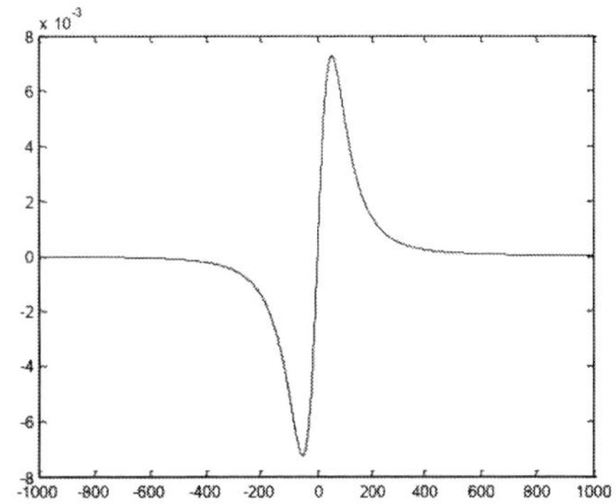

Figure 3. The derivative of the weight function.

4. The Mathematical Principle of the Proposed Algorithm

Because of the INS deficiencies, there are displacement and heading errors between the reference trajectory which is calculated by INS and the real trajectory. Figure 4 is the sketch of the trajectories.

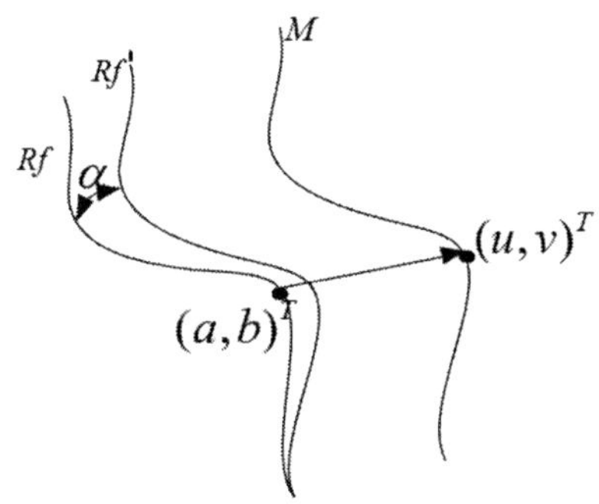

Figure 4. The sketch of the trajectories.

In Figure 4, Rf represents the trajectory which is calculated by INS and it is called the reference trajectory. M is the trajectory which is calculated by geomagnetic matching algorithm and it is called the matching trajectory. Rf' is parallel to M. α is the heading error between the reference trajectory and the matching trajectory. The displacement error consists of the longitude and latitude errors. The longitude error is denoted by Δx and the latitude error is denoted by Δy. $(a,b)^T$ represents a point on the reference trajectory, then the corresponding point on the matching trajectory is $(u,v)^T$. $(a_1,b_1)^T$ is the initial point on the reference trajectory. Therefore the relation equation between the matching trajectory and the reference trajectory can be represented by equation (3).

$$\begin{bmatrix} u \\ v \end{bmatrix} = \begin{bmatrix} \cos\alpha & \sin\alpha \\ -\sin\alpha & \cos\alpha \end{bmatrix} \begin{bmatrix} a-a_1 \\ b-b_1 \end{bmatrix} + \begin{bmatrix} a_1 \\ b_1 \end{bmatrix} + \begin{bmatrix} \Delta x \\ \Delta y \end{bmatrix} \tag{3}$$

A suitable choice for geomagnetic matching algorithm is the correlation criterion. The correlation criterion is selected as follows:

$$E = I(u,v) - I_t(a,b) \tag{4}$$

where $I(u,v)$ represents the geomagnetic field intensity of the matching point $(u,v)^T$. $I_t(a,b)$ is actual magnetic field intensity measured by magnetic sensor when the carrier is located at reference point $(a,b)^T$.

The matching trajectory is located nearby the reference trajectory. Thus the geomagnetic field intensity can be expanded in a Taylor series around the reference trajectory.

$$I(u,v) = I(a+(u-a), b+(v-b)) = I(a,b) + \frac{\partial I(a,b)}{\partial x}(u-a) + \frac{\partial I(a,b)}{\partial y}(v-b) + O_2 \tag{5}$$

where $I(a,b)$ is the geomagnetic field intensity of the reference point $(a,b)^T$ which can be obtained from geomagnetic map. $\partial I(a,b)/\partial x$ represents the geomagnetic field intensity gradient along the longitude at the point $(a,b)^T$. $\partial I(a,b)/\partial y$ is the geomagnetic field intensity gradient along the latitude at the point $(a,b)^T$. O_2 represents higher order term. Substituting equation (5) into equation (4) and neglecting the higher order term O_2, we can obtain equation (6).

$$E = I_x(u-a) + I_y(v-b) + I(a,b) - I_t(a,b) \tag{6}$$

where

$$I_x = \frac{\partial I(a,b)}{\partial x} \tag{7}$$

$$I_y = \frac{\partial I(a,b)}{\partial y} \tag{8}$$

Then substitute equation (3) into equation (6) and discretize it, thus equation (9) is acquired as follows:

$$E_n = I_{xn}(a'_n\cos\alpha + b'_n\sin\alpha - a'_n + \Delta x) + I_{yn}(-a'_n\sin\alpha + b'_n\cos a - b'_n + \Delta y) + I(a_n,b_n) - I_t(a_n,b_n) \tag{9}$$

where

$$I_{xn} = \left.\frac{\partial I(a,b)}{\partial x}\right|_{a=a_n, b=b_n} \tag{10}$$

$$I_{yn} = \left.\frac{\partial I(a,b)}{\partial y}\right|_{a=a_n, b=b_n} \tag{11}$$

$$a'_n = a_n - a_1 \tag{12}$$

$$b'_n = b_n - b_1 \tag{13}$$

and n represents the ordinal of the sampling points. a_n and b_n are the coordinates of the sampling point on the reference trajectory.

The robust target function is acquired by combining equation (1) with equation (9).

$$F = \sum_{n=1}^{p} \frac{E_n^2}{\sigma + E_n^2} \tag{14}$$

where p represents the total number of the sampling points and $p > 2$.

The geomagnetic matching problem is changed into searching for the values of Δx, Δy and α to make the robust target function minimum, therefore the first-order partial derivatives of F function to Δx, Δy and α are obtained as follows:

$$g_1 = \frac{\partial F}{\partial \Delta x} = \sum_{n=1}^{p} G_n I_{xn} \tag{15}$$

$$g_2 = \frac{\partial F}{\partial \Delta y} = \sum_{n=1}^{p} G_n I_{yn} \tag{16}$$

$$g_3 = \frac{\partial F}{\partial \alpha} = \sum_{n=1}^{p} [I_{xn}(-a_n' \sin\alpha + b_n' \cos\alpha) + I_{yn}(-a_n' \cos\alpha - b_n' \sin\alpha)] G_n \tag{17}$$

where

$$G_n = \frac{\sigma E_n}{(\sigma + E_n^2)^2} \tag{18}$$

Then nonlinear equations can be acquired as follows:

$$\begin{cases} g_1(\Delta x, \Delta y, \alpha) = 0 \\ g_2(\Delta x, \Delta y, \alpha) = 0 \\ g_3(\Delta x, \Delta y, \alpha) = 0 \end{cases} \tag{19}$$

Newton iteration [12] is employed to solve nonlinear equations in this paper. Then the solutions Δx, Δy and α of the nonlinear equations are substituted into equation (3), and the matching trajectory can be acquired.

It should be noted that the large initial errors can disturb the matching precision for neglecting the higher order term of the Taylor series. So the proposed algorithm can be iterated many times and the first iteration matching result is regarded as the second iteration reference trajectory. Because the first iteration matching result eliminates most of the displacement and heading errors, the second iteration will acquire the matching result which is closer to the real trajectory, and the matching precision will be further improved.

5. The Process of the Proposed Algorithm

By employing Newton iteration to solve the nonlinear equations mentioned earlier, the concrete process of the proposed algorithm based on robust estimation is as follows:

Step 1: Obtain the coordinates (a_n, b_n) of sampling points from INS and their corresponding actual magnetic field intensity $I_t(a_n, b_n)$ measured by magnetic sensor.

Step 2: Retrieve the geomagnetic field intensity $I(a_n, b_n)$ and the geomagnetic field intensity gradient I_{xn}, I_{yn} from geomagnetic map.

Step 3: Initialize the longitude, latitude and heading errors:

$$M = [\Delta x \quad \Delta y \quad \alpha]^T = [0 \quad 0 \quad 0]^T \tag{20}$$

Step 4: Calculate L and G as follows:

$$L=\begin{bmatrix} \frac{\partial^2 F}{\partial \Delta x^2} & \frac{\partial^2 F}{\partial \Delta x \partial \Delta y} & \frac{\partial^2 F}{\partial \Delta x \partial \alpha} \\ \frac{\partial^2 F}{\partial \Delta y \partial \Delta x} & \frac{\partial^2 F}{\partial \Delta y^2} & \frac{\partial^2 F}{\partial \Delta y \partial \alpha} \\ \frac{\partial^2 F}{\partial \alpha \partial \Delta x} & \frac{\partial^2 F}{\partial \alpha \partial \Delta y} & \frac{\partial^2 F}{\partial \alpha^2} \end{bmatrix} \tag{21}$$

$$G=\begin{bmatrix} \frac{\partial F}{\partial \Delta x} & \frac{\partial F}{\partial \Delta y} & \frac{\partial F}{\partial \alpha} \end{bmatrix}^T \tag{22}$$

Step 5: Calculate δM as follows:

$$\delta M=-L^{-1}\times G \tag{23}$$

Step 6: Update the longitude, latitude and heading errors:

$$M=M+\delta M \tag{24}$$

Step 7: If the process meets the end condition (the iterative number reaches the predetermined number or the 2-norm of δM is less than the predetermined value), go to Step 8, and otherwise go to Step 4.

Step 8: Substitute the values of Δx , Δy and α into equation (3), and output the matching result.

6. Simulation and Results

In the simulations, international geomagnetic anomaly data is regarded as the geomagnetic map. The reference trajectory is generated by the strap-down inertial navigation system (SINS) [13]. Table I shows the simulation parameters.

Table 1. Simulation parameters.

Parameters	Value
Initial velocity(m/s)	(40,35)
Initial position error(°)	(0.017,-0.018)
Initial heading error(deg)	7.3
Initial velocity error(m/s)	(0.1,0.25)
Accelerometer drift(ug)	80
Gyroscopic drift(°/h)	0.15
Simulation time(s)	1000
Geomagnetic sensor noise(nT)	3

In the simulations, the proposed algorithm in this paper is compared with traditional MSD algorithm[6] and the geomagnetic contour matching algorithm based on iterative method [8]. The different outliers are added on the real geomagnetic data in two simulations.

In simulation case 1, the outlier of 40nT is added on the eighth geomagnetic data measured by geomagnetic sensor. Figure 5 represents the matching trajectories, and Figure 6 shows the matching errors. It can be seen that the matching trajectory of the proposed algorithm is very close to the real trajectory and the matching error of the novel algorithm is the smallest. The maximum error of MSD algorithm is 0.100°. The maximum error of the geomagnetic contour matching algorithm based on iterative method is 0.042°. The maximum error of the proposed algorithm is 0.007°. Therefore, the proposed algorithm can effectively eliminate the influence of the outliers.

In simulation case 2, the outlier of 400nT is added on the eighth geomagnetic data measured by geomagnetic sensor. Figure 7 and Figure 8 show the results. It can be seen that the maximum error of the proposed algorithm is 0.017° while the other two algorithms fail to match. Therefore, the proposed algorithm still can realize the high precision matching when the outlier is big.

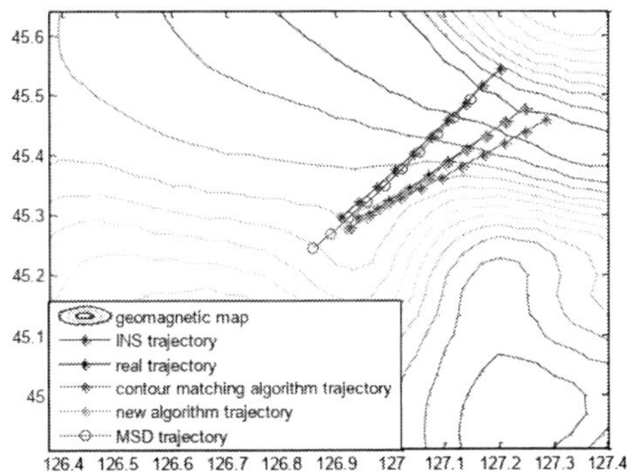

Figure 5. The matching trajectories with 40nT outlier.

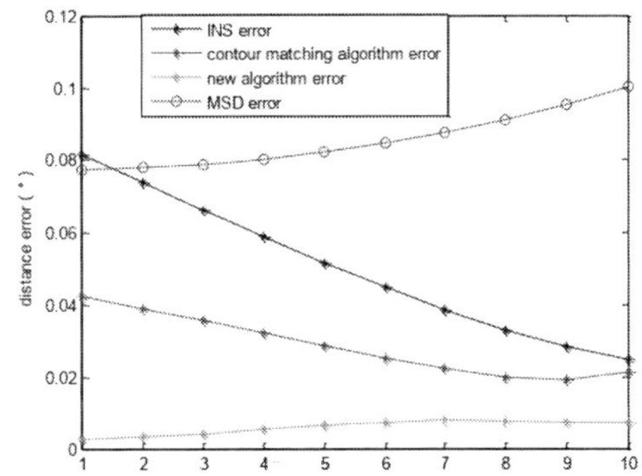

Figure 6. The matching errors with 40nT outlier.

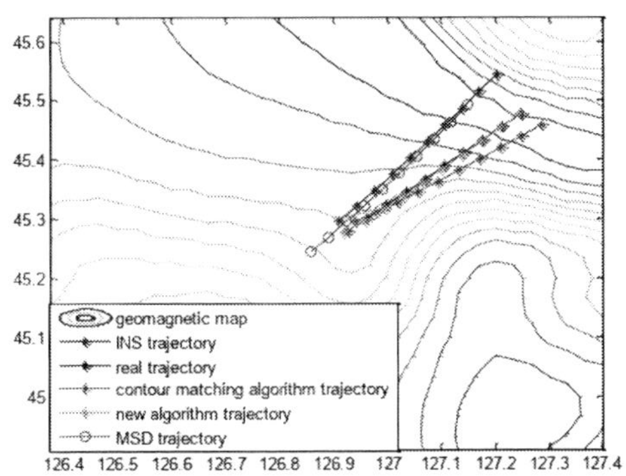

Figure 7. The matching trajectories with 400nT outlier.

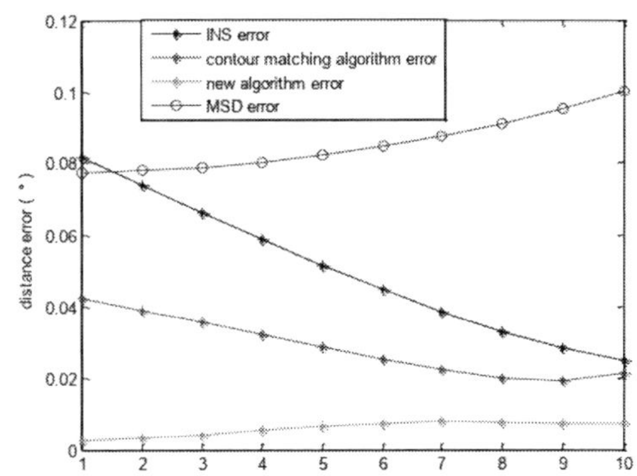

Figure 8. The matching errors with 400nT outlier.

7. Conclusion

A novel geomagnetic matching navigation algorithm which can eliminate the influence of the outliers is proposed in this paper. By constructing the robust target function, the robustness of the proposed algorithm is guaranteed. The matching result is calculated by Newton iteration. The simulation results indicate that the matching error of the proposed algorithm is decreased to 7.75% compared to the conventional MSD algorithm, and is decreased to 18.39% to the conventional iterative contour matching algorithm when the outlier is 40nT. Meanwhile, the maximum error of the proposed algorithm is 0.017° while the other two algorithms fail to match when the outlier is 400nT. Therefore, the proposed algorithm can effectively eliminate the influence of the outliers, and realize the high precision matching.

Further research will investigate better robustness of shape constraints against large inertial component drift.

Acknowledgments

This work is partially supported by the National Natural Science Foundation of China (61403095, 61375046, and 61640301).

8. References

[1] Inone J, Sanghyouk C, Joohwan C and Cheonsig S Robust cross-score algorithm against bit transitions for GPS interference suppression 2017 *IEEE Communications Letters***21**1325-28

[2] Xiaolin N, Mingzhen G, Jie Z, Jiancheng F and Gang LSolar oscillation time delay measurement assisted celestial navigation method 2017*Acta Astronautica***134** 152-158

[3] Ling Z, Xianghong C, Yixian Z and Yuan L 2015 Terrain aided navigation for long-range AUVs using a new bathymetric contour matching method *Proc. IEEE Int. Conf. on Advanced Intelligent Mechatronics* pp 249-254

[4] Caifa G, Anliang L, Hong C and Huabo Y 2011 Algorithm for geomagnetic navigation and its validity evaluation *Proc. IEEE Int. Conf. on Computer Science and Automation Engineering* pp 573-577

[5] Wei L, Zhitian W, Meiping W and Xiaoping H 2011 Geomagnetic matching technology based on iterative contour matching algorithm Proc. *The 10th Int. Conf. on Electronic Measurement & Instruments* pp 342-345

[6] Xianglei W, Fuping S and Po C Research on algorithm of geomagnetic matching navigation 2009 *Engineering of Surveying and Mapping***18** 37-39

[7] Weinan X, Zhenshen Q and Qinghua L 2013 A fast algorithm of the geomagnetic correlation matching based on MSD *Proc. 3rd Int. Conf. on Control, Automation and Systems Engineering* pp 59-62

[8] Weinan X, Qinghua L, Boqi X, Liping H and Changhong W Geomagnetic contour matching algorithm based on iterative method 2015 *Journal of Chinese Inertial Technology***23** 631-635

[9] Shitu L and Zhixin R Geomagnetic matching algorithm based on affine model 2010 *Journal of Chinese Inertial Technology***18** 462-465

[10] Yibin Y The design of weight function for robust estimation depending on equivalent analysis method 2001 *Engineering of Surveying and Mapping***10** 29-35

[11] Xiang L, Lingqun L. and Zhizhong G State estimation with the least robustness square method 2003 *Relay***31** 50-53

[12] Hui L Newton iteration method for nonlinear equation solutions and its application 2007 *Journal of Chongqing Institute of Technology***21** 95-98

[13] Liansheng L and Tao J 2011 Research on strap-down inertial navigation system simulation *Proc. 4th Int. Conf. on Intelligent Computation Technology and Automation* pp 1168-71

Design and realization of high voltage disconnector condition monitoring system

Jinrui Shi[1], Tianyang Xu[2], Shuixian Yang[3] and Buoyang Li[4]

[1]North China University of Water Resources and Electric Power, Zhengzhou, China, lily@ncwu.edu.cn
[2]Engineering Management Branch, the Yellow River Sanmenxia Mingzhu Group Co., Ltd., Sanmenxia, China, xty60830884@163.com
[3]North China University of Water Resources, and Electric Power, Zhengzhou, China, 315266468@qq.com
[4]North China University of Water Resources and Electric Power Zhengzhou, China, lby@ncwu.edu.cn

Abstract. The operation status of the high voltage disconnector directly affects the safe and stable operation of the power system. This article uses the wireless frequency hopping communication technology of the communication module to achieve the temperature acquisition of the switch contacts and high voltage bus, to introduce the current value of the loop in ECS, and judge the operation status of the disconnector by considering the ambient temperature, calculating the temperature rise; And through the acquisition of the current of drive motor in the process of switch closing and opening, and fault diagnosis of the disconnector by analyzing the change rule of the drive motor current, the condition monitoring of the high voltage disconnector is realized.

1. Introduction

High-voltage disconnector is the largest and most widely used high-voltage electrical equipment in power station, substation, and whole power system. But because of its main function of isolation, the manufacturer and user haven't paid enough attention to it even if its, structure is relatively simple and easy to make [1]. For a long time, the reliability of the high-voltage disconnector was significantly lower than that of other substation equipment. Therefore, high voltage disconnector has become the weakest link of electrical equipment in the running process, and its frequent failure has also become one of the important issues affecting the safe operation of the power grid [2].

For the traditional open disconnector, the two main categories of impact on its safe and reliable operation are : On the one hand, because of long-time operation, oxidation and dust laying, the current carrying part of high-voltage disconnector will constantly gain heat, what is worse is that too much heat may cause insulation aging, breakdown, or a short circuit, which may bring a major accident, and serious economic losses[7]; on the other hand, improper installation and adjusting of high voltage disconnector or mechanical clamping caused by corrosion problems may lead to the disconnector closing not in place, the operating torque increased and the decline in mechanical transmission parts strength, and further cause the overheat of conducting circuit, and a transmission mechanism of insulator rupture damage and deformation and other serious accidents [5].

IOP Conf. Series: Journal of Physics: Conf. Series **887** (2017) 012011 doi:10.1088/1742-6596/887/1/012011

2. System Structure

The high-voltage switch contact temperature and prediction system based on the wireless frequency-hopping communication technology is composed of two parts: the hardware temperature measuring device and the temperature analysis software [8].

2.1. The Hardware of the System

The temperature unit is arranged according to the number of contact points of the high voltage disconnector, at the same times measuring bus temperature. The online temperature monitoring system is composed of several wireless temperature measuring units [3, 4], wireless relay communication module, wireless gateway and monitoring center network system. The operating mechanism of disconnector selects the current acquisition module to measure the drive motor current [5], and uses the IO switch node to return to the disconnector state, using the network communication mode.

High voltage disconnector status monitoring system is shown in figure 1.

- Wireless temperature measurement module: Temperature sensing transmitter is composed of wireless transmitter module, temperature sensor and temperature layer. Contact of temperature sensor is according to the position and shape of the object to be measured, to achieve the temperature acquisition and signal emission.
- Ordinary temperature measurement equipment is used in the product of single channel communication, once the channel is occupied, the temperature data will be sent failed. Frequency hopping technology is used between wireless frequency module and wireless relay communication module. It will automatically switch to another channel communication when the channel is interfered. There are 64 available channel resources, and it will be effective to ensure the stability of data transmission, reliability and anti-jamming.

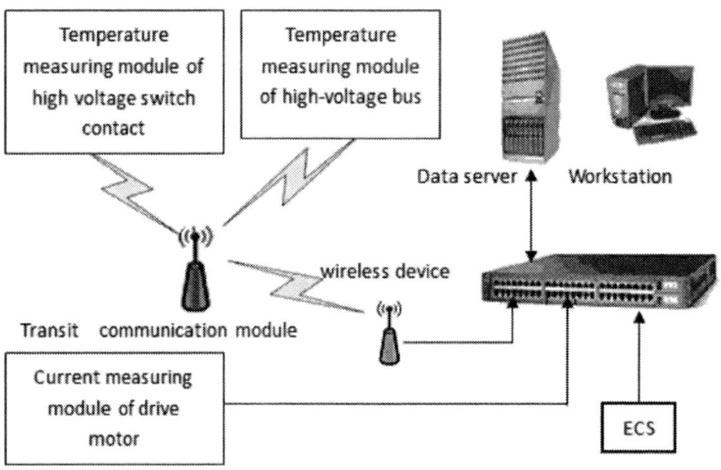

Figure 1. Illustration of status monitoring system

- Wireless transfer communication module: The wireless communication module identifies the data sent by the wireless temperature module, and the data transfer is sent to the wireless communication network computer to provide the data channel between the node and the host computer for the monitoring center computer.
- Current acquisition module: The current acquisition module is used to collect the three-phase current of the high-voltage disconnector, and the data is transmitted to the host computer through the optical fiber communication.
- Monitoring center network system: The network system is composed of a data server, a collection computer and a switch with a wireless communication device. It provides the hardware platform for the software and provides the ECS data channel.

IOP Conf. Series: Journal of Physics: Conf. Series **887** (2017) 012011 doi:10.1088/1742-6596/887/1/012011

2.2. The Structure of System Software

Wireless frequency hopping communication technology of high voltage switch contact temperature measurement and forecast system based on detection and monitoring system design, it is an important link of comprehensive detection and control of power system, and can run independently in the power plant and substation site.

Monitoring center software using Windows Server operating system, MS-SQL2008 database software, application software using B/S structure, and it selects Asp.net development tools. The software system manages all equipment running, the temperature signal of each temperature sensor to monitor regional identification reception, calculation, processing, alarm temperature detection, monitoring, display, overheating and fault judgment, and achieve temperature acquisition, data judgment, data preservation, system communication and other functions. The software system is shown in figure 2.

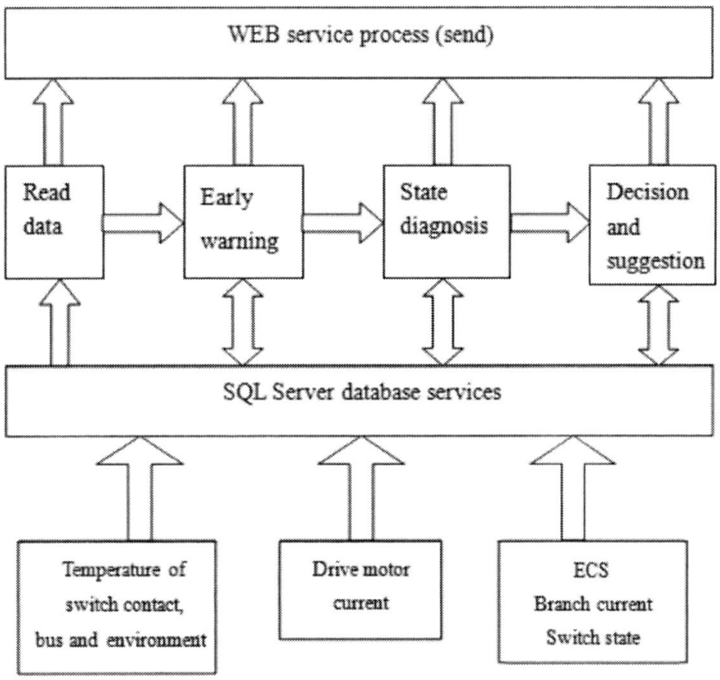

Figure 2. Diagram of software system

3. Realization of State Sonitoring Function

State monitoring function includes the disconnector contacts the normal operation of the temperature detection and disconnector operation process.

3.1. Temperature Monitoring and Alarm

The temperature acquisition module sends the data of the contact temperature, the bus temperature and the ambient temperature to the database according to the certain time interval, at the same time, according to the current value of each branch extracted from the ESC system, the temperature rise can be calculated:

$$\tau_{st} = \frac{I^2 R}{\alpha_t F}$$

In it:

τ_{st} : Stable temperature (°C)

IOP Conf. Series: Journal of Physics: Conf. Series **887** (2017) 012011 doi:10.1088/1742-6596/887/1/012011

α_t : Coefficient of total heat conductor (W/m^2.°C)

I: The current flowing through the conductor (A)

F: Heat conductor area (m^2)

R: Conductor resistance, the change of resistance with temperature is not necessarily consider.

After calculating the temperature rise, The management system calls the corresponding data in the database, the measured temperature of the switch contact or the high voltage bus is compared with the maximum allowable temperature and the temperature rise (according to the specification for electric materials: copper: 110 °C; aluminum: 100 °C; silver: 120 °C). When the value exceeds the allowable value, the system sends an alarm to alert the operator automatically, and the state of high voltage disconnector is diagnosed, in view of the problems, the corresponding suggestions are put forward, so as to provide reference for the operators and solve the corresponding problems as soon as possible, so as to avoid unnecessary economic losses. In addition, WEB service system also provides monitoring point temperature browsing, change curve, statistics, printing and other functions.

3.2. Monitoring of Mechanical Performance during Operations

There are many forms of disconnector operation faults. The severity and location of the faults are also different. However, it all show that the operating torque is increased, the transmission card is not smooth, and the motor current waveform changes compared with the normal situation. And the normal switching process of the high voltage disconnector generally requires 7 - 8s time. Switch status signal should be returned within 8s, or there may be a fault [2, 6].

Typical current waveforms are shown in Figure 3.

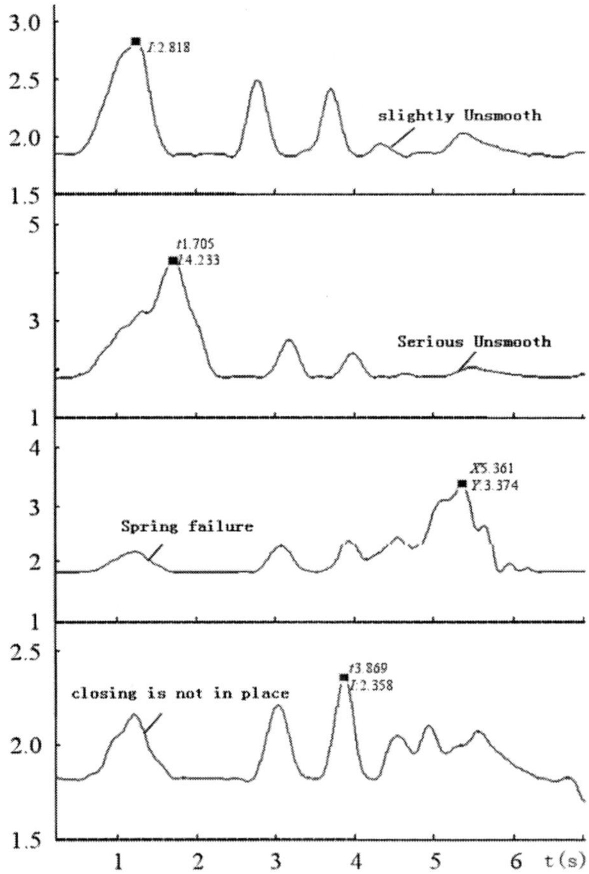

Figure 3. Motor current envelope waveform in different operating condition of the disconnector

We use the analysis of the current signal of the driving motor and the comparison of the closing time, and the mechanical faults of the high voltage disconnector can be effectively detected. The corresponding relationship between the mechanical properties and the motor current can be summarized as shown in table 1.

Table 1. The Corresponding Relation between the Main Mechanical Defects and the Current Characteristics of the Drive Motor

Mechanical state	Type of defect	Variation of motor current characteristic	Current peak period
Closing or Opening operation	Closing or Opening is not in place	In some cases the peak value of the current waveform is too large	5-7s
	Unsmooth operation	The peak value of the current is too large, The peak time is changed, The running time is longer	
	Power failure	The circuit failure No current signal Mechanical fault Abnormal current waveform	Indefinite
	Refuse separation Refuse to close	The peak of the starting current is serious	0-2s
Operrating Mechanical parts	Fastener fall off	The current value is too small And the waveform is stable	
	Balance spring failure	The current peak is large	5-6s

In the daily operation and maintenance management of substation, the drive motor current acquisition hardware system can be installed in the electric operating mechanism, and real time detection of the motor current signal in the operation of the disconnector is realized by the remote software system, and clamping fault diagnosis of disconnector can also be achieved. Compared with the normal working condition, the motor current signal in the closing operation can be used to judge the existence of the mechanical failure and the severity of the disconnector. Combined with the operation of the disconnector and the rotation mechanism, the location of the mechanical failure can be preliminarily judged, and the operation and maintenance measures can be taken in order to prevent the expansion of the accident to cause more damage.

4. The Features of System

The results are accurate and the transmission effect is stable. The system solves the problem of on-line temperature measurement of high voltage electrical equipment by traditional contact method; the wireless sensor is used to accurately monitor the temperature of the electrical equipment switch contact and the high voltage bus. Compared with the display wax sheet and infrared temperature measurement, the wireless sensor also has the advantages of quick response speed, small volume, easy installation, etc. The use of advanced wireless frequency hopping technology can successfully avoid the high voltage equipment in the electromagnetic field and channel interference, to ensure data integrity.

Remote monitoring and fire alarm. The system successfully achieved 24 hours online monitoring and remote monitoring, automatic early warning when the temperature exceeds the maximum allowable value, the location, name and temperature value of the measured object shown on the monitoring center computer screen, at the same time the voice alarm started, The system also provides advice and treatment method of characteristic curve of measured point temperature data to ensure the safe operation of power transmission and transformation equipment.

Convenient and flexible operation. The system is easy to learn and to enhance the ability of the transformer substation to cooperate with each other, and improve the comprehensive intelligence of substation and the flexibility of operating mode.

5. Conclusion

The state monitoring system of high voltage disconnector based on wireless frequency hopping communication technology can be used for 24 hours real-time monitoring of switch contacts and high voltage bus temperature. It uses the temperature sensor to ensure accuracy of data acquisition, to ensure the stability of data transmission by wireless frequency hopping technology. The measured location, name, number, and temperature curve will be shown on the computer of the control room. The temperature rise can be calculated according to the current value extracted from ECS to effectively ensure the accuracy of early warning system; by comparing the current signal of the driving motor and the disconnector under the normal condition, mechanical fault can be effectively detected. If there is any fault, the system can quickly issue a warning, and put forward the corresponding measures to solve the fault, avoid serious accidents, and provide a basis for the operation and maintenance to staff.

The system has been put into operation in a 220kV substation, the disconnector state detection, greatly facilitate the elimination of hidden dangers of staff, so as to prevent accidents especially fire, ensure the safety of power transmission equipment, ensure the safe and stable operation of power system .

6. References

[1] Li Da-wei, Zhang En-wei, Xu Guo-Guo, et al. Research on fault diagnosis method of disconnector based on motor current characteristic [J]. North China electric power technology, 2016,1:42-47.

[2] Qiu Zhi-bin, Ruan Jiang-Jun, Huang Dao-chun, etal. Mechanical fault diagnosis of high voltage disconnector based on motor current detection [J]. proceedings of the Chinese society of electrical engineering, 2015,35 (13): 3459-3465.

[3] Chen Jiang, Li Qing-min, Cong Hao-xi, et al. High voltage switchgear contact temperature monitoring technology research progress [J]. high voltage apparatus, 2015,51 (9): 146-151.

[4] Wu Nan. Research on wireless transmission technology for temperature detection of high voltage switchgear [D]. Shenyang: Shenyang University of Technology, 2016.

[5] Huang Yu-chen, Ruan Jiang-Jun, Qiu Zhi-bin, et.al Study on the relationship between the torque and current of the output shaft of the disconnector operating mechanism [J]. high voltage apparatus, 2015,51 (6): 139-144.

[6] Qiu Zhibin,Ruan Jiangjun, et.al Mechanical Faults Analysis of High Voltage Disconnectors and Review of Diagnosis Technology [J]. high voltage apparatus, 2015,51 (8): 171-179.

[7] Li Yanghua, Modification of temperature monitoring system for high voltage Voltage Disconnectors [J]. Science and Technology& Innovation, 2017-03,120-121.

[8] Yang Yi, Wang Zheng-xu, Zhang Bai-nian et.al Design of a real-time detection and warning system of high voltage switch cabinet hot spot temperature based on wireless infrared communication technology [J]. Journal of southwest University(Natural Science Edition), 2016.38(2):143-148

IOP Conf. Series: Journal of Physics: Conf. Series **887** (2017) 012012 doi:10.1088/1742-6596/887/1/012012

Study to the simulation for pedestrian-vehicle collision accident of the second crushing by PC-CRASH

Jiangyang Gu[1], Xiaoli Chao[2] and Xianguo Hu[3]

[1]Transportation safety applied technology synergy innovation center, AnHui SanLian University, Hefei, China, 943525232@qq.com
[2]Mathematics Department, Southeast University, Nanjing, China, 479183517@qq.com
[3]Mechanical Engineering Department, Hefei University of Technology, Hefei, China, 1621655046@qq.com

Abstract. In the pedestrian-vehicle accident, vehicle may make the second crushing for people after collision. Among the fatal accidents caused by this, there is necessary connection between the movement trail of vehicle and the position of pedestrian, therefore, this text compares with their position relationship through simulating the site of accident by PC-CRASH and the simulation between vehicle parameters and the position of pedestrian, analyzing the injury of each part in pedestrian and the condition of AIS. Analyze the whole process of second crushing and analyze accident cause to put forward scientific, effective solutions.

1. Introduction
At present, in the various accidents which cause the death of people, the accidents caused by motor vehicle continuously increase every year. The most fatal type is caused by second crushing of vehicle in pedestrian-vehicle collision accidents. Second crushing is that driver of motor vehicle re-crush suffer through backing the car or turning direction after colliding suffer to make him injured but not dead. This is the malignant activity to make suffer dead. Therefore, enhance the study to the simulation of pedestrian-vehicle collision accidents is good for just identification of accident liability. Now, the most mature technique to study simulation of pedestrian-vehicle collision accidents is to use PC-CRASH to simulate traffic accident, and it can judge the responsible party of accident through the data of simulation. In 2009, Hua Xiaomin put forward a type of research method to analyze the result error for simulation of pedestrian-vehicle collision accidents by PC-Crash, and this method introduces the relevant solutions. In 2014, Wang Chen made study to simulation of car-guard bar accident by PC-CRASH, which rebuilds field model of accident and specifically analyzes the preventive plan and the damage of driver caused in the process of collision. [1] However, China has no relevant experimental case before, so they have great research significance..

2. Introduction of PC-CRASH
In 1999, Austrian Doctor Hermann Steffan developed a type of new software based on the model Kudlich-Sliba, namely PC-CRASH, which is the basic software to study the simulation reappearance of traffic accident, and also the learning model for other relevant software. PC-CRASH, the typical software used for accident simulation through momentum and impulse momentum method, has very strong practicability. Later, in order to verify the applied accuracy of PC-CRASH in reality, Cliff and relevant people use the software data of JARI and RICSAC to make check analysis for this system, and the conclusion is that this software can automatically analyze result for reference of analyst. What is more, it can make the digital simulation and calculation for motor vehicle---motor vehicle, motor

vehicle---fixture, motor vehicle---pedestrian, motor vehicle---cylinder, turnover and other types of accident, and make complete three-dimensional animation display. [2] That means the final result output from analog computation can display through functional images.

3. The Establishment Of Second Crushing Accident Model

In the morning of Oct. 17th, 2013, at the gate of Wuhan Municipal Government, an accident happened that when one cyclist (female, 23 years old) crossed the road,. he was collided by the vehicle behind and then re-crushed by the vehicle finally leading to the cyclist gets damage of AIS51.30 grade. The vehicle in accident is Audi A4 produced in 2000, after accident, its front bumper is damaged, as shown in figure 1. Before accident, the distance between the cyclist and accident site is 1.8m, and the distance between after and before the collision is 22.2m. At that time, the weather is fine, view is wide, road surface is straight, broad and dry, and the rate-limiting of the road is 60 km/h. According to the description of parties, there is no braking before re-crushing, and the running speed of vehicle is about 56 km/h. [3]

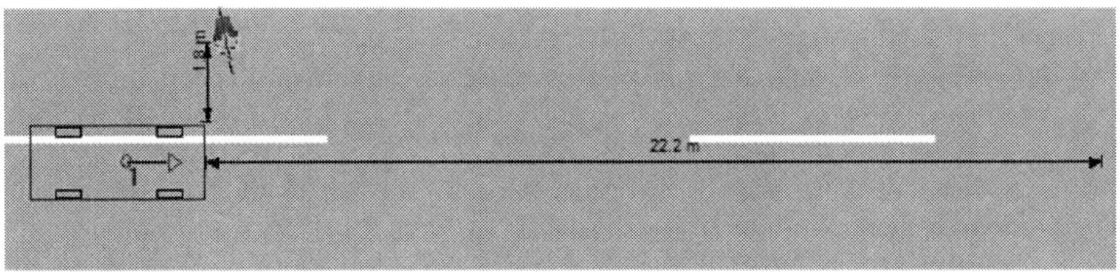

Figure 1. Schematic diagram for the site of accident

As shown in figures 2 and 3, generally, we can directly search vehicle information in the database of PC-CRASH, and also can set it through manual operation in relatively fuzzy parameters. In order to accurately obtain simulation data, we should maximum avoid generating information with large error when establish accident model [4]. We can obtain the specific parameters through using PC-RECT. As for cyclist model, we can replace it with using rigid-body model provided by inner software, and before replacement, we should input relevant height, weight, cycling vehicle length, weight and other detailed information. In the process of case simulation, the input information sees figure 1.

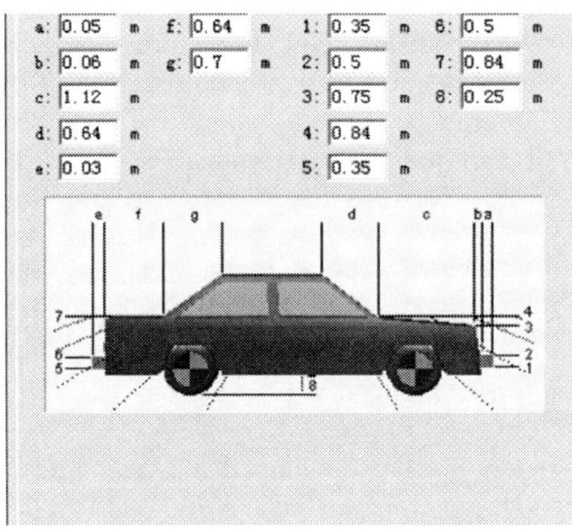

Figure 2. Contour parameter of vehicle in accident

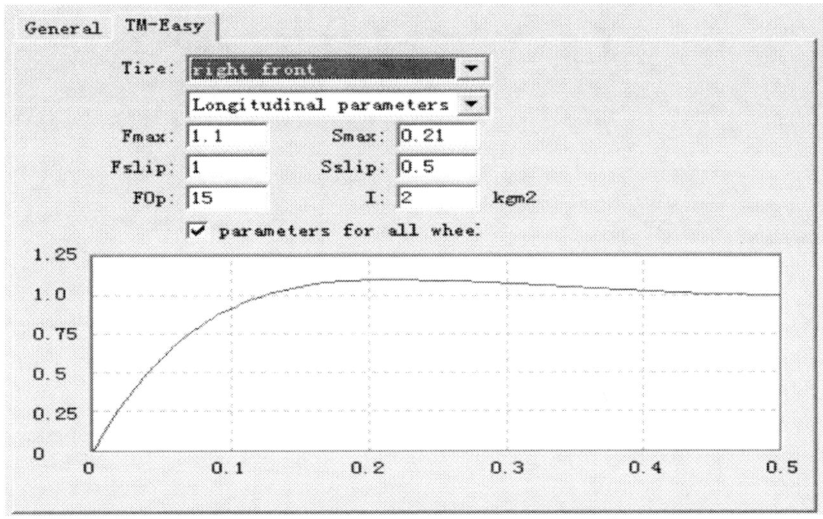

Figure 3. Tire longitudinal parameters

Table 1. Cyclist's Paraneter

Driver		Bike	
Height	Weight	Riding vehicle length	Weight
166cm	55kg	173cm	14kg

4. Simulated Test Of Second Crushing Accident

After the completion of parameter input and information input, it can begin to simulate second crushing accident. If it has inconsistent phenomenon with actual accident, belongs to normal range. Due to the site of accident has so many sundry factors, it needs fine adjustment through manual operation, that is, continuously adjust the angle of vehicle and cyclist in accident. In view of it is not very accurate for the bear load of vehicle and loaded speed of cyclist. Therefore, it needs to make proper adjustment for the motion parameter and outer contour parameter of vehicle and cyclist. Only up to the simulative parameters is consistent with the accident parameters, can adjustment stop. After the repeated fine adjustment, it finds that when speed of motor vehicle is 54km/h, coordinative time of braking system is 0.23s, friction coefficient between vehicle and road surface is 0.84, riding speed is 15km/h, friction coefficient between cycling and motor is 0.4, and friction coefficient between cyclist and road surface is 0.7, the simulation result is consistent with the information provided by accident.

Figure 4 shows simulation accident in the 0.13s, namely the first crushing; meanwhile, it provides the picture of relevant position relationship between cyclist and vehicle. From the figure, we can see that during the first crushing, the body of cyclist and the left tire of vehicle contact firstly, and then the first crushing happens. In this process, the cyclist also gets the certain damages from the crushing and scraping of bike.

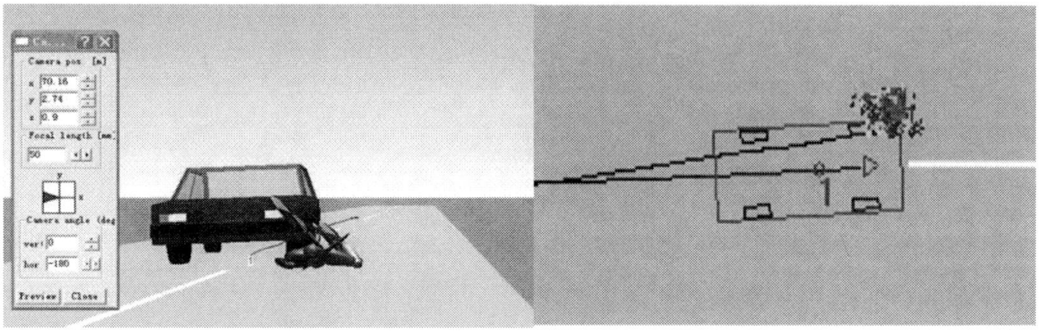

Figure 4. The information about participant's position happened in the first crushing in stimulation (the left side is three-dimensional view, the right side is two-dimensional view)

Form figure 5, we can see that after the first crushing, vehicle begins to reduce its speed, but at the same time, cyclist is relatively still, so at the moment of reducing speed, cyclist moves to the back of the vehicle. We also can see that after the 0.27s of collision, cyclist's foot begin to contact to the left rear wheel of vehicle, namely, the second crushing begins to happen.

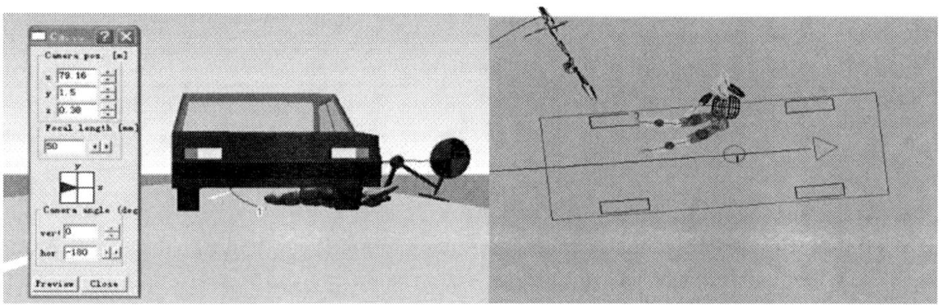

Figure 5. The relative position and location plan (0.27S) of accidental participant after the crushing first time

From figure 6, we can see that the second crushing is occurring, and the damaged part is cyclist's main body, which accords with the damage in actual accident. After the end of accident in 0.44s, the human body has crushing damages with total value of 1237.65N in the whole process, the accelerated speed of human body reaches to 47.7 m/s2, and the accelerated speed of main body of cyclist reaches to 143.3m/s2 .

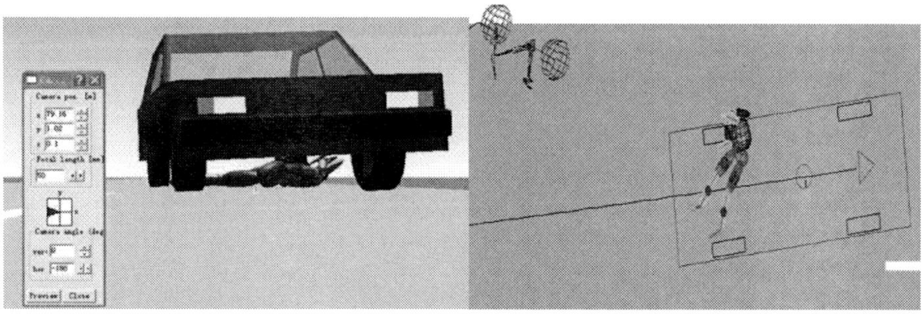

Figure 6. The information about participant's position happened in the second crushing in stimulation (the left side is three-dimensional view, the right side is two-dimensional view)

Figure 7 shows the maximum impact force for the head of cyclist in the whole accident, which means the cyclist's body injury mainly comes from the squeezing between tire and ground.

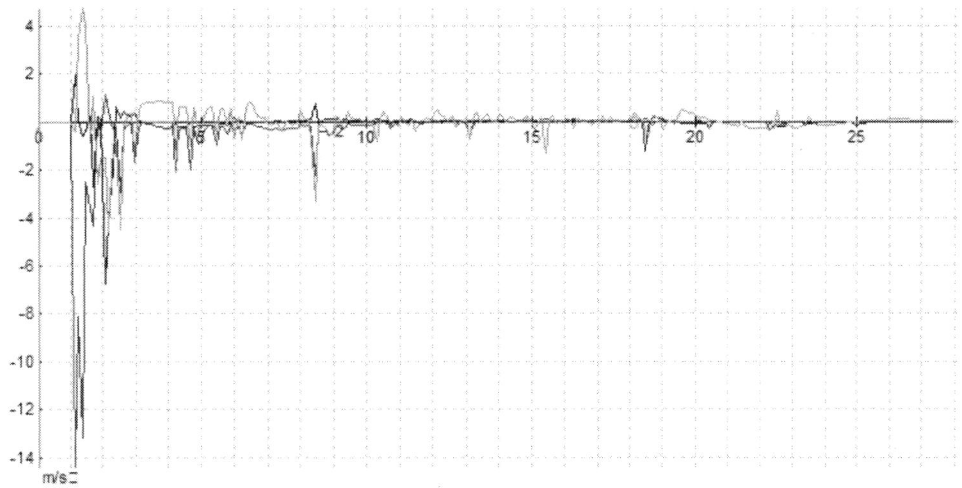

Figure 7. Graph about body affected by shearing force

5. Analysis to Simulation Data of Accident

After the accident simulation of second crushing, we can acquire the stress relations between cyclist and tire, as shown in figure 8.

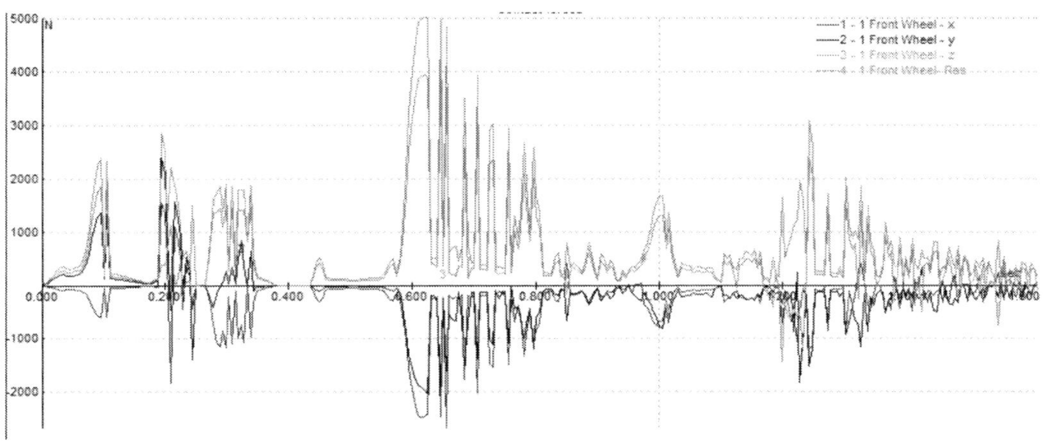

Figure 8. The schematic map about contact load between body and tire

In the first crushing, the shear force between road surface and tire for cyclist's body is less than the second crushing, and the maximum peak happens in the first half of second crushing. According to the data, we can speculate that when the second crushing begins, the speed of vehicle reduces, so the contact area between tires and cyclist increase, and during the simulation, the second crushing time is also more than the first time. Therefore, the fatal damage of cyclist happens in the first half of second crushing.

In the medical report, AIS grade damage of cyclist reaches to 51.30 in this case. According to the explanation of simulation data and mechanical model of TRL impactor, the stress in the first half of first crushing has already caused fatal damage for cyclist, so the fatal damage will not happen in the second crushing.

The result indicates that the damage is chest compression. From figure 9, we can see that the vertical stress time of left rear wheel in the second crushing is longer than the first time, but the instant rising speed rate of vertical stress in first time is larger than the second time, and appears explosive growth, so the damage for human body in short time can cause fatal damage.

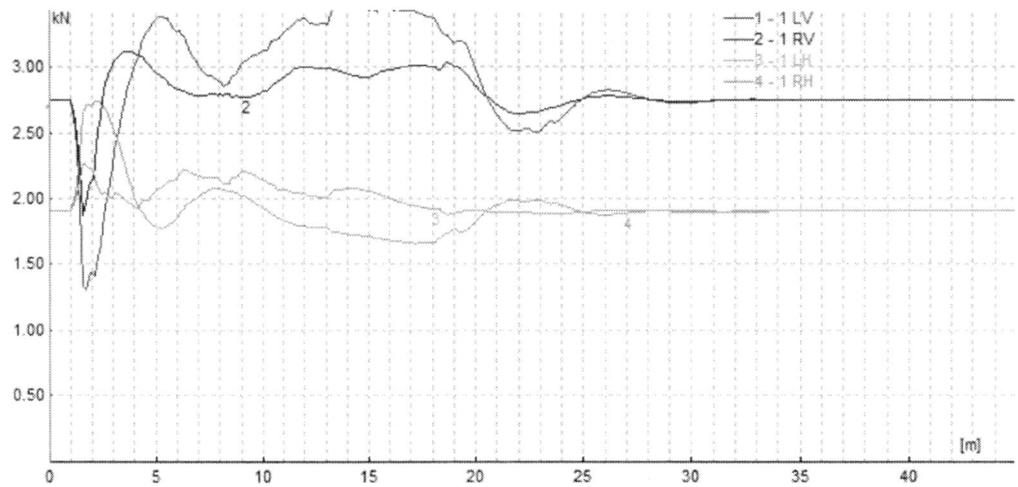

Figure 9. The graph about vertical stress of vehicle wheel

From the rising speed ratio of vertical stress, we can identify that the major fatal injury of cyclist in the second crushing is still generated from the first crushing, and the secondary fatal injury happens in the second crushing. Therefore, from the theoretical analysis, increasing the prevention for process of the second crushing plays very important role. However, if we want to fundamentally solve the fatal damage, should begin from preventing crushing. Form figure 10, we can see that if the cyclist gets the collision in this way, the head will have main injury, and the lower limbs will have secondary damage. So wearing helmet can avoid fatal injury.

Figure 10. The schematic diagram of contact collision between simulative cyclist and vehicle

6. Summary and prospect

This text uses PC-CRASH to simulate the typical accident of second crushing. It use lots of data to restore the process of accident, then analyze the lethality of accident through simulation data, finally obtain the conclusion that if it can avoid crushing, may prevent the fatal injury. Therefore, during the later prevention and control of accident, we can prevent the crushing through designing the front parts of vehicle. This study simulates the typical accident and provides prevention of accident with effective evidences and reference information.

7. References

[1] Xiaolong. Li and Jieyuan. Liu,"PC - CRASH simulation system in the traffic accident treatment courses teaching the application," Journal of course education research, vol. 34, Dec. 2014, pp. 205-206.

[2] Li. Shi, "Car collision theory elaboration and collision accident reappearance,"China mechanical, vol. 34, Mar. 2014, pp. 120-121.

[3] Qingbin. Deng, "Based on the energy absorption analysis of vehicle collision simulation and representation research,"Harbin industrial university, vol. 22, Mar. 2010, pp. 22-28.

[4] Cai. Siming., "Motor vehicles and non-motor vehicles and pedestrians traffic accident both sides of no-fault liability is briefly divided into,"Journal of shenyang university, vol. 23, Apr. 2011, pp. 22-24, doi: 10.3969/j.issn.1008-9225.2011.03.007.

[5] Tiefang. Zou, Zhi Yu and Ming Cai,"Based on Pc - people - car collision accident Crash software simulation research," Chinese journal of safety science, vol. 20, Jul. 2010, pp. 156-149, doi: 10.3969/j.issn.1003-3033.2010.02.009.

Word2vec and dictionary based approach for uyghur text filtering

Turdi Tohti, Yunxing Zhao and Winira Musajan

School of Information Science and Engineering, Xinjiang University, Faculty of Computer, Room 801, Science and Technology Building, No.666,Road Shengli, Urumqi 830046,P.R.China.
Email: turdy@xju.edu.cn

Abstract. With emerging of deep learning, the expression of words in computer has made major breakthroughs and the effect of text processing based on word vector has also been significantly improved. This paper maps all patterns into a more abstract vector space by Uyghur-Chinese dictionary and deep learning tool Word2vec, at first. Secondly, a similar pattern is found according the characteristics of the original pattern. Finally, texts are filtered using Wu-Manber algorithm. Experiments show that this method can get obvious filtering accuracy and recall of Uyghur text information improved.

1. Introduction

Information filtering refers to the technology of selecting user-wanted information and blocking unwanted or illegal information from large scale information flow. Information filtering has wide objectives such as text, audio, image and video formats in general, while usual (narrow sense) information filtering mainly focuses on text information.

With the rapid development of network technology, the more and more text data has been available for people and research issues. However, the openness, interactive and global characteristics of Internet made good and bad contents intermingled in spread. Many bad contented information such as violent, reactionary, pornographic or with obvious harmful emotional incitement brought troubles to user and the whole society, as well. Therefore, how to avoid bad information and ensure the healthy and sage networking environment has become an important research issue.

With increasing support to the development of western regions of China, the minority ethnic groups in this region also welcomed the fast economic, cultural and educational progress. Information processing technology for Uyghur and other minority languages in Xinjiang has stimulation under this excellent environment, too. In particular, Uyghur Web technology witnessed dramatic steps to make digitized text documents. Including the digitization of classic resources written on paper, quite a large volume of Uyghur text documents are being appeared in digital format [1]. Simultaneously, Uyghur information processing also faces the same or more problems that were encountered in Chinese and English information processing. Useful information is always hard to get from large, disordered information source, while the unhealthy information from webpages, text messages and other text carriers put serious threat on social order, national benefit in long run. Therefore, it is very important and urgent to find an efficient way of Uyghur text filtering.

Early text filtering techniques were mostly based on simple keyword matching and statistically filtering, which is simple in use and stable in effect. But these techniques fail to consider the overall content of document and diversity of word representations, only the documents which has the fixed key words are indexed, and often lead to mis-indexing or mis-filtering to some extent. The rise of deep learning let the representation of words in computer has new breakthrough. Deep learning in text filtering is also playing more and more important roles and providing better effects. Chinese and English text filtering technology has been matured, while Uyghur text filtering research is still in its infancy. The unique language features of Uyghur had many existing technologies can't be directly referenced, more studies still need to be further carried out.

This paper presents a Uyghur text filtering method combining the advantages of word2vec, Uyghur-Chinese dictionary and Wu-Manber filtering algorithm. Among them, word2vec is used for expanding the matching pattern, the transformation between the original and expanded patterns are realized by Uyghur-Chinese dictionary, and last filtering is conducted implementing Wu-Manber algorithm.

2. Related Works

2.1. Chinese and English Text Filtering

As early as 1958, Luhn proposed the idea of a "business intelligence machine". Then, in 1982, the concept of information filtering was first proposed by Denning who was able to identify emergent and regular mails in real-time e-mailing and thus expanded the scope of traditional information generation and information collection. In 1997, the information filtering was classified into adaptive filtering, batch filtering and shunt filtering in TREC1997, making the research on information filtering more specific [2].

It can be said that since 1950s, along with artificial intelligence, expert systems, machine learning and other fields of research, information filtering technology has acquired many experiences and breakthroughs. Information filtering for foreign languages such as English, already got very much advance. The concepts and techniques for each processing units are getting continuously improvements [3]. The study of Chinese text filtering started late in time. At present, most of them are working on feature extraction, similarity calculation, user template selection and related filtering algorithms and models in adaptive filtering system. Domestic research institutes including Tsinghua University, Fudan University, Chinese Academy of Sciences, Asian Research Institute of Microsoft and Harbin Institute of Technology first started their steps for text filtering and were gifted admirable push in this field [4].

2.2. Text Filtering Based on Pattern Matching

Multi-pattern matching received extensive attention in the field of string pattern matching. Common multi-pattern matching algorithms include Aho-Co-rasick algorithm, AC-BM algorithm, Wu-Manber algorithm, etc. Among them, Wu-Manber Algorithm is the most popular one [5] [6].

A Wu-Manber algorithm is the combination of the filtering with Boyer_Mooer algorithms. It uses character group to expand the transferring effect of bad characters and filter the pattern string through comparing with hash table, so that the matching time is reduced. Thus, we no longer use single character for investigation of matching, instead, a group of B characters is considered for every matching step, and the moving distance of pattern string is determined according to the comparing result[7][8].

Wu-Manber algorithm first conducts preprocessing on pattern string set and establishes three tables-Shift tables, Hash table, Prefix table which are used to filter text documents later [9].

This paper takes the left-right custom of Uyghur into consideration during whole procedures of preprocessing, table building and filtering, and made some modifications on Wu-Manber [10].

2.3. Word2vec

Since Hinton etc. put forward the concept of Deep Learning in 2006, it has been widely concerned for machine learning and text filtering. Deep learning uses the basic architecture of neural networks, automatically extracts high level and more abstract feature of samples through combination of multi-layer perceptron. The extracted high level features usually get better results with common machine learning techniques in last step. Deep learning got welcoming in many research fields due to the universality of neural networks that various kinds of features are easily manipulated and integrated.

The convolutional Neural Network-CNN in image processing, Deep Neural Network and Restricted Boltzman Machine in speech processing are already accepted as popular and recommended tools [11].

Thanks to Google that developed the deep learning based open-source language modeling tool Word2vec in 2012, in which a word is represented in form of numerical vector, and has been extensively concerned in the field of natural language processing. The tool implements a continuous bag-of-word model, and computes the skip-gram structure of the word vector. It takes the text set as input and generates the word vector for each word by training. These word vectors can be applied to other natural language processing problems as a feature of word. For example, you can calculate the degree of similarity between two words according to the word vector. Using word vector also can avoid the phenomenon of "curse of dimensionality" [12].

Word2vec is the realization of neural network based probability language model, is the combination of probabilistic model with artificial neural network to improve the n-gram model. The purpose of the n-gram model is to obtain the conditional probability distribution of the nth word w_n under the condition that the previous n-1 words w_1, w_{n-1} were given. In other words, if there are D words in the dictionary, N-gram is hoped to learn the $A^{n-1}D$ probability distribution P (wn| $w_1...w_{n-1}$), with context of $w_1... w_{n-1}$. Word2vec also has similar goal, but Word2vec assumes that $P(w_n \mid w_1... w_{n-1}) = f(w_1... w_{n-1}, w_n)$, where $f(w_1... w_{n-1}, w_n)$ is a function represented by a neural network.

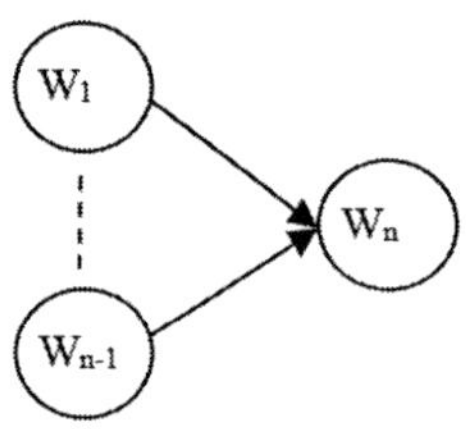

Figure 1. N-gram model.

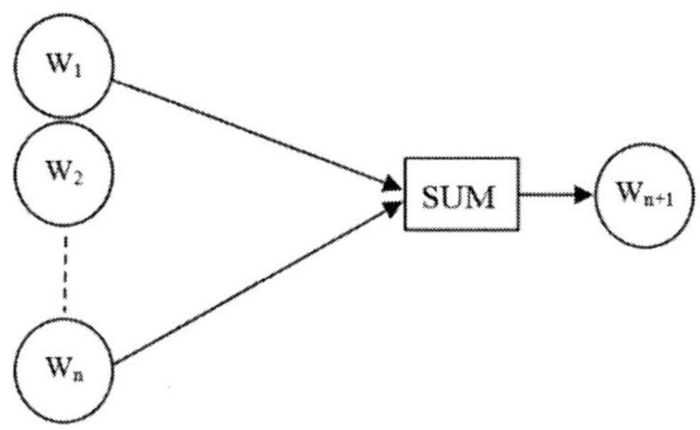

Figure 2. Continue bag of word mode.

It can be seen from Fig. 1 and Fig. 2 that for N-gram model, finding the probability of wn with known context w1 ..., wn-1 needs to visit the huge probability distribution table; While in continue bag-of-word model, the probability of this value can be simply calculated by the function f (w1 ...wn-1, wn). This function first maps the word as a word vector, then sums the word vectors, and finally calculates the value of f (w1 ..., wn-1, wn) [13][14].

In addition, Word2vec uses fewer parameters to represent model than N-gram does, that makes Word2vec able to avoid over-fitting problem to some extent. The most important thing is that each word after training has got the corresponding word vector, and it can be considered that the vector is the projection of the word in a semantic space

3. Word2vec and Dictionary Based Approach for Uyghur Text Filtering

The text filtering system proposed in this paper consists the modules of Chinese corpus preprocessing, Uyghur Chinese translation, Word2vec based training and mode expansion, and the filtering at last. According to Word2vec, words in corpus should be separated by space, and we found the Stanford Chinese word segmenting tool is good for this task in preprocessing. In order to avoid the inference of unnecessary words in corpus, the stop words has been filtered out from the word table using the special filtering tool from Harbin Institute of technology.

As word2vec does not support Uyghur text pattern directly at present, we used Chinese as intermediate language for the selected Uyghur patterns according to user's need. The Uyghur-Chinese dictionary developed by the Key Laboratory of Multi-lingual information processing of Xinjiang University is applied for translation between Uyghur and Chinese text patterns. The translated Chinese text patterns are processed with Word2vec, and the extended Chinese pattern is also translated into Uyghur for last filtering, simultaneously.

In the Word2vec based corpus training and pattern expanding module, the preprocessed Chinese text corpus is trained using Word2vec to obtain word vectors corresponding to the corpus. A pattern from user is extended using the word vectors to make more text patterns.

The extended patterns are processed by the modified Wu-Manber filter in last stage to select the appropriate patterns fit for user's need from the large Uyghur text corpus. The structure of whole filter system is shown in Fig.3.

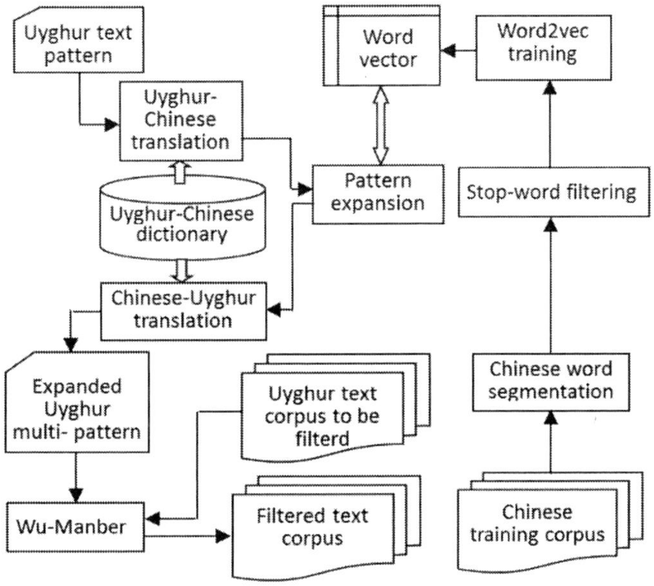

Figure 3. Flow of Text Filtering Process.

4. Experimental Results and Analysis

4.1. Preparation of the Experiment

The training corpus for Word2vec contains 10 years text documents collected from the websites of TianShan, Yaxin, with total volume of 4.4G including the categories of politics, economy and technology. The Uyghur text documents to be filtered are also from the websites of TianShan, Yaxin and Renmin, and has volume of 1.2G. The experiments are carried out on Lenovo desktop computer with 4G RAM and Win7 Operating system.

In this paper, we use Word2vec to train corpus, and then use the corresponding four word vector model file to expand pattern, and finally use the same pattern to filter the same Uighur text information.

4.2. Evaluation Methods

Recall and Precision are important terms always used in the field of natural language processing. Recall indicates the percentage of the detected relative items over the original ones, while precision refers the percentage of relative items over the detected ones in search. The former reflects how much useful items detected from total items should have been detected, and the latter indicates how much is useful among the detected ones [14]. The mathematical expression of Recall and Precision are as follows:

$$\text{Precision(p)}=\frac{\text{number of correct filtered text}}{\text{number of filtered text}} \tag{1}$$

$$\text{Recall(R)}=\frac{\text{number of correct filtered text}}{\text{number of should be filtered text}} \tag{2}$$

4.3. Experimental Results

The performance of the text filtering system was evaluated by calculating the Precision and Recall of the filtered Uyghur texts. In this experiment, we selected the patterns: كومپيۇتېر (computer), ۋىرۇس (virus) , ياغاچ ئات (trojan),تور بىخەتەرلىكى (network security), يوچۇق (holes).First, Word2vec is used to train the text corpus includes three categories such as with politics, economy and technology contents. Then, model expansion is done for the original model and filtering results are obtained using Wu-Manber algorithm. The comparisons of filter results with and without pattern expansion are given in Figure.4 and 5.

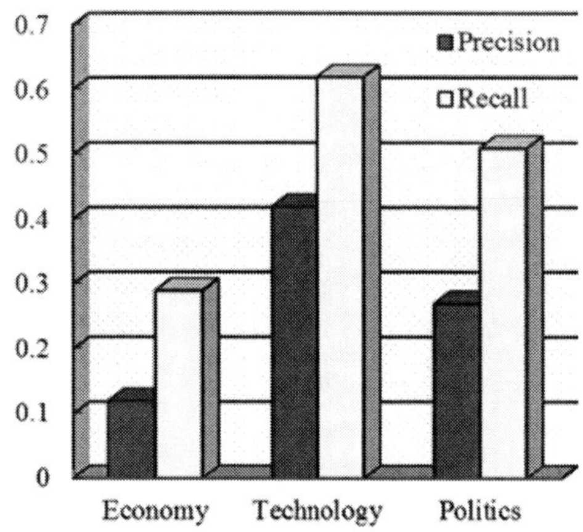

Figure 4. Experiment results of un-expanded patterns.

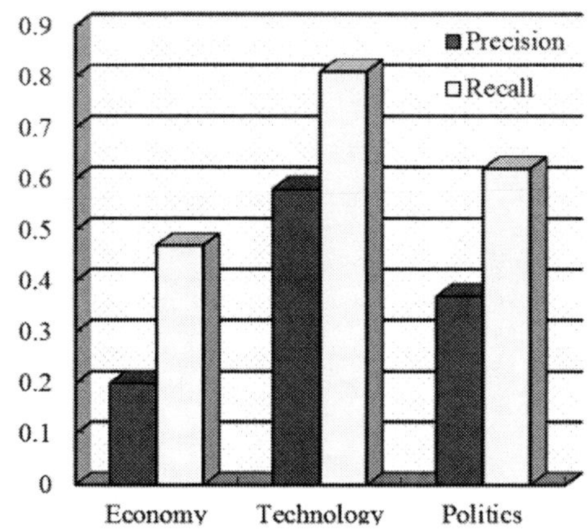

Figure 5. Experiment results of expanded patterns.

Through above experimental results we found that the proposed method has a large degree of improvement both in precision and recall than that of the traditional Wu-Manber text filtering. It should be noted that the precision and recall for technology category are higher than other two categories. Above results can let us to say that the proposed filtering system is superior to traditional Wu-Manber based text filtering? Also, larger correlation between the user-selected pattern and the training corpus yields better performance in filtering.

Finally, comparative experiments of different volume of corpus are carried out on the category of technology using the collected 3.5G Chinese text documents from Wangyi, Renmin, Sogou etc. The experiment environment, user-selected patterns, Uyghur text documents to be filtered is kept same. The 3.5G training corpus is randomly divided into 500M, 1G, 2G, the corresponding accuracy and recall rate are shown in Figure 6:

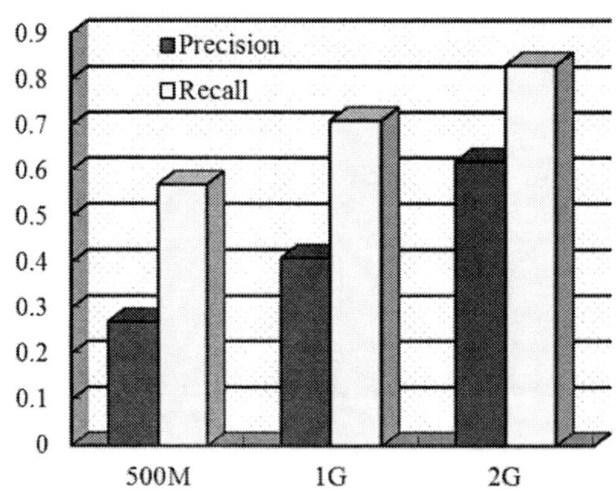

Figure 6. Experimental results with different volume of training corpus.

The experimental results indicate that, larger volume of training corpus can produce better filtering efficiency. This can be analysed as bigger corpus more possibly generates the extended pattern that is more similar to the original ones.

5. Conclusions and Prospect

The traditional pattern matching based text filtering has obvious disadvantages, though it is simple in implementation. First of all, user-selected pattern cannot accurately reflect the user's filtering requirements; Secondly, selected pattern is much fixed and easily misses the synonymous word during filtering, so the recall and precision of filtering are influenced.

The word vector based filtering method proposed in this paper can further improve filtering performance using larger volume of training corpus and the vocabulary of Uyghur-Chinese dictionary. Therefore, still more modification on filtering system can be implemented. 1). to train more accurate word vector with larger training corpus. 2) To eliminate even more unnecessary words using more effective word segmenting tool and stop-word filtering tool. 3) To increase the volume of Uyghur-Chinese dictionary, and establish a Uyghur-Chinese dictionary of better perfection and large vocabulary. 4) To divide the user selected patterns into more specific classes to expand the pattern using the word vectors generated from the corpus of same category and increase of volume and precision of extended patterns.5) To increase the number of layers of neural network to raise the semantic abstraction of word vectors and improve the Word2vec based language model.

6. Acknowledgments

This work has been supported by the National Natural Science Foundation of China (61562083, 61262062).

7. References

[1] Turdi Tohti et al. Semantics-based Feature Extraction and Its Application in Uyghur Text Classification. *Journal of Chinese information processing*, 28(4):140-144, 2014.
[2] Yalqin Almas et al. Uyghur Text Filtering Based on Vector Space Model. *Journal of Xinjiang University (Natural Science Edition)*, 32(2):221-226, 2015.
[3] Gao Yang et al. Pattern-based topics for document modelling in information filtering. *IEEE Transactions on Knowledge and Data Engineering*, 27(6):1629-1642, 2015.
[4] Yang Wenchuan et al. Study of an improved text filter algorithm based on trie tree. *IEEE International Symposium on Computer, Consumer and Control*, pages 594-597, 2016.
[5] Zhang Hongli et al. An efficient parallel algorithm for exact multi-pattern matching. *Security and Communication Networks*, 9(8): 1688-1697, 2015.
[6] V Bhardwaj et al. A Comparative Study of Wu Manber String Matching Algorithm and its Variations. *International Journal of Computer Applications*, 132(17): 34-38, 2015.
[7] L Zhang et al. Improvement on Wu-manber multi-pattern matching algorithm. *International Conference on Computer Science & Network Technology*, pages 608-611, 2014.
[8] YJ Wang et al. A Word-length Based Wu-manber Multi-pattern Matching Algorithm. *Journal of Chinese Computer Systems*, 34(7): 1650-1653, 2013.
[9] MO De-min et al. An Improvement of Wu-Manber Multiple Patterns Matching Algorithm. *Journal of Chinese Information Processing*, 23(1): 30-34, 2009.
[10] Maimaitiyiming Hasimu et al. Unique Character Based Statistical Language Identification for Uyghur, Kazak and Kyrgyz. *Journal of Chinese Information Processing*, 29(2): 111-117, 2015.
[11] R Collobert et al. Natural Language Processing (Almost) from Scratch. *Journal of Machine Learning Research (JMLR)*, 12(1): 2493-2537, 2011.
[12] EH Huang et al. improving word representations via global context and multiple word prototypes. *Proceedings of the 50th Annual Meeting of the Association for Computational Linguistics: Long Papers-Volume 1*, pages 873-882, 2012.
[13] T Mikolov et al. Linguistic regularities in continuous space word representations. *Proceedings of NAACL-HLT*, pages746-751, 2013.
[14] WANG Yi-pei et al. An Improved Wu-Manber Multi-pattern Matching Algorithm for Chinese Encoding. *Journal of Chinese Computer Systems*, 36(4): 778-781, 2015.
[15] WANG Zhen et al. A Parallel Multiple Pattern Matching Algorithm on Chinese/English Mixing. *Computer Engineering*, 40(4): 318-321, 2014.

IOP Conf. Series: Journal of Physics: Conf. Series **887** (2017) 012014 doi:10.1088/1742-6596/887/1/012014

SNMG: a social-level norm-based methodology for macro-governing service collaboration processes

Ji Gao[1,2,*], Hexin Lv[1], Zhiyong Jin[1] and Ping Xu[1]

[1]Zhejiang Shuren University, Hangzhou 310015, China
[2]Zhejiang University, Hangzhou 310027, China
*Corresponding author, Email: gaoji1@zju.edu.cn, +86-571-8799-6934

Abstract. In order to adapt to the accelerative open tendency of collaborations between enterprises, this paper proposes a Social-level Norm-based methodology for Macro-Governing service collaboration processes, called SNMG, to regulate and control the social-level visible macro-behaviors of the social individuals participating in collaborations. SNMG not only can remove effectively the uncontrollability hindrance confronted with by open social activities, but also enables across-management-domain collaborations to be implemented by uniting the centralized controls of social individuals for respective social activities. Therefore, this paper provides a brand-new system construction mode to promote the development and large-scale deployment of service collaborations.

1. Introduction

With progressing globalization and fiercely competitive market, the open tendency for collaborations between enterprises is accelerating. "Open" means that the enterprises participating in collaborations possess heterogeneous IT application systems which locate in independent management domains, and such across-management-domain collaborations often need to be created dynamically on demand. However, at present the application systems for open collaborations are still constructed according to the logically centralized control mode adopted by traditional software engineering. Moreover, open collaborations often involve the dynamical creation and change of social relationships between enterprises, and thereby belong to typical social application. However, this mode lacks the use of social-level abstract concepts, and therefore enables application systems only to be built by "low-level control flow abstraction" [1, 2]. As a result, this makes the reusability of system parts restricted seriously, and leads to unnecessarily complex software models and code, which in turn lead to expensive and painful software maintenance requirements.

In order to remove the predicament above, [3] proposed that the construction of social application should be grounded on basic social-level abstract concepts, and requires the social application languages and platforms for representing, computing, and reasoning those concepts. Since Social Dependences, called SDs, have been identified as basic social relationship, and popular social relationships, such as partner, friend, co-author, and so on, are only application-specific and are created on the top of SDs, SDs can be adopted as basic social-level abstract concepts. An SD means that some social individual (e.g., an enterprise participating in collaboration) depends on the assistance of another one to achieve an expectant goal, and the assistance is a social activity executed by the latter. For example, manufacturers depend on suppliers for parts, sellers depend on the bidders winning an auction for making the payment, and the invitees depend on the party sponsor for hosting the party. However, [3] have not specified how to use SDs to support the construction of those systems, and neither there is the subsequent achievements of this mode to be reported.

We propose that the key for innovation is to create the across-management-domain social-level macro-governance mode to replace the traditional centralized control mode limited in single management domains. Here, the macro-governance aims at regulating and controlling the social-level visible macro-behaviors of social individuals, which reflect how social sates are changed by social activities of those individuals, while the social-level norms for constraining those macro-behaviors are used as the governance basis. Because SDs arise in every social activity, formulating SD norms to constrain individual macro-behaviors enables macro-governance to be realized by the computation and reasoning of individuals for how to conform to those norms. Therefore, across-management-domain collaborations can be implemented by uniting the centralized controls of social individuals for respective social activities, and this enables the requirement for creating statically a physical or logical control center to be removed. Meanwhile, due to no need for concerning with the implementation detail of social activities, the macro-governance mode can reduce largely the difficulty for developing and maintaining open collaborations, and make the attention for developing open collaborations focused on designing architectures and solving challenging problems originated from openness.

This paper focuses the research of social-level macro-governance on open collaborations, and proposes a Social-level Norm-based methodology for Macro-Governing across-management-domain service collaboration processes, called SNMG, to support the open collaborations formed dynamically on demand. Here, "service collaboration" means that collaborations can be realized by creating the supply-demand relationship of business (i.e., software) services. Since SOC (Service-Oriented Computing) has grown into maturity, and the virtualization of services can remove the difficulty and complexity of the interaction between heterogeneous IT application systems, services have become the building blocks of collaborative business processes.

The layout of this paper is as follows. We first discuss the challenges and countermeasure for researching the macro-governance mode in section 2, and then the methodology SNMG is specified in section 3. In section 4 we introduce the implementation of SNMG. After discussing the relative work and comparison in section5, the conclusions (in section 6) are provided.

2. Challenge and Countermeasure

Open collaborations are confronted with the uncontrollability hindrance to social activities because the openness makes social individuals not controlled directly by outer management domains. The social-level macro-governance mode we proposes above can just be used as the general way for removing this hindrance: as long as the macro-behaviors of individuals for executing social activities conform to relevant norms, those activities are transformed into predicable and controllable ones. However, how to make individuals conform to norms is also the challenging problem which must be solved. For this we propose that social individuals should be constructed into rational ones which have both ability and desire for conforming to norms in order to ensure that their social activities can comply with the macro-behavior constraints prescribed by relevant norms. Of course, in order to restrain possible norm-violating behaviors, social individuals should register into a community so that the supervision system located in the community can urge those individuals to conform to norms and sanction the ones violating norms.

Due to the requirement for computing and reasoning how to conform to norms, the rational individuals participating in collaborations should have high autonomy and intelligence. Thus it is advisable adopting software agents as the brokers for enterprises to provide / consume business services through Internet.

3. Social-level Norm-based Macro-Governance Methodology SNMG

In order to achieve the macro-governance for a cross-management-domain service collaboration processes, **SNMG** consists of two levels of both standalone and complementary technologies: the social-level norm system for collaboration processes and the rational agents whose collaboration behaviors conform to norms. Although the implementation of **SNMG** also requires the assistance of the supervision system located in the agent community, the relevant introduction is omitted due to the limited space.

3.1. Social-Level Collaboration Process Norm System

The core for achieving social-level macro-governance is to create the social-level norm system for collaboration processes published in the agent community. An intuitive way may be to formulate, for each service collaboration instance, the SD norms which are used to constrain the macro-behaviors of the two parties participating in the instance and can be adopted by them. Since the process for implementing a collaboration instance often includes multiple social activities executed in succession, the norms for constraining those activities form the norm set called collaboration process pattern. However, the instance needs to be created dynamically on demand, and this makes creating the pattern from scratch by negotiation face the predicament that computation complexity and space-time consumption are too large. Therefore, we propose that the more convenient and effective approach should be: formulating a parameterized collaboration process template first, then instantiating dynamically this template, by negotiating and signing, into the shared pattern adopted by the two parties.

Define the social-level norm system $\mathcal{L}$ for collaboration processes as the set of norm systems for application domains: $\mathcal{L} = (\mathcal{L}_1, \mathcal{L}_2, \ldots, \mathcal{L}_w)$, and $\mathcal{L}_k$ (k=1, 2, …, w) is the set of collaboration process templates: $\mathcal{L}_k = \{M^{S1}, M^{S2}, \ldots, M^{Sn}\}$. Every template M^{Si} (i=1, 2, …, n) includes a group of SDNs (SD Norms) corresponding to a specific business service Si: $M^{Si} = \{SDN_{A1}{}^{Si}, SDN_{A2}{}^{Si}, \ldots, SDN_{Am}{}^{Si}\}$. Wherein, $SDN_{Aj}{}^{Si}$ is used to constrain social activity Aj (j =1, 2, …, m) and Aj may be an activity regular or for transacting abnormity. As an example, for "purchase" service (e.g. purchase a car), the collaboration process includes not only the regular activities executed alternately by the provider and consumer of this service: pay deposit, notice the pick-up of goods, pay for goods, provide goods, etc., but also the activities for transacting abnormity: complement deposit, defer the provision of goods, amerce, withdraw collaboration relationship, etc.

Define $SDN_{Aj}{}^{Si} = PSD_e{}^o(\rho \leq \delta | \sigma)$, which is used to represent the Parameterized SD (PSD), for social activity Aj, expected to maintain in the process performing bilateral (providing-consuming) collaboration oriented to service Si, and is endowed with the following meanings: under the condition that the current collaboration state satisfies σ and before the deadline δ is reached, party o depends on Aj' executer e (the other party) to transform this state into satisfying ρ.

Here, σ and ρ are also called the precondition for activating a norm and the postcondition for executing the norm respectively, and thereby make the execution of the norm possess the explicit start-up occasion and success-determining standard. An SDN defined in the description language corresponding to $PSD_e{}^o(\rho \leq \delta | \sigma)$ is shown in Figure 1.

```
(eio:SDNorm                //The norm for social activity "PayDeposit"
  NormPurpose:"Backbone" //This is the norm for driving a regular social activity
  NormNo:1
  Description:"Specify the performer, σ, δ, and ρ of this norm"
  NormTransactionType:"dmo:DMNormTransaction"
                           //The name of enumeration set of regular social activities
  NormTransaction:"dmo:DepositProviding" //The name of this activity in the set
  Performer:"InitiatingRole"     //This norm is performed by the service consumer
  Trigger:(@eio:ContractInuring ContractRegisterNo:?x Service:DataArrangement
              InuringDate:?y)     //?x and ?y are variables with "?" as their prefixes
  Deadline:(@eio:dateTimePeriod Type:RD BeginDate:?y Period:?nego_11)
  Postcondition: (@dmo:DepositStatus ContractRegisterNo:?x DepositLowLimit:
                   ?nego_12 Currency:"¥" DepositSatisfactionStatus:"Satisfying")
)
```

Figure1. The social dependence norm for social activity "PayDeposit" which can be activated by an event "ContractInuring", and should make the activity- ended time satisfy the deadline and the activity post-effect the postcondition.

Since SDNs only constrain the social-level visible Macro-behaviors of broker agents for executing social activities: activity-activated occasion, activity-ended time, and activity post-effect, this enables template M^{Si} to be formulated in the highly abstract form in which agent mental models and internal micro-opera-tors are invisible. Such a platform-independent description mode not only makes M^{Si} adopted and conformed to easily by heterogeneous agents, but also enables M^{Si} to be created on the basis of deep analysis and synthetic investigation. After created, M^{Si} is published in the agent community, becoming the negotiation foundation for creating collaboration relationships. Once the negotiation (which belongs to the scope of self-organization, and is not discussed here) successes, M^{Si} is instantiated into collaboration process pattern M_Y^{Si}, which is used to govern the current service collaboration instance Y, and the SDNs included in M_Y^{Si} is also specialized into appropriate to Y.

3.2. Rational Agents

Rational agents are configured into the individuals which have both ability and desire to make their social activities in collaboration processes accept the constraints from M_Y^{Si}, and therefore the outer observers (i.e., people or other agents) can believe those individuals reasonable and trustable. The key for achieving rationality is to install broker agents the local management policies for driving social activities in order to enable those agents to know how to execute those activities by using local micro-operators and decide how to conform to the constraints from M_Y^{Si}, including the norm-violating in an unavoidable situation. Since the SDNs included in M_Y^{Si} have already prescribed accurately the constraints agent social-level macro-behaviors should conform to, those constraints, as the basis for agents to compute and reason how to conform to norms, can be incorporated into the social activities driven by policies. Therefore, we propose creating the policy-driven Collaboration Process Management mechanism, called CPM, as the core mechanism for the rational agent running platform.

For collaboration instance Y oriented to business service Si, CPM is defined as the following multi-tuple:

CPM = (M_Y^{Si}, Policy, T-Event, C-State, M-State, B-Instruction, M-Operator, triggering, activating)

• M_Y^{Si}: the collaboration process pattern for constraining Y.

• Policy: the set of policies for managing the agent collaboration process oriented to Si. Those policies are divided into 4 types: P_{ea} (policies for examining collaboration state transition and activating norms), P_{rc} (ones for resolving the conflicts of multi-norm activation), P_{po} (ones for planning the micro-operators for executing social activities), and P_{ep} (ones for executing operator plans).

• T-Event: the set of events for triggering management policies. Those events are also divided into 4 kinds: E_{st} (events indicating collaboration state transformation), E_{cn} (ones indicating the conflicts of multi-norm activation), E_{ns} (ones indicating the single norm selected), E_{pg} (ones indicating the operator plan generated).

• C-State: the set of collaboration states occurring possibly in the collaboration process oriented to Y.

• M-State: the set of agent local micro-states occurring possibly in the collaboration process oriented to Y.

• B-Instruction: the set of local business instructions and transaction principles, oriented to Si, sent out by the owner of the agent statically, dynamically, or in dynamical initiation, such as the way for resolving the conflicts of multi-norm activation, the tolerance degree for QoS deviation, the tolerance degree for norm execution delay, and the constraints for replacing collaboration partners, etc.

• M-Operator: the set of agent local micro-operators, oriented to Si, for executing collaboration activities.

• triggering: T-Event → Policy, indicating that each e ∈ T-Event triggers a relevant policy.

• activating: C-State → $\mathbb{P}M_Y^{Si} \times (E_{cn} \cup E_{ns})$, which means: once a collaboration process starts or a previous social activity finishes, the current collaboration state activates one to multiple norms contained in M_Y^{Si}, by making the match examination with σ (the precondition for activating a norm), and an event $e \in (E_{cn} \cup E_{ns}) \subset$ T-Event is generated (note, the kind of e is decided according to the number of activated norms, and symbol $\mathbb{P}$ denotes power set).

A management policy p∈) Policy is represented in BNF as follows:

<p > ::= <trigger> <processing>

<Processing> ::= {< m-operator > | <rule> | <rule-group>}$^+$

<rule> ::= (→ <condition> {< m-operator>}$^+$)

<rule-group> ::= <selection-mode-for-activated-rules> {< rule >}$^+$

The main parts of a policy are its trigger and the transaction driven by it. The former is set as a trigger pattern, enabling the policy to be triggered by an event e (∈ T-Event). The latter is represented as a sequence of micro-operators, condition-action rules (conditional operators), and rule groups (for selecting activated rules), enabling an agent to make the decision analysis by detecting the activation of rules in order to drive appropriate local m-operators (⊂ M-Operator) for executing a social activity. Here, the decision analysis includes to examine the current collaboration state cs (∈C-State) and / or local micro-state ms (∈M-State), to inspect the relevant business instructions and transaction principles bis (⊂B-Instruction) from the agent owner, and to compute and reason how to conform to the current activated SD norm according to the constraints prescribed by this norm.

The policy-driven management for agent collaboration processes is illustrated in figure 2. The backbone of process management is divided into three stages: generating policy-triggered events, triggering the relevant management policies, driving appropriate local micro-operators. Note, the autonomic management for collaboration behaviors can be applied to drive both regular and abnormity transaction activities.

Figure 2. The agent collaboration process management Driven by policies (which is divided into three stages)

4. Implementation of SNMG

The implementation of methodology **SNMG** depends on the development of two-level technologies: social-level collaboration process norm system and rational agents. Because the key for creating the norm system is the formulation of SD norms, we have designed the norm description language corresponding to $SDN_{Aj}^{Si} = PSD_e^{o}(\rho \leq \delta | \sigma)$ and defines SD norms as the structured description called SDNorm, including a group of attribute slots, and an example for SD norms has been shown in Figure 1. The rational agents can be developed with the agent modelling and running platform, which are one of authors' previous research results: trusted autonomic service cooperation model IGTASC and application development framework [4].

We have established multiple simulation test instances oriented to service collaborations, such as small meeting arrangement, knowledge provision, data mining, and obtaining-employment management for graduates, to examine the feasibility and performance of **SNMG**. The test results have illustrated: formulating the collaboration process templates as the norms for constraining social-level macro-behaviors is the effective means for achieving autonomic coordination and maintenance of service collaboration processes.

5. Related Work and Comparison

Although [3] has proposed that the construction of social application should be grounded on basic social-level abstract concepts: SDs (social dependences), no feasible scheme and key implementation technology is specified. In contrast, we have proposed that the key for innovating the traditional software engineering is to create the across-management-domain social-level macro-governance to replace the logically centralized control limited in single management domains. In order to achieving the macro-governance, we have already created the norms for constraining the macro-behaviors of social individuals in the social level, and provided the support environment (rational agents plus the agent community supervision system) for promoting those individuals to comply with the norms. Besides, we have removed the limitation that the traditional theory for representing SDs doesn't support the representation for state change and dependence evolution [5, 6]. By using parameterized representation structures $PSD_c^o(\rho\leq\delta|\sigma)$ to accurately defines SDN_{Aj}^{Si}, agents are provided with the basis for computing and reasoning how to conform to norms.

Although NMASs (Normative Multi-Agent Systems) permits the sociology to be applied to MASs[7,8], the norms for constraining agent macro-behaviors visible in the social level are represented in the form of deontic logic [9], and thereby cannot represent the social relationships and their evolution intuitively. Besides, current NMASs are disjointed from conventional application software, making norms difficult to display their value for supporting real collaboration application. In contrast, the macro-governance methodology **SNMG** has not only established the solid foundation for developing high-performance social application of service collaborations, but also overcome the drawback of disjointing by incorporating SOC (Service-Oriented Computing) and policy-driven management technologies.

6. Conclusions

This paper has proposed the innovatory system construction mode for open collaborations: across-management-domain social-level macro-governance, and made macro-governance created on the foundation of basic social-level abstract concepts: SDs (social dependences). For implementing this mode, we have created the methodology for macro-governing collaboration processes **SNMG** to remove effectively the uncontrollability hindrance confronted with by open social activities and enable across-management-domain collaboration processes to be implemented by uniting the centralized controls of social individuals for respective social activities, forming a distributed control mode which is created dynamically and accepts the macro-governance. Finally, developing real application and solving the real-life problems will be our further research work.

7. Acknowledgment

We gratefully acknowledge the support of the National Science Foundation of China under Grant No. 61375071, the Zhejiang Province Science Foundation of China under Grant No. LZ15F020001, and the Zhejiang Province Science Foundation of China under Grant No. LY14F020008.

8. References

[1] Garlan D, Cheng S W, Huang A C, Schmerl B, and P. Steenkiste P, 2004, Rainbow: Architecture-based self-adaptation with reusable infrastructure, *IEEE Computer*, 37 (10), pp. 46–54.

[2] Zinnikus I, Cao X, Fischer K, 2013, Agent-supported collaboration and interoperability for networked enterprises: Modeling interactions and service compositions, *Computers in Industry*, 64 (2013), pp. 938–950.

[3] Chopra A, 2011, social computing: Principles, platforms, and applications, *RESC 2011*, pp. 26-29.

[4] Gao J and Lv H, 2012, Institution-governed cross-domain agent service cooperation: a model for trusted and autonomic service cooperation, *Applied Intelligence*, 37(2), pp. 223-238.

[5] Lau B, Singh A, Tan T, 2015, a review on dependence graph in social reasoning mechanism, *Artif Intell Rev*, 43(2), pp. 229-242.

[6] Caire P, van der Torre L, and Villata S, 2013, Argumentation Theoretic Foundations for Abstract Dependence Networks, Agreement Technologies, Lecture Notes in Computer Science, Volume 8068, pp. 180-194.

[7] Criado N, Argente E, and Botti V, 2011, Open issues for normative multi-agent systems, AI COMMUNICATIONS, 24(3), pp. 233-264.

[8] Chopra A and Singh M, 2016, Custard: Computing Norm States over Information Stores, *Proceedings of the 2016 International Conference on Autonomous Agents & Multiagent Systems (AAMAS '16), and 1096-1105.*

[9] Sergot M, 2010, Norms, Action and Agency in Multi-agent Systems, *Deontic Logic in Computer, Lecture Notes in Artificial Intelligence*, Volume 6181.

Determination method for aero-engine optimal operation performance interval

Xuyun Fu, Guangyao Zhang, Yapeng Tian and Shisheng Zhong

School of Naval Architecture and Ocean Engineering, Harbin Institute of Technology at Weihai, Weihai, China
Email: fuxuyun@hit.edu.cn

Abstract. Considering total cost with the aim to minimize operation cost and maintenance cost in unit time, an aero-engine optimal operation performance interval determination model is proposed in this paper. Compared with previous researches, this model is more reasonable, because the model takes into consideration both operation cost and maintenance cost. The lower limit of the optimal operation performance interval corresponds to maintenance occasion, and the upper limit corresponds to performance restoration degree. Taking an aero-engine as an example, the performance degradation, the operation cost and the maintenance cost are studied and the optimal operation performance interval is solved based on the proposed model. The results show that the proposed model can be applied to solving the aero-engine optimal operation performance interval and determining aero-engine maintenance occasion and maintenance scope.

1. Introduction

Aero-engine is a typical complex equipment, which has the characteristic of long running cycle and high maintenance cost. One of the most complicated problems in aero-engine operation is to determine the maintenance occasion and the maintenance scope. The problem can be decomposed into multiple sub problems such as maintenance occasion determination problem, maintenance objective determination problem, and maintenance scope determination problem based on single factor or multiple factors. These factors include performance degradation, life limited parts, hardware damage, and lubrication consumption. In these problems, one of the key problems is to determine the aero-engine maintenance occasion and the performance restoration degree based on performance degradation.

The performance is declining with the operation of equipment. To ensure the availability of equipment, a performance threshold will be set generally. When performance drops to the threshold, the equipment must be maintained to restore its performance. The present study mainly focuses on forecasting the time when equipment degrades to its performance threshold, that is, to determine maintenance occasion by performance degradation. The methods used to forecast performance threshold include regression analysis [7-8], neural network [9-11], support vector machine [12-14], pattern recognition [15], combination forecasting [16], etc. Because aero-engine performance degradation has significant influence on the maintenance cost, the maintenance occasion solved by above method is generally not the optimal one. The performance restoration degree is generally determined by the engineer's experience.

Obviously, the method for determining maintenance occasion and performance restoration degree considering both operation cost and maintenance cost was not proposed up to now. Equipment performance has significant influence on the equipment operation cost. In general, the better the performance is, the lower the equipment operation cost is, and vice versa. The equipment maintenance cost is affected deeply by performance restoration degree. The higher the restoration degree is, the higher the maintenance cost is. When the performance degradation pattern is known, determining the equipment maintenance occasion and performance restoration degree is equivalent to determine an operation performance interval. The lower bound of the optimal operation performance interval corresponds to maintenance occasion, and the upper bound corresponds to optimal performance restoration degree. How to solve the optimal operation performance interval considering the total cost is the main problem to be solved in this paper.

The optimal operation performance interval of aero-engine is connected with the performance degradation pattern, the effect of performance status on operation cost and the effect of performance restoration degree on maintenance cost. In this paper, an aero-engine optimal operation performance interval determination model considering the total cost is presented to minimize the operation cost and maintenance cost in unit time.

The paper is organized as follows. In section II, we describe the processing of the model establishment. In section III, we selected a case for validation analysis and got the results. In section IV, we summarized the conclusion.

2. Model Establishment

The aero-engine performance degradation is showed by the performance parameters degradation which can be expressed by a monotonic continuous function. The function is expressed by $p = f_{\mathrm{P}}(t)$, where p means performance parameter value and t means operation time.

The aero-engine operation performance interval can be expressed as $[p_2, p_1]$, corresponding to the operation time interval $[t_1, t_2]$. In this interval, initial value of the aero-engine performance parameter is p_1. With the increase of t, the performance degrades gradually. When the performance parameter falls to p_2, the aero-engine must be repaired to restore its performance parameter to p_1.

The constitution of aero-engine operation cost and maintenance cost is complicated, so this paper just study the operation cost and maintenance cost influenced by performance. Unit operation cost generally increases with the performance degradation. Denoting $f_{\mathrm{O}}(p)$ as unit operation cost, then the operation cost c_{O} in the interval $[t_1, t_2]$ can be defined as follows:

$$c_{\mathrm{O}} = \int_{t_1}^{t_2} f_{\mathrm{O}}(p)dt \tag{1}$$

The aero-engine maintenance cost is related with both the performance before maintenance and the performance after maintenance. For $[p_2, p_1]$, the performance before maintenance is p_2, and the performance after maintenance is p_1; then the maintenance cost is represented as $c_{\mathrm{M}} = f_{\mathrm{M}}(p_1, p_2)$.

To minimize the operation cost and maintenance cost in unit time from t_1 to t_2, and take t_1 and t_2 as decision variables, the aero-engine optimal operation performance interval determination model considering total cost can be expressed as equation (2).

$$\min\ c_{\mathrm{u}} = \frac{f_{\mathrm{M}}[f_{\mathrm{P}}(t_1), f_{\mathrm{P}}(t_2)] + \int_{t_1}^{t_2} f_{\mathrm{O}}[f_{\mathrm{P}}(t)]dt}{t_2 - t_1} \\ s.t. \begin{cases} p_{\mathrm{r,min}} \le f_{\mathrm{P}}(t_1) \le p_{\mathrm{r,max}} \\ p_{\mathrm{min}} \le f_{\mathrm{P}}(t_2) < f_{\mathrm{P}}(t_1) \end{cases} \tag{2}$$

where $p_{\mathrm{r,min}}$ is the minimum of performance restoration, $p_{\mathrm{r,max}}$ is the maximum of performance restoration, and p_{min} is the performance parameter threshold.

In general, equation (2) is a nonlinear programming model which can be solved by intelligent optimization algorithm such as particle swarm optimization, genetic algorithm and so on [17]. Equation (2) can be solved by interior point method when this equation can be proved that it is a convex optimization problem and that both the objective function and the constraint function are differentiable two times [18]. Besides, when the solution space scale is not big, the traversal search method can be used to solve this problem.

3. Case Description

In this section, the proposed model is applied to solving the optimal operation performance interval of an aero-engine. The unit of operation time is indicated as flight cycle considering that aero-engine performance degradation has higher correlation with flight cycle.

3.1. Performance Degradation

The parameter which is used to represent aero-engine performance in engineering application is exhaust gas temperature margin (EGTM). The higher the EGTM is, the better the aero-engine performance is, and vice versa. When the EGTM is lower than 0, the aero-engine core components would be damaged due to the high temperature, such as, turbine blades. Therefore, the aero-engine must carry out maintenance in this occasion. In order to ensure flight safety, EGTM threshold is generally set to a value greater than 0. In this paper, $p_{\min} = 0^{\circ}\mathrm{C}$.

A lot of EGTM data of multiple aero-engines in the same type is collected to obtain the EGTM recession tendency. Short term fluctuation of EGTM data is very large, but there is a clear downward trend for a long time. Linear regression model is adopted to match the EGTM data, and the result is shown in equation (3).

$$p = -4\times 10^{-3} t + 95 \tag{3}$$

From equation (3), the EGTM initial value of this type of aero-engine is 95, namely $p(0) = 95$.

According to the historical data, when this type of aero-engine is maintained, the minimum of performance restoration is 60, and the maximum of performance restoration is 95. Namely, $p_{\mathrm{r,min}} = 60^{\circ}\mathrm{C}$ and $p_{\mathrm{r,max}} = 95^{\circ}\mathrm{C}$.

3.2. Operation Cost

The aero-engine operation cost influenced by performance is mainly fuel cost. The main performance parameters related to fuel consumption are fuel flow (FF) and delta fuel flow (DFF). The actual measured value of FF which fluctuates larger is closely related to the flight environment and flight state. DFF is the percentage that FF standardized value deviates from the FF baseline value. Namely, DFF is the FF deviation percentage that the aero-engine in the standard flight environment compares to the healthy aero-engine in the same flight state. Compared with the actual measured value of FF, DFF can essentially reflect the aero-engine fuel consumption. Therefore, this paper presents a method to calculate the fuel consumption in the time interval $[t_1, t_2]$ based on DFF. Firstly, the DFF calculation equation is given as equation (4).

$$f_{\mathrm{d}} = \frac{f_{\mathrm{s}} - f_{\mathrm{b}}}{f_{\mathrm{b}}} \times 100 \tag{4}$$

where f_{d} is DFF; f_{s} is FF standardized value (kg/h); and f_{b} is FF baseline value (kg/h).

According to equation (4), the calculation equation of f_{s} can be obtained, which can be showed as equation (5).

$$f_{\mathrm{s}} = f_{\mathrm{b}} + f_{\mathrm{d}} f_{\mathrm{b}} / 100 \tag{5}$$

The FF average value of aero-engine in health condition is about 1600kg/h, then $f_{\mathrm{b}} = 1600\mathrm{kg/h}$. The flight hour cycle ratio of the aero-engine is 2.2, and let unit fuel price be u. The fuel cost per unit time in standard flight environment is shown as equation (6).

$$f_{\mathrm{O}}(p) = 2.2u \cdot \left(1600 f_{\mathrm{d}} / 100 + 1600\right) \tag{6}$$

To build the relationship between fuel cost and EGTM, the data of EGTM and DFF from multiple aero-engine in the same type is collected, as shown in figure 1. As can be seen from figure 1, although there is a big fluctuation between EGTM and DFF, but a clear trend can be found.

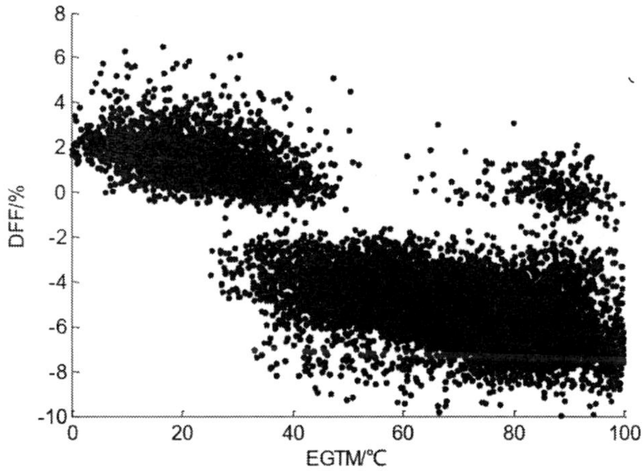

Figure 1. Data of EGTM and DFF

The relationship between EGTM and DFF can be fitted by two degree polynomial regression model, and the result is shown in equation (7).

$$f_d=1.5\times10^{-3}p^2-0.2646p+5.8924 \quad (7)$$

Bring equation (6) and equation (7) to equation (1), then the total fuel cost in interval $[t_1,t_2]$ can be calculated.

3.3. Maintenance Cost

The aero-engine maintenance cost influenced by performance is mainly the maintenance cost except the cost of life limited parts replacement. This part maintenance cost can be divided into the basic decomposition cost c_b that dividing whole engine into module and the cost c_p caused by module performance restoration. The maintenance cost can be expressed by equation (8).

$$c_M=c_b+c_p \quad (8)$$

According to the historical maintenance data, the basic decomposition cost c_b is about 160 thousand dollars. When the maintenance level of each module is performance restoration, EGTM can generally be restored to about 60 degree, and the corresponding maintenance cost c_p is about $1350000. When the maintenance level of each module is overhaul, EGTM can generally be restored to about 95 degree, and the corresponding maintenance cost c_p is about $2250000. Since the quantity limitation of the collected historical repair sample, the establishment of precise maintenance cost model is difficult. In order to simplify the maintenance cost model, assuming that c_p is independent of the EGTM before maintenance p_2, the EGTM after maintenance $p_1\in[60,95]$, and the relationship of c_p and p_1 is linear, then the calculation equation of c_p can be presented by equation (9).

$$c_p=25714.29p_1-192857.4 \quad (9)$$

3.4. Model Solution

Equation (3), Equation (6-9) is plugged into (2). The optimal operation performance interval determination model of this type of aero-engine is built and shown as (10).

$$\min c_u = v \cdot \frac{-102.86t_1 + 2.41\times10^6}{t_2 - t_1} + u \cdot \left[2.8\times10^{-7}\cdot\left(t_1^2 + t_1 t_2 + t_2^2\right) - 1.4\times10^{-3}\cdot\left(t_1 + t_2\right) + 3319.11\right]$$
$$s.t. \begin{cases} 0 \le t_1 \le 8750 \\ t_1 < t_2 \le 23750 \end{cases} \tag{10}$$

When $v=1$, is a fluctuate coefficient considering the maintenance cost affected by market, the maintenance cost does not change; and when $v>1$, the maintenance cost increases by $(v-1)\times100\%$. As $(v,u)\in\{1, 1.05, 1.1, 1.15, 1.2\}\times\{0.6, 0.8, 1, 1.2, 1.4, 1.6, 1.8, 2, 2,2, 2.4\}$, and the unit of u is dollars/kg. The standard particle swarm optimization algorithm (PSO) is used to solve equation (10). Some results are shown in Table 1 and figure 2.

Table 1. Partial optimization results

v	u /(dollars /kg)	t_1 /cycles	p_1 /°C	t_2 /cycles	p_2 /°C	c_u /(dollars /cycles)
1	0.6	0	95	20155.66	14.38	2162.35
1	0.8	0	95	18396.06	21.42	2841.50
1	1.0	0	95	17143.32	26.43	3517.98
1	1.2	0	95	16187.08	30.25	4192.66
1	1.4	0	95	15423.00	33.31	4866.03
1	1.6	0	95	14792.35	35.83	5538.39
1	1.8	0	95	14259.08	37.96	6209.95
1	2.0	0	95	13799.63	39.80	6880.86
1	2.2	0	95	13397.81	41.41	7551.23
1	2.4	0	95	13042.06	42.83	8221.13
1.2	0.6	0	95	21361.8	9.55	2185.57
1.2	0.8	0	95	19491.47	17.03	2866.93
1.2	1.0	0	95	18159.83	22.36	3545.27
1.2	1.2	0	95	17143.32	26.43	4221.57
1.2	1.4	0	95	16331.04	29.68	4896.38
1.2	1.6	0	95	15660.58	32.36	5570.04
1.2	1.8	0	95	15093.62	34.63	6242.79
1.2	2.0	0	95	14605.13	36.58	6914.79
1.2	2.2	0	95	14177.89	38.29	7586.18
1.2	2.4	0	95	13799.63	39.80	8257.04

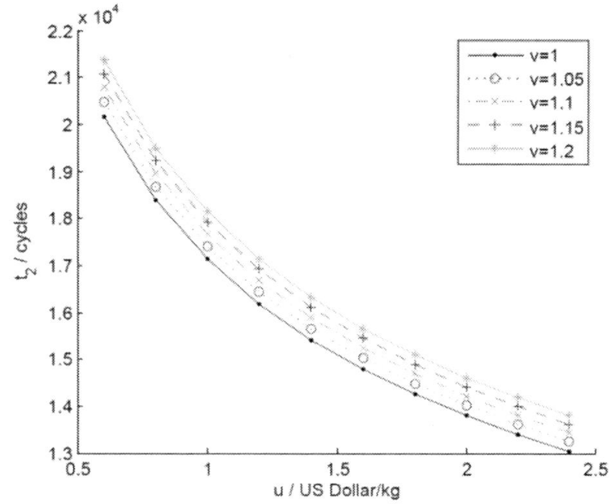

Figure 2. Relation for u, v, t_2

According to the Table I, when $v=1$ and $u=0.6$, the optimal operation performance interval of this type of aero-engine is $[14.38℃,95℃]$, the corresponding maintenance interval is 20155.66 flight cycles, and the operation cost and maintenance cost in unit time is $2162.35 per flight cycle.

Combined with Fig 2, when the maintenance cost (v) is constant and the fuel price (u) increases, the optimal operation performance interval gradually shorten, the corresponding maintenance interval also gradually shorten, but the operation cost and maintenance cost in unit time gradually rises. When $v=1$ and $u=2.4$, the optimal operation performance interval of this type of aero-engine is $[42.83℃,95℃]$, and the corresponding maintenance interval is 13042.06 flight cycles.

When the maintenance cost (v) increases and the fuel price (u) is constant, the optimal operation performance interval gradually enlarges, the corresponding maintenance interval also gradually enlarges, and the operation cost and maintenance cost in unit time gradually rises. When $v=1.2$ and $u=2.4$, the optimal operation performance interval is $[39.80℃,95℃]$, and the corresponding maintenance interval is 13799.63 flight cycles.

In conclusion, if the maintenance cost is higher and the fuel price is lower, the optimal operation performance interval is larger; vice versa. The result is consistent with the engineering practice.

The model optimization result is closely related to the performance degradation function, the operation cost function and the maintenance cost function. With the accumulation of the performance data samples and the historical maintenance samples, the optimization result will be more reasonable.

The fact that the maintenance occasion decided by the optimal operation performance interval only takes into account the performance factor deserves to be noticed. The factors which influence maintenance occasion include the replacement of multiple life limited parts, the mechanical failure and the execution of airworthiness directive. Therefore, the determination of aero-engine maintenance occasion needs to combine with the above factors. The determination method based on the optimal operation performance interval presented in this paper can be an important reference to determine the maintenance occasion.

4. Conclusion

To solve the problem of the maintenance occasion and the performance restoration degree based on the performance degradation, an aero-engine optimal operation performance interval determination model considering total cost that influenced by aero-engine performance is presented to minimize the operation cost and maintenance cost in unit time.

The optimal operation performance interval is solved based on the above model. This paper analyzes the aero-engine performance degradation, the operation cost and the maintenance cost. The results show that the established model is suited to solve the optimal operation performance interval, and the maintenance cost and fuel price have a great impact on the interval. When the maintenance cost increases and the fuel price decreases, the optimal operation performance interval of the aero-engine will become larger, which indicates the aero-engine maintenance interval become larger, vice versa.

In determining the aero-engine maintenance occasion and the maintenance workscope, in addition to considering the performance degradation, it is also necessary to consider other factors such as life limited parts, hardware damage, etc. Integrating the various factors to determine the optimal maintenance occasion and the maintenance workscope is a research direction.

5. Acknowledgments

Thanks to the National Natural Science Foundation of China (No. 51305096). The project was also funded by the Civil Aviation Administration of China (No. MHRD20150104), Shandong Independent Innovation and Achievements Transformation Fund (No. 2014CGZH1101) and the Fundamental Research Funds for the Central Universities (No. HIT.NSRIF.201704).

6. References

[1] J.Z. Sun, H.F. Zuo, K. Liang. Remaining useful life estimation method for the turbine blade of a civil aircraft engine based on the QAR and field failure data[J]. Journal of Mechanical Engineering, 2015, 51(23): 53-59. (in Chinese)

[2] S. M. Chun. Simulation of engine life time related with abnormal oil consumption[J]. Tribology International, 2011, 44(4): 426-436.

[3] G.R. Weckman, J.H. Marvel, and R.L. Shell. Decision support approach to fleet maintenance requirements in the aviation industry[J]. Journal of Aircraft, 2006, 43(5): 1352-1360.

[4] S.J. Joo. Scheduling preventive maintenance for modular designed components: A dynamic approach[J]. European Journal of Operational Research, 2009, 192(2): 512-520

[5] X. Y. Fu, Y. Chen, S.S. Zhong. Approach for civil aero-engine repair objective determination based on life limited parts[J]. Journal of Aerospace Power, 2014, 29(7): 1556-1561. (in Chinese)

[6] X.Y. Fu, S.S. Zhong, G. Ding. Civil aero-engine module workscope decision-making[J]. Journal of Aerospace Power, 2010, 25(10): 2195-2200. (in Chinese)

[7] J. Wang, T. Zhang. A degradation prediction method by use of autoregressive algorithm[C]//2008 IEEE International Conference on Industrial Technology. New Jersey: IEEE, 2008: 1-6.

[8] Y.X. Song, K.X. Zhang, Y.S. Shi. Research on aeroengine performance parameters fore-cast based on multiple linear regression forecasting method[J]. Journal of Aerospace Power, 2009, 24(2): 427-431. (in Chinese)

[9] B.R. Wu, Z.G. Tian, M.Y. Chen. Condition-based maintenance optimization using neural network-based health condition prediction[J]. Quality and Reliability Engineering International, 2013, 29(8): 1151-1163.

[10] G. Ding, X.Y. Fu, S.S. Zhong. Aero-engine performance parameters prediction based on process neural network[J]. Computer Integrated Manufacturing Systems, 2011, 17(1): 198-207. (in Chinese)

[11] G. Chen. Forecasting engine performance trend by using structure self-adaptive neural network[J]. Acta Aeronautica Et Astronautica Sinica, 2007, 28(3): 535-539. (in Chinese)

[12] V.T. Tran, T.P. Hong, B.S. Yang, et al. Machine performance degradation assessment and remaining useful life prediction using proportional hazard model and support vector machine[J]. Mechanical Systems and Signal Processing, 2012, 32,320-330.

[13] J.H. Hu, S.S. XIE, G.Q. LUO, et al. Study of support vector machines for aeroengine performance trend forecasting[J]. Journal of Propulsion Technology, 2005, 26(3): 260-264. (in Chinese)

[14] H.J. Wang, J.M. Zhang, X.L. Xu. Study on combination trend prediction technology for mechaninery system based on SVM[J]. Journal of Vibration Engineering, 2006, 19(2): 242-245. (in Chinese)

[15] X.Y. Fu, S.S. Zhong, M.N. Zhang. An improved method for aeroengine residual life prediction[J]. Journal of Harbin Institute of Technology, 2013, 45(5): 51-55. (in Chinese)

IOP Conf. Series: Journal of Physics: Conf. Series **887** (2017) 012015 doi:10.1088/1742-6596/887/1/012015

[16] D. Li, B.W. Li, K. Zhao, et al. Combined prediction of aero-engine performance index based on singular value trend decomposition and LS-SVR[J]. Journal of Propulsion Technology, 2013, 34(10): 1406-1413. (in Chinese)
[17] D.W. Wang, J.W. Wang, H.F. Wang, et al. Intelligent optimization methods[M]. Beijing, Higher Education Press, 2007. (in Chinese)
[18] B. Stephen, V. Lieven. Convex optimization[M]. Oxford, Cambridge University Press, 2009.

Improving prediction accuracy of cooling load using EMD, PSR and RBFNN

Limin Shen[1], Yuanmei Wen[2*] and Xiaohong Li[2]

[1]Department of Mechanical and Electronic Engineering, Guangzhou Institute of Railway Technology,100 Qinglong Middle Road, GZ, China.
Email: slmluck@126.com
[2] School of Information Engineering, Guangdong University of Technology, 100 Waihuan Xi Road, GZ, China.
Email: ym0218@gdut.edu.cn and eddy900812@163.com

Abstract. To increase the accuracy for the prediction of cooling load demand, this work presents an EMD (empirical mode decomposition)-PSR (phase space reconstruction) based RBFNN (radial basis function neural networks) method. Firstly, analyzed the chaotic nature of the real cooling load demand, transformed the non-stationary cooling load historical data into several stationary intrinsic mode functions (IMFs) by using EMD. Secondly, compared the RBFNN prediction accuracies of each IMFs and proposed an IMF combining scheme that is combine the lower-frequency components (called IMF4-IMF6 combined) while keep the higher frequency component (IMF1, IMF2, IMF3) and the residual unchanged. Thirdly, reconstruct phase space for each combined components separately, process the highest frequency component (IMF1) by differential method and predict with RBFNN in the reconstructed phase spaces. Real cooling load data of a centralized ice storage cooling systems in Guangzhou are used for simulation. The results show that the proposed hybrid method outperforms the traditional methods.

1. Introduction

District ice storage cooling systems use centrally-located chilling equipment to generate cooling in the form of chilled water so that it can be distributed to the nearby buildings for air conditioning. As energy prices increase, optimal control of district cooling systems has become more critical for energy cost reduction. Among issues of optimal control, accurate cooling load prediction for next-day demand is essential for the optimal scheduling of cooling system operating.

Various strategies have been developed to model cooling load demand. These attempts can be categorized in two approaches: physical-based software simulating methods and statistical methods. The physical-based software simulating methods are based on the thermal knowledge and physical equations of the buildings [1]. The statistical methods are based on the data collected in the cooling system. For district ice storage cooling systems, it is unrealistic to obtain all the building thermal behaviour. For this reason, statistical methods are more suitable for district cooling load prediction. The statistical methods include the single algorithm and the algorithm combinations. Linear-regression analysis [2], the artificial neural networks (ANN) [3], and the support vector machines (SVM) [4] are the primary ones of the single algorithm. The main feature of the algorithm combinations [5] is to combine more than one single algorithm to obtain prediction precision higher than that could be obtained from a single algorithm.

Regression analysis models formulate the cooling load prediction problem by an empirically mathematical function of the main parameters that affect the cooling load, such as temperature, humidity [6]. Common models reported in the literature include ARX, MLR, and ARMAX models. Generally, such a model is suitable for long-term predicting with large samples. For instance, Yoshida [7] applies this method to predict air-conditioning load based on an ARX model derived from building load simulation. Results show that its average prediction error is 29% in summer and 12% in winter. Li [8] applies four different methods including ARMAX, MLR, ANN and RC network to predict short term building cooling load demand, the results show that the MLR model and the ARMAX model are superior in prediction precision .Usually, regression models require a much shorter training time, but are less capable of modelling nonlinearity.

Artificial neural network and support vector machine has been proved to obtain prediction precision higher than the regression analysis models, due to its capable of modelling nonlinearity. ANN, SVM, modified ANN and SVM have been widely used to predicting, such as power load forecasting, natural gas usage trend forecasting, energy consumption in a building, and etc [9]. It is reported that, in forecasting a building air conditioning load, compared with back propagation neural network (BPNN), the root mean square error and the average relative error obtained by RBFNN are reduced by 36%. It is also reported that, for building air conditioning load prediction, the root mean square error obtained by SVM is about 50% lower than that obtained by BPNN [6].

Cooling load demand depends on many external and internal factors of cooling system, such as outdoor weather parameters, indoor equipment usage, and indoor human activities. Nonlinear interaction of these factors in cooling system leads to chaos of cooling load demand. The chaos especially the self-similarity and non-stationary of cooling load should be considered in cooling load prediction. EMD, which is data-driven technique that represents nonlinear and non-stationary data as a sum of a finite zero-mean components referred to as IMF, can be employed to transform the self-similarity time series into the short-related components. Ying [10] used EMD to forecast an hour-ahead wind speed and power, and Yang [11] integrated EMD, Chaos-based neural network and PSO for financial time series forecasting. EMD has also been used to predict networking traffic [12], price of power [13], and etc. To improve the prediction accuracy for the next-day Cooling load demand of a centralized ice storage cooling systems in Guangzhou, we propose the hybrid EMD-PSR-RBFNN model.

The rest of this paper is organized as follows. Section 2 is a collection of cooling load data, including tests. This serves to confirm that the data are chaos time series. In section 3, we introduce the proposed EMD-PSR-RBFNN model. Section 4 consists of simulations and results of the proposed model and other relevant models. Finally, Section 5 concludes this paper and summarizes the discussion.

2. Data Collection and Proofing of Chaos

This paper adopted the real cooling load time series of a centralized ice storage cooling systems in Guangzhou, the data that is collected a total of 121 days from July 1, 2012 to October 31, 2012, in which the former 106 data are taken as the training set and the remained 15 data as the testing set.

In order to investigate the chaotic nature of the cooling load time series, firstly, we applied mutual information method [14] and Cao method [15] to calculate the delay time and the embedding dimension respectively. Time delay is the delay time between the adjacent two coordinates of reconstructed phase space vector. Mutual information method utilizes the first minimum value of mutual information function to determine the optimum delay time, while the mutual information function is a measure of general random association between two random variables. The delay time of the cooling load time series in this paper is calculated to be 1.

Takens embedding [16] theorem deems that: For infinite and noiseless scalar time series with d-dimensional chaotic attractors, an m-dimensional phase space embedded can always be found in the sense of topological invariant, as long as dimensions $m \geq 2d+1$. By Cao method, the embedding dimension m of the cooling load time series is calculated to be 5.

Based on the calculated delay time and embedding dimension, Wolf method [17] is applied to calculate the maximum Lyapunov exponent. The maximum Lyapunov exponent of the experimental data is 0.0676, which indicates that the cooling load series has chaotic nature. Phase space reconstruction method of chaotic series is then applied to predict the cooling load demands in this paper.

3. Method

Chaos is a complex behaviour manifested by the low-order deterministic nonlinear dynamical system, which has the characteristic of nonlinearity, self-similarity and etc. In this paper, the proved self-similarity cooling load chaotic series is decomposed with EMD, which broken the self-similarity time series into several short-related IMFs. An IMF combining method is proposed to balance the computation complexity and prediction accuracy, and the differential method is used to smooth the high frequency component of EMD (IMF1) to improve the prediction accuracy further. Then reconstruct the phase space for the combined series and predict with RBFNN algorithm.

3.1. Emd and Combination of IMFs

EMD in Hilbert-Huang transform (HHT) [18] is mainly applied to decompose the original signal into several IMFs and a residue. The essence is to determine the basic shock mode of valid signal in the series by experience. The IMFs which contain local characteristics of cooling load in different time must satisfy the following two conditions: ①in the whole data set, the number of extreme and the number of zero-crossings must either be equal or differ at most by one. ②at any point, the mean value of the envelope defined by local maxima and the envelope defined by the local minima is zero. N.E .Huang, who was the founder of HHT, explained the meaning of these two conditions in the literature [19]. The first condition stipulates that the signal form of IMF is similar to the traditional narrow band requirements for a stationary Gaussian process, which can be characterized as a form of $a(\mathrm{t})\mathrm{e}^{j\phi(t)}$, in which $a(t)$ represents the envelope of the signal, $\phi(t)$ represents the phase of the signal .The second condition guarantees the symmetry of IMF waveform, and ensures that the instantaneous frequency does not fluctuate after Hilbert transform.

The process of EMD is as follows: Firstly, identify all the local extreme points of cooling load series $h(t)$. Secondly, connect all the local maxima points and the local minima points by a cubic spline as the upper envelope and the lower envelope, the upper and the lower envelopes should cover all the series between them, the mean of upper and lower envelope value is designated as $l(\mathrm{t})$. Thirdly, set $\mathrm{m}(t)=h(t)-l(t)$, judge whether $m(t)$ is an IMF component ,if $m(t)$ satisfies the conditions of IMF, $m(t)$ is the first IMF of $h(t)$, otherwise $m(t)$ is treated as $h(t)$, and repeat the above processes. Suppose after k iterations, the difference value between signal and the mean of the envelope is $m_{k,1}(t)=m_{k-1,1}(t)-l_{k-1,1}(t)$, $m_{k,1}(t)$ is a new series which removes the low frequency, after repeating up to $k-1$ times , the difference value is $m_{k-1,1}(t)$. In order to reduce the computation, when the equation (1) is satisfied, $m_{k,1}(t)$ is considered as the first IMF, where

$$SD=\frac{\sum[m_{k,1}(t)-m_{k-1,1}(t)]^2}{\sum[m_{k-1,1}(t)]^2}\leq\varepsilon \quad (1)$$

The threshold value ε is set according to the actual needs. The smaller the threshold is, the closer to the real IMF component $m_{k,1}(t)$ which satisfies equation (1) is. In this paper, the threshold is set to 0.3. Assuming $m_{k,1}(t)$ is the first IMF, $d(t)=h(t)-m_{k,1}(t)$ will be considered as $h(t)$, repeating the above processes. When the residual component is a monotonic function or its amplitude less than a predetermined value, just stop calculating. So we can get several IMFs, these components can be designated as $c_i(t)$, and the final residual component is $r(t)$, then

$$h(t)=\sum_{i=1}^{n}c_i(t)+r(t) \quad (2)$$

The original cooling load series is decomposed as the superposition of a variety of different feature fluctuation series with EMD. The IMFs have different time scale and can be ordered from highest to lowest frequency. In order to highlight the local detail of the high-frequency IMFs which mainly affect the cooling load prediction accuracy, this paper keeps the high-frequency IMFs and the residual unchanged with the low-frequency IMFs combined together. The details are as follows,

① Calculate the means of IMF components $c_1(t), c_2(t), \cdots, c_k(t)$ in turn; ② Determine the first component $c_i(t)$ which the mean deviates from zero significantly; ③ Keep the former $i-1$ IMFs $c_1(t), c_2(t), \cdots, c_{i-1}(t)$ (the high-frequency IMFs) and the residual $r(t)$ unchanged, combine the last $k-i+1$ IMFs (the low-frequency IMFs) as a low-frequency component according to the equation

$$\bar{c}_{\text{low}}(t) = \sum_{n=i}^{k} c_n(t) \quad (3)$$

3.2. Phase Space Reconstructing

Takens theorem [16] considers that the evolution of a component is decided by the interaction of other components in the system, and the information in these related components are implied in the development process of any component. Therefore, phase space reconstructing [20] only need to inspect a component, and reconstruct an equivalent phase space by computing the observing values of certain constant delay points. In section 2, the delay time τ and the embedding dimension m have been obtained to be 1 and 5 respectively. We calculated that the number of sample series after phase space reconstructing is 117.

Assuming the high-frequency IMFs, the combined low-frequency component and the residual forming a time series set $D = \{x(i), i = 1, 2, \cdots, N\}$, after reconstructing the phase space, we can get the series set $\overline{D} = \{X(t), Y(t)\}, t = 1, 2, \cdots, M$ in phase space domain, in which matrix X and Y represent the input and output set respectively, where

$$X(t) = \begin{bmatrix} x(1) & x(1+\tau) & \cdots & x(1+(m-1)\tau) \\ x(2) & x(2+\tau) & \cdots & x(2+(m-1)\tau) \\ \vdots & \vdots & \vdots & \vdots \\ x(M) & x(M+\tau) & \cdots & x(M+(m-1)\tau) \end{bmatrix} \quad Y(t) = \begin{bmatrix} x(2+(m-1)\tau) \\ x(3+(m-1)\tau) \\ \vdots \\ x(M+1+(m-1)\tau) \end{bmatrix} \quad (4)$$

3.3. Rbfnn

In the reconstructed phase space, RBFNN algorithm is utilized to predict the cooling load demand. There are three parameters need to be found for RBFNN simulation: the centres of base functions, the variance and the weights between the hidden layer and the output layer. In this paper, the centres of base functions were selected by self-organized method, and Gaussian function is chosen as the radial basis function. Therefore, the activation function of RBFNN is expressed as

$$R(x_p - c_i) = \exp\left(-\frac{1}{2\delta^2}\|x_p - c_i\|^2\right) \quad (5)$$

In equation (5), $\|x_p - c_i\|$ is Euclidean norm, c_i is the centre of Gaussian function, δ is the variance of Gaussian function. The output of RBFNN is

$$y_i = \sum_{i-1}^{k} w_{ij} \exp\left(-\frac{1}{2\delta^2}\|x_p - c_i\|^2\right) \quad j = 1, 2, \cdots, n \quad (6)$$

In equation (6), $x_p = (x_1^p, x_2^p, \cdots, x_m^p)^T$ is the p-th input sample; $p = 1, 2, 3, \cdots, P$, P is the total number of samples; c_i is the node centre of network hidden layer; w_{ij} is the weight between the hidden layer and the output layer; $i = 1, 2, 3, \cdots, h$, i is the number of hidden layer nodes; y_i is the actual output of the j-th output node corresponding to the input samples.

Assuming d is the desired output, the variance of base functions can be expressed as $\delta = \frac{1}{P}\|d_j - y_j c_j\|^2$

In this paper, k means clustering algorithm [21] is used to calculate the centre points c_i , and the least squares method [22] is used to calculate the connection weights w_{ij} .

3.4. Hybrid Prediction Model Based on EMD-PSR-RBFNN

This paper proposes a hybrid method based on EMD, PSR and RBFNN (EMDPSRRBFNN) for short-term cooling load demand predicting. The steps are as follows:

Step1: Pre-process the original cooling load series, including data cleansing and data restructuring.

Step2: Decompose the cooling load series with EMD, and then get k IMFs and a residual.

Step3: Combine the IMFs according to the IMF combination methods described in section 3.1.

Step4: Reconstruct the combined time series set $D=\{x(i), i=1,2,\cdots,N\}$ into the series set $\overline{D}=\{X(t), Y(t)\}, t=1,2,\cdots,M$ in the phase space.

Step5: Normalize data sets $\overline{D}=\{X(t), Y(t)\}, t=1,2,\cdots,M$ of phase space domain, and process the IMF1 by differential method.

Step6: Set up the RBFNN prediction models in the reconstructed phase space for IMF1, IMF2, IMF3, the combined low-frequency component (call it IMF4-IMF6 combined) and the residual respectively, then anti-normalize the predicted values.

Step7: Add the predicted values above together to achieve the final prediction result of the cooling load demand.

4. Simulation and Result

This section discusses the simulation works and results of the proposed EMD-PSR-RBFNN model in predicting the cooling load demand. Works begin with evaluation criteria in section 4.1. The results in section 4.2 verified that EMD model acts as a good filter in non-stationary and self-similarity cooling load time series systems. Simulation results of each EMD component using the PSR-RBFNN are described in section 4.3. Both are displayed in fig.3 and table 3. Following to this, it is the determination of high-frequency components, low-frequency components of EMD. Predicting comparison of different IMFs combination methods are shown in section 4.4. Section 4.5 serves to improve the prediction accuracy of IMF1. Section 4.6 conducts to demonstrate the performance of the proposed EMD-PSR-RBFNN model in predicting cooling load demand. We used models of SVM, LSSVM, RBFNN and SVM, LSSVM, RBFNN based on EMD-PSR for predicting comparison.

4.1. Evaluation Criteria

The root mean square relative error (RMSRE) and the mean relative error (MRE) are used as the evaluation criteria to assess the prediction accuracy, where in equation (7), y_i' is the actual load result, y_i is the predicted load result, n is the total number of the predicted result,

$$E_{RMSRE}=\left(\frac{1}{n}\sum_{i=1}^{n}\left(\frac{y_i'-y_i}{y_i}\right)^2\right)^{1/2}\times 100\% \quad E_{MRE}=\frac{1}{n}\sum_{i=1}^{n}\left|\frac{y_i'-y_i}{y_i}\right|\times 100\% \tag{7}$$

4.2. EMD and Analyzing of IMFs

4.2.1. Empirical mode decomposing. After EMD to the original cooling load time series, we got six IMFs and a residual, which are shown in figure 1. In figure 1, the horizontal axis represents the samples (days) and the vertical axis represents the cooling load demand (W).

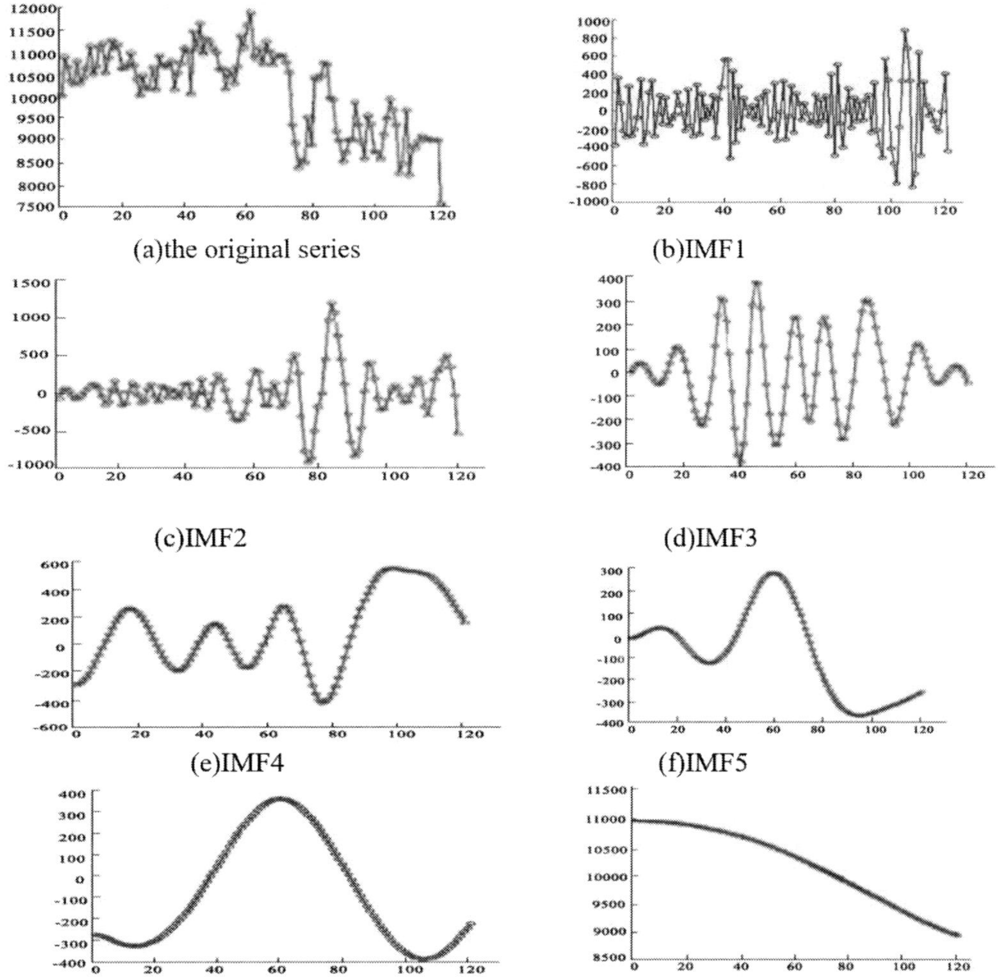

(a)the original series (b)IMF1 (c)IMF2 (d)IMF3 (e)IMF4 (f)IMF5 (g)IMF6 (h)residual

Figure 1. Original series and EMD components.

4.2.2. Comparing of stationary. In order to verify the influence of EMD on the stationary of cooling load series, the Hysteron Proteron method [23] is used to test the stationary of the original series and the EMD components, The computing results of the inverse number A and λ value are shown in table 1.

Table 1. The inverse number A and λ value.

Parameters	original series	IMF1	IMF2	IMF3	IMF4	IMF5	IMF6	residual
inverse number A	12	23	31	30	36	18	25	0
λ	-2.3355	-0.6228	0.6228	0.4671	1.4013	-1.4013	-0.3114	-4.2039

According to the Hysteron Proteron method, if the value of λ is between -2 to 2, the series should be stationary; otherwise, the series should be non-stationary. As shown in table 1, the original time series is non-stationary and the IMF1 to IMF6 are stationary respectively while the residual is non-stationary. That's to say, after EMD, the non-stationary original series has been transformed into several stationary series except the residual.

4.2.3. Comparing of self-similarity. In order to verify the influence of EMD on the self-similarity of cooling load series, the variance-time plots method [24] is used to estimate the Hurst index H for the original cooling load series and each EMD component, the estimation results are shown in table 2.

Table 2. The Hurst index H.

Parameter	Original series	IMF1	IMF2	IMF3	IMF4
H	0.9131	0.7327	0.3567	0.553	0.7493

The Hurst index H is the parameter to characterize the self-similarity. When $H \in (0.5, 1)$, the series is self-similarity. The greater the H is, the stronger the self-similarity is. As shown in table 2, after EMD, the self-similarity of IMF1~IMF4 reduced significantly. It indicates that EMD reduced the self-similarity of the original cooling load time series. From figure 2 in section 4.2.1, we can see that the waveforms of IMF5, IMF6 and residual approach sinusoidal waveforms gradually which resulting to the random reduced greatly. The calculation error of its Hurst index H is large and we will not discuss this topic in this paper.

Normally, a self-similar random process is also a non-stationary random process. Long-range dependence time series has the disadvantages of the high model algorithm complexity and poor precision. EMD is an effective method to break the self-similarity time series down into several short-related components.

4.3. RBFNN Prediction for Each EMD Component

Figure 2 shows the EMD components of the original cooling load series and their predicting results with PSR and RBFNN modelling. In figure 2, the horizontal axis represents the samples (days) and the vertical axis represents the cooling load demand (W).

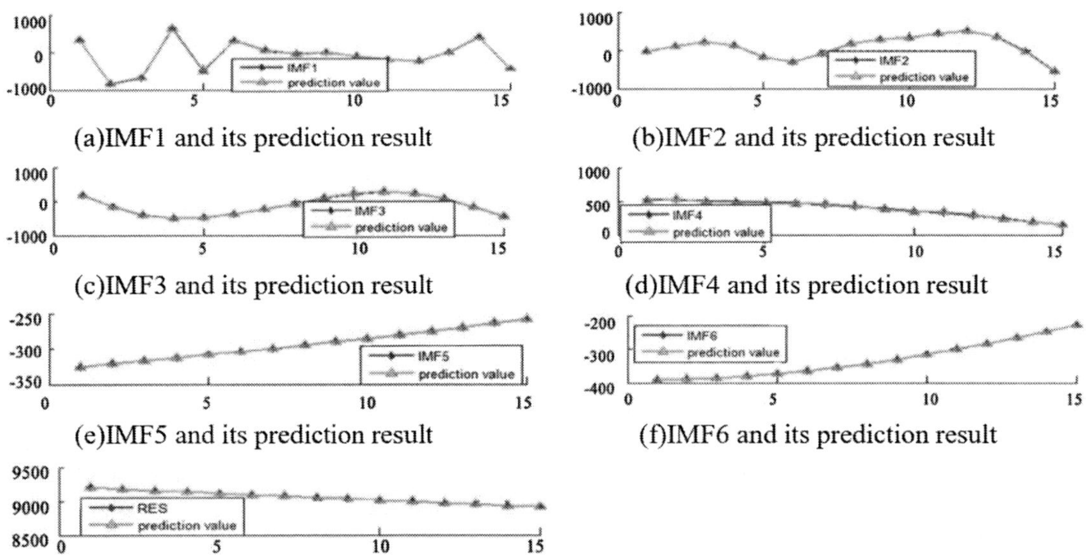

(a)IMF1 and its prediction result (b)IMF2 and its prediction result

(c)IMF3 and its prediction result (d)IMF4 and its prediction result

(e)IMF5 and its prediction result (f)IMF6 and its prediction result

(g)the residual and its prediction result

Figure 2. EMD components and their prediction results of the next 15days by PSR-RBFNN.

Prediction errors corresponding to each EMD component are shown in table 3.As can be seen in table 3 and figure 2, there exists larger prediction errors for the former IMFs (especially the IMF1) when RBFNN modelling, while the prediction errors are smaller for the later EMD components. In section D, we will compare different combination ways of the EMD components to balance the computation complexity and the prediction accuracy.

Table 3. The prediction error of each EMD component with RBFNN modelling

Evaluation criteria	IMF1	IMF2	IMF3	IMF4	IMF5	IMF6	residual	reconstruction result
RMSRE(%)	2.262	1.571	0.115	0.3816	4.33E-02	3.00E-02	8.19E-03	2.218
MRE (%)	1.540	0.874	0.103	0.3165	3.81E-02	1.76E-02	5.04E-03	1.516

4.4. Comparing of Prediction Errors for Different IMF Combination Methods

The means of IMF1 to IMF6 are calculated to be -9.9246, 6.4726, -2.9915, 100.369, -83.4171 and -86.3867 respectively. The first IMF whose mean deviates from zero significantly is IMF4. It indicates that IMF1, IMF2 and IMF3 are high-frequency EMD components and represent the influence factor of significant events. In the meanwhile, the IMF4, IMF5, IMF6 are low-frequency EMD components and represent the normal fluctuations factor of cooling load. The residual represents the long-term trend of the cooling load demand.

To illustrate why combining the EMD components as mentioned in section 3.1, this section gives the prediction error of two different combination ways of the EMD components respectively in table 4 and table 5, then compares the results of table3, table 4 and table 5. In addition to the different combination ways of the EMD components, the combined components are all predicted respectively with RBFNN model after PSR.

According to the IMFs combination method mentioned in section 3.1, keep the IMF1, IMF2,IMF3 and the residual unchanged, and superimpose IMF4, IMF5, IMF6 into a new low-frequency component (call it IMF4~IMF6 combined). The prediction error of each combined component is shown in table4. We can find that the predicting accuracy of the new combined low-frequency component and the residual are excellent, but the prediction errors of the high-frequencies, especially the IMF1, are so great.

Table 4. The prediction error after combining IMF4-IMF6.

Evaluation criteria	IMF1	IMF2	IMF3	IMF4~IMF6 combined	residual	reconstruction result
RMSRE(%)	2.267	1.571	0.115	0.3889	8.19E-03	2.221
MRE(%)	1.539	0.874	0.103	0.3333	5.04E-03	1.517

The predicting error of another IMFs combination method is shown in table 5. In which, IMF1~IMF3 are superimposed into a new high-frequency component, IMF4~IMF6 are superimposed into a new low-frequency component, and keep the residual unchanged as before. Table V shows that the prediction error of superimposing IMF1~IMF3 is so high that the final error (RMSRE) of the cooling load prediction runs up to 3.695%.

Table 5. The prediction error after combining IMF1~IMF3 and combining IMF4~IMF6.

Evaluation criteria	IMF1~IMF3 combined	IMF4~IMF6 combined	residual	reconstruction result
RMSRE(%)	3.8561	0.3889	8.19E-03	3.695
MRE(%)	2.2082	0.3335	5.04E-03	2.112

By comparing table 3 and table 4, we find that the prediction accuracy are almost the same for the two IMFs combination methods, in which table 3 shows the prediction errors of keeping all the EMD components alone and table 4 shows the prediction errors of combination method mentioned in section 3.1, the RMSRE are 2.221% and 2.218%, the MRE are 1.516% and 1.517% respectively. In other words, whether the low-frequency EMD components (IMF4~IMF6) combining or not, the prediction errors are similar.

Comparing table 4 and table 5, we find that the RMSRE of the former method is 2.221% and the later method is 3.695%. That is to say, when combining the high-frequency EMD components (IMF1~IMF3), the prediction accuracy will get worse.

Therefore, to take the prediction accuracy and the computation complexity into account, this paper selects the combination method of EMD components which is proposed in section 3.1.

4.5. Improving Prediction Accuracy for High Frequency Component of EMD
Figure 3(a) and table 3 in section 4.3 show that the stochastic of IMF1 leads to the poor prediction accuracy. Differential [25] is found as a good method for pre-processing the IMF1 before RBFNN prediction.

Table 6 shows the RESRE and MRE of the IMF1 prediction with RBFNN after zero to six order differential operation. It is evident that the RESRE and MRE of the IMF1 prediction after one order differential operation is the minimum, which are 1.9% and 1.3% respectively. Obviously, one order differential pre-process to IMF1 is the first choice.

Table 6. The prediction error of IMF1 with RBFNN after 0~6 order differential operation.

Differential order	RMSRE(%)	MRE(%)
0	2.27	1.54
1	1.90	1.30
2	2.46	2.02
3	5.97	4.81
4	11.61	9.4327
5	23.13	18.55
6	47.25	40.11

Comparing to table 4, the only change to table 7 is that the IMF1 is pre-processed by one order differential operation. Table 7 shows that the RESRE and MRE of the IMF1 prediction with RBFNN are decreased from 2.267% to 1.903% and 1.539% to1.3035% respectively. And the final prediction errors are decreased from 2.221% to 1.794% and 1.517% to1.288% respectively. The differential operation to IMF1 improved the prediction accuracy indeed.

Table 7. The RBFNN prediction error of each component after combining IMF4~IMF6 and pre-processing IMF1 with differential.

Evaluation criteria	IMF1 after differential	IMF2	IMF3	IMF4~IMF6 combined	residual	reconstruction result
RMSRE (%)	1.903	1.571	0.115	0.3889	8.19E-03	1.794
MRE (%)	1.3035	0.874	0.103	0.3333	5.04E-03	1.288

4.6. Comparison with Different Prediction Methods
In order to discuss the influence of different prediction methods on the prediction accuracy, three single prediction methods and three hybrid prediction methods are compared. The single prediction methods are LSSVM (least square support vector machine), SVM, and RBFNN. The hybrid prediction methods are EMD-PSR-SVM, EMD-PSR-LSSVM, and EMD-PSR-RBFNN (the proposed method). The process of hybrid prediction methods are: decomposing the original cooling load series with EMD, then combining the IMFs as mentioned in section 3.1, reconstructing the phase space for the combined EMD components, predicting with different single algorithms finally. The IMF1 of the three hybrid prediction methods are all pre-processed with differential method. The RMSRE and MRE of different prediction methods are shown in table 8.

Table 8. Prediction Error of Different Models.

Evaluation criteria	LSSVM	SVM	RBFNN	EMD-PSR-SVM	EMD-PSR-LSSVM	EMD-PSR-RBFNN
RMSRE(%)	5.589	5.736	5.093	3.10	3.762	1.794
MRE(%)	4.924	4.455	3.251	2.53	3.511	1.288

Table 8 shows that RESRE and MRE of the three hybrid prediction methods are smaller than the three single prediction methods. Among the hybrid prediction methods, the prediction accuracy of the proposed EMD-PSD-RBFNN is the maximum.

5. Conclusion

Predicting non-stationary and nonlinear cooling load demand for next day is very challenging, since there are many correlated independent variables involved. This paper introduced a new model comprised the EMD algorithm, the PSR, and the RBFNN. While the EMD functions as a filter for self-similarity and non-stationary cooling load time series, the RBFNN algorithm predicts the EMD-PSR components.

We proved the chaotic nature of the real operation cooling load time series of a district ice storage cooling systems in Guangzhou. The results in section 4.2 verified that EMD model acts as a good filter in non-stationary and self-similarity cooling load time series systems. In terms of the EMD component combination, different EMD component combination methods results to different prediction accuracy. To balance the computation complexity and prediction accuracy, the combination method is excellent.

We found that the prediction accuracy of high-frequency EMD component is lower than the prediction accuracy of low-frequency EMD components in section 4.3. The results in section 4.5 verified that differential method can be used to improve the prediction accuracy of high frequency EMD component.

Moreover, we compared the accuracy of the cooling load prediction using 'the proposed EMD-PSR-RBFNN' method with other algorithms. The results show that the hybrid prediction EMD-PSR models outperform the models not processed by EMD-PSR, and the proposed EMD-PSR-RBFNN model outperforms the other two hybrid prediction methods.

6. References

[1] Qu Y, PanY Q and Huang ZZ 2008 Heating Ventilating & Air Conditioning 10 9-17
[2] Li ZW and Huang GS 2013 Energy and Buildings 62 442-9
[3] SolmazO,Ozgoren M and Aksoy MH 2014 Energy Conversion & Management 82 177-87
[4] Yu J and Lee BB 2014 Multimedia Tools and Applications, 71(1), 293-307
[5] Hou ZJ 2006 Applied Energy 83(9) 1033-46
[6] Wen Y M and Chen Y Y 2014 Journal of Software 8(9) 2093-2104
[7] Yoshida H and Kumar S 1999 Energy Conversion & Management 40 1191-1206
[8] Li Z W and Huang G S 2013 Energy and Buildings 62 442-9
[9] Mahmoud MOM, Mhamdi F and Jaidane-Saidane M 2009 Proc. Int. Conf. on Power Tech (Bucharest: Romania /IEEE) p1–6
[10] Hong Y T, Yu Y S and Liu C Y 2013 Energies 6 6137-52
[11] Yang H L, Lin H C 2015 Economic Computation and Economic Cybernetics Studies and Research 49(1)
[12] Dai L F, Yang W G and Gao S X 2014 Proc. Int. Conf. on Mobile Ad Hoc and Sensor Systems, (Philadelphia: PA/ IEEE) p 539-44
[13] Kurbatsky V, Tomin N, Sidorov D and Spiryaev V 2010 Proc. Int. Conf. on Environment and Electrical Engineering (Prague: Czech Republic /IEEE) p 381-3
[14] Fraser M and Swinney H L 1986 Physical Review a 33(2) 1134-40
[15] Cao L Y 1997 Physica D 110 43-50
[16] Takens F 1981 Detecting strange attractors in turbulence Dynamical Systems and Turbulence (Springer Berlin-New York) 898 366-81
[17] Awrejcewicz J, Krysko AV, Zagniboroda NA, Dobriyan VV and Krysko VA 2015 Nonlinear Dynamics 79(1) 11-29
[18] Huang N E, Long S R and Shen Z 1996 Adv. Appl. Mechanics 32 59-111
[19] Huang N E, Shen Z and Long S R 1998 Proc. R. Soc. Lond. A 454 903–95
[20] Rajeev Sharma and Ram Bilas Pachori 2015 Expert Systems with Applications 42 (3) 1106-17
[21] Gao G Q, Li M 2014 Proc. Int. Conf. on Sensing and Control (Miami: FL/ IEEE) 451-6
[22] Ph. Wahl. 1979 Biophysical Chemistry 10(1) 91-104
[23] Weisenburger S 1992 Texas Studies in Literature and Language, 87-105

[24] Zhang H F, Shu Y T and Yang O 1997 Proc. Pacific Rim Conf. on Communications, Computers and Signal Processing Vol 2 (Victoria: Canada/IEEE) 883-6
[25] Enge P K, Kalafus RM and Ruane M F 1988 Communications Magazine, IEEE 26(7) 48-60

A social-level macro-governance mode for collaborative manufacturing processes

Ji Gao[1,2,*], Hexin Lv[1], Zhiyong Jin[1], Ping Xu[1]

[1]Zhejiang Shuren University, Hangzhou 310015, China
[2]Zhejiang University, Hangzhou 310027, China
*Corresponding author, Email: gaoji1@zju.edu.cn, Phn: +86-571-8799-6934

Abstract. This paper proposes the social-level macro-governance mode for innovating the popular centralized control for CoM (Collaborative Manufacturing) processes, and makes this mode depend on the support from three aspects of technologies standalone and complementary: social–level CoM process norms, CoM process supervision system, and rational agents as the brokers of enterprises. It is the close coupling of those technologies that redounds to removing effectively the uncontrollability obstacle confronted with by cross-management-domain CoM processes. As a result, this mode enables CoM applications to be implemented by uniting the centralized control of CoM partners for respective CoM activities, and therefore provides a new distributed CoM process control mode to push forward the convenient development and large-scale deployment of SME-oriented CoM applications.

1. Introduction

Faced with progressing globalization and intense market competition, Collaborative Manufacturing (CoM, for short) becomes the promising approach for small and medium size enterprises (SMEs, for short) to win the market competition. Meanwhile, along with the development of SOC (Service-Oriented Computing) and Cloud Computing technologies, the mechanisms for supplying manufacturing (mfg, for short) services through Internet such as ASP (Application Service Provider) [1], CMfg (Cloud Manufacturing) [2, 3], etc. have occurred. Those mechanisms provide the effective support for SMEs to develop CoM applications, and facilitate the research of SME-oriented CoM application development architectures, such as Dynamic Manufacturing Network[4], collaborative business ecosystem[5], cluster supply chain[6], non-hierarchical manufacturing networks[7], and so on.

However, a common characteristic of those architectures is to focus only on the dynamic on-demand formation of CoM applications, but the execution and supervision of CoM processes after the formation still adopts the centralized control mechanism dependent on artificial static customization. Therefore, though the optimization of CoM partners and enterprise alliances achieved by the Dynamic on-demand formation can promote the market competition ability of SMEs, the manpower and time consuming for customizing CoM processes severely weakened the ability for SMEs to quickly respond to the changes of market environments and client requirements. Besides, the centralized control mechanism is also confronted with the obstacle of uncontrollability for cross-management-domain CoM activities since collaborative partners usually work in different management domains while pursuing respective benefits. Moreover, the characteristic of "dynamic on-demand formation" aggravates the difficulty for removing this obstacle.

In order to eliminate these defects, this paper proposes the social-level macro-governance mode for innovating the centralized control mechanism implemented by leading SMEs (i.e., the SMEs which sponsor CoM and enterprise alliance) and limited in single management domains. Here, macro-governance aims at regulating and controlling the social-level macro-behaviors of CoM participators, that is, the social-level visible effects of their CoM activities, and uses the social-level CoM process norms as the governance basis. Evidently, as long as CoM participators make their macro-behaviors for providing and consuming mfg services conform to relevant norms, their CoM activities are transformed into predicable and controllable ones. Therefore, the macro-governance mode not only redounds to removing the uncontrollability obstacle, but also enables cross-management-domain CoM applications, though CoM activities executed by CoM participators are distributed, to be implemented by uniting the centralized control of those participators for respective CoM activities, and thereby removes the requirement for creating the centralized control mechanism. Due to no need for concerning with the implementation detail of CoM activities, the macro-governance mode can reduce largely the difficulty for developing and maintaining CoM applications, and thereby facilitates effectively the convenient development and large-scale deployment of CoM applications.

The layout of this paper is as follows. We first discuss the countermeasure for realizing the macro-governance mode in Section 2. Then in Sections 3 the three support technologies for this mode are specified. Finally, the evaluation and conclusions are given in Section 4.

2. Countermeasure for Realizing Macro-Governance Mode

As mentioned above, the social-level macro-governance mode redounds to removing the obstacle of uncontrollability for cross-management-domain CoM activities. However, how to make enterprises conform to social-level CoM process norms is also the challenging problem which must be solved. Therefore, we propose that enterprises should be constructed into the rational individuals which have both ability and desire for conforming to norms in order to ensure that their CoM activities can comply with the macro-behavior constraints prescribed by relevant norms. Meanwhile, enterprises should register into a CoM community in order to accept the assistance and supervision from this community, and thereby restraining possible norm-violating behaviors.

In order to enable enterprises to have high autonomy and intelligence to compute and reason how to conform to norms, it is advisable adopting software agents as brokers for enterprises to provide / consume mfg services through Internet. Moreover, in order to promote those agents into rational individuals, called rational agents, social-level norms should specify accurately the constraints for agent social-level macro-behaviors so that how to make the CoM activities of agents conform to norms can be determined by inspecting whether CoM states satisfy those constraints.

In human society, the popular CoM applications are the enterprise alliances created dynamically on demand. Such an alliance often includes multiple bilateral collaborations for obtaining different mfg services, called multi-service collaboration, so that the enterprise sponsoring the alliance can compose those services, by using a local business process, to accomplish the mfg tasks dynamically occurring and beyond the enterprise's own ability, and thereby responds successfully to market changes and wins the competition. Because a multi-party collaboration can be transformed into multiple bilateral collaborations and nested alliances can be created for suiting complicated collaboration demand, we focus our research work on the dynamic alliances which are oriented to the multi-service collaborations and whose members are the agents, without loss of generality. Obviously, bilateral collaborations oriented to single services should become the core for achieving the macro-governance

3. Social-Level Macro-Governance Mode

According to the analysis in Section 2, the realization of this mode demands the support from three aspects of technologies standalone and complementary: social–level CoM process norms, CoM process supervision system, and rational agents as the brokers of enterprises (Figure 1). The former two technologies form the management system configured in the CoM community in order to support effectively the macro-governance for the social-level visible behaviors of enterprises participating in CoM while the latter enables the norms to be exerted effectively to CoM processes and makes the CoM activities executed by those enterprises accept the supervision of this community.

3.1. Social-Level CoM Process Norm system

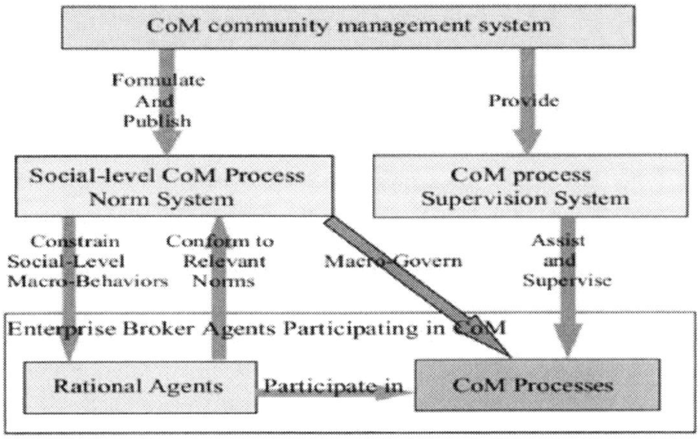

Figure 1. Social-level norm-governance mode and its support technologies

Because social activities are the means for maintaining social relationships, the social-level norms for governing those activities should be formulated based on the abstract concepts oriented to social relationships. Since Social Dependence (SD, for short) is a basic social relationship in human society, and occurs in every social activity [8-10]. We propose to represent CoMprocess norms in terms of SDs. An SD means that some social indivi-dual (e.g., an enterprise participating in CoM) depends on the assistance of another one to achieve an expectant goal, and the assistance is a social activity (e.g., a CoM activity) executed by the latter. Therefore formulating norms in the form of SDs, called SD norms, redounds to describing deeply and in the round social relationships and their evolution which should be complied with in CoM activities.

Since the CoM process for a mfg service often includes multiple CoM activities executed in succession, the norms for constraining those activities form the norm set called CoM process pattern for macro-governing the CoM process. However, every CoM instance needs to be created dynamically on demand, and this makes creating, by negotiation, the pattern for an instance from scratch face the predicament that computation complexity and space-time consumption are too large. Therefore, we propose that the more convenient and effective approach should be: formulating the parameterized CoM process template for a mfg service first, then instantiating dynamically this template, by negotiating and signing, into the shared pattern adopted by CoM partners for this instance (i.e., the two broker agents for providing / consuming this service).

Define the CoM process norm system $\mathcal{L}$ for a mfg area as the set of CoM process templates: $\mathcal{L} = \{M^{S1}, M^{S2} \dots M^{Sn}\}$. Every template M^{Si} (i=1, 2... n), corresponding to a specific mfg service Si, includes a group of SDNs (SD Norms): $M^{Si} = \{SDN_{A1}{}^{Si}, SDN_{A2}{}^{Si} \dots SDN_{Am}{}^{Si}\}$. Wherein, $SDN_{Aj}{}^{Si}$ is used to constrain CoM activity Aj (j =1, 2... m) and Aj may be an activity regular or for transacting abnormity. As an example, for a mfg service "CPUSupply" (defined in the mfg area "Computing Equipment"), the CoM process includes not only the regular activities executed alternately by the provider and consumer of this service: pay deposit, notice the pick-up of goods, pay for goods, provide goods, etc., but also the activities for transacting abnormity: complement deposit, defer the provision of goods, amerce, withdraw CoM relationship, etc. By formulating relevant SDNs, M^{Si} provides CoM participators with the constraint and guidance for the macro-behaviors in those activities, becoming the basis for macro-governing CoM processes from the social level.

Define $SDN_{Aj}{}^{Si} = PSD_e{}^o(\rho \leq \delta | \sigma)$, which is used to represent the Parameterized SD (PSD) of CoM activity Aj expected to maintain in the process performing bilateral (providing-consuming) collaboration oriented to service Si. $PSD_e{}^o(\rho \leq \delta | \sigma)$ is endowed with the following meanings: under the condition that the current collaboration state satisfies σ and before the deadline δ is reached, party (agent) o depends on Aj' executer e (the other party) to transform this state into satisfying ρ. Here, σ and ρ are also called the trigger for activating a norm and the postcondition for executing the norm respectively, and thereby make the execution of the norm possess the explicit start-up occasion and success-determining standard.

"Parameterized" means that o and e are played by the provider and consumer of service Si respectively, but bound to the agents as their brokers only when the CoM process of a CoM instance starts up. And ρ, δ, and σ are all the collaboration state patterns described in logic expressions, which are the logical combination of predicate expressions, relation expressions, and true functions. Therein, the parameters of a predicate expression may be variables whose value range can limited by relation expressions and true functions while the concepts, attributes, and relations referred to by predicate expressions are defined in the ontologies of mfg areas. Therefore, SDN_{Aj}^{Si} prescribes accurately the constraints which the social-level macro-behaviors of CoM activity Aj should conform to, and thus can be used as the basis for computing and reasoning how to comply with norms.

Since SDNs only constrain the social-level visible Macro-behaviors of broker agents for executing CoM activities: activity-activated occasion, activity-ended time, and activity post-effect, but does not prescribe the execution details, this enables template M^{Si} to be formulated in the highly abstract form in which agent mental models and internal micro-operators are invisible. Such a platform-independent description mode not only makes M^{Si} adopted and conformed to easily by heterogeneous agents, but also enables M^{Si} to be created on the basis of deep analysis and synthetic investigation for CoM objectives and possible benefit relations or conflicts. Moreover, the SDNs contained in M^{Si} can express explicitly the expected collaboration state transition and reveal the SD relationship evolution happening along with the successive execution of CoM activities, this redounds to abstracting and describing, in systematic mode, the benefit-transferred relationships between CoM participators and the changes, and thereby makes creating perfect M^{Si} become possible.

After created, M^{Si} is published in the CoM community, becoming the negotiation foundation for creating CoM relationships. Once the negotiation (which belongs to the scope of self-organization, and is not discussed here) successes, M^{Si} is instantiated into CoM process pattern M_Y^{Si}, which is used to govern the current CoM instance Y, and a contract for the two parties of collaboration to adopt M_Y^{Si} is signed. Meanwhile, the SDNs included in M_Y^{Si} is also specialized into appropriate to Y: o and e in the definition of SDN_{Aj}^{Si} are enacted by the agents participating in Y while the variables occurring in M_Y^{Si} are instantiated into constants or the limit of variable values are increased.

3.2. Rational Agents

Enterprise broker agents are configured into the rational agents which have both ability and desire to make their CoM activities accept the constraints from M_Y^{Si}, and therefore the outer observers (i.e., people or other agents) can believe those agents reasonable and trustable. The key for achieving rationality is to install those agents the local management policies for driving CoM activities in order to enable those agents to know how to execute those activities by using local micro-operators and decide how to conform to the constraints from M_Y^{Si}, including the norm-violating in an unavoidable situation. Since the SDNs included in M_Y^{Si} have already prescribed accurately the constraints agent social-level macro-behaviors should conform to, those constraints, as the basis for agents to compute and reason how to conform to norms, can be incorporated into the CoM activities driven by policies. A management policy p is represented in BNF as follows:

<p > ::= <trigger> <processing>
<processing> ::= {< micro-operator > | <rule> | <rule-group>}$^+$
<rule> ::= (→ <condition> {< m-operator>}$^+$)
<rule-group> ::= <selection-mode-for-activated-rules> {< rule >}$^+$

The main parts of a policy are its trigger and the transaction driven by it. The former is set as a trigger pattern (it and the condition part of a rule are all represented as the logical expressions similar to the collaboration state patterns mentioned in section 3.1), enabling the policy to be triggered by a state-change event. The latter is represented as a sequence of micro-operators, condition-action rules (conditional operators), and rule groups (for selecting executed rules), enabling an agent to make the decision analysis by detecting the activation of rules in order to drive appropriate local micro-operators for executing a CoM activity.

Since template M^{Si} and the SD norms included in it has been published in the agent community, this enables the desire for rational agents to conform to norms to be embodied by setting relevant management policies while the local micro-operators embedded into the policy transaction parts and the executing software for implementing those operators make those agents possess the abilities for performing the norms constraining CoM activities (include executing and monitoring those activities), in order to achieve the governance effect of those norms for these activities.

3.3. CoM Process Supervision System

The supervision system is set in the CoM community and constructed by coupling mechanisms of social commitment, commitment supervision, and reputation management in order to enforce and assist the enterprise broker agents as community members to perform the social commitment for complying with CoM process patterns. This system is configured with five supervision roles: community manager, service matchmaker, CoM arbitrator, CoM monitor, and reputation manager (new roles may be added on requirement), and creates a group of supervision agents enacting those roles. Meanwhile, a set of CoM assistance and supervision services are configured to those roles in order to support the work of those mechanisms.

3.3.1 Social Commitment. This mechanism promises broker agents to create the social commitment for conforming to a CoM process pattern, say M_Y^{Si}, by registering into the community and signing the contract for adopting M_Y^{Si}. Note, once the negotiation for creating CoM relationships successes, broker agents can sign such contracts by invoking the signing transaction provided by relevant mfg services. As a result, this brings on the Parameterized representation form of norm SDN_{Aj}^{Si} in M_Y^{Si}: $PSD_e^o(\rho \leq \delta | \sigma)$ to be transformed into social commitment $SCommit_e^o(\rho \leq \delta | \sigma)$ / social trust $STrust_e^o(\rho \leq \delta | \sigma)$. Such a transformation has following semantics: the executer e of CoM activity Aj commits the CoM partner o to / o trust e for that under the condition that the CoM state satisfies σ and before the deadline δ is reached, e transforms this state into satisfying ρ.

3.3.2 Commitment Supervision. By executing the supervision services provided by the community, such as CoM process monitoring, norm-violation arbitration, norm-violation analysis, norm-violation punishing, reputation evaluation, norm-violation sanction, etc., and the CoM supervision policies of norm-violation punishing, reputation reduction, and norm-violation sanction published in the community, the commitment supervision mechanism, based on social commitments represented as $SCommit_e^o(\rho \leq \delta | \sigma)$, aims at supervising CoM processes, arbitrating conflicts of norm-violation identification, and sanctioning norm-violation agents in order to generate the deterrent force for driving agents to perform social commitments and make the equitable transaction for norm-violation events (Figure 2). Evidently, this force can drive CoM participators to comply with social commitments, reduce greatly the CoM indeterminacy, and thereby make the CoM activities executed in CoM processes transform into being predicable, controllable, and then trustable, and in turn restrain the requirements for transacting abnormity. Although CoM process patterns contain the norms for norm-violation transaction activities, it is also necessary asking the community, as the third party, to arbitrate conflicts of norm-violation identification and sanction norm- violation agents.

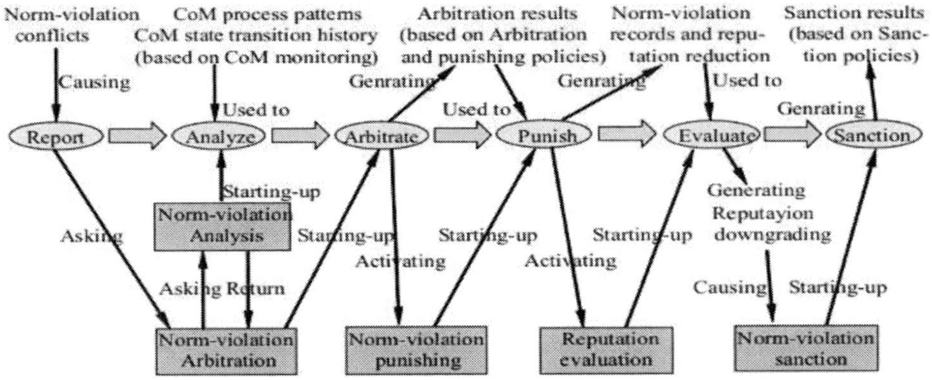

Figure 2. Commitment supervision mechanism (square frames indicate supervision services, double arrows the order of supervision activities)

3.3.3 Reputation Management. The reputation management mechanism aims to inspect, evaluate, and maintain the reputation for CoM participators to perform social commitments by executing supervision services: norm-violation analysis, norm-violation punishing, reputation evaluation, reputation enquiry, etc., and the punishing policies for decreasing reputation levels. Because the broker agents with bad reputation would lose the chance for participating in CoM, the reputation evaluation itself also generates the deterrent force for driving broker agents to perform social commitments, and therefore can be used as the important means for commitment supervision. Meanwhile, selecting the providers with good reputation as the CoM partners can also prevent the occurrence of norm-violation events effectively.

4. Evaluation and Conclusions

This paper proposes the social-level macro-governance mode for innovating the popular centralized control for CoM processes, and makes this mode depend on the support from three aspects of technologies standalone and complementary: social–level CoM process norms, CoM process supervision system, and rational agents as the brokers of enterprises.

We have already created an architecture prototype based on the social-level macro-governance mode, which is divided into two layers: CoM community for providing social–level CoM process norms and CoM process supervision and enterprise broker agents participating in CoM. Meanwhile, we have designed a representation language for SD norms and the relevant visualized editing interfaces to support the formulation of CoM process templates for mfg services, and provided the developing and running platform of rational agents [11]. Then, several experimental applications have been designed to validate the feasibility of this macro-governance mode.

Here we briefly describe one of them: the CoM application for the mfg area "Computing Equipment". A computing equipment manufacturer plans to develop a new product P which consists of mainframe, CUP, Memory, operation software, and photography parts. The manufacturer produces the mainframes and operation software of P by itself and completes the equipment assembly while the parts of CUP, Memory, and photography depend on the supply from outer enterprises, and the performance test of P is deputed to an external professional test center (those are abstracted into the mfg services provided by outer enterprises). Therefore, this manufacturer needs to dynamically create an enterprise alliance including the selected enterprises for providing the parts of CUP, Memory, and photography and the performance test, and configure a local business process to schedule the multi-service collaboration for composing those local and outer services.

Note, as long as the CoM process oriented to each outer service is governed reliably, it is not difficult to compose mfg services (how to compose is beyond the scope of this paper). In order to support the execution of CoM processes, the development work for this application has included to formulate and publish, in the CoM community, the CoM process norm subsystem for this mfg area, and to create the rational agents as the enterprise brokers for providing / consuming mfg services. Then. let the manufacturer agent sponsor and perform several dynamic enterprise alliances based on multi-service collaborations.

The results for running this case demonstrate: the macro-governance effect of social-level norms can be exerted reliably into CoM processes by rational agents. Therefore, not only the uncontrollability obstacle confronted with by cross-management-domain CoM processes can be removed, but also the control for CoM processes can be implemented by uniting the centralized control of CoM partners for respective CoM activities. This, plus automatically creating CoM process patterns by instantiating relevant templates, can remove the requirement for customizing statically and artificially the centralized control mechanism, and thereby promote the ability for SMEs to quickly respond to the changes of market environments and client requirements.

The norm-based governance originates from the research area of NMAS [12, 13] (Normative Multi-Agent System). However, NMASs are focused on the autonomic formation of collaborations, and disjointed from conventional application software, thereby difficult to display their value for supporting real collaboration application. Moreover, the norms for constraining social-level visible macro-behaviors of agents are represented in the form of deontic logic [14] (including obligation, permission, and prohibition), but cannot describe intuitively the social relationships and their evolution occurring in social activities. In contrast, this paper proposes to adopt the basic social concept "social dependence" to formulate the norms for constraining social-level macro-behaviors[8-10], redounding to describing deeply and in the round social relationships and their evolution which should be complied with in social activities. Meanwhile, using parameterized SD (social dependence) $PSD_e^o(\rho\leq\delta|\sigma)$ to accurately defines SD norms can not only remove the limitation that traditional structures for representing SDs do not accommodate state transition and SD evolution[10,15], but also provide the reliable basis for agents to compute and reason how to conform to norms. Besides, by integrating the technologies of "policy-driven" and SOC (Service-oriented Computing), the developing and running platform of rational agents mentioned above has already achieved the seamless connection between agent intelligence and conventional software.

The limitation of our research work is that the feasibility and performance of the social-level macro-governance mode is examined only based on the simulation test case. Although the key characteristics and main circumstance of CoM applications can be simulated well, many real-life problems required to be solved have still been ignored yet. Moreover, the completeness verification of CoM process templates also is a both important and challenging task. Therefore, developing real applications and solving the verification and other real-life problems will be our further research work.

5. Acknowledgment

We gratefully acknowledge the support of the National Science Foundation of China (Grant 61375071), the Natural Science Finds of Zhejiang Province, China (Grant LZ15F020001), and the Natural Science Finds of Zhejiang Province, China (Grant LY14F020008).

6. References

[1] Smitha D, Rupp W T, 2002, Application service providers (ASP): moving downstream to enhance competitive advantage, Information Management and Computer Security, 10(2/3): 64-72.

[2] Li B, Zhang L, Wang S, et al. 2010, Cloud manufacturing: a new service-oriented networked manufacturing mode, Computer Integrated Manufacturing Systems, 16 (1): 1-16.

[3] Xu X, 2012, from cloud computing to cloud manufacturing, Robotics and Computer-Integrated Manufacturing, 28 (2012): 75–86.

[4] Papakostas N, Georgoulias K, Koukas S, and Chryssolouris G, 2015, Organisation and operation of dynamic manufacturing networks, International Journal of Computer Integrated Manufacturing, 28 (8): 893–901.

[5] CardosoT and Camarinha-Matos L M, 2013, Pro-Active Service Ecosystem Framework, International Journal of Computer Integrated Manufacturing, 26 (11): 1021-1041.

[6] Xue X, Liu Z Z, and Wang S F, 2016, Manufacturing service composition for the mass customised production, International Journal of Computer Integrated Manufacturing, 29 (2): 119–135.

[7] Andres B and Poler R, 2016, Models, guidelines and tools for the integration of collaborative processes in non-hierarchical manufacturing networks: a review, International Journal of Computer Integrated Manufacturing, 29 (2): 166–201.

[8] Chopra A, 2011, Social computing: Principles, platforms, and applications, 1st International Workshop on Requrements Engineering for Social Computing (RESC 2011), pp. 26-29.

[9] Mokom F, 2011, Social Dependence-Based Methods for Coalition Formation in Multi-Agent Systems, http://cs.uwindsor.ca/richard/cs510/survey_felicitas_mokom.pdf.

[10] Lau B, Singh A, Tan T, 2012, A review on dependence graph in social reasoning mechanism, Artif Intell Rev, 16 (November): 1-14.

[11] Gao J and Lv H, 2012, Institution-governed cross-domain agent service cooperation: a model for trusted and autonomic service cooperation, Applied Intelligence, 37(2): 223-238.

[12] Criado N, Argente E, and Botti V, 2011, Open issues for normative multi-agent systems, Ai Communications, 24(3): 233-264.

[13] Boella G, Pigozzi G, and van der Torre L, 2009, normative systems in computer science – ten guidelines for normative multiagent systems, in: Normative Multi-Agent Systems, Dagstuhl Seminar Proceedings, No. 09121.

[14] Sergot M, 2010, Norms, Action and Agency in Multi-agent Systems, Deontic Logic in Computer, Lecture Notes in Artificial Intelligence, Volume: 6181.

[15] Caire P, van der Torre L, and Villata S, 2013, Argumentation Theoretic Foundations for Abstract Dependence Networks, Agreement Technologies, Lecture Notes in Computer Science, Volume 8068, 180-194.

IOP Conf. Series: Journal of Physics: Conf. Series **887** (2017) 012018 doi:10.1088/1742-6596/887/1/012018

Design of agricultural product quality safety retrospective supervision system of Jiangsu province

Kun Wang

Taizhou Agricultural Internet of things Engineering Technology Center, Jiangsu Agri-animal Husbandry Vocational College, 225300, Taizhou, Jiangsu, China

Email: jswangkun@yeah.net

Abstract. In store and supermarkets to consumers can trace back agricultural products through the electronic province card to query their origin, planting, processing, packaging, testing and other important information and found that the problems. Quality and safety issues can identify the responsibility of the problem. This paper designs a retroactive supervision system for the quality and safety of agricultural products in Jiangsu Province. Based on the analysis of agricultural production and business process, the goal of Jiangsu agricultural product quality safety traceability system construction is established, and the specific functional requirements and non-functioning requirements of the retroactive system are analyzed, and the target is specified for the specific construction of the retroactive system. The design of the quality and safety traceability system in Jiangsu province contains the design of the overall design, the trace code design and the system function module.

1. Introduction

As the quality and safety of agricultural products directly affect the health of the public, social harmony and the healthy development of the national economy, so the problem to the government and the public attention. Agricultural standardization refers to standardization activities targeted at agriculture [1]. It is a kind of unified activities for the development and implementation of standards for all kinds of objects in agricultural economy, technology, science and management activities. A comprehensive technical basis for agricultural modernization is agricultural standardization. The quality of agricultural standards is not perfect, is not conducive to ensuring the quality and safety of agricultural products, improve the market competitiveness of products [2]. The establishment of agricultural enterprises agricultural product traceability system, the use of agricultural information technology, is conducive to promoting the process of agricultural information; promote agricultural standardization [3]. The traceability of fruit and vegetable agricultural products refers to the information from the downstream of the supply chain, which is traced back to the orchards and farmers located in the upper reaches of the supply chain [4]. The process information such as planting, processing and transportation of the fruits and vegetables products is obtained by recording the identification. Consumers or sellers package products to its base production and sales of the status of the traceability. Therefore, the retrospective can not only be used to identify product quality problems, but also can learn other excellent products management production technology.

Traceability system is a kind of product history, use or location and other related information to record storage quality assurance system [5]. In the whole process of product supply, the quality traceability system can quickly and effectively query the problems in the raw materials or processing problems in the event of product quality problems, so that when necessary, product recall can be carried out, or the implementation of targeted penalties, and thus to ensure product quality.

2. System Design

Traceability of the system needs to be achieved from the production site to the sales of each link can be traced each other, that is, the need to establish a database record agricultural production, business records. Information on producers, base environments, the use of agricultural inputs, field management, processing, packaging and other information, and open to consumers, so that producers, consumers and distributors through the coding and query related to the quality and safety related information. The process of traceability of agricultural products is carried out around the life cycle of agricultural production management. The life cycle includes planting, picking, processing, transportation and other production processes and activities. The life cycle of agricultural production management is shown in Fig.1.

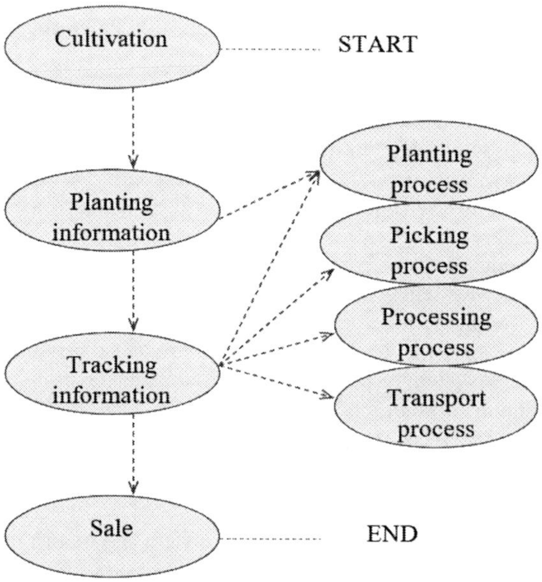

Figure 1. Jiangsu agricultural production management of the life cycle diagram

Jiangsu agricultural product quality and safety traceability system is mainly through the Internet to build products traceability system. The information on the picking, processing and transportation of agricultural products is entered for retrospective and inquiry. Consumers, wholesalers through the tracing code, the product cultivation, picking, processing and transportation related information to query, the quality of the existence of complaints.

Ease of use requirements are one of the important non-functional requirements that this system is designed to consider. System interface is the user and the server data exchange between the media. As the system users for the cooperative management staff, joined the farmers and consumers, computer operating level is low, so the interface design to facilitate the operation, highlighting the ease of use. Functional plate boundaries clear, so that users of the plate can easily to use it.

3. Implementation of Specific Processes

The non-functional requirements of the system refer to the specific requirements of the software products in addition to the necessary functional requirements, such as the system application environment, hardware environment, etc., which are important basis for the impact of non-functional requirements. The non-functional requirements of this system focus on ease of use, security, and implementation requirements. Picking the information management system, the picking navel orange picking time, plots, quantity, traceability and pesticide residue detection corresponding to the entry, so that the future use of traceability. Picking information / management business process shown in Figure 2.

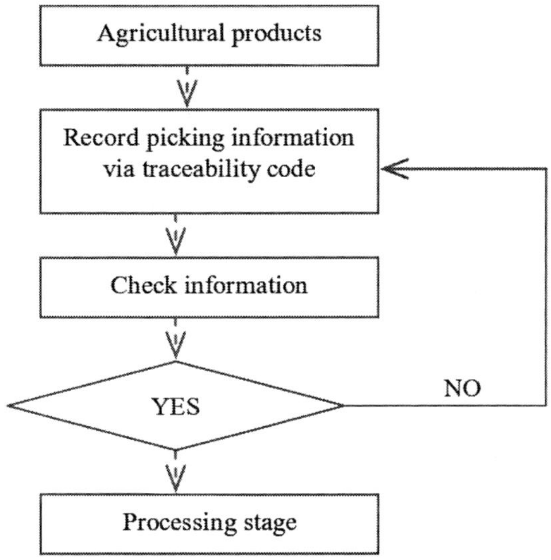

Figure 2. Flow chart of picking information management

The system use the telecom center room hosting server, the establishment of data center, the server bandwidth of 100 trillion, in addition to the establishment of data collection center in the planting base. Through ADSL, wireless networks, etc. access to the Internet to access the server. At the same time with the Internet and GPRS network, through the touch screen, website, mobile source terminal, etc., to provide consumers with agricultural product quality and safety traceability. The server hardware configuration is shown in Table 1.

Table 1. Server Hardware Configuration

Accessories	Specific parameters
processor	CPU Model: Xeon E5335
	CPU frequency: 2GHz
	Standard CPU Quantity: Standard 1 xeon E5335 processor
	Maximum number of CPUs: 2 stars
	L2 cache: 2 * 4MB
	Bus specification: FSB 1333MHz
Motherboard	Motherboard chipset: Intel 5000p
	Expansion slots: 4 optional PCI-Express slots
RAM	Memory type: FB-DIMM
	Memory capacity: 2GB
	Maximum memory capacity: 32GB
Storage	Hard disk interface type: SAS / SATA
	Standard hard disk capacity: standard does not provide
	Hot-swap disk: support hot-swappable
	RAID mode: 256MB HP Smart Array P400 controller (RAID 0/1/1 + 0/5)
	CD-ROM: optional DVD, DVD / CD-RW combination drive
	IDE controller: ATA 100

4. Coding and Implementation of the Trace Chain

Based on the analysis of the actual production and processing process of agricultural products, this study draws out the Jiangsu agricultural production and processing business model: from the initial planting, after picking, and then enter the basis of the agricultural production traceability. First, warehouse, in the production and processing, from the warehouse to extract Jiangsu agricultural products into the production workshop, second, a series of production processes, by the quality inspection staff during the inspection, and then the finished product packaging, and finally the circulation of the transport and final for sale.

Based on the research process of Jiangsu agricultural production process, the paper designs and designs the Jiangsu agricultural product quality tracking and traceability coding system with batch as the core, and links the agricultural production process information in Jiangsu through various links: picking link, processing link, transportation link and sales link of the batch number to connect, the next link contains information on the previous link, through the batch number, can be easily traced back. And finally designed to "product trace code sales batch number, transport batch processing batch number picking batch number origin code" for the chain of the retroactive program, which not only increases the consistency of information, but also easy to understand, At the same time in Jiangsu agricultural products traceability can quickly access all aspects of the quality and safety information, the specific process shown in Figure 3.

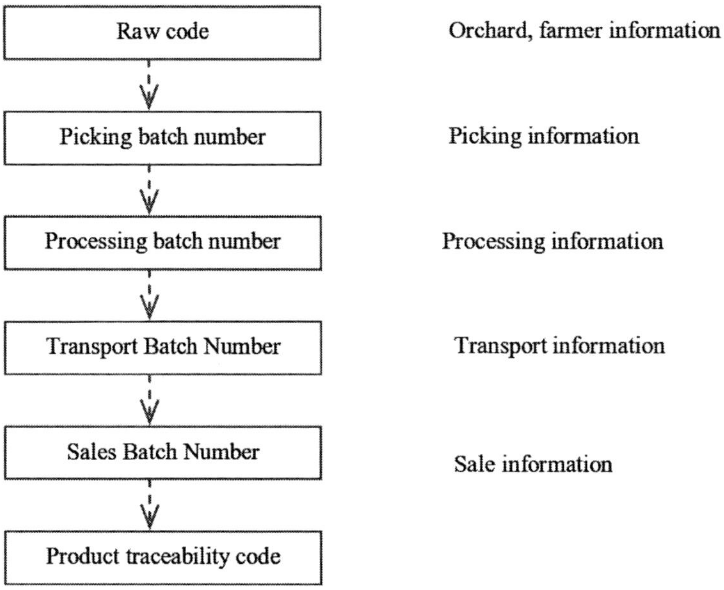

Figure 3. The retroactive process based on the batch number

Which is based on the data centre as the core of the retrospective link, the data centre mainly stores the whole process of Jiangsu agricultural products such as the use of inputs, production, processing, storage, transportation, sales and other aspects of quality and safety information, with the Internet and GPRS network, Website, mobile tracing terminal, etc., to the consumer, Jiangsu agricultural products processing enterprise management personnel, and quality supervisors through the Jiangsu agricultural product quality and safety traceability code, access to the product quality and safety information to the data centre as the core of the retrospective program as shown in figure.4.

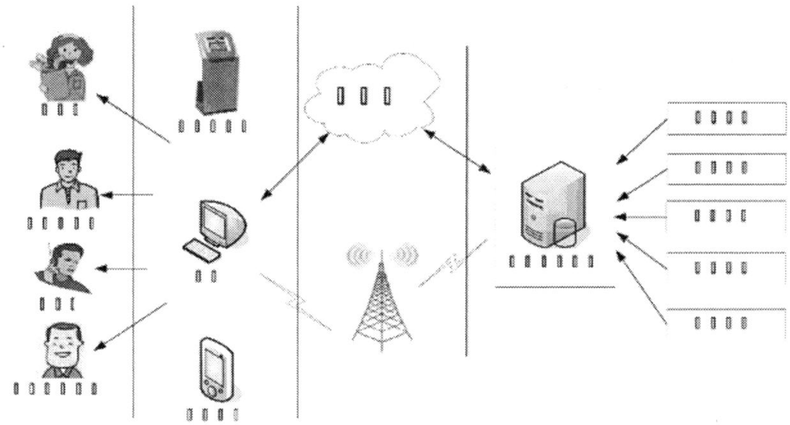

Figure 4. The retrospective based on the data

System design is the core of system development; the process is to use the results in the system requirements analysis, the target system of the logical model into a physical model. According to the basic function of Jiangsu agricultural product quality safety traceability system, the retroactive system design basic information management subsystem, fertilization information management subsystem, drug information management subsystem, agricultural information management subsystem, picking information management subsystem, processing information management subsystem , Transport information management subsystem, user information management subsystem, retroactive query subsystem, a total of nine subsystems. Among them, the product number design as shown in Figure 5:

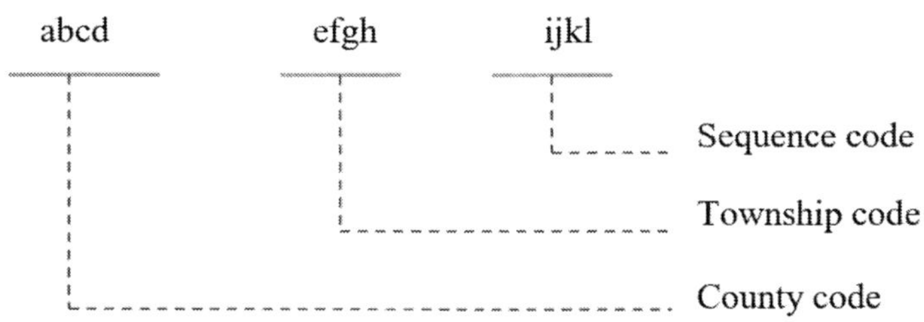

Figure 5. Design of agricultural products number

5. Conclusions

The establishment of quality and safety traceability system is the agricultural enterprises to participate in international market competition, breaking the trade barriers, enhance the brand value of the inevitable trend. The construction of agricultural enterprises through information construction, especially the traceability system of agricultural products, can avoid the bottleneck of development after increasing the operation cost of agricultural enterprises to a certain scale due to the increase of farmer and production area. At the same time, through the retrospective system and the Internet and other information means to promote the standardization of agricultural production, improve product quality and enhance the competitiveness of agricultural enterprises in the market.

This paper designs and realizes the traceability system of Jiangsu agricultural product quality and safety based on B / S structure. Based on the current situation and the business of Jiangsu agricultural products, the system establishes the construction goal of Jiangsu agricultural product quality safety traceability system, analyzes the functional requirements and non-functioning requirements of the system, and designs the functions on the basis of the demand analysis, including the overall design , System function module design, retrospective chain coding system design and other aspects of the design, in order to achieve its Jiangsu agricultural products in the cultivation, picking, processing and transportation and other aspects of the quality and quality of traceability, and through the results back to the farmers to achieve the process of planting for standardized management.

6. Acknowledgments

Authors wishing to acknowledge Fund Project: Agricultural products quality and safety special fund project in Jiangsu Province (2130109).

7. References

[1] Li J, Ma C, Wang JJ 2012 *Advanced Materials Research* 550-**553** 1952
[2] Marlies de K, Renzo A, Martin G, Jacqueline M. B, Jack G. A. J 2017 *European Journal of Operational Research* **262** 535
[3] Jonathan R. M, Wallace E. H 2017 *Food Policy* **69** 82
[4] Minghua Z, Michael R. Z, Shu G 2015 *Journal of Integrative Agriculture* **14** 2340
[5] Tianjin C, Gang C, Shuming Y, Yan Z, Zhihua Y 2017 *Food Control* **72** 306

IOP Conf. Series: Journal of Physics: Conf. Series **887** (2017) 012019 doi:10.1088/1742-6596/887/1/012019

A novel approach to describing and detecting performance anti-patterns

Jinfang Sheng[1], Yihan Wang[1], Peipei Hu[1] and Bin Wang[1,*]

[1] School of Information Science and Engineering, Central South University, Changsha, 410083, China

* Corresponding author

Abstract. Anti-pattern, as an extension to pattern, describes a widely used poor solution which can bring negative influence to application systems. Aiming at the shortcomings of the existing anti-pattern descriptions, an anti-pattern description method based on first order predicate is proposed. This method synthesizes anti-pattern forms and symptoms, which makes the description more accurate and has good scalability and versatility as well. In order to improve the accuracy of anti-pattern detection, a Bayesian classification method is applied in validation for detection results, which can reduce false negatives and false positives of anti-pattern detection. Finally, the proposed approach in this paper is applied to a small e-commerce system, the feasibility and effectiveness of the approach is demonstrated further through experiments.

1. Introduction

As an extension of the concept of the pattern, the anti-pattern describes a commonly used bad solution, which will have a negative impact on the application system [1]. Anti-pattern detection as a bridge between design patterns and inexperienced personnel emphasizes the identification of harmful software structures that repeatedly appear in multiple projects. Anti-patterns are likely to be at all stages of the software life-cycle. As a software defect, anti-patterns can cause serious damages, especially in the software project management [2].

The presence of anti-patterns in the system can lead to many issues such as poor maintainability, scalability, reusability, and low performance. As a kind of anti-pattern, the existence of performance anti-pattern may have a serious impact on system performance. Therefore, it is very important to study the effective method of anti-pattern detection.

2. Related Work

Through the analysis of the current situation of anti-pattern detection, this paper holds that the current anti-pattern detection methods have the following problems.

1) The research on the detection of performance anti-pattern is relatively less.

2) The description capability of anti-pattern is insufficient.

3) No result process of performance anti-pattern detection.

4) No evaluation on the impact of the performance anti-pattern on the system.

In view of the shortcomings of these methods, this paper summarizes the 14 performance anti-patterns in literature [3-8], and selects the more common performance anti-patterns which have relatively high performance impact: Empty Semi Trucks [3] as an example for detailed analysis and discussion. This paper adopts first order predict method to describe anti-patterns based on the overall consideration of anti-patterns' manifestations and symptoms against deficiencies. In addition, this paper proposes a verification method of detection results based on Bayesian classifier, which can effectively reduce underreporting and misinformation. Finally, we apply this method to a small e-commerce system to further validate the feasibility and effectiveness of it.

3. Anti-Pattern Description Method Based on First-Order Predicate

Description1 Determine the measurement used to describe anti-patternsM_i, $M_i \in \sigma, \sigma$ is a set of all measurements, that is $\sigma = \{M_1, M_2, \cdots M_n\}$;

Description2 Measurement conditions:

$$P_i = f(M_i), R$$

Among them,P_i is a proposition which represents the comparison between $f(M_i)$ and R, $f(M_i)$ is a function to measure M, R represents the threshold of $f(M_i)$,a number of P_i constitute the conditions used to describe anti-patterns.

Description 3 AP is the name of anti-pattern, C_i represents the composition relationships between measurements, they are expressed as below:

$$AP = C_1 V C_2 V \cdots V C_n$$

$$C_i = P_1 \wedge P_2 \wedge \cdots \wedge P_n$$

When AP is true, it means the conditions of anti-pattern detection is met.

4. The Anti-Pattern Detection Method Based on Bayesian

4.1. The Probability Calculation of Candidate Anti-Patterns

4.1.1. Extremum Setting and Interval Division. Extremum is when measurement reaches it we can clearly determine if this pattern is anti-pattern. As shown in figure 1, the horizontal axis shows measurement and the vertical axis shows the probability of candidate anti-pattern whose range is from 0 to 1. L is the pre-set threshold, which is used to determine some rules shaped like $t_{value} \leq L$. The probability of L is set by specialists.

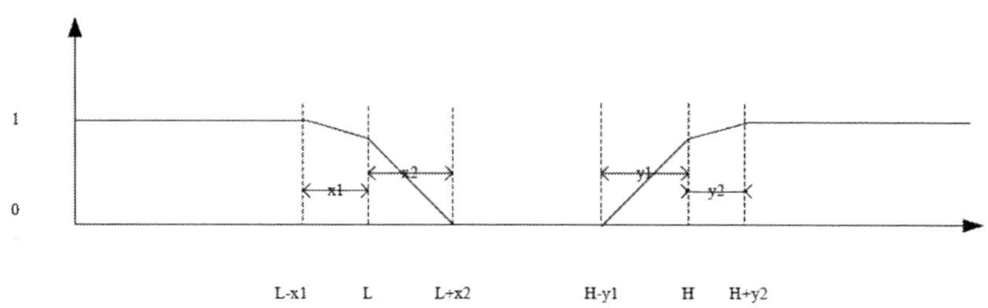

Figure 1. The probability setting interval of candidate anti-patterns

H is used to determine rules shaped like $t_{value} \geq H$. The probability of H is set by specialists.

4.1.2. The Probability Calculation of Intervals. The probability calculation of intervals can be divided into the following three types:

(1) The probability is calculated according to different intervals. As shown in equation 1.

$$p = \begin{cases} 1, \ t_{value} \leq L - x_1; \\ p_i + \frac{L - t_{value}}{x_1}(1 - p_i), \ L - x_1 < t_{value} \leq L; \\ p_i - \frac{t_{value} - L}{x_1}(1 - p_i), \ L < t_{value} \leq L + x_2; \\ 0, \ t_{value} > L + x_2; \end{cases} \tag{1}$$

(2) The probability is calculated according to different intervals. As shown in equation 2.

$$p = \begin{cases} 0, \ t_{value} < H - y_1; \\ p_{j-}\frac{H-t_{value}}{y_1}(1-p_j), \ H - y_1 \leq t_{value} < H; \\ p_j + \frac{t_{value}-H}{y_2}(1-p_j), \ H \leq t_{value} \leq H + y_2; \\ 0, \ t_{value} \geq H + y_2; \end{cases} \tag{2}$$

(3) When we need to determine rules shaped like L<BP<H, we can represent it as BP>L and BP<H and calculate its probability by combining equation2 and equationr3.

4.1.3. The Probability Calculation of Synthesis Rules. (4)For those synthesized with some conditions, shaped like A^B, the probability can be expressed as the multiplication of all these conditions' probability. As shown in equation 3 in which p_i is the probability of a single condition.

$$p = \prod_{i=1}^{n} p_i \tag{3}$$

4.1.4. The Probability of Combined Rules. For those combined with some conditions, shaped likeA ∨ B, the probability can be calculated by equation4

$$f(x) = \begin{cases} 1, \ \forall p_i, \exists p_i = 1 \\ 1 - \sum_{i=1}^{n}(1-p_i), \ \forall p_i x \neq 1 \end{cases} \tag{4}$$

p_i is the probability of the single rule which is anti-pattern. If $p_i = 1$, the probability of this condition being anti-pattern is 1. In other case, the probability can be calculted by equations.

4.2. The Verification of Candidate Anti-Patterns

Combining the Bayesian classification process, the process of anti-pattern detection based on Bayesian is defined as follows.

1) Suppose an anti-pattern can be represented by an n-dimension eigenvector $X = \{x_1, x_2, \cdots x_n\}$ and x_i describes n attributes of it.

2) The detection results of anti-pattern can be divided into two kinds:$c = \{c_1, c_2\}$. c_1 means the detection result is anti-pattern , on the contrary, c_2 means it is not.

3) The calculation of prior probability.

4) According to equation7:

$$p(C_1/x) = (p(x/C_1)p(C_1))/(p(x))$$
$$p(C_2/x) = (p(x/C_2)p(C_2))/(p(x)) \tag{5}$$

If considering system contexts (distributed applications? &expert knowledge) during anti-pattern detection, then $p(x/c_i)$ can be expressed as equation8

$$P\left(\frac{x}{c_i}\right) = \prod_{j=1}^{n} p\left(\frac{x_j}{c_i}\right) \tag{6}$$

Combining equation7 and8, we can know:

$$p(x) = \sum_{k=1}^{m} \prod_{j=1}^{n} p\left(\frac{x_j}{c_m}\right) p(c_m) \tag{7}$$

So according to equation7 and 9, we can know the probability of belonging to c_i is $p(c_i/x)$, which can be expressed as equation 10.

$$p\left(\frac{c_i}{x}\right) = \frac{\prod_{j=1}^{n} p\left(\frac{x_j}{c_i}\right)}{\sum_{k=1}^{m} \prod_{j=1}^{n} p\left(\frac{x_j}{c_m}\right) p(c_m)} p(c_i) \tag{8}$$

5) Respectively calculate $p(c_1/x)$ and$p(c_2/x)$. If $p(c_1/x) > p(c_2/x)$, this sample is anti-pattern, otherwise not.

5. Experiment

5.1. Sample Scenario Description
This paper applies the method mentioned earlier to the research object Java Pet Store1.3.2 [9] for anti-pattern detection. Java Pet Store1.3.2 is a simulated pet shop which simulates the whole process of buying pets by users. By analyzing the source code of Java Pet Store1.3.2, we find the anti-pattern Empty Semi Trucks appears during the period of viewing orders by users.

5.2. Results and Analysis
In order to complete the detection of Empty Semi Trucks, we firstly set the threshold of this anti-pattern' parameters based on experience. The sequence diagrams of system calls at runtime can be acquired according to the analysis of source code, so that the observed value corresponding to these parameters can be obtained. The anti-pattern detection starts after getting required information and completing threshold setting. Anti-pattern detection is divided into the following four steps:

1) Extremum setting and interval partition

Set the extremum and interval of each parameter.

2) The probability calculation of candidate anti-patterns at intervals

Combine equation 2 and 3 to calculate the probability of observed value being anti-pattern.

3) The probability calculation of synthetic rules and combined rules

Combine equation 4 and 5 to calculate the probability of candidate anti-patterns under various combined conditions.

4) The verification of detection results

In order to explain the application of Bayesian to anti-pattern detection, we simulate the following results aimed at Empty Semi Trucks. The experiment is divided into 3 groups, the data of group a meet the threshold conditions, that of group B is near the threshold and that of group C dissatisfy the threshold. Experimental data is illustrated in table 1.

Table 1. Simulated data

No.	A	B	C
1	$Th_{maxRemMsgs}=18$ $Th_{maxRemInst}=6$	$Th_{maxRemMsgs}=12$ $Th_{maxRemInst}=5$	$Th_{maxRemMsgs}=11$ $Th_{maxRemInst}=4$
2	$Th_{maxRemMsgs}=18$ $Th_{minNetUtils}=0.28$	$Th_{maxRemMsgs}=13$ $Th_{minNetUtils}=0.28$	$Th_{maxRemMsgs}=10$ $Th_{minNetUtils}=0.28$
3	$Th_{maxGetMethNum}=6$	$Th_{maxGetMethNum}=4$	$Th_{maxGetMethNum}=2$
4	$Th_{maxRemMsgs}=15$ $Th_{minNetUtils}=0.28$ $Th_{maxRemInst}=6$	$Th_{maxRemMsgs}=13$ $Th_{minNetUtils}=0.31$ $Th_{maxRemInst}=4$	$Th_{maxRemMsgs}=10$ $Th_{minNetUtils}=0.35$ $Th_{maxRemInst}=3$
5	$Th_{maxRemMsgs}=15$ $Th_{maxRemInst}=6$ $Th_{maxGetMethNum} = 4$	$Th_{maxRemMsgs}=11$ $Th_{maxRemInst}=4$ $Th_{maxGetMethNum} = 3$	$Th_{maxRemMsgs}=10$ $Th_{maxRemInst}=4$ $Th_{maxGetMethNum} = 2$

Conduct the anti-pattern detection by using the method mentioned in this paper and an ordinary method respectively.

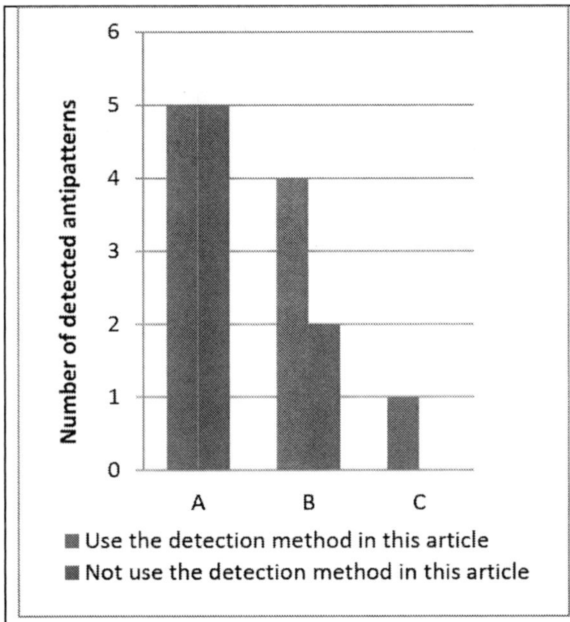

Figure 2. Results after verification based on Bayesian classifier

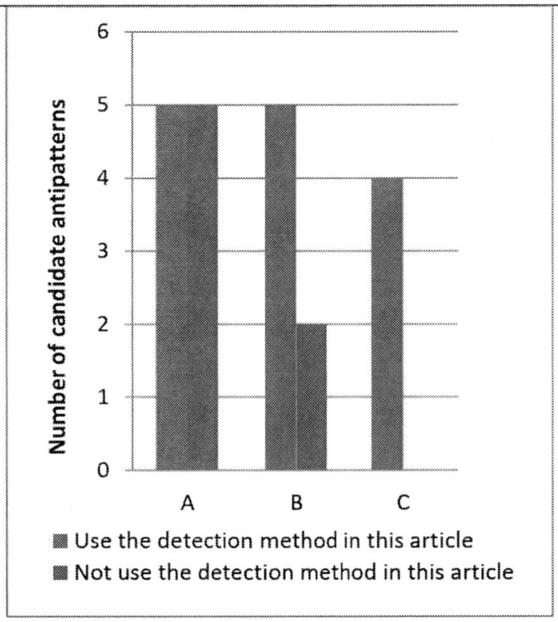

Figure3. Calculations of candidate anti-pattern

From the above-mentioned results, for conditions meeting the threshold, anti-patterns that are detected by an ordinary method can also be detected by the method in this paper; for conditions fluctuating across the threshold, the underreporting of anti-patterns can be effectively reduced; for those dissatisfying the threshold, the anti-patterns that are not detected by an ordinary method can be detected partly by the method in this paper, which can highly reduce the misinformation of anti-pattern detection.

6. Conclusion

The description method of anti-pattern based on first order predicate combines the manifestation with symptoms of anti-patterns to improve the reliability of description. This method is independent of specific rules templates and analytical models, so that it can modify and expand at any time. That is why the scalability of this method is superior to those with specific templates and models. Furthermore, this method is easy to implement without limitation to specific analytic language or inference engine. Meanwhile, adopting the Bayesian classifier with the existing testing information can effectively reduce the misinformation of detection results when introducing the concept of candidate anti-patterns. In addition, introducing the automated verification method based on Bayesian classifier before manual verification can also reduce the workload of artificial work so that improve the efficiency of manual verification. The results show that the description and detection method in this paper has obvious advantages.

7. References

[1] Keck P, Van Hoorn A, Okanović D, et al. Antipattern-Based Problem Injection for Assessing Performance and Reliability Evaluation Techniques[C]//Software Reliability Engineering Workshops (ISSREW), 2016 IEEE International Symposium on. IEEE, 2016: 64-70.

[2] Khomh F, Vaucher S, Sahraoui H. BDTEX: A GQM-based Bayesian approach for the detection of antipatterns [J]. Journal of Systems & Software, 2011, 84(4):559-572.

[3] Smith C U, Williams L G. More new software performance antipatterns: Even more ways to shoot yourself in the foot[C]//Computer Measurement Group Conference. 2003: 717-725.

[4] Crasso M, Zunino A, Moreno L, et al. JEETuningExpert: A software assistant for improving Java Enterprise Edition application performance [J]. Expert Systems with Applications, 2009, 36(9): 11718-11729.

[5] Smith C U, Williams L G. New software performance antipatterns: More ways to shoot yourself in the Foot[C]//Int. CMG Conference. 2002: 667-674.

[6] Williams L G, Smith C U. PASA (SM): An Architectural Approach to Fixing Software Performance Problems[C]//Int. CMG Conference. 2002: 307-320.
[7] Dugan Jr R F, Glinert E P, Shokoufandeh A. The Sisyphus database retrieval software performance antipattern[C]//Proceedings of the 3rd international workshop on Software and performance. ACM, 2002: 10-16.
[8] Marin M, Van Deursen A, Moonen L. Identifying aspects using fan-in analysis[C]//Reverse Engineering, 2004. Proceedings. 11th Working Conference on. IEEE, 2004: 132-141.

IOP Conf. Series: Journal of Physics: Conf. Series **887** (2017) 012020 doi:10.1088/1742-6596/887/1/012020

Application of harmonic detection technology in methane telemetry

Yuehua Huo[1] and Weiqiang Fan[2]

1. Center of Modern Education Technology, China University of Mining & Technology, Beijing. Beijing 100083, China
2. School of Electromechanical and Information Engineering, China University of Mining & Technology, Beijing. Beijing 100083, China
Email: huoyh@cumtb.edu.cn

Abstract. Methane telemetry plays a vital role in ensuring the safe production of coal mines and monitoring the leakage of natural gas pipelines. Harmonic detection is the key technology of methane telemetry accuracy and sensitivity, but the current telemetry distance is short, the relationship between different modulation parameters is complex, and the harmonic signal is affected by noise interference. These factors seriously affect the development of harmonic detection technology. In this paper, the principle of methane telemetry based on harmonic detection technology is introduced. The present situation and characteristics of harmonic detection technology are expounded. The problems existing in harmonic detection are analyzed. Finally, the future development trend is discussed.

1. Introduction

Methane telemetry can not only promote coal mine safety production, but also can improve the accident prevention and emergency response capacity. At present, the existing mine infrared methane sensor in the application of coal mine gas monitoring in China has the advantages of full-scale methane concentration monitoring, long calibration cycle in use, fast response time measurement and long life of sensing element, but the temperature, Pressure, humidity and others in the measurement environment impact on the measurement accuracy and higher cost of equipment and other issues [1]. At the same time the type of sensor is difficult to reach the upper corner of the mining face, tunneling face and the high-risk locations [2]. Therefore, with the development of high-performance laser generator and harmonic detection technology, a new type telemetry device of laser methane for the rescue team ahead of harmful gas detection, mined-out gas detection has become one of the research hotspots in the application of coal mine gas monitoring of China.

Harmonic detection technology as the core of the methane telemetry device, through the photoelectric receiver to detect the information containing methane concentration of the reflected laser signal, the use of phase shifter containing methane gas concentration signal to extract the various harmonics. It will be a good solution for the existing problems that based on the technology of methane telemetry device can make up for the current lack of mine gas monitoring, coal mine to achieve a full range of methane concentration monitoring[3].

In this paper, the principle of methane telemetry based on harmonic detection technology is introduced. The key techniques of gas absorption law, harmonic function, second harmonic measurement and harmonic ratio measurement in harmonic detection are introduced. Harmonic detection technology in the application of methane telemetry status and characteristics, and finally from the application of the status quo and characteristics of the current point of view of some of the existing problems and trends.

2. The Principle of Methane Telemetry Based on Harmonic Detection

In the current gas telemetry, harmonic detection is used to improve the detection accuracy, and the central wavelength of the laser is swept through the gas absorption peak by wavelength modulation spectroscopy [4]. The high frequency sine wave is used to modulate the low frequency sawtooth wave related to the current amount. The frequency and power of the modulated wave are as follows

$$\upsilon=\upsilon_0+\upsilon_f \sin(\omega t) \tag{1}$$

$$I(\upsilon,t)=I_0(\upsilon,t)\left[1+n\sin(\omega t)\right] \tag{2}$$

Formula: υ_0 :laser light source center frequency; υ_f :frequency modulation amplitude; I_0 :laser is not modulated before the optical power; $I(\upsilon,t)$:laser output modulation optical power; n :light intensity modulation coefficient.

When the laser emitted from the laser is passed through the open space of the methane-containing gas, it is absorbed by the methane gas, and the absorption relation in the methane telemetry conforms to the Lambert-Beer gas absorption law [5].

$$I_r=\eta K(R,A)I(\upsilon,t)\exp\left[-\alpha(\upsilon)cL\right] \tag{3}$$

Formula: I_r : photodetector receives optical power; η :reflectivity of the telemetry reflector; $K(R,A)$: collection efficiency (the ratio of the received optical power to the transmitted power when no gas is absorbed and the diffuse reflectance is 1), which is related to the detection distance R and the receiving lens area A ; $\alpha(\upsilon)$:the gas absorption coefficient, which is related to air temperature and pressure; L :laser telemetry of the optical path.

There is $\alpha(\upsilon)cL << 1$, when the methane gas is a low concentration, the absorption coefficient ($\alpha(\upsilon)$) is relatively small, and the absorption optical path (L) is not very large. Formula (3) is developed by a series expansion method, and a linear part is taken

$$I_r=\eta K(R,A)I(\upsilon,t)\left[1-\alpha(\upsilon)cL\right] \tag{4}$$

The gas absorption line is represented by a Lorentz linear function, and the intensity and frequency are both time functions [6], the linear function is brought into (4) and then propagated by Fourier series

$$I_r=\sum_{n=0}^{\infty}A_n(\upsilon)\cos n\omega t \tag{5}$$

Formula: $A_n(\upsilon)$:fourier coefficients of c-th harmonic components of absorption coefficient.

After the normalization of the Lorentz linear function, the second harmonic coefficient is calculated [7].Using the lock-in amplifier to extract the different harmonic components， which based on the formula (5). The gas concentration is proportional to the magnitude of the even harmonic component.

$$I_f=\eta K(R,A)nI_0 \tag{6}$$

$$I_{2f}=\eta K(R,A)k\alpha_0 LcI_0 \tag{7}$$

Formula: I_f : first harmonic amplitude; I_{2f} : second harmonic amplitude.

It can be seen from (6) and (7) that the first harmonic signal is proportional to the average power of the initial light intensity; the second harmonic signal is not only related to the initial light intensity, but also to the modulation coefficient, telemetry distance and gas concentration The Therefore, the concentration of the second harmonic amplitude or the second harmonic amplitude and the first harmonic amplitude can be used to invert the concentration of methane gas.

3. Analysis of Harmonic Detection

The input laser through the methane gas is effectively absorbed, and the output signal will be distorted in the methane telemetry study [3]. As the methane gas concentration increases, the output harmonics of the output signals increase. As the harmonic signal contains methane concentration information, so in the methane concentration telemetry theory of harmonic signal has been one of the focus of the problem [8].

At present, the methane telemetry device uses the detection signal to reflect the laser back, and through the gas absorption law and Lorentz linear function simulation of gas absorption coefficient normalization and Fourier series expansion, demodulation harmonic signal to reverse the concentration of gas [9]. At present, the harmonic detection is mainly through the measurement of low-order harmonic signal components, through the signal processing to calculate the open chamber methane concentration. Therefore, the focus of methane telemetry is how to accurately extract the methane absorption spectrum of the harmonic signal components. For the harmonic signal detection using band-pass filter or lock-in amplifier two options.

In the telemetry methane gas concentration, the laser signal received after receiving the detection target is limited. Therefore, the harmonic signal component is very weak. In the experiment, the band-pass filter can filter the out-of-band noise signal. However, in practice, this method should filter out the white noise and other noise in the harmonic signal as much as possible. This requires that the pass-band width is very narrow, since the quality factor Q of the band-pass filter is usually not more than 100, the pass-band width can't be narrowed indefinitely. Therefore, the amplitude-frequency characteristic curve is not flat enough in the pass band or the phase-frequency characteristic curve changes nonlinearly in the pass-band. [10] In order to compensate for the shortcomings of the band-pass filter in the weak signal extraction, the ambient light effects of the lock-in amplifier to remove the external noise and extract high-precision harmonic signal. Figure.1 shows that the principle of phase lock.

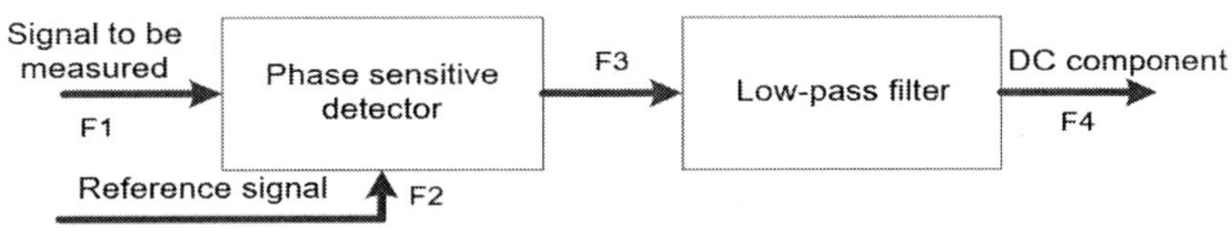

Figure 1. Phase lock principle

The phase-locked amplifier extracts the harmonics signals by using and referencing the reference signal of the signal at the same frequency. This technique shows excellent performance in terms of the density signal extraction that is small and submerged in the noise signal, greatly improving the detection signal-to-noise ratio for weak signals [11]. Therefore, the relevant areas of scientific research are widely used in phase-locked amplifier.

At present, analog and hybrid phase locked amplifiers are used in the detection of trace signals, because the temperature and phase shift of analog technology, the error of harmonic amplitude signal extracted by analog phase locked amplifier is larger [12]. With the rapid development of DSP technology, digital lock-in amplifier can convert analog signals to the digital domain for processing. The method has the advantages of small measurement error, large dynamic range, no temperature drift and bias effects [13]. Therefore, the methane telemetry is more important to use digital lock-in amplifier to demodulate each harmonic.

According to the ideal condition, the amplitude of the harmonic signal is proportional to the concentration of the methane gas. It is feasible to extract the second harmonic component separately in the harmonic detection. In the system, the temperature of the laser driver is and the current changes, resulting in laser light intensity and wavelength fluctuations [9]. This has a great impact on the measurement accuracy, the measured gas concentration is based on laser fluctuations and the existence of time-varying. The first harmonic is at the center frequency of 0, the second harmonic at the center frequency of the maximum absorption peak, by the formula (6), (7) analysis, the second harmonic than the value of the first harmonic signal and the concentration of methane gas is proportional to the value. Therefore, in order to overcome the second harmonic detection due to the use of various devices to introduce the noise signal and light source fluctuations caused by the impact [3]. At present, the more mature algorithm is to use the second harmonic signal and the first harmonic signal ratio method, through the ratio of the ratio of the relationship between the concentration value, accurate measurement of open environment in the methane concentration.

At present, in order to quickly extract the first harmonic and the second harmonic signal, accurately reflect the gas concentration and eliminate the measured signal and the reference signal between the phase difference, the direct acquisition of the target signal amplitude and phase [9]. Related research institutions designed a new orthogonal dual-channel digital lock-in amplifier, and has become a current research hotspot.

Through the methane telemetry signal generator for the lock-in amplifier reference channel with the measurement signal with the frequency of the sine, cosine two orthogonal reference input signal. The two reference signals are multiplied by the measured signal and then extracted by low-pass filter DC component, and then by square sum, square start operation, get the second harmonic signal [10]. Figure.2 shows that the quadrature digital phase lock amplifier.

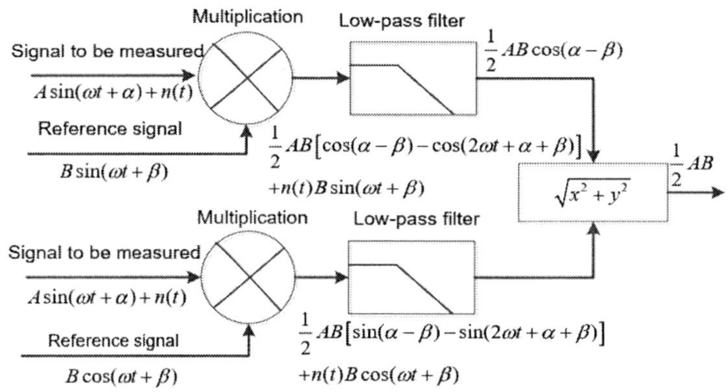

Figure 2. Orthogonal digital lock-in amplifier

4. Problems in Current Harmonic Detection

At present, the gas concentration measurement based on harmonic analysis mostly uses the gas absorption law series to start, taking its linear part approximation [14]. However, in the study of the current trace gas measurement, there is no deep study of the relationship between the three variables in the approximation condition of the above equation (4), which leads to a large error in the accuracy of the measurement of methane after telemetry. When the methane gas is telemetered in engineering use, the change of the surrounding medium will cause the diffuse reflectance to change and affect the accuracy of the gas received signal intensity and concentration inversion results.

At the same time in the methane concentration inversion test, due to the temperature and pressure range of large changes in the calculation of a certain type of line will bring the error. However, most of the current research is based on the Lorentz linear function, and does not explore the mechanism of harmonic linear function. So at different temperatures and pressures, should be based on the corresponding linear broadening of the specific circumstances to choose the appropriate line [15]. Therefore, it is very important to study the relationship between the line profile of methane absorption line and the change of ambient temperature and pressure. By learning the temperature and pressure knowledge of the linear function [15], the Lorentz line is mainly due to the collision between the particles when the gas pressure is large, which is related to the ambient pressure and the collision cross section. The Doppler effect caused by environmental temperature changes [16]. Therefore, the absorption mechanism of the gas is different under different conditions, and should be described by different linear functions [15].

In the published literature, there is little influence on the influence of ambient temperature and pressure variation on the optimal modulation coefficient and modulation depth in the concentration inversion error [14]. At present, the test is to set a fixed frequency modulation coefficient and absorption spectrum half width at wavelength modulation, and calculate the modulation depth by Arndt normalization [8]. The harmonic optimal modulation coefficient k is obtained from the modulation depth m by the formula 7 in this paper. At present, after setting the optimal modulation coefficient under certain temperature and pressure, the inversion of methane concentration according to the empirical formula can't meet the requirement of high precision measurement. When the environmental change of methane telemetry is large, it is necessary to calculate the methane concentration according to the preset optimal modulation coefficient inversion. Therefore, it is necessary to consider the optimal modulation coefficient in the test process.

At present, when the harmonic signal absorbed by methane is extracted based on wavelength modulation, the spectral intensity is also modulated during the modulation process, resulting in a fixed phase angle between the modulation frequency and the modulation intensity. However, most of the current studies only consider the linear part of the light intensity modulation and ignore the non-linear part of the residual amplitude modulation (RAM), resulting in the detection signal line and system noise to a certain extent [8]. Considering the linear and non-linear effects of light intensity modulation, the signal modulation coefficient has a large degree of influence on the linearity and noise level of the harmonics to be detected. In a certain range, as the modulation coefficient increases, the asymmetry of the linearity increases and the baseline offset increases to bring about greater noise [14]. Therefore, the selection of the appropriate modulation factor is particularly important for improving the detection accuracy, usually with the maximum signal to noise ratio as the modulation amplitude of the choice to suppress the RAM noise [8]. Although the use of high harmonics for gas detection, the laser residual amplitude modulation RAM is small [17], but the current high harmonic signal is weak, extraction technology is difficult. Laboratory uses a lock-in amplifier with low-frequency detection function, in the harmonic analysis can only be directly used to extract a lock-in amplifier, the second harmonic.

At present, the second harmonic and first harmonic ratio method is adopted in the research of harmonic detection. This method effectively solves the influence of light intensity fluctuation and the removal of noise in the second harmonic method, but the method needs to use more Channel-locked amplifiers extract primary and secondary harmonics signals, which increases the complexity of the system and reduces the stability of the system [18]. In practice, due to the changes in the temperature and current of the laser, the output wavelength of the laser produces a drift, and the actual frequency produces a phase difference with respect to the central absorption peak of the methane. Therefore, in practice, the amplitude of the first harmonic at the center frequency Not 0, that is, containing gas absorption concentration information. When using the second harmonic and the first harmonic signal ratio to reflect the gas concentration caused by the harmonic ratio method will bring measurement error.

In the process of inversion of methane concentration based on harmonic analysis, the change of ambient temperature and pressure change on the output harmonic amplitude of telemetry [15] is not perfect, and most of the research is based on the change of ambient temperature and pressure Discussion on the Influence of the Second Harmonic Peak. When the measurement of environmental conditions change more intense, often take to establish an equation to eliminate the temperature and pressure on the measurement results [15], through the simulation test to obtain part of the methane gas concentration, the least squares method to obtain empirical formula [19]. In the study of harmonic ratio method, the influence of environmental temperature and pressure variation on the first harmonic peak and the dynamic correction function are used in the inversion of methane gas concentration. So how to ensure the measurement accuracy in telemetry system is one of the main research directions.

5. Conclusions

Harmonic detection is a key technology of methane telemetry device, but the ambient temperature, pressure and other effects resulting insufficient of harmonic detection accuracy, higher price of digital orthogonal lock-in amplifier in harmonic detection, and it is difficult to extract the high-order harmonic. These factors limit the widespread use of methane telemetry devices in mines. With the progress & development of science & technology, we can build a different temperature and pressure gas concentration inversion algorithm database and coal mine methane gas monitoring platform and the methane telemetry device based on harmonic detection technology gradually extended to the oil , Chemical, environmental testing and other fields, making it the future development of gas telemetry an important direction.

6. Acknowledgements

This work was supported by the Foundation for National Key R&D Program of China plan under Grant 2016YFC0801800.

7. References

[1] Sun J P 2015 Coal mine monitoring new technology and new equipment *J. Industry and Mine Automation.*, 41 (**1**): 1-5.
[2] Sun J P 2016 New coal mine information automation technology and development J. *Coal Science and Technology.* 44 (**1**): 19-23.
[3] Zhou H K 2015 Technology of laser methane based on TDLAS J. *Coal Mine Safety.*, 46 (**7**): 113-115.
[4] Li B, Liu H F, and He Q X, etc 2016 Methane detection system based on 1654nm distributed feedback laser J. *Spectroscopy and Spectral Analysis.* , 36 (**1**): 20-26.
[5] Zhang Z, and Zhang L Y 2011 Model telemetry technology based on TDLAS *J. Journal of Changchun University (Social Science Edition)* .,21 (**6**): 9-11.
[6] Wang X Y 2011 Research on laser telemetry methane concentration based on TDLAS D. *Tianjin: Tianjin University.*
[7] Zhang Z R, Sun P S, and Xia S, etc 2015 Modulation parameters of tunable semiconductor laser absorption spectroscopy and its optimal selection J. *Acta Photonica Sinica.* 44 (**1**): 2-9.
[8] Hu Y J, Zhao H, and Zhang R, etc 2013 Study on the Influence of Light Intensity Modulation on the Second Harmonic Line Type in Wavelength Modulation *J. Acta Optica Sinica,.* 33 (**11**): 1130002.
[9] Chen S W, Sun T, and Tang D L 2017 Study on Hydrogen Sulfide Gas Sensor Based on Harmonic Detection J. *Technical Journal of Sensors and Actuators.* 30 (**1**): 31-34.
[10] Nan X F, Zhang Y, and Yu M , etc 2016 Laser gas telemetry test system automation test software design *J. Optical instruments.*, 37 (**3**): 210-213.
[11] Cong M L, Sun D D, and Wang Y D 2017 Application of Logarithmic Transformation-Wavelength Modulation Spectroscopy in Gas Detection *J. Infrared and Laser Engineering,.* 46 (**2**): 1-6.
[12] GIUSEPPE D M. and F, ARNALDO D A, etc 2012 A fully-analog lock-in amplifier with automatic phase alignment for accurate measurements of ppb gas concentrations *J. IEEE Sensors Journal.* 12(**5**):1377-1383.
[13] Liu H F, Bin Li, and He Q X, etc 2016 Development of digital orthogonal lock-in amplifier and its application in methane detection *J. Acta Photonica Sinica* ,. 45 (**4**): 0423004.
[14] Liu Y, Wu J N, and Chen M M, etc 2016 Trace methane gas detector based on TDLAS-WMS *J. Spectroscopy and Spectral Analysis.*, 36 (**1**): 279-282.
[15] Jia L Q, and Wenqing Liu 2014 The influence of temperature and pressure variation on the second harmonic inversion results *J. Chinese Journal of Lasers,.* 41 (**12**): 1215004.
[16] Rui Z 2014 Research on High Precision Gas Detection Techniques of Tunable Diode Laser Absorption Spectroscopy *D. Tianjin: Tianjin University.*
[17] Li H, and Liu J G, etc 2013 Simulation and analysis of the second harmonic signal of tunable diode laser absorption spectroscopy *J. Spectroscopy and Spectral Analysis.* 23 (**4**): 881-884.
[18] Zhang K K, Qi Y, and Fu X, etc 2014 Measurement of carbon dioxide temperature based on TDLAS first harmonic *J. Optoelectronic Technology.* 34 (**4**): 300-307.
[19] Zhang W J, Hou Y B, and Zhang W T, etc 2017 Study on Denoising of Coal Mine Gas Data Based on GA-LSSVR *J. Mining Safety and Environmental Protection,.* 44 (**1**): 45-48.

Design of vehicle overload detection system based on geophone

Siquan Hu[1, 2, a], Min Kong[1, b] and Chundong She[3, c]

[1] School of Computer and Communication Engineering, University of Science and Technology Beijing, Beijing, 100083, China.
[2] Chengdu Visual Microcosm Technology Ltd. Chengdu, 610072, Sichuan, China.
[3]School of Electronic Engineering, Beijing University of Posts and Telecommunications, Beijing, 100876, China
Email: [a]husiquan@ustb.edu.cn, [b]AMay5478@163.com, [c]shurcd@vip.sina.com

Abstract. A vehicle overload detection system is proposed based on geophone. Under normal circumstances, when overloaded vehicles and ordinary vehicles pass through the road, the amplitude of the ground vibration will be different, and the geophone sensor can detect tiny vibrations of the ground. The system includes information acquisition module, signal conditioning module and wireless transmission module. The collected vibration data is transmitted through the wireless transmission module to the background, and the SVM algorithm is used to classify the information and determine whether the vehicle is overloaded. Experiments show that the system can detect overload accurately.

1. Introduction

With the rapid development of current logistics industry, the question of transport vehicles overloading is becoming more and more serious. Overloaded vehicles will cause extensive damage to the road and huge economic losses to society, which seriously threaten traffic safety[1]. According to statistics, 70% of traffic accidents are caused by vehicles overloading. Therefore, the overloading of transport vehicles has become one of the crucial issues of traffic management.

In the field of overload detection, traditional static weighing system needs to intercept a part of vehicles from the normal traffic flow and weigh them in a static field [2], it takes a long time and affects the normal traffic. There are also many unsolved problems in the dynamic weighing technology, such as the small measuring speed range and the large system, which is inconvenient for installation and maintenance. This paper presents a non-intrusive dynamic overload detection method based on the basic observation that overloaded vehicles on the road will cause different vibration from normal vehicles. The proposed system can be used to detect the overload within a short time, and also can be used to make an early warning in the dynamic situation before forcing the suspected vehicles weighed statically by the road administration. At the same time, it has a small size, and the installation and maintenance are very convenient.

In this paper, vibration sensor is used to collect vibration data, according to the varied sizes of vibration data to detect overload. Raspberry pi is used as the core of data collection equipment. A signal conditioning circuit board is designed independently to filter and amplify original vibration signals, and convert analog signals to digital signals. The embedded nodes are connected by wireless network. Classic SVM model is used to classify the data and determine the vehicle status.

2. Overall Design and Proposed Embedded Node

2.1. System Architecture

As shown in figure 1, the whole detection system consists of detection nodes, transmission networks and the back-end servers. Detection nodes are used for sampling, the data is transmitted to the PC through a wireless router so that the waveform generated by ground vibration can be observed in real time. Data is stored in the database, and processed by the algorithm in the server. When the overload is detected, the server will send out a warning.

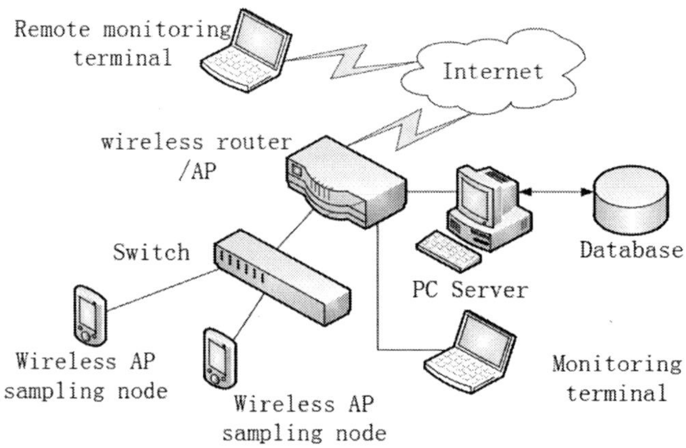

Figure 1. System Architecture

In order to meet the requirements of high precision and multiple nodes vibration data acquisition, the Wi-Fi wireless networking mode is used in system. EDUP EP-N8508GS is used as Wi-Fi module, which is a wireless AP with USB interfaces. In the case of a small range, star network with a single center can be used in wireless network, and all wireless sampling nodes are connected to the central switch. Between the slave computer and the host computer, TCP/IP protocol is used to ensure the reliability of data. In case of a wide range, due to the characteristics of Wi-Fi protocol, several wireless routers with WDS function [3] should be used.

2.2. Detection Node Design

Detection nodes need to be designed after the system design is completed, including selection of suitable control boards and sensors, as well as circuit design.

Since the USB Wi-Fi module provides the Linux driver, a control board with Linux environment is adopted. The raspberry pi is used in this paper, it has rich interfaces [4], which can well meet the need of development. The raspberry pi is connected with the USB Wi-Fi module, also connected with the vibration sensor through the signal conditioning board. The signal conditioning board sets aside SPI interface for raspberry pi. The read speed and time interval of SPI are adjusted through the SPI interface by raspberry pi, to control the sampling frequency.

In this paper, CDJ28HZ is selected as the geophone sensor to collect vibration data, it is mainly used for low frequency detection [5]. It has advantages that the output voltage signal is in proportional to the speed of the vibration signal, and the application field can be compatible with high frequency, intermediate frequency and low frequency; it has low output impedance, high signal-to-noise ratio. Therefore, it is very suitable to use geophone sensor to detect weak ground vibration. Feedback circuit is used to achieve the sensor's low-impedance output, coupled with A/D conversion circuit, the digital signal can be output. The A/D conversion circuit design is shown in figure 2.

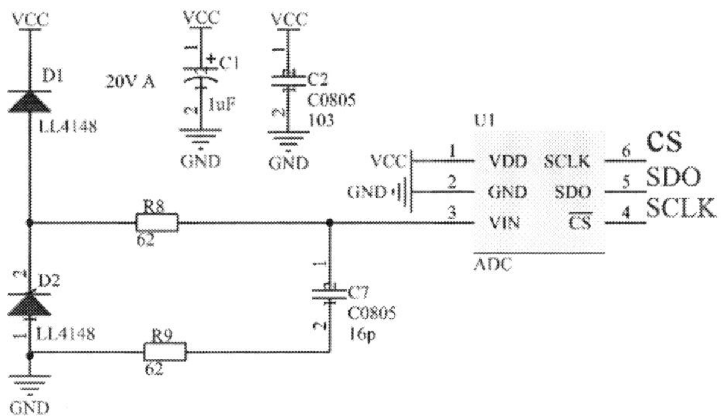

Figure 2. AD conversion circuit of CDJ28HZ

3. SVM Classification Algorithm

In order to distinguish overloaded vehicles from ordinary vehicles, classification algorithm is required. In the field of machine learning, classification algorithm requires that the system to analyze unknown pattern of input, in order to predict unknown categories.

3.1. Support Vector Machine(SVM)

The support vector machine is a two-class model, of which purpose is to find a hyper-plane to split the sample, and to separate positive and negative examples. Therefore the interval between positive and negative examples is the largest and classification results are more credible, and there would be better forecasting abilities for unknown new samples [6].

In the overload detection system, the sample is derived from the processed vibration data, the sample set is in format:$(x_{1,}y_1),\dots,(x_{i,}y_i), x \in R^n, y \in \{+1,-1\}$, x_i represent multidimensional feature vectors, and y_i represent classification markers with values of -1 and 1. -1 is used to show the normal situation while 1 is used for the vehicle overloading situation. The equations of the hyper-plane can be written as:

$$(\omega x) + b = 0, \omega \in R^n, b \in R$$

After being normalized:

$$y_i\big((\omega x_i) + b\big) \geq 1, i = 1,\dots,l$$

The classification interval is $\frac{2}{\|\omega\|}$. To maximize sorting interval, $\|\omega\|^2$should be minimized.

To optimize the above formula, the Lagrange theorem and KKT conditions are used for derivation, then the following form can be got:

$$Max \quad Q(\alpha) = \sum_{i=1}^{l} \alpha_i - \frac{1}{2}\sum_{i=1}^{l}\sum_{j=1}^{l} \alpha_i\alpha_j d_i d_j x_i^T x_j$$

$$s.t. \quad \sum_{i=1}^{l} \alpha_i d_i = 0, \alpha_i \geq 0, i = 1,2,\dots,l$$

This is the famous QP problem, the decision plane equation is $\sum_{i=1}^{l} \alpha_i^* d_i x_i^T x + b^* = 0$ and α_i^* is the optimal solution to this problem.

Since there may be accidental errors while collecting data, the slack variable is used to optimize the problem [7]. Using ξ_i to represent the slack variable, the above problem is translated into the following optimization problem:

$$Min \quad \frac{1}{2}\omega^T\omega + C\sum_{i=1}^{l}\xi_i$$

$$s.t. \quad d_i(\omega^T x_i + b) \geq 1 - \xi_i \;\; i = 1,2,\dots,l \quad \xi_i \geq 0 \;\; i = 1,2,\dots,l$$

In the above formula, C is a penalty factor, which is a coefficient specified by user that represents the penalty for adding points to the wrong point. The objective function can be finally got:

$$Max \quad Q(\alpha) = \sum_{i=1}^{l} \alpha_i - \frac{1}{2}\sum_{i=1}^{l}\sum_{j=1}^{l} \alpha_i \alpha_j d_i d_j x_i^T x_j$$

$$s.t. \quad \sum_{i=1}^{l} \alpha_i d_i = 0, 0 \leq \alpha_i \leq C, i = 1,2,\ldots,l$$

To make the result of overload detection more accurately, the parameters in the classification model need to be adjusted. From the view of loss function, above problem can be understood as regularization term + loss term, and there is a C parameter to balance two parts. For the regularization term, there are L2-regularized and L1-regularized. With L2, the limitation of model space can be achieved, and over-fitting can be avoided. For loss term, hinge loss (for soft margin SVM) is commonly used, it can be subdivided into hinge loss (L1 loss) and squared hinge loss (L2 loss).

3.2. Algorithm Application

In the overload detection system, data obtained after being filtered is stored in the file in libSVM data format. The eigenvalues are separated by spaces in LibSVM data, +1 and -1 represent categories. In this system, the format is as bellow.

+1 1:71.8461538462 2:133.538461538 3:0 4:13.1538461538 5:44.1538461538

-1 1:0 2:0 3:0 4:0 5:0

The function of reading data: *y, x = svm_read_problem('data file location')*

The function of training model: *model = train (svm_problem, svm_parameter)*

The trained model is saved to the specified location. In following experiments, the model and the predict function can be used to predict experimental data.

4. System Testing and Analyzing

4.1. Experimental Testing

To identify differences between of overloaded vehicles and normal vehicles more intuitively, the wxPython module is used to construct the interface and display the vibration waveform in real time. After filtering, the contrast of vibration and vibration-free waveform is showed in figure 3.

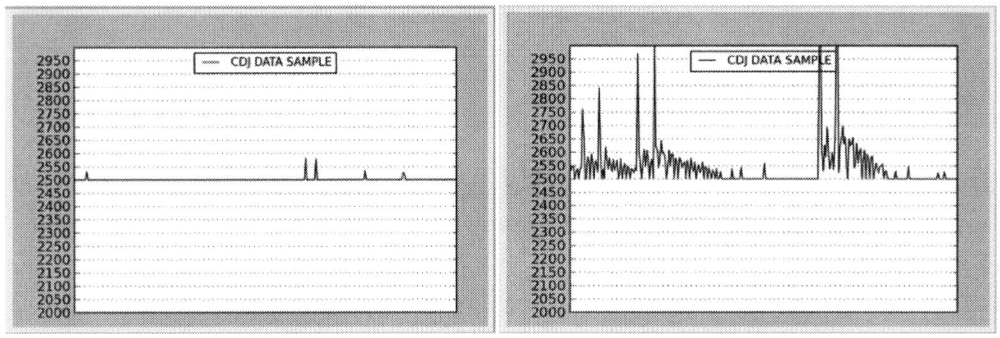

Figure 3. The contrast of vibration and vibration-free waveform

After filtering, it can be distinguished by naked eye that whether there is vibration or not. Then it can be determined that, after using SVM algorithm, the large vibration caused by overloading and normal vibration can also be very well classified. Field testing is carried out at a bus station. When there are no passengers and some passengers on the bus, the waveform contrast is showed in figure 4.

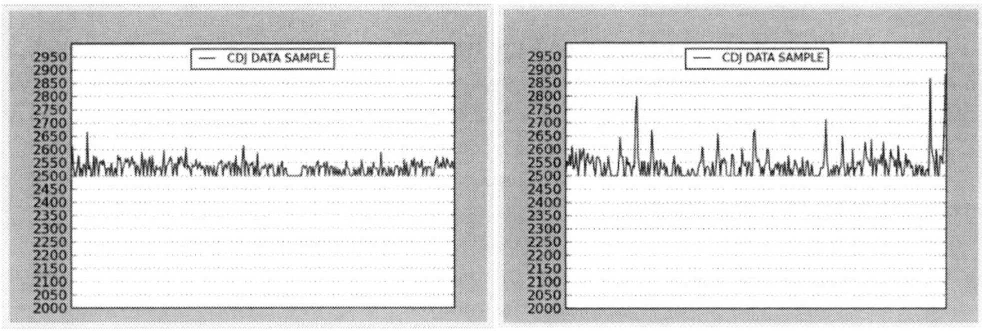

Figure 4. The waveform contrast between empty and crowded buses

The waveform comparison of the the full load and overload buses is showed in figure 5.

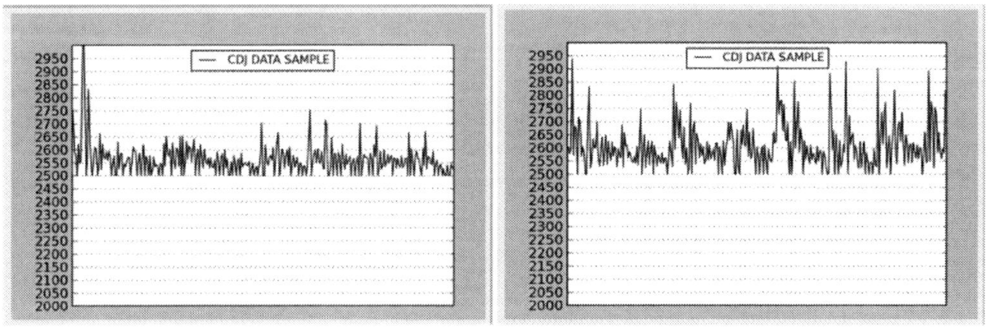

Figure 5. The waveform contrast between full-load and overload buses

4.2. Training Model Process and Application Result

This section focuses on adjusting of two important parameters--- loss type (-s) and cost (-c), to find the most appropriate parameters for prediction. Some data measured at the bus station are used as a training set, sample dimension is 5, and 156 samples are trained. Among two parameters, -s took 1(L2-regularized L2-loss support vector classification) and 3(L2-regularized L1-loss support vector classification), while -c rises from the minimum. The comparison is shown in the table below:

Table 1. Accuracy of training set when adjusting parameters

Value of s	Value of c	Accuracy	Value of s	Value of c	Accuracy
1	0.001	50	3	0.001	50
1	0.01	65.3846	3	0.01	52.5641
1	0.1	94.8718	3	0.1	82.0513
1	1	96.1538	3	1	95.5128
1	2	100	3	2	100
1	3	100	3	3	100

From the above table, with the increasing number of training iterations, the training set error decreases and finally converge. The training set and sample dimensions are relatively small, so it is easy to reach the convergence and when parameter C is set to 2 or more, the accuracy of the training set can reach 100%. In the same case, when the parameter s is set to 1, the accuracy is clearly higher. Therefore, in the final implementation, the parameter s is set to 1 and the parameter c is set to 2.

After being fully trained, the model is used in the actual detection. When an overloaded vehicle passes by, the output value of the model is 1 and the overload is detected. Due to the complexity in real traffic, false judgment may occur sometimes, and further improvement will be needed.

IOP Conf. Series: Journal of Physics: Conf. Series **887** (2017) 012021 doi:10.1088/1742-6596/887/1/012021

5. Conclusion

In this paper, based on the geophone sensor and SVM algorithm, a new type of vehicle overload detection system is established, which make it possible to detect the overloaded vehicle and transfer the relevant data to the background after the vibration signal is obtained. Then further processing can be done. The system has advantages of small size, high sensitivity, and it can adapt to different test environments. Relevant system tests showed that overloaded vehicles can be accurately detected by this system, which has a very good practical prospect.

6. Acknowledgments

The study was supported by the National Natural Science Foundation of China (No. 91438120) and the Sichuan Science and Technology Support Program (No.2016GZ0339).

7. Reference

[1] Zhong H J, Huo Y and Deng C C 2013 *Computer Measurement & Control* **4** 75-77.

[2] Huang P Y 2016 *Journal of Qiqihar University (Natural Science Edition)* **1** 9-12.

[3] Brown A, Mortier R and Rodden T 2012 *Acm Sigcomm Conference on Applications* **42 (4)** 275-76.

[4] Richardson M and Wallace S 2012 *Sebastopol: O'Reilly Media.*

[5] Luo F L, Yi B J and Luo L B 2005 *Equipment for Geophysical Prospecting* **15(1)** 6-14.

[6] Hsu C W, Chang C C and Lin C J 2003 *Taipei: Department of Computer Science and* Information *Engineering, University of National Taiwan* 67(5).

[7] Xia S Y, Wang Y and Zhang Q 2012 *Computer Applications and Software* **29(4)** 121-124.

Research on SEU hardening of heterogeneous Dual-Core SoC

Kun Huang, Keliu Hu, Jun Deng and Tao Zhang

Sichuan Institute of Solid-State Circuits, China Electronics Technology Group Corp, Chongqing 400060, P. R. China
Email: hkof11@126.com

Abstract. The implementation of Single-Event Upsets (SEU) hardening has various schemes. However, some of them require a lot of human, material and financial resources. This paper proposes an easy scheme on SEU hardening for Heterogeneous Dual-core SoC (HD SoC) which contains three techniques. First, the automatic Triple Modular Redundancy (TMR) technique is adopted to harden the register heaps of the processor and the instruction-fetching module. Second, Hamming codes are used to harden the random access memory (RAM). Last, a software signature technique is applied to check the programs which are running on CPU. The scheme need not to consume additional resources, and has little influence on the performance of CPU. These technologies are very mature, easy to implement and needs low cost. According to the simulation result, the scheme can satisfy the basic demand of SEU-hardening.

1. Introduction

With the rapid development of space technology, the fault tolerance of integration circuits is increasingly important. In the harsh space environment, errors introduced by SEU occupy more than half of total radiation related errors [1]. More and more researches have focused on SEU effect. At process level, the SOI processing can make a chip single-event latch immune [2]. At circuit level, dual interlocked storage cell (DICE) achieves decent immunity against SEU effect [3]. At system level, Error Correcting Code (ECC) [4, 5] and Triple-Modular-Redundancy (TMR) [6, 7] are very helpful. At software level, control-flow checking by software signature and error detection by duplicated instructions can be a very suitable candidate [8-10].

These techniques are mature. However, there are still many shortcomings when the difficulty and the cost of implementation are considered. TMR is the most popular adopted solution, but the area cost is too high. The work [6, 7] had added TMR into the synthesized netlist, which was very inflexible. The area of DICE is smaller than TMR, but DICE requires a special digital cell library, which lead to a library to be developed afresh and result in consumption of a large number of manpower and resources. SOI is naturally immune to SEL, but the tape-out cost is expensive. The work [8] had presented a control-flow checking by software signature which required a special compiler, and would influence the performance of CPU. The work [10] adopted a redundant core to check the software signature, which still affect the performance of CPU when the current software signature is generated.

This paper proposes a novel SEU-hardening scheme for heterogeneous dual-core SoC (HD SoC). It consists of the following three techniques. First, the automatic TMR technique is adopted to harden the register heaps of the processor and the instruction-fetching module. Second, to harden the random access memory (RAM), Hamming codes are used. Last, a software signature technique is applied to check the programs running on CPU. This scheme can correct soft errors at circuit level and monitor soft errors at software level. The biggest benefit of this technology is that the implementation is very easy and the effect can satisfy the basic demand of SEU-hardening.

IOP Conf. Series: Journal of Physics: Conf. Series **887** (2017) 012022 doi:10.1088/1742-6596/887/1/012022

2. Heterogeneous Dual-Core SoC Hardening

It is common that a SoC in embedded application utilizes a DSP processor to enhance the capacity of arithmetic computing, thus a heterogeneous dual-core architecture has been established, and both performance and efficiency demand could be matched. An example of dual-core architecture is shown in Figure 1.

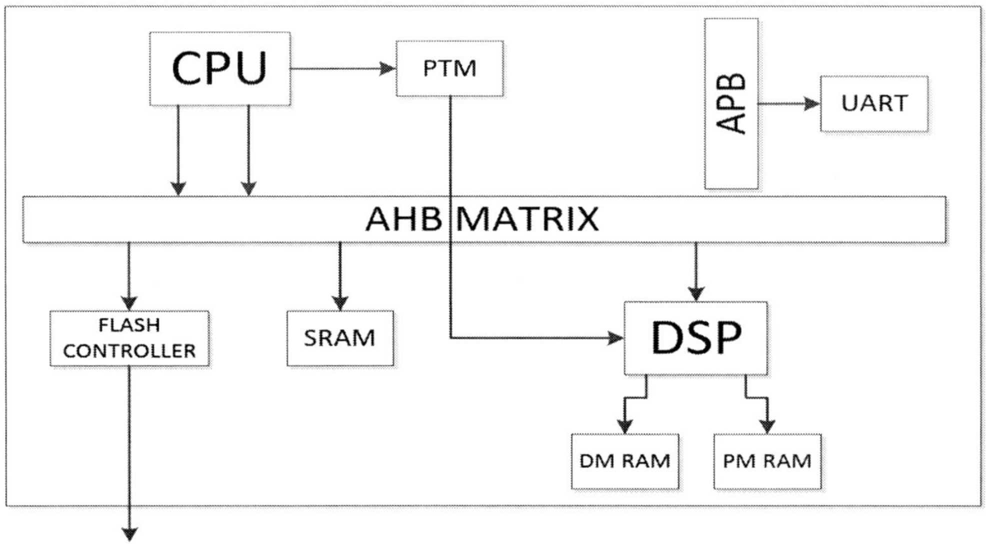

Figure 1. Arch of HD SoC

The two processers in HD SoC collaborate in a master-slave manner: CPU is the master processer which is in charge of control of the whole system, which is also a master on AHB bus matrix. DSP is a slave on AHB bus matrix and plays as a slave processer role to accomplish the data processing. The DSP has individual program ram and data ram which is also CPU-accessible. In addition, there is a program tracing module (PTM) which is used for software signature checking.

The hardening technique of HD SoC is comprised of three aspects: TMR to harden the CPU and DSP cores, ECC to protect the embedded memory, and PTM to monitor errors in a software style.

TMR is commonly used to protect sequential circuits, or storage elements from SEU, nevertheless, in the current situation, only the critical part of CPU and DSP should be TMR-hardened, in consideration of the area-consuming feature of TMR.

ECC is a widely-used method in radiation hardened memory design, and among the ECCs, Hamming code may be a simple and effective choice. In this paper, Hamming encoder and decoder will be added to fix the errors in the embedded SRAMs induced by SEU.

The function of PTM is tracing program, which records every PC value of branch instructions when CPU is running. DSP subsequently reads these messages to check the software signature, estimate whether if error occurs and run the error-fixing routine if necessary.

3. Automatic Design for TMR

TMR technique will make three copies of D flip-flops (DFF) whose outputs are connected to a voter. The structure is shown in Figure 2.

The function of the voter is:

$$Q = AB + BC + AC \quad (1)$$

If one value of (A, B, C) changes, the value of Q will not change. This makes DFF to be SEU immune.

This paper proposes a program, which can add TMR into the design of RTL automatically. It is shown in Figure 3.

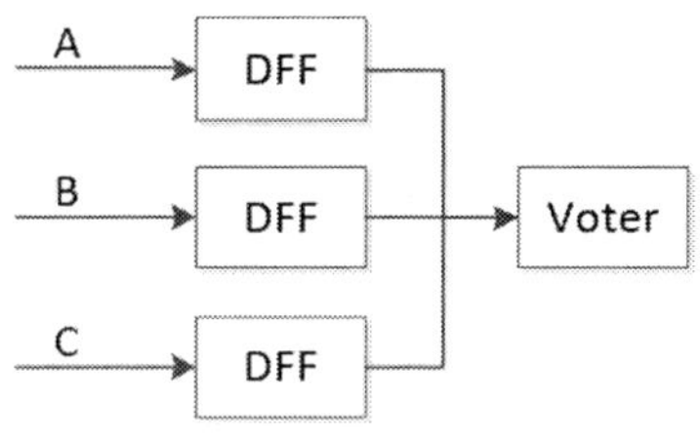

Figure 2. TMR structure

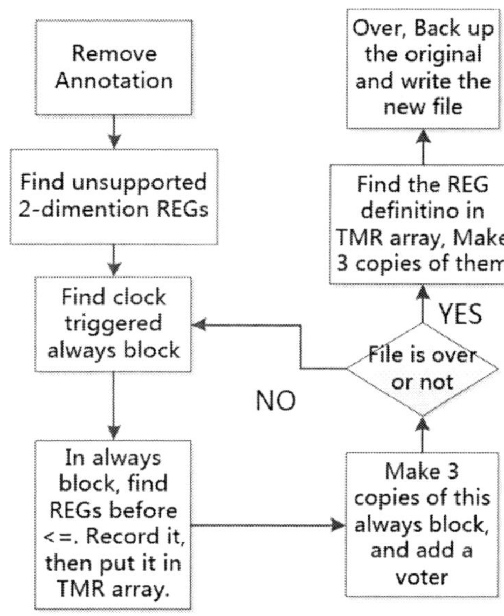

Figure 3. Automatic TMR Design

Before the automatic program to be applied, all the annotations should be removed. If 2-dimention REG arrays make 3 copies, the increase in area would not be acceptable. All 2-dimention REG arrays should be ignored. The key of this automatic algorithm is find and handle the 'always' block which is triggered by clock edge as this kind of block will be synthesized into DFF. Three copies of REG should be added when a find is hit, and the REG will be putted into a TMR array. The last step is to change the definition of REG in TMR array. All TMR REGs should be recorded, which will be used in synthesis.

4. Error-Correcting Code

Error-Correcting Code originates from Channel Coding Theory in communication, and is now widely used in radiation hardened memory design. There are many kinds of ECCs, such as Hamming code, BCH, RS (Reed-Solomon), LDPC, etc. In this paper, Hamming code is selected to apply the static RAM protection.

Hamming codes were proposed by Richard Hamming in 1950. With a k-bit message, Hamming codes generate an r-bit checksum through a matrix, and during the decoding, the checksum can be used to detect and correct 1-bit error in the message, in case it happened. The number k and r should match (2).

$$2^r \geq k + r + 1 \tag{2}$$

In this HD SoC, only the program memory of CPU and DSP is hardened, and the bit width is 32 and 24 respectively. For the 32-bit ram, choose (39, 32) Hamming codes. For the 24-bit ram, use (13, 8) and (22, 16) Hamming codes as a combination. The Hamming encoder and decoder will be added to the ram writing and reading circuit, a diagram for the circuit is shown in Figure 4.

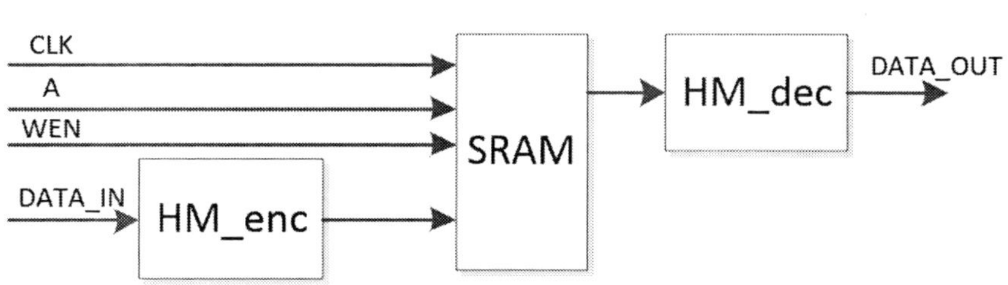

Figure 4. SRAM with Hamming Code Protection

5. Control-Flow Checking by Software Signatures

Control-flow checking by software signature is used to monitor the control-flow errors. The basic idea is as follows. The assemble source program is divided into basic block. Each block has no branch instruction except the last instruction. A unique signature is given to each block in a pre-defined way. All signatures are recorded. When the program is running into a new block, it generates a new signature and is compared with the pre-defined signature. If equal, it means the control-flow is correct. If not, it means error occurred. A special situation should be considered. When there are many blocks running into one block, Run-time Adjusting Signature should be included.

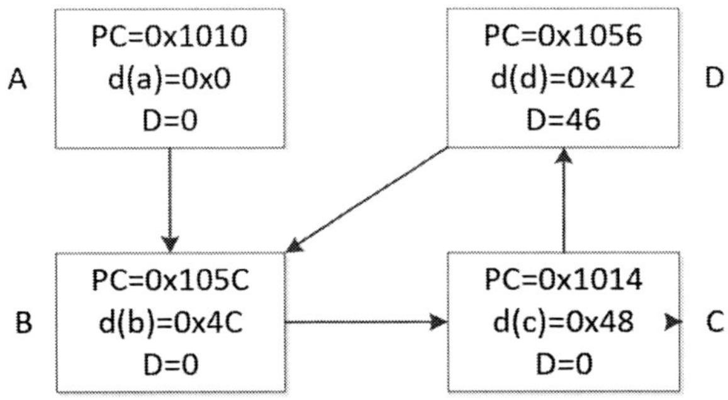

Figure 5. Example of software signature checking

In Figure 5, there are 4 program blocks, represented by node A to D. The PC of last branch instruction is the node signature. And D represents the Run-time Adjusting Signature. d(i) represents the distance (xor distance) from the source node to the destination node. For example,$d(b) = PC(B)xorPC(A)$ = 0x1010 $\oplus$ 0x105C = 0x4. d (i) is pre-defined and is restored in each node. When the program jumped from A to B, $PC(A) \oplus d(b)$ will produce the signature of node B. Then compare the signature with the PC value of current node (node B). If equals, it's right. If not, error occurs.

If the program jumped from D to B, then$PC(D) \oplus d(b) = 0x101A$. This value is not equal to the PC of node B (PC=0x105C). Node B has two source nodes, A and D. So Run-time Adjusting Signature $D = PC(A) \oplus PC(D)$ should be included. When program jumped from D to B, the signature will be$PC(D) \oplus d(b) \oplus D = 0x105C$. It is equal to the PC value of node B.

In [10], it uses a redundant CPU to check the signature of other CPU. This paper adopts a similar method. DSP is used to check the signature. The difference is how to produce the current signature, this paper use the program trace module instead of CPU which generate the signature by itself. The PTM is used to trace the signature of the current program. The PC value of the last branch instruction is defined as the signature of the block. The PTM needs to record all the PC of the branch instruction value when the program is running.

In order to record all the PC of branch instruction, it needs to find out all the branch instructions in the executing stage of pipeline and the current PC value. All these information should be organized as branch states. If a branch occurs, the PC value will be putted into a FIFO. Then the DSP reads out these PC value in the FIFO to check the software signature. The PTM is shown in Figure 6.

Figure 6. Program Trace Module

The algorithm of software signature checking is shown as following:

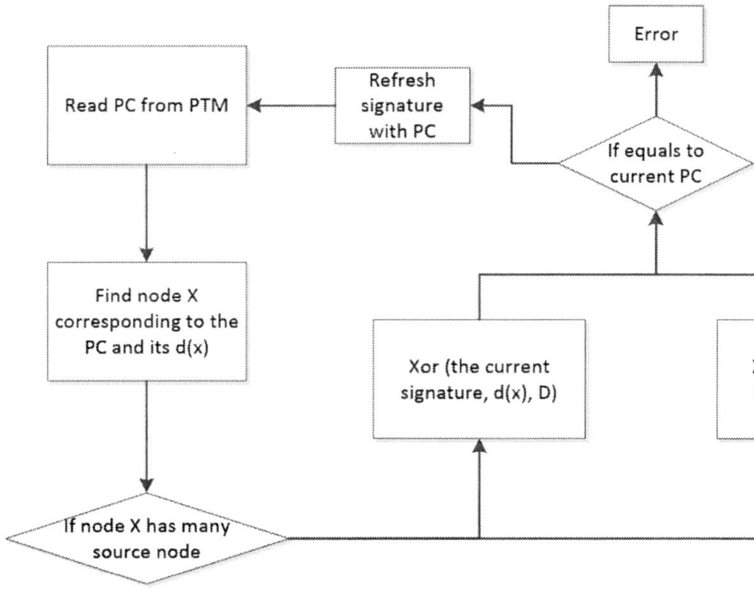

Figure 7. Algorithm of software signature checking

Since the current signature PC is generated by PTM instead of CPU itself, the performance of CPU will not be affected. And the program running on CPU doesn't need to be changed.

6. Soft Error Inject Simulation

The performance of this SEU-Hardening technique is evaluated by injecting soft errors to the whole SoC during simulation. Since SEU affects DFF and SRAM, this paper only inject errors to the registers. If soft errors are injected manually, it is not efficient since the SoC is very complicated. The soft error injecting procedure is shown as follows:

- With the openkdb tools in Debussy, the full name of all REGs in SoC can be extracted by the REG triggered InstanceBased mode.
- A perl script is written to select some REGs randomly. Then a Verilog stimulus should be generated which makes the selected REGs to turn over (thus to emulate SEU errors) at predetermined time.
- CPU runs a 3x3 matrix multiplication. During this time, soft errors should be injected by the stimulus and then the result could be observed.

The running time of 3x3 matrix multiplication is 12.96us. Errors should be injected during the time. N REGs is selected randomly and errors are injected. This action is repeated 10 times.

Table 1 shows the result. When the error inject number is below 15, the program function is right which means the soft error is corrected by TMR or Hamming Code. When the error inject number is between 15 and 30, most errors could be detected which means the soft error can't be corrected but can be monitored. When the error inject number is above 30, the program run out frequently. The SEU-hardening technique is effective which meets the basic SEU-hardening requirements.

Table 1. Simulation Result

inject error num	function is right	error detected	program run out
10	10	0	0
15	8	2	0
20	7	2	1
25	7	3	0
30	6	3	1
35	4	4	2
40	4	1	5

7. Conclusion

This paper presents a novel SEU-hardening scheme for heterogeneous dual-core SoC. This scheme contains TMR tech, Hamming Code tech, and software signature checking tech. Advantages of the scheme are not only practical and uncomplicated implementation, but also little effect on the performance of CPU, as no changes are applied to the software program. The simulation result shows that this scheme can meet the basic SEU-hardening requirements.

8. Reference

[1] Changhe W. 1998 *The Influence with Reliability of Motional Satellite by the Single-Event Phenomena(Semiconductor Information vol 35)*pp 1-8

[2] Rathod S S, Saxena A K and Dasgupta S.2011*Alpha-particle-induced effects in partially depleted silicon on insulator device: with and without body contact[J](IET Circuits, Devices & Systems vol 5)*pp 52-58

[3] Jing S and Haiwei X. 2016*Design of Radiation Hardened SRAM Based on DICE(Electronics & Packaging vol 3)* pp 26-30

[4] Zhixiong S and Haixia X. 2014 *Implementation of Hamming code encoder and decoder based on VHDL(Microcomputer & Its Applications vol 12)*pp72-77

[5] JiangtingX and RuiZ. 2011 *Improved of Hamming Code and Circuits Realized in Memory (Electronics & Packaging vol 5)*pp20-22

[6] Ranran X and Hai-bo M.2014 *Triple modular redundancy design for VLSI gate level netlist(Computer Engineering & Science vol 12)*pp2356-2360

[7] Hui X and Wei T. 2015*Research of Radiation Harden Based on Triple Modular Redundancy for ASIC(MICROPROCESSORS vol 5)*pp1-4

[8] Oh NShirvani and P PMcCluskey.2002*Control-flow checking by software signatures(IEEE Transactions on Reliability vol 51)*pp111-122

[9] Oh NShirvani and P PMcCluskey.2002*Error detection by duplicated instructions in super-scalar processors(IEEE Transactions on Reliability vol 51)*pp63-75

[10] Liu T and Zhangqin H. 2014 *Fault detection approach for MPSoC by redundancy core(Journal of Computer Applications vol 1)*pp41-45

IOP Conf. Series: Journal of Physics: Conf. Series **887** (2017) 012023 doi:10.1088/1742-6596/887/1/012023

The prediction of the impact of climatic factors on short-term electric power load based on the big data of smart city

Yunfei Qiu[1], Xizhong Li[1], Wei Zheng[1], Qinghe Hu[2], Zhanmeng Wei[1] and Yaqin Yue[3]

[1] National Grid Yingkou Company, East 40, Bohai Street, Zhanqian Destrict, Yingkou 115000, China.
[2] Corresponding Writer. College of Information Science and Enginering, Northeastern University, No. 3-11, Wenhua Road, Heping District, Shenyang 110819, China. Email: huqinghe@ise.neu.edu.cn
[3] Software College, Northeastern University, No.195, Chuangxin Road, Hunnan District, Shenyang, 110169, China.

Abstract. The climate changes have great impact on the residents' electricity consumption, so the study on the impact of climatic factors on electric power load is of significance. In this paper, the effects of the data of temperature, rainfall and wind of smart city on short-term power load is studied to predict power load. The authors studied the relation between power load and daily temperature, rainfall and wind in the 31 days of January of one year. In the research, the authors used the Matlab neural network toolbox to establish the combinational forecasting model. The authors trained the original input data continuously to get the internal rules inside the data and used the rules to predict the daily power load in the next January. The prediction method relies on the accuracy of weather forecasting. If the weather forecasting is different from the actual weather, we need to correct the climatic factors to ensure accurate prediction.

1. Introduction

With the continuous development of China's economy, people rapidly increase their demand for electric power. The electric power system is the foundation of a country. Saving electric power is of great significance for the country. The climate changes have great impact on the residents' electricity consumption [1], so the study on the impact of climatic factors on electric power load is important. It is the power sector's job to predict power load. Different prediction method has different effect. Effective forecasting method can greatly improve the electricity efficiency.

At present, there are regression analysis method, chaotic forecasting method and time series method for the prediction of climatic factors to power load. But these methods cannot meet the needs of modern society, especially unsuitable for China [2] [3] [4]. With the continuous efforts of many scholars, the researches on neural network have drawn people's attention. Due to the late research on neural network to power load prediction in China, the techniques are not mature. At present, more scholars use Error Back Propagation Algorithm for power load prediction. It takes the effects of the historical load as input, the interrelation among neural cells form the functional relations from input to output and from output to input. The difference between output and actual data meeting requirements determines the cycle to terminate or not [5].

With the continuous construction of smart cities, the corresponding big data, such as the population data, legal data, economic data and other big data of people's daily lives are open and shared nowadays. Because the power load has some certain periodicity and similarity, we can forecast the power load in a future period of time based on climatic factors of the big data of smart city. The continuous changing of the climate makes climatic factors and future power load uncertain to some degree. People's living environment is composed of the humidity, temperature, wind, rainfall, sunshine and a series of other factors. As we know that the temperature of human body is constant. So, in order to keep their body's temperature constant, people need to exchange of energy with the around the surrounding to balance the needed energy and the absorbed energy [6]. Among the climatic factors, the temperature, rainfall, and wind have obvious impact on human being, so they are the major factors causing the short-term fluctuation of power load. So, it is of significant to study the effects of temperature, rainfall and wind on the short-term load.

In this paper, the authors predict the impact of climatic factors on short-term electric power load based on the big data of smart city. Power load forecasting with big data is a very important task, it is important for residents to save power and improve power supply efficiency. Effective forecasting can be helpful for balancing power supply and saving resource consumption, to provide good foundation for reasonable power supply among industries and construction of power grid. With the continuous development of China's economy, people's quality of life is getting better and better, and people are using more and more household electrical appliances, such as air conditioners, heaters, refrigerators and other high-power appliances. These appliances take great proportion of the total electricity consumption [5]. The usage of the appliances is due to the climatic changing. People are paying more and more attention to the impact of climatic changing to the electric power load. Because climatic changing is uncertain, it is difficult to have a concrete way to establish the relation between climate and power load. While Matlab neural network has obvious advantages in dealing with this problem, because it doesn't need concrete model but some input and output data as samples, which the neural network will train the samples to approximate a function to set up a model. The trained model can effectively forecast the short-term power load in the future.

2. Prediction Algorithm

Please follow these instructions as carefully as possible so all articles within a conference have the same style to the title page. This paragraph follows a section title so it should not be indented. In this paper, the effects of the data of temperature, rainfall and wind of smart city on short-term power load is studied to predict power load. The authors studied the relation between power load and daily temperature, rainfall and wind in the 31 days of January of one year. In the research, the authors used the Matlab neural network toolbox to establish the combinational forecasting model [7]. The authors trained the original input data continuously to get the internal rules inside the data and used the rules to predict the daily power load in the next January. The prediction method relies on the accuracy of weather forecasting. If the weather forecasting is different from the actual weather, we need to correct the climatic factors to ensure accurate prediction.

When BP neural network is taken for learning and training, input data should be the 31 days' climatic factors, e.g. temperature, rainfall and wind, from the big data of a smart city. But the factors cannot be input just directly, they must be quantified. For example, as for rainfall, it can be classified as sunny, cloudy, overcast, drizzle, snow and other weather conditions, and sunny is set as 1, cloudy is set as 0.9, overcast is set as 0.8 and so on. Thus, climatic factors can be qualified to better use BP neural network. In BP neural network, the output represents the daily electric power load of the 31 days. A model is set up via BP neural network, and is configured with the daily climatic factors of the next January to predict the daily electric power load of the month.

2.1. BP Neural Network Structure

In general, neural network is composed of input layer, hidden layer, and output layer. Hidden layer can be one-layered or multi-layered. Input layer and output layer is determined by given data, most of them are one-layered. BP neural network is an error corrective learning algorithm composed of forward propagation and back propagation. Signals are input from the input layer, processed in the hidden layer and transferred to the output layer. The states of each layer's neural cells effect only those of the next layer. If the expected output cannot be got from the output layer, the back propagation of the error signal is input, the correlation weight among the neural cells of each layer is modified to minimize the error [8]. Researches proved that 3-layered BP neural network can realize complex non-linear mapping. BP neural network is shown as Figure 1.

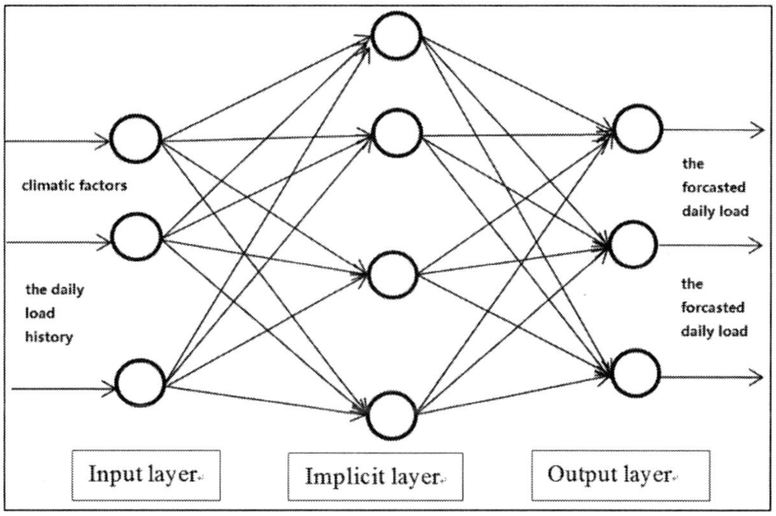

Figure 1. The structure of BP neural network.

The algorithm flow of BP neural network is described as follows.

```
begin
  given data of training samples
  initialize weight and threadhold
  given input and expected output
  calculate the output of hidden layer and output layer
  calculate the error E between the actual output and the expected output
  if (the error E meets the condition) then
    if (all of the errors meet the conditions) then
      go to end
    end if
  else
    calculate the error of hidden layer
    calculate error gradient
    learn and correct weight and threshold
  end if
end
```

2.2. Qualify Climatic Factors

2.2.1. Qualify Temperature. The temperature in January usually varies between -10 and 10 degree. The qualified average daily temperature is set between 0 and 1. The qualified temperature is shown in Table 1.

Table 1. The qualified temperature

Temperature	-10	-8	-6	-4	-2	0	2	4	6	8	10
Qualified temperature	0	0.1	0.2	0.3	0.4	0.5	0.6	0.7	0.8	0.9	1

2.2.2. Qualify Rainfall. Normally, rainfall is classified as sunny, cloudy, overcast, drizzle, snow and so on. The qualified rainfall is shown in table 2.

Table 2. The qualified Rainfall

Rainfall	Qualified Rainfall	Rainfall	Qualified Rainfall	Rainfall	Qualified Rainfall
sunny	0	cloudy	0.1	overcast	0.2
drizzle	0.4	light rain	0.3	moderate rain	0.4
heavy rain	0.5	rain and snow mixed	0.6	light snow	0.7
moderate snow	0.8	heavy snow	0.9	blizzard	1

2.2.3. Qualify Wind. Wind can be classified into calm, light air, light breeze and so on. The qualified wind to be the input is listed as in table 3.

Table 3.The Qualified Wind

Wind	calm	light air	light breeze	gentle breeze	moderate breeze	......
Qualified Wind	0	1	2	3	4	......

2.3. The Transfer Function of BP Neural Network

The temperature in January usually varies between -10 and 10 degree. The qualified average daily

In the BP neural network, the output is the daily total power load. Hyperbolic tangent function is used for the transfer layer as shown in (1). In the hyperbolic tangent function, the input scope is mapped into (-1, 1) from real number range. The input of the to-be-trained samples is mapped into (-1, 1), which can reflect the accurate relation of the neural network.

$$y = f(x) = \tan sig(x) = \frac{e^{x} - e^{-x}}{e^{x} + e^{-x}} \tag{1}$$

2.4. Set up Neural Network

2.4.1. Input & Output. An approximate model is got by simulated training with the daily weather, temperature and wind of last January as the input of the neural network and the daily output as the output. Current month's power load is predicted via the trained neural network taking the daily weather, temperature and wind as the input.

For the simulation, a 3-layered BP neural network is constructed with 3 nodes of input and 1node of output.

2.4.2. Determine the number of the nodes of hidden layer. When neural network is used to predict electric power load, the determination of the number of the nodes of the hidden layer can be made based on the below formula.

$$\begin{aligned} &h = \sqrt{n+m} + a^{*} \qquad 1 < a^{*} < 10 \\ &0.02m < h < 4m \\ &h = \log_2 n \end{aligned} \tag{2}$$

Therein, h is the number of the nodes in the hidden layer, m is the number of nodes in the input layer, n is the number of the nodes in the output layer.

From the aforementioned neural network, we can get that m=3, n=1. So, h can be between 3-12, let's say h=8, i.e., the number of hidden layers is 8. S-shaped tangent function tansig is taken as the activation function of the neural cells of hidden layer, and linear function purelin is taken as the activation function of the neural cells of the output layer.

2.4.3. Set Training Parameters. Our testing showed that we can get the best output when the learning rate is 0.035. The period of showing intermediate result is 50; the maximum iteration is 2000; the target error is 1e-3.

3. Prediction Result Analysis

The climatic factors and daily electric power load of last January from the database of smart city are taken as input and output data. Network error training curve is predicted via BP neural network shown as Figure 2. The curve proves that BP neural network not only reaches the set objective error, but also reaches the set training steps.

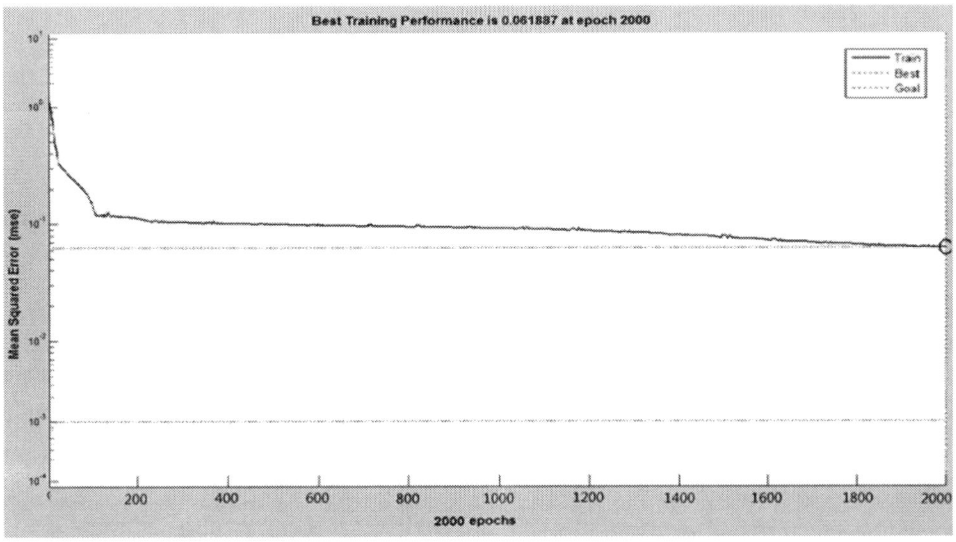

Figure 2. Network Training Error Curve

Figure 3 presents the power load output by the neural network and the actual power load, and shows that the errors between the training output and the actual data is not big and acceptable. Once a neural network is trained, the trained model can take the qualified climatic factors of the 31days of the next January to get the daily power load of the 31 days of the next January, like shown in Figure 4.

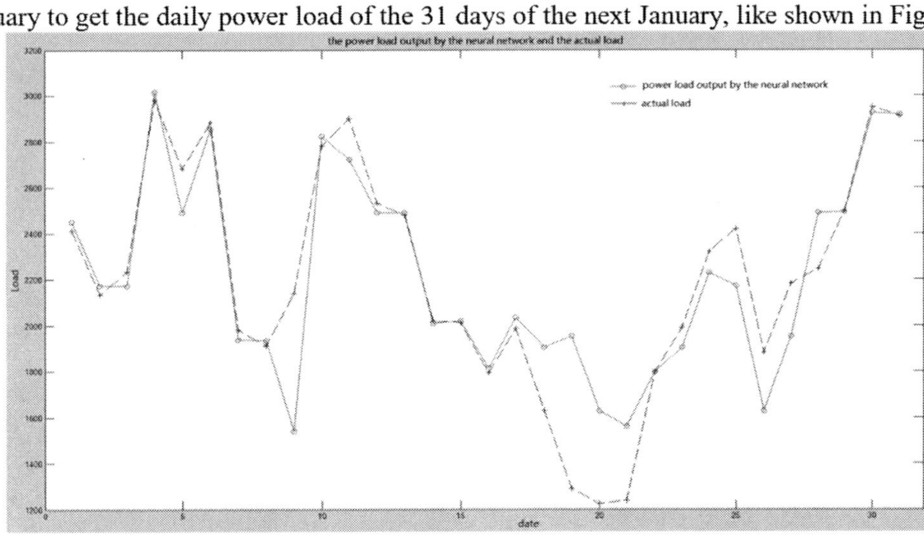

Figure 3. The curve of the power load output by the neural network and the actual load

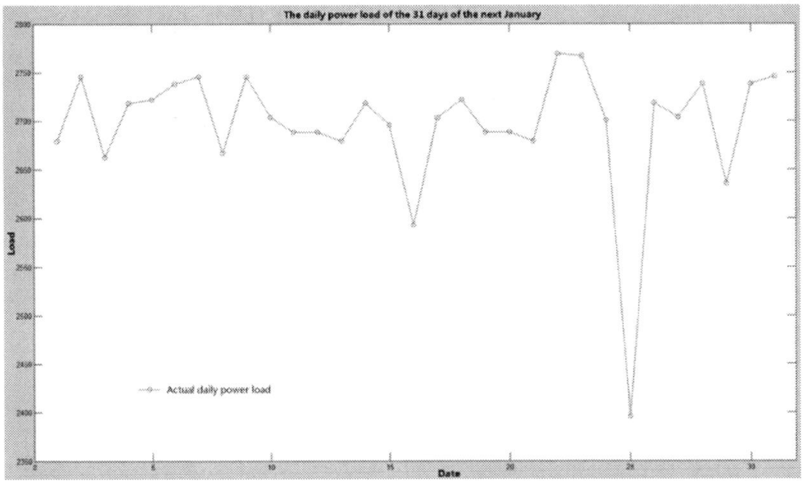

Figure 4. The daily power load of the 31 days of the next January

4. Conclusion

The prediction of the impact of climatic factors on short-term electric power load based on the big data of smart city id the foundation of electric power system. Accurate and effective prediction is of great significance for national power sector. With the increasing of people's quality of life and development of smart cities, the power load by household electrical appliances takes more and more proportion of total load. Climatic changes influence the variation of power load. Difference areas have different climate and geographic factors. And the fitness of persons to the climate also impacts power load indirectly. So, when the neural network is designed, the local climatic factors and the actual power load must be considered to predict the impact of climatic factors on electric power load.

In the paper, BP neural network is adopted to set up a prediction model; and the climatic variation and daily electric power load from the big data of a smart city of the last January is used to accurately train the relationship between temperature, rain, wind and power load. Experiment data proves that the BP neural network based prediction model is an effective prediction method.

5. References

[1] Fu Guiqin, Li Yunzong. Influence Analysis of Meteorological Conditions on Electric Loads [J]. Meteorological Science and Technology, 2008, 36(6) PP 795-800

[2] KANG Chongqing, ZHOU Anshi, WANG Peng, ZHENG Guangjun, LIU Yi. Impact Analysis of Hourly Weather Factors in Short-Term Load Forecasting and Its Processing Strategy [J]. Power System Technology, 2006, 30(7) PP 5-10

[3] JR Jervis, JK Pringle, A Study of the Effect of Seasonal Climatic Factors on the Electrical Resistivity Response of Three Experimental Graves [J], Journal of Applied Geophysics, 2014, 108 (9):53-60

[4] VB Tupov, AA Taratorin, Influence of Climatic Factors and the Ground Surface on the Required Noise Abatement from Power Equipment [J], Thermal Engineering, 2013, 60 (7): 493-498

[5] ZHENG Gang, LIU Bin, ZHOU Yong, LIU Ding, MU Guoqiang. Short-Term Load Forecasting Based on Neural Network[J].Journal of Xi'an University of Technology,2002,18(2) PP 126-130

[6] Fu Guiqin, Cao Xin, Shi Qingge. Short-Term Power Load Forecast Model Based on the Meteorological Condition [J]. Hebei Province meteorology science and technology service center Shijiazhuang 050021 PP 66-71

[7] LIU Shuang, YANG Lixi, WANG Zhigang, JIA Defeng, CHEN Genyong. Combined Power Load Forecast Model Based on Matlab Neural Network Toolbox [J]. Electric Power Automation Equipment, 2003,23(3) PP 59-61

[8] Zhao Juanjuan. Short-Term Load Forecasting Based on BP Neural Network in Power System [D]. Harbin University of Science and Technology, 2013

Research on optimal driver steering model based on Multi-Point preview

Jun Gu[1] **and Aijing Ma**[2]

[1] College of Information and Mechanical Electrical Engineering, Jiangsu Open University, Nanjing, Jiangsu 210019, China
Email:gujun_nj@126.com
[2] College of Energy and Power Engineering, Nanjing University of Aeronautics and Astronautics, Nanjing Jiangsu 210016, China

Abstract. In this paper, multi-point preview control algorithm is applied to driver steering control model. This paper builds multi-point preview road model in the form of state shift register. Based on the linear quadratic regulator (LQR) optimal control theory it optimizes driver steering control model with multi-point preview. Meanwhile, in the Matlab \ Simulink environment, based on vehicle system dynamics and optimal control theory, different preview points and weighted coefficients are simulated to study the influence of driver steering model. The simulation results show that the multi-point preview control mode has excellent driving performance. And in this paper, the main parameters affecting the preview control algorithm such as speed, preview weighted coefficients and the number of preview points and so on are discussed.

1. Introduction

The driver steering model describes the driver's steering behaviour characteristics in the form of mathematical model. It is good for the study of driver steering control system and real vehicle test validation, analysis and aided design. Therefore, the driver steering model research has received extensive attention of the researchers [1-2]. With the deepening of the study there has been a large number of different control methods of the driver steering model [3-6]. At present, the most widely used model is the optimal preview model proposed by MacAdam [7], and the model optimizes the input with lateral deviations. The preview optimal curvature model proposed by Academician Guo, which establishes the relationship between vehicle steering characteristic parameters and driver characteristic parameters [8]. And Sharp optimizes both the lateral deviation and the yaw angle deviation [8]. Then, for the shortcomings of the specific implementation process of the steering angle, Cole et al. proposed a driver steering control model based on Neuromuscular System (NMS) [9].

This paper builds multi-point preview road model in the form of state shift register. Based on the linear quadratic (LQR) optimal control theory it optimizes driver steering control model with multi-point preview. Meanwhile, this paper chooses different vehicle speed, weighted coefficients and preview points' number to simulate, respectively, to study the driver's steering behaviour characteristic, which provides experience and reference for better establishing driver steering model.

2. Single Track Vehicle Model

Referring to single track vehicle model shown in figure 1 and ignoring the influence of the nonlinearity of the suspension and tire this paper directly uses the wheel steering angle as input. Select state variables: vehicle lateral speed v_y, yaw angular speed ω, lateral displacement y, and yaw angle ψ. The following

IOP Conf. Series: Journal of Physics: Conf. Series **887** (2017) 012024 doi:10.1088/1742-6596/887/1/012024

four-dimensional linear vehicle model is established, as shown in Equation (1) and the specific parameter values are shown in table 1.

$$\begin{bmatrix} \dot{v}_y \\ \dot{\omega} \\ \dot{y} \\ \dot{\psi} \end{bmatrix} = \begin{bmatrix} \frac{C_f + C_r}{v_x m} & \frac{l_f C_f - l_r C_r}{v_x m} - v_x & 0 & 0 \\ \frac{l_f C_f - l_r C_r}{v_x I_{zz}} & \frac{l_f^2 C_f + l_r^2 C_r}{v_x I_{zz}} & 0 & 0 \\ 1 & 0 & 0 & v_x \\ 0 & 1 & 0 & 0 \end{bmatrix} \begin{bmatrix} v_y \\ \omega \\ y \\ \psi \end{bmatrix} + \begin{bmatrix} \frac{-C_f}{m n_{rsw}} \\ \frac{-l_f C_f}{I_{zz} n_{rsw}} \\ 0 \\ 0 \end{bmatrix} \theta_{sw} \quad (1)$$

Where $\theta_{sw} = n_{rsw}\delta$, and θ_{sw} is the wheel steering angle.

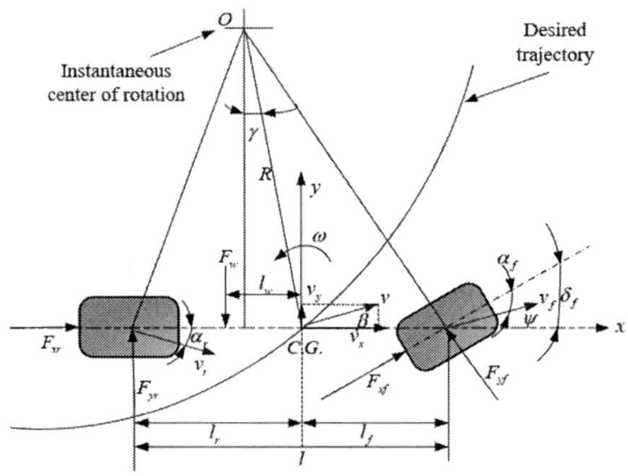

Figure 1. Force analysis structure diagram of vehicle.

The above state-space Equation (1) can be converted into the following standard form

$$\begin{aligned} \dot{x} &= Ax + Bu \\ y &= Cx + Du \end{aligned} \quad (2)$$

Assume that the input is the zero-order hold state and transfer it into a discrete system by the function c2d in Matlab.

$$\begin{aligned} x(k+1) &= A_d x(k) + B_d u(k) \\ y(k) &= C_d x(k) + D_d u(k) \end{aligned} \quad (3)$$

Where k denotes time stride, $\boldsymbol{x(k)} = [v_y(k) \quad \omega(k) \quad y(k) \quad \psi(k)]$.

Table 1. Single track vehicle model parameters.

symbol	name	value	unit
I_{zz}	rotational inertia	2550	kgm^2
M	vehicle weight	1673	kg
l_r	rear wheelbase	1.73	m
l_f	front wheelbase	0.913	m
C_f	front wheel cornering stiffness	-2*88310	N/rad
C_r	rear wheel cornering stiffness	-2*64076	N/rad
n_{rsw}	gear ratio	16	-

3. Multi-Point Preview Model for Road

3.1 Road Model

Taking the center line of the test road as the ideal road track that drivers desire to track, the concrete ideal road parameter settings are shown in table 2. Double change road is shown in figure 2.

Table 2. Ideal road curve setting.

Longitudinal displacement (x/m)	lateral displacement (y/m)
<321	0
363~398	3.5
>433	0

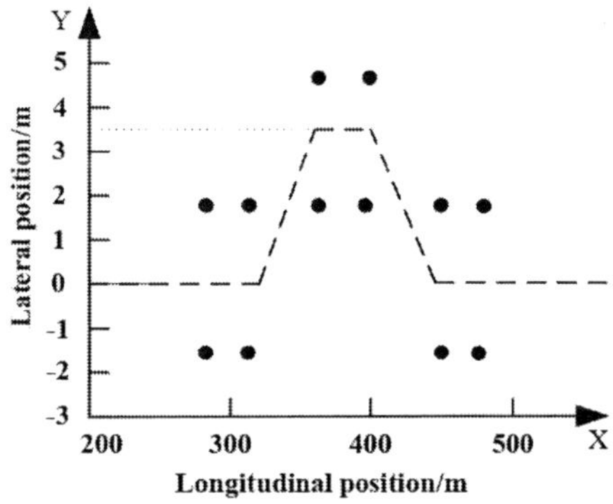

Figure 2. Two-shift road setting.

3.2 Multi-Point Preview Model for Road

When the vehicle is moving, the driver firstly obtains the front road information through the preview, and then select a plurality of road model preview points in the road model as the pre-road input in the case of suitable road condition.

When establishing a road model, the linear form of road model is converted into a discrete form by sampling. Its concrete conversion process is similar to "queue": choose N_p+1 discrete points, and then take them as system state variables. New discrete points continue to enter the system, first in first out, and repeat it in this way until all road sampling points gradually entering into the system. As time goes by, these points will be near the vehicle gradually (shown below)

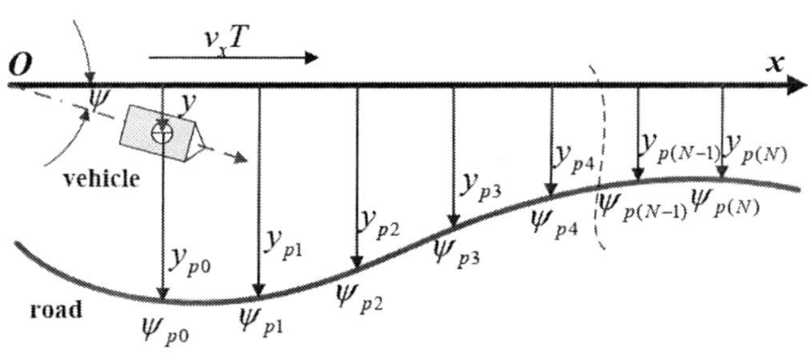

Figure 3. Vehicle and preview road structure chart in geodetic coordinate system.

Where v_x is vehicle speed, and T is the simulation stride. y_{p0}, y_{p1}, y_{p2} … $y_{p(N)}$ denotes vehicle lateral displacements, and ψ_{pi} denotes road heading angle. Assume that the road ahead is divided into N_p+1 equidistant points, then this dynamic road model can be represented as a mathematical model as follows

$$y(k) = Dy(k) + Ey_p(k) \tag{4}$$

Where $\boldsymbol{y_p(k)} = [y_{p0}(k) \quad y_{p1}(k) \quad y_{p2}(k) \quad ...... \quad y_{p(N-1)}(k) \quad y_{p(N)}(k)]^T$ denotes road input. The road model system matrix defined as

$$D = \begin{bmatrix} 0 & 1 & 0 & \dots & 0 \\ 0 & 0 & 1 & \dots & 0 \\ \vdots & \vdots & \vdots & & \vdots \\ 0 & 0 & 0 & \dots & 1 \\ 0 & 0 & 0 & \dots & 0 \end{bmatrix} \quad E = \begin{bmatrix} 0 \\ 0 \\ \vdots \\ 0 \\ 1 \end{bmatrix} \tag{5}$$

4. Control Algorithm

4.1. Control Method

Using LQR theory, taking full account of road information and establishing optimal-preview driver model can integrate vehicle and road into a whole. The "road-vehicle" system is shown in figure 3. It can be written as a form of Equation (6)

$$\begin{bmatrix} x(k+1) \\ y_p(k+1) \end{bmatrix} = \begin{bmatrix} A_d & 0 \\ 0 & D \end{bmatrix} \begin{bmatrix} x(k) \\ y_p(k) \end{bmatrix} + \begin{bmatrix} 0 \\ E \end{bmatrix} y_{p(N+1)}(k) + \begin{bmatrix} B_d \\ 0 \end{bmatrix} \theta_{sw}(k) \tag{6}$$

According to the loss function Equation (7), this paper optimizes the input of the system by using the linear quadratic optimal control theory.

$$J = \sum_{k=0}^{\infty} \begin{pmatrix} q_y (y(k) - y_{p0}(k))^2 + q_\Psi (\psi(k) \\ -\psi_{p0}(k))^2 + R_2 (\theta_{sw}(k))^2 \end{pmatrix} \tag{7}$$

Weighted matrix defined as

$$R_1 = C^T Q C, \ R_2 = \begin{bmatrix} 1 & 0 \\ 0 & 1 \end{bmatrix}$$

$$C = \begin{bmatrix} 0 & 0 & 1 & 0 & -1 & 0 & 0 & 0 & \dots & 0 \\ 0 & 0 & 0 & 1 & 1/v_x T & -1/v_x T & 0 & 0 & \dots & 0 \end{bmatrix}, \ Q = \begin{bmatrix} q_y & 0 \\ 0 & q_\Psi \end{bmatrix}$$

Where q_y and q_Ψ are lateral position and heading Angle error loss function weighted coefficient, respectively. And the two together generate the weighted matrix R_1. R_2 is a scalar that controls the weight of the input.

By employing DLQR function in Matlab, considering R_1, R_2 and "road-vehicle" model, the state feedback gain matrix K_p will be obtained when the loss function J reaches the minimum.

$$\boldsymbol{K_p} = [k_1 \quad k_2 \quad k_3 \quad k_4 \quad k_{p1} \quad k_{p2} \quad \cdots \quad k_{pn}]$$

Where k_1 to k_4 denote state gains ($\left[v_y(k) \quad w(k) \quad y(k) \quad \psi(k)\right]^T$) associated with vehicle status, respectively. Then k_{p1} to k_{pn} denote preview gains associated with road information, respectively. Thus, producing the optimal control input

$$\hat{\theta}_{sw}(k) = -\boldsymbol{K_p} \left[x(k) \quad y_p(k)\right]^T$$

4.2. *Vehicle Coordinate System Transformation*

In the actual case, the information used by the driver when making decisions is observed in the vehicle coordinate system, as shown in figure 4. Then the road information obtained by the driver is the vehicle coordinate system observation values: $[y_{r0}, y_{r1}, y_{r2}, \cdots, y_{r(N)}]$, namely, the driver uses the relative error between vehicle and the preview road, rather than the observation values in the geodetic coordinate system $[y_{p0}, y_{p1}, y_{p2}, \cdots, y_{p(N)}]$.

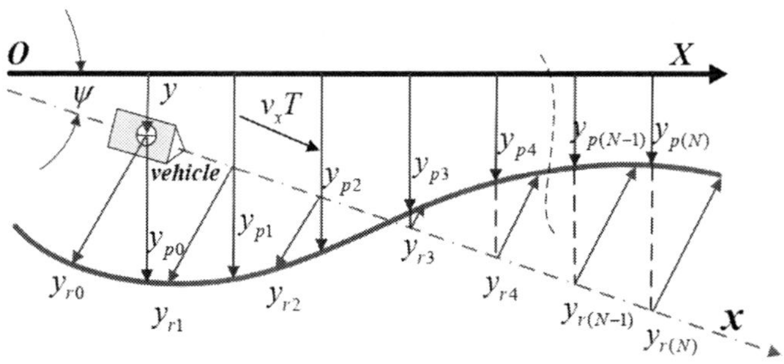

Figure 4. Vehicle and the preview road diagram in the vehicle coordinate system.

Assuming that yaw angle is very small, in the geodetic coordinate system, vehicle lateral displacement defined as

$$y_{pn}(k) = y_{rn}(k) + y(k) + nv_xT\psi(k) \tag{8}$$

The optimal steering angle in the geodetic coordinate system is

$$\begin{aligned} \hat{\theta}_{sw}(k) &= k_1 v_y(k) + k_2 w(k) + k_3 y(k) \\ &+ k_4 \Psi(k) + \sum_{n=0}^{N} k_{pn} y_{pn}(k) \end{aligned} \tag{9}$$

The optimal steering angle in the vehicle coordinate system is

$$\hat{\theta}_{sw}(k) = k_1 v_y(k) + k_2 w(k) + \sum_{n=0}^{N} k_{pn} y_{rn}(k) \tag{10}$$

The optimal steering angles in the two coordinate systems are variable, so Equation (9) and Equation (10) must be equal.

$$k_3 = -\sum_{n=0}^{N} k_{pn} \, , k_4 = -\sum_{n=0}^{N} k_{pn} n v_x T \tag{11}$$

5. Simulation Results and Analysis

The multi-point preview control algorithm is related to the parameters such as vehicle preview distance, the number of preview point and weighted coefficient, etc. In order to study the validity of the multi-point preview algorithm, in the Matlab / Simulink software environment, this paper simulates and analyses different preview distance, the number of preview point and preview weighted coefficient, respectively.

5.1. Influence of Preview Distance and the Number of Preview Point on Steering Performance

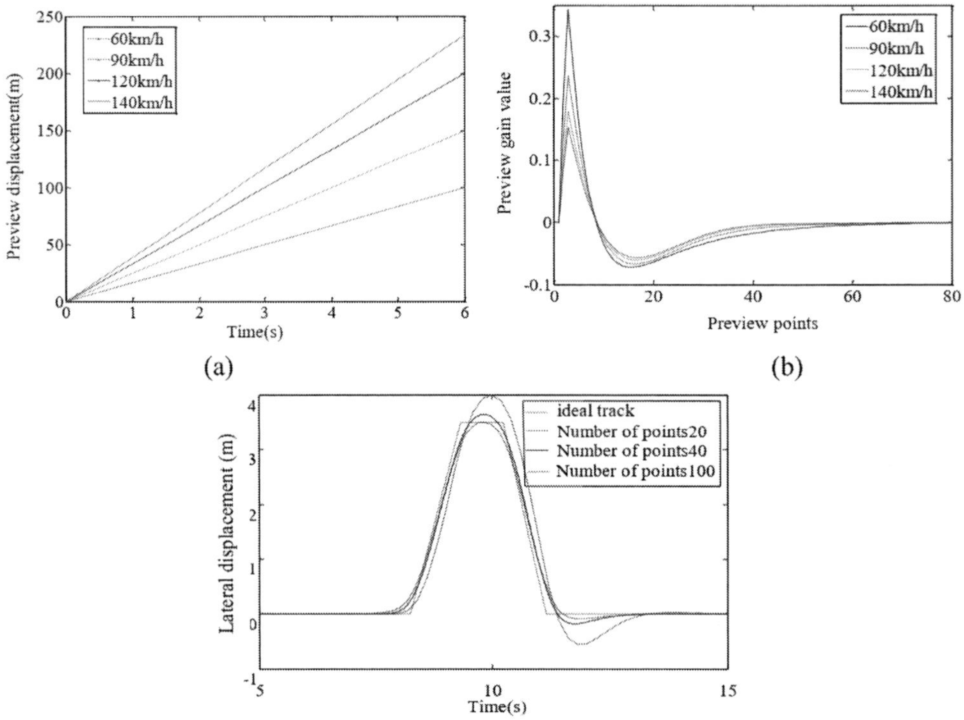

(c)

Figure 5. Influence of preview distance and the number of preview point on steering performance. (a) Relationship between preview distance and speed; (b) Relationship between preview gain and speed; (c) The movement track of different preview points, speed is 140 km/h.

Figure 5(a) & (b) are the simulation results of different speeds when the number of preview points is invariant. When the preview time is constant, as the speed increases, the preview distance increases. And the farther the driver preview, the closer the preview gain value to zero. When the distance exceeds the driver's physiological limits, then the preview feedback gain value is zero, and the preview point in a far distance has no effect on the steering performance. It can also be seen from figure 5(c) that in the case of multi-point preview, when the number of preview points is less or the preview distance is too close, the prediction effect of the trajectory is deteriorated, which results in the decrease of tracking accuracy. The more the number of points, the better tracking result is. Therefore, in essence, the preview part is associated with the whole system, and the driver can control the vehicle only when the driver is within the physiological limits. Otherwise, the vehicle will be out of control.

5.2. Influence of Different Weighted Coefficients on Steering Performance

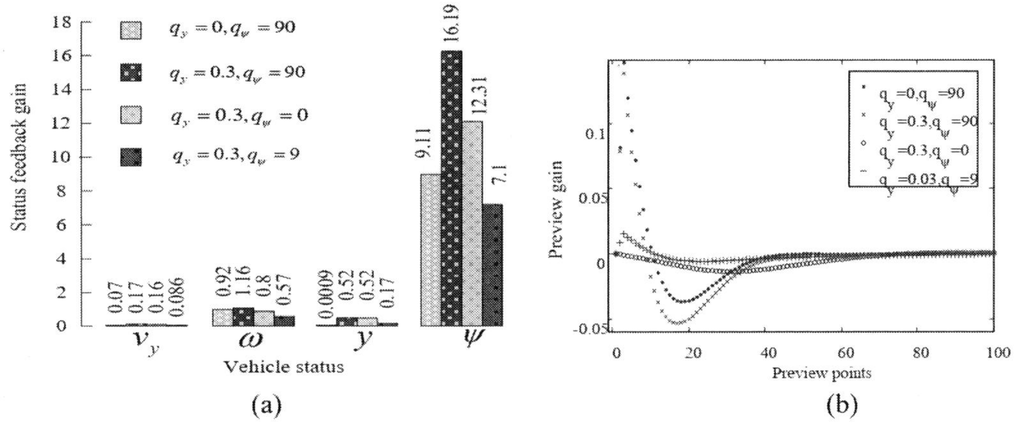

Figure 6. Status and preview gain. (a) Status gain; (b) Preview gain.

Figure 6 shows the control trajectory following characteristics of different weighted coefficients and the gain. With the change of the weighted coefficients, both the trajectory following characteristic and the gain values all change. In the status gain values, the yaw rate changes greatly, which has a great influence on the driver steering control, namely, the driver has a greater dependence on the yaw rate. At the same time, obviously, the preview is more dependent on weight. As the preview distance increases, the preview gain gradually approaches zero, and the preview gain at infinity is zero, which means that the point at which the driver's preview distance is closer has an effect on the driver steering control. However, with the increase of the preview distance, the influence of the far points on the steering control is getting smaller and smaller.

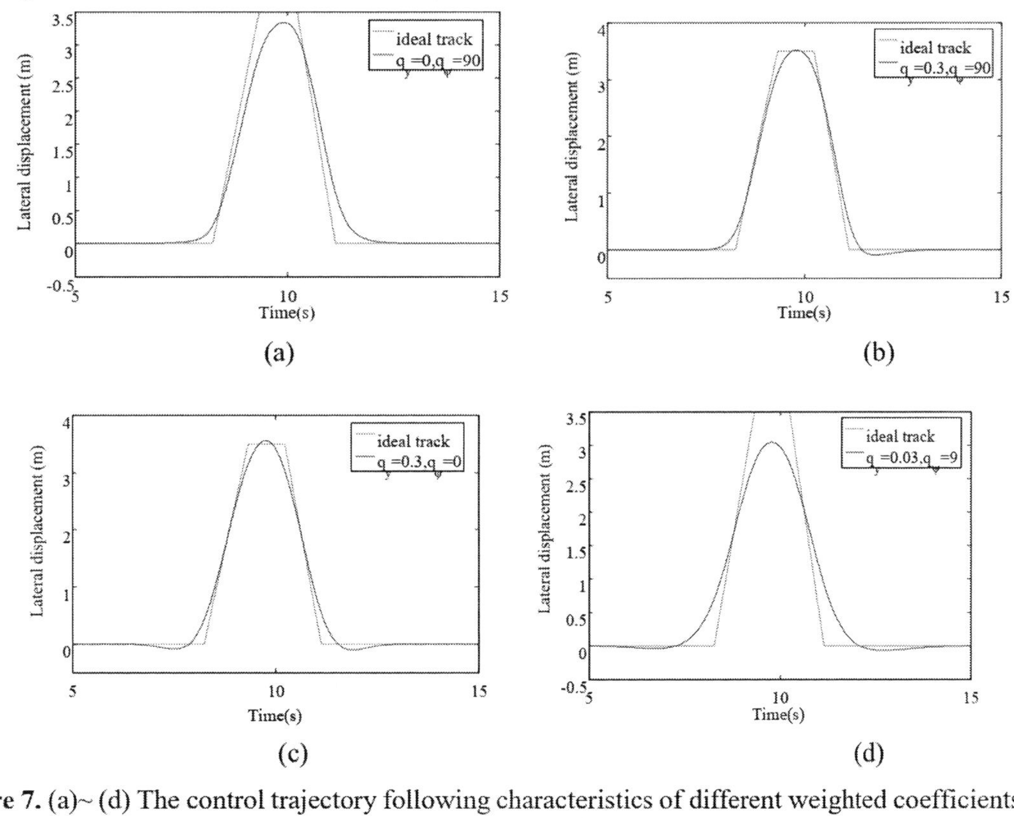

Figure 7. (a)~ (d) The control trajectory following characteristics of different weighted coefficients. (a) Controller 1; (b) Controller 2; (c) Controller 3; (d) Controller 4.

The trajectory following characteristics of each controller are shown in figure 7. The controller 1 shows that when the lateral position deviation is very small and the heading angle error is very large the road trajectory following characteristic is remarkably poor. As the weight of the lateral position error increases, the trajectory following control characteristics gradually improve, as shown by the controller 2. When the controller 3 shows that the heading angle error is small, the driver turns to a certain amount of rotation in the reverse direction, and the reverse steering operation is related to the vibration at the time of steering input by the driver. When the lateral position error and the heading angle error weight are small (controller 4), the steering shear angle often occurs. For different weight conditions, there are different trajectory following characteristics. Even if different drivers in the same road conditions, there will be a different trajectory following result, so it's necessary to seek the best control strategy to the driver steering control model design.

6. Conclusion

Closed-loop simulation results show that the multi-point preview of the driver steering model has a good trajectory following effect. The optimal driver steering model can be obtained by selecting different weighted coefficients and the number of preview points, which can be used as a reference for the future research of various driver steering models.

7. Acknowledgments

This work was supported by the Natural Science Foundation of the Jiangsu Higher Education Institutions of China [Grant Number 13KJB510005].

8. References

[1] MacAdam C C. Understanding and modeling the human driver [J]. *Vehicle System Dynamics*, **40**(1-3):101-134.

[2] Plochl M,Edelmann J. Driver models in automobile dynamics application[J]. *Vehicle System Dynamics,* 2007, **45**(7):699-741.

[3] Kondo M. Directional stability (When Steering is Added) [J]. *Journal of the Society of Automotive Engineers of Japan (JSAE)*, 1953, **7**(5–6): 1-28.

[4] Salvucci D D, Rob G. A Two-Point visual control model of steering [J]. *Perception*, 2004, **33**(10): 1233-1248.

[5] Shen H, Ling R, Li S M. Large curvature steering control method based on preview optimal curvature model [J]. *China Mechanical Engineering*, 2012, **23**(17): 2111-2116.

[6] Duan L F, Gao Z H, Wang D P. Driver's adaptive control behavior modeling for vehicle direction [J]. *Chinese Journal of Mechanical Engineering*, 2011, **47**(8): 121-125.

[7] MacAdam C C. Application of an optimal preview control for simulation of Closed-Loop automobile driving [J]. *Transactions on Systems, Man, and Cybernetics*, 1981, SMC-**11**(6): 393-399.

[8] Guo K H. Preview optimal curvature model controlled by Drivers -Vehicle Close-Loop system [J]. *Automotive Engineering*, 1984, **6**(3): 1-16.

[9] Sharp R S. Driver steering control and a new perspective a car handling qualities [J]. *Proceedings of the Institution of Mechanical Engineering, Part C: Journal of Mechanical Science*, 2005, **219**(10): 1041-1051.

[10] David J. Cole D J. A Path-Following Driver–Vehicle model with neuromuscular dynamics, including measured and simulated responses to a step in steering angle overlay[J]. *Vehicle System Dynamics*, 2012(4): 573–596

IOP Conf. Series: Journal of Physics: Conf. Series **887** (2017) 012025 doi:10.1088/1742-6596/887/1/012025

Medium- and long-term electric power demand forecasting based on the big data of smart city

Zhanmeng Wei[1], Xiyuan Li[1], Xizhong Li[1], Qinghe Hu[2], Haiyang Zhang[3] and Pengjie Cui[4]

[1] National Grid Yingkou Company, East 40, Bohai Street, Zhanqian Destrict, Yingkou 115000, China.
[2] Corresponding Writer. Northeastern University, No. 3-11, Wenhua Road, Heping District, Shenyang 110819, China. Email: huqinghe@ise.neu.edu.cn
[3] Northeastern University, No. 3-11, Wenhua Road, Heping District, Shenyang 110819, China.
[4] Software College, Northeastern University, No.195, Chuangxin Road, Hunnan District, Shenyang, 110169, China.

Abstract. Based on the smart city, this paper proposed a new electric power demand forecasting model, which integrates external data such as meteorological information, geographic information, population information, enterprise information and economic information into the big database, and uses an improved algorithm to analyse the electric power demand and provide decision support for decision makers. The data mining technology is used to synthesize kinds of information, and the information of electric power customers is analysed optimally. The scientific forecasting is made based on the trend of electricity demand, and a smart city in north-eastern China is taken as a sample.

1. Introduction

Electric power demand forecasting plays an important role in the operation, planning and control of electric power system. It is an important part to ensure the safe operation of power system, to meet the scientific management and the same scheduling of the joint power grid, and to plan the commercial operation. However, the traditional prediction introduces fewer factors as the variables [1], by considering only the power system's internal data, such as the maximum load, the average regional electricity consumption, and by using the traditional data mining and statistical methods, based on limited sampling data, proprietary data in the field, such as marketing systems, distribution management systems, production management systems, energy management systems, equipment inspection and monitoring systems, customer service systems and so on. Therefore, the prediction accuracy of power information is very limited, and it is urgent to propose a new approach to forecast power demand. That's to say data preparation has become particularly important.

The development of smart city brings opportunities for cross-sectoral data acquisition. Powerful data analysis based on smart city, integrates both internal data and external data of power enterprises, especially government based public data. By exploring the data association and using as many categories of data as possible, analysis based on big data and a simple algorithm can be more effective than a small number of data and a complex algorithm. We included types of data, and brought meteorological information, geographic information [2], population information [3], business information, economic information [4-7] and other external data into the big database.

Based on the big data inside power supply companies and inside the smart city, we devised an improved algorithm to forecast the medium and long term power demand. Simulation results proved that the big data of smart city can lead to more accurate power demand forecasting.

2. Research Content

(1)Power Big Data Analysis Model

When electric power information is forecasted, the focus is shifted from complex algorithms to big data preparation. Big data and a simple algorithm is more effective than a small amount of data and a complex algorithm, so data preparation becomes particularly important, i.e. to absorb as many data indicators as possible, to extract the correlation between the data characteristics, and to predict the target data based on the associated data. The larger the total amount of data, the more categories of data, the more comprehensive the data feature, the more accurate the analysis is. For example, the annual electricity consumption, if we include the economic factors into the short-term load forecasting to consider meteorological factors, the accuracy of prediction degree can be greatly improved. Table 1 shows the comparison of the two analysis approaches.

Table 1. Comparison of the two analysis approaches

Comparison Item	Analysis based on traditional power supply data [8]	Analysis based on big data of the smart city
Data amount	very limited	huge
Data sources	power enterprise production data only	from the public basic database of both electric power enterprises and government departments
Variable	power grid operation and equipment testing or monitoring data, power enterprise marketing data, etc.	The variables used by traditional data analysis methods, the type of regional population, weather meteorology, national policy, industrial legal persons, large customers, etc..
Algorithm	traditional algorithms and intelligent algorithms	correlation analysis and feature extraction
Algorithm rate	large convergence rate, long computation time	mergers and acquisitions of the data can speed up rate
Algorithm accuracy	accuracy of the original data not high	may fall into the local minimum

(2) By extracting the characteristic relationship between the total electricity consumption and the population, the GDP, the per capita GDP, the number of industrial enterprises above designated size and the total amount of the import and export, and so on. Combined with the internal data of the power enterprises, medium and long term electric demand forecasting model is established.

Data type is a big feature of the big data. Power consumption is affected by multiple factors, so we took as many data sources and data categories as possible as associated objects, and forecast power information by extracting associated features. We made full use of the big data from the smart city basic database outside power enterprises, and did the demand forecasting and evaluation basing on the variation of the associated objects and features.

3. Algorithm Implementation

3.1. Functional Structure Modeling

According to this paper, a model of the functional structure which relies on basic database of smart city is established. Firstly, a data database is built, then the relevant data are extracted, the data cleaning and format conversion is carried out. Finally, the mining results are analysed. The effectiveness of customer analysis depends heavily on the selection and measurement of customer evaluation criteria. The mining method aiming for the actual needs is the key of the whole model.

Power consumption is affected by many factors. This paper not only takes into account the internal data of electric power enterprises but also makes full use of the information provided by the smart city database. The specific names of the data indicators are as follows [9].

(1) The speed of economic development (GDP)

(2) Changes in industrial structure (adjustment of industrial structure, the proportion of high energy consuming industries in industry)

(3) Population and living standards (total population, living standards of residents)

(4) Temperature and climate change (temperature, precipitation, natural disaster and so on)

(5) Demand management measures (shift peak load Valley, energy storage equipment and so on)

(6) Prices of other alternative energy (It mainly refers to the price of natural gas)

(7) Policy factors (macro industrial structure adjustment, energy saving)

(8) Supply factors (power supply capacity and wind power integration)

(9) Large number of complex data information.

The architecture of the forecasting model is shown in Fig.1.

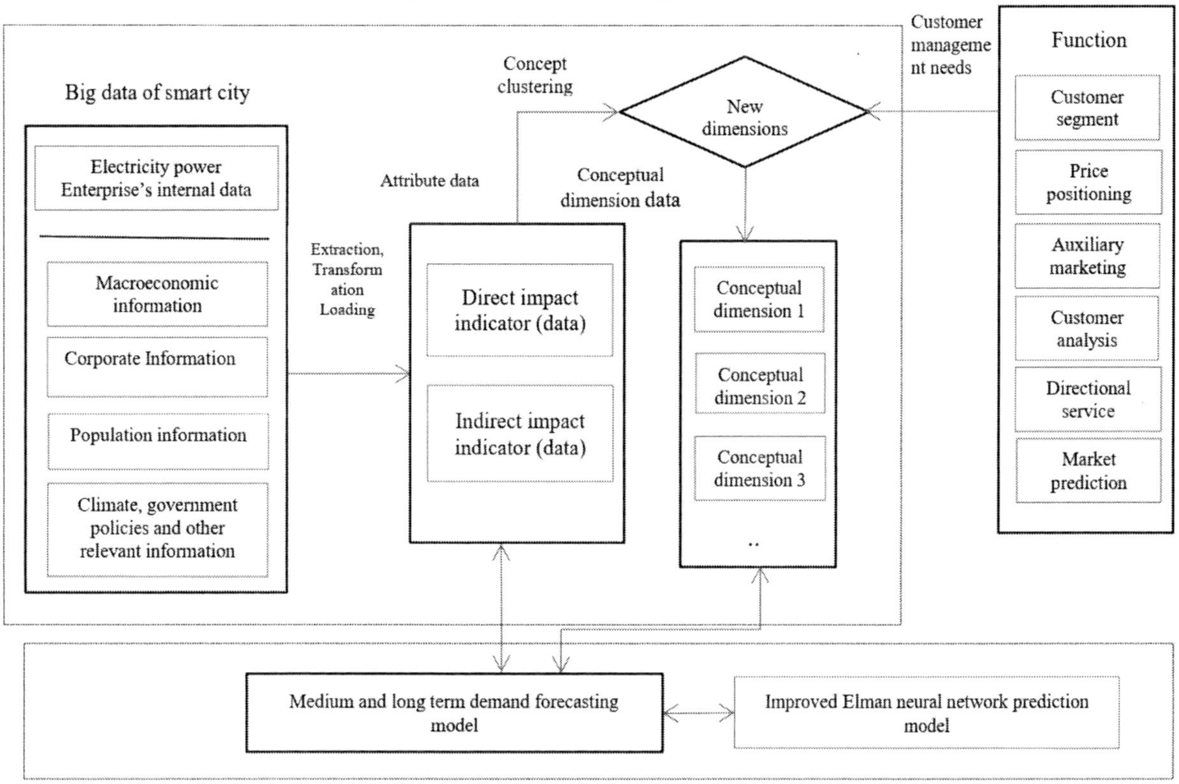

Figure1. The Architecture of the Forecasting Model

3.2. Elman Neural Network Structure

We include an undertake layer to record the input of middle layer (i.e. hidden layer) to increase the sensitivity to historic data on purpose. The adding of internal feedback network increases the network's ability of self-processing dynamic information. Thus, a dynamic model is achieved.

The proposed Elman neural network structure is as shown in Fig. 2.

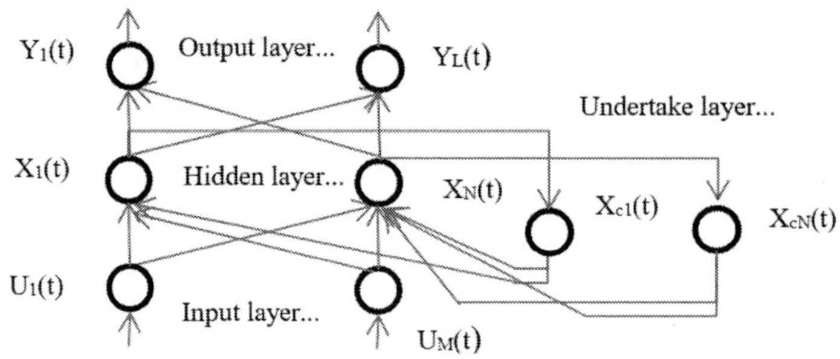

Figure 2. Elman neural network

3.3. Improved Algorithms

3.3.1. Improved Learning Algorithm. In the process of learning, the traditional Elman neural network often shocks and converges slowly, and falls into local optimal solution. In order to solve the above problems, the nonlinear damping least square method is used to optimize it. The nonlinear damping least square method is an improved method based on quasi-Newton method. The equation of weights and thresholds of the network is as follows.

$$W(k+1) = W(k) - [J^T J + \mu I]^{-1} J^T e \tag{1}$$

Wherein, W is the network parameter set that will be adjusted, and k is the times of iteration. Where J is the Jacobi matrix of the error function and contains the weights and threshold derivative. Wherein, μ is the learning rate, and I is the unit matrix. And e is the error vector of network. When μ is 0, equation (2) is quasi-Newton method. When μ is larger, it is gradient descent with smaller stride. The quasi-Newton method can approach the minimum error quickly and accurately. And μ decreases continuously after each iteration. Then the algorithm is close to Quasi-Newton. If error performance increases after iteration, μ should be enlarged. Therefore, the performance of the iterative error has been reduced.

3.3.2. Improved Excitation Function. The excitation function [10] of Elman neural network model is generally a sigmoid function *f(x) =1/ (1+exp (−x))*. However, it is easy to have the model to converge slowly and to fall into local minimum. The improved function is as follows.
Its function is as equation (2).

$$f(x) = m + \frac{1}{1+\exp(-h(x+n))} \tag{2}$$

The derived function of equation (2) is as equation (3).

$$\begin{aligned} f'x &= \frac{h[\exp(-h(x+n))]}{1+\exp(-h(x+n))^2} \\ &= h\left\{\frac{1}{1+\exp(-h(x+n))} - \frac{1}{1+\exp(-h(x+n))^2}\right\} \\ &= -h\left[\frac{1}{1+\exp(-h(x+n))} - \frac{1}{2}\right]^2 + \frac{h}{4} \end{aligned} \tag{3}$$

The authors get equation (4) from equation (2).

$$\frac{1}{1+\exp(-h(x+n))} = f(x) - m \tag{4}$$

By substituting equation (4) into equation (3), the authors get equation (5).

$$f\acute{x} = -h\left[f(x)-(m+\frac{1}{2})\right]^2+\frac{h}{4} \tag{5}$$

Wherein, *m*, *n* is a constant and *h* is a slope. Set *m*, *n* value, so that the function moves vertically and horizontally. In the model, the values of m, n and h arse corrected with the error function. When the function value *f(x)* is closer to *m+1/2*, the larger the derivative value is, the faster the function converges. In addition, the value of *f'(x)* is positively related to the learning rate. And the higher the value is, the faster the learning rate is. By adjusting the *m*, *n* and *h* values, the convergence speed and prediction accuracy of the model can be optimized.

4. Case Analysis

In order to verify the effectiveness of the proposed methods of load forecasting, the power grid from 2006 to 2015 in a certain region of the historical annual electricity sales data is used, as shown in Table 2. It shows that over the past years the electricity sales data has an overall growth trend, but a fluctuation in 2009 which has a great influence on the load forecasting precision. This paper mainly considers the load forecasting of the coming month to the coming several years which is medium- and long-term electric power demand forecasting for dispatching department and planning department. In Table 2, years from 2006 to 2015 are numbered by 1~10, and the corresponding sale data is taken as the original input of the model [11].

Based on the traditional data of the internal data of the electric power, the Elman neural network prediction model of the large and medium data of the smart city is selected and the improved data of the internal and external data of the smart city are obtained by using the Mathlab with raw data from year 2006 to 2015 as the input. Elman neural network prediction model is used to fit the load series from year 2006 to 2015, and the three models are used to predict the data of year 2014 and 2015, and the historical sales volume in year 2014 and 2015 is taken as the model testing samples.

The load model of year 2006 to 2015 is normalized by the established prediction model. The fitting results and the relative error ratios are shown in Table 3.

Table 2. Data of electricity sold from 2006 to 2015

Serial number	Particular year	Electricity sales/(TWh)	Serial number	Particular year	Electricity sales/(TWh)
1	2006	26	6	2011	37
2	2007	27	7	2012	46
3	2008	31	8	2013	52
4	2009	32	9	2014	54
5	2010	34	10	2015	56

The results of the analysis show that the maximum prediction error ratio of the traditional prediction model based on the data of internal power system is 12.91%, and the minimum error ratio is 4.01%. As for Elman neural network model based on big data of smart city, the maximum error ratio is 11.78%, and the minimum error ratio is 1.93%. As for the proposed Elman neural network model based on big data of smart city, the maximum error ratio is 6.70%, the minimum error ratio is 1%. It can be concluded that it is possible to achieve better power forecasting by referring to the big data inside and outside the power company, and the improved Elman neural network model has higher prediction accuracy.

Table 3. Comparison between the forecasting data and actual data

Serial number	Actual charge value (TWh)	Traditional forecasting model based on internal data of power company		Elman neural network model based on big data of smart city		Improved Elman neural network model based on big data of smart city	
		calculated value (TWh)	relative error ratio %	calculated value (TWh)	relative error ratio %	calculated value (TWh)	relative error ratio %
1	26	——	——	——	——	——	——
2	27	24.7765	8.24	26.1558	3.13	26.5483	1.67
3	31	26.9980	12.91	27.3493	11.78	28.9244	6.70
4	32	35.8791	12.12	34.3442	7.32	34.1335	6.67
5	34	36.0213	5.94	35.0225	3.01	34.8756	2.58
6	37	39.4652	6.66	38.6643	4.50	38.1243	3.04
7	46	47.8455	4.01	46.8871	1.93	46.4554	1.00
8	52	55.6752	7.07	53.1062	2.13	52.9876	1.90

The number of neurons in the input and output layers of the neural network is determined by the problem itself. In this paper, the number of neurons in the Elman neural network is set as 8. And in the experiment, the neurons in the hidden layer are gradually increased and decreased. The count of the units in hidden layer is 9, and the count of neurons in the output layer is 1. With the original load data as the output data *T*, the output layer uses Logsig as the output function, the training function is Trainlm, and the training is 1000-round. The predicted load data by the forecasting model is input into the training-ready network, and the simulation result is just the forecasting result, as shown in Table 4.

Table 4. Prediction results

Year	Actual charge value (TWh)	Traditional forecasting model based on internal data of power company		Elman neural network model based on big data of smart city		Improved Elman neural network model based on big data of smart city	
		calculated value (TWh)	relative error (TWh)	calculated value (TWh)	relative error (TWh)	calculated value (TWh)	relative error (TWh)
2014	54	59.8034	10.75	57.9767	7.36	55.3216	2.45
2015	56	61.0043	8.94	58.1062	3.76	57.4061	2.51

5. Conclusion

Based on the big data of smart city, the authors include more indicators and consider the relationship among the data. Compared with traditional forecasting, the proposed algorithm is better on prediction accuracy and stability. And, by using improved Elman neural network prediction model, the proposed model has better network computing, time-varying adaptability, error controlling and network learning ability. The forecasting results of a practical case proved both the effectiveness of the smart city data for medium- and long-term power demand forecasting and the availability and superiority of the improved algorithm.

6. References

[1] Muhammad Qamar Raza, Zuhairi Baharudin, Perumal Nallagownden, et al. A Comparative Analysis of PSO and LM Based NN Short Term Load Forecast with Exogenous Variables for Smart Power Generation [C]. 2014 5th International Conference on Intelligent and Advanced Systems (ICIAS), 2014, 1-6.

[2] Franklin L.Quilumba, Wei-Jen Lee, Heng Huang, et al. Using Smart Meter Data to Improve the Accuracy of Intraday Load Forecasting Considering Customer Behavior Similarities [J], IEEE Transactions on Smart Grid, 2014, 6(2): 911-918.

[3] Shu Fan, Methapraynoon K, Wei-Jen Lee. Multiregion Load Forecasting for System with Large Geographical Area [J]. Industry Applications, IEEE Transactions on, 2009, 45(4): 1452-1459.

[4] Lee Chien-Chiang, Chang Chun-Ping, Structural Breaks. Energy Consumption and Economic Growth Revisited: Evidence from Taiwan [J]. Energy Economics, 2005, 27(6): 857-872.
[5] Paresh Kumar Narayan, Russell Smyth. Energy Consumption and Real GDP in G7 Countries: New Evidence from Panel Co-Integration with Structural Breaks [J]. Energy Economics, 2008, 30(5): 2331-2341.
[6] Nicholas Apergis, James E, Payne. Energy Consumption and Economic Growth in Central America: Evidence from a Panel Co-Integration and Error Correction model [J]. Energy Economics, 2009, 31(2): 211-216.
[7] Hwang D B K, Gum B. The Casual Relationship between Energy and GNP: the Case of Taiwan [J]. The Journal of Energy and Development, 1992, 16: 219-226.
[8] Wu Jinwei. Big Data Technology in the Application of Electricity Demand Analysis [D]. Jiangsu: Southeast University, 1-65 (2016).
[9] ZHANG Ai-hui. Study on Medium and Long Term Electric Demand Forecasting Based on Multi-index Model [D]. Hunan: Changsha University of Science and Technology, 1-52 (2013).
[10] Chen Ya, Li Ping. Study on Short-Term Load Forecasting Based on Neural Network [J]. Electrical Technology, (1): 26-29 (2017).
[11] Zhang Jianmei, Zhou Shuxiang, Lin Nan, Zhang Qin, Chen Jie. Gray Elman Neural Network Medium and Long Term Load Forecast [J], 25 (4): 146-149 (2013).

Research progress of control of condensate depression for condenser

Ying Liu [1], **Run Liang** [2] **and Fengyu Li** [3,*]

[1]College of Electric Engineering, Naval University of Engineering;
[2]First Cadet Brigade twelfth team, Naval University of Engineering;
[3]Naval University of Engineering, Wuhan 430033 China, lifengyu905@163.com

Abstract. It is introduced that significance and structure of the condensate depression control system. In accordance with controller devised procedure, we analyze and elaborate how to construct the lumped parameter and dynamic mathematical model which possesses distinct physics significance. Neural network model being called black-box model is also introduced. We analyze and contrast the control technique of condensate depression as conventional PI control, fuzzy PI control and fuzzy control. It is indicated that if the controller of condensate depression were devised inappropriate, while the steam discharged of turbine varying by a large margin, would result in the rotation rate of cooling water circulating pump accelerating at a great lick even to trigger the galloping danger which is less impressive for the units operating safely.

1. Introduction

The condenser is a significant portion of steam turboset. The thermal economy of the whole of set is influenced immediately by condenser, the same as operational reliability. And the condensate depression is a significant economic indicator to evaluate the condenser [1]. If the condensate depression exceeded that would bring about increase dissolved oxygen of condensation water. And that would exacerbate the corrosion of low tension regenerative system. On the contrary, minor condensate depression stand for cold source is minor, and the thermal economy of the set is better [2]. Nonetheless, condensate depression cannot be too small. If condensate depression is lesser than 0°C would result in supercharged jet pump coming into cavitation erosion.

Therefore, it is extremely necessary to control condensate depression. The control method of condensate depression must be chose felicitously which is of momentous significance.

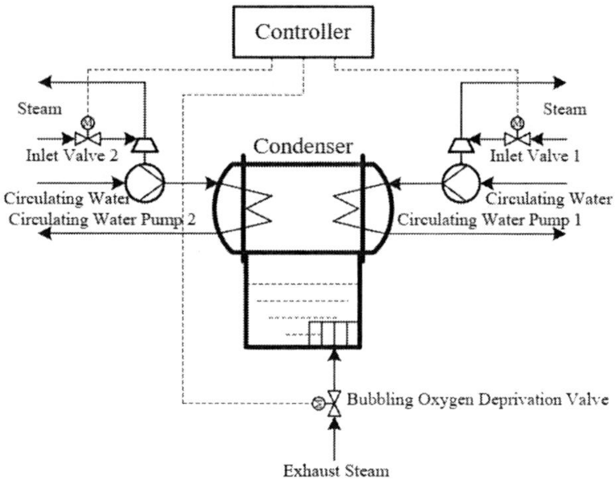

Figure 1. The structure chart of condensate depression control system

The function of condensate depression control system is as followed. On the premise of ensuring the vacuum of condenser, to enable the degree of condensed water approach the saturated temperature of current pressure. Condensate depression control system dominates cooling quantity of condensed water by adjusting the rotation rate of cooling water circulating pump. And it dominates exhaust steam heating condensed water by manipulating the opening of the bubbling oxygen deprivation valve. The rotation rate of circulating pump is regulated by adjusting the opening of the inlet valve. The structure chart of condensate depression control system is as Fig.1.

2. Mathematical Research in Modeling During the Controller of Condensate Depression Designed

There are various types of analysis methods for condensate depression control system. It is researched mainly on account of different mathematical model of condenser.

Condensate depression is defined as the magnitude of condensed water over cooling in condenser's hot well. Under the absolute static pressure in condenser shell, it is the difference between condensed water temperature t_{cw} and the saturation temperature of steam in condenser t_s [3]. As followed:

$$\Delta t = t_s - t_{cw} \tag{1}$$

The mathematical model of condenser is founded on principle of condenser devices. And it is the pivotal factor for researching and simulating the controlling means of condensate depression. As to design the controller of condensate depression, it is extremely essential to establish the lumped parameter and dynamic model of condenser. There are quiet a few elements affect the function of system. As an example, bubbling oxygen deprivation, steam resistance, leakage of condenser, flash evaporation of hot well water and swept volume of air ejector can be chose to take into consideration. And it should be made its choice on the basis of actually condition.

But for the sake of founding the mathematical model of condenser explicitly, quiet a few matters are assumed on the basis of simplification principle of mathematical model [4].

(1)The steam and non-condensed gas can be regarded as ideal gas as a result of the low pressure and density in condenser conform to the requirement of ideal gas.

(2)The system simulation takes the main parameters of condenser into consideration. The internal microcosmic variation is not keynote of research. Hence we can presume that working medium in shell-side is in equilibrium and saturation state except condensed water in condenser's hot well. Parameters of working medium are selected amongst saturation value, and the equation is built at equilibrium point ignoring the air content.

(3)Circulating cooling water is flowing at pipe-shell, and its velocity distribution is a parabola. In the process of simulation, we adopt its average velocity and suppose its current velocity and mass flow rate distributing uniformly.

(4)The temperature variation of circulating cooling water is calculated on the basis of its average temperature, and we presume it is uniformly.

(5)We presume that the mass of metal of circulating cooling water tube is distributed uniformly, same as its specific heat capacity and temperature gradient.

There are numerous researches of lumped parameter and dynamic model of condenser. Xue Ruojun [5] presented a real-time model describing the dynamic behavior of a nuclear power shell-and-tube condenser. Normal condition and accident condition are simulated base on simulation platform, and it is appropriate for the whole simulation range. DING Yan et. [6] Described the working process of the condenser by a series of differential equations, thus presenting the dynamic mathematical model of the condenser. And several simulation result of different condition of condenser in 300MW are analyzed. That presents a more intuitive analysis of the problem, but no solution is given. WANG Wu-Chao [7] employed a new moving boundary method to develop the dynamic simulation mathematical model of condenser. The simulation model reflects the variation of length of different phase region condenser during working, and could be simplified to a set of ordinary derivative equation finally. It could be solved more easily, both the efficiency and accuracy. The model can provide detailed performance parameters of the condenser. It studied the dynamic response of condenser in the step change of different control parameters. ZHANG Yong-sheng [8] presented the simplified lumped parameter mathematic model of a marine condenser according to its structure and its working principle in order to effectively control the vacuum and condensate depression of the condenser. The control system of the vacuum and condensate depression for the condenser is designed. Matlab /Simulink software is used for the simulation of the control system. The operating condition of decreasing load is calculated.With the support of general simulation platform *Minis*, NI He et. [9] Established a dynamic mathematical model of condenser by a series of difference equations on basic principium of mass and energy conservation. And the model possesses a better static accuracy.

Although those models using different methods and emphasis, basically is to start a research on the basis of the working principle of the condenser.

3. Research into Control Method

There are a multitude of control methods of condensate depression, for instance, conventional PI control, fuzzy PI control, fuzzy control et.

On the basis of improving traditional controller, the conventional PI control, SUN Jian-hua [10] used the neural network method to establish the model of the condenser. The local hierarchical fuzzy control system is also designed. The effectiveness of the control system is verified by the simulation result. WAN Hua-qing [3] presented humanoid logical control and fuzzy control of condenser vacuum and condensate depression, on the basis of analysis the mechanism of condensate depression. The simulation results prove the validity of the control. LIU Wei-ting [11] studied condenser vacuum and condensate depression control system in an incisive way, and elaborated design principles, design model and realization method. Digital signal processing chip is adopted by the system hardware for ensuring the real-time of system. The fuzzy-neural network algorithm is adopted by the system software for conquering the no determinacy of system model. Results show that the designed system work stable and reliable, and possesses strong robustness.

3.1. Conventional PI control system

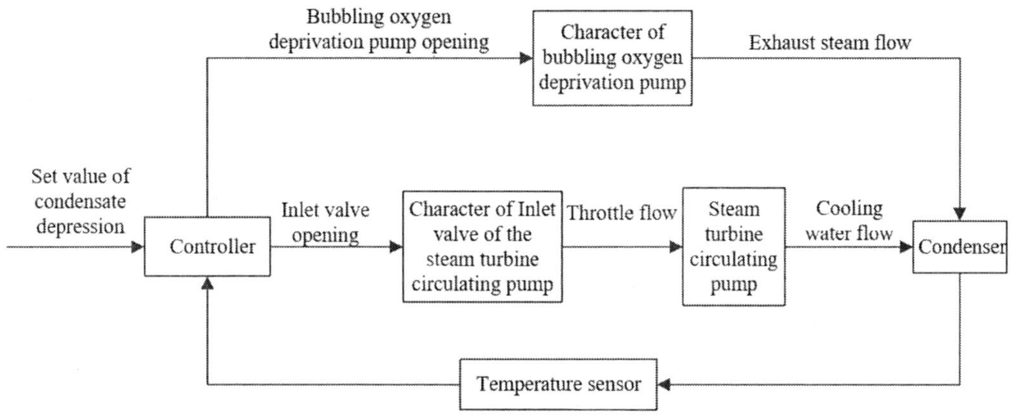

Figure 2. The functional block diagram of condensate depression control system

The functional block diagram of condensate depression control system is shown in Fig.2. The figure shows that the controller measures the temperature of condensed water though the temperature sensor. We can calculate the condensate depression of the condenser and then to compare it with its set value. The circulating pump of circulating water is dominated by D-value signal to modify the flow of circulating water. That can make the condensate depression of the condenser in a certain range [8].

3.2. Fuzzy PI control System

The fuzzy PI control system is the incorporation of fuzzy control and traditional PI control. Its design philosophy is to find out the fuzzy control relation between parameters of the PI controller and deviation *E* & deviation change rate *EC*. The parameters *E* and *EC* are constantly testing in the process of operation. And the parameters K_p and K_i are real-time modified according to the principle of fuzzy control. Thus, that could content the diverse demand of different *E&EC* to the controller parameters. Finally the controlled object possesses a positive dynamic and static performance [12]. The structure of control system is shown in Fig.3.

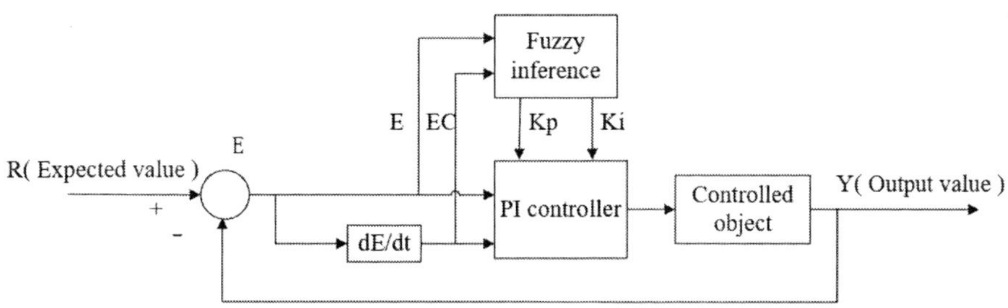

Figure 3. The structure of fuzzy PI control system

According to the design philosophy of the fuzzy PI control system, first of all we should define that the deviation E and the deviation change rate EC between the expected value of condensate depression and the actual value of condensate depression are input parameters. The output parameters are ΔK_p and ΔK_i the correction of control parameters of PI controller K_p and K_i. After that the universeoffuzzysets of input and output value are defined. In the test the control effect of condensate depression of the fuzzy PI controller is better than the conventional PI controller. The concrete performances are less volatile and shorter setting time. As the condenser operate in changing different condition changing, the control performance of the fuzzy PI controller is better than the conventional PI controller. That can content the control requirement across the condensate depression control while the steam discharged of turbine varying by a large margin, but the improving is not well.

3.3. Fuzzy Control System

The fuzzy control is been used to design the controller as for condensate depression control system. It is the fuzzy controller that been designed as a combined fuzzy controller. It can be seen as the integration of traditional digital controller and fuzzy controller.

As shown in Fig.4, the system can switch between the fuzzy controller and the conventional PI controller based on the changed working condition of turbine. When the perturbation varying drastically, the fuzzy controller will be disconnected and switch to traditional controller. And then the rotation rate of circulating pump and the opening of the oxygen deprivation valve of exhaust steam will be adjusted according to the working condition of steam turbine corresponding experience. When condensate depression basically stable near the experience value, the traditional controller should be disconnected and switch to the fuzzy controller. After that, the fuzzy controller will be fine-tuned according to the vacuum error *e*, the vacuum error derivative *ec,* the condensate depression before bubbling oxygen deprivation and the condensate depression before bubbling oxygen deprivation. Finally, the condensate depression will reach regulation control performance index requirements.

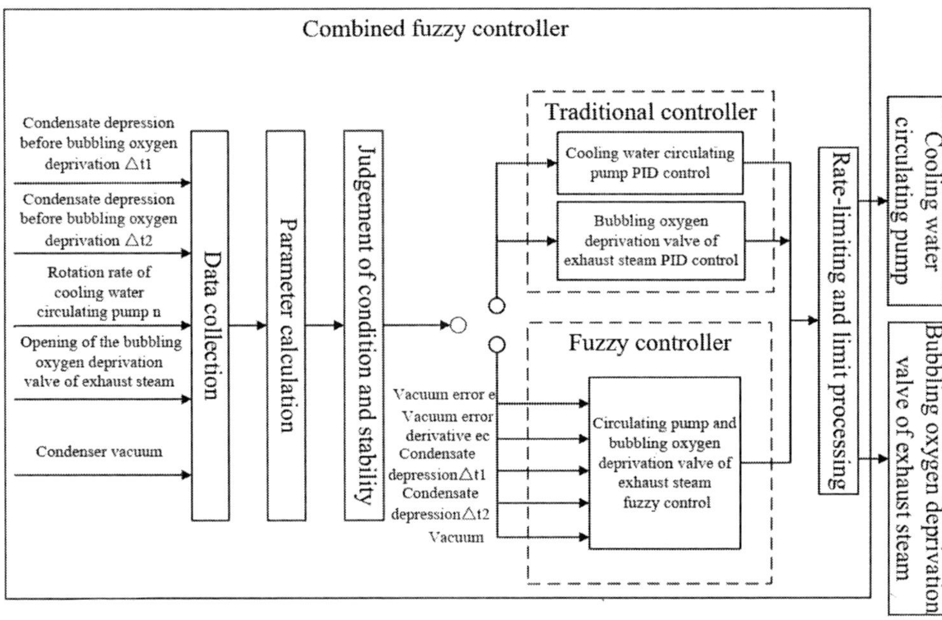

Figure 4. The functional block diagram of the condensate depression fuzzy control system

4. Conclusion

The condensate depression control is a significant portion of condenser, and to construct the lumped parameter and dynamic mathematical model which possesses distinct physics significance is the foundation in designing the condensate depression controller of the condenser. It is worth taking care that if the controller of condensate depression were devised inappropriate, while the steam discharged of turbine varying by a large margin, would result in the rotation rate of cooling water circulating pump accelerating at a great lick even to trigger the galloping danger which is less impressive for the units operating safely. It is central to study of condensate depression control method which has important practical significance according to the different condensing system operating characteristics.

5. References

[1] LIU Yu-kuai, PAN Yong-mei, CHEN Zhen-hong, ZHANG Yan-bin. Condensate Supercooling Degree and the Influence on the Unit [J]. Applied Energy Technology, 2008(11): 39-41.
[2] LU Xu-xiang, LI Lu-ping, WANG Yun-min. Analysis of Condensate Depression and its Effects on Performance of the Unit [J]. Turbine Technology, 2002, 44(2): 117-119.
[3] WAN Hua-qing, YAO Yong-tao, PAN Yan, SUN Jian-hua. Research of condensate depression and its control system [J]. Ship Science and Technology. 2008, 30(1): 80-83.
[4] XUE Ruo-jun. Research on Real-time Simulation of Dynamic Characteristics of Marine Condensation-steam Equipment [D]. Harbin Engineering University, 2002.
[5] XUE Ruo-jun., TIAN Zhao-fei, ZHAO Qiang. Real-Time Simulation of Nuclear Power Condensers [J]. Nuclear Power Engineering, 2008, 29(5): 119-123.
[6] DING Yan, GU Chang, FANG Qiong. The Dynamical Mathematical Model and Simulation of Condenser in 300MW Unit [J]. Turbine Technology, 2004, 46(5): 327-332.
[7] WANG Wu-Chao, ZHAO Jing-Quan. Dynamic Simulation of Condenser [J]. Journal of Engineering Thermophysics, 2005, 26(4): 631-634.
[8] ZHANG Yong-sheng,MA Yun-yi. Research on the control system of vacuum and condensate subcooling degree for the marine condenser [J]. Ship Science and Technology, 2011, 33(3): 62-64.
[9] NI He, CHEN Gang, SUN Feng- rui. Application of a Dynamic Mathematic Model of a Naval Condenser [J]. Computer Simulation, 2007, 24(7): 337-341.
[10] SUN Jian-hua, QIU Zhi-qiang. [J]. Research of Host Condenser Vacuum and Condensate Depression Intelligent Control [J]. Ship Science and Technology, 1999, 21(6): 15-19.
[11] LIU Wei-ting, ZHANG Bing, ZHU Zhi-yu. Fuzzy Control System Design of Host Condensator[J]. Journal of University of Electronic Science and Technology of China,
[12] LUO Ma, SUN Jian-hua, WANG Wei. Condenser Supercooling Degree Control System Based on Fuzzy PI Control [J]. Ship Eletronic Engineering, 2013, 33(4): 153-155. 2006, 35(3): 356-358.

Analysis and improvement of face detection based on surf cascade

Siquan Hu[1, 2], Caihong Zhang[1, 3] and Lei Liu[1, 4]

[1]School of Computer and Communication Engineering, University of Science and Technology Beijing, 100083, China
[2]Email:husiquan@ustb.edu.cn
[3]Email:s20150063@xs.ustb.edu.cn
[4]Email: liulei2776@gmail.com

Abstract. This paper aims to study limitations of the commonly employed boosting cascade framework. We focus on the factors like data, feature, weak classifier and stages. A set of novel experiments were done to show the relationship. The model contains three key points: SURF feature, weak classifier based on logistic regression and AUC-based cascade learning algorithm. This paper adds cross validation in logistic regression creatively which improves accuracy and speeds up convergence greatly. Eventually only five stages and about 100 weak classifiers are needed. The frontal face detector improves reject rate to 99% for the first three stages, decreases number of false positive greatly and achieves comparable performance among non-CNN techniques on FDDB dataset.

1. Introduction

Object detection is the basis of object recognition and tracking, and is widely used in security and other fields. The method based on machine learning consists of two steps: feature extraction and classifier design [1].Commonly used features are Hog, HAAR, Edgelet and so on. The classifier is mainly based on SVM, adaboost. Dalal proposed a pedestrian detection method based on HOG and SVM [2].Viola and Jones [3] used 3 types of 4 forms of HAAR for face detection. M.Pietikäinen [4] proposed LBP for texture feature extraction. Bo Wu [5] proposed edgelet. Dollar [6] proposed integral channel feature.

Object detection algorithms based on deep learning, like Faster RCNN [7] and YOLO [8], have achieved a high detection rate. But because of complex computing process and the requirement of GPU, it is difficult to achieve real-time demand.

2. Analysis of Boosting Cascade Model

The whole model is obtained by cascading the strong classifiers obtained by boosting a number of weak classifiers. We study all factors as below that influence model capability. We did quantities of comparison experiments to illustrate how to choose these parameters.

Data. The degree of difficulty of the samples varies greatly. Data augmentation like flip is used.

Feature. We should consider feature dimension and computational complexity. Then how to select the best feature becomes a difficult point for training.

Weak classifier. We should consider the calculation time and accuracy, and select the best method. Common methods are SVM, Logistic regression, etc.

Cascade. The number of weak classifiers and stages always influence final detection rate.

3. Details of Model

In our model, we use SURF feature, logistic regression with cross validation, AUC-based algorithm, and train in cascade structure. We will describe these below.

3.1. Feature

SURF [9], a local feature descriptor, reflects the shape and texture of feature points. It is invariant and has a great improvement in computing speed. Here, we use SURF descriptor to extract gradient information, which is robust to the rotation of faces. As shown in Figure1, We set the detection widow size 40×40, and the size of feature rectangle gradually increases from 8×8 to 40×40 and slides in the 40×40 detection window, resulting in thousands of candidate features. Each feature is divided into 2×2 or 1×4 image blocks, and we use [-1, 0, 1] to compute the gradient information of the horizontal, vertical and diagonal dimensions of the pixel. And we compute $|d_x| \pm d_x$. In this way, we get 8 dimensional information. Then, the 8 dimensional gradient information of each image block is summed $\sum(|dx| + dx), \sum(|d_x - d_x|)$. So we get 32 dimensional vectors for every feature.

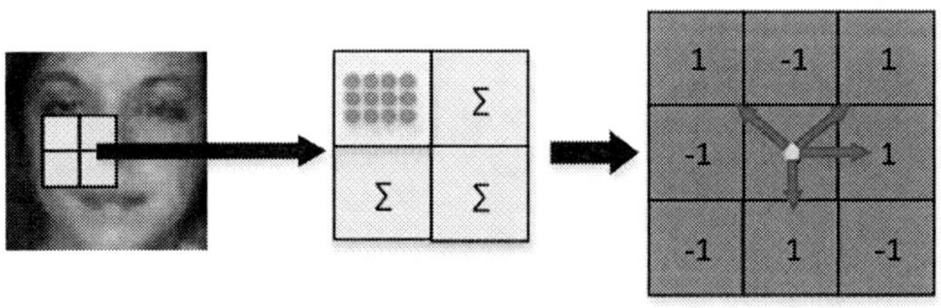

Figure 1. The flowchart of feature selection and calculation.

Compared to the 180 thousand of HAAR in 24×24 detection window, the number of candidate candidates for SURF features in this paper is greatly reduced. There is only 2143 candidate feature. There are usually three ways to select best feature from our candidate features. As in equation (1) the best feature can be found via minimizing the sum of the absolute values of the error weights of all samples, via minimizing the sum of squared errors of all samples as in equation (2).

$$\sum_{i=1}^{N} w_i |h_i - y_i| \tag{1}$$

$$\sum_{i=1}^{N} w_i (h_i - y_i)^2 \tag{2}$$

As in equation (3), the maximum value of AUC which combines the classifiers obtained in the previous training is used to find best feature.

$$J(H^{i-1} + h_j(x, w)) \tag{3}$$

3.2. Weak Classifier

The weak classifier is used to classify the result of a single feature, and the common methods are SVM, logistic regression. In this paper, we choose logistic regression, and the optimized processing method of [10], greatly improving the accuracy of weak classifier. Given samples $\{x_i, y_i\}_{i=1}^{N}$, the output of samples is define as P. The process of solving W is translated into minimization of L.

$$P = \frac{1}{1+exp(-yw^T x)} \tag{4}$$

$$L = \sum_{i=1}^{N} lg(1 + exp(-yw^T x)) + Cw^T w \tag{5}$$

The parameter C adjusts the proportion of the error and the regular term, including L1, L2.Table1 shows the great change of C. In this paper, we add cross validation to improve accuracy, and set the range of C from 0.001to 1024 and select best C which gets the minimum false positive rate. Figure 2 shows that the change of the weak classifier promotes the convergence quickly using means of cross validation.

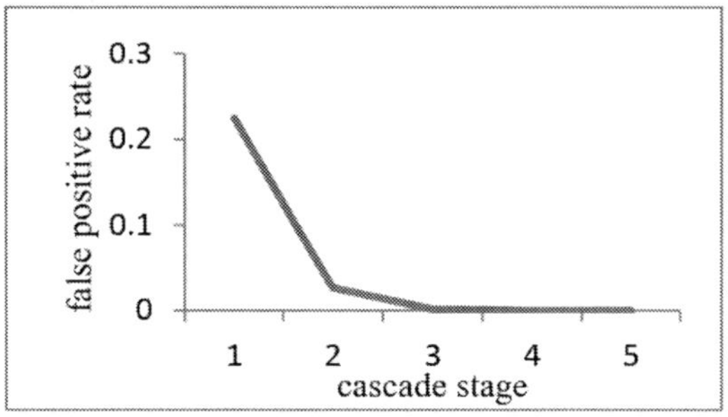

Figure 2. Accumulated false positive rate.

Table 1. Compare of parameter C.

Param.C	False positive rate
0.1	0.514
1	0.3114
10	0.2044
cross validation	0.2097

4. Experiments

The training and detection experiments were done on a personal workstation with 2.6GHZ Core-i5 and 8GB RAM. We evaluate our models on the public dataset FDDB, including 2,845 images with 5,771 faces, about 80% of which are frontal faces. We get ROC to show TPR (true positive rate) and number of FP (false positive). We only focus on four points on the ROC, the TPR of FP=0, FP=100, FP=250 and the max of FP. Frontal faces are collected from GENKI dataset, FERET dataset, totally 12,000 faces cropped and resized to 40×40.The negative samples are mainly from images download from network by tools and Caltech dataset etc. Finally, we collect 24,000 images without faces. All experiments bellow are trained with these data and evaluated on FDDB dataset.

4.1. Data

It is impossible to collect all the samples owing to the samples diversity. Data augmentation techniques are used to improve the capabilities of the model. We get other 12,000 faces with mirror transform, and 12,000 faces by random perspective transforming face image within [-10, 10] degree. The number of false positive in the model decrease 130 but true positive rate decrease 1% compared with model without data augmentation. The model with data augmentation has more stages (5 vs 8) because of more complex data.

4.2. Feature

Figure 3 shows the features of the first three stages, which are clearly focused on the eyes, nose, mouth. The number of candidate feature is restricted by growth step size and moving step size in the detection windows, which has a great impact on memory usage and training time.Table2 shows the different results on number of candidate feature.

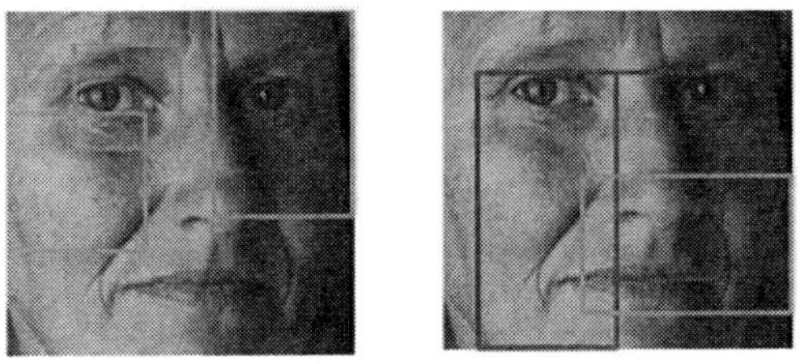

Figure 3. The feature of first three stages.

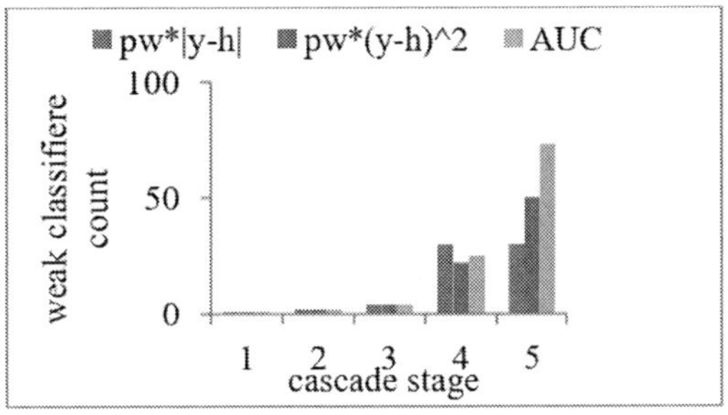

Figure 4. The number of weak classifier.

Table 2. The result of models with different number of candidate feature on FDDB.

candidate feature	TPR/FP=0	TPR/FP=100	TPR/FP=250	TPR/FP-max
482	0.392	0.705	0.742	0.762/506
1100	0.397	0.715	0.748	0.765/458
1631	0.358	0.724	0.749	0.764/414
2143	0.481	0.725	0.755	0.766/421

As the number of candidate feature increases, the true positive rate of the model is gradually improved, and the number of false positive has dropped. However, considering memory usage and training time, compared with number-2143, number-482 is a better choice. As mentioned in 3.1, there are three methods on selecting best feature. All these can rapidly converge to the FPPW 1e-6. But the AUC-based is stricter, and Figure 4 shows that the number of weak classifier of AUC-based is slightly higher. Table 3 shows that AUC-based model has smaller number of false positive. Thus we chose AUC-based method.

Table 3. The result of models with three methods of selecting best feature on FDDB.

select-feature	TPR/FP=0	TPR/FP=100	TPR/FP=250	TPR/FP-max
|pw|	0.471	0.726	0.752	0.771/606
pw^2	0.54	0.727	0.755	0.768/493
auc	0.365	0.725	0.75	0.761/ 401

4.3. Weak Classifier
Here we compare two classification algorithms, logistic regression and linear SVM with solver means with dual and original. Table 4 shows the results of SVM and LR are almost same. With smaller memory usage and faster computation speed, LR with solver means original is more suitable for the weak classifier in this paper.

Table 4. The result of models with different solver means on FDDB.

solver	TPR/FP=0	TPR/FP=100	TPR/FP=250	TPR/FP-max
L2R_LR	0.392	0.705	0.742	0.762/506
L2R_LR_DUAL	0.436	0.716	0.740	0.763/552
L2R_L1LOSS_SVC_DUAL	0.318	0.722	0.746	0.764/489

4.4. Cascade

Threshold of every stage is mainly depends on min_TPR (minimum of true positive rate). To retain more faces, we set high minimum true positive rate, resulting more weak classifiers of every stage and more stages.

Table 5. The result of models with different min_TPR on FDDB.

min_TPR	TPR/FP=0	TPR/FP=100	TPR/FP=250	TPR/FP-max
99.0%	0.159	0.687	0.743	0.746/271
99.3%	0.445	0.708	0.741	0.756/429
99.5%	0.392	0.705	0.742	0.762/506
99.6%	0.465	0.714	0.743	0.769/738
99.7%	0.609	0.724	0.749	0.779/926

Table5 shows the set of min_TPR has great influence on the TPR. Considering the final TPR and the total number of FP, we usually choose min_TPR=99.5% or 99.6%.

For an image with 450×431 pixels, we set scale to 1.1, step size to 4, and minimum of face size to 20×20, so that 200 thousand of the 40×40 detection windows need to be classified. Usually almost windows are rejected on the first few stages.

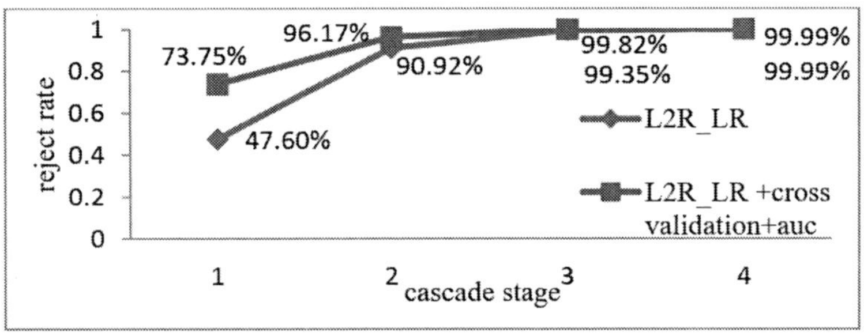

Figure 5. The accumulated reject rate of the first few stages of three different models

Table 6. The recall of models on FDDB.

Models	all2stages	all3stages	all4stages	all stages
L2R_LR	95.20%	92.88%	83.63%	75.99%
L2R_LR+crossvalidaiton+auc	95.57%	90.24%	80.58%	78.18%

Figure 5 and Table 6 show that our model with cross validation and AUC-based has higher reject rate in the first few stages and higher final TPR. The 99% reject rate shows our model has stronger learning ability.

Each parameter setting is a trade-off between detection rate and number of false positive. Compared with HAAR cascade model, as in figure 6, our model has improved a lot on the point FP=0, FP=100.And our model only has 5 stages with 105 weak classifiers. HAAR cascade model has 24stages with 2916 weak classifiers. The true positive rate of our frontal face detector is 77.9%, only missing 2% of frontal faces.

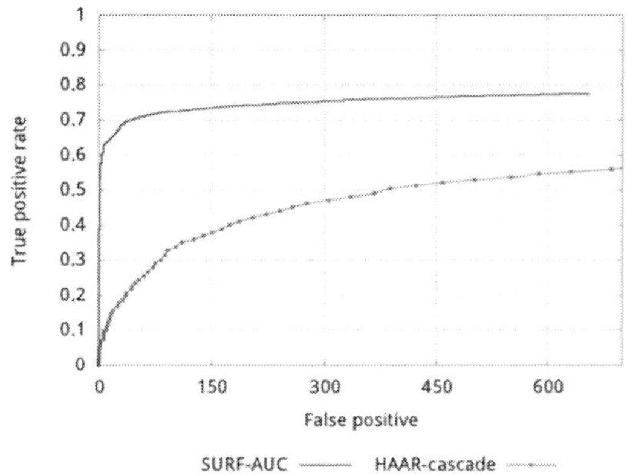

Figure 6. Compare of SURF cascade model and HAAR cascade model on FDDB dataset

5. Conclusions

This paper studies the cascade detection model based on SURF feature. The main contributions are three points. First, we analyze factors restricting the model, including data, feature, weak classifier, and cascade. And we did groups of experiments to make explicit the effect of these parameters. Second, we add cross validation in logistic regression creatively to improve accuracy, greatly speeding up convergence. Third, frontal face detector can achieve results comparable to state-of-the-art detectors using traditional machine learning means. The true positive rates are 77.9% and 60.9% when false positives are max and 0. Our detector decreases number of false positive greatly and outperforms than HAAR-cascade frontal face detector. Future work may consider combine convolutional network with cascade structure on multi-view face detection.

6. References

[1] Zhang C and Zhang Z 2010 *Technical Report* **66** 1-17

[2] Dalal N and Triggs B 2005 *Conf on Computer Vision and Pattern Recognition* vol1 (San Diego) pp886-893

[3] Paul V and Michael J 2001*Proc of the 2nd Int Workshop on Statistical and Computation* Theories of *Vision Modeling Learning Computing and Sampling* vol57 (Vancouver) p 87

[4] Ojala T and Pietikäinen M 2002 *Transactions on Pattern Analysis & Machine Intelligence* **24** 971-87

[5] Wu B and Ramakant N 2005 *Tenth IEEE International Conference* vol1 (Beijing) pp 90-97

[6] Dollár P et al 2009 *British Machine Vision Conf* vol 2(London) p3

[7] Ren S et al 2015*Transactions on Pattern Analysis & Machine Intelligence* 91-99

[8] Z Redmon J et al 2016 *Conf on Computer Vision and Pattern Recognition* (Las Vegas) pp 779-788.

[9] Li J et al 2012 *Int Conf on Computer Vision Workshops* vol21 pp2183-90

[10] Fan RE et al 2008 *Journal of Machine Learning Research* **9** 1871-74

The enhancement of TextRank algorithm by using word2vec and its application on topic extraction

Xiaolei Zuo[1], Silan Zhang[1, 2] and Jingbo Xia[2, 3,4*]

[1]College of Science, Huazhong Agricultural University, Wuhan, China
[2]Institute of Applied Mathematics, Huazhong Agricultural University, Wuhan, China
[3]College of Informatics, Huazhong Agricultural University, Wuhan, China
[4]Hubei Key Laboratory of Agricultural Bioinformatics, Wuhan, China
*Corresponding author, mailto: xiajingbo.math@gmail.com

Abstract. TextRank is a traditional method for keyword matching and topic extraction, while its drawback stems from the ignoring of the semantic similarity among texts. By using word embedding technique, Word2Vec was incorporated into traditional TextRank and four simulation tests were carried on for model comparison. The results showed that the hybrid combination of Word2Vec and TextRank algorithms achieved better keyword/topic extraction towards our testing text dataset.

1. Introduction

With the booming development of new media and the Internet, the text data of unstructured or semi-structured news is proliferating. Extracting effective information from the complex and irregular texts, which can certainly improve daily reading efficiency, is of great significance.

PageRank is a sorting algorithm for webpages [1]. The algorithm assigns a heavier weight to webpage which is more frequently cited by other webpages. In another word, the importance metric of a webpage relies on the amount of the linking resources. Assume $V = \{v_1, v_2, \cdots v_n\}$ is the set of webpages, m_j is the weight for each webpage, $In(V_i)$ refers to set of webpages that link to V_i, and the importance metric of webpage V_i is regarded as $p(v_i)$:

$$p\left(v_i\right) = \frac{1-d}{n} + d \sum_{v_j \in In(V_i)} m_j \times p\left(v_j\right) \tag{1}$$

where d is a factor, which normally is set as 0.85. A drawback of this algorithm is that an outlier will draw significant effect to the result, e.g., a dramatic change of one webpage in $In(V_i)$ will bring dramatic value change. Henceforth, several enhancement were made to enhance PageRank [2, 3].

In light of PageRank, Tarau and Mihalcea proposed TextRank in 2004 [4]. In TextRank, article is divided into basic text units, i.e., words or phrases. As treated as webpage in PageRank, text unit maps to vertex in graph, and edge between vertexes refers to the link between text units.

The research of this paper is to introducing semantic similarity into text unit of sentences so as to achieve better topic representation of a text. Word2Vec [5] are used for the purpose of semantic embedding. Four simulation tests are carried on after the algorithm design and coding. The data and codes are available in GitHub, https://github.com/zuoxiaolei/TextRankPlus.

2. Material and Method

2.1. Material

Literature abstracts are extracted from www.sciencedirect.com. Four type of topics are chosen randomly for keyword evaluation. As shown in Table 1.

Table 1. Literature sampling for evaluating the keywordk extraction

Topic for literature	Filed	Amount of the literature	Keywords from expertise's view
Biomedical natural language processing	Computer science [6]	5	NLP, text mining
Intrusion detection system	Computer science [7]	5	IDS, classifier, evaluation metric
Bioinformatics method for predicting thermophilic proteins.	Bioinformatics [8]	5	Prediction, Pseudo AAC, thermophilic
Thue equation	Mathematics [9]	5	Thue equation

The reason of the topics selection comes from authors' expertise in these wide-range disciplinary fields, based on which we released sufficient amount publications, thus made it accurate to define the relevant keywords/topics for each texts. From the view of our expertise, the keywords for each topic are safely pre-set, as shown in the last column in Table 1, and all of the topics will be tested in our simulation test in the Result section. The text data for evaluation is with proper length that make it sufficient to test the accuracy of keyword/topic extraction. The data is also delivered in Github for free downloading (https://github.com/JingboXia/Enhancement_of_TextRank).

2.2. Method

2.2.1. TextRank algorithm for keywords ranking. TextRank [4] is built in a graph-based unsupervised learning frame, and it has been widely used in keywords extraction and automatic abstracting. The core of TextRank come from vertex voting, where the voting action equals to an edge between two vertexes. The keyword is mapped with higher value, if the vertex it represents has higher relevance with the rest vertexes. In our research, the idea of TextRank is used for keywords ranking among four types of scientific abstracts.

PageRank algorithm for page ranking.

In the beginning, each vertex is assigned with equal weight, and afterwards a recurrent calculation update the weight thought voting. Here, $G=(V,E)$ is the graph, with V being the vertex set, E being the edge set. The importance metric of each vertex is as shown in the formula:

$$W(V_i)=(1-d)+d*\sum_{j\in In(V_i)}\frac{w_{ji}}{\sum_{V_k\in out(V_j)} w_{jk}}W(V_j) \tag{2}$$

where $In(V_i)$ is the set of subscripts for vertexes (text units) which share a common window with V_i in linear order in sentence, $Out(V_j)$ is the set of vertexes which share a common window with V_i, d is a damping factor, and its default value is 0.85. Normally, weight assigned to V_j to V_i, w_{ji}, is counted by calculating the chances of two text units co-occurred in a text window with fixed size, and the usual size equals to 2.

In the initialization step, weight of each text unit is one, and all of the weights reach consistency after recurrent calculation by formula (2). The text units in the top ranking list are considered to be keywords of the text.

The flowchart of the algorithm is shown in Figure 1.

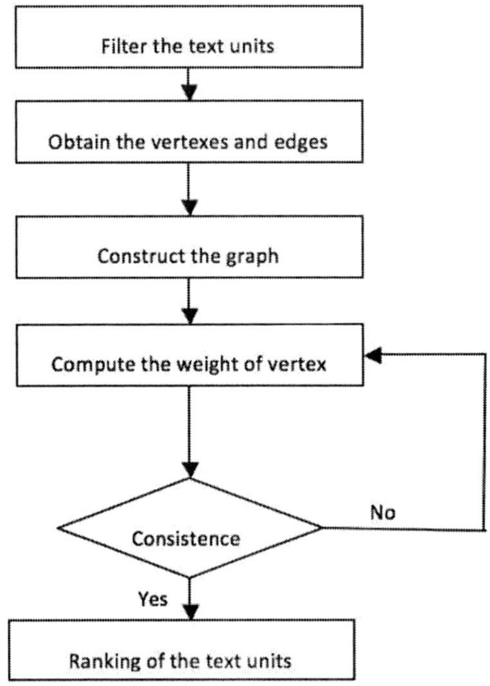

Figure 1. Classic TextRank algorithm workflow

The advantage of TextRank is that it is an unsupervised learning algorithm in no need of huge corpus for training. It make it easy to be adopted for handling other text resources in an efficient way.

2.2.2. Word embedding (CBOW and Skip-gram). The disadvantage of TextRank is that it omit the keywords which has lower chance to appear though being meaningful in context. The natural way to enhance TextRank is to use the semantic similarity of words and avoid the miss selection of vital keywords.

Word2vec is a word embedding algorithm proposed by Google in 2013, which have two varities: CBOW and Skip-Gram. The main idea of Word2vec is to find numerical vector representation of word by using neural networks.

The idea of Word2vec comes from the probability calculation of Bayesian occurrence estimation, Let $T = w_1, w_2, \cdots w_n$ be sentences including n words, the probability of occurrence of the sentence T is:

$$P(T) = \prod_{i=1}^{n} p(w_i \mid w_{i-n-1} w_{i-n-2} \cdots w_{i-1}). \tag{3}$$

Similarly, the Bayesian estimation of the occurrence chance of the i-th word is:

$$p(w_i \mid w_{i-n-1} w_{i-n-2} \cdots w_{i-1}) = \frac{C(w_1 w_2 \cdots w_n)}{C(w_1 w_2 \cdots w_{n-1})}$$

where $C(w_1 w_2 \cdots w_n)$ is the probability of the sentence $w_1 w_2 \cdots w_n$ in the corpus.

In CBOW, context information of each word $w \in W$ is concerned within a ws - width window. The purpose of CBOW training is to maximize the probability of $(w, C_w) \in T$ and minimize the probability of $(w, C_w) \notin T$. Here, the probability of the occurrence of w based on C_w is:

$$p(w \mid c_w) = \frac{1}{1 + e^{-v_{c_w} W_w}}$$

where W_w is the weights matrix connecting the hidden layer and softmax layer, v_{c_w} refers to the sum of the numerical vectors which are in the flank side of the target word:

$$v_{c_w} = \sum_{w \in c_w} v_w$$

The likelihood function of the model is

$$OBJ_{cbow} = arg \max_{v_w, W_1} \left(\prod_{(w,c_w) \in T} log(p(w \mid c_w)) \cdot \prod_{(w,c_w) \notin T} (1 - log(p(w \mid c_w))) \right) \tag{4}$$

Negative sampling strategy is used to obtain $(w, c_w) \notin T$, that make a quicker implementation. And gradient descent algorithm is used for parameter optimization as well.

Similar as CBOW, Skip-gram is to predict the neighbor words, and its objective function is to calculate the greatest average logarithm:

$$\frac{1}{T} \sum_{t=1}^{T} \sum_{-c \le j \le c, j \ne 0} log\, p(w_{t+j} \mid w_t) \tag{5}$$

where c refers to the window width. As shown in Figure 2, CBOW and Skip-Gram own similar ideas and both were considered in our research.

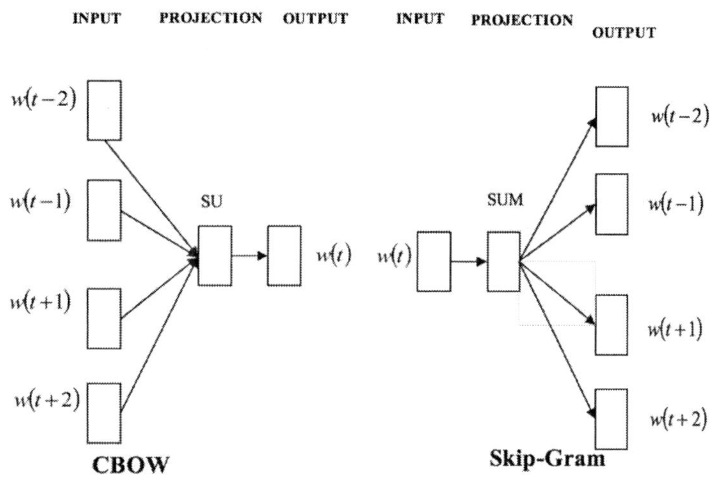

Figure 2. CBOW and Skip-Gram model structures

2.2.3. Enhanced TextRank algorithm. The classic TextRank algorithm only considered the occurrence info of text units in sentences, but omit the semantic understanding of them. Furthermore, the same initial weights is assigned for each text unit, without differentiating the semantic representation of text unit. Henceforth, the word appeared more often is readily be selected to be keyword, regardless its importance. In order to solve this drawback, Word2Vec and TextRank are combined so as to form our proposed algorithm. In this way, lower-rank numerical vector is assigned for each word, and semantic similarity between text units are remained.

For implementation, popular python package Gensim is used for Word2Vec training and model construction (https://radimrehurek.com/gensim/), and Wikipedia Corpus is selected as training corpus. Here, Wikipedia corpus is a domain-free corpus which ensures a model with better generalization ability. Built-in Word2Vec computation in Gensim treate each text unit as a vertex of graph, and the similarities distribution among text units are calculated as edge between vertexs. Finally, in the integration part of TextRank and Word2Vec, the egde value between words are set as semantic similarity. In detail, to replace formula (2) in TextRank, weight value w_{ji} is reset by using semantic similarities counted by Word2Vec distance.

3. Result and Conclusion

3.1. Results of Comparison

The result of comparison among four selected topics are shown in Table 2.

Table 2. Comparison of performance for testing dataset with selected topics

Topics for testing text	TextRank	Our algorithm
Topic 1: Biomedical natural language processing	1. system-oriented evaluation 2. structured information 3. extraction performance 4. underlying biomedical 5. hypothesis generation 6. classification 7. 5-10 8. Enormous	1. biomedical text mining (√) 2. next 5-10 3. undertake user-oriented 4. great promise 5. PubMed search (√) 6. published biomedical 7. past year 8. text mining (√)
Topic 2: Intrusion detection system	1. detection evaluation 2. taxonomy 3. accurate 4. particular 5. knowledge encoding 6. statistical 7. general-purpose intrusion-detection (√) 8. extensive	1. DARPA Intrusion Detection Evaluation (√) 2. Mining Audit Data 3. real-time intrusion-detection expert (√) 4. general-purpose intrusion-detection expert (√) 5. Current IDSs pose 6. collaborative-based wireless IDPS 7. novel R2L 8. Intrusion Detection (√)
Topic 3: Bioinformatics method for predicting thermophilic proteins	1. reliable classifier 2. important sequence 3. thermostability (√) 4. classification 5. sequence 6. Decision 7. predictive successful 8. Matthews correlation	1. pseudo amino acid composition (√) 2. classification rule generator 3. amino acid composition (√) 4. amino acid distribution (√) 5. % non-thermophilic (√) 6. 10-fold cross-validation

		7. 5-fold cross-validation
		8. Decision tree
Topic 4: Thue equation	1. Logarithms	1. f r = g |
	2. u201cThue equationu201d (✓)	2. deg g n · max|
	3. well-known corollary	3. u201cThue equationu201d f (✓)
	4. defined	4. deg f I
	5. practical general	5. + deg g
	6. Theorem	6. f I
	7. integral	7. x =
	8. u201cintegralu201d	8. n =

As can be seen from the table, our proposed algorithm achieved better performance in the simulation test. For the first topic, terms including “Biomedical text mining”, “PubMed Search”, and “text mining” are highlighted in the result. For the second topic, “DARPA intrusion detection evaluation”, “real-time intrusion-detection expert”, “general-purpose intrusion-detection expert”, “Intrusion Detection” were selected as topic keyword. The results are accurate. Comparatively, the original TextRank algorithm mainly picked “general-purpose intrusion detection” as the keyword. In the third simulation test, “PSEAAC”, “AAC”, “amino acid distribution” were correctly extracted from the texts, and these terms referred to exact methodology used in these articles. “Non-thermophilic” was also chosen, which was a exact term to address the research target. Finally, for the topic 4, “Thue equation” were extracted in both our algorithm and TextRank. In this case, the topic extraction is challenging, as the texts are consists of un-regular mathematic symbols that shuffle the order of the natural language representation in sentences.

3.2. The Conclusion of the Result

The result obtained from the simulation result showed that semantic similarity is reliable aid for enhancement of keyword extraction in terms with TextRanking. In light of this effect, further attempt of computational linguistic method will be worthy for our further research.

4. Acknowledgement

This research is funded by the National Natural Science Foundation of China (Grant no. 61202305) and the Fundamental Research Funds for the Central Universities (Project No. 2013PY120).

5. References

[1] Page L, Brin S, Motwani R et al. The PageRank citation ranking: Bringing order to the web[R]. Stanford: Stanford InfoLab, 1999.

[2] Wen C, Huang TZ, Shen ZL. A note on the two-step matrix splitting iteration for computing PageRank. Journal of Computational and Applied Mathematics. 2017 May 1; 315:87-97.

[3] Liu Q, Xiang B, Yuan NJ, Chen E, Xiong H, Zheng Y, Yang Y. An Influence Propagation View of PageRank. ACM Transactions on Knowledge Discovery from Data (TKDD). 2017 Mar 21; 11(3):30.

[4] Mihalcea R, Tarau P. TextRank: Bringing Order into Texts [J]. *UNT*, 2004, 90:404-411.

[5] Mikolov T, Chen K, Corrado G et al. Efficient Estimation of Word Representations in Vector Space [J]. Computer Science, 2013, 103:1301-1309.

[6] Kevin Cohen, Jingbo Xia, Christophe Roeder, Lawrence Hunter. Reproducibility in natural language processing: A case study of two R libraries for mining PubMed/MEDLINE. Workshop on Research Results Reproducibility and Resources Citation in Science and Technology of Language, Collocated with LREC2016 – 10th Language Resources and Evaluation Conference. Portorož, Slovenia. 2016. Pp.6-12.

[7] Li Yinghui, Xia Jingbo, Zhang silan, Yan Jiakai, Ai Xiaochuan, Dai Kuobin. An Efficient Intrusion Detection System Based on Support Vector Machines and Gradually Feature Removal Method. Expert Systems with Applications. 2012, 39(1): 424-430.

[8] Wang De, Yang Liang, Fu Zhengqi, Xia Jingbo. Prediction of Thermophilic Protein with Pseudo Amino Acid Composition: An Approach from Combined Feature Selection and Reduction. Protein & Peptide letters, 2011, 18(7): 684-689.

IOP Conf. Series: Journal of Physics: Conf. Series **887** (2017) 012028 doi:10.1088/1742-6596/887/1/012028

[9] Xia Jingbo, Zhang Silan, Chen jianhua. Simplifying method on Algebraic approximation of certain algebraic numbers. Mathematical notes. 2012, 92(3):383-395. Published in Russian in Matematicheskie Zametki, 2012, Vol. 92, No. 3, pp. 426–439.

A monocular vision system based on cooperative targets detection for aircraft pose measurement

Zhenyu Wang[1], Yanyun Wang[1], Wei Cheng[1], Tao Chen[2] and Hui Zhou[1]

[1]School of Control and Computer Engineering, North China Electric Power University, Beijing 102206, P. R. China
[2]School of Business, Central South University, Changsha 410083, P. R. China
Email: zywang@ncepu.edu.cn

Abstract. In this paper, a monocular vision measurement system based on cooperative targets detection is proposed, which can capture the three-dimensional information of objects by recognizing the checkerboard target and calculating of the feature points. The aircraft pose measurement is an important problem for aircraft's monitoring and control. Monocular vision system has a good performance in the range of meter. This paper proposes an algorithm based on coplanar rectangular feature to determine the unique solution of distance and angle. A continuous frame detection method is presented to solve the problem of corners' transition caused by symmetry of the targets. Besides, a displacement table test system based on three-dimensional precision and measurement system human-computer interaction software has been built. Experiment result shows that it has a precision of 2mm in the range of 300mm to 1000mm, which can meet the requirement of the position measurement in the aircraft cabin.

1. Introduction

The aircraft pose measurement system is an important subject of aircraft monitoring and control. Rich information can be obtained by visual measurement methods, which has a good performance in short distance. Targets detection in visual measurement can be divided into cooperative targets and non-cooperative targets. When measuring based on a cooperative target, a target made with a specific size and shape is used, together with a single camera to require pictures of it. The key problem of monocular vision measurement is target detection and pose solution [5, 7, 8]. Target detection is to detect the feature points for the later calculation. In the aspect of target detection, this paper uses the checkerboard as the target to detect the four corners of the boundary, which transforms the problem into similar P4P problem with four points as the characteristic point[2]. In terms of pose solution, a four-point fixed method based on spatial fixation is adopted.

In order to verify the measurement range and precision of the whole measurement system, a stepper motor is used to test the distance and angle of X, Y and Z in three directions. The reliability and measurement of the algorithm accuracy is verified as well at the same time. The human-computer interaction interface based on MFC (Microsoft Foundation Classes) is developed, and the location information can be observed in real time and the intuitive real-time dynamic curve is displayed. Experiment shows that the chessboard target has more features and is easy to extract the feature reliability, and the extracted feature pixel is sub-pixel level, which can meet the requirement of high precision calculation within 300mm ~ 1000mm.

IOP Conf. Series: Journal of Physics: Conf. Series **887** (2017) 012029 doi:10.1088/1742-6596/887/1/012029

2. System Design

2.1. Hardware System Design

The system consists of simulated aircraft, a specific target, the lens, the camera, the light source, the computer the installation of fixed parts. The camera uses an industrial camera model whose image size required is 1286×962 and pixel size is 3.2μm×3.2μm. The interface is GigE (Gigabit Ethernet). The camera is selected as the computer H0514-MP2 with the focal length of 5mm.Light source is the natural light; I3 processor is chosen for the computer. The hardware system can be used in the determination of the spacecraft's own position. A series of chessboard targets can be set up in the cabin to determine the location of its own aircraft. Hardware build platform is shown in Figure 1.

Figure 1. Hardware build platform

2.2. Software System Design

A block diagram of the software system is shown in Figure 2. In this paper, MFC operating interface is designed in the environment of Windows7 operating environment and Visual Studio2010.As shown in Figure 3, the window for the collection of video images is on the right side and the real-time operating curve is on the left side. The bottom of the interface is the accurate data output value and the results are displayed in the corresponding dialog box. All of these can give an intuitive response to the location of the system's dynamic information. The operation interface is shown in Figure 3.

2.3. Implement Error Analysis

The reason for the error may be that as follows: 1. the accuracy of the target design error that cannot fully meet the size and shape of the system; 2. Estimate the error when extracting the coordinates of the feature points; 3. the error of the displacement table's movement.

2.3.1. Position Error Estimation. According to the error propagation theory, the horizontal measurement error caused by the pixel extraction error δ_{x_u} and the focal length error δ_f is as follows:

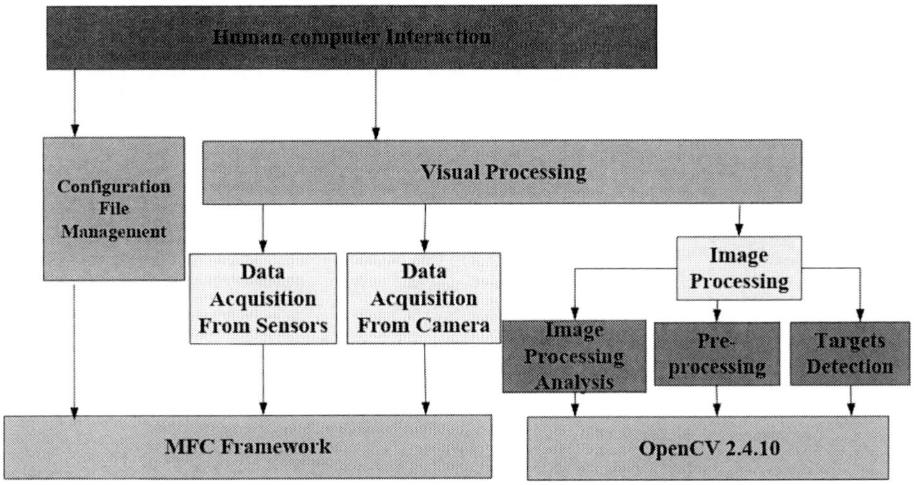

Figure 2. Block Diagram of Software System

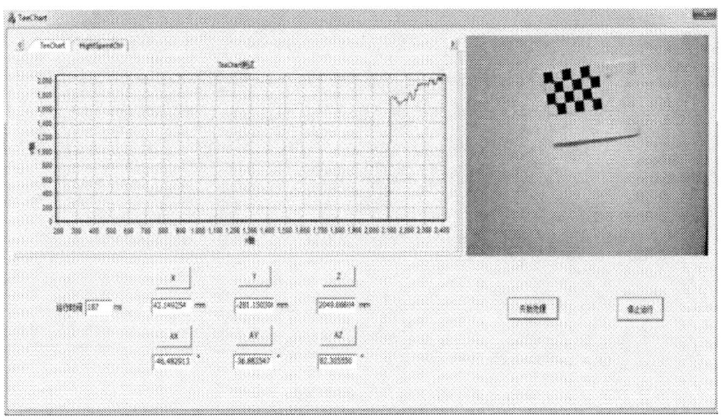

Figure 3. MFC operation interface

$$\delta_{X_c} = \sqrt{(\frac{\partial X_c}{\partial_{x_u}})^2 \delta^2_{x_\mu} + (\frac{\partial X_c}{\partial f})^2 \delta^2_f} = \frac{L}{f}\sqrt{\delta^2_{x_\mu} + \frac{x^2_u}{f^2}\delta^2_F} \quad (1)$$

Likewise, the position error on the Y_C axis is:

$$\delta_{Y_c} = \sqrt{(\frac{\partial Y_c}{\partial_{y_u}})^2 \delta^2_{y_\mu} + (\frac{\partial Y_c}{\partial f})^2 \delta^2_F} = \frac{L}{f}\sqrt{\delta^2_{y_v} + \frac{y^2_v}{f^2}\delta^2_f} \quad (2)$$

The propagation error on the Z_C axis is:

$$\delta_{Z_c} = \frac{L}{f}\sqrt{2\frac{L^2}{R^2}\delta^2_{x_\mu} + \delta^2_f} \quad (3)$$

Where L represents the objects distance.

2.3.2. Angle Error Estimate. Taking the roll angle as an example, the roll angle error estimate is as follows:

$$\delta\phi = \sqrt{(\frac{\partial\phi}{\partial x_u})^2 \delta^2_{x_u} + (\frac{\partial\phi}{\partial f})^2 \delta^2_f} = \sqrt{\frac{L^2}{D^2 f^2 - x^2_u L^2}\delta^2_{x_u} + \frac{x^2_u L^2}{D^2 f^2 (D^2 f^2 - x^2_u L^2)}\delta^2_f} \quad (4)$$

3. Measurement Principle

(1) Known that all control points P_0, P_1, P_2, P_3 are on the plane target π, the corresponding points on the image plane are q_0, q_1, q_2, q_3, $P_0P_1P_2P_3$ are rectangular, where $P_0P_1 \perp P_0P_3$, $P_0P_1 // P_2P_3$, $|P_0P_1| = d_1$, $|P_2P_3| = d_2$;

(2)Spatial coordinates of P_0, P_1, P_2, P_3 in the target coordinate system are known, and the pixel coordinates of q_0, q_1, q_2, q_3 are known as well;

(3) The internal parameter matrix $A = \begin{bmatrix} \alpha_x & \gamma & u_0 \\ 0 & \alpha_y & v_0 \\ 0 & 0 & 1 \end{bmatrix}$ of the camera imaging model is known. Find the geometric transformation between the target coordinate system and the camera coordinate system.

4. Experiments

4.1. Camera Calibration

Camera calibration is the basis for studying visual measurement system. In this paper, Matlab2010a is used to verify the validity of the camera calibration method based on 2D planar target [1, 13]. A custom-made chessboard plate with an edge length of 20mm is used as the 2D planar target. 20 images are taken for each orientation of the target plane. Each image's size is 1286 * 962. The calibration results of the camera parameters are shown in Table 1.

Table 1. Parameter calibration results in the camera

α_x	α_y	γ	u_0	v_0	k_1	k_2
1338.84	**1345.02**	**0**	**618.15**	**475.73**	**-0.1116**	**0.1242**

4.2. Experimental Platform and Measurement Method

In order to verify the accuracy of the pose measurement method, a pose calculation platform is constructed. The test target is fixed on a 6-DOF rotary displacement table and the camera is fixed on the other side of the horizontal displacement table. Move the horizontal displacement table, the value of this position is the true value of this position. Operate of the displacement table to move a certain distance or rotation of an angle, which can be treat as the true value of pose changes. The distance test is performed by the horizontal displacement table. The angle test is accomplished by rotating the displacement table to rotate the target. Calculate the result of the posture output after moving or rotating, and compare the difference between the two calculated values with the displacement table or the rotation true value to obtain the measurement accuracy of the pose measurement system.

Figure 4.1. Target Operation Platform Figure

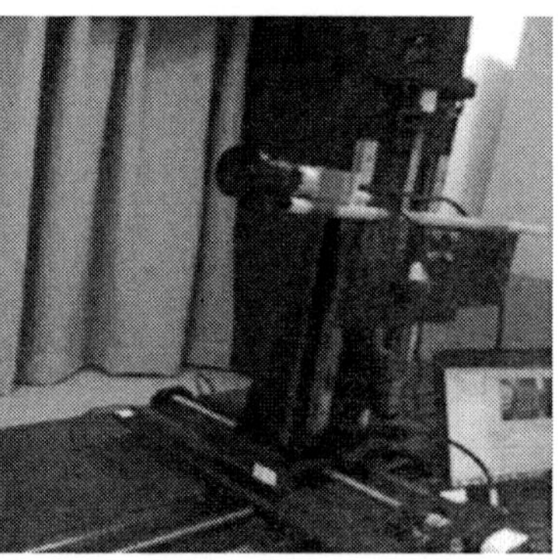

Figure 4.2. Lens Mount Platform

4.3. Experimental Process

According to the above-mentioned distance accuracy measurement method, the Z-direction distance test result is collated and the line chart is drawn as shown in Fig. 5; the Z-direction angle test result is collated and plotted as shown in Fig. 6

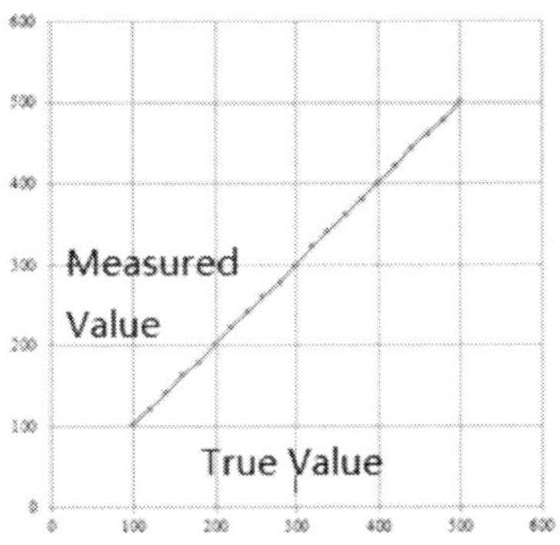

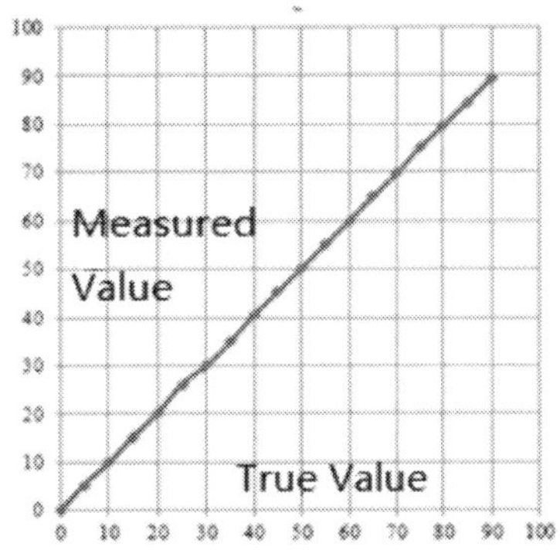

Figure 5.Z-direction distance test line chart (mm) **Figure 6.** Z-direction angle test line graph (°)

According to the experimental method above, five representative test pictures are selected, respectively, to obtain the distance and angle measurements and the true value of the comparison, the distance shown in Table 2, the angle comparison shown in Table 3:

Table 2. Target position relative to the camera data (mm)

	Actual Value			Measured Value			Piston Error		
Image	X_C	Y_C	Z_C	X_C	Y_C	Z_C	X_C	Y_C	Z_C
1	10	10	200	11.8	9.2	201.2	1.8	-0.8	1.2
2	50	50	300	50.4	51.3	300.8	0.4	1.3	0.8
3	80	80	350	78.4	81.2	350.3	-1.6	1.2	0.3
4	120	120	400	121.2	120.8	400.2	1.2	0.8	0.2
5	150	150	500	150.1	151.6	501.1	0.1	1.6	1.1

Table 3. Position and orientation data of target relative to camera (o)

	True Value			Measured Value			Gesture Value		
Image	θ	φ	ψ	θ	φ	ψ	θ	φ	ψ
1	1	10	10	1.3	9.5	9.1	0.3	-0.5	-0.9
2	65	15	15	65.5	16.8	16.1	0.5	1.8	1.1
3	80	25	25	80.1	25.7	25.6	0.1	0.7	0.6
4	90	50	50	90.5	50.3	50.8	0.5	0.3	0.8
5	120	60	60	119.8	60.9	61.4	-0.2	0.9	1.4

5. Conclusions

In this paper, the error of this specific monocular vision measurement system target can reach less than 2mm in the range of 300mm~1000mm for distance positioning; the error of positioning of the angle, Z-axis direction can reach the error of less than 0.5 ° accuracy. The error of X, Y direction can reach the accuracy of less than 2 ° accuracy. All of the precision can meet the requirements of an application scenario. In this paper, the distance and attitude test have been carried out in the experiment and reached the desired index results; this system has been able to achieve real-time measurement of the target now. In practical application, the chessboard can be installed in the vehicle environment, and image information can be also available to capture at any time to determine its location.

IOP Conf. Series: Journal of Physics: Conf. Series **887** (2017) 012029 doi:10.1088/1742-6596/887/1/012029

6. Acknowledgment
This work is supported by National Natural Science Foundation of China (61573139).

7. References

[1] Tang Jian-liang. Some of the solutions for the P4P problem [J] - Journal of Mathematics, 2006,26(2).
[2] Liu Chang, Zhu Feng,Xia Renbo. Monocular vision localization based on coplanar two point line feature [J]. Computer application research,2012,08:3145-3147.
[3] Zhao Lianjun. Research on monocular vision position and attitude measurement based on target feature [D]. Graduate University of Chinese Academy of Sciences (Institute of Opto Electronics Technology),2014.
[4] Zhou Fuqiang.2002. Research on Key Technologies of binocular stereo vision inspection. Postdoctoral Research Report of Beihang University.
[5] Zhu Daixian. Research and grasping based on stereo vision positioning [J]. Computer measurement and control, 2011, 19(1): 92·94.
[6] Feng Chun. Research on target recognition and localization based on monocular vision [D]. Nanjing University of Aeronautics & Astronautics,2013.
[7] Sheng Zunbing. Research on Key Technologies of machine vision image detection and localization system [D]. Harbin Institute of Technology,2009.
[8] Yu Xiaoqiao, Hu Zhanyi. Research and development of camera self-calibration method. Journal of automation, 2003, 29(1): 110-124.
[9] Ponce, J., Hebert, M., Schmid, C., Zisserman, A.(eds). Toward Category-Level Object Recognition [M]. New York: Springer,2006. 3-28.
[10] Rothganger, F., Lazebnikl, S., Schmid, C., and Ponce, J. 3D objective modeling and recognition using local affine-invariant image descriptors and multi-view spatial constraints [J]. International Journal of Computer Vision, 2006,66(3): 231-259.
[11] Ferrari,V., Tuytelaars, T., Van, Gool, L. Simultaneous object recognition and segmentation from single or multiple model views[J]. International Journal of Computer Vision, 2006,67(2):159-188.
[12] Sivic, J, Zisserman, An efficient visual search of videos cast as text retrieval [J]. IEEE Transactions on Pattern Analysis and Machine Intelligence, 2009, 31(4): 591-606.
[13] Mikolajczyk, K, Schmid, C. A performance evaluation of local descriptors [J]. IEEE Transactions on Pattern Analysis and Machine Intelligence,2005,27(10): 1615-1630.

IOP Conf. Series: Journal of Physics: Conf. Series **887** (2017) 012030 doi:10.1088/1742-6596/887/1/012030

Maturity stage distinction of pear based on visible/near infrared spectroscopy technology

Weiwei Xu [1], Min Lin [1,*], Yongmei Huang [1], Huijun Liu [1] and Xinqi Zhou [2]

[1]College of Metrology and Measurement Engineering, China Jiliang University Hangzhou, China
[2]Focused Photonics(Hangzhou),Inc.,Hangzhou 310052,China
*Corresponding author, e-mail address: linm@cjlu.edu.cn

Abstract. This paper selected pear as the research object, established a model for distinguishing different maturity stages of pear which bases on visible/near-infrared spectroscopy, discussed the influence of four types of spectral pretreatment, such as the Savitzky-Golay (SG) smoothing, first derivative correction, second derivative correction, and multiplicative scatter correction(MSC) on the model, and verified that competitive adaptive reweighted sampling(CARS) algorithm can effectively extract the effective information from the spectrum. Maturity stage distinction model of pears was established based on Visible/Near Infrared Spectroscopy Technology by using partial least squares discriminant (PLSDA). It analysed four different commen pretreatment methods to improve the model effect. The result shows that the first derivative correction carries the best predicted accuracy, and the recognition rate is up to 86.11%.The information from the spectrum was filtrated from the pretreated spectrum by adopting CARS algorithm, to optimize the maturity stage distinction model of pears which was established before. The results demonstrates that the CARS algorithm can actually extract effective information, and the recognition rate is up to 97.22%.

1. Introduction

The pear which is used as the research object is mature in mid-august every year. If the picking is premature, the maturity is not enough, the fruits' quality will be poor, on the contrary, it will make against the storage and transportation. Both of them will result in the unsatisfactory economic income, so timely picking can maximize the economic benefit. Accurately judging the maturity stage of the fruit can not only be used to arrange the optimal time of picking, but also make for cultivating good seeds, storage, transportation and further processing[1].

Traditional fruit maturity differentiation method is roughly divided into two kinds [2]. One is that the practitioner judge based on the experience. The other is the physical and chemical index detection instrument. The former has the advantages of quick and intuitive, while accuracy is the biggest drawback. The latter has the advantages of the former, also with high accuracy, small volume, light weight, easy to carry, but still not perfect. Currently, hardness tester and the sugar meter are generally used to discriminate the fruit maturity stage, both of the two kinds of physical and chemical indicators testing instruments will damage fruit samples. Therefore, it can only take random sampling method, and the result has the occasionality, rather than universal because of the destructive.

This paper puts forward a pear maturity near-infrared detection method based on CARS algorithm, which is a fast and intuitive, nondestructive, high accuracy and universal assessment method of fruit maturity stage [3]. It not only includes the advantages of the traditional assessment method of maturity, but also overcomes the disadvantages.

2. Theory

2.1. Introduction of Spectral Pretreatment

This paper adopts four pretreatment methods, which is first derivative correction, second derivative correction, the Savitzky-Golay smoothing, and multiplicative scatter correction. Signal smoothing is a common method to eliminate noise, which can improve the signal-to-noise ratio (SNR) of high frequency noise, and the selection of filter width is very important [4]. First derivative correction and second derivative correction both can effectively eliminate interference from other background, distinguish overlapping peaks, and improve resolution and sensitivity [5]. Multiplicative scatter correction is used to eliminate the influence of scattering on the spectrum.

2.2. Competitive Adaptive Reweighted Sampling

Near infrared spectrum usually consists of a large number of data points. Wavelength points is far more than sample number, and the collinearity of the spectrum is very serious. The effective information is very weak, and the noise level is not consistent. Therefore, wavelength variable screening is a necessary pretreatment method. CARS algorithm is a new method of variable selection which is based on Darwin's survival of the fittest theory. The method can pick out the maximum absolute value of regression coefficient through each cycle constructed model by using the exponential decay function, and get the subset which has the minimum root-mean-square error of cross-validation (RMSECV) among N subsets by using cross validation. It is defined as optimal variable subset. This method can not only remove the non-information variables in the original spectral data, but also compress and remove the collinear variables, which can effectively be used in the selection of high spectral data variables [6].

2.3. Partial Least Squares Discrimination Analysis

PLSDA is a multivariate statistical analysis method for discriminant analysis. Discriminant analysis is a common statistical analysis method based on observation or measurement to determine how to classify the object. PLSDA can reduce the influence of multiple collinearity among the variables.

3. Experiment

3.1. Technological Process

The research process of this paper has these following steps: preparing the pear samples, obtaining the physical and chemical indicators, collecting spectra, selecting the best pretreatment method, extracting the characteristic wavelength of the pretreated spectrum by using CARS algorithm, and establishing the calibration model. That is the way to identify the maturity stages of the unknown pear samples.

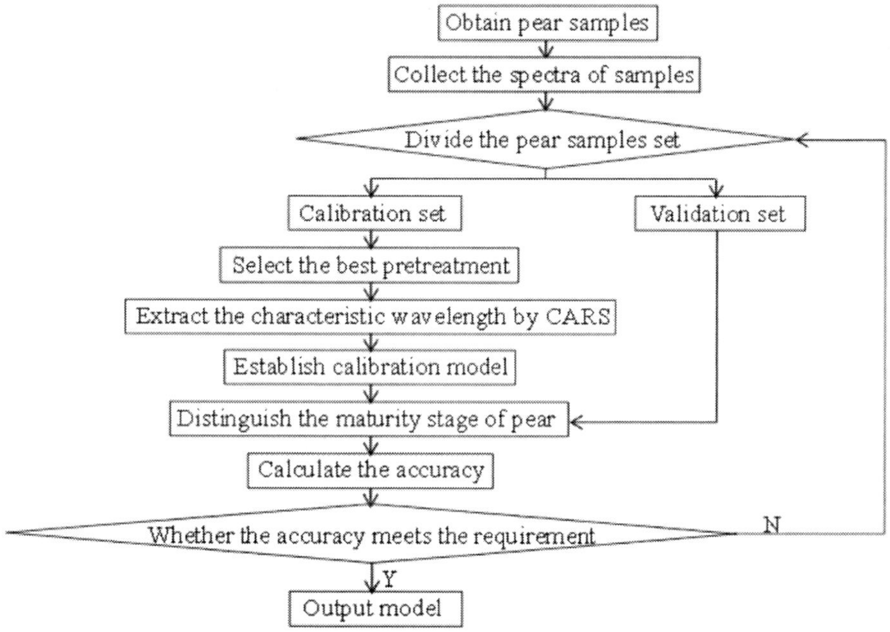

Figure 1. The flowchart of establishing a maturity stage discrimination model of pears

3.2. Sample Collection

The pear samples for experiment is from Hangzhou binjiang fruit industry co., LTD. There are totally 144 pear samples which should be numbered with 108 randomly selected as the calibration set and the rest of the 36 as the prediction set. The pear samples are divided into three maturity stages. As shown in Figure 2, it is the diffuse reflectance spectra Comparison of pear with three different maturity stages. The spectral range of the pear diffuse reflection spectra of the three maturity levels is 450nm - 1650nm.

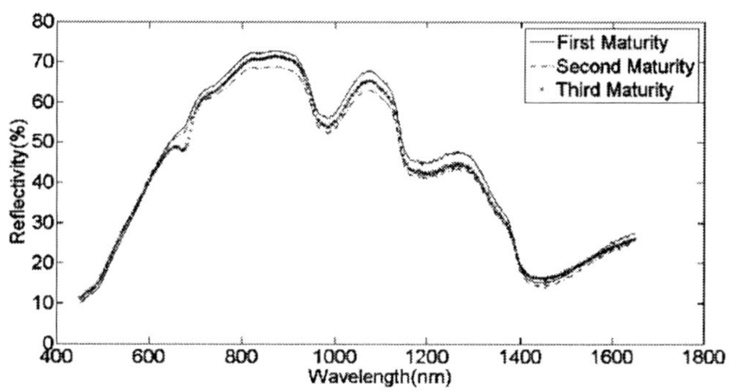

Figure 2. The diffuse reflectance spectra comparison of pear with three different maturity stages

3.3. Spectra Collection

The spectra collection instrument is a German Zeiss spectrometer with MCS600, and the spectrum collection area is 190nm - 2150 nm. The experiment is conducted at room temperature. In the experiment, each sample was scanned 12 times at the equator, and Ba_2SO_4 was used as the reference sample to take the mean spectrum. To avoid stray light interference, a 1.5mm gasket was used to protect the spectrum. The spectrum acquisition software is Aspect Plus.

3.4. Spectra Pretreatment

To reduce the noise in the spectrum and the useless information, this subject adopts four different spectral preprocessing methods, respectively, for the Savitzky-Golay smoothing, first derivative correction, second derivative correction, and multiplicative scatter correction. The calibration model of the maturity discrimination of pear was established by using the Partial least squares discrimination analysis method, and 36 pear samples of the forecast set are used to test the effect of the model.

During the experiment, it is found that combining the original pretreatment method with autoscale will influence the model effect. Therefore, the results of the model prediction of single pretreatment mode and further autoscale processing will be discussed together. In addition, two kinds of multiple scattering correction are used to pretreat the spectrum: MSC (mean) and MSC (median). The former uses the average spectrum, while the latter uses the intermediate value spectrum. Partial modeling results are shown in table 1.

Table 1. The model effect of different pretreatment modes

Pretreatment mode	Filter width	Principal divisor number	Accuracy (%)
First derivative correction	11	6	80.56
First derivative correction &Autoscale	9	4	86.11
Second derivative correction	17	10	86.11
Second derivative correction &Autoscale	19	3	80.56
SG Smoothing	19	3	80.56
MSC(mean)	25	9	75.00
MSC(mean)&Autoscale	\	4	69.44
MSC(median)	\	7	75.00
MSC(median)&Autoscale	\	9	75.00

Table 2. The prediction result of the pear maturity identification model when pretreatment is the first derivative with filter width with 9 combined with autoscale

	Actual Class		
	Class 1	Class 2	Class 3
Predicted as Class 1	15	1	0
Predicted as Class 2	1	10	1
Predicted as Class 3	0	2	6

According to Table 1, the model effect of SG smoothing is not ideal in general. The recognition rate is low, and the model effect of first derivative correction pretreatment is significantly better than that of SG smoothing. The pretreatment method that combines first derivative correction with autoscale is superior to the single first derivative correction treatment, and the model effect of second derivative correction is inferior to the first derivative correction, but better than the SG smoothing. The effect of autoscale on second derivative correction pretreatment is negligible. However, for multiple scattering correction, further autoscale processing can be optimized. To sum up, the best pretreatment method is first derivative correction with filter width of 9 combined with autoscale, the predicted results are shown in Table 2, and Class1 is the first maturity stage, Class2 is the second kind of maturity, and Class3 is the third kind of maturity.

3.5. Characteristic Wavelength Extraction

The spectra of the first derivative correction (filter width of 9) with autoscale pretreatment are extracted and the characteristic wavelength will be picked out by the CARS algorithm. According to figure 3, with the increase of the number of running times, the number of wavelengths retained has decreased, indicating that the useless information in the spectrum has been removed, and it is a selection process. When the number of runs was 24, the RMSECV is the smallest, 0.4062%, and the retention wavelength number is 33. When the number of runs is greater than 24, the RMSECV becomes larger. It means that the valid information in the spectrum is also eliminated.

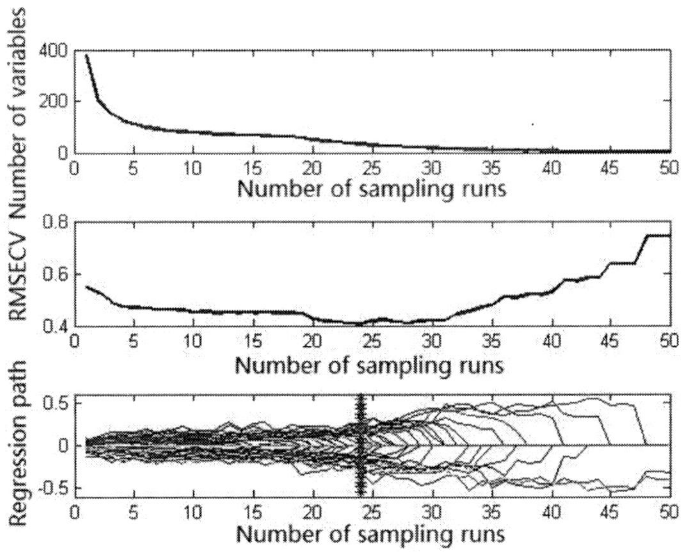

Figure 3. The variables selection of CARS with the first derivative correction (filter width of 9)

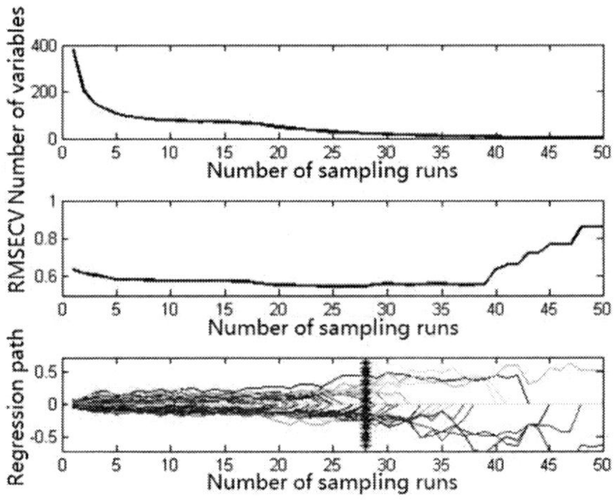

Figure 4. The variables selection of CARS with the first derivative correction (filter width of 17)

In addition, the model prediction result with the pretreatment method which is first derivative correction (filter width of 17) combined with autoscale is also good. Therefore, it is also carried out by the CARS algorithm to optimize the model, and the results are shown in figure 4. When the number of runs is 28, the RMSECV is the minimum, which is 0.5470%, and the number of retention wavelength is 21. The retention wavelength is less than that of the first derivative (filter width of 9), and the value of RMSECV is also greater.

3.6. Establish and Verify the Model

By using the CARS algorithm, the characteristic wavelength of the pretreated spectrum was screened and the model was reestablished. The pretreatment was no longer considered because the number of wavelengths was reduced. It can verify the model effect through forecast set,and when the pretreatment is first derivative correction (filter width of 9) combined with autoscale, the pear maturity identification model recognition correct rate is 97.22%. There is only a second stage pear sample is mistaken for the third, which is shown in table 3.

Table 3. The model prediction effect of the first derivative correction (filter width of 9) by CARS

	Actual Class		
	Class 1	Class 2	Class 3
Predicted as Class 1	16	0	0
Predicted as Class 2	0	12	0
Predicted as Class 3	0	1	7

However, the correct recognition rate of the pear maturity test model of the first derivative (filter width of 17) combined with autoscale is only 88.89%, which is shown in table 4.

Table 4. The model prediction effect of the first derivative correction (filter width of 17) by CARS

	Actual Class		
	Class 1	Class 2	Class 3
Predicted as Class 1	14	0	0
Predicted as Class 2	2	11	0
Predicted as Class 3	0	2	7

4. Conclusion

Difference of harvest period can cause noteworthy effect to the economic benefit. In order to improve the economic benefit of pears, this paper proposes a pear distinction method of near infrared detection which is based on CARS algorithm. It can help to arrange the best harvest of pear, as much as possible to guarantee the quality of the pear.

This paper compares the influence of four kinds of commonly used pretreatment methods on model, and concludes that the first derivative correction (filter width of 9) combined with autoscale pretreatment's effect is the best. Then it establishes a pear maturity stage identification model with good effect by using the pretreated spectrum, CARS algorithm and partial least squares discriminant analysis.

Contrast to the PLSDA model with full spectrum which does not extract characteristic wavelengths, those simplified by CARS algorithm has a greater accuracy of pear maturity identification, which raises from the original 86.11% to 97.22%. This shows that the CARS algorithm optimizes the model of the project.

5. Acknowledgment

This work is supported by National major scientific instruments and equipment development projects (2014YQ470377). The authors awfully thanks for anonymous reviewers who made constructive comments.

6. Reference

[1] Zhou R and Li Y 2015 *J. Transactions of the CSAE.* **23** 255-259
[2] He Y, Wei Y and Zheng Y 2010 *J. Northern Horticulture.* **33** 208-212

[3] Ying Y 2006 *J. Spectroscopy and Spectral Analysis.* **26** 63-66
[4] Li M 2013 *J. Journal of Green Science and Technology.* **50** 215-218
[5] Xia J, Lu Y and Su Y 2015 *J. Acta Tabacaria Sinica.* **23** 19-22
[6] Wu J and Xu Y 2011 *J. T Chin Soc Agric Mach.* **42** 162-166

Hybrid feature selection for supporting lightweight intrusion detection systems

Jianglong Song, Wentao Zhao*, Qiang Liu and Xin Wang

College of Computer, National University of Defense Technology, Changsha, China
Email: songjl@nudt.edu.cn, wtzhao@nudt.edu.cn, qiangliu06@nudt.edu.cn, dywangxin@foxmail.com

Abstract. Redundant and irrelevant features not only cause high resource consumption but also degrade the performance of Intrusion Detection Systems (IDS), especially when coping with big data. These features slow down the process of training and testing in network traffic classification. Therefore, a hybrid feature selection approach in combination with wrapper and filter selection is designed in this paper to build a lightweight intrusion detection system. Two main phases are involved in this method. The first phase conducts a preliminary search for an optimal subset of features, in which the chi-square feature selection is utilized. The selected set of features from the previous phase is further refined in the second phase in a wrapper manner, in which the Random Forest(RF) is used to guide the selection process and retain an optimized set of features. After that, we build an RF-based detection model and make a fair comparison with other approaches. The experimental results on NSL-KDD datasets show that our approach results are in higher detection accuracy as well as faster training and testing processes.

1. Introduction

Intrusion detection is the process of observing the events occurred in the network or computer systems and examining them for signs of possible events that violate the security policies or security standard practices [1]. Intrusion Detection Systems (IDS) have become important and widely-used tools for ensuring network security. Since the amount of audit data that an IDS needs to examine is very large even for a small network, analysis work becomes difficult for harder detection of suspicious behaviour pattern with extraneous features [2]. IDS must reduce the amount of data to be processed to improve detection efficiency, which desires a lightweight detection system. Feature selection is one of the important and frequently-used techniques in data preprocessing for intrusion detection systems [3]. It reduces the number of features, removes irrelevant, redundant or noisy data, and brings the immediate effects for applications.

Many literatures have tried to figure out important features in order to minimize overhead of detection models and maximize detection rates. Dong et al. [4] proposed a feature selection method based on Genetic Algorithm (GA) and Support Vector Machine (SVM). Hai et al. [5] proposed an automatic feature selection procedure based on Correlation Feature Selection (CFS). Zhu et al. [6] a binary feature selection (BFS) framework in kernel spaces, where each feature is projected into kernel spaces and a binary classification task is constructed in this space. Ambusaidi et al. [7] proposed a mutual information based algorithm that analytically selects the optimal features for intrusion detection. Tama et al. [8] combined particle swarm optimization and correlation-based feature selection (PSO-CFS) with tree-based classifiers for detection task. Thaseen et al. [9] designed an intrusion detection model using fusion of chi-square feature selection and multi class SVM. Dong et al. [10] proposed a new approach to build lightweight intrusion detection system based on RF with a measure of feature importance. Besides this, some PCA (Principal Component Analysis) and ICA (Independent Component Analysis) approaches also have been proposed to decrease overhead of IDS and increase the detection rates [11].

Those methods can fall into two categories: filter and wrapper. Filter methods utilize the underlying characteristics of the training data to evaluate the relevance of the features by some independent measures such as distance measure, correlation measures and consistency measures [12][13]. It is more lightweight than wrapper method in terms of computation time and overheads since it is performed independent of classification algorithms. However, their results are not always acceptable. Due to the lack of interaction between the classifier and the dependence among features, filter methods might fail to choose the best available subset or might select redundant features [14]. Wrapper method exploits a machine learning algorithm as a fitness function to evaluate the goodness of features. It provides better performance of selecting suitable features since it employs performance of learning algorithm as an evaluation criterion. In comparison with filter methods, wrapper methods are argued to be more accurate. However, the wrapper approaches are often computationally more complicated when dealing with large sets of features than the filter approaches [15].

To cope with the aforementioned drawbacks, we propose to adopt the principle of chi-square feature selection and RF algorithm to build a lightweight intrusion detection model in a hybrid manner. The proposed approach is able not only to significantly decrease training and testing time while retaining high detection rates but also to figure out important features simultaneously. Then, we design an RF-based detection model framework with the proposed hybrid feature selection method and conduct several experiments on intrusion detection benchmark. The experimental results on NSL-KDD datasets indicate the feasibility of our method.

This paper is organized as follows. In Section 2, the proposed hybrid feature selection approach is introduced. Experiments and results are presented in Section 3. Finally, the paper is concluded in Section 4.

2. Hybrid Feature Selection Method

In this section, we introduce the proposed hybrid feature selection method which utilizes both filter and wrapper methods. The framework of the approach is depicted in figure 1. It consists two stages: the first stage at which the chi-square statistics is used for feature ranking and elimination, and the second stage which determines the optimal subset using RF algorithm, and contributes maximum classification accuracy on training dataset.

2.1. Filter Method for Feature Pre-Selection

The χ^2 method [16] evaluates features individually by measuring their chi-square statistic with respect to the class labels. The χ^2 value of a feature is defined as:

$$\chi^2 = \sum_{i=1}^{t}\sum_{j=1}^{l}\frac{(A_{ij}-E_{ij})^2}{E_{ij}} \tag{1}$$

where

$$E_{ij} = R_i * L_j / S \tag{2}$$

t is the number of different values in a feature, l is the number of class labels, A_{ij} is the number of samples in the i-th feature value and j-th class, R_i is the number of samples in the i-th feature value, L_j is the number of samples in the j-th class, S is the total number of samples and E_{ij} is the expected frequency of A_{ij}.

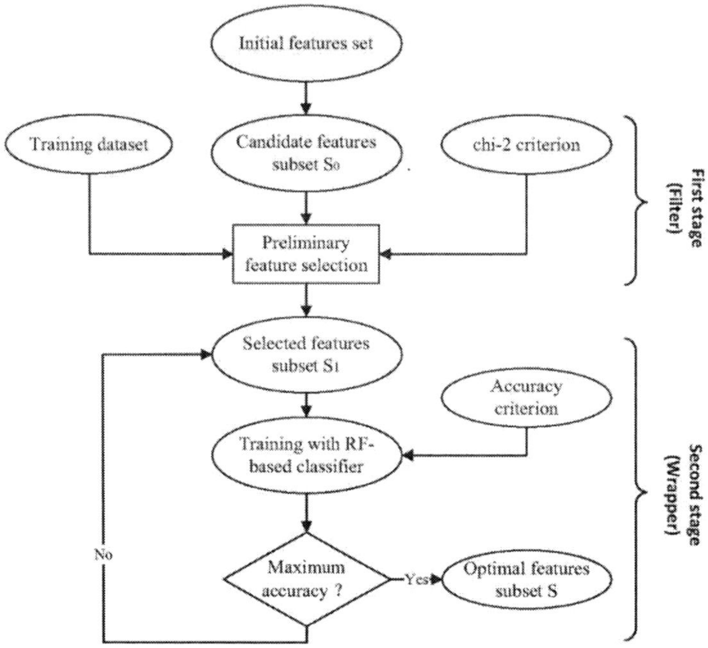

Figure 1. The framework of hybrid feature selection.

After calculating the χ^2 value of all considered features in each sample, these values can be sorted with the largest one at the first position, as the larger the χ^2 value, the more important the feature is. This will provide an ordered ranking of features, and a threshold needs to be established. The filtering process is applied to eliminate irrelevant and redundant features from the initial set. This helps the wrapper method-based feature selection to narrow the searching range from the entire original feature space to the pre-selected features.

2.2. Wrapper Feature Selection Using RF

Random Forests builds an ensemble of CART (Classification and Regression Tree) tree classifications using bagging mechanism [17]. By using bagging, each node of trees only selects a small subset of features for the split, which enables the algorithm to create classifiers for high dimensional data very quickly. This somewhat counterintuitive strategy turns out to perform very well, compared with the state-of-the-art methods in classification and regression. Also, RF produces additional facilities, especially the feature importance by numerical values [18].

Once the filter method finishes its task, the second stage evaluates the candidate feature subsets, using wrapper scheme, to determine the optimal subset of feature that can produce the best classification performance. To do so, RF and the classification accuracy are employed. If the performance reaches the best accuracy rate, the selection process is completed while outputs the last optimal subset of features. The procedure is specified as follows:

(i) Delete one input attribute from the data.
(ii) The resultant data are used for training.
(iii) The results of the classifier are analyzed using the performance metrics.
(iv) The rules are used to rank the attribute by its importance level.
(v) Repeat the steps (i) to (IV) for each of the features.

2.3. RF-Based Intrusion Detection Framework

In this paper, for demonstrating the efficiency and effectiveness of our approach, we build an intrusion detection model based on RF using the features selected by our proposed hybrid feature selection method. The overall architecture of the model is depicted in figure 2.

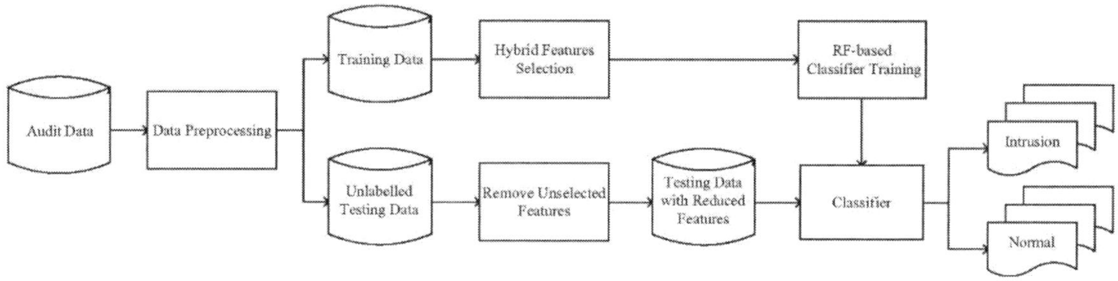

Figure 2. The framework of the RF-based intrusion detection model.

3. Experiments and Results

All experiments were performed in a Win7 OS having configurations Intel(R) Core(TM) i7-4790K CPU@4.00GZ, 16GB RAM. We have used RF version in an open machine learning library scikit-learn using Python programming language [19] [20].

3.1. Description of the Benchmark Dataset

The NSL-KDD dataset was proposed by Tavallaee et al. [21], which is a new revised version of the KDD Cup 99 dataset. The important deficiency in the KDD Cup 99 dataset is the huge number of redundant records that will cause learning algorithms to be biased towards the more frequent records, and thus prevent it from learning infrequent records. The existence of these repeated records in the test set, on the other hand, will cause the evaluation results to be biased by the methods which have better detection rates on the frequent records [21]. The provided NSL-KDD dataset does not suffer from any of the mentioned problems. Furthermore, the number of records in the train and test sets is reasonable. This advantage makes it affordable to run the experiments on the complete set without the need to randomly select a small portion. Consequently, evaluation results of different research work will be consistent and comparable.

The total number of records in the training dataset is 125973, where 67343 records are normal data and the rest indicate attacks. The total number of features is 41, which include numerical, nominal, and binary features. This dataset consists of five different classes, where one shows normal behavior and the rest indicate attacks. Attacks are categorized as DoS, Probe, R2L, and U2R. The test set consists of 22544 records. In this paper, 20% percent of the full training data is applied to the proposed method. Furthermore, to validate the performance of our hybrid feature selection, we test our RF-based detection model with the proposed hybrid feature selection on KDDTest-21 dataset and make a comparison with other detection model.

3.2. Performance Metrics

In this paper, we classify the dataset into two main classes: normal and abnormal. In order to classify them, we use RF to train a binary classifier. Several experiments have been conducted to evaluate the performance of the proposed method. For this purpose, the accuracy rate, detection rate, and F-measure metric are applied, and they are defined by:

$$Accuracy = \frac{TP+TN}{TP+TN+FN+FP} \tag{3}$$

$$DetectionRate = \frac{TP}{TP+FN} \tag{4}$$

where True Positive (TP) is the number of actual attacks classified as attacks, True Negative (TN) is the number of actual normal records classified as normal ones, False Positive (FP) is the number of actual normal records classified as attacks, and False Negative (FN) is the number of actual attacks classified as normal.

The F-measure metric is a statistical technique for examining the accuracy of a system by considering both precision and recall of the system. F-measure used in this paper assigns the same weights to both Precision and Recall, and is given by:

$$F-measure = \frac{2(Precision * Recall)}{Precision+Recall} \tag{5}$$

The Precision is the ratio of items correctly classified as X to all items classified as X. A higher value of precision means a lower false positive rate and vice versa. The Recall is another important value for measuring the performance of the detection system and to indicate ratio of items correctly classified as X to all items that belong to class X [22].

$$Precision = \frac{TP}{TP+FP} \tag{6}$$

$$Recall = \frac{TP}{TP+FN} \tag{7}$$

3.3. Experimental Results and Analysis

We compare the results of the RF-based detection model using the proposed hybrid feature selection method with the detection model using all 41 features and the chi-square filter algorithm only. Features selected by the aforementioned three feature selection methods are listed in Table 1. Table 2 summarizes the classification results of the different selection methods with respect to detection rate, accuracy and F-measure. Through Table 2, we can see that our detection model with the proposed hybrid method achieve the highest accuracy rates of 80.07%. In addition, the proposed approach achieve detection rate of 71.21%. In general, in terms of the F-measure results for all methods, the detection model with the proposed hybrid feature selection enjoys higher rates.

Table 3 shows the average training and testing time of the detection model with the proposed hybrid feature selection, compared with using only filter method and those using all 41 features. It can be observed that the proposed approach illustrates the best average time of building and testing processes.

Table 1. Feature importance with three different feature selection methods.

Algorithm	#	Feature ranking
Proposed hybrid method	7	*f5, f11, f4, f3, f2, f29, f35*
Filter method	30	*f4,f5,f33,f11,f3,f1,f34,f38,f2,f31,f22,f35,f27,f32,f29...*
All features	41	*f3,f4,f28,f5,f33,f29,f25,f2,f36,f31,f1,f24,f22,f11,f32,f35,f23...*

Table 2. Performance of classification on the KDD Test+ dataset.

RF model with:	Detection rate	Accuracy	F-measure
Proposed hybrid method	**71.21**	**80.07**	**80.27**
Filter method	64.49	78.57	77.41
All features	65.71	79.23	78.28

Table 3. Average time of building and testing processes on the KDD Test+ dataset.

	RF model with:	Accuracy time
Building time(ms)	Proposed hybrid method	**110**
	Filter method	156
	All features	172
Testing time(ms)	Proposed hybrid method	**160**
	Filter method	310
	All features	330

In order to demonstrate the performance of the RF-based model with our hybrid feature selection, experiments have been conducted to make comparisons with some state-of-the-art detection model on the KDD Test+ dataset. The results illustrates in figure 3 strongly indicate that the proposed detection model shows promising results compared with other models. Regarding the results obtained by [21], it can be seen that the proposed approach enjoys the best accuracy and our proposed method can significantly improve the classification performance (up to 12.38 % improved) compared with other six competitors.

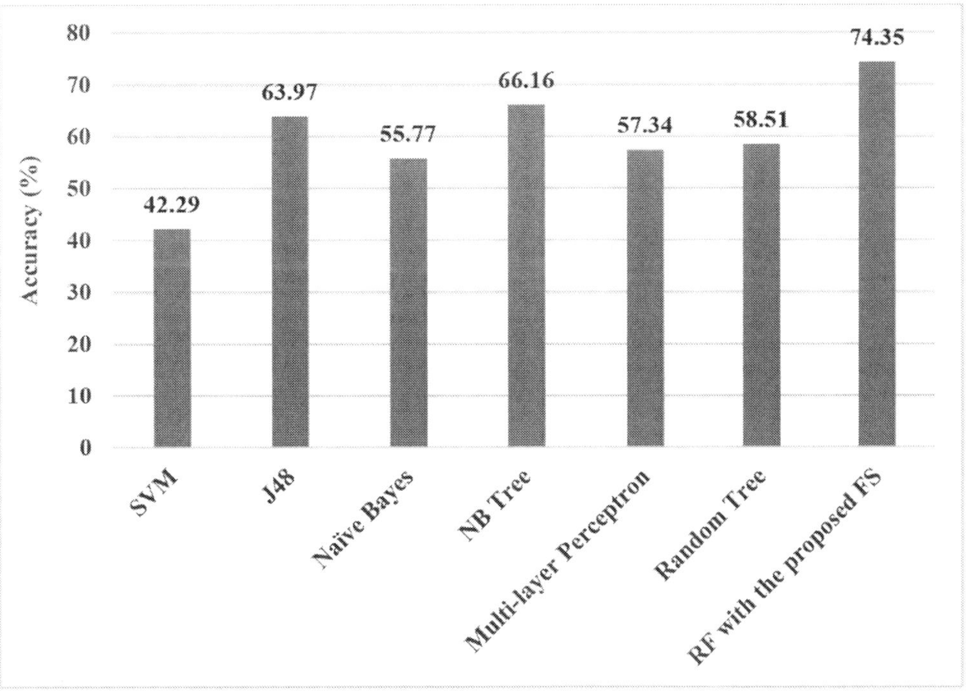

Figure 3. The detection accuracy of different models on the KDDTest-21 dataset

4. Conclusion
In this paper, for building a lightweight intrusion detection model, a hybrid feature selection approach combining the filter and wrapper selection processes is proposed to achieve more accurate detection as well as fast training and testing process. It contains two stages: (1) chi-square filter feature eliminating phase; and (2) wrapper feature selection using RF. Experiments on the NSL-KDD Test+ dataset exhibit promising results in terms of classification accuracy, low computational cost and F-measure. In addition, compared with those models that have been evaluated on the NSL-KDD Test-21 dataset, the RF-based detection model with the proposed hybrid method shows good a comparable result in terms of detection accuracy.

5. Acknowledgments
This work is partly sponsored by the National Defense Key Laboratory Fund Project under Grant No.1602068614001.

6. References
[1] Barry B I A and Chan H A 2011 *Intrusion Detection Systems* (Berlin, Heidelberg: Springer)
[2] Lee W, Stolfo S J and Mok K W 1999 *Proc. of the 1999 IEEE Symp. on Security and Privacy (Oakland)* vol 00 (Los Alamitos: IEEE Computer Society) pp 120–132
[3] Chen Y, Li Y, Cheng X Q and Guo L 2006 *Current Medical Research and Opinion* **44** 1309–18
[4] Dong S K, Nguyen H N, Ohn S Y and Park J S 2005 *Advances in Neural Networks - ISNN 2005, 2nd Int. Symp. on Neural Networks (Chongqing)* vol 3498 (Berlin, Heidelberg: Springer) pp 415–420
[5] Nguyen H T, Franke K and Petrovic S 2011 *Int. J. Mob. Comput. Multimed. Commun.* **3** 21–34

[6] Zhu C, Liu X, Zhou S and Liu Q 2014 *Int. Joint Conf. on Neural Networks* (Beijing: IEEE) pp 4190–97
[7] Ambusaidi M A, He X, Nanda P and Tan Z 2016 *IEEE Transactions on Computers* **65** 2986–98
[8] Tama B A and Rhee K H 2015 *Advances in Computer Science and Ubiquitous Computing* vol 373 ed Park D S, Chao H C and et al (Singapore: Springer) pp 489–495
[9] Thaseen I S and Kumar C A 2016 *Journal of King Saud University - Computer and Information Sciences* (2016)003
[10] Dong S K, Sang M L and Park J S 2006 *Lecture Notes in Computer Science* **3973** 224–230
[11] Cataltepe Z, Genc H and Pearson T 2007 *15th European Signal Processing Conf.* (Poznan: IEEE) pp 970–974
[12] Dash M, Liu H and Motoda H 2000 *4th Pacific-Asia Conf. on Knowledge Discovery and Data Mining, Current Issues and New Applications* vol 1805 ed Terano T, Liu H and et al (Berlin, Heidelberg: Springer) pp 98–109
[13] Almuallim H and Dietterich T G 1994 *Artificial Intelligence* **69** 279–305
[14] Peng Y, Wu Z and Jiang J 2010 *Journal of Biomedical Informatics* **43** 15–23
[15] Huang J, Cai Y and Xu X 2006 *18th Int. Conf. on Pattern Recognition* vol 2 (Hong Kong: IEEE) pp 618–621
[16] Liu H and Setiono R 1995 *Proc. of 7th IEEE Int. Conf. on Tools with Artificial Intelligence* vol 00 (Los Alamitos: IEEE Computer Society) pp 388–391
[17] Duda R O, Hart P E and Stork D G 2000 *Pattern Classification (2nd Edition)* (New York: Wiley-Interscience)
[18] Cutler A, Cutler D R and Stevens J R 2012 *Ensemble Machine Learning: Methods and Applications* ed Zhang C and Ma Y (Boston: Springer) pp 157–175
[19] Pedregosa F and et al 2011 *Journal of Machine Learning Research* **12** 2825–30
[20] Buitinck L and et al 2013 *ECML PKDD Workshop: Languages for Data Mining and Machine Learning* vol 238 pp 108–122
[21] Tavallaee M, Bagheri E, Lu W and Ghorbani A A 2009 *Proc. of the 2nd IEEE Int. Conf. on Computational Intelligence for Security and Defense Applications* (New York: IEEE) pp 53–58
[22] Buczak A L and Guven E 2016 *IEEE Communications Surveys Tutorials* **18** 1153–76

The research of emergency managers' risk attitude based on genetic neural network

Chen Pan[1], Chuanliang Jia[2], Jie Su[2] and Junlin Chen[2]

[1]Department of Management Science and Engineering, Central University of Finance and Economics, Faculty of Management Science and Engineering, room343, Student apartment Building 5, Beijing, China.
[2] Department of Management Science and Engineering, Central University of Finance and Economics, Faculty of Management Science and Engineering, room415, Tupei Building, Beijing, China.

Abstract: Since the 21st century, the sudden emergencies occurred around the world. The decision which emergency managers made in the emergency environment is actually a risk decision-making. And decision-makers will be affected by its risk attitude in the risk decision-making. The paper puts forward a kind of forecasting method of risk attitudes of emergency managers, which are influenced by various factors such as their own gender, age, training experiences, treatment experiences, actual average accuracy and self-confidence and so on. Based on the questionnaire results from Kunming and Wuhan, the above six factors were chosen as the input variables of the BP neural network model and the risk attitude values as the output variable. Genetic algorithm was used to optimize the weights and thresholds of the neural network and establish BP Neural network prediction model of emergency managers. The results of risk attitude value prediction show that the GA-BP is a more effective and accurate method to predict emergency managers' risk attitude, which can provide the reference for the risk managers' risk attitude prediction and more efficient management.

1. Introduction

In recent years, some of the sudden public events that are well known to the public are frequently erupted, which is no time to remind the human that sudden public events bring a lot of potential threats to people, and more and more emergency managers are in need to make appropriate decision in a short time. But the decision-making process is largely dependent on the individual judgment of the emergency manager. The decision-making of public emergencies is one of the important steps to deal with emergencies. It has special characteristics such as time urgency, high degree of uncertainty, social impact and centralized decision-making. These characteristics determine the decision-making behavior of emergency managers, especially decision-making attitude will be affected by a variety of factors.

In the genetic neural network established in this paper, the output variable is the risk attitude of the emergency manager. In general, the risk attitude is the choice of people when faced with uncertainty. In Kahneman and Tversky's Prospect Theor, three basic principles are proposed: (1) Most people are risk-averse when faced with "gains"; (2) Most people are risk-taking when faced with "loss" (3) People are sensitive to "loss" than "gains"[1]. Moreover, the characteristics of risk perception and risk behavior have been found in risk decision research. Weber, Blais, and Betz point out the significant differences in risk decisions in financial, health, entertainment, morality, and social, which is clearly reflected in the DOSPERT scale designed by Weber [2].

Many researchers at home and abroad are committed to the impact of risk attitude factors of decision-makers, that is, the input variables in the network. Hartog et al. (2002) found that there was a correlation between risk attitudes and personal traits [3]. Hu Xiao-xi and other studies have shown that women's executives decision-making behavior compared with men are more conservative and cautious [4]. The results of Croson and Gneezy's 2004 study also found that men and women were significantly different in risk-taking behavior [5]; According to Hambrick et al. (1984), age is an important factor influencing executives' risk attitudes [6]; Carlsson and Karlsson (1970), Vroom and Pahl (1971) have shown that older managers tend to taking less risky decisions [7]; Bantel and Jackson (1989) argue that higher qualifications of executives is more favorable to the company's strategic changes, etc.[8]. Bao Xing's (2008) research shows that the risk attitude and the ability of the manager in the emergency management will affect the emergency cost and the operation cost, and there is an unclear correlation between them [9]. Russo and Schoemaker pointed out that experienced bridge athletes are more confident when bidding, but less inexperienced athletes often cannot win the bidding that they can win [10]. Grifin and Tversky (1992) find that people tend to be overconfident when answering questions that are extremely difficult [11].

Because the risk attitude of emergency managers is related to many factors, but not necessarily linear relationship, BP neural network model has a good effect in analyzing and dealing with nonlinear relationship, which can effectively show the relationship between input and output. However, in practical applications, the BP neural network shows the shortcomings of global network search ability, slow iterative convergence and partial optimal. By using the genetic algorithm to optimize the connection weight of the BP neural network model, the training speed of the BP neural network can be improved, and the network can overcome being trapped into the partial optimum.

2. Data Source and Index Selection

2.1. Data Source and Analysis of Questionnaire Reliability

The data sample of this paper is a questionnaire issued to emergency managers in Kunming and Wuhan. A total of 104 questionnaires were received, of which 98 were completed with all the experiments, and 3 questionnaires missing the second part of the questionnaire of the self-confidence measure of the experiment, 3 questionnaires missing the third part of the risk attitude coefficient of the experiment.

In order to ensure the reliability of the questionnaire and the consistency of the survey, this paper uses SPSS to analyze the reliability of the questionnaire. The data processing of the results is shown in Table 1 below:

Table 1. Reliability statistics

Cronbach's Alpha	Cronbach's Alpha based on standardized items	Number of items
0.744	0.605	11

Through its reliability analysis we can infer that the internal consistency coefficient of the questionnaire is 0.744, and the internal consistency coefficient based on the standardized item is 0.605, which indicates that the questions of the questionnaire can basically express the original intention and will not cause the ambiguity and the obtained data basically effective.

Table 2. Total statistics

	Removed average scale	removed scale variance	Corrected total correlation	The square of multiple correlations	removed scale Cronbach's Alpha values
Gender	10.3226	29.110	0.210	0.078	0.766
Age	8.5065	23.792	0.400	0.293	0.725
Marital status	11.2112	24.588	0.830	0.826	0.705
Education level	10.3511	13.672	0.926	0.969	0.605
Years of work	10.2889	13.041	0.926	0.969	0.608
Unit nature	11.0842	22.466	0.783	0.702	0.683
Job level	11.1231	23.309	0.847	0.821	0.690
Regularly attend the training	10.0402	27.885	0.014	0.081	0.762
Whether have been handled such incidents	9.8822	29.835	0.297	0.142	0.780
Actual average correct rate	11.1813	28.175	0.029	0.041	0.754
Self-confidence	10.7112	28.814	0.203	0.104	0.759

In this reliability analysis, we also found that the Cronbach's Alpha values after each variable have been deleted are above 0.6 that are greater than the critical of 0.5, indicating that the set of questions in the questionnaire has good internal consistency and can reflect the degree of measurement of the influence factors of decision-making quality of emergency managers and the relationship between self-confidence level and risk attitude. Therefore, this paper considers the design and results of the questionnaire to be effective.

2.2. Sample Description and Index Selection

The questionnaire consists of three parts, each of which tests the different aspects of the subject. In the first part, the basic information of the subjects' personal characteristics is tested. According to the relevant research at home and abroad, combined with the availability of data, this paper focuses on the subjects' gender, age, and whether they have received training. The second part tests subjects' decision-making behavior in four different types of emergencies. Emergency management field covers a wide range. In general, it can be divided into social security, natural disasters, public health, and accident disaster class. From the different sub-areas, whether the subjects have experiences in dealing with emergency, and the level of manager's self-confidence as well as the actual average accuracy can be obtained through the test. The third part tests the risk attitude of the subjects. With reference to Weber scale, this paper designed a table based on the emergency fields. In the test, γ is the risk attitude coefficient, and the value is in the range of (0,1). γ is equal to 1 indicates that the risk is neutral. The smaller the γ is, the stronger the degree of risk aversion.

In the three parts of the questionnaire, the previous two parts of the questionnaire served as input value of genetic neural network, and the third part of the questionnaire results served as output value for effective prediction.

3. Basic Principles of Genetic Neural Network

BP neural network optimized by genetic algorithm can be divided into three steps: determination of BP neural network's structure, weights and thresholds optimized by genetic algorithm, BP neural network training and prediction. Among them the topology of the BP neural network is determined by the number of input/output parameters of the sample, so that the number of parameters optimized by genetic algorithm can be determined. Then the coding length of the population can be determined. When the parameters of BP neural network are optimized by genetic algorithm, the weights and thresholds are known as long as the network structure is known. And the weights and thresholds of the neural network are usually determined by random initialization [-0.5, 0.5] interval of the random number. The initialization parameters has a great impact on the network training, but cannot be accurately obtained. For the same initial weight value and threshold, the network training results are the same. The introduction of genetic algorithm is to optimize the best initial weights and thresholds.

Genetic algorithm optimization of flow BP neural network is shown as below:

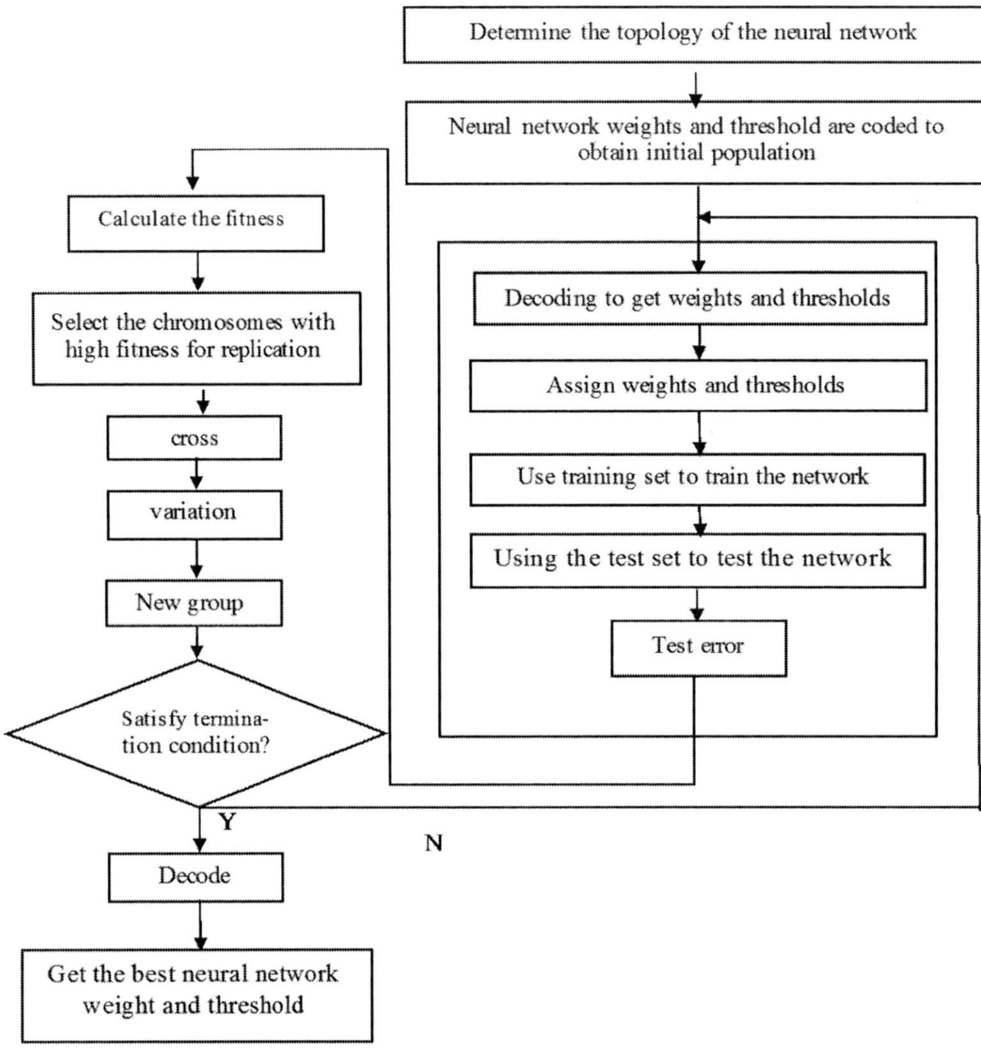

Figure 1. The flow chart

4. Predictive Examples Based on Genetic Neural Networks

In this example, we select the complete 208 datum, and the first 188 data are chosen as a training set, the last 20 data are served as a test set to predict. With gender, age, whether to participate in training, whether there is treatment experience, self-confidence level and the actual average accuracy is used as the influencing factor of the risk attitude of the emergency manager, and the risk attitude value is taken as the output value of the genetic neural network.

The establishment of the network topology is reasonable or not, which directly affects the accuracy, objectivity and adaptability of the network model. The determination of network topology generally includes network layer number, input layer, hidden layer, output layer and so on. In the GABP model established in this paper, the designing of network structure is based on the experience to determine the number of layers of the neural network, usually three layers.

(1) The number of input layers

Through the previous analysis, 6 evaluation indicators were identified, which were the gender, age, training experiences, treatment experiences, self-confidence level and the actual average accuracy of six. And each index corresponds to the neural network input layer of neurons, so the number of neurons in the neural network input layer is 6.

(2) The number of output layers

The final requirement of this paper is to predict the risk attitude of the emergency manager. The result of the final model output is the risk attitude value of the model. Therefore, the number of nodes in the output layer is 1.

(3) The number of hidden layer nodes

In the three-layer network, there are approximate relationships between the number of hidden layer neural networks and neurons in the input layer:

$$n_2 = 2 \times n_1 + 1 \tag{1}$$

In this case, there are six input parameters, so here is a value of 13, that is, there are 13 hidden layers.

(4) Other parameters set

The BP neural network's structure is 6-13-1, that is, the input layer has 6 nodes, the hidden layer has 13 nodes, and the output layer has 1 node, a total of 6 * 13 +13 * 1 = 91 weight, 13 + 1 = 14 thresholds, so the number of parameters optimized by genetic algorithm is 91 + 14 = 105. In other network parameter settings, the training times epochs = 1000, training goal = 0.01, learning rate lr = 0.1. The parameters of the genetic algorithm needed in the genetic neural network are shown in the following table:

Table 3. Other parameters set

NIND	MAXGEN	PRECI	GGAP	PX	PM
40	50	10	0.95	0.7	0.01

Running the GABP code, the resulting evolution curve and the training error curve are shown as below:

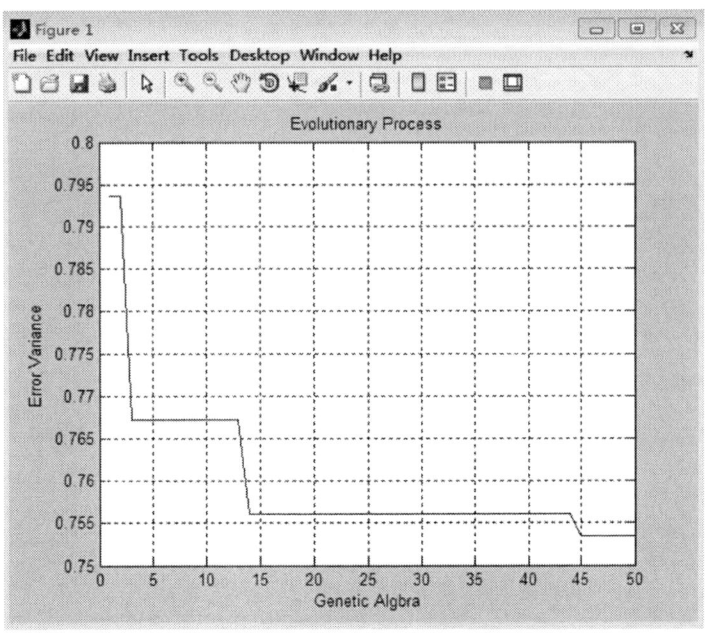

Figure 2. Evolution curve

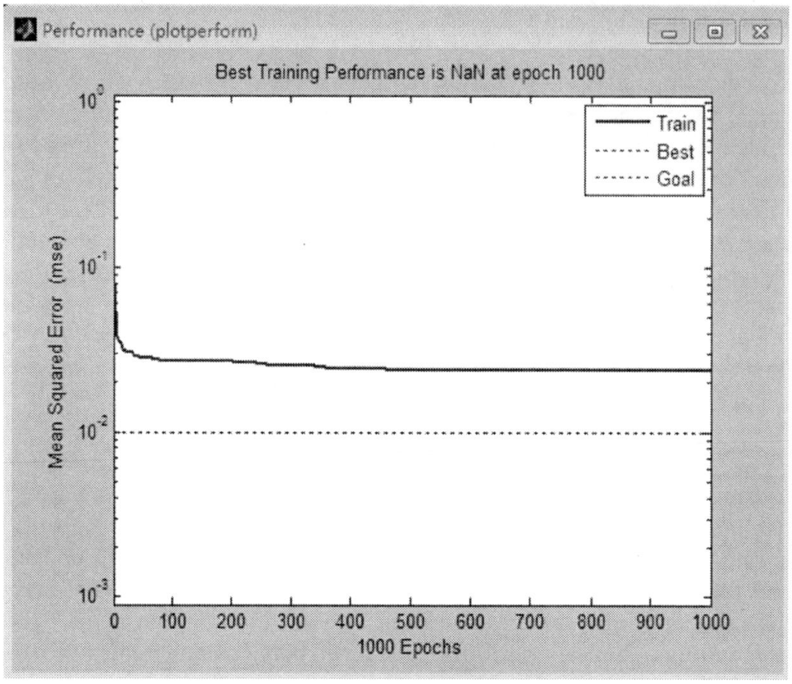

Figure 3. Training error curve

```
1、The random weights and thresholds
Prediction of test sample:

Y1 =

  Columns 1 through 11

    0.6083    0.4791    0.5761    0.6467    0.6398    0.4620    0.4791    0.6529    0.4828    0.2266    0.7191

  Columns 12 through 20

    0.6448    0.3515    0.6448    0.6448    0.4465    0.4568    0.6801    0.7827    0.6945

The simulation error of test sample:1.0352
The simulation error of training sample:3.0422
```

Figure 4. Test results using random weights and thresholds

```
2、The optimized weights and thresholds
Prediction of test sample:

Y2 =

  Columns 1 through 11

    0.8198    0.7871    0.7926    0.6342    0.4866    0.8290    0.7871    0.6561    0.4603    0.0328    0.5542

  Columns 12 through 20

    0.4439    0.3569    0.4439    0.4439    0.5172    0.5802    0.5609    0.7925    0.5071

The simulation error of test sample:0.75345
The simulation error of training sample:2.1596
```

Figure 5. Test results using optimized weights and thresholds

From the fitness function curve and the training error curve, it can be seen that the fitness function is not changed after the evolution of the individual 14 times, and the training error curve tends to be gentle soon. It can be also seen that the genetic algorithm improves the convergence speed and error precision of the BP neural network. The effect is outstanding.

From figure 4 and figure 5, it can be seen from the comparison of the two prediction results that the error of the test sample after optimization of the initial weight and the threshold is reduced from 1.0352 to 0.75345, and the error of the training sample is reduced from 3.0422 to 2.1596. The test results of BP network training and test sample have been relatively improved.

5. Conclusion

In the field of emergency, a risk prediction model based on genetic neural network is proposed to explore the influencing factors of current managers' risk attitude and its prediction. The neural network has strong self-learning, adaptive and fault tolerance, as well as connection weights and thresholds optimized by genetic algorithm, which can weaken the human factors in the evaluation, improve the accuracy and authority.

The prediction example also shows that the error of both training set and test set are produced by using genetic neural network to operate, indicating that the use of genetic neural network can significantly improve the neural network prediction accuracy. And the fitting effect is better than neural network, but also can solve the neural network exists into the partial optimal point and the slow convergence of the problem.

The prediction of the risk attitude of the genetic algorithm for emergency managers will effectively improve the management quality and efficiency of emergency managers. Because the gender, age training experiences, treatment experiences are easy to obtain, and self-confidence level and the actual accuracy can also be obtained through simple testing, so it can be predicted by genetic neural network. According to the forecast results we can understand the risk attitude of each manager and according to the different results we can select the appropriate positions and departments for each manager, so that people can do their best and give full scope to the talents.

6. Reference

[1] Tversky A and Kahneman D 1985 the Framing of Decisions and the Psychology of Choice. *Springer US*, **211**453-458

[2] Weber E U, Blais A R and Betz N E 2002 a domain-specific risk-attitude scale: measuring risk perceptions and risk behaviors *Journal of Behavioral Decision Making*, **15**263–290

[3] Hartog J, Ferrer-I-Carbonell A and Jonker N 2002 Linking Measured Risk Aversion to Individual Characteristics *Kyklos*, **55**3-26

[4] Hu Xiaoxi 2015 Female executives involved in enterprise management will affect the earnings management of listed company's decision-making it? *China Certified Public Accountants*, **5** 66-71

[5] Croson R and Gneezy U 2004 Gender differences in preferences. *Journal of Economic Literature*, **47**448-474

[6] Hambrick D.C. and Mason P.A. 1984 Upper Echelons:the Organization as a Reflection of Its Top Managers. *Academy of Management Review*, **9**193-206

[7] Carlsson G, Karlsson K 1970 Age,Cohorts and the Generation of Generations *American Sociological Review* **35**710-718

[8] Bantel K.A and Jackson S.E 1989 Top management and innovations in banking: Does the composition of the top team make a difference? *Strategic Management Journal* **10**107-124

[9] Bao Xing, Ji Jianhua 2008 Strong Management of Large Service Operation System Capability Considering Manager's Risk Attitude *Journal of Systems Management,* **18**555-561

[10] Russo J.E and Schoemaker P.J.H 1992 Management Overconfidence *Sloan Management Review/ Winter* **33**7-17

[11] Griffin D and Tversky A1992 The weighing of evidence and the determinants of confidence *Cognitive Psychology*, **24**411-435

An underwater acoustic beacons positioning method using single hydrophone

Jia Liu, Feng Xu, Qian Zhang and Lijun Jiang

Institute of Acoustics, Chinese Academy of Sciences, Beijing, China
liujia@mail.ioa.ac.cn

Abstract. When an aircraft crashed the sound emitted by the beacon would help to detection and location the black box. While in the deep sea research, it is hardly to use the USBL, SBL and LBL as usual. In this paper a new method was introduced. In this method, the beacon could be located with only one hydrophone. The time difference of the hydrophone at different place was used to located the beacon. To verify the performance of this method, the simulation was processed. The results show that this method is effectively.

1. Introduction

When an aircraft crashed the sound emitted by the beacon (pinger) would help to detection and location the black box for prompt recovery and analysis. Because the Operating Life of the Beacons is limited, it is very important to search and location the beacon as quickly as we can [1, 2].

It is well known that the traditional technique for localization acoustic beacons is Long Baseline, Short Baseline and Ultra Short Baseline. USBL uses phase differencing measurements between closely spaced (< 1 m) hydrophones to estimate the direction and elevation angle of an incoming signal. These are combined with travel time measurements to estimate range for pinger positioning. SBL is based on a similar concept, except that the hydrophones are more widely spaced (5 - 20 m) and instead of phase differencing, time difference of arrival is used. Both USBL and SBL systems are commonly mounted on vessel hulls, but can also be moored on the bottom or mounted in buoys at the surface. LBL systems measure travel-times between an array of transponders and a moving pinger. With multiple travel time, or range measurements, a least squares algorithm can be used to estimate the position of the pinger using either a hyperbolic or spherical solution. However sometimes only have one hydrophone especially in a deep ocean search. The attitude of the hydrophone array can't be obtained easily in a deep ocean search. Therefore, the common methods were hard to be used at this time [3-6].

In this paper, we will propose a method to locate the acoustic beacon with only single hydrophone. This paper is organized as follows: In the section 2, we introduce the location method. In section 3, we present solution of the method.

2. Basic Principle

The acoustic beacon is at $\mathrm{M}(x, y)$. The receive is a single hydrophone, which goes along A_1 A_2 A_3. Each position and time of A_1 A_2 A_3 are known.

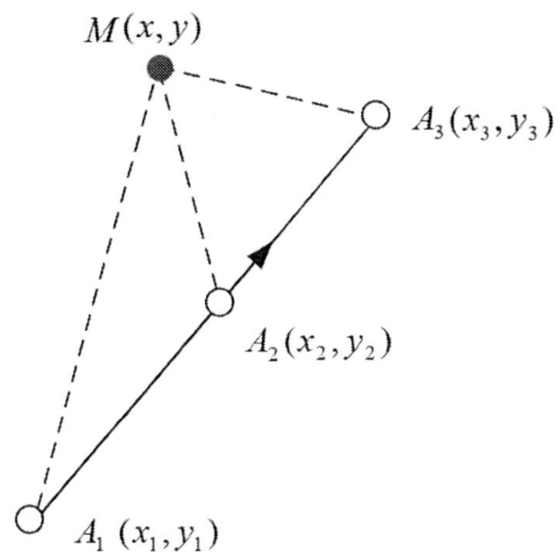

Figure 1. The location model of the method

The pulse length of the beacon is 10ms. The pulse repetition period is 1s.

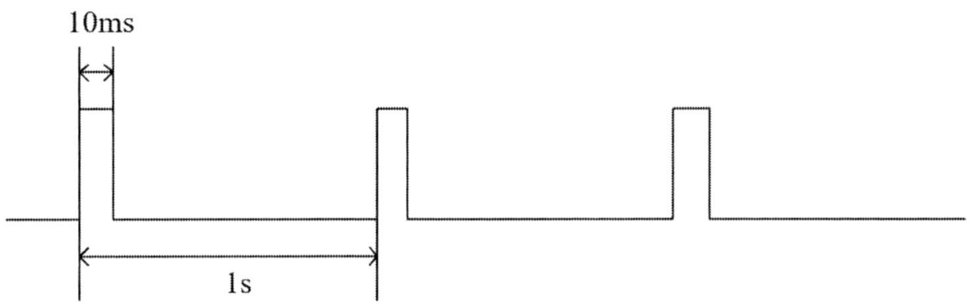

Figure 2. The signal of the acoustic beacon

The receive hydrophone spends t_s to go from A_1 to A_2. In this time the receiver has received n pluses, which is timed t_n .And, there is a time difference $\Delta t = t_s - t_n$. So, the difference of distance between M A_1 and M A_2 is

$$\Delta t_{12}(c) = \sqrt{(x_1 - x)^2 + (y_1 - y)^2} - \sqrt{(x_2 - x)^2 + (y_2 - y)^2} \tag{1}$$

And The difference of distance between M A_1 and M A_3 is

$$\Delta t_{13}\,c = \sqrt{(x_1 - x)^2 + (y_1 - y)^2} - \sqrt{(x_3 - x)^2 + (y_3 - y)^2} \tag{2}$$

Solved the equation, then the positon of M can be obtained.

3. The Solution

Set the A_1 to be the original point. Where, $x_1 = 0, y_1 = 0$, then (1) can be write as

$$x_2^{\ 2} + y_2^{\ 2} - 2x_2 x - 2y_2 y = \Delta t_1^{\ 2} c^2 + 2\Delta t_1 c\sqrt{x^2 + y^2} \tag{3}$$

Where , $r_2 = \Delta t_{12} c$, so we can get

$$\frac{x_2^2+y_2^2-r_2^2}{2r_2}-\frac{x_2x}{r_2}-\frac{y_2y}{r_2}=\sqrt{x^2+y^2} \tag{4}$$

Where, $F_{2r}=\frac{x_2^2+y_2^2-r_2^2}{2r_2}$, so we can get

$$F_{2r}-\frac{x_2x}{r_2}-\frac{y_2y}{r_2}=\sqrt{x^2+y^2} \tag{5}$$

As the same one, we can get

$$F_{3r}-\frac{x_3x}{r_3}-\frac{y_3y}{r_3}=\sqrt{x^2+y^2} \tag{6}$$

$$F_{3r}-\frac{x_4x}{r_4}-\frac{y_4y}{r_4}=\sqrt{x^2+y^2} \tag{7}$$

Then the equation can be re-write as

$$A_1x+B_1y+D_1=0 \tag{8}$$

$$A_2x+B_2y+D_2=0 \tag{9}$$

Where,

$$A_1=(\frac{x_2}{r_2}-\frac{x_3}{r_3}) \quad A_2=(\frac{x_2}{r_2}-\frac{x_4}{r_4})$$

$$B_1=(\frac{y_2}{r_2}-\frac{y_3}{r_3}) \quad B_2=(\frac{y_2}{r_2}-\frac{y_4}{r_4})$$

$$D_1=F_{3r}-F_{2r} \quad D_2=F_{4r}-F_{2r}$$

The equation of (1) and (2) can be solved, as

$$x=\frac{B_1D_2-B_2D_1}{A_1B_2-A_2B_1} \tag{10}$$

$$y=\frac{B_1D_2-B_2D_1}{A_1B_2-A_2B_1} \tag{11}$$

4. Simulation Results

To verify the performance of this method, the simulation has been processed. In the simulation, the time error is less than 0.1s, and the location error of the hydrophone is less than 0.1m. The simulation results (the location errors of the beacon) are shown in Fig.3. The error of locating is less than 500m in 6km.

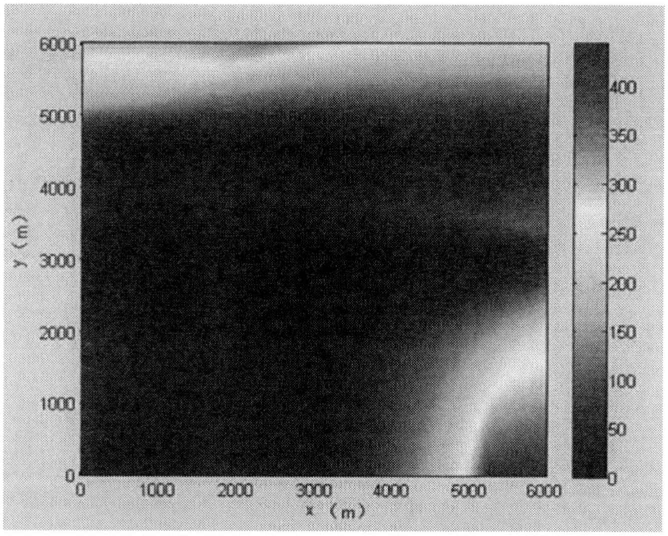

(a) The location error of coordinate X

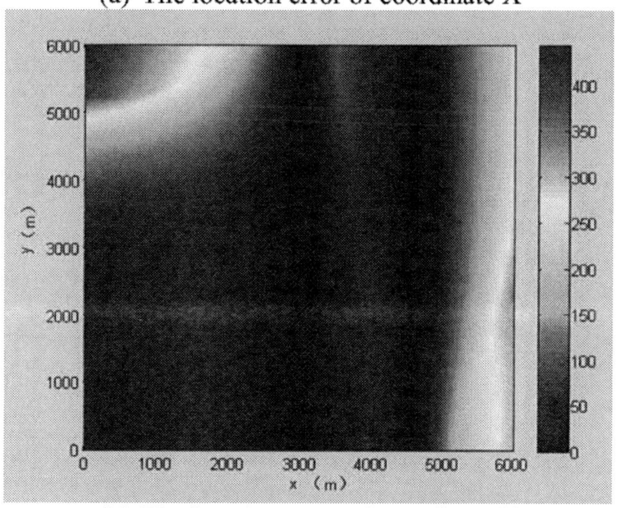

(b) The location error of coordinate Y

Figure 3. The location error of this method

5. Summary

In this paper, we tackles the problem of localization the underwater acoustic locating beacons of the "black box". In the deep-sea research, it is hardly to use the USBL, SBL and LBL. So a new method is introduced. In this method, the beacon was located with only one hydrophone, by the time difference of the hydrophone at different place. To verify the performance of this method, the simulation has been processed. The error of locating is less than 500m in 6km in the simulation. The results show this method is effectively.

6. Acknowledgment

National Natural Science Foundation of China NO.11404365 supported this work.

7. References

[1] Commonwealth of Australia, "MH370 – Definition of UnderwaterSearch Areas," Australian Transportation Safety Bureau, June 2014.

[2] Justin Manley; Joe Borden; Dale Green; Russ Blake; Tom Altshuler, Advanced Technologies for Undersea Object Location, 2014 Oceans - St. John's Pages: 1-5, 2014.

[3] D. Phillip, "An evaluation of USBL and SBL acoustic systems and the optimisation of methods of calibration - part 1," The Hydrographic Journal, no. 108, pp. 18-25, April 2003.

[4] M. Wikstrom, U. Ahnstrom, J. Falk, and P. Handel, "Implementation ofan acoustic location-finding system for TDOA measurements," in Nordic Matlab Conference 2003, October 2003.
[5] R. M. Eustice, H. Singh, and L. L. Whitcomb, “Synchronous-Clock, One-Way-Travel-Time Acoustic Navigation for Underwater Vehicles,” journal of field robotics, vol. 28, no. 1, SI, pp. 121–136, JAN-FEB,2011.
[6] B. Bingham, “Precision autonomous underwater navigation,” Ph.D. dissertation, Massachussetts Institute of Technology, Massachussetts, May 2003.

IOP Conf. Series: Journal of Physics: Conf. Series **887** (2017) 012034 doi:10.1088/1742-6596/887/1/012034

Analysis and optimization of calibration method of digital energy meter

Songxue Hou[1], Yuyou Liu[2,*], Yunying Xu[1], Shunchao Wang[1] and Dan Xu[1]

[1]Wuhan Electric Power Technical College, Wuhan, 430070, China
[2]Huazhong University of Science and Technology, School of Electrical and Electronic Engineering, Wuhan, 430074, China
[*]Email: liuyuyou1037@foxmail.com

Abstract. At present, the short of calibration of digital energy meter include deficiency of standard, inefficiency, and the lack of proper methods. In this paper, based on four commonly used methods(Watt-second method, standard digital meter method, standard digital power source method, standard analog meter method), the factors that cause the error of these calibration methods are introduced, and relevant measures of improvement and optimization are put forward.

1. Introduction
With the rapid development of intelligent substation, digital energy meter has gradually replaced the traditional electronic energy meter, becoming an important part of the energy measurement and trade settlement. At present, the commonly used methods of error calibration are standard digital meter method, standard digital source method, standard analog meter method[1-4] and watt-second method, which are still imperfect. Calibration experiments using analog sources can be used to trace the digital meter, but introduce new system errors such as A/D conversion, signal conditioning[5]and so on. Compared with the analog source method, the digital source method has the advantages of shorter calibration time and less error introduction[6]. This paper introduces the four commonly used calibration methods, and provides the technical reference for the basic error test scheme of digital energy meter to the standardization, systematization and specialization.

2. Error Analysis of Calibration Methods for Digital Energy Meter

2.1. Watt-Second Method
The watt-second method calculates the power of the calibrated meter by detecting the fixed number of electric pulses m and the fixed measurement time interval t, and compare it with the power P_0 set by the standard digital power source to get the error. Its principle is as Figure 1. According to the principle of detection, the error expression is as formula (1).

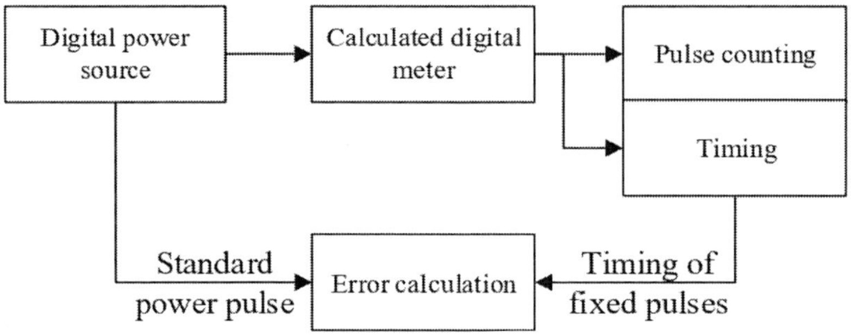

Figure 1. Process of Watt-second calibration method

$$\gamma = \frac{P_L - P_0}{P_0} = \frac{\frac{m}{K'C_0 t} - P_0}{P_0} \tag{1}$$

C_0 is the standard meter pulse constant and P_L is the power value of the calibration meter when it is converted to a digital source. K' is the coefficient. When the calibration meter works in the primary power mode, $K'=1$; When the calibration meter works in the secondary power mode, $K'=10^3/(K_I K_U)$, in which K_I and K_U are the virtual current ratio and the virtual voltage ratio of the digitized electric energy meter.

According the formula (1), we can see the source of error introduced from the watt-second method has three aspects: measurement time, the overflow error and power calculation in non-full period.

According to the formula (1), $PL = m / K'C_0 t$ is obtained, and it can be seen that the accuracy of the measurement time interval t will affect the measurement result of the watt-second method. Assuming that the actual measurement time is $t + \Delta t$, the calibrated meter is accurate under ideal conditions, that is, $m / K'C_0 t = P_0$, we can get the relative error of this method is as follows:

$$\varepsilon = \frac{\frac{m}{K'C_0(t+\Delta t)} - P_0}{P_0} = \frac{m}{K'C_0 P_0(t+\Delta t)} - \frac{m}{K'C_0 P_0 t} \approx -\frac{\Delta t}{t} \tag{2}$$

The measurement time error Δt is a fixed value related to the hardware circuit and the timing of the high frequency pulse. In general, the measurement time error Δt will not exceed 100us [7]. Therefore, when t>1s, the introduction error of the measurement time interval is less than 0.01%, so the introduction error brought by the uncertainty of the measurement time can be ignored.

Digital energy meter output power pulse by the way of the energy pool [8]. The energy meter calculates the corresponding power based on the data obtained $E_t(k)=U(k)I(k)\Delta t$, and then put the $Et(k)$ into the energy pool, when the energy pool is full, it emits a power pulse. But when the energy meter emits the last power pulse, the energy pool may "overflow". This part of the overflow of energy will eventually be remnants to the energy pool and cause system errors. And because $\Delta t = 1 / fc$, so the overflow error expression is:

$$\Delta y = \frac{P(k)}{f_c P_0 t} = \frac{U_m I_m \sin(\omega t + \alpha)\sin(\omega t + \alpha + \varphi)}{f_c \frac{U_m I_m}{2}\cos(\varphi) t} = \frac{\cos(\varphi) - \cos(2\omega t + 2\alpha - \varphi)}{f_c \cos(\varphi) t} \tag{3}$$

Therefore, from the formula (2), we can see that when $t = \varphi/2w\text{-}\alpha$, the maximum relative error is as follows:

$$\triangle y_{\max} = \frac{\cos(\varphi)+1}{f_c \cos(\varphi)t} \tag{4}$$

It can be seen that the relative system error of the digital energy meter is inversely proportional to the sampling rate f_c, measurement time interval t and the offset angle φ. The f_c in the digital meter is 4 kHz generally. If $\cos\varphi = 0.5$, the relative error calculated by the formula (3) is 0.01% when the measurement time t>7.5s, so the introduction error can be ignored.

When digital energy meter is calibrated, the time length corresponding to the calculated data frame is most likely not the entire period, resulting in a systematic error in the calculation of the average power [9]. The instantaneous power of the sine wave output from the standard power source is as formula (5). In time t, the theoretical value of the energy in the full period is $P=UIt\cos\varphi$, so the energy error obtained by the non-periodic method is as formula (6).

$$\begin{aligned} \mathrm{P}(t) &= 2UI\sin(\omega t)\sin(\omega t\text{-}\varphi) \\ &= \mathrm{UI}(\cos(\varphi)-\cos(2\omega t-\varphi)) \end{aligned} \tag{5}$$

$$\begin{aligned} \varepsilon' &= \frac{\int_{t_0}^{t+t_0} \mathrm{P}(t)dt - Pt}{Pt} \\ &= \frac{-\int_{t_0}^{t+t_0} UI\cos(2\omega t+2\alpha-\varphi)dt}{UIt\cos(\varphi)} \end{aligned} \tag{6}$$

Where φ, α, t *and* t_0 respectively represent the power factor angle, the initial phase angle, the actual calculation time and the time when the timing is started. The formula (6) can be further reduced to the formula (7) as follows:

$$\varepsilon' = \frac{\sin(2\omega t_0\text{-}\varphi)-\sin(2\omega t+2\omega t_0\text{-}\varphi)}{2\omega t\cos(\varphi)} \tag{7}$$

Because of $|\sin(2\omega t_0-\varphi)-\sin(2\omega t+2\omega t_0-\varphi)|\leq 2$, the greatest error caused by single-phase non-periodic power calculation is:

$$\varepsilon'_{\max} = \frac{1}{2\pi t f_N \cos(\varphi)} \tag{8}$$

In equation (8), f_N is the fundamental frequency. Let $\cos(\varphi)$=0.5 by equation (8) can be introduced, in $U=U_N$, under the condition of $I=120\%I_N$, when t>63.66s, the error caused by non-full period power calculation is less than 0.01%, so it can be ignored, but when the load current is 5% of the rated current, calibration time is extended by 24 times. Therefore, the error in the calculation of the non-full period is the main error of the watt-second method in the single phase calibration.

When the three-phase symmetrical calibration is practiced, the error induced by the calculation of the non-periodic power is:

$$\varepsilon^{(3)}=\frac{\int_{t_0}^{t+t_0}\cos(2\omega t-\varphi)+\cos(2\omega t-\varphi+\frac{2\pi}{3})+\cos(2\omega t-\varphi+\frac{4\pi}{3})dt}{3UIt\cos(\varphi)} \tag{9}$$

$$=0$$

It can be seen from the formula (9) that the introduction error of non-periodic calculation is 0 when the three-phase symmetrical calibration is practiced. Therefore, it is only necessary to extend the calibration time to eliminate the overflow error and test the time interval error. Therefore, it is more efficient to use the watt-second method to perform three-phase symmetrical calibration.

From the above error analysis, it can be seen that the error effect caused by the non-periodic sampling is better than that of the digital energy meter and the error due to the inaccurate measurement time interval (which needs the longest measurement time to reduce the error effect). According to the error synthesis formula, the total relative error is:

$$\varepsilon_z=\sqrt{\varepsilon^2+\varepsilon'^2+\Delta y^2} \tag{10}$$

According to the error theory and the definition of small error, when the system error introduced by the test method is less than the error limit of 1/10, then the error introduced by the test method can be ignored [10]. By the formula (2), (3), (8) we can see that all of the errors are related to the measurement time interval *t*, so the most effective way to reduce the system error is to increase the measurement time interval. When the time *t* is large enough, the introduction error can be ignored.

2.2. Standard Digital Meter Method

The digital power source can send data packets containing voltage and current signals. It transmitted to the standard meter and the calibrated meter through the switch. The standard digital meter sends the high frequency pulse *m* according to the received data message. Pulse N converted to the standard digital meter for the m_0 [11]. The principle is as Figure 2.

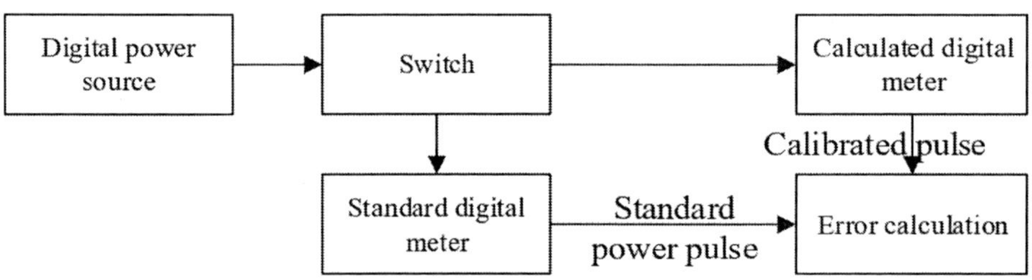

Figure 2. Process of Standard digital meter method

The error characteristics are as formula (11).

$$\gamma=\frac{\mathrm{m}-\mathrm{m}_0}{m}\times 100\% \tag{11}$$

According to the principle of digital power source and standard meter method, there is overflow error and pulse counting error, the maximum value of pulse count error is ±1. So the introduction error of the method due to the pulse counting calibration is as follows:

$$\varepsilon_n=\frac{\mathrm{m}\pm 1-\mathrm{m}_0}{m\pm 1}-\frac{\mathrm{m}-\mathrm{m}_0}{\mathrm{m}}\approx\frac{1}{\mathrm{m}}=\frac{1}{\mathrm{c}_0\mathrm{P}_0\mathrm{t}} \tag{12}$$

It can be seen from the formula (11) that the greater the pulse constant, the longer the measurement time interval, the smaller the error of this test method. For a given standard meter, the pulse time constant C_0 is constant, so only by increasing the measurement Time t to increase the number of calculation of the number of pulses to reduce the relative reference error of the standard meter method. The relationship between precision of standard meter method and the number of calculated pulses is as follows:

Table 1. The relationship between the accuracy class and the pulse number

Accuracy level	Pulse Number
0.05	50000
0.1	20000
0.2	10000
0.3	6000

It can be seen from Table 1 that this method is very demanding on the number of pulses. When the number of calculated pulses reaches 10k, it can reach the precision level. The pulse frequency of the standard digital meter is generally not more than 1 kHz, so its detection demands long time, especially in the case of small load current, the test time will be extended.

2.3. Standard Digital Power Source Method

The standard digital power source sends out the high frequency pulse m proportional to the electric energy, meanwhile the calibrated meter outputs the low frequency pulse N, and then converts N into the pulse number m_0 of the standard source output using the conversion formula. Its principle is as Figure 3.

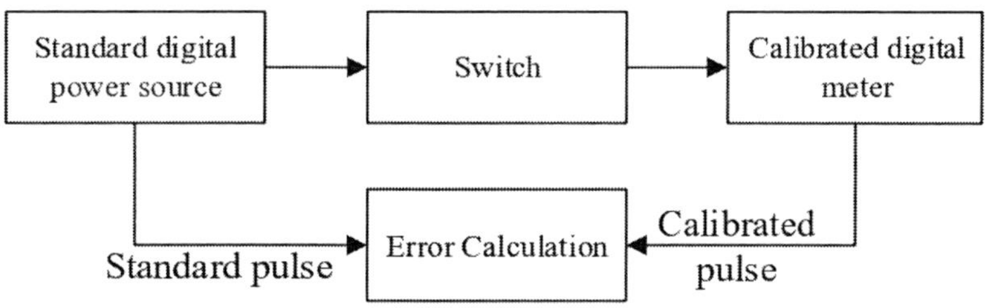

Figure 3. Process of standard digital power source method

The error characteristic is as formula (13).

$$\gamma = \frac{m - m_0}{m} \times 100\% \tag{13}$$

The main reference error of standard power source method is still overflow error and pulse counting error. Using it to test the accuracy of 0.2-level energy meter needs at least 10s, of which overflow error can be ignored, and using standard power source method, the test time can be adjusted for the pulse constant C_0 is adjustable, so that the credibility of calibration can be ensured in short calibration time. But when the power pulse constant C_0 set large, there will be a certain impact in the test system. When the overflow error $\Delta y_{max} > 3\varepsilon$, the overflow error becomes the main introduction error, and the synchronization pulse error is small and can be ignored. Similarly, when $\varepsilon_n > 3\Delta y_{max}$, the synchronization pulse error dominates, and the overflow error can be ignored. In other cases, the overflow error and the synchronization pulse error should be considered.

2.4. Standard Analog Meter Method

Analog signal source output standard sinusoidal signal, after A/D conversion, the data frame sent to the calibrated energy meter, and then it compares the power pulse output from calibrated meter and the power pulse converted by the is standard sinusoidal signal, to analyze the error characteristics of the calibrated meter.

The standard analog meter has a mature technical system and a clear means of traceability. It is a commonly used method to calibrate the digital energy meter at present. But compared with the standard digital source method, it introduces error ε_s of the signal conditioning and A/D conversion equipment, and calibration method of pulse output will cause overflow error Δy, and the error ε_n caused by pulse counting. The synthetic error is:

$$\varepsilon_{z}' = \sqrt{\Delta_{y}^{2} + \varepsilon_{n}^{2} + \varepsilon_{s}^{2}} \tag{14}$$

As the A/D conversion circuit and other error modules can be controlled by technical means to reduce errors, and the using of analog source method can complete the traceability of digital energy meter, so this method currently is still more commonly used.

3. Optimization of Calibration Methods for Digital Energy Meter

For the watt-seconds method, its source of error is measurement time interval error, overflow error and non-full period sampling error. To improve the efficiency, we can increase the accuracy of the measurement interval with high-frequency pulse for the time counting, and the fixed pulse counting method is changed to the fixed time range to detecting the pulse number and pulse time interval, so that the working condition interference can be avoided. It applies to three-phase symmetric conditions.

For the standard digital meter method, its source of error is overflow error and sync pulse error. To improve the efficiency of calibration, interpolation can be used for standard meter to improve the pulse output frequency. Energy meter with larger pulse frequency constant can be used as standard energy meter. Its applicable condition is rated conditions with larger load current.

For the standard digital source method, its source of error is overflow error, synchronization pulse error and introduction error of standard digital power source. To improve the efficiency, we can increase the frequency of the pulse according to the actual working condition, and in the three-phase symmetry test, the average power method is used. Its applicable condition is rated conditions with larger load current

For the standard analog meter method, its source of error is overflow error and synchronization pulse error and error caused by A/D conversion and other signal debug module. To improve the efficiency, we can select the analog standard meter with smaller error for comparison and reduce the system error form A/D conversion, signal conditioning and protocol group frame. It can be used in the requirement for traceability.

The principle of standard digital source method is basically the same as the standard digital meter method, so the introduction error of each is the same. Since the standard digital source method can adjust the pulse frequency constant, the standard digital source method is more efficient. Standard analog meter method is a mature method of calibration system, but because it contains modules of the introduction of error, in general, it is inferior to digital power source method. Watt-second method introduced a new non-integer period sampling error, in the case of single-phase meter calibration, Watt-second method required a longer time. Therefore, Watt-second method is most suitable for calibration in three-phase symmetrical working condition.

4. Conclusion

In this paper, the error analysis and comparison of four commonly used methods of calibrating digital energy meter are carried out. According to the error characteristics of each calibration method, some relevant improvement measures are given. Based on the above analysis, the choice of calibration scheme should be based on its error characteristics to take appropriate measures to improve the test accuracy and reduce error in the introduction of the error. In order to solve the problems such as slow calibration speed and incomplete traceability chain, improvements are made to accurately and quickly calibrate the measurement accuracy of digital energy meter. It is important to promote the improvement of digital energy meter detection technology.

5. References

[1] Liu Yu, Establishment *of calibration standard for digital meter and evaluation of uncertainty* [J]. Measurement and measurement technology, 2010, S2:40-43.

[2] Wu Yuanfang, Qi Yanshou, Yang Jun, et al. *New efficient standard source for calibrating electric energy meter* [J]. Qinghai Electric Power, 2015, S2:1-3.

[3] Liu Huiying, Yin Xin, Cui Xu, et al. *Research and application of calibration technology for digital energy meter* [J]. Technology and enterprise [J].2015, 21:111.

[4] Zhu Xiaohong, Misunderstanding and Countermeasures of watt second method for checking single-phase watt hour meter [J].Electrical Measurement & Instrumentation, 2010, S2:98-101.

[5] Yuan Hui, Zhang Chaojun, *Research on detection method of digital electric energy meter*[A].Proceedings of the Third Symposium on the academic development direction of electromagnetic measurement and instrumentation[C], 2012

[6] Wang Hanjie, *Progress of inspection technology for electric energy meter* [J]. Electrical Measurement & Instrumentation, 2004 06:28-30.

[7] Li Qibing, *Research and design of high precision multi-function time calibrator*[J].Journal of Sichuan University of Science and Engineering (NATURAL SCIENCE EDITION),2009,05,89-92

[8] Xie Dong, Wu Tao, Wu Bohua, et al. *Discussion on Accuracy Class Issue for Digital Interface Electricity Meter* [J]. Electrical Measurement & Instrumentation, 2011, 08:87-91.

[9] Pan Feng, Sun Weiming, Zhao Wei, et al. *Error Analysis of Numerical Integration Algorithm in Digital Power Measurement* [J]. Electrical Measurement & Instrumentation, 2012, 06:23-28.

[10] Zhang Shuhan, *Measurement error theory and electrical measuring instrument technology* [J]. Electrical Measurement & Instrumentation, 1983, 05:7.

[11] Q/GDW 11111-2013, *Standardization of digital energy meter calibration* [S]

IOP Conf. Series: Journal of Physics: Conf. Series **887** (2017) 012035 doi:10.1088/1742-6596/887/1/012035

Research on backbone node deployment for Wireless Mesh Networks in dynamic environments

Meiyi Li[1] and Shengling Cao[2]

[1]College of Information Engineering, Xiangtan University, Xiangtan, China, 1356695709@qq.com
[2]College of Information Engineering, Xiangtan University, Xiangtan, China, shenglingcao@qq.com

Abstract. Wireless Mesh Network is a type of wireless networks in which demands of bandwidth for users has mobility. The backbone node placement of wireless mesh networks in a dynamic scenario is investigated, and the TSDPSO algorithm is used to adapt the dynamic environment, which updates node deployment location to adapt to changes in demand if it detects environmental changes at the beginning of the cycle time. In order to meet the demands of bandwidth for users and network connectivity, particle swarm optimization algorithm is employed to select the gateway location, then nodes to the backbone network is added constantly until all requirement is covered. The experimental results show that algorithm could get effective solution in dynamic environment.

1. Introduction

Wireless Mesh Network (WMN) is a type of multi-hop wireless network, which is different from traditional wireless network. In the era of wireless network, it has received the attention of a large number of researchers at home and abroad. Mesh Router (MR), Mesh Gateway (MG) and Mesh Client (MC) constitute the basic framework of WMN. MR has access and forwarding function in the network, it provides network services for MC within the coverage, and forwards data to MG for other MR in multi-hop networks. MG is a special kind of MR, it not only has the function of MR in wireless Mesh network, also assume the connection task with Internet by wired connection. MC is the terminal device that accessing network by MR, such as notebooks, mobile phones, etc. The backbone node deployment for WMN in dynamic environment is mainly introduced.

The literature [2] proposes a WMN router deployment greedy algorithm named NF-Greedy, which is based on network flow. In order to minimize the number of deployed MR, the algorithm selects the node with the largest weight in the MR candidate location set by iterative method, and uses the network flow method to solve the problem. In the literature [3], the problem of WMN gateway node placement is abstracted as the problem of geometric K- center, and an adaptive particle swarm optimization algorithm is proposed to solve the problem of gateway node deployment in order to minimize the path length between nodes and gateways. The literature [4] is for the WMN backbone network deployment optimization problems, in order to meet the condition of the minimum number of MR users which network bandwidth demand as the goal, reference MG and MR separately position deployment plan, proposed in MG and MR solution in the condition of mutual influence.

At present, the approach of finding the optimal solution to WMN backbone network node deployment problem in the static environment has been relatively mature. But in real life, WMN is always in a dynamic environment, MR and MC has mobility. Therefore, the deployment of WMN backbone network nodes in dynamic environment is more meaningful. In order to meet the requirement of network connection and coverage in the dynamic environment, a particle swarm optimization algorithm is proposed to implement the deployment of MR by the literature [5]. However, the study assumes that the deployment location of MR can be moved freely, which is too costly to implement in real life. A new research to solve the deployment problem of WMN backbone nodes in dynamic environment should be proposed.

2. Problem Definition

In real life, WMN is in a dynamic environment. And in different time, MC will be in different requirements and geographical location. After the change, the user demand is stable. Therefore, when the demand changes are detected, the location of the backbone nodes is adjusted to meet the demand changes. The environment will be detected when the new cycle begins. The deployment of the backbone node location can be seen as the deployment of nodes in a two-dimensional plan. It is assume that there are enough suitable backbone node deployment locations, and the user's demand for the network has been discretized into the user needs.

2.1. Definitions and Variables

It is assumed that the wireless network communication is reliable, and the communication between nodes does not interfere with each other, then set the following variables.

R_c: Node coverage radius. A circular region centered on the node with R_c radius. The node can provide services for MC in this area.

R_t: Node communication radius. A circular region centered on the node with R_t radius. The node can communicate with other nodes in this area.

Cap: Node access capacity, Refer to that the node provides the maximum access bandwidth for the MC covered by it.

H : Maximum hop. Refer to that to ensure the quality of communication, the maximum number of hops to the gateway communication path.

T:Cyc. Refer to that the cycle which be seted by the actual situation.

In the two-dimensional plane, WMN corresponds to a topological graph $G=(V,E)$, where $V=\{v_1,v_2,\cdots,v_n\}$ represents a set of nodes in the network, Adjacency matrix $E=\{e_{i,j}|i,j\in\{1,2,\cdots,n\}\}$ represents the relationship between nodes. MR Candidate (MRC), refers to the pre-selected candidate positions to deploy MR in the two-dimensional plan.

- User Demand Node (UDN), refers to user demand point after the user's bandwidth demand has been discretized, reflect the distribution of user demand in the deployment area.
- Actual coverage radius, refers to the maximum distance that a backbone node can actually provide for its coverage UDN.
- Environmental change times, refers to the number of cycles that are currently running time, i.e. the number of environmental changes.

2.2. Mathematical Model

Based on the above definitions and variables, a formal description of the problem has been presented. Optimization goal as follow.

$$\min \sum_{i=1}^{n} x_{i,t} \tag{1}$$

It is meant to minimize the number of deployed MR after each environmental change which on the premise of satisfying the user's bandwidth requirement and network connectivity. Constraint condition as follow.

$$\sum_{k=1}^{n} e_{i,k} \neq 0, \forall x_{i,t} = 1, \forall i \in \{1,2,\cdots,n\} \tag{2}$$

$$\sum_{j=1}^{m} d_{i,j} \neq 0, \forall x_{i,t} = 1, \forall i \in \{1,2,\cdots,n\} \tag{3}$$

$$d_{i,j} = 1, \forall y_{i,j} \neq 0, \forall i \in \{1,2,\cdots,n\}, \forall j \in \{1,2,\cdots,m\} \tag{4}$$

$$x_{i,t} = 1, \forall y_{i,j} \neq 0, \forall i \in \{1,2,\cdots,n\}, \forall j \in \{1,2,\cdots,m\} \tag{5}$$

$$\sum_{i=1}^{n} y_{i,j} = b_{j,t}, \forall j \in \{1,2,\cdots,m\} \tag{6}$$

$$r_{i,t} \geq 0, \forall i \in \{1,2,\cdots,n\} \tag{7}$$

$$h_{i,t} \leq H, \forall i \in \{1,2,\cdots,n\} \wedge x_{i,t} = 1 \tag{8}$$

$$p_{i,j}^{k} \leq x_{i,t}, \forall i,j,k \in \{1,2,\cdots,n\} \tag{9}$$

$$p_{i,j}^{k} \leq x_{j,t}, \forall i,j,k \in \{1,2,\cdots,n\} \tag{10}$$

$$p_{i,j}^{k} \leq e_{i,j}, \forall i,j,k \in \{1,2,\cdots,n\} \tag{11}$$

$$\sum_{i=1}^{n} p_{i,j}^{k} = x_{k,t}, \sum_{j=1}^{n} p_{i,j}^{k} = 0, \forall k \in \{1,2,\cdots,n\} \tag{12}$$

$$\sum_{i=1}^{n} p_{i,j}^{k} - \sum_{j=1}^{n} p_{i,j}^{k} = 0, \forall k \in \{1,2,\cdots,n\}, i \in \{1,2,\cdots,n\} \setminus \{k\} \tag{13}$$

$$\sum_{j=1}^{n} p_{i,j}^{k} \leq x_{k,t}, \forall k \in \{1,2,\cdots,n\}, i \in \{1,2,\cdots,n\} \setminus \{k\} \tag{14}$$

$$\sum_{i=1}^{n}\sum_{j=1}^{n} p_{i,j}^{k} = x_{k,t}, \sum_{i=1}^{n}\sum_{j=1}^{n} p_{j,i}^{k} = 0, \forall k \in \{1,2,\cdots,n\} \tag{15}$$

The formula (2) indicates that the selected MRC must be able to connect to the gateway via a multi hop network. Formula (3) indicates that a UDN must be present in the selected MRC coverage area. Formula (4) and (5) show that MRC provides network services for UDN if and only if UDN is within MRC coverage and MRC has deployed MR. Formula (6) indicates that all user requirements in the entire network must be satisfied. Formula (7) shows that each backbone node for its coverage of the bandwidth of the UDN cannot exceed its access capacity. The formula (8) shows that the minimum hop number of each backbone node connected to the MG through the multi hop network cannot exceed the maximum hop count. The formula (9), (10) and (11) indicate that the nodes can be connected to each other only if the nodes and nodes are selected to deploy MR and the two can communicate with each other through the path to the MG. The formula (12), (13), (14) and (15) indicate that the node must satisfy the current conditions in order to have a multi hop path connected to the MG.

3. Backbone Node Deployment Algorithm in Dynamic Environment

3.1. Backbone Node Deployment Algorithm

The deployment of MR approach is that after determining the location of MG, The node is added to the backbone network node collection from the deployable MRC collection in an iterative manner. The deployment status will be updated until all UDN bandwidth requirements have been met. According to this method, the key point is how to deploy the MR from the deployable MRC collection. In each iteration, the neighbor nodes of the current backbone network will be selected from the non-deployed MRC collection, according to the principle of maximum coverage and minimum coverage radius, the node is added to the backbone network. In order to improve the efficiency of the algorithm, cross linked list is used to store data. The steps are as follows.

Step 1. Initializes the deployed MR, the covered the UDN and other placement variables, according to the UDN and MRC coordinates to compute the demand coverage table, then update the MG collection.

Step 2. Update the deployed MR, the covered the UDN and demand association tables, calculate the maximum bandwidth requirement of the current network, determine whether the current network is equal to the total flow of traffic, if it is equal to the total flow then go to step 5.

Step 3. Compute node residual bandwidth and the minimum hop that from deployed MR to MG, then update the shortest path of backbone node and node coverage requirements table. Add node to backbone network according to above method, update available MRC collection.

Step 4. Select a node with the largest weight in the adjacent MRC collection and add it to the deployed MR collection. If there is no available adjacent MRC, select the path with the largest weight, then go to step 2.

Step 5. Evaluate deployment results, end algorithm.

The above MR deployment method is deployed at the MG location, but the location of the MG and MR is mutually affected. BPSO algorithm is used to solve the problem of MG deployment, initialize a particle swarm composed of particles, searching extreme value in D dimensional binary space. According to the optimal solution and the global optimal solution, the particle swarm is modified to obtain the optimal solution. The formula for update velocity and position of particle swarm as follow.

$$v_i^{k+1} = wv_{i,d}^{k} + c_1 r_1 (x_{pbest,d}^{k} - x_{i,d}^{k}) + c_2 r_2 (x_{gbest,d}^{k} - x_{i,d}^{k}) \tag{16}$$

$$z_i^{k+1} = \begin{cases} 1 & \rho < Sigmoid\left(v_i^{k+1}\right) \\ 0 & Others \end{cases} \tag{17}$$

$$Sigmoid\left(v_i^k\right)=\frac{1}{1+\exp\left(-v_i^k\right)} \tag{18}$$

$$Fitness\left(Z_i\right)=\frac{1}{\sum_{i=1}^{n} x_i} \tag{19}$$

According to formula (17), when v_i^{k+1} becomes closer to $v_{\max}$, it is more likely for c_i to be chosen as MG. According to the formula (16) to update the particle velocity and position, the results will not meet the constraints of the number of MG. Make improvements that updating the location by the stability of sorting, which means to start the update from the candidate position the lowest stability, if the position change, delete the candidate position from the sequence, or skip the candidate position, repeat the operation until the number of MG meet the target number. To sum up, the algorithm steps are as follows.

Step 1. Randomly generate a particle size of $PopSize$, initialize the position and velocity of discrete particles.

Step 2. Deployment of MR based on MR deployment algorithm, according to formula (19) to calculate the fitness of the particle swarm.

Step 3. If there is a better fitness value, update the best location and the best position.

Step 4. Update particle velocity and position according to formula (16).

Step 5. To determine whether the maximum number of iterations, or whether the results meet the requirements, if not go to step 2.

Step 6. Output particle swarm optimal results (MG, MR position and number).

3.2. TSDPSO Algorithm in Dynamic Environment

In order to solve the problem in dynamic environment, TSDPSO algorithm has been presented. The current deployment solution will be checked that whether it can meet the network connectivity and user needs after the cycle. In the D dimension space, $pbest^t$ represents the particle's current optimal solution, $gbest^t$ represents the optimal solution of the particle in the population, x_i represents the position vector of the particle itself, and v_i represents the velocity vector of the particle itself. At the beginning of the algorithm, the population is divided into two parts, core-subgroup update the velocity and position of the particle according to the speed and position evolution formula (20) and formula (21) of the standard PSO algorithm. In order to maintain the exploration ability of the population, the differential evolution strategy is applied to the aux-subgroup, and the aux-subgroup is updated according to the following formula (21) and formula (22).

$$v_{i,d}^{t+1}=wv_{i,d}^{t}+c_1r_1(x_{pbest,d}^{t}-x_{i,d}^{t})+c_2r_2(x_{gbest,d}^{t}-x_{i,d}^{t}) \tag{20}$$

$$x_{i,d}^{t+1}=x_{i,d}^{t}+v_{i,d}^{t+1} \tag{21}$$

$$v_{i,d}^{t+1}=wv_{i,d}^{t}+\operatorname{sgn}\left(rand(\ )-0.5\right)\left(c_1r_1(x_{pbest,d}^{t}-x_{i,d}^{t})+c_2r_2(x_{pbest,d}^{t}-x_{i,d}^{t})\right) \tag{22}$$

$$\operatorname{sgn}=\begin{cases}-1 & t<0\\ 0 & t=0\\ 1 & t>0\end{cases} \tag{23}$$

In the evolutionary process, the particle of core-subgroup always attracted by the global optimal particle, the aux-subgroup of the differential evolution strategy is adopted to keep the attraction and repulsion of each particle with the probability of 50%.

The current deployment plan will be evaluated after environmental changes, determine whether the deployment plan to meet the network connectivity, MR can be calculated to meet the user's bandwidth demand, If the target is not met, reset deployment plan with symmetric displacement mapping strategy. Symmetric displacement mapping strategy steps are as follows.

Step 1. The difference between the particle position and the local optimal position is calculated on every dimension.

Step 2. Count the number of positive and negative values of the position difference in each dimension, positive number is denoted as $n_{1d}(d=1,2,\cdots,D)$, negative number is denoted as $n_{2d}(d=1,2,\cdots,D)$.

Step 3. On the one side of a large number of particles, the $\mathrm{int}\left(\frac{|n_{1d}-n_{2d}|}{2}\right)$ particles are selected from the position of the local optimum particle.

Step 4. The one side of a less number of particles update particle positions in accordance with the one side of a large number of particles by the formula (24), update the particle position.

$$x_{ijnew}=\begin{cases} g_p+|x_{ij}-g_p| & x_{ij}-g_p<0 \\ g_p-|x_{ij}-g_p| & x_{ij}-g_p>0 \end{cases} \tag{24}$$

So the algorithm is as follows.

Step 1. Randomly initialize the particle swarm, the particle swarm is divided into two sub groups, initialize particle velocity and location.

Step 2. Deploy backbone nodes according to the method described before, the current optimal solution and the global optimal solution are evaluated by comparing the fitness values of the main and auxiliary subgroups.

Step 3. Update the velocity and position of main and auxiliary subgroups.

Step 4. To determine whether the environment changes after the cycle, evaluate whether the current deployment solution can meet the network connectivity and user requirements, if not go to step 2 after core-group take the space of symmetric displacement mapping.

Step 5. To determine whether the algorithm to achieve the number of times, if not go to step 2.

Step 6. The end of the algorithm, record the output value.

4. The Analysis of Simulation and Experimental Results

According to the network parameters of Roofnet experimental network platform built by Massachusetts Institute of Technology, Set the node coverage radius to 150, the node communication radius to 250, node access capacity to 54, UDN bandwidth demand to 10, the maximum number of hops to the gateway to 4, simulation user demand set of real population distribution which meet the normal distribution. Set the cycle to 15 minutes, the number of UDN changes in each environmental change is a random integer between 0 and udn-count/8. The coordinates of each UDN move randomly in the scene. The results of static and dynamic deployment in normal distribution are shown in table 1.

Table 1. Static and dynamic deployment results in a normal distribution scenario

Scene	**Static and dynamic deployment results in a normal distribution scenario**		
	Scene Establishment	***Static deployment results***	***Dynamic deployment results***
1	200*200,10,1,15	2.00	2.21
2	300*300,20,1,25	4.00	4.59
3	400*400,40,2,45	7.00	7.31
4	600*600,80,3,80	12.50	14.70
5	800*800,150,4,110	17.05	19.12
6	1000*1000,200,8,140	20.55	24.25
7	2000*2000,450,16,360	68.15	69.10

The results of static and dynamic deployment in the average distribution are shown in table 2.

Table 2. Static and Dynamic Deployment Results in an Average Distribution Scenario

Scene	**Static and dynamic deployment results in an average distribution scenario**		
	Scene Establishment	***Static deployment results***	***Dynamic deployment results***
1	200*200,10,1,15	2.00	2.38
2	300*300,20,1,25	4.00	4.24
3	400*400,40,2,45	7.00	7.60
4	600*600,80,3,80	12.80	14.42
5	800*800,150,4,110	18.05	19.00
6	1000*1000,200,8,140	21.55	23.50
7	2000*2000,450,16,360	69.80	70.05

The experimental results show that this algorithm in uniform distribution and normal distribution, the number of MR deployments in the MR deployment scheme that in dynamic environment is close to the static deployment results, which proves the effectiveness of the proposed algorithm.

5. Acknowledgment

In this work we propose an optimal algorithm for the deployment of backbone nodes in wireless Mesh networks. In order to meet the user bandwidth requirement and the network connectivity, using particle swarm optimization algorithm to select gateway location, then to minimize the number of routers as the goal to gradually add the maximum weight of the adjacent nodes to complete the deployment. This paper, using TSDPSO algorithm to adapt to dynamic environment. At the beginning of each new cycle, the algorithm detects the change of the environment, and updates the location of the deployment of the nodes to meet the demand changes. The experimental results show that the algorithm can get an effective deployment plan in a dynamic environment.

6. References

[1] D. Benyamina, A. Hafid, M. Gendreau. "Wireless Mesh Networks Design——A Survey". IEEE Communications Survey & Tutorials, vol. 14, pp. 299-310, 2012.

[2] Wen-jia WU, Ming YANG, Jun-zhou LUO. "A Bandwidth-Aware Router Placement Scheme for Wireless Mesh Networks". Chinese Journal of Computers, vol. 37, pp. 344-355, 2014.

[3] Shu-qiang HUANG, Gao-cai WANG, Zhi-guang SHAN, et al. "Node Deployment Optimization of Wireless Network in Smart City". Journal of Computer Research and Development, vol. 51, pp. 278-289, 2014.

[4] Quan LIN, Mei-yi LI. "Research on backbone node deployment algorithm in Wireless Mesh Networks". Computer Engineering, vol. 41, pp. 147-152, 2015.

[5] Chun-cheng Lin. "Dynamic Router Node Placement in Wireless Mesh Networks: A PSO Approach with Constriction Coefficient and Its Convergence Analysis". Information Sciences, vol. 232, pp. 294-308, 2013.
[6] F. Xhafa, C. Sanches, L. Barolli. "Genetic Algorithms for Efficient Placement of Router Nodes in Wireless Mesh Networks", Proceedings of the 24th IEEE International Conference on Advanced Information Net-working and Applications. Perth, Australia, 2010, pp. 465-472.
[7] J. Kennedy, R.C. Eberhart. "Particle Swarm Optimization", Proceeding of IEEE International Conference on Neural Networks, 1995, pp. 1942-1948.
[8] Electronic Publication: Digital Object Identifiers (DOIs): /Eberhart R C, Shi Yu-hui. Tracking and Optimizing Dynamic Systems with Particle Swarms[C]//Proceeding of Congress on Evolutionary Computation. New York, USA, 2001, pp. 94-97.
[9] Sheng-Xiang Yang. "A Clustering Particle Swarm Optimizer for Locating and Tracking Multiple Optima in Dynamic Environment". IEEE Transactions on Evolutionary Computation, vol. 14, pp. 959-974, 2010.
[10] S. Sakamoto, E. Kulla, T. Oda, et al. "A Comparison Study of Simulated Annealing and Genetic Algorithm for Node Placement Problem in Wireless Mesh Networks". Journal of Mobile Multimedia, vol. 9, pp. 101-110, 2013.

IOP Conf. Series: Journal of Physics: Conf. Series **887** (2017) 012036 doi:10.1088/1742-6596/887/1/012036

Design and implement of mobile equipment management system based on QRcode

Runze Yu, Xiaohui Duan and Bingli Jiao

Center of Wireless Communication and Signal Processing, Peking University, Beijing, China, {yurunze, duan, jiaobl}@pku.edu.cn

Abstract. A mobile equipment management system based on QRcode is proposed for remote and convenient device management. Unlike conventional systems, the system here makes managers accessible to real-time information with smart phones. Compared with the conventional method, which can only be operated with specific devices, this lightweight and efficient tele management mode is conducive to the asset management in multiple scenarios.

1. Introduction

With the rising variety of equipment in various fields, the need for equipment management is becoming increasingly intense. In the traditional way, the use of manpower and paper for the equipment management is a waste of resources [1]. Most of the current equipment management systems use specific device to operate and manage, for instance card [2] [3] and RFID [4] [5] Reader. Lars el al. had presented the architecture of medical equipment service management and details of integration to develop a service program which is not a web service [6]. Hu Bin had developed a laboratory equipments management system in university which uses a RF-SIM card Reader to identify the RFID tags on the equipment [7]. S. Nutdanai had developed a medical equipment management in hospital which uses a bar code reader device to recognize the bar code on the medical equipment [8] [9]. They all have significant contributions to the equipment management in specific scenarios.

However, if using the smart phone for telemanagement, the system will enhance the user experience and broaden the application scenario. This paper develops a mobile equipment management system (MEMS) using the telemanagement mode based on mobile internet. In this system the user operates and manages only requiring to scan the QRcode on the equipment with a smart phone. This lightweight and efficient system is conductive to the asset management in multiple scenarios.

The rest of paper is organized as follow: Section II describes the design of the system. Section III details the implementation of the system. In section IV, the system operation and evaluation results are demonstrated. Section V concludes the paper.

2. System Design

2.1. System User Scenario

The user scenario of the MEMS is shown in Figure 1. Every equipment is affixed with a unique label made by the QRcode printer. Then the user can send a request of borrowing or returning to the cloud server after scanning the QRcode. Meanwhile, the web server makes manager accessible to real-time request information. Once the manager has made a check, the information will be fed back to the user and recorded in the database.

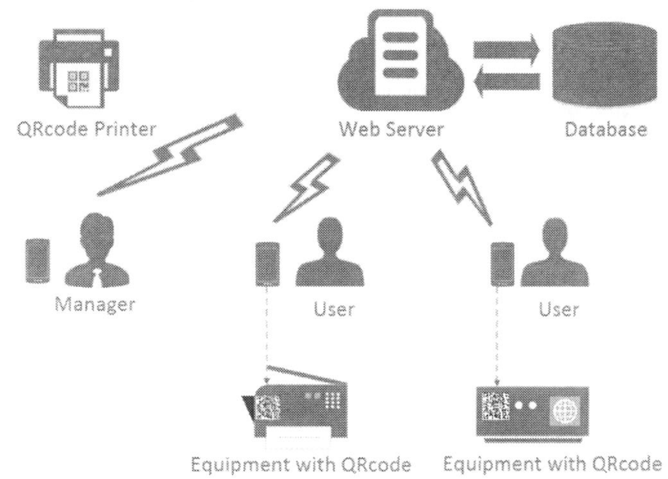

Figure 1. The user scenario of MEMS. The user borrows equipment through scanning QRcode. The web server makes manager accessible to real-time request and synchronize database information.

2.2. System Function Design

Figure 2 shows the block diagram of system functions. The function of the MEMS is divided into two main parts, namely user management and the equipment management. Two parts are subdivided into seven specific functions. They are detailed as follow:

- User Query: The manager can query user information, such as the history, current status, telephone, etc.
- User Remind: Once the user borrows or returns equipments, the system will remind the manager of the audit via SMS. In addition, the user can also be reminded to return the equipment on schedule.
- Equipment Registration: After the identity authentication, the manager is able to register the equipment with a smart phone.
- Equipment Borrowing: After the manager passes the audit, the user can borrow the equipment
- Equipment Returning: The user can return the equipment after the manager has audited.
- Equipment Query: According to the equipment name or the equipment ID, the manager can query the information of equipment, such as the history, status, images and etc.
- Equipment Inventory: The manager can investigate the equipment by scanning the QRcode on the equipment.

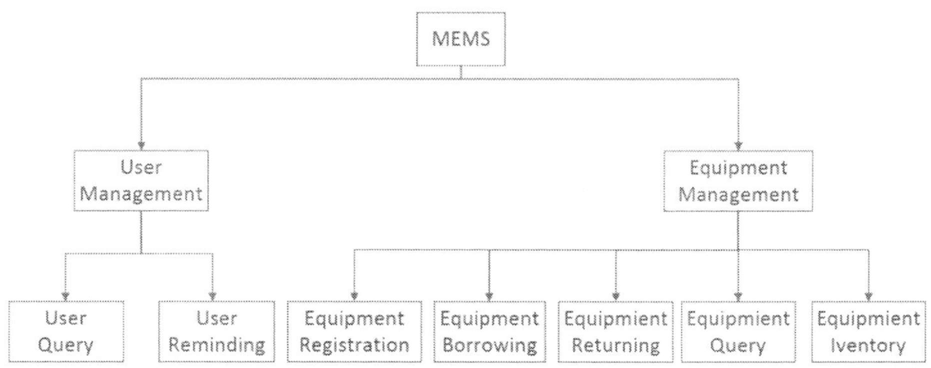

Figure 2. The block diagram of MEMS functions. The system function is divided into two main parts,user and equipment management. They are subdivided into seven specific functions such as equipment registration, borrowing, query and etc.

2.3. System Architecture Design

Based on the user scenario and function, we designed the system architecture. It can be seen from Figure 3 that the MEMS is web-based system. The developing environment is based on LAMP [10] (Apache + MySQL + PHP under Linux), and the server uses the MVC framework [11], which separates the input, processing procedure and output [12]. This framework eases the management and development procedure.

In our system, the user's smart phone communicates with the web server via http protocol in real time. The browser sends the request to the Controller, then the Controller updates the Model in line with requests. Hereafter, the Model modifies the data, synchronizes to database and responses the Controller. Finally, the Controller renews the View, then the View feeds the updated the interface and information to the browser.

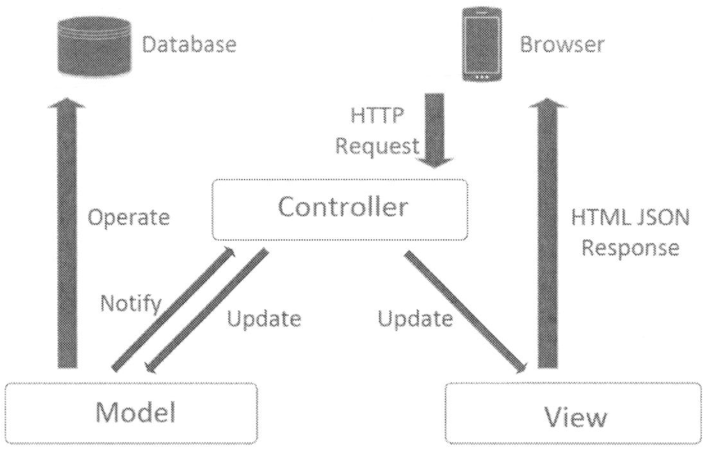

Figure 3. The MEMS architecture. The system uses MVC framework. The Controller receives requests from browser, then updates the Model and the View. The Model synchronizes database information. The View displays interfaces.

3. System Implementation

3.1. System Function Implementation

The system functions have been implemented according to the system functional design. We condense the function into three pivotal parts, namely the equipment registration, the equipment borrowing and the information query.

- The equipment registration is premised. The process is shown in Figure 4. Initially, manager enters the system through scanning QRcode, meanwhile interface sends request to server for details. Then the server retrieves information in the database. Once finding the corresponding data, the server will return data and update pages. Next, the manager photographs equipment and himself. Finally, the manager sends registration request. The server updates data and return results.

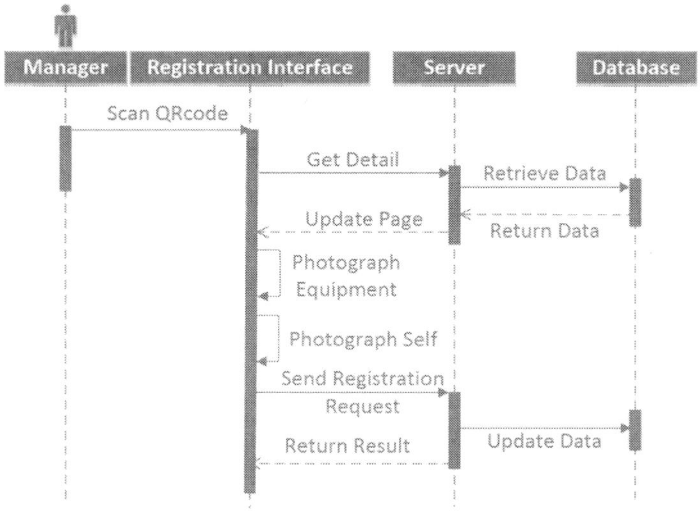

Figure 4. Sequence diagram of equipment registration. The manager registers equipments through scanning QRcode, photographing equopemnt and self, send registration requests.

- Only the equipment has been registered, the user can borrow. The process of equipment borrowing is illustrated in Figure 5. Firstly, the user scans QRcode and then enters the borrow interface. Then the server updates information and page. After photographing the equipment and self, the user sends request to the server for borrowing. When receiving the request, the server calls API to remind the manager of check via SMS. Once the manager has made feedback, the server will update data and return results.

The process of equipment returning is the same as above.

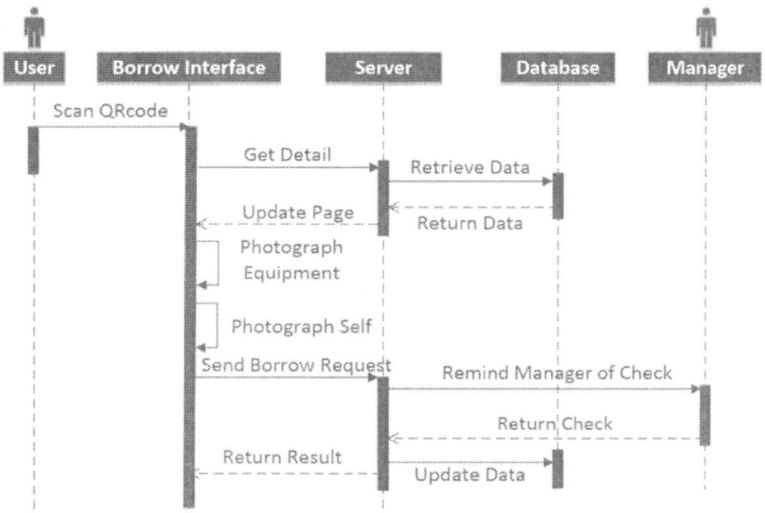

Figure 5. Sequence diagram of equipment borrowing. User borrows equipment through five steps, namely scanning QRcode, photographing equipment and self, sending borrow request.

- The information query can be operated at any time. As is shown in Figure 6, the manager logs in the background management interface after verifying identity. Then the manager can query the current state and history of users and equipments. If necessary, the manager can also export the data.

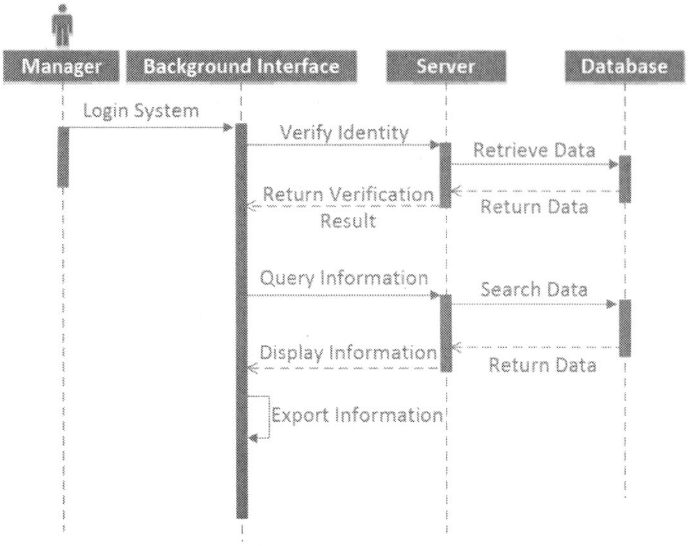

Figure 6. Sequence diagram of information query. After the authentication, the manager can query and export information in the background interface.

3.2. Database Creation

The background database of the MEMS stores various types of data and related parameters, such as the basic information and parameters of equipment, user's parameter and record of operation. According to system needs, we create database and tables. Data tables are linked to others and some of tables are built by users during the operation. Part of the table structure and field information is shown in Fig. 7 as follow.

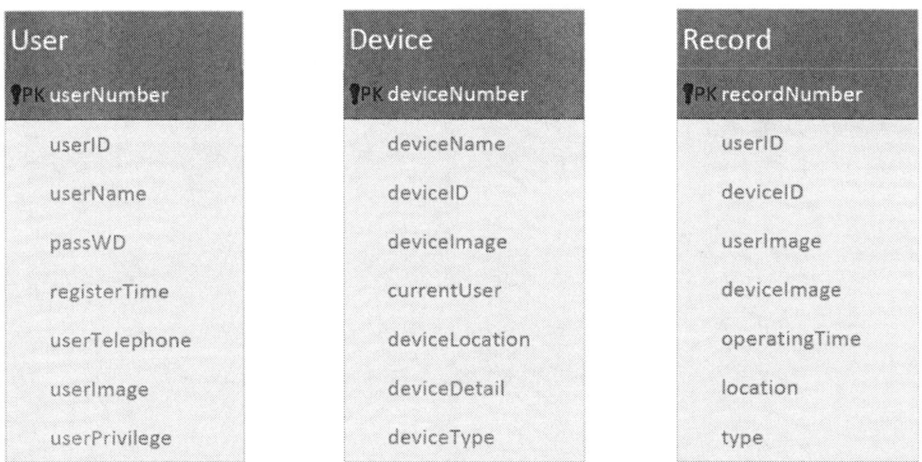

Figure 7. Part of the database structure. The user and device table stores basical information and parameters. The record table stores operation history

4. System Results

According to the user scenario, we built a test environment. As is shown in Figure 8, the system includes a QRcode printer (GPRINTER GP-3150TIN), an oscilloscope used as an example (RIGOL DS1052E), a user mobile device (iPhone 6) for borrowing or returning, a manager mobile device (iPhone 5S) for checking requests and a management software based on a PC browser (MacBook Pro).

In addition, the mobile browser is Safari and the PC browser is Chrome. Then we tested the system from two aspects.

Figure 8. The test environment. a: The QRcode printer generates label. b: The oscilloscope with QRcode label is used as an example of the equipment. c: The user mobile for borrowing. d: The manager mobile for check. e: The management software based on a PC browser

4.1. Test Procedure

The system procedure is tested according to the system function sequence diagram. Firstly, the QRcode printer generates a QRcode shown in Fig. 9 for the equipment. Next, the manager scans the QRcode and registers the equipment. Then the user sends a request for borrowing and the manager checks the information. This part corresponds to the Figure 10. In addition, all of the above information can be queried and exported in the background management interface.

Figure 9. The label generated by QRcode printer. Every equipment has a specific QRcode label.

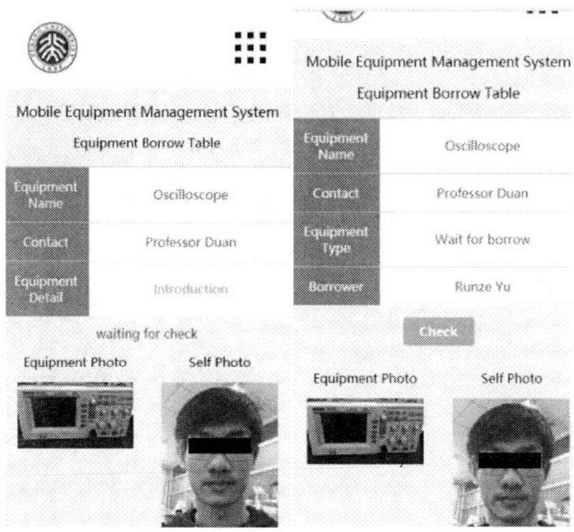

Figure 10. The actual system interface. (a) The user sends request for borrowing and waits for check. (b) The manager checks the borrow request.

4.2. Test Performance

The transmission delay of information is mainly due to the size of photo. In order to reduce the delay, we compress the files before transmission. Next we measure transmission delay of information 50 times through the data capture package method. The average delay is about 47ms.

5. Conclusion

In this paper, we designed and implemented the MEMS, which can receive requests, process data and display information in real time. Moreover, the system can be remotely operated and monitored based on mobile Internet. The system assessment contains robustness and latency. Practical application results show the MEMS is a promising solution for lightweight and high-confident asset management.

6. Acknowledgment

This work was supported by the Joint Seeding Grant for MI of Peking University (2016-MI-04). The authors would like to thank the support from Engineering Research Center for Mobile Digital Hospital System Ministry of Education, CN.

7. References

[1] Liu Yajing, Tian Guie, "Design and Implement of University Laboratory Equipment Network Management Platform Based on C/S and B/S Model", Industrial Control and Electronics Engineering (ICICEE), 2012 International Conference, 23-25 Aug. 2012

[2] Bao-Xun He, Ke-Jun Zhuang, "Research on the Intelligent Information System for the Multimedia Teaching Equipment Management", Information System and Artificial Intelligence (ISAI), 2016 International Conference, 24-26 June 2016

[3] N. Saleh, S. Rosati, A. Sharawi, M. Abde, G. Balestra, "Application of quality function deployment and genetic algorithm for replacement of medical equipment", Biomedical Engineering Conference (CIBEC) 2014 Cairo International, 11-13 Dec. 2014.

[4] Jingyu Wang, Sumei Li, Chungui Liu, "Design and Implementation of Smart Equipment Management System Based on RFID", Wireless Communications, Networking and Mobile Computing (WiCOM), 2011 7th International Conference, 23-25 Sept. 2011

[5] Yinghui Huang, Guanyu Li, "Descriptive models for Internet of Things", Intelligent Control and Information Processing (ICICIP), 2010 International Conference, 13-15 Aug. 2010

[6] L. Lars, T Mario, B. Thomas, S. Oliver, "Integration of medical equipment into SOA — Enabling technology for efficient workflow management", Emerging Technologies & Factory Automation (ETFA) 2011 IEEE 16th Conference, 5-9 Sept. 2011.

[7] Hu bin, "The Design and Implementation of Laboratory Equipments Management System in University Based on Internet of Things", Industrial Control and Electronics Engineering (ICICEE), 2012 International Conference, 23-25 Aug. 2012
[8] S. Nutdanai, A. Sanpanich, "A development of medical equipment registration and spare part module in WepMEt system for medical equipment management in Thai hospital", Biomedical Engineering International Conference (BMEiCON), 2014 7th, 26-28 Nov. 2014
[9] S. Nutdanai L. Pornthip A. Sanpanich, "Development of an information system for medical equipment management in hospitals" Biomedical Engineering International Conference (BMEiCON), 2016 9th, 7-9 Dec. 2016
[10] S.-J. S. Tsai, C. C. Luo, "Synchronized Power-Quality Measurement Network With LAMP", IEEE Transactions on Power Delivery, Volume: 24, Issue: 1, Jan. 2009 , pp. 484-485, doi: 10.1109/TPWRD.2008.2005361
[11] Edward Curry, Paul Grace, "Flexible Self-Management Using the Model-View-Controller Pattern", IEEE Software, Volume: 25, Issue: 3, May-June 2008, pp. 84-90, doi: 10.1109/MS.2008.60
[12] Badurowicz Marcin, "MVC architectural pattern in mobile web applications", Actual Problems of Economics, vol. 6, no. 120, pp. 305, January 2011.

A novel low-scattering and wideband monopole antenna based on artificial magnetic conductor

Pei Yao[1] , Binzhen Zhang[2],* Junping Duan[3] and Qiong Bai[4]

[1]Science and Technology on Electronic Test & Measurement Laboratory, School of Instrument and Electronics, North University of China, Taiyuan, China, 1621388471@qq.com

[2]Science and Technology on Electronic Test & Measurement Laboratory, School of Instrument and Electronics, North University of China, Taiyuan, China ,zhangbinzhen@nuc.edu.cn

[3]Science and Technology on Electronic Test & Measurement Laboratory, School of Instrument and Electronics, North University of China, Taiyuan, China , duanjunping@nuc.edu.cn

[4]Science and Technology on Electronic Test & Measurement Laboratory, School of Instrument and Electronics, North University of China, Taiyuan, China, 825702367@qq.com

Abstract. A novel low-scattering monopole antenna based on the phase cancellation between perfect electric conductor (PEC) and artificial magnetic conductor (AMC) is designed, the bandwidth of RCS reduction is 9.3 GHz~11.3 GHz, and the maximum RCS reduction value reaches 26.2 dB at 10.1 GHz. Besides, a wideband range of 10 dB return loss is achieved for the proposed design, compared with the original reference antenna, of 3.3 GHz~12.2 GHz (114.84%) and the radiation pattern is also kept in good condition. Finally, a prototype of the proposed new low-scattering monopole antenna was manufactured and measured, the experimental results coincide well with simulated counterpart.

1. Introduction

Nowadays, the metamaterials has been widely penetrated into aerospace, wireless communication [1], military equipment [2-4] and other related fields. The anti-radar stealthy technology based on metamaterials [5-8] has been applied to military equipment to decrease the radar cross section (RCS), which is characterized the scattering echo of target with the incidence of normal electromagnetic wave. The antenna in the stealthy platform, as a combination of scattering and radiation source, make a great contribution to the overall RCS, and therefore the first step to realize electromagnetic stealth of the whole platform is to reduce the RCS in antenna.

Several methods have been proposed in literatures to reduce RCS as well as keep the radiation performance of antenna [9,10], such as adopting radar absorbing material [11] changing the shape of the radiation patch [12], and destructive cancellation between different cells. The metallic perfect electric conductor (PEC) reflect the incident waves with a 180° phase change while AMC cells present a 0° phase change to the reflected wave within its working frequency, as in [13], the destructive interference between PEC and AMC is first applied to achieve RCS reduction, the 10 dB RCS reduction is obtained within the work frequency of AMC (15 GHz~16 GHz). Keeping the radiation performance of proposed antenna is a crucial issue, in [14], the octagonal ring arrangement composite surface is placed around waveguide slot antenna as the ground, the bandwidth of 10 dB RCS reduction is 5.24 GHz~5.92 GHz, besides, the gain is improved by 5 dB to obtain an improvement on radiation performance.

To make further efforts on broadening the impedance bandwidth of monopole antenna [15] and improving its scattering performance [16], a design approach aimed at reducing the RCS and increasing the bandwidth is proposed in this paper, details of the proposed ultra-wideband monopole antenna with lower RCS are describe and experimental results of the radiation performance are presented and analyzed.

2. Design of Proposed AMC Cell

Full numerical analysis of the AMC cells has been carried out by means of the software HFSS using a unit cell with Floquet port and appropriate periodic boundary conditions, the prototype of proposed AMC cell is shown as Fig.1. A 2 mm Polytef substrate with dielectric constant ε_r=2.65 and tangent of 0.003 is employed, on the top layer, a periodic element with unit length D=5 mm is designed, the effect of side length $L1$ of the periodic patch on the reflection phase condition is simulated and shown in Fig.2. According to the model of RLC equivalent circuit [17], an equation relationship among the resonant frequency f, equivalent inductance L and the equivalent capacitance C is given as (1):

$$f = \frac{1}{2\pi\sqrt{LC}} \tag{1}$$

With the side length of patch is increased, the gap between the metallization edge is reduced, and the equivalent capacitance C is enhanced, and then the resonant frequency is decreased, from the Fig.2, it is observed that the resonant frequency is decreased with the increase of the side length $L1$, the simulation results indicate that the resonant frequencies f at 9.6 GHz, 8.8 GHz, 7.7 GHz with bandwidths, defined for the distance between reflection phase ±90°, of about 7.9 GHz~11.7 GHz, 7.4 GHz~10.5 GHz, 6.6 GHz~9.0 GHz, respectively.

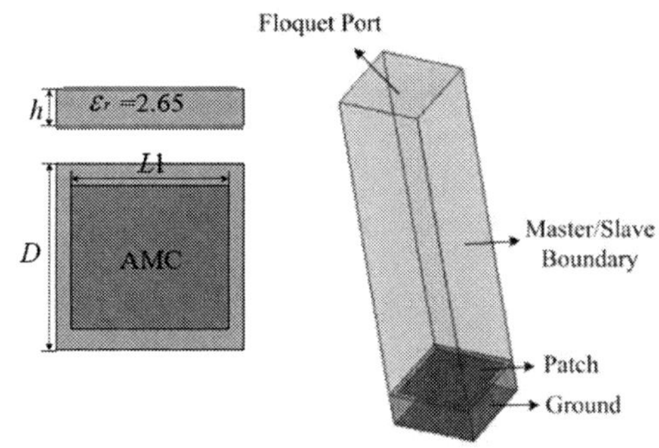

Figure 1. Prototype of the proposed AMC unit cell

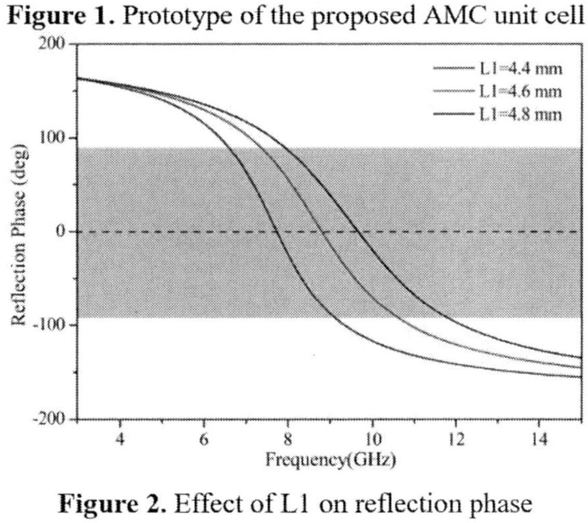

Figure 2. Effect of L1 on reflection phase

3. Design and Analysis of Proposed Antenna

The schematic configuration of the proposed low-RCS wideband planar monopole antenna with AMC reflector is shown in Fig.3 (b), and the reference antenna without loading AMC reflector is depicted in Fig.3 (a). The square radiation patch has a side length of Lp=8 mm and is printed on a Polytef dielectric substrate of length W=50 mm, thickness h=2 mm and relative permittivity ε_r=2.65, which is fed by the coplanar waveguide of length L=18.1 mm, Lc=20.8 mm, g=0.5 mm respectively. The AMC block composed by 3×10 unit cells whose side length $L1$=4.6 mm is loaded upon the substrate, with the corresponding ground on the bottom.

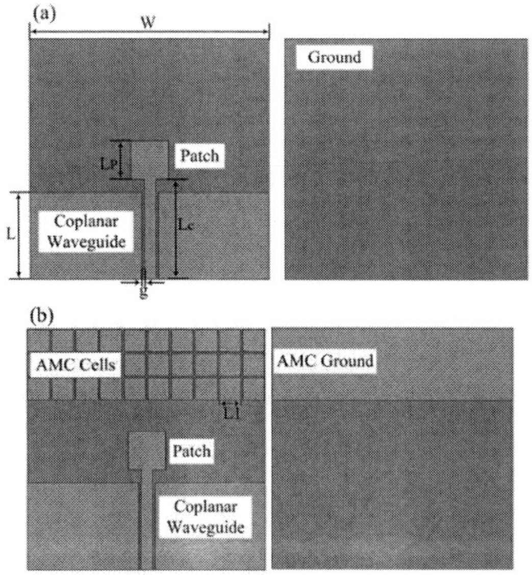

Figure 3. Proposed antenna: (a) reference; (b) after loading the AMC block

The preservation of the reference antenna radiation properties is a vital target for the proposed antenna design. The return loss curve after loading the AMC cells is shown as Fig.4, the original reference antenna shows multi frequency characteristic for 10 dB return loss bandwidths on 3.9 GHz~5.8 GHz, 7.3 GHz~9.7 GHz and 11.2 GHz~12.1 GHz, respectively, corresponding to an impedance bandwidth of 40.43%, 26.67% and 7.69% with respect to the resonant frequencies. However, the bandwidth of proposed monopole antenna loaded with the AMC reflector cover 3.3 GHz~12.2 GHz (114.84%), that is due to the surface current distribution of reference antenna is changed after loading the AMC block, following with the improved impedance matching. Besides, the radiation patterns at resonant frequencies is depicted in Fig.5, loading the AMC reflecting screen hardly has any significant impact at the lower resonant frequency, the gain is enhanced in certain directions at the third resonant frequency, we could conclude that the novel low-RCS monopole antenna work well.

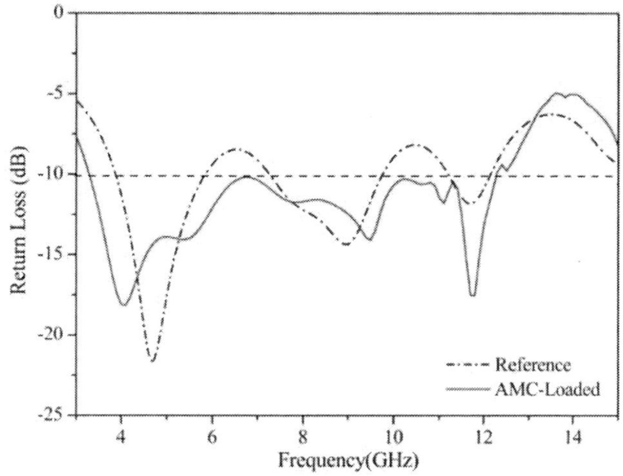

Figure 4. Simulated return loss curve of reference and proposed antenna

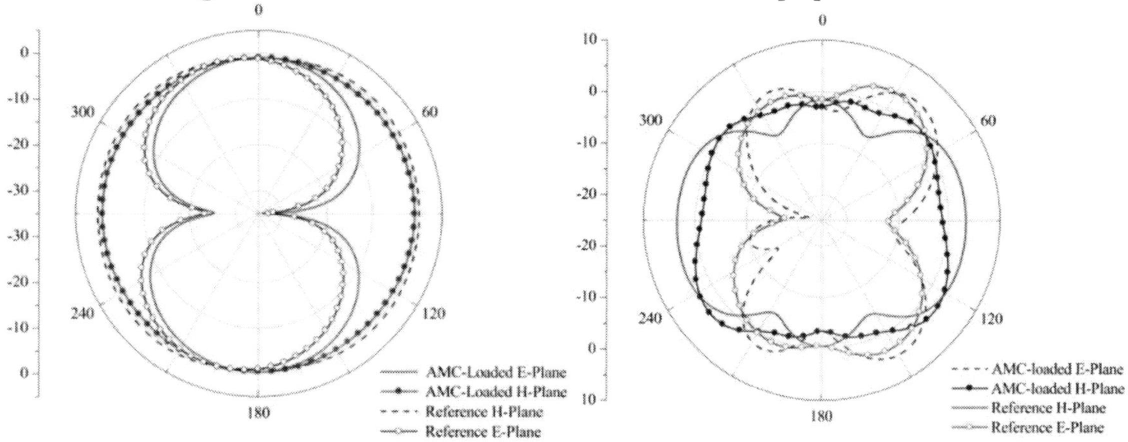

Figure 5. Comparison of radiation patterns: (a) the first resonant frequency; (b) the third counterpart

The main idea of this paper is aimed at the design of a surface that reflects the incident plane wave in phase and counter-phase simultaneously, and achieve a null in specular direction, the power of incident wave will be reflected over other directions. The simulated results of monostatic RCS is depicted in Fig.6, the RCS is reduced over the 9.3 GHz~11.3 GHz compared with the reference antenna, meet the pervious theory, and obtained the maximum RCS reduction value 26.2 dB at 10.1 GHz.

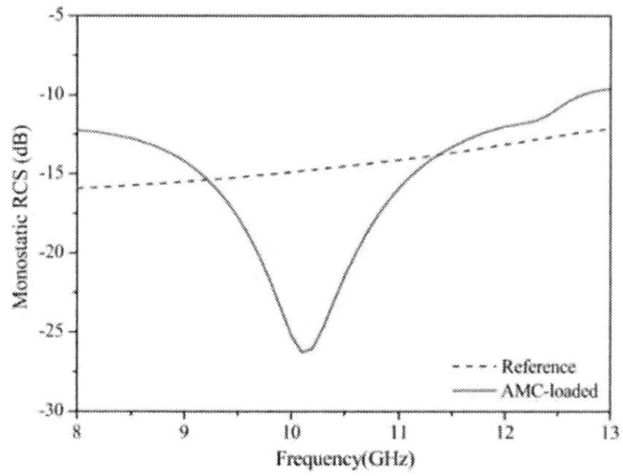

Figure 6. Simulated results of monostatic RCS

IOP Conf. Series: Journal of Physics: Conf. Series **887** (2017) 012037 doi:10.1088/1742-6596/887/1/012037

4. Fabrication and Measurement

Once the structure of the proposed antenna is confirmed, a prototype was fabricated, as depicted in Fig.7. The return loss against frequency for the proposed monopole antenna was measured by using the Agilent N5224A vector network analyzer and is presented in Fig.8, the 10 dB return loss bandwidth is extended to 3.4 GHz~12.4 GHz (113.92%), coincidence with the simulated counterpart, and thus the proposed low-RCS antenna has an ultra-broad bandwidth.

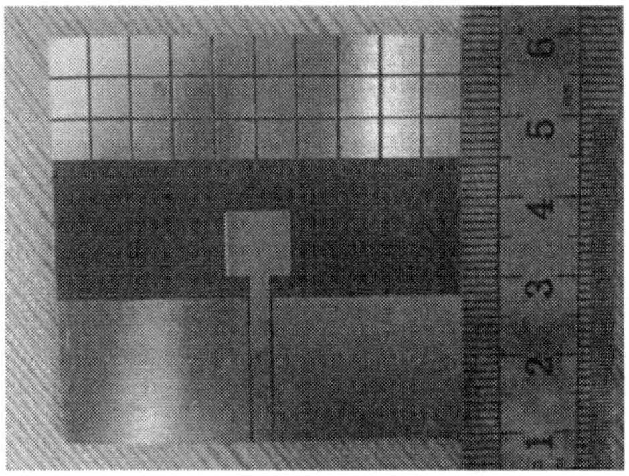

Figure 7. Prototype of proposed low-RCS wideband monopole antenna

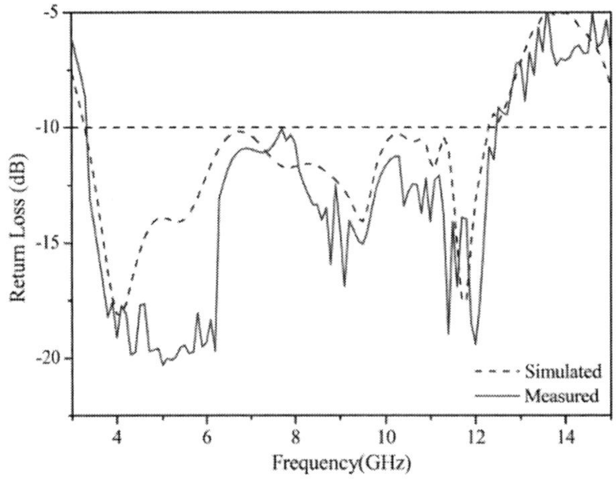

Figure 8. Experimental results of return loss

The monostatic behavior of the proposed structure has been evaluated by means of the standard horn antenna reflection coefficient characterization in an anechoic chamber. And from the obtained results in Fig.9, it can be seen that RCS reduction is reached within 8.9 GHz~11.2 GHz, match with the simulated results.

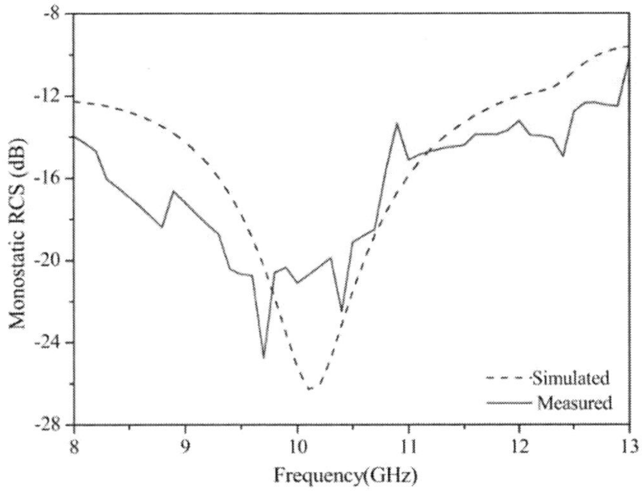

Figure 9. Experimental results of monostatic RCS

5. Conclusion

The principle of destructive interference between PEC and AMC for RCS reduction in monopole antenna has been validated both theoretically and experimentally, reasonable agreement between the simulation and measurement is achieved. In addition, the radiation performance is also improved by loading the AMC reflector on the monopole antenna, the measured bandwidth of new antenna cover 3.4 GHz~12.4 GHz (113.92%), and the radiation pattern is preserved compared with the original antenna, and thus obtain a low-RCS and wideband monopole antenna. In addition, the bandwidth of 10 dB RCS reduction is still insufficient, for further work, it is aimed at extending the bandwidth of 10 dB RCS reduction by improving the proposed structure.

6. Acknowledgment

Project supported by the Joint Funds of the National Natural Science Foundation of China (Grant No. U1637212), the National Natural Science Foundation of China (Grant No. 61401405, 51475438), Natural Science Foundation of Shanxi Province, China (Grant No.2014011021-4), and the Shanxi Provincial Foundation for Returned Scholars (Main Program), China (Grant No. 20150603ZX).

7. References

[1] Q. Fu, C,L Fan, S.J. Li, G. Wang, and X.Y. Cao, "Ultra-Broad Band Radar Cross Section Reduction of Waveguide Slot Antenna with Metamaterials," Radioengineering, vol.25 Jun. 2016, pp. 241-246, doi:10.13164/re.2016.0241.

[2] Y.H. Liu and X.P. Zhao, "Perfect Absorber Metamaterial for Designing Low-RCS Patch Antenna," IEEE Antennas Wireless Propag. Lett., vol.13, Jul. 2014, pp. 1473-1476, doi: 10.1109/LAWP.2014.2341299.

[3] A. Ghosh, T. Mandal, and S. Das, "Design of Triple Band Slot-Patch Antenna with Improved Gain Using Triple Band Artificial Magnetic Conductor," Radioengineering, vol. 25 Sept. 2016, pp.442-448, doi: 10.13164/re.2016.0442.

[4] Y.W Zhong, G.M Yang, and L.R Zhong, "Gain enhancement of bow-tie antenna using fractal wideband artificial magnetic conductor ground," Electron. Lett., vol.51, Feb. 2015, pp. 315–317, doi: 10.1049/el.2014.4017.

[5] H. Malekpoor and S. Jam, "Improved Radiation Performance of Low Profile Printed Slot Antenna Using Wideband Planar AMC Surface," IEEE Trans. Antennas Propag., vol. 64, Nov. 2016, pp. 4626-4638, doi:10.1109/TAP.2016.2607761.

[6] Y.J. Zheng, J. Gao, X.Y. Cao, Z.D. Yuan, and H.H. Yang, "Wideband RCS Reduction of a Microstrip Antenna Using Artificial Magnetic Conductor Structures," IEEE Antennas Wireless Propag. Lett., vol.14, Mar. 2015, pp.1582-1585, doi: 10.1109/LAWP.2015.2413456.

[7] S.J. Li, J. Gao, X.Y. Cao, Y. Zhao, Z. Zhang, and H.X. Liu, "Loading metamaterial perfect absorber method for in-band radar cross section reduction based on the surface current

distribution of array antennas," IET Microw. Antennas Propag, vol.9, Oct. 2015, pp. 399-406, doi: 10.1049/iet-map.2014.0490.

[8] A. Kush, Nasimuddin, and A. Arokiaswami, "Wideband Circularly Polarized AMC Reflector Backed Aperture Antenna," IEEE Trans. Antennas Propag., vol. 61, Mar. 2013, pp. 1456-1461, doi: 10.1109/TAP.2012.2227446.

[9] S. H. Esmaeli and S.H. Sedighy, "Wideband radar cross-section reduction by AMC," Electron. Lett. vol. 52, Jan. 2016, pp. 70–71, doi: 10.1049/el.2015.3515.

[10] M. Mighani and G. Dadashzadeh, "Broadband RCS reduction using a novel double layer chessboard AMC surface," Electron. Lett. vol. 52, Jul. 2016, pp. 1253–1255, doi:10.1049/el.2016.1214.

[11] S. M. Rouzegar, A. Alighanbari, and O. M. Ramahi, "Wideband Uniplanar Artificial Magnetic Conductors Based on Curved Coupled Microstrip Line Resonators," IEEE Microw. Wireless Compon. Lett., vol. 27, Apr. 2017, pp.326-328, doi: 10.1109/LMWC.2017.2678434.

[12] W. Jiang, Y. Liu, S.X. Gong, and T. Hong, "Application of Bionics in Antenna Radar Cross Section Reduction," vol. 8, Nov. 2009, pp.1275-1278, doi: 10.1109/LAWP.2009.2037168.

[13] M. Paquay, J. Iriarte, I. Ederra, R. Gonzalo, and P. Maagt, "Thin AMC Structure for Radar Cross-Section Reduction," IEEE Trans. Antennas Propag, vol.55, Dec. 2007, pp. 3630-3638, doi:10.1109/TAP.2007.910306.

[14] X. Liu, J. Gao, X.Y. Cao, Y. Zhao, W.Q. Li, S.J. Li, and N. Li, "A High-gain and Low-scattering Waveguide Slot Antenna of Artificial Magnetic Conductor Octagonal Ring Arrangement," Radioengineering, vol. 25, Apr. 2016, pp. 46-52, doi: 10.13164/re.2016.0046.

[15] [C. M. Dikmen, S. Çimen, and G. Çakır, "Planar Octagonal-Shaped UWB Antenna With Reduced Radar Cross Section," IEEE Trans. Antennas Propag., vol. 62, Jun. 2014, pp.2946-2953, doi: 10.1109/TAP.2014.2313855.

[16] J.J. Zhang, J.H. Wang, M. Chen, and Z. Zhang, "RCS Reduction of Patch Array Antenna by Electromagnetic Band-Gap Structure," IEEE Antennas Wireless Propag. Lett., vol. 11, Aug. 2012, pp.1048-1051, doi: 10.1109/LAWP.2012.2215832.

[17] Y. Liu, Y.W. Hao and S.X. Gong, "Low-profile high-gain slot antenna with Fabry-Pérot cavity and mushroom-like electromagnetic band gap structures," Electron. Lett.,vol. 51, Feb. 2015, pp. 305–306, doi: 10.1049/el.2014.4190.

IOP Conf. Series: Journal of Physics: Conf. Series **887** (2017) 012038 doi:10.1088/1742-6596/887/1/012038

The design and implementation of network attack and defense platform based on cloud desktop

Zining Yan[1], Yunhan Wang[2], Shijiao Shao[3] and Boquan Li[4]

[1]Embedded System Lab. College of Computer Science and Technology, Harbin, China, E-mail: 954277966@qq.com
[2]Embedded System Lab. College of Computer Science and Technology, Harbin, China, E-mail: 2224206409@qq.com
[3]Embedded System Lab. College of Computer Science and Technology, Harbin, China, E-mail: 769351769@qq.com
[4]Institute of Information Engineering Chinese Academy of Sciences; School of Cyber Security, University of Chinese Academy of Sciences
Key Laboratory of Network Assessment of Technology, Chinese Academy of Sciences, Beijing, China, E-mail: liboquan@iie.ac.cn

Abstract. This paper discusses the cloud desktop technology, virtualization technology and penetration testing technology used in the network attack and defense training platform, and introduces the design and implementation process of the network attack and defense training platform. This paper focuses on the cloud desktop construction scheme based on B/S structure, and aims to enhance the flexibility and convenience of network attack and defense training platform.

1. Introduction

Network security incidents occur frequently so that the importance and key of network security are self-evident, to enhance the network attack and defense technology is the most effective means to prevent and solve the problem of network security. In order to improve the network attack and defense technology, the country has opened the research and construction of network shooting field, the United States has launched "cyberspace Manhattan Plan" with a budget up to 30 billion dollars, the US Defense Advanced R & D Program launched the "National Cyber Range" [1], to provide training and training field for the US network attack and defense. For China, network security has gradually become an important issue associated with national military and economic development. Therefore, to construct the network attack and defense exercise system and to enhance security technician' network attack and defense capabilities, is an important foundation for the construction of China's network information security.

Though we have done much research in the field of network attack and defense exercise platform at home and abroad, there are still some ubiquitous problems among the existing platform, like too elaborate to operate, excessive construction costs, maintenance difficulties and etc. On account of the above problems, this paper uses the XenServer system as the underlying virtualization server system to build the cloud desktop based on the B / S structure and realizes the network attack and defense exercise platform based on the cloud desktop. The user can operate the remote desktop through browser so as to reduce resources, and enhance the flexibility and convenience of the offensive and defensive exercise platform.

Section 1 of this paper introduces the research background and significance of this subject. In Section 2, the key technologies such as penetration testing technology, virtualization technology, and cloud desktop technology are studied. Section 3 introduces the remote control technology of cloud desktop based on B / S Structure, analyzes the construction of virtual environment based on XenServer, and the realization process of cloud desktop platform based on B / S Structure. Section 4 describes the process of the design and implementation of the exercise system, Section 5 analyzes the platform function and performance tests.

2. Key Technical Analysis

2.1. Penetration Testing Technology

Penetration testing exercise system is the most common way of network attack and defense exercise. The network penetration test is a method of evaluating the security of a computer network system by simulating a malicious hacker's attack method. The security technician detects the network security from the attacker's point of view· [2] and takes the initiative to analyze the system vulnerabilities, defects, and other weaknesses. A penetration test can effectively check out the existence of network security problems, help to block the attack that the network may be suffered, so that network security is more secure.

The process of penetration testing is similar to the hacker attack process and requires vulnerabilities or technical flaws in exploiting the system. Testers need to exploit vulnerabilities or defects to carry out penetration testing while ensuring the least destructive nature of the network environment and target systems. Penetration testing is a step-by-step process with a certain rule [3].

A penetration test can not only detect whether the protective measures of the target system are perfect but also can restore the security problems existing in the system. Testers can solve the problem in time according to the test results and adjust the repair mode in time to provide a reliable basis for the safety protection mechanism of the target system.

2.2. Cloud Desktop Technology

2.2.1. Cloud Desktop Technology Introduction. Cloud desktop technology is the user's desktop environment associated with the terminal equipment, the complete desktop environment stored in the server. Users can access remote desktops with other end devices that can connect to the network [4]. Cloud desktop technology will put all the resources processing into the background server, the system maintainers through the background allocation center manage a variety of desktop unify, this process reduces the system maintenance time and workload, while control management of the desktop set enhances system security. Cloud desktop technology schematic diagram is shown in Figure 1.

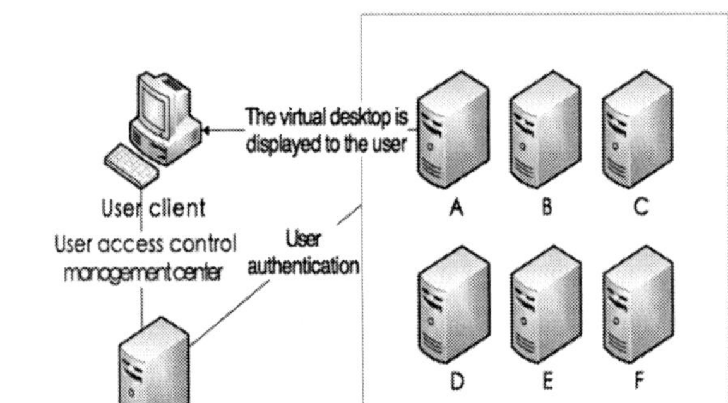

Figure 1. Cloud desktop technology schematic

As can be seen from the figure, the principle of cloud desktop technology is mainly: the user through the terminal equipment use the network connection to the configuration management center, configuration management center will verify the identity of the user, if verified through, the user will access the data center server to access and control Remote cloud desktop [5].

2.2.2. VNC remote control tool. When a user accesses a remote cloud desktop, the client needs to support the Remote Desktop Protocol for the purpose of controlling cloud desktop access. VNC (Virtual Network Computing) protocol is a remote control to achieve the remote desktop protocol, which provides a remote management function that displays the entire desktop of the remote server on the client side. VNC remote control tool based on C / S (Client / Server) architecture, made up of the VNC Server and VNC Viewer two parts. The schematic diagram is shown in Figure 2.

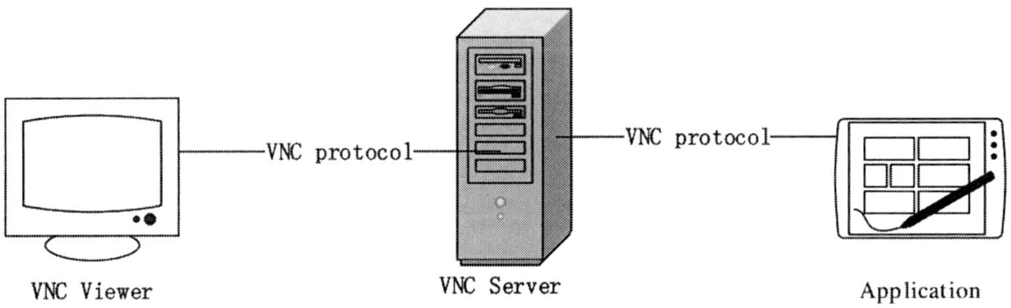

Figure 2. Remote control tool schematic

As can be seen from the above figure, VNC Server allows the application in the system to run on the desktop of the VNC Server through the X protocol, so that the user client installed VNC Viewer can access and control the remote desktop through the VNC protocol. The nature of VNC Viewer is a monitor program, through the VNC protocol, the user client's monitor program can communicate with the remote server , and the server connect to a number of hardware devices and run a number of applications, so Users can monitor and operate the hardware and applications of remote desktops with VNC Server [6].

2.3. Xenserver Server Virtualization Solution

Server virtualization technology is the process of abstracting the physical resources of a server into logical resources, isolating a server into multiple virtual servers, or merging multiple servers into a single server [7]. Server virtualization technology to achieve the CPU, memory, I / O, disk, and other hardware dynamic management, improve the utilization of resources, so that the system management is more flexible and simpler.

XenServer is a server virtualization solution of standalone virtualization [8]. The underlying layer is based on Linux, so XenServer is also a special Linux system. XenServer has the advantages of compatibility and security, source code opening and free features [9], so it is widely used by companies or research organizations for critical applications or secondary development.

XenCenter is a tool for managing XenServer that can operate easily the creation, start, stop, and generation of virtual machine lifecycles. The graphical interface interacts well with users. The XenServer virtualization solution efficiently completes the goals that server virtualization needs to meet, allowing enterprise, individuals or any user to use.

3. Cloud Desktop Platform Based On B/S Structure

3.1. Cloud Desktop Remote Control Technology based on B/S Structure

The remote control technology of cloud desktop based on B/S structure is mainly realized by Guacamole project. Guacamole is a WEB application based on HTML 5.Users can access the Web application in the server through a browser and access and control the remote desktop in conjunction with the Remote Desktop Transport Protocol (VNC or RDP) or Remote Login Session Protocol (SSH).Its working principle is shown in figure 3.

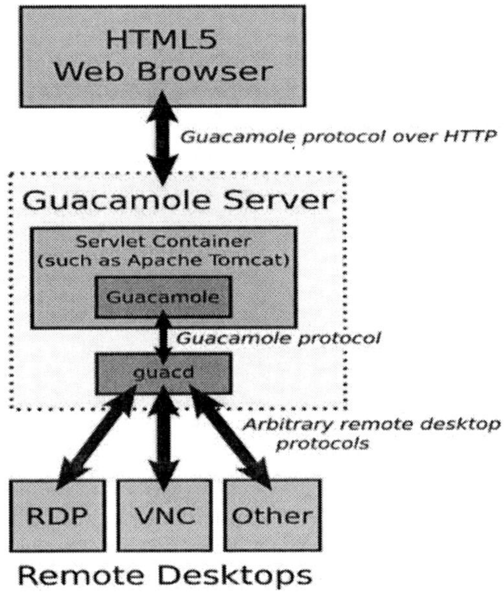

Figure 3. Guacamole working schematic

As can be seen from Figure 3, when the users visit Guacamole server through the browser, Guacamole server with WEB application server (such as Apache Tomcat) provides services to users. Guacd helps to deploy the application in the Guacamole server to parse the Guacamole protocol and helps to identify which desktop transport protocol (RDP, VNC, etc.) is used for remote desktop applications. When each step goes smoothly, the user can communicate with the server via HTTP.

In the Guacamole project, the Guacamole protocol is a remote display and interactive protocol that is responsible for displaying and rendering messages, web applications deployed in the Guacamole server allow the application itself to support other desktop transport protocols (RDP, VNC, etc.) by parsing the Guacamole protocol. WEB application is written in Java, front-end is realized by html5 and java script implementation, the WEB application cannot achieve any desktop protocol by itself, but rely on the entire project core Guard (plug-in dynamic support Remote Desktop Protocol) to resolve the Guacamole protocol and identify other desktop transport protocol.

The desktop transform protocol used in this topic is the VNC protocol. Guacamole replaces the VNC Viewer in the VNC remote control tool with a browser, allowing users to access remote desktops with any operating system or even mobile devices.

3.2. The Construction of Cloud Desktop Platform

3.2.1. Server virtualization environment. The network attack and defense exercise platform need to provide each user with the same desktop environment so that each user can conduct network attack and defense exercises in the same environment. Server virtualization technology achieves to divide a server into a number of virtual servers and assign to each user a virtual machine. While using XenServer virtualization solutions for virtual machine cloning and porting and other functions, a well-deployed exercise environment is cloned into a number of exercise environment to achieve the same conditions for each user's exercise condition. The server virtualization environment architecture is shown in Figure 4.

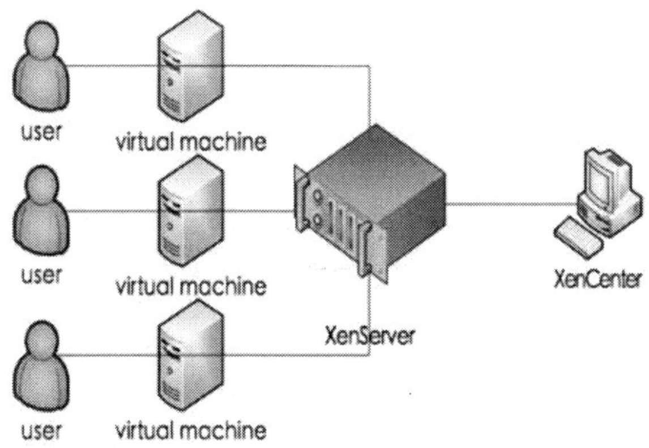

Figure 4. Server Virtualization Environment Architecture

3.2.2. WEB application server. A key link of the construction of the cloud platform based on the B/S structure is to realize the remote cloud desktop access through the browser. The cloud desktop is evolved into a WEB application, through accessing to WEB applications to indirectly access the remote cloud desktop. This topic uses Guacamole to connect the browser with VNCServer and uses Guacamole Server as the medium to access and control the remote cloud desktop through VNC remote desktop protocol.

WEB application server store the application and data files for the network attack and defense exercise platform, providing online information browsing service, which is the system and user interaction hub, but also for the entire network attack and defense exercise platform design and implementation laid the foundation.

4. Design and Implementation of Exercise System

4.1. Vulnerability Database

Vulnerabilities are the root of the system and application security issues. Therefore, the construction of the exer-ncise system vulnerability database is the basis of system development. The vulnerability database should contain the basic information, characteristics, hazards and protection programs of the security vulnerabilities. The vulnerability database has some reference value to the research of the penetration test technology and the network attack and defense technology. At the same time, it can help the attack and defense technician to master the characteristics of the vulnerabilities better and develop the corresponding repair plan according to the characteristics of the vulnerabilities, and also can have some guiding effect on the system protection.

Based on the investigation of the main security vulnerabilities involved in the current penetration test, this system has constructed a database of typical vulnerabilities including brute force, command execution vulnerability, cross-site request forgery, SQL injection, file upload vulnerability, cross-site scripting attack, etc. The vulnerability database contains two parts, the document, and program. The documents and programs in the vulnerability database will be directly applied to the development of the exercise environment, which make the system's development and operation more flow and formal. It is the basis of the system exercise function. The construction of the vulnerability database is a continuous process, with the update and development of applications, vulnerabilities should continue to add the newly discovered vulnerabilities.

Through the collection, programming, writing documents and other processes to build typical vulnerabilities in the exercise process, and develop the corresponding exercise environment for each vulnerability by using PHP language, the program, and documentation contained in the vulnerabilities database will be applied directly to the development of exercise system and exercise environment. The main vulnerabilities currently contained in this project are shown in Table 1.

Table 1. Network vulnerability typical vulnerability information table

Vulnerability name	Vulnerability description
Brute force	Test continuously the system password or command and other verification information for several times, until you use the correct authentication information to log on to the system.
Command execution vulnerability	Through the browser to the function filter does not submit the implementation of the order, causing the attacker by changing the system path or other program environment.
Cross station request forged	The use of the site to the browser's trust, the temptation of users in the WEB application has been registered on the implementation of non-intended operation.
SQL injection	Enter the malicious SQL statement in the Web form to get the data in the server database.
File upload vulnerability	Upload files that contain malicious code and be executed by the server by using the Upload File feature.

4.2. Penetration Test Exercise Environment

Penetration test exercise environment includes six typical vulnerabilities: Brute force, command execution vulnerability, cross-site request forgery, SQL injection, file upload vulnerability, and cross-site scripting. Each vulnerability achieves low-level and advanced two exercise environments according to the degree of filtering. Each vulnerability is specifically designed and implemented as follows:

The main way of brute force is to guess the password for the system and other authentication information and try a number of logins until you enter the correct authentication information log on the system or get important data. The exercise environment of brute force vulnerability is a basic form that contains only the username and password. When the user enters the correct username and password, the user is prompted passing, the system will give a prompt when the username and password do not match. Users can help brute force through the system's pre-provided Brute tool.

The exercise environment of brute force sets the difficulty based on the degree of filtering of the characters entered in the form. The main code for filtering user names and passwords for special characters is as follows:

```
$user = mysql_real_escape_string( $user );
$pass = mysql_real_escape_string( $pass );
```

The system uses the mysql_real_escape_string () function to escape special characters in the string used in the SQL statement, which filters out the special characters to increase the difficulty of brute force.

The exercise principle for the command execution vulnerability is through the browser to submit an execution command to the server that does not filter functions, which cause the attacker to change the system path or other program environments. An exercise environment is a form that prompts for an IP address. The system feeds back related information to users based on user-submitted commands.

In command execution vulnerability exercise environment, the input coment is not filtered by low difficulty level .with the difficulty level increasing "&&", ";" and other characters are be filtered out. The main code for filtering the input characters is as follows:

```
$substitutions = array ('&&' => '', ';' => '',);
$target = str_replace (array_keys ($substitutions), $substitutions, $target);
```

The system will add special characters to the blacklist, when the user input sensitive characters, it will automatically replace the sensitive characters so that the input command of the system will not be able to perform.

The main principle of cross-site requests forgery vulnerabilities is to take advantage of the site's trust in the browser and entice the user to perform unintended operations on the logged-in web application. An exercise environment is a form that prompts you to change the password, and the system feedbacks the correct password to the user according to the correctness of the user's password.

The exercise environment of cross-site request forged vulnerabilities, any judgment for the input content is not made by the exercise environment at the low difficulty level, the password can be changed whether the original password input is right or not. The environment at the high level using the specific functions to determine the source of the request. The main source of the judgment of request source is as follows:

```
If (eregi ("127.0.0.1", $ _SERVER ['HTTP_REFERER'])) {}
```

High-level difficulty using $ _SERVER ['HTTP_REFERER'] function judge the source for the user request, so that users need a certain means to modify the password.

The SQL injection vulnerability is done by entering a malicious SQL statement in the Web form to get the data in the server database. An exercise environment is a form that prompts the user to enter an ID, where the system feedback the information to the user according to the ID entered in the database.

Similar to the command execution vulnerability, the SQL injection exercise environment also sets the difficulty by filtering special characters. The main code for filtering characters is as follows:

```
$id = mysql_real_escape_string ($id);
```

The system through the MySQL built-in escaping function filter special characters, the user needs to bypass the escape function in a way for SQL injection attacks.

The main way of file upload vulnerability exercise is uploading the file, which contains malicious code and can be executed by the server, to the server through the way that application upload files. The file is executed by the application to achieve the purpose of malicious attacks. The file upload vulnerability exercise environment provides an entry for uploading files and submitting, and the system stores the files uploaded by the user to the server.

Difficulty is set by way of whitelist filtering to verify the properties of the uploaded picture or file. The main code for white list filtering is as follows:

```
$uploaded_name = $_FILES ['uploaded'] ['name'];
$uploaded_type = $_FILES ['uploaded'] ['type'];
$uploaded_size = $_FILES ['uploaded'] ['size'];
If (($uploaded_type == "image/jpeg") && ($uploaded_size < 100000)) {}
```

The system first gets the file name, file format, and file size of the uploaded file. If the uploaded file meets the requirements, the file can be uploaded, otherwise the file will not be uploaded.

Cross-site scripting attacks are divided into two types: reflection and storage. The main principle of storage-type cross-site scripting attacks is to enter the malicious executable code to the web form, which makes users access the page that is, execute malicious code, through the browser .while reflective cross-site scripting attacks have no database connection, the environment is in the form that can directly return the input .after the system execute the malicious code, the successful attack is proved by the pop-up window.

Reflective cross-site scripting exercise environment is a form that prompts for a name. The system will directly feedback the input content to the user in the form.

In a reflection-type cross-site scripting attack, the system uses the str_replace () function to filter out keywords and symbols that may appear in the popup window. Filter Keyword The main code is as follows:

Str_replace ('<script>', '', $ _GET ['name'])

Storage-type cross-site scripting attack environment is a message board. The system will store the name and message content entered by the user in the database.

Similar to the reflective cross-site scripting attack, the system also uses the str_replace () function to filter out keywords and symbols that may appear in the user's name, and the message form uses a relatively complex filtering function. The user needs to first select the form of the attack and try to choose a breakthrough form to attack.

5. System Function Test and Case Analysis

5.1. Functional Testing of Cloud Desktop Platform Based on B/S Structure

This section examines the functionality and performance of the cloud desktop platform with personal computers (desktop or laptop), tablet PCs, and smartphones, and analyzes the test results.

Personal computer access to the cloud desktop test with the Firefox browser. The test results are shown in figure 5.

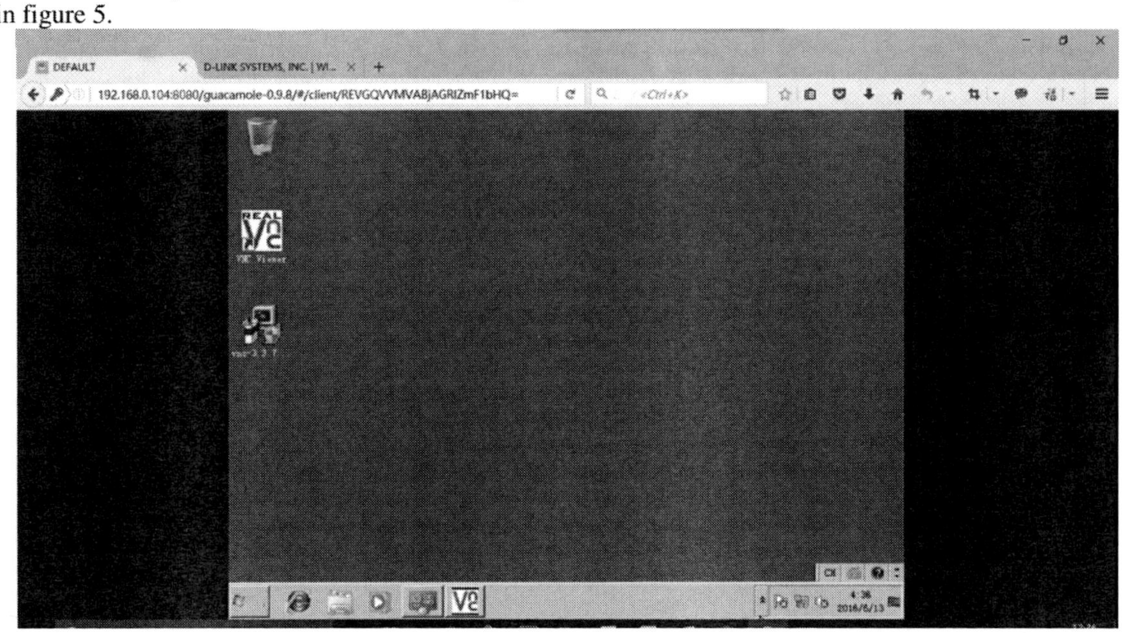

Figure 5. Personal computer visit the cloud desktop interface through the browser

In the test of Tablet PC visiting the cloud desktop, this test with the Tablet PC comes with Safari browser, the test results shown in Figure 6.

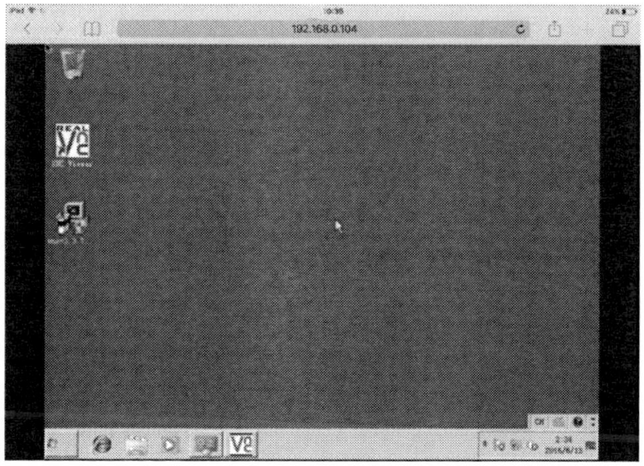

Figure 6. Tablet PC access the cloud desktop interface through the browser

In the smartphone access to the cloud desktop test, the same with the smartphone Safari browser for testing. The test results are shown in Figure 7.

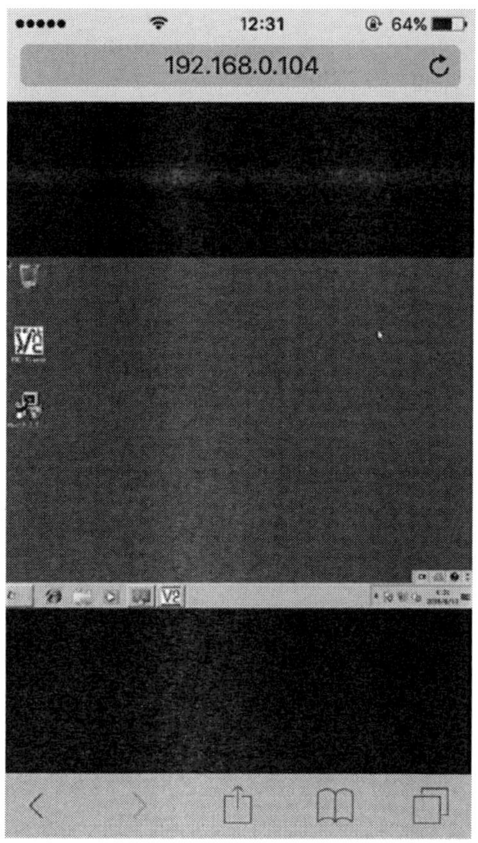

Figure 7. Smartphone through the browser to access the remote cloud desktop

Through the test results can be seen, the cloud desktop platform can be stable on the server to run, and through personal computers, tablet PCs, mobile phones and other devices to access the remote cloud desktop browser, the screen is stable and clear, which achieve the desired results and demand.

5.2. Functional Testing and Case Analysis of the Exercise System

The exercise system is a dynamic website of the B/S structure developed by using the combination of Apache + MySQL + PHP. Its function is mainly to drill up to set the difficulty of the drilling environment and select the appropriate type of vulnerabilities to drill. This section will select the storage-type cross-site scripting vulnerability as an example, to drill the identity of the user login training system, the difficulty set to primary, complete a complete exercise process. The exercise environment is shown in Figure 8:

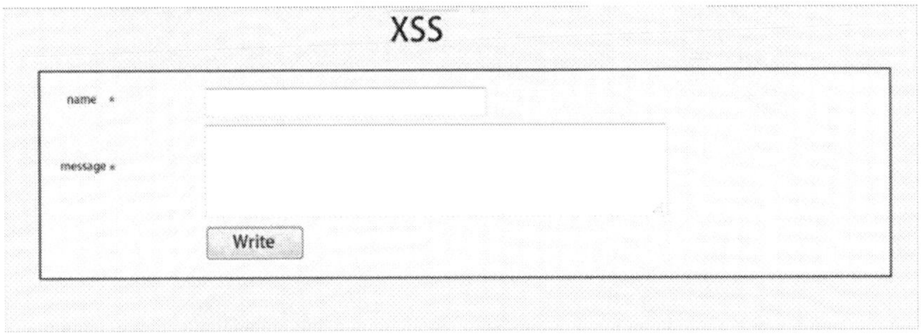

Figure 8. Storage type XSS attacks interface

Submit the following malicious code to the form:
<<SCRIPT>alert ("XSS") ;//<< /SCRIPT>
JavaScript code of the original system is modified, cross-site scripting attacks successfully executed. Successful attack system interface is shown in Figure 9.

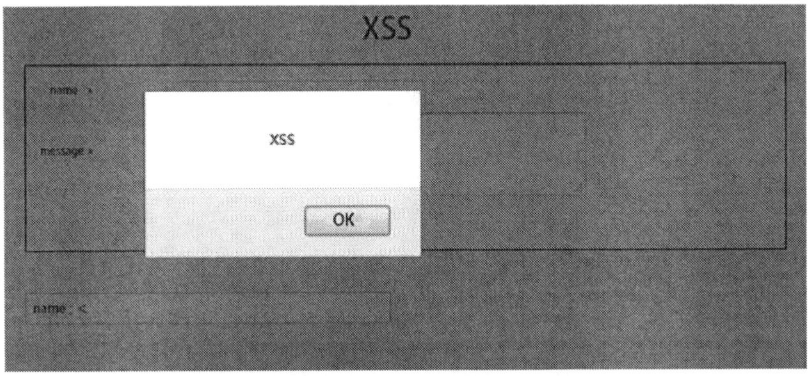

Figure 9. Cross-site scripting attack successful interface

Cross-site scripting attacks is a kind of Web applications attacks, an attacker trying to inject malicious script code into a trusted site to perform malicious operations. In a cross-site scripting attack, malicious code is executed on the browser side of the affected user and has an impact on the user.

This example interface appears pop-up, indicating that malicious script has been stored in the database, and successfully executed in the browser to achieve the desired results.

6. Conclusion and Prospect

In the process of the platform design and implementation, using XenServer as the underlying virtualization server system can improve the resource utilization.

With the help of the Guacamole project to build a cloud desktop platform based on B/S structure, the system's convenience and flexibility are improved. The good interaction between the exercise system and the user is achieved. The vulnerability database has some reference value to the research of the penetration test technology and the network attack and defense technology and can help the security technicians to improve the network attack and defense ability. Although XenServer virtualization solutions are used to manage resources efficiently, they still need to install XenCenter on a Windows-based computer to manage them.In the future, the XenServer management client based on the B / S architecture can be developed for the features of XenCenter, allowing the entire drill system to be managed only by the browser, making the operation more flexible. At the same time, the system only for the current penetration test drills in the typical loopholes in the exercise environment development, need to continue to vulnerabilities to add other vulnerability information. In the future, the above shortcomings can be further improved.

7. References

[1] Information Security Vulnerability Report (2015) [EB / OL]. Http: //www.cnnvd. Org.cn/news/jklbz/id/CNNVD Situation Report.
[2] Yan-Guang L I.Design and Realization of Network Attack and Defense Simulation System Terminal Subsystem [J] .Computer & Modernization, 2014 (3): 169-172,185.
[3] Chang Yan, Wang Guan. Network security penetration test research [J]. Information Network Security, 2012 (11): 3-4.
[4] Divorra Escoda O, Espinola Torrent D.CLOUD DESKTOP SYSTEM WITH MULTI-TOUCH CAPABILITIES: US, US 20130047095 A1 [P] .2013.
[5] Design and Implementation of Virtual Desktop Platform Based on Cloud Computing [D]. Fudan University, 2012.Kalje R A,Kosbatwar S P.Android Based Remote Control of Mobile Devices Using VNC System[J].International Journal of Computer Science & Information Technolo,2014.
[6] Kaur L,Kakkar H.A Review On optimization technique in Server Virtualization[J].International Journal of Computer Science & Information Technolo,2014.
[7] Machida F,Dong S K,Trivedi K S.Modeling and analysis of software rejuvenation in a server virtualized system with live VM migration[J]. Performance Evaluation, 2013, 70(3):212-230.
[8] Shirinbab S,Lundberg L,Ilie D.Performance Comparison of KVM, VMware and XenServer using a Large Telecommunication Application[J].2014.
[9] Guacamole [EB/OL]. http://guacamole.incubator.apache.org/. Woo Yun- 2011- 02330[EB/OL]

A channel estimation scheme for MIMO-OFDM systems

Chunlong He, Chu Tian, Xingquan Li, Ce Zhang, Shiqi Zhang, and Chaowen Liu

College of Information Engineering, Shenzhen University, Shenzhen, China, {chunlonghe@163.com}

Abstract. In view of the contradiction of the time-domain least squares (LS) channel estimation performance and the practical realization complexity, a reduced complexity channel estimation method for multiple input multiple output-orthogonal frequency division multiplexing (MIMO-OFDM) based on pilot is obtained. This approach can transform the complexity of MIMO-OFDM channel estimation problem into a simple single input single output-orthogonal frequency division multiplexing (SISO-OFDM) channel estimation problem and therefore there is no need for large matrix pseudo-inverse, which greatly reduces the complexity of algorithms. Simulation results show that the bit error rate (BER) performance of the obtained method with time orthogonal training sequences and linear minimum mean square error (LMMSE) criteria is better than that of time-domain LS estimator and nearly optimal performance.

1. Introduction

Multiple input multiple output (MIMO) systems have been discussed indispensable for the transmission improvement capacity. Orthogonal frequency division multiplexing (OFDM) has been considered as an important technique in the broadband wireless communication systems. Their combination, MIMO-OFDM technology, has attracted great interests in these recent years for enhancing the capacity, energy efficiency and improving the link reliability for the future fourth generation (4G) mobile communication systems [1-3]. However, in order to get the promised theoretical capacity and to achieve maximum gain of diversity, MIMO-OFDM systems require accurate channel state information (CSI) at the receiver [4]. The knowledge of the CSI at the receiver is crucial for coherent detection and space-time decoding. In OFDM systems, channel estimation has been successfully used to improve the system performance [5-6]. It has been demonstrated that coherent demodulation can be used instead of differential de-modulation a 3-4dB gain of signal-to-noise ratio [7-8]. However, the method of channel estimation in OFDM can't be used to MIMO-OFDM systems directly because the antenna interfere with each other. There are some literatures on channel estimation for MIMO-OFDM systems. The minimum mean square error (MMSE) channel estimators in [5] and the time-domain Least square (LS) channel estimators in [6] are presented, both of which have good performance at the cost of high complexity of calculating the pseudo-inversion of the matrix. Hence, several reduced complexity channel estimation methods are proposed in [9-14], whereas introducing performance degradation and limitations of application.

In this paper, we obtain a novel reduced complexity channel estimation method for MIMO-OFDM systems which is applicable to the slow-fading channel. This approach can transform the complexity of MIMO-OFDM channel estimation problem into a simple SISO-OFDM channel estimation problem and therefore there is no need for large matrix pseudo-inverse. As demonstrated by computer simulations, the proposed method provides better performance and lower complexity.

The rest of this paper is organized as follows. The system model is presented in the second section. Section III introduces the time-domain LS channel estimator. A novel reduced complexity channel estimation method is derived In Section IV. In Section V, the simulation environment and results are described. Section VI concludes the paper.

Notation: In this paper, Bold upper and low letters denote matrices and column vectors, respectively; $(.)^T$, $(.)^*$, $(.)^H$ denote transpose, conjugate, and Hermitian transpose, respectively; $(.)^+$ and $|.|$ denote matrix pseudo-inversion and absolute value, respectively.

2. System Model

A MIMO-OFDM system with two transmitter and two receiver antennas is shown in Fig.1 [8]. At a transmission time n, a binary data block $\{\mathbf{b}[n,k]: k=0,1...K-1\}$ is coded into two different signals, $\{\mathbf{t}_i[n,k]: k=0,1...K-1, i=1,2\}$, where K, k, and i are respectively the number of sub-channels of the OFDM systems, sub-channel index, and antenna index. Each of these signals forms an OFDM block. At the receiver, the discrete Fourier transform (DFT) of the received signal at each receiver antenna is the superposition of two distorted transmitted signals. The received signal at the j-th receiver antenna can be expressed as in [8]

$$\mathbf{r}_j[n,k]=\sum_{i=1}^{2}\mathbf{H}_{ij}[n,k]\mathbf{t}_i[n,k]+\mathbf{w}_j[n,k] \tag{1}$$

where $\mathbf{H}_{ij}[n,k]$ is the channel frequency response for the k-th sub-channel at the time n, corresponding to the i-th transmitter antenna and j-th receiver antenna; $\mathbf{w}_j[n,k]$ is the additive complex Gaussian noise, with zero mean and variance δ_n^2. The noise is uncorrelated for different n, k, or j [8].

The channel is assumed to be a wide-sense stationary (WSS) channel model and unchanged over the duration of one OFDM symbol. The LS estimator for the current OFDM symbol is not related to the previous or the following OFDM symbols. The goal of MIMO-OFDM demodulation is to recover the transmitted signal from the received signals. From (1) we can obtain the frequency response at the k-th tone of the nth block corresponding to the i-th transmitter antenna can be shown as [8]

$$\mathbf{H}[n,k]=\sum_{l=0}^{K_0-1}\mathbf{h}[n,k]\mathbf{W}_K^{kl} \tag{2}$$

where K_0 is the number of multi-path, $\mathbf{W}_K=\exp(-j(2\pi/K))$.

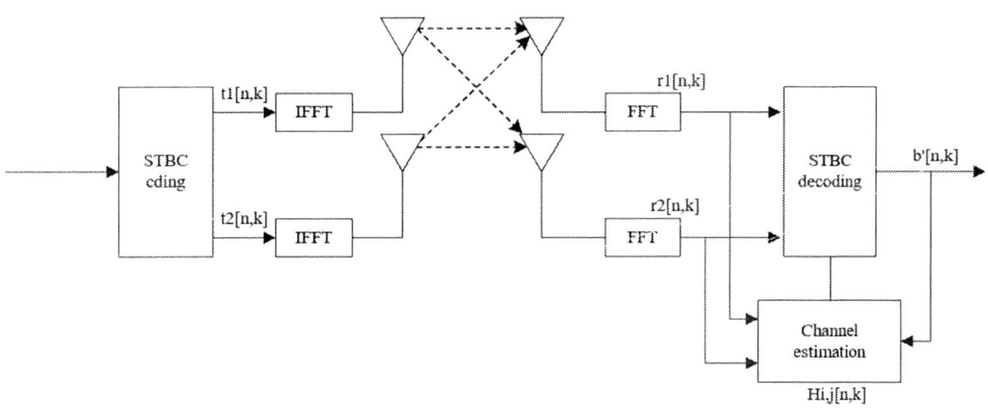

Figure 1. Block diagram of a STBC-OFDM system with two transmitter and two receiver antennas

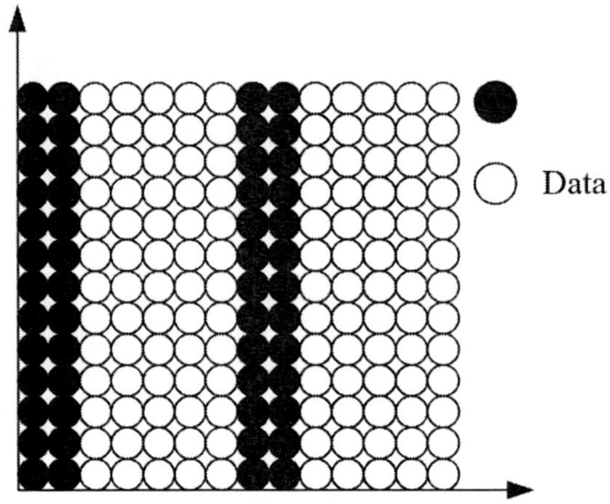

Figure 2. Block-type pilot sub-carrier arrangement

3. The Time-Domain Ls Channel Estimator

For an MIMO-OFDM systems, the received signal is overlapped signals transmitted from different transmit antennas, which makes the channel estimation difficult. Fortunately, channel parameters for different sub-carriers of each channel are correlated and a channel estimator has been developed in [8] based on this correlation. From (2), we only need to estimate $\mathbf{h}_i[n,k]$, then $\mathbf{H}_{ij}[n,k]$ can be obtained by FFT. If the transmitted signals $\{\mathbf{t}_i[n,k]: k=0,1...K-1, i=1,2\}$ are known, the temporal estimation of $\mathbf{h}_i[n,k]$, can be found by [8]

$$\begin{bmatrix} \mathbf{h}_1(n) \\ \mathbf{h}_2(n) \end{bmatrix} = \begin{bmatrix} \mathbf{Q}_{11}(n) & \mathbf{Q}_{21}(n) \\ \mathbf{Q}_{12}(n) & \mathbf{Q}_{22}(n) \end{bmatrix}^{-1} \begin{bmatrix} \mathbf{p}_1(n) \\ \mathbf{p}_2(n) \end{bmatrix} \tag{3}$$

Where,

$$\mathbf{p}_j[n,l] = \sum_{k=0}^{K-1} \mathbf{r}[n,k]\mathbf{t}_j^*[n,k]\mathbf{W}_K^{-kl} \tag{4}$$

$$\mathbf{q}_{ij}[n,l] = \sum_{k=0}^{K-1} \mathbf{t}_i[n,k]\mathbf{t}_j^*[n,k]\mathbf{W}_K^{-kl} \tag{5}$$

$$\mathbf{Q}_{ij}[n] = \begin{pmatrix} \mathbf{q}_{ij}[n,0] & \mathbf{q}_{ij}[n,-1] & \dots & \mathbf{q}_{ij}[n,-K_0+1] \\ \mathbf{q}_{ij}[n,1] & \mathbf{q}_{ij}[n,0] & \dots & \mathbf{q}_{ij}[n,-K_0+2] \\ \dots & \dots & \dots & \dots \\ \mathbf{q}_{ij}[n,K_0-1] & \mathbf{q}_{ij}[n,K_0-2] & & \mathbf{q}_{ij}[n,0] \end{pmatrix} \tag{6}$$

$$\mathbf{p}_i[n] = (\mathbf{p}_i[n,0], \mathbf{p}_i[n,1], ..., \mathbf{p}_i[n,K_0-1])^T \tag{7}$$

From (6), we can see that $2K_0 \times 2K_0$ matrix inversion is required to get the temporal estimation of $\mathbf{h}_i[n,k]$, which requires intensive computations. Next we will propose a novel method to reduce the computation complexity, with this method no matrix inverse is required.

4. The Novel Reduced Complexity Channel Estimation Method

Assuming that the channel is quasi-fixed over two consecutive OFDM symbols and every two neighboring pilots are grouped as one cluster, and these clustered pilot groups are equi-spaced as show in Fig. 2. The pilot signal is assigned to a particular OFDM block at each transmit antennas, which is sent periodically in time-domain. Let's the first cluster pilot OFDM for example, letting

$$\mathbf{H}_{11}[1] = \mathbf{H}_{11}[2]$$

$$\mathbf{H}_{21}[1] = \mathbf{H}_{21}[2]$$

From (1), the received signals can be expressed as

$$\mathbf{r}_1[1] = \mathbf{H}_{11}[1]\mathbf{t}_1[1] + \mathbf{H}_{21}[1]\mathbf{t}_2[1] + \mathbf{w}_1[1] \tag{8}$$

$$\mathbf{r}_1[2] = \mathbf{H}_{11}[2]\mathbf{t}_1[2] + \mathbf{H}_{21}[2]\mathbf{t}_2[2] + \mathbf{w}_1[2] \tag{9}$$

From (8) and (9), we have

$$\begin{aligned}
\mathbf{Z}_{11} &= \frac{\mathbf{r}_1[1]}{\mathbf{t}_2[1]} - \frac{\mathbf{r}_1[2]}{\mathbf{t}_2[2]} \\
&= \frac{\mathbf{H}_{11}[1]\mathbf{t}_1[1] + \mathbf{H}_{21}[1]\mathbf{t}_2[1] + \mathbf{w}_1[1]}{\mathbf{t}_2[1]} - \frac{\mathbf{H}_{11}[2]\mathbf{t}_1[2] + \mathbf{H}_{21}[2]\mathbf{t}_2[2] + \mathbf{w}_1[2]}{\mathbf{t}_2[2]} \\
&= (\frac{\mathbf{H}_{11}[1]\mathbf{t}_1[1]}{\mathbf{t}_2[1]} - \frac{\mathbf{H}_{11}[2]\mathbf{t}_1[2]}{\mathbf{t}_2[2]}) + (\frac{\mathbf{w}_1[1]}{\mathbf{t}_2[1]} - \frac{\mathbf{w}_1[2]}{\mathbf{t}_2[2]}) \\
&= \mathbf{H}_{11}[1](\frac{\mathbf{t}_1[1]}{\mathbf{t}_2[1]} - \frac{\mathbf{t}_1[2]}{\mathbf{t}_2[2]}) + (\frac{\mathbf{w}_1[1]}{\mathbf{t}_2[1]} - \frac{\mathbf{w}_1[2]}{\mathbf{t}_2[2]}) \\
&= \mathbf{H}_{11}[1]\mathbf{T1} + \mathbf{W1}
\end{aligned} \tag{10}$$

Where

$$\mathbf{T1} = \frac{\mathbf{t}_1[1]}{\mathbf{t}_2[1]} - \frac{\mathbf{t}_1[2]}{\mathbf{t}_2[2]} \tag{11}$$

$$\mathbf{W1} = \frac{\mathbf{w}_1[1]}{\mathbf{t}_2[1]} - \frac{\mathbf{w}_1[2]}{\mathbf{t}_2[2]} \tag{12}$$

Similarly

$$\mathbf{Z}_{21} = \frac{\mathbf{r}_1[1]}{\mathbf{t}_1[1]} - \frac{\mathbf{r}_1[2]}{\mathbf{t}_1[2]} = \mathbf{H}_{21}[1]\mathbf{T2} + \mathbf{W2} \tag{13}$$

$$\mathbf{Z}_{12} = \mathbf{H}_{12}[1]\mathbf{T1} + \mathbf{W1} \tag{14}$$

$$\mathbf{Z}_{22} = \mathbf{H}_{22}[1]\mathbf{T2} + \mathbf{W2} \tag{15}$$

Where

$$\mathbf{T2} = \frac{\mathbf{t}_2[1]}{\mathbf{t}_1[1]} - \frac{\mathbf{t}_2[2]}{\mathbf{t}_1[2]} \tag{16}$$

$$\mathbf{W2} = \frac{\mathbf{w}_1[1]}{\mathbf{t}_2[1]} - \frac{\mathbf{w}_1[2]}{\mathbf{t}_2[2]} \tag{17}$$

So the channel responses corresponding to different transmit antennas are decoupled. And the complexity of MIMO-OFDM channel estimation problem transformed into a simple SISO-OFDM channel estimation problem. The LS [14-15] estimators for pilot signals channel responses are given by

$$\hat{\mathbf{H}}_{11LS} = \mathbf{T1}^{-1}\mathbf{Z}_{11} = (\frac{\mathbf{Z}_{11}[1,0]}{\mathbf{T1}[1,0]}, \frac{\mathbf{Z}_{11}[1,1]}{\mathbf{T1}[1,1]} \cdots \frac{\mathbf{Z}_{11}[1,K-1]}{\mathbf{T1}[1,K-1]}) \quad (18)$$

$$\hat{\mathbf{H}}_{21LS} = \mathbf{T2}^{-1}\mathbf{Z}_{21} = (\frac{\mathbf{Z}_{21}[1,0]}{\mathbf{T2}[1,0]}, \frac{\mathbf{Z}_{21}[1,1]}{\mathbf{T2}[1,1]} \cdots \frac{\mathbf{Z}_{21}[1,K-1]}{\mathbf{T2}[1,K-1]}) \quad (19)$$

$$\hat{\mathbf{H}}_{12LS} = \mathbf{T1}^{-1}\mathbf{Z}_{12} = (\frac{\mathbf{Z}_{12}[1,0]}{\mathbf{T1}[1,0]}, \frac{\mathbf{Z}_{12}[1,1]}{\mathbf{T1}[1,1]} \cdots \frac{\mathbf{Z}_{12}[1,K-1]}{\mathbf{T1}[1,K-1]}) \quad (20)$$

$$\hat{\mathbf{H}}_{22LS} = \mathbf{T2}^{-1}\mathbf{Z}_{22} = (\frac{\mathbf{Z}_{22}[1,0]}{\mathbf{T2}[1,0]}, \frac{\mathbf{Z}_{22}[1,1]}{\mathbf{T2}[1,1]} \cdots \frac{\mathbf{Z}_{22}[1,K-1]}{\mathbf{T2}[1,K-1]}) \quad (21)$$

And the mathematical representation for LMMSE [15] estimators of pilot signals channel responses as follows:

$$\hat{\mathbf{H}}_{11LMMSE} = R_{hh}[R_{hh} + \sigma_n^2(\mathbf{X}_1{}^H\mathbf{X}_1{}^{-1})]^{-1}\hat{\mathbf{H}}_{11LS} \quad (22)$$

$$\hat{\mathbf{H}}_{21LMMSE} = R_{hh}[R_{hh} + \sigma_n^2(\mathbf{X}_2{}^H\mathbf{X}_2{}^{-1})]^{-1}\hat{\mathbf{H}}_{21LS} \quad (23)$$

$$\hat{\mathbf{H}}_{12LMMSE} = R_{hh}[R_{hh} + \sigma_n^2(\mathbf{X}_1{}^H\mathbf{X}_1{}^{-1})]^{-1}\hat{\mathbf{H}}_{12LS} \quad (24)$$

$$\hat{\mathbf{H}}_{22LMMSE} = R_{hh}[R_{hh} + \sigma_n^2(\mathbf{X}_2{}^H\mathbf{X}_2{}^{-1})]^{-1}\hat{\mathbf{H}}_{22LS} \quad (25)$$

Where,

$$\mathbf{X}_1 = diag(\mathbf{T1}[1,0], \mathbf{T1}[1,1], \ldots, \mathbf{T1}[1,K-1]) \quad \mathbf{X}_2 = diag(\mathbf{T2}[1,0], \mathbf{T2}[1,1], \ldots, \mathbf{T2}[1,K-1])$$

R_{hh} is the auto-covariance matrix of **h**. Then, the data sub-carriers can be estimated by linear interpolating neighboring pilot sub-channels.

5. Simulation Results

For the simulations, two transmit and two receive antennas MIMO-OFDM system with 64 sub-carriers and a cyclic extended guard interval of length 16 is considered. The channel bandwidth B=20 MHz, the sampling rate fs=20 MHz, which is shown in Table I and the carrier frequency fc=3.5GHz are selected similar to the HiperLAN/2 standard [16]. The channel is modeld as Rayleigh fading with parameters shown in Table II. QPSK is used for the bit error rate (BER) simulations, and the time-domain channel response length $K0$ is equal to 11 samples. The BER performance is measured by averaging over 10000 OFDM blocks and perfect synchronization between the transmitter and the receiver is assumed.

Table 1. Simulation Parameters

Parameters	Specifications
FFT Length	64
Sub-carriers	64
Guard Interval	16
Pilot Interval	4
Modulation Level	QPSK
Channel Model	Rayleigh Fading

Table 2. Rayleigh Fading Parameters

Path Number	Average Power(dB)	Delay(ns)
1	0	0
2	-6	100
3	-12	200
4	-18	300
5	-24	400
6	-30	500

The BER performance curves versus Eb/N0 for the time-domain LS estimation, the proposed estimation method and the ideal estimation in slow Rayleigh fading channel are shown in Fig.3-4. Fig.3 shows that the time-domain LS estimation exhibits 2~2.5dB BER gain compared with the obtained method under the LS criteria when the pilot sequences are non-orthogonal. When the pilot sequences are orthogonal, the obtained method has the same good performance with the time-domain LS estimation. Fig.4 shows that the performance of the obtained method under the LMMSE criteria is 1~1.5dB better than that of the time-domain LS estimation and nearly optimal performance when the pilot sequences is orthogonal. So we can get the conclusion that the obtained method with time orthogonal training sequences and LMMSE criteria has better estimation performance compared with the time-domain LS estimator. The calculation complexity of the obtained method is significant reduced and it is easy to implement in practical systems.

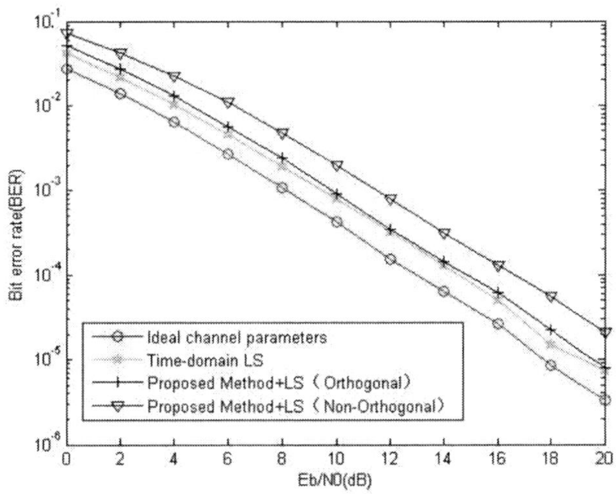

Figure 3. The BER performance versus Eb/N0 with LS criteria (vehicle speed= 15 km/h)

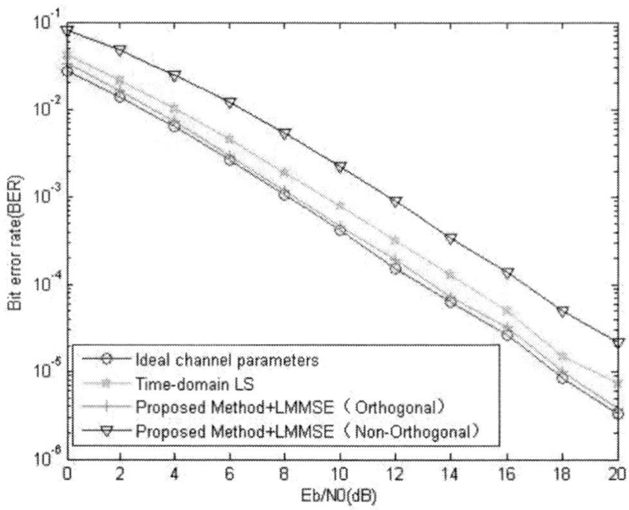

Figure 4. The BER performance versus Eb/N0 with LMMSE criteria (vehicle speed= 15 km/h)

6. Conclusions

In this paper, a reduced complexity channel estimation method is obtained for MIMO-OFDM system. Since the matrix inversion is avoided, significant complexity reduction is obtained while achieving an accurate frequency domain channel estimates. Compared with the time-domain LS estimator, better performance and low complexity can be obtained.

7. Acknowledgment

This work was supported by the Science and Technology Innovation Commission of Shenzhen under Grants JCYJ20150324140036835, Natural Science Foundation of China under grants 61601300, the O pen Research Fund through the National Mobile Communications Research Laboratory, Southeast Uni versity, Nanjing, China, under Grant 2017D10, Shenzhen University start-up funding No.2016053.

8. References

[1] Sampath H , Talwar S , Tellado J , et al. "A fouth-generation MIMO-OFDM broadband wireless system: design ,performance and field trail results," *IEEE Communications Magazine* , 2002 , pp: 143-149.

[2] H. Bolcskei, D. Gesbert, A. J. Paulraj, "On the Capacity of OFDM Based Spatial Multiplexing Systems," *IEEE Trans. Commun.*, vol. 50, no. 2, pp. 225–34, Feb. 2002.

[3] Yang H W. "A road to future broadband wireless access: MIMO-OFDM-based air interface," *IEEE Communications Magazine*, 2005, 43(1), pp:53-60.

[4] V. Pohl, P. H. Nguyen, V. Jungnickel, and C. von Helmolt, "How often channel estimation is needed in MIMO systems," in Proc. *IEEE Global Telecommun. Conf.*, San Francisco, Calif, USA, 2003, vol. 2, pp. 814–818.

[5] S. Coleri, M. Ergen, A. Puri. "Channel estimation techniques based on pilot arrangement in OFDM systems," *IEEE Trans. Broadcasting*, vol. 48, pp. 223-229, Sept. 2002.

[6] B. Song, L. Gui, and W. Zhang, "Comb type pilot aided channel estimation in OFDM systems with transmit diversity," *IEEE Trans. Broadcast.*,vol. 52, pp. 50–57, Mar. 2006.

[7] Runhua Chen and K. B. Letaief, "Channel Estimation for Space Time Coded OFDM Systems in Non-Sample-Spaced Multipath Channels," in Proc. *IEEE WCNC 2002*, pp: 61-66.

[8] Ye Li, Seshadri N , Ariyavisitakul S. "Channel estimation for OFDM systems with transmitter diversity in mobile wireless channels," *IEEE Journal Select Areas Communication*, 1999, 17 (1) , pp: 461-471.

[9] Ye Li. "Simplified channel estimation for OFDM systems with multiple transmit antennas," *IEEE trans. on wireless communications*. 2002 , 1(1), pp :67-75.

[10] Hlaing Minn, Dong In Kim, "A Reduced complexity channel estimation for OFDM systems with transmit diversity in mobile wireless channels," *IEEE Transaction communications*, May 2002, vol.50, pp: 799-807.
[11] Liu Gang, Guo yi, Ge jianhua, "An efficient channel estimation of MIMO-OFDM system", Journal of Jilin university, 2004,vol.6, pp: 29-32.
[12] C. L. He, B. Sheng, P.C. Zhu, X. H. Yu and Li, G. Y. "Energy-and spectral-efficiency tradeoff for distributed antenna systems with proportional fairness', *IEEE Journal on Selected areas in Communications*, Vol. 31,No. 5, pp.894-902,2013.
[13] C. L. He "Comparison of Three Different Optimization Objectives for Distributed Antenna Systems" *International Journal of Electronics and Communications*, vol. 70, pp. 442-448, 2016.
[14] B. Song, L. Gui, and W. Zhang. "Comb type pilot aided channel estimation in OFDM systems with transmit diversity," *IEEE Trans. Broadcast.*,vol. 52, pp. 50–57, Mar. 2006.
[15] C. L. He, Y. Li, F. Zheng and X. You. "Power Allocation Criteria for Distributed Antenna Systems" *IEEE Transactions on Vehicular Technology*. vol.64, no.11, pp.5083-5090, Nov. 2015.
[16] ETSI,"Broadband radio access networks (BRAN); HIPERLAN type 2; physical (PHY) layer," TSI TS 101475 v1.1.1, Apr. 2000.

Statistics-based email communication security behavior recognition

Junkai Yi[1], Yueyang Su[2, 3] and Xianghui Zhao[4]

[1]College of Information Science and Technology, Beijing University of Chemical Technology, professor, Beijing, China.
[2] College of Information Science and Technology, Beijing University of Chemical Technology, Beijing, China.
[3]Email:Suyy1225@163.com.
[4] China information technology security evaluation center, Beijing, China.

Abstract. With the development of information technology, e-mail has become a popular communication medium. It has great significant to determine the relationship between the two sides of the communication. Firstly, this paper analysed and processed the content and attachment of e-mail using the skill of steganalysis and malware analysis. And it also conducts the following feature extracting and behaviour model establishing which based on Naive Bayesian theory. Then a behaviour analysis method was employed to calculate and evaluate the communication security. Finally, some experiments about the accuracy of the behavioural relationship of communication identifying has been carried out. The result shows that this method has a great effects and correctness as eighty-four percent.

1. Introduction

With the development of the Internet, e-mail can be found in all aspects of people's lives due to its effectiveness, immediacy and low cost. Meanwhile, social security of emails is more and more valuable, information and evidence involved in network crimes are increasingly significant. So it is a new challenge to analyse email information quickly and effectively, as well as the content and relationship of emails accurately, so as to specify communication behaviour of email contacts.

Currently there are mainly two methods below to classify emails at home and abroad: one is based on analysis of text content and the other one is based on analysis of email attachments such as texts, pictures, etc. The ultimate purpose of classification is to recognize and filter spam, among which Bayesian Classification Algorithm is the most common method [1-3]. Put forward by Pearl in 1988, it is a model composed of a directed acyclic graph and a set of probability distribution. Naive Bayesian classification algorithm, via structure expansion, is indeed a very powerful classifier and has outstanding performance in the practical application, it has achieved great success in many areas, especially in spam classification and disease diagnosis. It assumes that the features are independent of each other, but in real life, this assumption is simply not established. However, it is still very good, even the assumption is not established. Especially in the case of small-scale samples, it has a solid mathematical foundation and a relatively stable classification efficiency. At the same time, the Naive Bayesian classification model is less sensitive to data loss, with less parameter estimation and digestible algorithm.

In the field of text classification research, in the early 1950s researcher H. P. Luhn firstly put forth the theory based on frequency that characteristics of the entries appear in the text to calculate classification information. After that, some scholars put forth a new text model, called Vector Space

Model (short for VSM). It can be represented by vector space composed of a series of characteristics, which reduces the complexity of text representation greatly and improves the efficiency of text classification. With the development of research, machine learning algorithm replaced the previous classification algorithm gradually. There are multiple classification learning methods used in message text classification, such as Naive Bayes, Neural Network, Support Vector Machine (SVM), etc. Naive Bayes, K-nearest Neighbor Algorithm and SVM are more commonly used. Text Feature Selection is the key step of automatic text classification. It is a process that applying computer technology, under the predefined classification system, according to the document content to be classified, attributes it to one or more categories. Automatic text classification technology research began in the 1950s, but now many classification models based on different theories have appeared, among which VSM is used to represent documents. For example, the letter T presents vocabulary collection contained by the document. Each word and its weight in the text are regarded as feature items, so the document can be expressed as the vector d = (t1, t2 ... tm) (ti ∈ T, 1 ≤ i ≤ m), and then according to the document vector and category vector, calculate the similarity, so as to determine document category.

This paper proposes a new analysis method of email communication security behavior. Communication behavior is a behavior relationship between both sides while communicating via email. Naive Bayesian classification principle in applied statistics is to classify communication behavior between the sender and the receiver of email [4-5]. The fundamental principle of this method is firstly to analyze all documents in both email and its attachments such as text content and pictures via relatively mature file analysis technology, for example, steganalysis, as well as the set of characteristic attribute of email content generated by technology. Take the characteristic attribute of email content as input set of Naive Bayesian Classifier, and the communication behavior relationship between the sender and the receiver as classified results. Then the classifier will be generated by training samples. Finally, this classifier will be applied to classify communication behavior relationship of the sender and the receiver of new email.

2. Naive Bayesian Classification Principle

Naive Bayesian classification is an easy and effective method, and applied widely. With excellent performance compared with Decision Tree and Neural Network Classification Algorithm, in some cases it is even better than other classifiers. Naive Bayes fundamental principle is to calculate the probability that all types appear on the condition of items to be classified. The item with the biggest probability will be regarded as that type. The definition of Naive Bayesian classification is as follows: set X={a1, a2, ... , am} as an item to be classified; every a is the characteristic attribute of X, category set C={y1, y2, ... , yn}, then calculate P(y1|x), P(y2|x), ... , P(yn|x). If P(yk|x)={P(y1|x), P(y2|x), ... , P(yn|x)}, then x∈yK, among which the key step is to calculate every conditional probability of P(y1|x), P(y2|x), ... , P(yn|x).

To calculate conditional probability, firstly find out a known classification set to be classified, called training sample set; then count conditional probability of characteristic attribute in every category, P(a1|y1), P(a2|y1), ... , P(a1|y1); P(a1|y2), P(a2|y2), ... , P(am|y2); ... ; P(a1|yn), P(a2|yn), ... , P(am|yn). If all characteristic attributes are in conditional independence, then it can be deduced according to Naive Bayes:

$$P(y_i \mid x) = \frac{P(x \mid y_i)P(y_i)}{P(x)}$$

Denominator is always constant, so just maximize the member. Since all characteristic attributes are in conditional independence, hence it can be drawn:

$$P(x \mid y_i)P(y_i) = P(a_1 \mid y_i)P(a_2 \mid y_i)...P(a_m \mid y_i)P(y_i) = P(y_i)\prod_{j=1}^{m} P(a_j \mid y_i)$$

It can be reached that the key step of Naive Bayesian classification is to calculate every conditional probability P(a|y). When characteristic attribute is discrete value, counting frequency of appearing in every category among training samples, then P (a|y) can be estimated. Naive Bayesian classification process is shown in Figure 1:

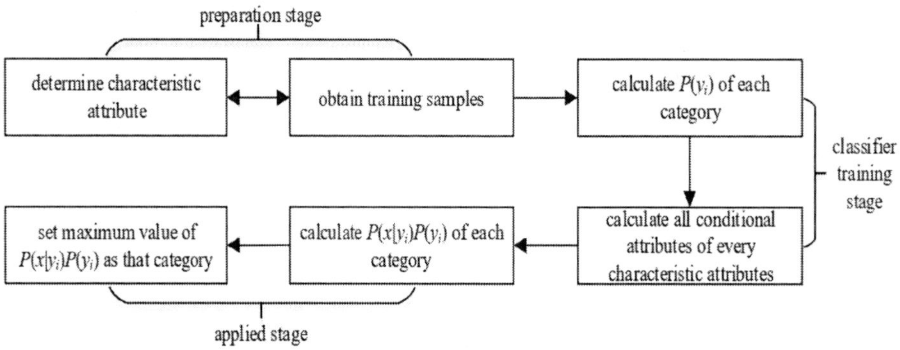

Figure 1. Naive Bayesian classification process.

3. Data Collection

3.1. Date Collection

Some of email data from Enron email data sets and one security company, and some emails collected by ourselves at spare time, are selected 10,000 randomly as the test set, among which are selected 1000 as training samples to classify. See the classification of communication behavior relationship of the sender and the receiver according to characteristic attribute in Table 1. There are 730 normal emails, 192 emails that one of two sides belongs to malicious attack, 44 emails that both sides belong to leaked relationship, 34 emails that both sides have no relationship.

Table 1. Classification results of sample training emails.

	Normal Email	Malicious Email	Leaked Email	Email With No Characteristics
Total	730	192	44	34

3.2. Data characteristic of Pre-extraction and Processin

Characteristic attributes of email contents will be extracted and divided before classifying communication behavior relationship between sender and receiver of email. Firstly, texts and attachments in email are extracted, and characteristic attributes come the second by means of steganalysis. This method is to extract keywords from text, and to analyze hidden information such as pictures and videos in the attachments. The process of analysis on emails and extraction of characteristic attributes is shown in Figure 2.

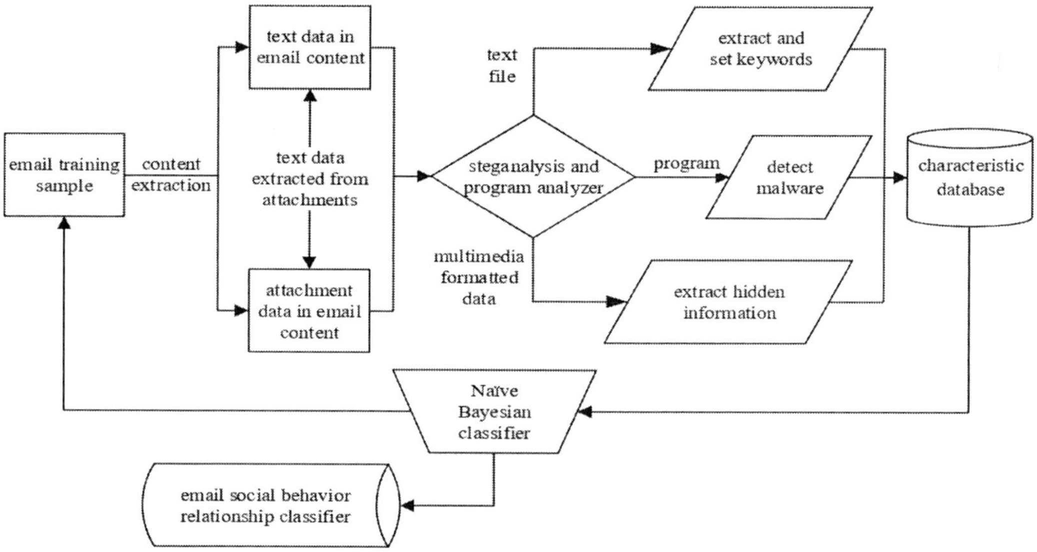

Figure 2. Process of analyzing emails and extracting characteristic attributes.

The detection process of texts and attachments in email will apply specific steganalysis and malware detection technology. Both of them have been mature, and own high reliability. The results are shown in Table 2. However, when malware is detected in attachments, it may not be a real one. Sometimes it may be an important file the receiver needs. Therefore, under such circumstance, other characteristic attributes should be analyzed together, so as to judge whether it is a malicious email or not.

Table 2. Feature extraction results of sample training emails.

Email NO	Number of keywords (a1)	Malware	Number of leaked information (a3)	Classification
1	6	None	6	Leaked email Normal email
2	2	None	1	Normal email
3	1	Exist	0	Normal email
4	7	Exist	8	Malicious email
5	2	None	1	Email with no characteristics
6	0	None	1	Normal email
...	...	...	...	...

4. Classify Email Behavior Relationship Based On Naive Bayes

4.1. Extraction and Classification of Characteristic Attributes

4.1.1. Characteristic of times that keywords appear. It is important to analyze email communication relationship via specific keywords in email. Corresponding keywords can be set at will and filtered in the detection process of email contents. In the experiment, unified keywords set has been used to avoid influencing results, and times that they appear are taken interval partition as 3 intervals: $(0, 6)$, $[6, 10)$, $[10, +\infty)$.

4.1.2. Characteristic of whether malware exists. Sometimes malware may exist in the email attachments, but it depends on the receiver system environment. It may be a real one but at times it may be an important file the receiver needs. Therefore, under such circumstance, other characteristic attributes should be analyzed together, so as to judge whether it is a malicious email or not. But anyway, it is an important parameter to identify communication relationship. {Exist, None} is to differentiate emails in the experiments.

4.1.3. Characteristic of times that leaked information appears. Emails can be used to pass on secret messages via multimedia files like pictures and videos, as well as other office files. There are many methods of file encryption such as hidden information in file storage structure, adding vital messages to postfix of file name, etc. During the detection process of the whole email content, analysis software can identify and find out the file with hidden information effectively, which can help judge entity relationship of both sides in email communication better. In the experiment, times that leaked information appears are also taken interval partition as 3 intervals: (0, 3), [3, 6), [6, +∞).

4.2. Classifier Construction of Email Behavior Relationship

If tests the set manually, it will cost lots of manpower and material resources, and it is poor efficient. On the contrary, if automatic monitoring mechanism can be introduced, work efficiency will be dramatically increased. Consequently, in order to classify email communication behavior relationship, classification types should be given: C=0 stands for normal communication; C=1 stands for one side belonging to malicious attack; C=2 stands for both sides belonging to leaked relationship; C=3 stands for both sides being in sensitive area. According to last part, 3 characteristic attributes are selected: a1: times that keywords appear in file; a2: whether malware exists in file; a3: times that leaked information appears. Divide them in accordance with data characteristics: a1: {0<a1<6, 6<=a1<10, 10<=a1}, a2: {a2=0 (None), a2=1 (Exist)}, a3: {a3<3, 3<=a3<6, 6<=a3}.

Take 500 emails extracted randomly as samples, and calculate frequency of every classification:

P(C=0) = 730/1000 = 0.73, P(C=1) = 192/1000 = 0.192, P(C=2) = 44/1000 = 0.044, P(C=3) = 34/1000 = 0.034. Then, calculate the frequency classified by characteristic attribute under every class condition:

P (a1<5|C=0) = 0.65, P (5<=a1<10|C=0) = 0.22, P (a1>10|C=0) = 0.13, P (a1<5|C=1) = 0.63,
P (5<=a1<10|C=1) = 0.28, P (a1>10|C=1) = 0.09, P (a1<5|C=2) = 0.09, P (5<=a1<10|C=2) = 0.19,
P (a1>10|C=2) = 0.72, P (a1<5|C=3) = 0.15, P (5<=a1<10|C=3) = 0.25, P (a1>10|C=3) = 0.60,
P (a2=0|C=0) = 0.84, P (a2=1|C=0) = 0.16, P (a2=0|C=1) = 0.23, P (a2=1|C=1) = 0.77,
P (a2=0|C=2) = 0.75, P (a2=1|C=2) = 0.25, P (a2=0|C=3) =0.43, P (a2=1|C=3) = 0.57,
P (a3<2|C=0) = 0.66, P (2<=a3<5|C=0) = 0.21, P (a3>5|C=0) = 0.13, P (a3<2|C=1) = 0.45,
P (2<=a3<5|C=1) = 0.27, P (a3>5|C=1) = 0.28, P (a3<2|C=2) = 0.07, P (2<=a3<5|C=2) = 0.15,
P (a3>5|C=2) = 0.78, P (a3<2|C=3) = 0.08, P (2<=a3<5|C=3) = 0.23, P (a3>5|C=3) = 0.69,

Now, classifier construction has mainly been completed.

5. Experimental Results and Analysis

5.1. Sample training

Based on the previous conclusion, applied Naive Bayesian classification method, 1000 emails selected randomly taken as training samples, Naive Bayesian classifier will be established. See characteristics and details of email entity relationship in Table 3, see part of sample training data in Table 4. Finally, train sample data and establish basic Bayes classifier model.

Table 3. Classification of sample data.

Name	Number of keywords (a1)	Malware	Leaked place (a3)	Classification
	a_1<6	Exist	a_3<3	Normal communication
	6< a_1<10	None	3< a_3<6	Malicious attack
Classify	a_1>10		a_3>6	Leaker email
				Email with no characteristics

Table 4. Part of sample training data.

Email No	Number of keywords (a1)	Malware	Leaked place (a3)	Classification
1	$a_1<6$	None	$a_3<3$	Normal email
2	$a_1<6$	Exist	$a_3>6$	Attack email
...	...	...	...	...
999	$a_1<6$	Exist	$a_3<3$	Normal communication
1000	$a_1>10$	None	$a_3>6$	Leaked email

5.2. Experimental Results and Analysis

Based on classifier model, our system will classify the following 3,500 emails data, and monitor the accuracy of classifier. With the increasing of email data examples continuously, the classification results are shown in Table 5 below:

Table 5. Assessments of classification results.

Number of email	Number of correct classification	Rate
1,000	*883*	88.3%
1,500	*1285*	85.67%
2,000	1687	84.35%
2,500	*2106*	84.24%
3,000	*2520*	84.0%
3,500	*2937*	83.91%

It is shown in figure 3 below that with the increasing of testing data, the classification accuracy is gradually to be steady as over 84%, which means classification results by classifier is reliable.

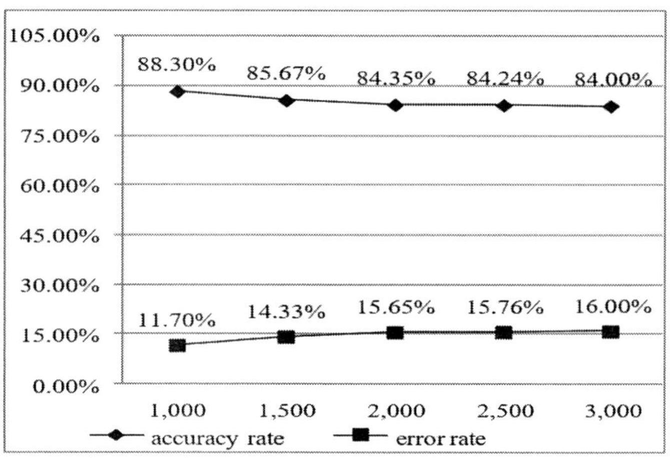

Figure 3. Line graph of Naive Bayesian Classification accuracy and error rate.

6. Conclusion

This paper aims to propose a new email communication security behavior analysis method by theoretical analysis and experimental data verification. This method can effectively analyze the relationship between the two sides of e-mail communication. At present, one simple system of classification system has been mainly completed, and has gained good impact in the field of email communication behavior. The method can also be improved by optimizing the Bayesian classification.

Meanwhile, in the future practical applications, more characteristic attributes can be extracted in terms of requirements, and classified in more detail.

7. Acknowledgments

This work has been supported by Projects U1536116 and U1636208 funded by National Natural Science Foundation of China (NSFC).

8. References

[1] Yoo J Y, Yang D. Classification Scheme of Unstructured Text Document using TF-IDF and Naive Bayes Classifier [J]. 2015

[2] Banday M T, Sheikh S A. Multilingual email classification using Bayesian filtering and language translation[C]//Contemporary Computing and Informatics (IC3I), 2014 International Conference on. IEEE, 2014: 696-701

[3] Tang B, He H, Baggenstoss P M, et al. A Bayesian classification approach using class-specific features for text categorization [J]. IEEE Transactions on Knowledge and Data Engineering, 2016, 28(6): 1602-1606

[4] Boryczka U, Probierz B, Kozak J. A New Algorithm to Categorize Email Messages to Folders with Social Networks Analysis [M]// Computational Collective Intelligence. Springer International Publishing, 2015

[5] Yoshinaga N, Itaya S, Tanaka R, et al. Content Propagation Analysis of Email Communications [C]// Ieee/wic/acm International Conference on Web Intelligence and Intelligent Agent Technology. 2010:79-82

[6] Ma Y, Yu Z, Ding J. A Method of User Recommendation in Social Networks Based on Trust Relationship and Topic Similarity [M]// Social Media Processing. Springer Berlin Heidelberg, 2014:240-251

[7] Chandrasekar P, Qian K. The Impact of Data preprocessing on the Performance of a Naive Bayes Classifier[C]//Computer Software and Applications Conference (COMPSAC), 2016 IEEE 40th Annual. IEEE, 2016, 2: 618-619.

[8] Yang T, Qian K, Lo D C T, et al. Spam filtering using Association Rules and Naive Bayes Classifier[C]//Progress in Informatics and Computing (PIC), 2015 IEEE International Conference on. IEEE, 2015: 638-642

[9] Li L, Li C. Research and Improvement of a Spam Filter Based on Naive Bayes[C]//Intelligent Human-Machine Systems and Cybernetics (IHMSC), 2015 7th International Conference on. IEEE, 2015, 2: 361-364.

The intelligent technologies of electronic information system

Xianyu Li

Department of System Collectivity, Tian Chen Information Technology and System Engineering Institute, Beijing, China.
Email: 919wh_cn@sina.com

Abstract. Based upon the synopsis of system intelligence and information services, this paper puts forward the attributes and the logic structure of information service, sets forth intelligent technology framework of electronic information system, and presents a series of measures, such as optimizing business information flow, advancing data decision capability, improving information fusion precision, strengthening deep learning application and enhancing prognostic and health management, and demonstrates system operation effectiveness. This will benefit the enhancement of system intelligence.

1. Introduction

Electronic information system is a distributed man-machine system that effectively organizes related resources, conducts information processing and actualizes business activities. Service-oriented architecture can translate function units of different applications into information services among that interface and protocols are favourably defined, and the information services are the integration basis of electronic information system. Along with the continuous advancement of information technology, electronic information system need further satisfy the requirements of dynamic function evolvement and realization, and meantime the practices and development of information service system can provide available technical approaches for it [1][2].

Intelligence support and intelligent technology is necessary for electronic information system to improve performance level and advance decision advantage, and it has been an important technology emphasis during system development. System intelligence generally comprises computation intelligence, perception intelligence and cognition intelligence. Perception intelligence area is typical of the capability of image comprehension, identification and language translation. Computation intelligence area is characterized by the core of formalization or regularization operation such as scientific operation, logic processing and statistic inquiry. Cognition intelligence area emphasizes thinking and decision capability with the core of comprehension, reasoning and decision-making [3].

Intelligence comparison of electronic information system is drawn in table 1.

Table 1. Intelligence comparison of electronic information system.

Category	Objective	Function	Present State
Cognition intelligence	Comprehension, reasoning and decision-making	Decision support (to increase cognition speed)	Inferior
Computation intelligence	Computation, statistical analysis and inquiry	Elementary business operation	Relatively preferable
Perception intelligence	Audition, seeing, speaking, action and response	Man machine interface	Better and hopeful to achieve a real breakthrough

2. Information Service Development

Information service attributes generally consist of the function attribute of kernel metadata and message format, and the non-function attribute of service state, service quality and user-defined [4]. Kernel metadata means the descriptive information that pictures the service characteristics of status, content and copyright; message format means the communication protocol that serves when service applicant transfer related services; service state is meant to describe whether present service is available; service quality means the capability of information service to meet user requirements or complete stated tasks, and includes the defined static parameters of robustness, scalability and security, and the supervised dynamic indexes of accuracy, delay and reliability; and user-defined is for the purpose of extending the description of information service.

Information service is the encapsulation entity of system function units, which comprises three logic layers such as description layer, entity layer and interface layer, as is shown in figure 1.

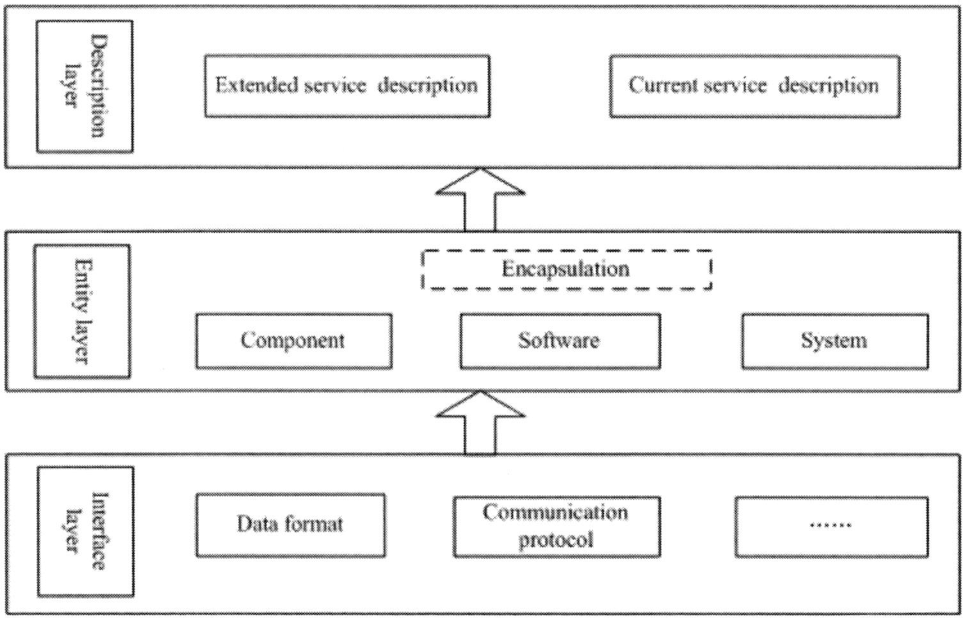

Figure 1. The logical organization of information service.

- Interface layer specifies the outer interfaces that comprise communication protocol and message format, and actualizes the separation between service interfaces and functions.
- Entity layer presents the related service applications, and is composed of function realization and resources encapsulation.
- Description layer provides the description information needed for the service discovery and transfer.

Information service system contains various integrative management and business services that sustain the related infrastructure on the basis of service oriented technology [5]. Service oriented electronic information system is substantially based on various information services distributed in a network in light of mission requirements, and integrated into a mutual operational system by means of service combination and transfer, alternation and cooperation.

The distributed functions and resources are illustrated and encapsulated into independent information services with unitive and feasible interface standards; the alternation and communication among services are devised via service bus topology to raise system data and function sharing; and the information services are recombined and aggregated according to certain operational business flow, and eventually transferred, alternated and collaborated so as to accomplish the desired intelligent and dynamic system.

3. Intelligent Technology Framework

Based upon information services, we follow the overall thread of resources-service-application, and set up intelligent technology framework. Intelligent technologies of electronic information system generally comprise intelligentized application technology, common service technology, kernel service technology, service engine technology and resources supervision technology [5] [6]. Intelligent technology framework of electronic information system is depicted in figure 2.

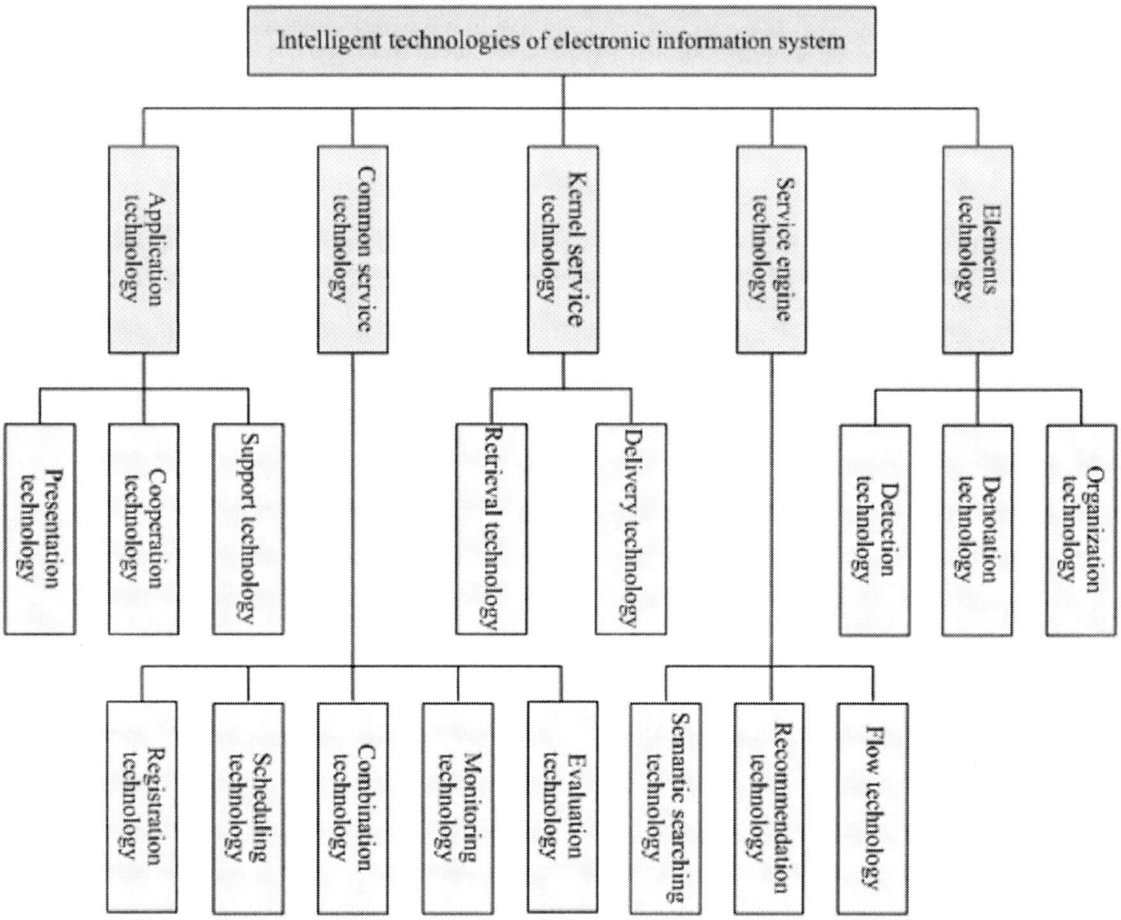

Figure 2. Intelligent technology framework of electronic information system.

- Intelligent application technology is to apply various services and resources to system functions and processes, and comprises intelligent presentation, cooperation and support technology.
- Intelligent common service technology is to reorganize and encapsulate distributed intelligent resources, investigate and construct required service models, and comprises registration, scheduling, combination, monitoring and evaluation technology.
- Intelligent kernel service technology is to achieve proper sharing, reuse and collaboration, and comprises intelligent retrieval and delivery technology.
- Intelligent service engine technology is to carry out entity modeling and semantic modeling, and chiefly consists of semantic searching engine, recommendation engine and flow engine technology.
- Intelligent elements technology mainly deals with the storage, management and maintenance of intelligent resources, is the basic of service and application, and includes detection, denotation and organization technology.

IOP Conf. Series: Journal of Physics: Conf. Series **887** (2017) 012041 doi:10.1088/1742-6596/887/1/012041

4. System Improvement Measures

According to the overall scheme and associated standards, we need take a series of practicable measures to smoothly accelerate system service capacity and improve system intelligence grade [6] [7] [8], and the related measures are primarily as follows.

4.1. Actualizing Business Flow Optimization

We should optimize business information flow with respect to the related processes of information collecting, transferring, processing, feedback and control in the light of new situation and requirements, make use of various technology means of parallel programming mode, distributed resource management, and carry out system architecture design, quality enhancement of software design and realization, and improvement of program algorithm efficiency so as to implement the intelligent engine and agile operation.

4.2. Advancing Data Decision Capability

Big data analysis, online analysis and data excavation technologies should be properly applied to databases from multi-sources so that decision information and decision knowledge can be produced to improve model database and knowledge database, and decision report is established to devote to business decision making, while business information is fed back to business department for further demand and to related database for new knowledge. Composite data structure can meet the application needs of field decision making, and trend analysis and pattern identification can supply mature data product and enhance decision service capacity.

4.3. Improving Information Fusion Precision

Information fusion technology is to synthesize redundancy information from various types and isomorphic and isomerous data sources, get hold of accurate evaluation of concerned object, and present proper judgement and decision. Information fusion consists of data level, character level and decision level, and continually obtains finer outcome by means of a series of optimized algorithms on the basis of method base, model base and knowledge base. It can enhance information dimensionality and dependability, and improve system performance and robustness.

4.4. Strengthening Deep Learning Application

Deep learning approach roots in traditional nerve cell network. It is to simulate the mechanism for simulating brain analysis and study by establishing multi connotative layer, and carry out independent study of brain character via rules that are set up based on a great deal of data and experience. Deep learning can implement abstract and comprehension of concept and law layer after layer by means of training for multilayer nerve network, and solidify the comprehension in a form of nerve cell network so as to intelligently advance system perceiving and operation capability.

4.5. Enhancing Prognostic and Health Management

Prognostic and health management is mainly to administer system state in real time, deal with and examine the collected information, and forecast system operation state and regulate maintenance work. We shall further investigate the related intelligent perceiving, data transfer, fault diagnosis, fault prognostics, decision support and interface technologies, and utilize performance parametric monitoring method and physics of failure model-based prediction method so as to enhance the self-support capability and intellectual level of electronic information system.

5. System Operation Effectiveness

Based upon the integration of intelligent technologies and associated measures by means of information service units, the operation capabilities of electronic information system are evidently increased as compared to the ordinary system, and it is mainly reflected in the three aspects of business processing delay-time, information sharing capacity and service availability [9][10][11]. And performance contrast of ordinary and intelligent electronic information system is drawn in table 2.

Table 2. Performance contrast of ordinary and intelligent electronic information system.

Category	Business Processing Delay time	Information Sharing Capacity	Service Availability
Ordinary System	11.331 s	20	0.9971
Intelligent System	3.0820 s	100	0.9995

6. Conclusions

We novelly establish intelligent technology framework of electronic information system, and carry out effective improvement with respect to various system functions via intelligent technology utilization and information service integration. In the light of the new demands of electronic information system, we should further promote the application of information services and intelligent technologies to the system integration and development. In addition, we should continually optimize business information flow, advance data decision capability, improve information fusion precision, strengthen deep learning application and attach importance to prognostic and health management. It is the complete scheme, systemic technologies and appropriate standards of man machine intelligence and information services that would fully ensure system intelligent and collective capabilities.

7. Acknowledgment

I am grateful to the experts and secretaries for their insight, guidance and patience; they are a reliable source of support and encouragement.

8. References

[1] Chuanqing Cai, Guangqi Huang 2011 *Command Information System and Technology* **2** 53
[2] Wei Jung, Zifeng Cui 2011 *Command Information System and Technology* **4** 10
[3] Xiaofeng Hu, Shengming Guo, Xiaoyuan He 2016 *Command information System and Technology* **3** 1
[4] P Y Zhang, B Huang, Y M Sun 2007 *Journal of System Simulation* **2** 2872
[5] Zhen Shu and so on 2010 *Fire Control & Cornrnand Control* **9** 53
[6] Xianyu Li, Hua Wang, and so on 2015 *Proceedings of ICCMCEE* p 848
[7] Hui Peng, and so on 2015 *Journal of Equipment Academy* **6** 94
[8] Lei Cao and Yongliang Zhang 2015 *Operations Research and Systems Engineering* **3** 69
[9] Yushi Lan 2010 *Command Information System and Technology* **1** 1
[10] Deyi Li and so on 2006 *Military Science* **3** 111
[11] Shuying Dong, Ting Li 2016 *Journal of Ordnance Engineering College* **3** 43

IOP Conf. Series: Journal of Physics: Conf. Series **887** (2017) 012042 doi:10.1088/1742-6596/887/1/012042

On construction method of shipborne and airborne radar intelligence and related equipment knowledge graph

Ruizhe Hao and Jian Huang

[1]College of Mechatronic Engineering and Automation, National University of Defense Technology, Changsha, China, E-mail: Richard_HRZ@163.com
[2]College of Mechatronic Engineering and Automation National University of Defense Technology, Changsha, China, E-mail: 13973166586@163.com

Abstract. Knowledge graph construction in military intelligence domain is sprouting but technically immature. This paper presents a method to construct the heterogeneous knowledge graph in the field of shipborne and airborne radar and equipment. Based on the expert knowledge and the up-to-date Internet open source information, we construct the knowledge graph of radar characteristic information and the equipment respectively, and establish relationships between two graphs, providing the pipeline and method for the intelligence organization and management in the context of the crowding battlefields big data.

1. Introduction

The capability of radar intelligence collection and analysis is a vital aspect of national military comprehensive strength, and it is significant for strategic situation evaluation to timely and accurately collecting and analyzing radar intelligence. Under the circumstances of information extracted from mass data overloading during the operation, a problem surfaced that it is essential to converting the information advantage to decision strength, which requires accumulating the intelligence superiority by flattening the information sharing and convergence [1].

Knowledge base is the fundamental techniques of artificial intelligence and intelligent service applications [2]. Since the Google's proposition of building the knowledge graph utilizing the big data in 2012, it has raised wide attention as a new type of knowledge representation and service technology for both academic and industry sphere [3]. The research sphere of knowledge graph includes information extraction, knowledge representation, knowledge reasoning, database management, etc. It is a burgeoning domain, originated from information retrieval, now applied in the search engine, medical research, bioinformatics as well as finance analyzing. However, study of it in the military intelligence domain is still scarce. Jiang Kai [4] proposes an intelligence based military knowledge search framework. Yijun Yan [5] designs a reconnaissance situation synthesis system to deal with the intelligence acquired from multiple source. The most successful application case of knowledge graph related domain in intelligence analyzing is Palantir's Dynamic Ontology, which helped international antiterrorist campaign in the Middle East.

The association of radar intelligence and equipment information benefits battlefield knowledge management, effectively reducing the information barrier, connecting the signal layer, information layer and decision layer together and better realizing the integrative joint operations. Knowledge graph's strong semantic query capabilities provide richer expression and more accurate results, so that it can be more widely used in the lack of expert knowledge combat environment. The Decision Support System (DDS) roots in the knowledge graph and assists commanders to plan and solve various actions.

2. Graph Modeling

The knowledge graph is a semantic network that describes the various entities, concepts, and relationships between them in the real world. The difference between the knowledge map and the expert system is that the expert system is more concerned with the rules building and reasoning, but the bottlenecks of knowledge and the portability of the system are obvious while knowledge graph focus more on the existing entities and relations [6]. By integrating multi-source heterogeneous data into the form of graph, taking good advantage of some unstructured or isolated data becomes possible, and a good purpose for multi-data type fusion decision will achieves. In the light of this, the radar intelligence graph pays more attention to the association information mining, rendering knowledge representation and knowledge reasoning a significantly stronger ability [7].

2.1. Knowledge Graph Framework

According to the coverage of knowledge range, knowledge graph includes general type and domain type, and the construction methods are also slightly different for them. The general knowledge graph is constructed from a bottom-up approach, which facilitates the discovery of new patterns of knowledge. The domain graph focuses on the hierarchical structure of knowledge, usually need top-down approach to build the graph, and then expanding the entity network [8].

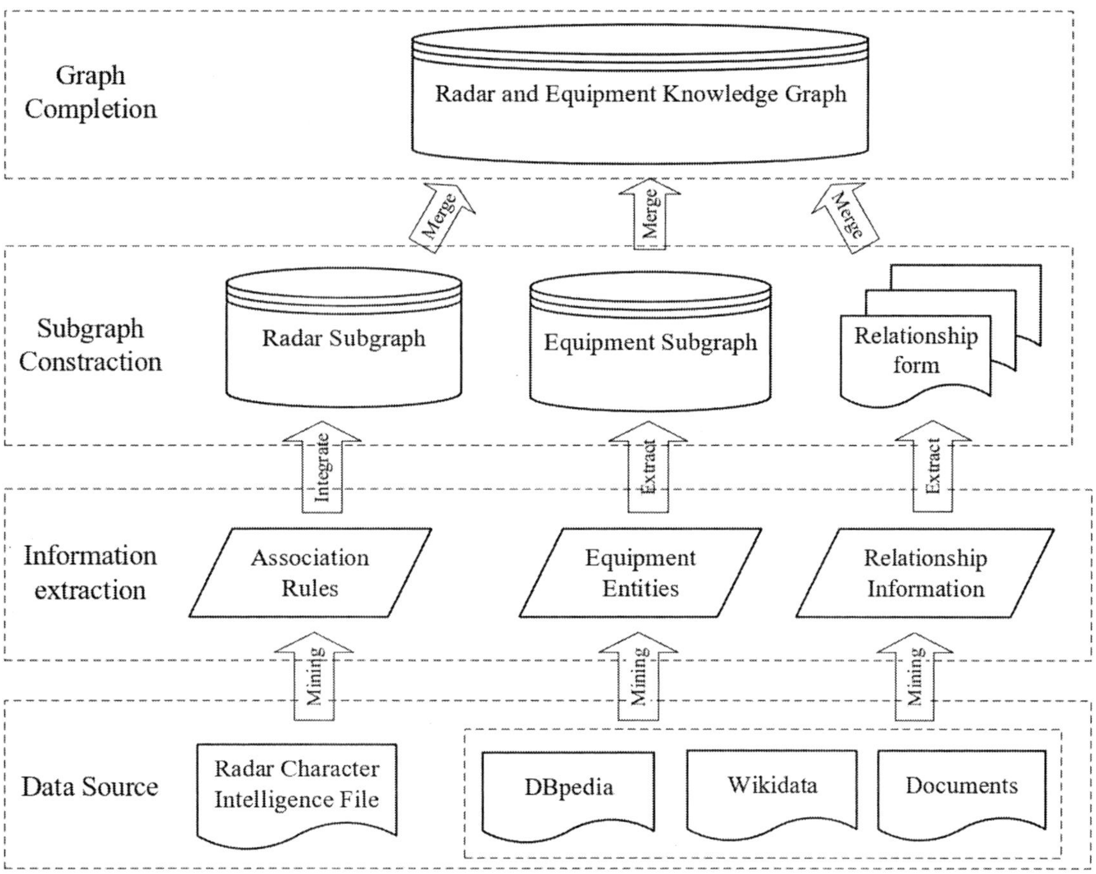

Figure 1. Radar Intelligence and Equipment Knowledge Graph Construction Framework

We combine two approaches mention above that design two subgraphs, one is radar intelligence graph, and the other is equipment knowledge graph and then using expert knowledge as well as open source knowledge to establish relations between the entities of them and on the top of them, we design a concept ontology to organize the whole graph. The former subgraph contains the shipborne and airborne radar signal features and the co-occurrence association rules, the task of which is to establish a semantic web model to represent, visualize and store association rules. The latter one is comprised of equipment entities, the task of which is extracting equipment knowledge from open source projects and integrating new confidence entities into them, including the tasks of entity extraction, entity resolution and link prediction, etc. In the process of knowledge accumulation, we need to deal with the tacit knowledge and explicit knowledge. The explicit knowledge is extracted from big data and it's a long-term process to gradually accumulate. The tactic knowledge, embedded in the human brain, cannot expressed easily and require the collaborative work. These make the radar knowledge graph modeling more specifically in domain, a long-term progress and build up technically.

The main source of intelligence knowledge is the association rules mined from frequent radar signal item, and the equipment knowledge is mainly from the Internet open source domain knowledge base and some specific intelligence documents including academic papers and intelligence magazines such as Jane's Defense Weekly. The framework of knowledge graph is as Figure 1.

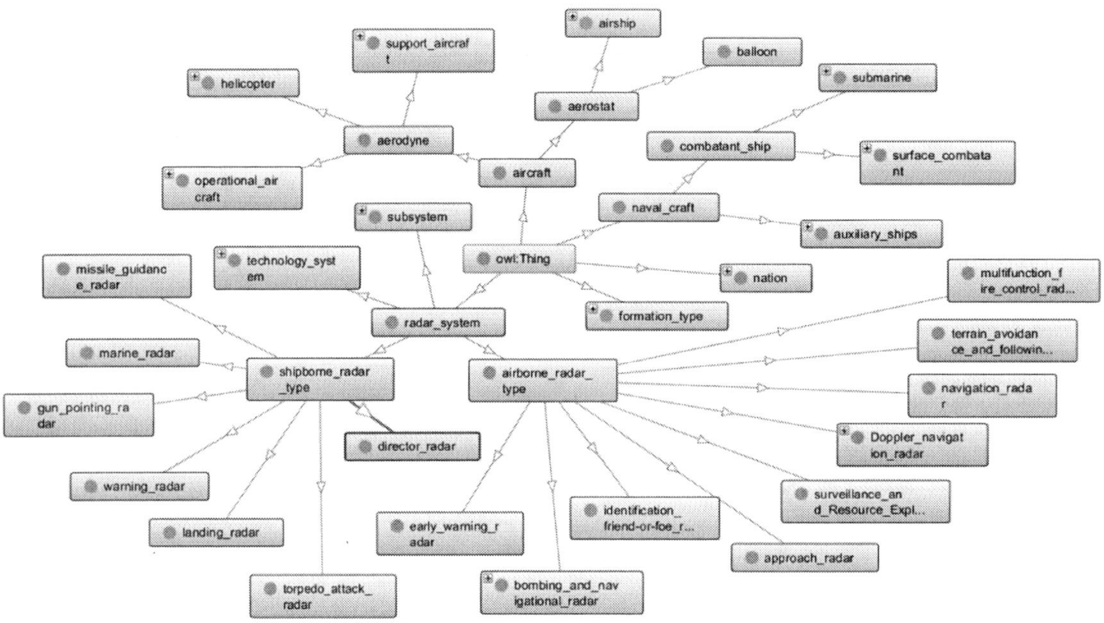

Figure 2. Up-layer Ontology Concept Model

2.2. Model Terminology

Our ontology modeling methods employs Ontology Development 101 and comprehensively refers to the Radar Handbook, World Airborne Radar Manual, and Airborne Radar Handbook, Jane's Defense Weekly and various academic papers and military encyclopedias to define the terminology. Figure 2 is the up-layer ontology model.

By means of top-down approach we build the hierarchy of classes, and based on this the properties are essential to enrich the semantic information of ontology layer of knowledge graph. We have four main top layer classes:

Country, formation, aircraft, radar system and the naval craft. The object properties are designed as follows:

1) subclassof: Describing the child relationship of concept;

2) armament: Describing the relationship between radar and equipment;

3) cooperatewith: Representing the cooperative relation between combatant equipment;

4) hasfunc: Describing the properties radar type has;

5) ispartof: Describing the dependency of equipment. *ispartof:* Describing the dependency of equipment.

6) operator: Describing the relation that the equipment is the armament of a country.

3. Construction Techniques

3.1. Radar Intelligence Subgraph Construction

This paper focuses on the construction of radar intelligence association rules in the format of semantic web. Graph is an intuitive pipeline to visualize the complex association rules of feature signals and the semantic web provides expressive and convenient way for querying. The data source of the radar intelligence graph is the rule mined from radar intelligence information, which is born with confidence and good operability. We propose two semantic relationship in the subgraph: "isfeatureof" and "meantime". The "isfeatureof" describes the relationship between radar and its feature signals, and the "meantime" shows the co-occurrence relationship between signals.

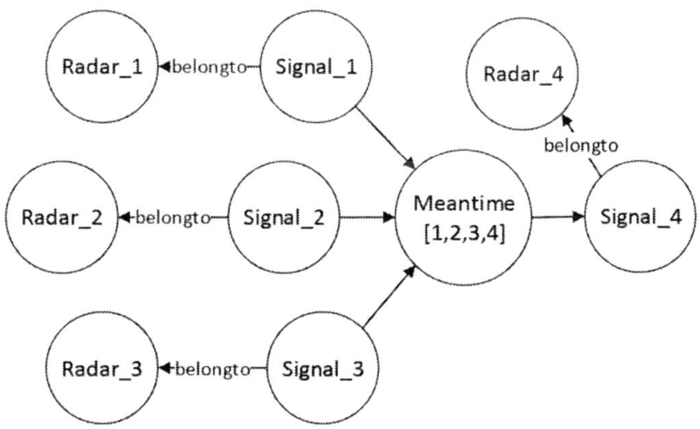

Figure 3. Radar Intelligence Subgraph Structure

As for those multiple-to-one rules, we introduce new node into the graph to express the meaning "and". For example, the rule "Meantime (1, 2, 3) → (4)" is represented as figure 3. Every node in the graph is allocated with a Uniform Resource Identifier (URI) and the intermediate node is named after "Meantime" plus the identifier of signal.

3.2. Equipment Subgraph Construction

The equipment subgraph is a vertical knowledge graph that focus on the depth of knowledge and the overall hierarchy. It is rational to reuse some existing knowledge projects to help better build knowledge graph. Linked Open Data (LOD) projects such as *Dbpedia* and *Wikidata* achieve quite well on the general knowledge graph construction so we reuse, partition and merging them to extract those domain knowledge we need. The *Jane's Defence Weekly*, military publications and academic papers provide the entities of equipment subgraph.

This paper proposes an incremental construction method called "one foundation and multi sources" to build up the equipment subgraph. We reuse the domain knowledge of *DBpedia* as the initial knowledge source and adding new knowledge from *Wikidata*, intelligence documents and publications. The recursive build-up process is illustrated as figure 4.

The specific steps are as follows:

1) *Using the SPARRQL query language and Protégé ontology editor to partition the military domain knowledge;*
2) *Mapping those equipment class concepts to the Wikidata dump, and find all the leaf nodes within 2-length in the Wikidata dump;*
3) *Using Jena tool to write the new individuals from dump into the knowledge graph;*
4) *Adding new entities extracted from web, publications and papers;*
5) *Check the consistency of knowledge graph;*
6) *Repeat step(2) to step(5).*

During the process of dealing with *Wikidata*, we simplify the property of entities for the reason that the data properties are not uniform. We save the *Wikidataid*, aliases, label and description of all the properties. The object properties *Wikidata* uses are identifiers in the format of "P001", so we retrieve the real name on the internet and substitute it.

The process of entity matching is achieved by defining the distance measurements for the string. If the string similarity is less than the threshold value, the name is the entity mention. The measurements we use are Levenshtein Distance [9] and Jaccard Similarity [10]. The similarity degree of two strings is the bigger score of two measurement.

If the machine could not figure out the corresponding entity of the concept, then the mention would be marked to wait for manual processing.

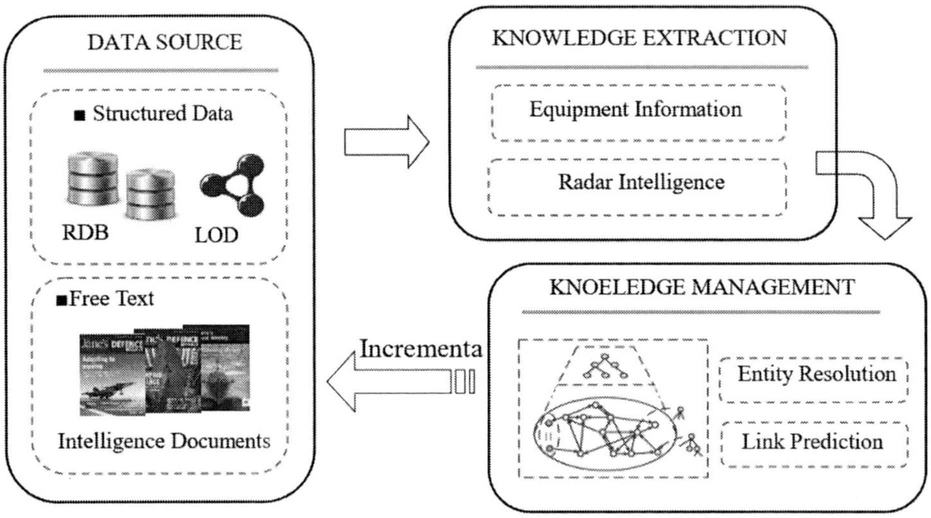

Figure 4. Equipment Knowledge Graph

3.3. Equipment Relationship Extraction

In this part we focus on the relation between two subgraphs, specifically the relation between radars and equipment.

In Wikipedia, military radar information is under the category of "Military radars", but it is hard to find the structured information about radar and equipment, and this sort of information is hidden in the unstructured text. The web page crawling and cleaning work is done by the python package "Beautiful soap". The extraction steps are listed as below:

1) *Crawling the web pages and extract each text under a theme into a "txt" file;*
2) *Taking advantage of Stanford NLTK natural language processing tool to parse the text, tagging the part of speech;*
3) *Looking for those sentences with the named entities in the concept graph and match the pattern sentences containing "suitable", "equipped", "installed" , "combined", "used in", "destroyer", "carrier", "frigate", etc.*
4) *Use the pattern to match the relationship. If the one sentence has named entities belonging to distinguished category of "radar" and the category of "aircraft" or "naval craft", saving them into the relationship form.*

Table 1 shows an example that extracting information from the Wikipedia page "EL/M-2052" and "APAR".

The EL/M-2052 has direct relationship to the equipment of aircrafts while APAR only connects to the frigate type, so it requires more work to detail the information. Under this circumstance, we retrieve the key words in the Wikipedia and find the subclass of the equipment and add them into the relationship form.

Under the light of these methodology the shipborne and airborne radar intelligence and equipment knowledge graph gradually grows bigger and the circle of construction still continues.

Table 1. Relationship Form of Two Kinds of Radars

Radar type	Equipment type	Radar type	Equipment type
EL/M-2052	F-15	Active Phased Array Radar (APAR)	De Zeven Provinciën class frigates
	Mirage 2000		F124 Sachsen class frigates
	MiG-29		Royal Danish Navy Ivar Huitfeldt class frigates
	Sukhoi Su-27/30		

4. Consistency Evaluation and Application Scenario

The consistency evaluation is to check the consistency of the knowledge graph and the up-layer ontology. Protégé furnishes some reasoning plugins such as Pellet, HermiT and FaCT++. We use Pellet inference engine to evaluate the consistency and gives an American's Nimitz aircrafts carrier formation instance. After executing the reasoning machine, the Pellet engine will return the inference result if the ontology is consistent. Otherwise, it will pop up an error window. The reasoning result is shown as Figure 5, the yellow part of which is the inference information.

The shipborne and airborne radar intelligence and equipment knowledge graph can be widely implemented in the target recognition task. The knowledge graph is the decision layer for the situation evaluation. The feature intelligence, analyzed through several steps, is send to the decision layer to identify which kind of target it belongs to by matching the characters defined in the ontology. After identifying the enemy radar, it can figure out the potential enemy armament and formation information as well as other intelligence of the radar and equipment, which rendering power for commanders to make operational decisions.

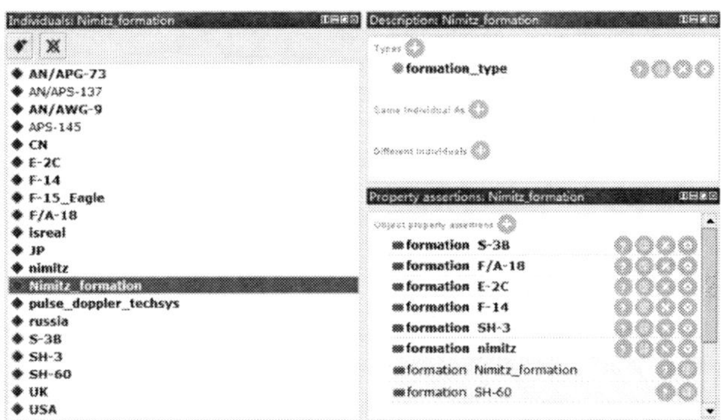

Figure 5. Consistency Evaluation and Formation Reasoning

5. Conclusion

The knowledge graph construction is a part of knowledge engineering so it is more like a project. This paper proposes a methodology to construct multi-source heterogeneous shipborne and airborne radar intelligence and equipment knowledge graph and verify the feasibility of the method. This method provides a new idea of semantic representation of radar intelligence rules and integrates open source information and expert knowledge, which applies the ontology engineering method and large data semantic network construction technology. Besides this, the graph construction also requires reasoning to enrich the knowledge capacity to meet the consistency and completeness, so this is a process of constantly updating and iteration. The application of the method facilitates improving the level of automatic processing of our military intelligence and lays the foundation for situation evaluation and threat discovery.

6. References

[1] Kim, D., Xie, L., & Ong, C. S. (2016). Probabilistic Knowledge Graph Construction: Compositional and Incremental Approaches. In Proceedings of the 25th ACM International on Conference on Information and Knowledge Management (pp. 2257-2262). ACM.

[2] Lin Hailun, Wang Yuanzhuo, Jia Yantao, Wang Weipeng, (2017), "Network Big Data Oriented Knowledge Fusion Methods: A Survey," *Chinese Journal of Computers*, vol.40, pp. 1-27

[3] Nickel, M., Murphy, K., Tresp, V., & Gabrilovich, E. (2015). A review of relational machine learning for knowledge graphs. Proceedings of the IEEE, 104(1), 11-33.

[4] Jiang Kai, Qian Kui, heng Xuan. (2016). Military Information Retrieval Technology Architecture Based on Knowledge Graph. Command Information System, 7(1), 47-52.

[5] Yi Junyan. (2006), "Research and Design of Situation Fusion System of Reconnaissance Intelligence Situation," doctoral dissertation, CHN: Southeast University.

[6] Dong, X., Gabrilovich, E., Heitz, G., Horn, W., Lao, N., & Murphy, K., et al. (2014). Knowledge vault: a web-scale approach to probabilistic knowledge fusion. ACM.

[7] Usunier, N., & Usunier, N. (2016). Constructing and Mining Web-scale Knowledge Graphs. International ACM SIGIR Conference on Research and Development in Information Retrieval (pp.1195-1197). ACM.

[8] Bollacker, K., Evans, C., Paritosh, P., Sturge, T., & Taylor, J. (2008). Freebase:a collaboratively created graph database for structuring human knowledge. ACM SIGMOD International Conference on Management of Data, SIGMOD 2008, Vancouver, Bc, Canada, June (pp.1247-1250). DBLP.

[9] Levenshtein, V. I. (1966). Binary codes capable of correcting deletions, insertions and reversals. Soviet Physics Doklady, 10(1), 707-710.

[10] Kosub, S. (2016). A note on the triangle inequality for the jaccard distance.

Link prediction in social network based on local information and attributes of nodes

Yingying Liang[1, 2], Lan Huang[1,2,3] and Zhe Wang[1,2,*]

[1]College of Computer Science and Technology, Jilin University, 130012, Changchun, China,
[2]Key Laboratory of Symbolic Computation and Knowledge Engineering (Jilin University), Ministry of Education, 130012, Changchun, China
[3]Zhuhai College of Jilin University, 519041, Zhuhai, China
[*]Corresponding Author:wz2000@jlu.edu.cn

Abstract. Link prediction is essential to both research areas and practical applications. In order to make full use of information of the network, we proposed a new method to predict links in the social network. Firstly, we extracted topological information and attributes of nodes in the social network. Secondly, we integrated them into feature vectors. Finally, we used XGB classifier to predict links using feature vectors. Through expanding information source, experiments on a co-authorship network suggest that our method can improve the accuracy of link prediction significantly.

1. Introduction

Link prediction plays an important role in the field of data mining [1].The traditional methods are based on the topological information, which have been widely used in various fields [1], like social recommendation, information retrieval and other fields. It is also very helpful to solve the problem of information overload [2][3]. The traditional methods can be divided into two categories, the first category is based on the local information like CN [4], and the other one is based on the global information like Katz [5].

Centrality is used to describe importance of nodes in the network [6][7]. If the centrality of a node is higher than others, it is easier for it to attract attention and construct new links with other nodes in the network, so researchers improve accuracy of link prediction methods with the help of centrality[8].Some researchers construct feature vectors using characteristics including centrality of nodes to predict links[9][10],and some researchers use machine learning methods to improve performance of link prediction methods[11].

It has become more convenient for people to communicate with others using social media websites which produced a huge amount of information. In order to improve the accuracy of link prediction, researchers begin to pay attention to the relevance between content and structure of the social network, and then design new methods to predict links effectively [12][13]. Many researchers used some characteristics of social networks to improve accuracy of link prediction, and got good results[14][15].According to [12], people belong to the same social network tends to form a local community, so they defined a new index: CAR which can improve the effect of classical CN index significantly .CAR is also better than AA index [14] and RA index [15].

The relationship in social networks is dynamic with the change of time and network context [9]. In a co-authorship network, if two authors have no co-authorship at the moment, how can we predict whether they will construct relationship in the future? In this paper, we focus on topology and context of network and consider attributes of nodes, then design feature vectors to describe characteristics of nodes. In addition, this kind of framework is suitable for most social networks since the calculation of features is very easy. Our method is based on classification using XGB classifier, and experiments on a part of DBLP (Digital Bibliography Library Project) show that our method can predict links effectively.

2. Related Definitions and Concepts

In order to formalize characteristics of the social network, researchers usually use graph theory to model it. For a social network, each individual is considered as a point, and the relationship between them is considered as an edge, here we will give some basic definitions.

Definition 1. Social network. $G = (V, E)$ represents a social network, where V represents the set of users, E represents the set of edges, and $e_{i,j} \in E$ represents an edge between nodes v_i and v_j.

Definition 2. CN index. For any network G, if there is no direct link between v_i and v_j at the moment, then $CN(v_i, v_j)$ index(Common Neighbors)is defined as in equation(1), where $\Gamma(v_i)$ represents the set of neighbors of v_i ,and $\Gamma(v_j)$ represents the set of neighbors of v_j.

$$CN(v_i, v_j) = \Gamma(v_i) \cap \Gamma(v_j) \tag{1}$$

Definition 3. CAR index. If nodes v_i and v_j have no direct edges at the moment, then CAR index [12]is defined as in equation(2), where $LCL(v_i, v_j)$ represents the number of edges between nodes which belong to $CN(v_i, v_j)$. CAR index can improve the accuracy of DC index effectively without increasing too much computation.

$$CAR(v_i, v_j) = CN(v_i, v_j) \times LCL(v_i, v_j) \tag{2}$$

Definition 4. DC index. DC index (Degree Centrality) represents centrality of a node using the number of neighbors of it.Taking an example of v_i, $DC(v_i)$ is defined as in equation (3), where $k(v_i)$ represents the number of neighbors of v_i, and N is the total number of nodes in the network, so $N-1$ represents the maximum number of neighbors of a node in the network.

$$DC(v_i) = \frac{k(v_i)}{N-1} \tag{3}$$

We choose XGB classifier [16]to predict links. Babajide [17] proposed an algorithm to predict molecular activities based on XGB and experiments show that it is more accurate than some popular algorithms at present, like support vector machine, stochastic forest, naive Bayesian and so on.

3. Methods and Experiments Setup

Common neighbors of two users often have an impact on the construction of new links between strangers, so we propose the concept of relevant force to describe influence of CN nodes at first and design the modified CN index by combining relevant force. Then we design some other features to describe attributes of nodes. Finally, we construct feature vectors using those features to predict links.

3.1. Our Method

3.1.1. Relevant force (RF). RF calculates CN nodes' influence according to specific context, if there is a link between v_i and v_z ,then RF from v_z to v_i is defined as in equation (4):

$$RF(v_i,v_z)=\frac{1+CN(v_i,v_z)}{D(v_i)+\sum_{m=1}^{D(v_i)}CN(v_i,v_m)}(v_z\in CN(v_i,v_j),v_m\in Nei(v_i)) \tag{4}$$

where $\sum_{m=1}^{D(v_i)}CN(v_i,v_m)$ represents the number of common neighbors between v_i and its neighbors, and $D(v_i)$ represents degree of v_i.

3.1.2. MCN index. We assume that there is no direct link between v_i and v_j,and v_z is one of nodes of $CN(v_i,v_j)$,then RF index is defined as in equation (5):

$$MCN(v_i,v_j)=\sum_{v_z\in\Gamma(v_i)\cap\Gamma(v_j)}((RF(v_i,v_z)+RF(v_j,v_z))\times DC(v_z)) \tag{5}$$

For any node belongs to $CN(v_i,v_j)$, there are two factors affect the influence from it to v_i and v_j: its importance in local network and the trust degree from v_i and v_j to it. We measure them through DC and RF respectively. The higher value of MCN means that there is greater possibility to construct new links.

3.2. Experiments Setup

DBLP is a collection of English literatures in the field of computer science which lists information of authors and papers. Link prediction on DBLP refers to predicting co-author relationship between authors in the future based on information within specific period. The author of the paper corresponds to a node, and the co-author relationship corresponds to the edge. We pay attention to authors' research fields, and integrate co-authorship information into MCN.Here are some definitions based on DBLP.

3.2.1. Similarity of research area. In the field of computer science, different international academic conferences focus on different research areas. We select 15 international conferences in DBLP and divide them into four areas. The details as follows: 1) Data mining: KDD, ICDM, PAKDD; 2) Database: SIGMOD, VLDB, ICDE, PODS; 3) Information retrieval: SIGIR, ECIR, WWW, CIKM; 4) Machine learning: ICML, NIPS, AAAI, CVPR.

We assume that if the author published a paper at an international conference which belongs to a specific research area, there will be a link between the author and the research area. Then we regard the number of links between authors and research areas as a feature which expresses author's interests. We use cosine similarity of feature vectors to represent the similarity between authors. We regard ***A*** = $(A_1,A_2,..,A_n)$ and ***B*** = $(B_1,B_2,..,B_n)$ as the authors' feature vectors, then the $CosSim(Cosine\ Similarity)$ is defined as in equation(6).

$$CosSim(A,B)=\frac{\sum_1^n(A_i\times B_i)}{(\sum_1^n A_i^2)^{1/2}\times(\sum_1^n B_i^2)^{1/2}} \tag{6}$$

We define the vector ***Confer*** to represent author's interests, so similarity of research areas is defined as $ConfSim$ which is calculated as in equation (7):

$$ConfSim(v_i,v_j)=CosSim(Confer(v_i),Confer(v_j)) \tag{7}$$

3.2.2. Modified MCN index. In DBLP, authors who have not direct links may construct a link if they can communicate with each other through a common neighbor. Researchers often construct new links with each other due to similar research topics. We define $CoPub$ to describe the times of co-author between authors, and then define modified MCN index: $MMCN$ using $CoPub$ and RF . $MMCN$ is defined as in equation (8):

$$MMCN(v_i,v_j) = \sum_{v_z \in CN(v_i,v_j)} DC(v_z) \times \sum_{k=i,j} (RF(v_k,v_z) \times CoPub(v_k,v_z)) \tag{8}$$

Unlike the MCN index, $CoPub$ and $ConfSim$ are the indexes which are designed according to the characteristics of DBLP. According to the specific conditions, we can extract different kinds of domain knowledge from different fields, and integrate them to the feature vectors, all of which will be helpful to improve the accuracy of link prediction methods.

4. Results and Discussion

After obtaining the information such as network topology and attributes of nodes, we can calculate the CAR and other indexes, and then construct the feature vector to predict links. Let the feature vector as ***Features***$=\{f_1, f_2,..., f_n\}$ where f_i represents various specific indexes 1CAR, DC, MCN and so on. We use the indexes presented in this paper to construct a model of feature vector and test it on DBLP.

4.1. Data Set

We select a part of DBLP and define it as DBLP_fourarea which contains 1500 authors and 6303 co-author relationships. It includes information of authors and papers published in 15 international conferences mentioned in 3.2.1 during the period from 2004 to 2010.Then we select information of DBLP_fourarea from 2004 to 2008 as the train set, and regard the information from 2009 to 2010 as the test set. Our aim is to predict co-author relationships which are included in train set but not exist in test set.

4.2. Evaluation Standards

In this paper, we choose F-value and AUC to evaluate the performance of our method. F-value combines precision and recall, which is the harmonic mean of precision and recall, so it is more reliable. The traditional F-value is defined as in equation (9):

$$F-value = 2 \times \frac{precision \times recall}{precision + recall} \tag{9}$$

The area under the curve (AUC) describes the ability of the classifier to classify the patterns submitted correctly. It is defined as in equation (10):

$$AUC = \frac{1}{2} \sum_{i=1}^{m-1} (x_{i+1} - x_i) \times (y_i + y_{i+1}) \tag{10}$$

where x_i and y_i represent the X-axis and Y-axis of the ROC curve respectively.

4.3. Experiments and Analysis

The experimental results of the single index based on the DBLP_ fourarea are shown in Table 1.

Table 1. Results of DBLP_ fourarea based on the single index

Feature	F-Value	AUC
CAR	0.0962	0.5491
MCN	0.0831	0.5419
MMCN	0.1095	**0.5567**
ConfSim	**0.5299**	0.4162

We can see that the performance of single *CAR* index is not that good. *MMCN* and *ConfSim* indexes take consider of domain knowledge, so their performance are better than *CAR* and *MCN* indexes according to AUC and F-value respectively, which shows that the domain knowledge is a significant factor for the improvement of link prediction methods. Experiments of *MMCN* index are better than *CAR* index, which illustrates the effectiveness of our algorithm. Table 2 shows experimental results of the feature vector based on combination of indexes, where *CAR* & *DC* represents that ***Features***= {CAR, DC}.

Table 2. Results of DBLP_ fourarea based on the combination of indexes.

Feature Vector	F-value	AUC
CAR	0.0962	0.5419
CAR&DC	0.6493	0.4903
CAR&MMCN	0.1831	0.5491
CAR&ConfSim	0.4238	0.5097
CAR&DC&MMCN&ConfSim	**0.6513**	**0.8839**

As can be seen from Table 2, the feature vector which combines some indexes has improved performance of link prediction methods dramatically, which indicates that expanding information source is efficient. In addition, it can be seen that the feature vector combining with attributes of nodes has a certain improvement comparing with the single index, so we should utilize attributes of nodes and network context to enhance the accuracy of link prediction.

5. Conclusion

We propose a new method to predict links in social network. We calculate some indexes to represent topological information and attributes of nodes to construct feature vectors, and predict links based on XGB classifier. Experiments show that our method improves accuracy of link prediction effectively. In addition, components of the feature vector are easily calculated, so it can be applied to many types of networks and allows us pay more attention to the domain knowledge.

6. Acknowledgements

This work was supported by National Natural Science Foundation of China (NSFC Grant No. 61373051, No.61472159 and No.61572227), and the key support of e-commerce technology engineering laboratory innovation project of Jilin Province.

7. References

[1] Lü Linyuan and T Zhou. "Link prediction in complex networks: A survey." Physica A Statistical Mechanics & Its Applications390.6 (2011):1150-1170.

[2] Wang, Peng, et al. "Link prediction in social networks: the state-of-the-art." Science China Information Sciences 58.1(2015):1-38.

[3] Liben-Nowell, David and J Kleinberg. "The link prediction problem for social networks." Journal of the Association for Information Science & Technology 58.7(2007):1019–1031.

[4] Lin Dekang. "An Information-Theoretic Definition of Similarity." 15th Int. Conf. on Machine Learning Morgan Kaufmann Publishers Inc. 1998:296-304.

[5] Katz,Leo."A new status index derived from stoichiometric analysis. Psychometrika 18.1(1953):39-43.

[6] Freeman, Linton C. "Centrality in social networks conceptual clarification." Social Networks 1.3(1978):215-239.

[7] Sabidussi G. "The centrality of a graph." Psychometrika31.4 (1966):581.

[8] Liu Haifeng, et al. "Hidden link prediction based on node centrality and weak ties." Epl 101.1(2013):18004.

[9] Li Yuhua, et al. "Research of Dynamic Link Prediction Method Based on Link Importance." Journal of Computer Research & Development (2011).

[10] Scellato, Salvatore, Anastasios Noulas and Cecilia Mascolo. "Exploiting place features in link prediction on location-based social networks." Proc. of the 17th ACM SIGKDD Int. Conf. on

Knowledge discovery and data mining. ACM, 2011.
[11] Hasan and a Mohammad. "Link Prediction using Supervised Learning."Proc. of Sdm Workshop on Link Analysis Counterterrorism & Security30.9 (2006):798-805.
[12] Cannistraci, Carlo Vittorio, G Alanislobato and T Ravasi. "Erratum: From link-prediction in brain connectomes and protein interactomes to the local-community-paradigm in complex networks." Sci Rep3.4 (2013):1613.
[13] Valverde-Rebaza J and A D A Lopes. "Structural link prediction using community information on Twitter." Int. Conf. on Computational Aspects of Social Networks IEEE, 2012:132-137.
[14] Adamic, Lada A and E Adar. "Friends and neighbors on the Web."Social Networks 25.3(2003):211-230.
[15] Zhou Tao, L Lü and Y C Zhang. "Predicting missing links via local information." European Physical Journal B 71.4(2009):623-630.
[16] Chen Tianqi and Carlos Guestrin. "Xgboost: A scalable tree boosting system." Proc. of the 22nd acm sigkdd Int. Conf. on knowledge discovery and data mining. ACM, 2016.
[17] Babajide Mustapha, Ismail and Faisal Saeed. "Bioactive molecule prediction using extreme gradient boosting." Molecules 21.8 (2016): 983.

IOP Conf. Series: Journal of Physics: Conf. Series **887** (2017) 012044 doi:10.1088/1742-6596/887/1/012044

Review on open source operating systems for internet of things

Zhengmin Wang, Wei Li and Huiliang Dong

Information Center, China tobacco ZheJiang industrial CO.,LTD., room 217, building No.1, No.77 Zhongshan South Road, Hangzhou, Zhejiang, China,Email:wangzm@zjtobacco.com

Abstract. Internet of Things (IoT) is an environment in which everywhere and every device became smart in a smart world. Internet of Things is growing vastly; it is an integrated system of uniquely identifiable communicating devices which exchange information in a connected network to provide extensive services. IoT devices have very limited memory, computational power, and power supply. Traditional operating systems (OS) have no way to meet the needs of IoT systems. In this paper, we thus analyze the challenges of IoT OS and survey applicable open source OSs.

1. Introduction

In recent few years, the interest in Internet of Things (IoT) is growing from industrial field and academic field for its applications. The development of several technologies such as sensor technology, network communication technology and cloud computing made it possible to achieve the vision of IoT. We are moving towards an automated era, where we have smart planet, smart cities, smart homes, all equipped with intelligent IoT capable of carrying out tasks by themselves. IoT is an integrated system of uniquely identifiable communicating devices which exchange information in a connected network to provide extensive services.

There are a large number of embedded devices in IoT. Compared with the traditional embedded devices, IoT devices generally have scarcity of energy and memory resources. They are generally small and battery-operated. Furthermore, they should be safe and reliable. IoT devices need to support a variety of communication protocols and protocol conversion between the layers and have the ability of cloud computing in application layer. In software aspect, the software of IOT devices is more complex than traditional embedded device software, and it also puts forward higher requirement for embedded operating system. In order to meet this requirement, IoT operating system came into being.

This paper is organized as follows. First, we analyze the structure of IoT and challenges of IoT OS. Then, we survey the main open source operating systems for IoT and we emphasize to compare the network stacks in them. Finally we forecast the trend of IoT operating systems.

2. Structure of IoT

The Internet of things (IoT) is the inter-networking of physical devices, buildings, and other items embedded with electronics, software, sensors, actuators, and network connectivity which enable these objects to collect and exchange data. The classical structure of one IoT system is showed as the following Fig.1. Embedded system is the core of the IoT and it is responsible for managing all the sensors, actuators, drives, and need to connect the network. For a large number of networking needs, the embedded operating system due to the closed features cannot be directly applied to the Internet of things equipment, mobile iOS and Android operating system due to the larger, high power consumption and other reasons cannot be very good fit networking terminal, therefore the operating system of IoT needs a new round of development and innovation.

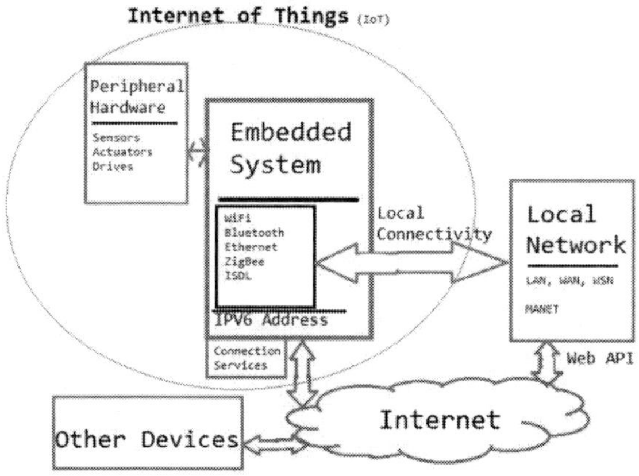

Figure 1. Classical structure of IoT system

3. Challenges of IoT OS

The design of an operating system includes many aspects, such as architecture, scheduling algorithm, memory management, network protocol, security measures, etc. In this section, we just analyze three very important aspects of OS for IoT devices: network connectivity, support for heterogeneous hardware, memory management.

3.1. Network Connectivity

In the foreseeable future, it is expected that trillions of devices will be connected to the Internet. Internet connectivity is a fundamental requirement of IoT devices. IoT system must support low power, highly reliable and internet-enabled communication stack. The stack must be flexible, so that with minimal changes it can be configured to meet the demands of a wide range of IoT applications. The support to Ipv6 is mandatory in IoT systems to have unique identities in tremendously large networks. A standardized approach, using latest developments in the IEEE 802.15.4 and IETF working groups, has been proposed [1, 2]. Maria etc. describe in detail power-efficient IEEE 802.15.4-2006 PHY layer, the power saving and reliable IEEE 802.15.4e MAC layer, the IETF 6LoWPAN adaptation layer enabling universal Internet connectivity, the IETF ROLL routing protocol enabling availability, and finally the IETF CoAP enabling seamless transport and support of Internet applications[1]. The IoT OS should cater for multiple network stacks and for continuous network stack evolution.

3.2. Support for Heterogeneous Hardware

There are various kinds of Internet of things nodes in the market, such as Arduino [3], OpenMote nodes [4], and TelosB nodes [5]. IoT devices are based on various microcontroller architectures and families, such as 8-bit (e.g., Intel 8051/52), 16-bit (e.g., TI MSP430), 32-bit (e.g., ARM7). They have different sizes of RAM and ROM and they can be equipped with a wide variety of communication technologies. Thus the IoT OS should support this heterogeneity in hardware architectures.

3.3. Memory Management

Memory management is an important function of the operating system. IoT devices usually have small resources (KB level RAM and MB level Flash). One of the requirements for IoT OS is to fit within such memory constraints. IoT OS should provide a set of optimized libraries (potentially cross layer) providing common IoT functionality, and efficient data structures.

Sections, Subsections and Subsubsections

The use of sections to divide the text of the paper is optional and left as a decision for the author. Where the author wishes to divide the paper into sections the formatting shown in table 2 should be used.

4. Open Source OS for IoT

4.1. F Contiki

Contiki [6] is a flexible and portable OS.

For communication, Contiki provides powerful low-power Internet communication. Contiki supports fully standard IPv6 and IPv4, along with the recent low-power wireless standards: 6lowpan, RPL, CoAP. With Contiki's ContikiMAC and sleepy routers, even wireless routers can be battery-operated.

Contiki has microIP (uIP) [7] and Rime stacks. Rime stack comprises a set of custom light-weight protocols. uIP stack is a light-weight stack which supports IPv6. uIP stack uses protocols especially developed for low-power and wireless memory-constrained real-time sensor devices. Another application layer protocol uXMPP (Micro eXtensible Messaging and Presence Protocol) [8] exists for event-driven Contiki for web services such as short messaging, publish subscribe, basic authentication mechanisms etc.

In wireless networks, nodes may need to relay messages from others to reach their destination. With Contiki, even relay nodes, so-called routers, can be battery-operated thanks to the ContikiMAC radio duty cycling mechanism which allows them to sleep between each relayed message. Some call these sleeping routers, we call it sleepy routers.

In situations when bandwidth is at a premium or where the full IPv6 networking stack is overkill, Contiki provides a tailored wireless networking stack called Rime. The Rime stack supports simple operations such as sending a message to all neighbors or to a specified neighbor, as well as more complex mechanisms such as network flooding and address-free multi-hop semi-reliable scalable data collection. Everything runs with sleepy routers to save power.

For memory, Contiki is designed to run in small amounts of memory. A typical system with full IPv6 networking with sleepy routers and RPL routing needs less than 10 k RAM and 30 k ROM. The OS supports dynamic memory allocation but there is no Memory Protection Unit (MPU). Contiki is therefore highly memory efficient and provides a set of mechanisms for memory allocation.

For heterogeneous hardware, Contiki runs on a range of different hardware platforms from 8-bit to 64-bit and is designed to be easy to port to new hardware.

At present, contiki has become the most important choice in the field of IoT.

4.2. TinyOS

Together with Contiki, TinyOS is the most prominent OS for WSN applications, targeting very constrained 8-bit and 16-bit platforms and is known for its sophisticated design [9, 10].

For communication, TinyOS has Berkeley Low-power Internet Stack (BLIP) [11] which consists of TCP, UDP, ICMPv6 and IPv6 and supports 6LoWPAN, RPL and CoAP. The OS uses Hydro routing protocol to ensure reliable unicast communication within an IPv6 sub network with lossy links.

For memory, TinyOS uses compile-time static memory allocation. There are no heaps, function pointers or virtual memory concepts of dynamic memory allocation. In version 2.1 onwards [12], Safe TinyOS has MPU.

For heterogeneous hardware, compared to Contiki, there are only a few hardware supported by TinyOS.

4.3. RIOT

RIOT [13] is developed separately in two parts [14], hardware-dependent and hardware-independent code with well-defined interface. This helps in independent configuration of hardware-specific code to exploit the underlying platform. The kernel and system libraries can remain unchanged in case hardware specific features are modified.

For memory, it doesn’t have MMU or Floating Point Unit. It has both memory allocation provisions: static and dynamic. It uses tickless scheduler to reduce power consumption. The system enters in sleep mode by switching to idle thread whenever there are no pending tasks and remains in the mode until woke up by an interrupt.

For communication, RIOT has modular Internet stack. Modular stack is more flexible than layered stack as modules can be separately maintained. The protocols 6LoWPAN and RPL are provided with support to TCP, UDP and IPv6.

For heterogeneous hardware, RIOT can run on 8-bit platforms (e.g. Arduino Mega 2560), 16-bit platforms (e.g. MSP430), and on 32-bit platforms (e.g. ARM).

4.4. FreeRTOS

FreeRTOS [15] is written mostly in C with a few assembly functions. It uses Nabto [16] to provide communication between peer IoT devices. It supports two privilege modes of operation: kernel mode and user mode. The programming model is based on tasks and routines. Co-routines are light-weight tasks having smaller shared stacks. Generally tasks are scheduled preemptively, each having a stack of its own and co-routines follow cooperative scheduling amongst themselves. A task can preempt a co-routine. Priority based preemptive scheduling of OS supports real-time systems.

For memory, FreeRTOS supports dynamic memory allocation with two heaps. Fast internal RAM is used by task stack, while the other heap stores application data in slow external RAM. FreeRTOS has FreeRTOS-MPU to protect the kernel from failures. It has optional tickless mode to save energy.

For communication, it supports FreeTCP/IP (based on uIP stack, still under development) and lwIP networking stacks which are based on IPv6 and other less power consuming protocols like 6LoWPAN, CoAP etc.

For heterogeneous hardware, a variety kinds of MCUs, such as Atmel, Freescale, ST, and TI and so on, offer official RTOS ports for FreeRTOS.

4.5. MbedOS

MbedOS [17] was developed by the ARM, specifically for the operation of the ARM processor device design and networking equipment. It includes all the features you need to develop a connected product based on an ARM Cortex-M microcontroller, including security, connectivity, an RTOS, and drivers for sensors and I/O devices.

For communication, mbedOS supports IPv6 and 6LoWAN, and takes SSL and TLS, the standard protocols for securing communications on the internet, and allow the users to include them in their mbed project with a simple API.

4.6. UClinux

The original uClinux [18] was a derivative of Linux 2.0 kernel intended for microcontrollers without Memory Management Units (MMUs). However, the Linux/Microcontroller Project has grown both in brand recognition and coverage of processor architectures. Today's uClinux as an operating system includes Linux kernel releases for 2.0 2.4 and 2.6 as well as a collection of user applications, libraries and tool chains. It has full TCP/IP stack which can be replaced with a light-weight stack like uIP and lwIP [7]. uClinux has higher footprint compared to other OSs.

4.7. µC/OS

µC/OS [19] is a full-featured embedded operating system. Features support for TCP/IP, USB, CAN bus, and Modbus. Includes a robust file system, and graphical user interface

µC/OS has a compact, reliable, high-performance TCP/IP protocol stack designed for embedded applications, featuring dual IPv4 and IPv6 support. It allows multiple interfaces and multiple IP addresses per interface. It has multicast IPv6, ICMPv6, TCP and UDP. Some application layer protocols like FTP, POP3 etc., are not compatible with IPv6. The Ethernet device controllers provide hardware abstraction for easy porting to a variety of Ethernet devices.

4.8. Lite OS

Huawei LiteOS [20] is a Unix-like OS. It has Unix based programming environment. The OS consists of three subsystems: LiteFS, LiteShell and the kernel.

For memory, LiteFS file system provides support to file and directory operations. The concurrent programming model is thread-based with some support to events via callback functions. Whenever an already registered event occurs, callback function is invoked. Callback functions reduce the number of active threads and improve system efficiency. The memory allocation to applications is done dynamically through system calls using malloc and free API functions. LiteOS has an event trace logger mechanism to provide system insight. Every running application has a buffer to register triggered events. When the buffer runs off the space, it is emptied in a file in external flash. This mechanism can help in protecting memory.

For communication, LiteOS supports plug-and-play routing stack, it does not have any built-in networking protocols.

4.9. Other OSs

In addition to the above, there are many other IoT OS, such as Android Things [21], OpenWSN [22], NutOS [23].

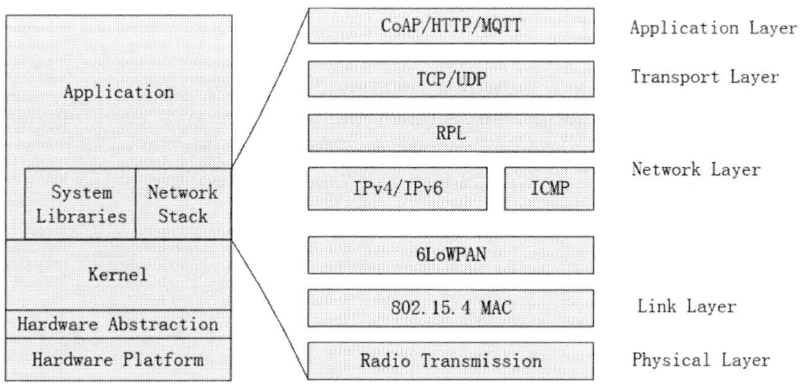

Figure 2. Classical structure of IoT OS

In this section, we list the predominant open source OSs targeting for IoT devices. There is a continuously growing long chain of OSs with a diverse set of features. According to analysis of existing open source OSs, they all have the classical structure as Figure 2.

5. Conclusion

IoT devices have some special features as compared to conventional computing devices. Therefore, the OS is to be designed according to the distinct requirements of IoT devices and target application areas. In this paper, we briefly introduce the challenges of IoT OS and briefly review the open source operating systems for IoT and make a simple comparison on network stack of the state-of-the-art OS. The paper would help researchers to understand the IoT and choose the appropriate OS according to their requirements.

6. References

[1] M. R. Palattella et al., "Standardized Protocol Stack for the Internet of (Important) Things," in IEEE Communications Surveys & Tutorials, vol. 15, no. 3, pp. 1389-1406, Third Quarter 2013.

[2] D. Dragomir, L. Gheorghe, S. Costea and A. Radovici, "A Survey on Secure Communication Protocols for IoT Systems," 2016 International Workshop on Secure Internet of Things (SIoT), Heraklion, 2016, pp. 47-62.

[3] "Arduino due," [Online]. Available: http://arduino.cc/en/Main/ arduinoBoardDue

[4] OpenMote. "OpenMote-CC2538," [Online]. Available: http://www. openmote.com/hardware/openmote-cc2538-en.html

[5] MoteIV Corporation. "Telos—Ultra Low Power IEEE 802.15.4 compliantwireless sensor module," Datasheet [Online]. Available: http://www.willow.co.uk/html/telosb_mote_platform.php

[6] "Contiki operating system," [Online]. Available: http://www.contikios.org.

[7] Dunkels A. Full TCP/IP for 8-Bit Architectures. In Proc. Mobisys, 2003.

[8] A. Hornsby, and E. Bail. XMPP: Lightweight Implementation for Low Power Operating System Contiki. In Proc. ICUMT, 2009.
[9] E. Baccelli, M. Gunes, and T.C. Schmidt. Operating Systems for the IoT Goals, Challenges, and Solutions. In Proc. IPSN, 2008.
[10] [Online]. Available: https://github.com/tinyos
[11] K. N. Nithin (2010).BLIP: An Implementation of 6LoWPAN in TinyOS. Available: http://bit.ly/1rD5dbd.
[12] TinyOS Alliance. TinyOS 2.1: Adding Threads and Memory Protection to TinyOS. In Proc. Sensys, 2008.
[13] RIOT [Online]. Available: www.riot-os.org.
[14] E. Baccelli, and O. Hahm. RIOT: One OS to Rule Them All in the IOT. In Ineria Research Report n 8176, 2013.
[15] FreeRTOS [Online]. Available: www.freertos.org.
[16] Nabto [Online]. Available: http://nabto.com/#intro.
[17] mbedOS [Online]. Available: www.mbed.com
[18] uClinux [Online]. Available: http://www.uclinux.org
[19] uCOS [Online]. Available: https://www.micrium.com/rtos/
[20] LiteOS [Online]. Available: http://www.huawei.com/minisite/liteos/cn/
[21] Android Things [Online]. Available: https://developer.android.google.cn/things/index.html
[22] OpenWSN [Online]. Available: https://openwsn.atlassian.net/wiki/
[23] NutOS [Online]. Available: http://www.ethernut.de/en/firmware/nutos.html

IOP Conf. Series: Journal of Physics: Conf. Series **887** (2017) 012045 doi:10.1088/1742-6596/887/1/012045

Hybrid employment recommendation algorithm based on Spark

Zuoquan Li[1], Yubei Lin[2]*and Xingming Zhang[1]

[1] School of Computer Science and Engineering, South China University of Technology, Higher Education Mega Centre, Panyu District, Guangzhou, China
[2] School of Software Engineering, South China University of Technology, Higher Education Mega Centre, Panyu District, Guangzhou, China
Email: yupilin@scut.edu.cn

Abstract. Aiming at the real-time application of collaborative filtering employment recommendation algorithm (CF), a clustering collaborative filtering recommendation algorithm (CCF) is developed, which applies hierarchical clustering to CF and narrows the query range of neighbour items. In addition, to solve the cold-start problem of content-based recommendation algorithm (CB), a content-based algorithm with users' information (CBUI) is introduced for job recommendation. Furthermore, a hybrid recommendation algorithm (HRA) which combines CCF and CBUI algorithms is proposed, and implemented on Spark platform. The experimental results show that HRA can overcome the problems of cold start and data sparsity, and achieve good recommendation accuracy and scalability for employment recommendation.

1. Introduction

With the improvement of communication infrastructure and the development of Internet-related technologies, the amount of network information is exploding, which greatly satisfies the growing information needs of people. However, it makes the user cannot find the desire information in time. Therefore, people put forward recommendation algorithms, which can quickly find the desire information by analysing people's historical behaviour data, and provide an optimal recommendation result. At the same time, the continuous development of the network has also greatly broadened the channels of employment. Both job seekers and companies have begun to use the network for employment. However, there is increasing difficulty to deal with massive job-related information. Therefore, a good employment recommendation algorithm is so important that it is worth doing further research.

Recommendation algorithm, which is the core of recommendation system, has become a hot topic since the research of recommendation system was begun at the end of the last century. In 1994, user-based collaborative filtering algorithm was applied for news recommendation [1]. In 2003, item-based collaborative filtering algorithm (IBCF) was put forward [2]. In the same year, Slope one collaborative filtering algorithm was proposed. It is simple and scalable. However, it cannot solve the problems of cold start and data sparsity [3]. To overcome the cold-start problem, content-based recommendation algorithm (CB) was proposed [4]. In recent years, machine learning algorithms, such as clustering, association rules and matrix decomposition, have been implemented for recommendation [5]. In [6] a collaborative filtering algorithm based on user interest change was presented. A graph recommendation algorithm based on attention relation and multi-user behaviours was introduced in [7]. A factor decomposition machine algorithm based on high order deviation was developed in [8].

Some scholars introduced recommendation algorithms into the field of human resources. The implicit semantic model was used to conduct two-way recommendation of talent and position in [9]. The technology of machine learning was implemented to predict the future career change of applicants in [10]. In [11] the ontology-based network recruitment recommendation was proposed. In [12] PageRank algorithm based on random walk model for job recommendation was introduced. Models of user interest and job preference were built to improve the collaborative filtering recommendation algorithm in [13].

For employment recommendation, not only requirements of companies, but also the job seekers' conditions and preferences should be considered. Moreover, with the continuous expansion of network information, a good recommendation system must be powerful enough to deal with the massive data. However, the algorithms mentioned above were almost designed as stand-alone calculation models. Therefore, in this paper a hybrid recommendation algorithm (HRA) base on Spark, which is a fast and versatile computing engine designed for large-scale data processing, is proposed. First a clustering collaborative filtering algorithm (CCF) is developed. The predictive scores of users can be obtained by CCF. The highest five predictive scores of each user are filled into the original matrix of user-item rating. Then this new rating matrix serves as the input data of a content-based algorithm with users' information (CBUI), which takes job seekers' information into account to improve the users' preference vectors. Finally, better recommendation results can be achieved.

2. HRA Based on Spark

2.1. Data Preprocessing

The experimental data of this paper are collected from a research project, including the applicants' personal information, jobs' basic information and the historical behaviour of applicants. The quality of the experimental data will greatly affect the final recommendation result. Since the collected data cannot be used directly in the recommendation algorithm, it is necessary to conduct data pre-processing as follows.

1) Some original data have duplicate values. For example, some companies repeatedly publish the same job. Only the latest information of this job will be retained after data pre-processing.

2) Some original data have missing or illegal values. In data pre-processing, default values will be set for these data. For example, the average wage of all jobs will be set as the default value of salary.

3) The salaries in the original data can be paid monthly, weekly or daily, which makes the wage-related attribute very confusing. We need to convert these data to a union form.

4) Some of the attributes in the original data, such as applicant's telephone number, job title, contact telephone number, and so on, are not necessary for our recommendation algorithm. Therefore, we should delete these attributes in data preprocessing.

2.2. CCF Based on Spark

The traditional IBCF has to find the neighbours of the target item in all items, which results in a lot of time spent on searching. We can reduce the time for finding neighbours by introducing clustering to the algorithm. Specifically, the clustering algorithm is used to divide the item set into clusters. Then we can find the nearest neighbours of the target item in each cluster. Finally, IBCF is used in the neighbour set of the target item. At the same time, because a cluster set is much smaller than the entire set of all items, we can find most of the neighbours of target item in a smaller range rather than the entire item space, and effectively improve the speed of the recommendation algorithm. The flow chart of CCF is shown in figure 1.

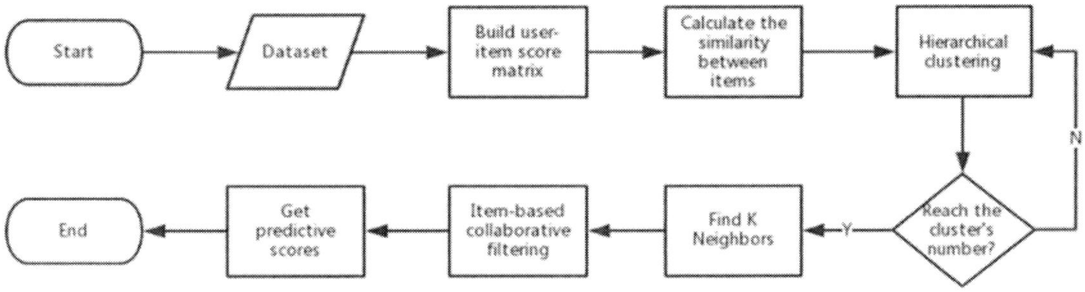

Figure 1. Flow chart of CCF.

As can be seen in figure 1, the algorithm mainly includes the following steps:

1) Construct the matrix of user-item score from the processed data set. In this paper, data cleaning, data conversion, attribute reduction, data aggregation and other methods are conducted in data pre-processing. After data pre-processing, the matrix of user-item score can be constructed by using the applicants' historical score data.
2) Calculate the similarity scores between items. Assume that vectors of item $\boldsymbol{i}$ and $\boldsymbol{j}$ are expressed by $\{r_{i1}, r_{i2}, \ldots, r_{in}\}$ and $\{r_{j1}, r_{j2}, \ldots, r_{jn}\}$ respectively, where r_{ix} and r_{jx} are the scores which user x awards to the items $\boldsymbol{i}$ and $\boldsymbol{j}$ respectively. The similarity score between items $\boldsymbol{i}$ and $\boldsymbol{j}$ defined as $\mathrm{sim}(\boldsymbol{i}, \boldsymbol{j})$ can be calculated by the following equation:

$$\mathrm{sim}(\boldsymbol{i}, \boldsymbol{j}) = \frac{\sum_{x=1}^{n}(r_{ix} \times r_{jx})}{\sqrt{\sum_{x=1}^{n} r_{ix}^2}\sqrt{\sum_{x=1}^{n} r_{jx}^2}} \tag{1}$$

3) Conduct hierarchical clustering on items. After calculating the similarity scores between items, the items can be clustered according to the similarity scores. Once the clustering is complete, we can define a cluster centre for each cluster. The new item needs to be compared with each cluster centre. Then the neighbour in the closest cluster can be found to improve the search speed.
4) Get the predictive score. After the K-nearest neighbours of the target item in the clustering result are obtained in the previous step, IBCF can be applied in the neighbour set to obtain the user's predictive score.

To improve the scalability and computing performance of CCF, the algorithm is implemented on Spark platform. In this platform, parallelism is achieved primarily through Resilient Distributed Datasets (RDD). The overall RDD conversion diagram of CCF is shown in figure 2.

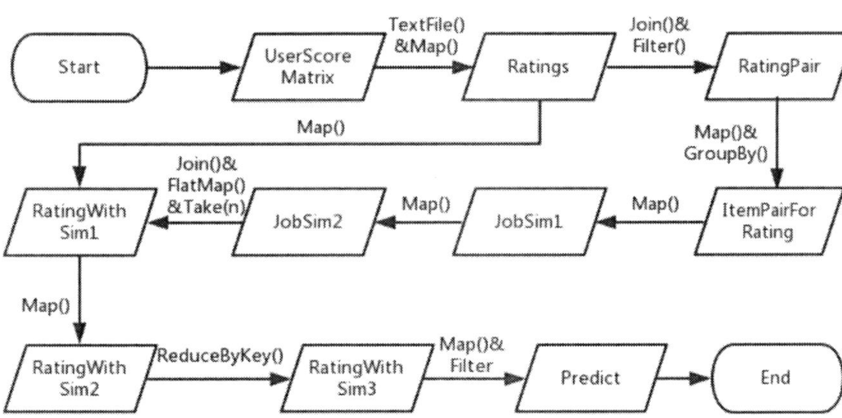

Figure 2. RDD conversion diagram of CCF.

It is noted that in figure 2, *UserScoreMatrix* is the input data of this algorithm, *Ratings* is the score matrix built from *UserScoreMatrix*, *RatingPair* is the collection of items rated by the same user, *ItemPairForRating* is a pair of items rated by the same user, *JobSim1 and JobSim2* represent the degree of similarity between each pair of items, *RatingWithSim1*, *RatingWithSim2* and *RatingWithSim3* are the records of *n* neighbouring items with the highest similarity to the target item, and the final predictive score is saved in the RDD named *Predict*.

2.3. CBUI Based on Spark

In Section 2.2, we study the clustering collaborative filtering algorithm, which is based on the user rating to cluster the items and improve the real-time performance of the employment recommendation algorithm. However, it only considers the users scores. Neither users' information nor items' information is taken in account. Moreover, collaborative filtering recommendation algorithm has the problem of cold start. To address these problems, an approach based on the traditional content-based recommendation algorithm is proposed, in which users' information, such as age, gender, educational level, and so on, is considered to find the users' preferences. The workflow of CBUI is as follows:

1) Construct the attribute vectors of items. The attributes of the items, including gender requirement, age requirement, academic requirements, job salary and job benefits, can be used to build the attribute vectors.
2) Mix the user's basic information and rating data, and construct the preference vectors of users. On one hand, we can obtain the user's favourite items from the user's rating information, and use the items' attributes obtained in the previous step to select these items' worktime, job benefits, job salary and other attributes. On the other hand, we can select some attributes from the user's basic attributes, such as gender, age, degree of education, working life, as part of the attributes of the user's preference vector. In this way, once the number of the user's rating is limit, the user's preference vector can be constructed according to the user's basic information.
3) Get Top-N recommendation results. After obtaining the attribute vectors of the item $\boldsymbol{i}$ and the user's preference vector $\boldsymbol{u}$, the similarity score between these two vectors defined as $\mathrm{sim}(\boldsymbol{u}, \boldsymbol{i})$ can be calculated by the following equation:

$$\mathrm{sim}(\boldsymbol{u}, \boldsymbol{i}) = \frac{|U \cap I|}{|U \cup I|} \tag{2}$$

where U and I are the attribute sets of $\boldsymbol{u}$ and $\boldsymbol{i}$ respectively. After obtaining the similarity scores, we can sort them and recommend a group of items with the highest similarity score.

Likewise, to improve the scalability and computational performance of CBUI, this algorithm is implemented on Spark. The parallel processing for CBUI mainly consist of constructing the item's attribute matrix, constructing the user's preference vector, calculating the similarity scores and obtaining the Top-N recommendation result. The RDD conversion diagram of CBUI is shown in figure 3.

It is noted that in figure 3, the RDD named *DataSet* is the input data of the algorithm, and we built the items' attribute matrix which is named *JobAttrs*, the users' rating matrix which is named "Ratings" and the users' attribute matrix which is named *UserAttrs* from *DataSet*. The other created RDDs are introduced as follows:

- *Ratings2*: the top *n* items rated by each user.
- *ItemWithAttrs2*: the set of jobs' attributes grouped by user.
- *Profile PartA*: the preference attribute of top *n* items with highest rating by each user.
- User Profile: the preference attribute of user.
- *Profile With Job Attrs*: the cartesian set of *UserProfile* and *JobAttrs*.
- *FilterItem*: rated items.
- *FilterResult*: result without rated items.
- *Recommend Result*:top *n* recommendation results.

2.4. HRA
In this section, a HRA based on CCF and CBUI can be implemented as follows. First, users' predictive scores are obtained by CCF. Then the highest five scores of each user are filled into the original user-item rating matrix. Finally, the new rating matrix serves as the input data of CBUI, which can improve the user's preference vector and achieve better recommendation results. The overall process of HRA is shown in figure 4.

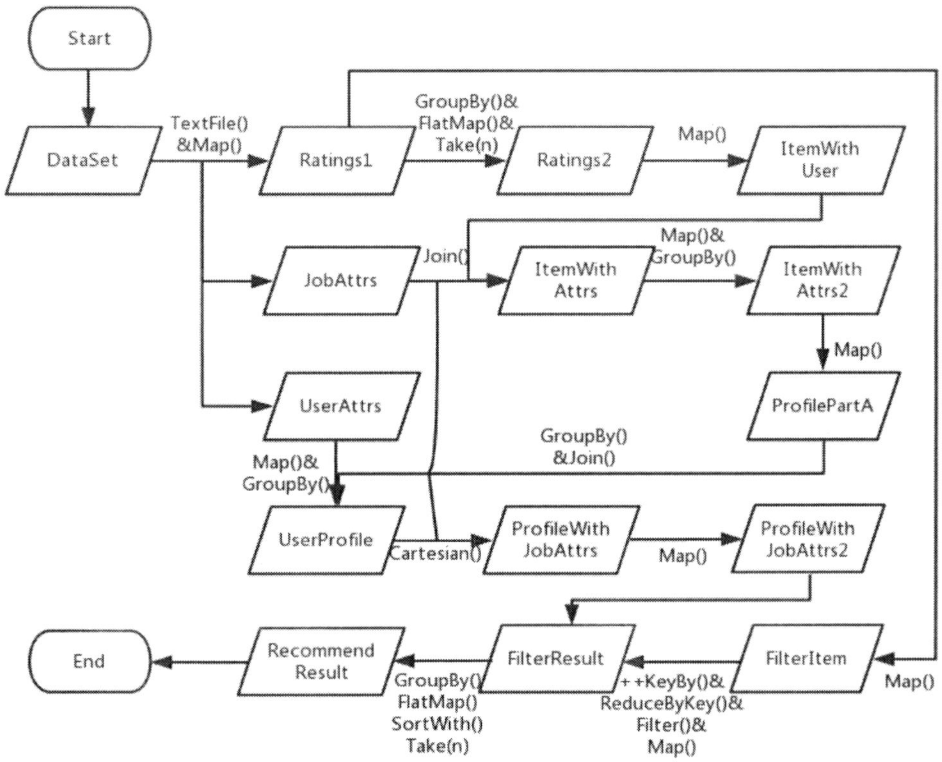

Figure 3. RDD conversion diagram of CBUI.

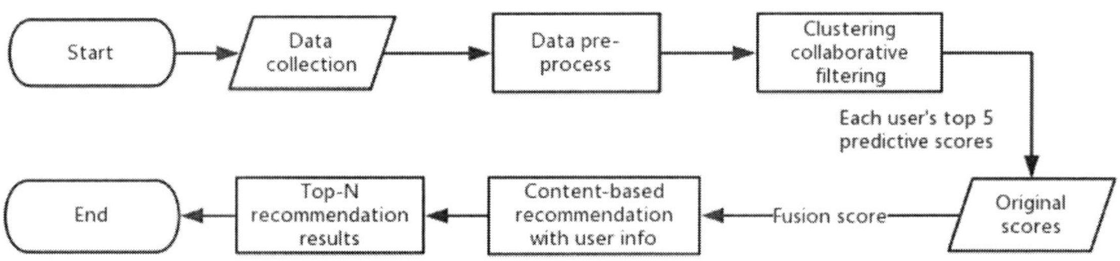

Figure 4. The overall process of the proposed HRA.

3. Experimental Setting and Results

3.1. Experimental Setting
Dataset & Metrics. In our experiment, 4692 users' information, 15000 jobs' information and 170844 user-job score data are involved. In addition, to verify the experimental results, accuracy rate P_u defined as the following equation (3) is used as the experimental indicator.

$$P_u = \frac{|R_u \cap I_u|}{R_u} \tag{3}$$

where R_u is the collection of the recommended items, and I_u is the set of items selected by the users in the test set.

Experimental Environment. In the experiment, we build a Spark cluster with four virtual machines. One of the virtual machines serves as the master node of the cluster, and the other three virtual machines serve as worker nodes. The configuration of each virtual machine is as follows: 8G memory, 4 cores 1.799GHZ CPU, 80G disk space, centos 6.7 operating system, 1.6.0 version of Spark, 2.6.0 version of Hadoop and 2.10.4 version of Scala.

3.2. Results and Analysis

Predictive Accuracy. CBUI is based on the traditional CB, so we firstly test these two algorithms on the same data set for the sake of comparison. The recall rate of these two algorithms are shown in figure 5.

As shown in figure 5, CBUI is superior to the traditional CB, which means that the users' basic information is used rationally so that when the ratings of users are less or even missing, users' preferences vectors can be built according to users' information. It can effectively alleviate the problem of cold start and improve the recommended results.

In addition, in the experiment, CCF, CBUI and HRA are tested on the same data set. When the number of clusters is 90 and the number of neighbour items is 40, the accuracy rate of these three algorithms are shown in figure 6.

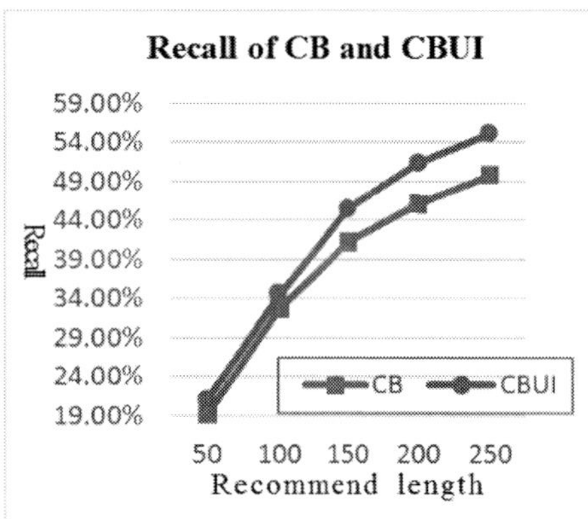

Figure 5. Recall of CB and CBUI.

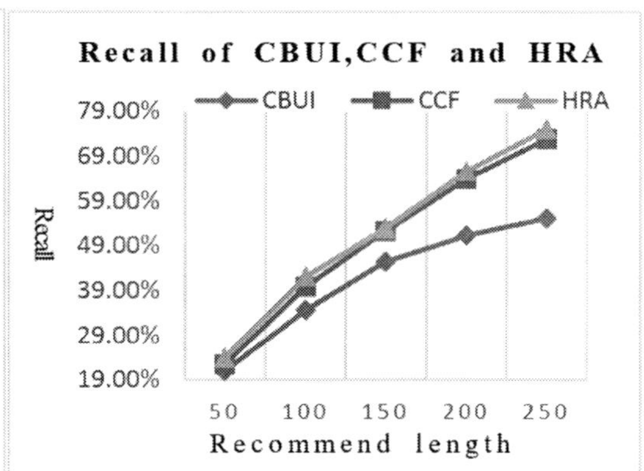

Figure 6. Recall of CBUI, CCF and HRA.

As shown in figure 6, with the increment of the recommendation list's length, the recall rates of these three algorithms are increased, while the rising speed is gradually slowed down. When the recommendation list's length is 250, the recall rates of these three algorithms are 75% (HRA), 72.8 % (CCF) and 55.1% (CBUI) respectively. Since the predictive score of CCF can alleviate the sparsity of the users' scores and improve the users' preferences vectors, the recall rate of HRA is much higher than that of CBUI, which shows that the proposed hybrid strategy is effective.

Parallelism. The parallelization of HRA is implemented on Spark. Speedup is defined as equation (4) to measure the scalability and performance of the parallel algorithms.

$$Speedup = \frac{T_1}{T_p} \tag{4}$$

where T_1 is the operation-time in the single-processor system and T_p is the operation-time in the p-node-processor system. In the ideal case, the parallel algorithm's Speedup should be linearly incremented as the number of nodes increases. In the Spark cluster, T_1 refers to the algorithm operation time in pseudo-distributed mode.

To validate the *Speedup* of the parallel HRA, the number of nodes in the cluster is gradually increased, and the actual operation time required for the task of the parallel HRA is recorded. The

Speedup of the parallel algorithm is shown in figure 7. As can be seen from it, with the increment of the number of data nodes, the *Speedup* curve of HRA basically conforms to the linear trend, which shows that the algorithm is scalable and can shorten the time by increasing the nodes. The *Speedup* curve of the algorithm is different from the ideal curve "Y = X", mainly because the data transmit between nodes takes time, and the problem of data tilt will also affect the completion time.

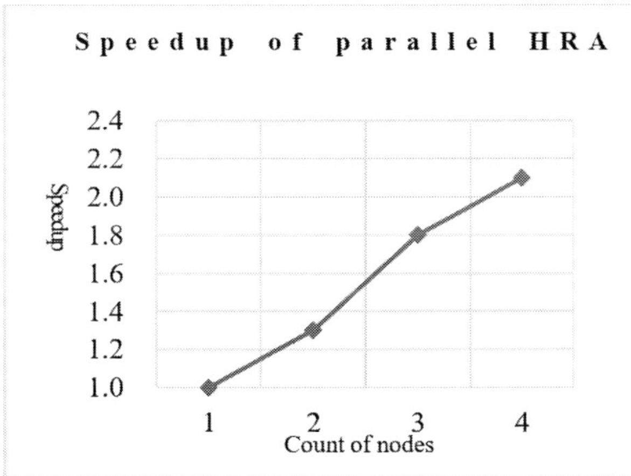

Figure 7. Speedup of parallel HRA.

4. Conclusion

The application of the employment recommendation algorithms, including CCF and CBUI, are studied in this paper. In addition, aiming at addressing the shortcomings of the single recommendation algorithm, a hybrid recommendation algorithm based on these two algorithms is proposed, which effectively solves the problem of cold start and alleviates the influence of data sparsity of the traditional CB. At the same time, the parallelization of the proposed algorithm is designed and implemented on Spark, which improves the parallelism and expansibility of the proposed algorithm. However, it is found that the parallelization efficiency of HRA has a certain gap with the ideal value, and it still has the space for improvement. Therefore, we will carry out an in-depth study on Spark to further improve the efficiency of the proposed algorithm.

5. Acknowledgement

This work is financially supported by Guangdong Provincial Science and Technology Plan (No. 2016B030308002).

6. References

[1] Goldberg D, Nichols D, Oki B M and Terry D 1992 Using collaborative filtering to weave an information tapestry Comm. ACM 35(12) 61-70

[2] Sarwar B, Karypis G, Konstan J and Riedl J 2001 Item-based collaborative filtering recommendation algorithms Int. Conf. on World Wide Web (Hong Kong) vol 4 (ACM) p 285

[3] Lemire D and Maclachlan A 2008 Slope one predictors for online rating-based collaborative filtering Comput. Sci. 21-3

[4] Longfei C and Xue Z 2013 The summary of personalized recommendation technology J. Hebei Normal Univ. Sci. Techno. 57-60

[5] Defu L, Cong Z, Xing X, Guangzhong S, Enhong C and Yong R 2014 GeoMF: joint geogra-phical modeling and matrix factorization for point-of-interest recommendation ACM SIGKDD Int. Conf. on Knowledge Discovery and Data Mining (New York City) (ACM) p 831

[6] Daoguo L, Dijiang H and Lianjie L 2017 Collaborative filtering recommendation algorithm based on user interest change Productivity Res. 1 19-21+55.

[7] Mengjuan L, Wei W, Yangxi L, Xucheng L and Zhiguang Qin 2017 AttentionRank ~ +: a graph recommended algorithm based on attention relation and multiuser behaviour J. Comput. 3 634-48

[8] Zihao Wang and Guiqiong Xu 2017 Recommended algorithm for factor decomposition machine based on high order deviation Application Res. Comput. 2 339-42
[9] Malinowski J, Keim T, Wendt O and Weitzel T 2006 Matching people and jobs: a bilateral recommendation approach Hawaii Int. Conf. on System Sciences (Hawaii) vol 6 (IEEE) p 137.3
[10] Paparrizos I, Cambazoglu B and Gionis A 2011 Machine learned job recommendation Proc. of the 5th ACM Conf. on Recommender Systems (Chicago) (ACM) p 325
[11] Qingqing S 2009 User modelling research of personalized network recruitment recommended system based on the ontology Nanjing Univ. Aeronautics and Astronautics 1 81-4
[12] Di W, Lijuan Z and Hongfei Lin 2011 Research and implementation of employment recommended system based on random walking 2011 J. Guangxi Normal Univ. (Natural Sci. Edition) 29(1) 179-85
[13] Hongjiao C 2014 The design and implementation of employment recommendation system for the graduate student based on context-aware technology Huazhong Normal Univ.

IOP Conf. Series: Journal of Physics: Conf. Series **887** (2017) 012046 doi:10.1088/1742-6596/887/1/012046

Near infrared spectroscopy detection of the content of wheat based on improved deep belief network

Wenwen Li[1], Min Lin[1,*], Yongmei Huang[1], Huijun Liu[1] and Xinqi Zhou[2]

[1]College of Metrology and Measurement Engineering, China Jiliang University Hangzhou, China
[2]Focused Photonics (Hangzhou), Inc. Hangzhou 310052, China
*Corresponding author, e-mail address: linm@cjlu.edu.cn

Abstract. In order to solve the complicated problem of traditional detection of the content of Wheat, a method for predicting the content of wheat components by near infrared spectroscopy based on improved deep belief network is proposed. In this paper, wavelet transform is used to preprocess near infrared spectroscopy of wheat, and then a quantitative analysis model of wheat's moisture, protein and ash content is established by using deep belief network. And combined with the random hidden algorithm, the network model is sparse processed, and the sparse network is obtained. So as to improve the accuracy and stability of the network. The experimental results show that using the improved deep of belief network to establish the quantitative analysis model on the content of wheat, and the correlation coefficient of moisture, protein and ash content were 0.9978, 0.9928, 0.9920, standard error of prediction were 0.0069, 0.0628,0.0535. Compared with the traditional deep belief network (DBN) and the traditional shallow learning BP neural network algorithm, the prediction results have been significantly improved.

1. Introduction

Wheat is one of the main sources of people's daily food. Wheat is one of the important sources of protein in people's diet. Protein is the nutrients that the body must take every day. Too much or too little intake can affect the body. Therefore, the quality and safety of wheat is directly related to the health of consumers. However, in recent years, adding excessive whitening agent and lime in wheat flour with the purpose of weight gain was repeatedly exposed, which not only has a huge negative impact on society, but also harm the health of consumers and food safety seriously. With the rapid, accurate, non-destructive testing technology is adopted to detect the content of wheat in line with modern high quality of life needs. Near infrared spectroscopy nondestructive testing technology has the advantages of fast, high efficiency, no pollution and no need of sample pretreatment. At present, it has been widely used in food quality analysis [1], fruit quality analysis [2] and many other fields. The establishment of an analytical model for moisture, protein and ash in wheat has a certain practical significance for wheat quality testing.

At present, the new methods are emerging, and the popular methods are mainly shallow learning methods such as PLS [3], neural network [4] and Support Vector Machine [5] (SVM). A common feature of these networks is that they all use the structure of no more than three layers to transform the original input signal into a feature space. There is no doubt that shallow learning is very effective to solve simple problems, in solving complex practical applications, there are many problems such as insufficient expression of function. However, in the same number of nodes, the deep learning network is usually more expressive than the shallow network.

In this paper, a method based on improved DBN (deep belief network) is proposed for the determination of wheat component content by near infrared spectroscopy. Wavelet transform can remove the spectral information with the noise in the data and realize effective compression of hyperspectral data and reduce the amount of computation model. The pretreated data and the corresponding analytical data were used as input to the network model, and the content of wheat components was used as the network output. A wheat network model for predicting the component content of wheat was established.

2. Deep Belief Network

Deep belief network model is composed of a plurality of restricted Boltzmann machine (RBM) and a layer of BP neural network [6]. First, unsupervised training, the first RBM to get the initial weight, and then the training of the output as the next RBM input, layer by layer training, until the training of all the RBM.

2.1. Gauss Restricted Boltzmann Machine

In this paper, the Gauss restricted Boltzmann machine is used instead of the restricted Boltzmann machine. And then the Gauss restricted Boltzmann machine model is obtained. The energy model is as follows.

$$E(v,h)=\frac{1}{2}\sum_{i}\frac{(v_i-c_i)^2}{\sigma_i^2}-\sum_{i,j}\frac{v_i}{\sigma_i}W_{i,j}h_j-\sum_{i}c_iv_i-\sum_{j}b_jh_j \tag{1}$$

In the upper form, c is bias at the input layer node, b is bias of hidden layer nodes, W is the connection weight between the input layer node and the hidden layer node, σ is the standard deviation of the input spectral data. All parameters are real numbers.The corresponding conditional probability distributions of the visible and hidden layers are as follows.

$$\begin{aligned} p(h_j=1\mid v)&=sigmoid(c_j+\sum_{i}\frac{v_i}{\sigma_i}W_{i,j}) \\ p(v_i=1\mid h)&=N(b_i+\sigma_i\sum_{j}W_{i,j}h_j,\sigma_i^2) \end{aligned} \tag{2}$$

In the upper form, $sigmoid=1/(1+e^{-x})$ N is Gauss distribution.

2.2. The Training of Deep Belief Networks

The training of deep belief network consists of two stages. The training process is as follows:

(a)Unsupervised pre training: physicochemical corresponding near infrared spectra of cotton polyester blended data and Analysis on its value as the input of Gauss restricted Boltzmann machine of each layer is trained using unsupervised learning method, and the output of each Gauss restricted Boltzmann machine as the next Gauss restricted Boltzmann machine input. Layer training Gauss restricted Boltzmann machine, until the end of the top.

(b)Supervised fine tuning: the parameters obtained by pre training are used as initial values of the network, and the whole network is trained with a supervised BP network to tune parameters.

2.3. Random Hidden Algorithm

The random hidden algorithm is an implicit model fusion algorithm, which can effectively prevent the overfitting of the model by changing the network structure. In the course of training and training, the random concealment algorithm changes the probability parameters of the network layer, which causes the network to hide some hidden layer nodes randomly until the end of training.

In this paper, the random hidden algorithm is implemented by changing the network structure. The nodes in the dashed line indicate that they are not involved in training and testing, and the weight is zero, similar to the hidden layer noise reduction method. The random hidden algorithm will change the probability parameters of the network layer, which will cause the network to randomly discard some hidden neurons. When the model is trained, the weights of certain hidden layers are changed randomly, so that the weights of all nodes in the hidden layer are 'fused'.

3. Experiment

3.1. Experimental Data

In this paper, the spectra of wheat samples were obtained by using SupNIR-1520 near infrared spectrometer(Focused Photonics Inc), and the spectral area is 1000~2500nm. The collected spectral data were processed by MATLAB2015a software. A total of 400 wheat samples were collected, and their moisture, protein and ash contents were determined by quantitative chemical analysis. The near infrared spectroscopy of wheat are shown in Figure 1.

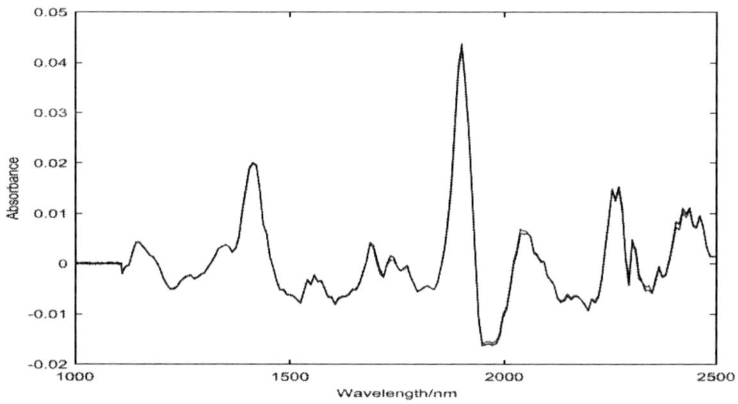

Figure 1. The original near infrared spectroscopy of Wheat

3.2. Preprocessing of the Original Spectrum

The original spectrum contains irrelevant information, a lot of redundant information in the spectrum will increase the complexity of the model, and even affect the prediction accuracy. In order to reduce the influence of various interference factors on the performance of the correction model, it is necessary to preprocess the original spectral data. The wavelet transform is used to process the original spectral data. Wavelet transform can separate noise and real spectral signals, and highlight spectral information, and compress the spectral data. The method entails using db3 wavelet at 3 levels decomposition to process original spectral data. The compressed near-infrared spectrum data matrix is obtained, and the compressed spectral data points are reduced to 13% of the original data, the energy retention rate is 99%. The spectrum after the wavelet transform is shown in Figure 2.

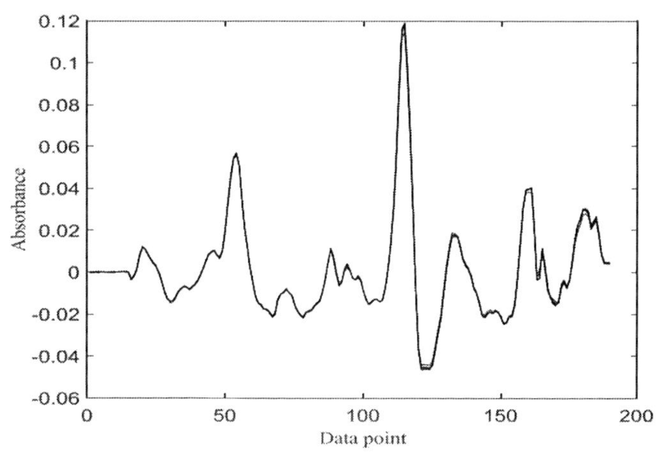

Figure 2. Near infrared spectroscopy of wheat after wavelet transformation

3.3. Partition of Sample Set

The establishment of quantitative analysis of near infrared spectra requires the selection of reasonable and representative sample sets. The training set sample data limits the applicable range and prediction accuracy of the model. Therefore, this paper used SPXY algorithm according to the ratio of 3:1 to divide 400 samples. The method can guarantee the uniform distribution of the spatial distance of the sample spectral data, and make the samples evenly distributed in the training set and the prediction set, so as to ensure the rationality and representativeness of the sample set. The training set is used to establish the analytical model, and the prediction set is used to test the reliability and stability of the model. The training set has 300 samples, and the prediction set has 100 samples.

3.4. Determination of Network Model

The random hidden algorithm changes the structure of the network model by hiding the hidden layer nodes, thus improving the stability and prediction accuracy of the network. Set the probability of random concealment, the initial value is 0.05, and then increase to 1 by 0.05 for the step size, then train under the probability parameters, and compare the prediction results. The experimental results show that the network model with stable and higher prediction accuracy can be obtained when the random concealment parameter is set to 0.6.

4. Results of Experiment and Discussions

300 training set samples and 100 prediction set samples of wheat near infrared spectrum data were collected as input data of improved DBN.

The improved DBN method for the determination of the content of wheat components is as follows.

(a) Inputing the spectral data of training set to the RBM in the underlying layer. The learning rate of the connection weight between the hidden layer and the visual layer, the learning rate of the visual layer and the hidden layer bias are initialized to 0.1.

(b) The output of the bottom RBM is used as the input of the DBN, and the weights and offsets of the previous RBM are kept unchanged. Using the state of the hidden layer neuron to train the next RBM, and then output the result of the parameter vector.

(c) The connection weights of hidden layer nodes are obtained by pre training to initialize corresponding nodes. And the probability ratio of random hidden algorithm is 0.6, the learning rate is 0.1. The network is trained by the function of neural network, and the hidden layer nodes are discarded randomly.

(d) Using stochastic gradient descent method to update the weights, and train them many times until convergence.

(e) The parameters that are iterated by random concealment algorithm are used as the optimal values, and the predicted data is input into the trained network for prediction. The experiment was repeated several times and analyzed and compared with the simulation results. The experimental results were obtained.

In accordance with the improved DBN model established in this paper, the NIR spectra of wheat were predicted. The results are compared with the traditional deep belief network model and the traditional shallow learning BP neural network, the results are shown in table 1.

Table 1. The prediction results obtained by different modeling methods.

	BP neural network		Deep belief network		Improved Deep Belief Network	
	R	RMSEP	R	RMSEP	R	RMSEP
Moisture	0.9322	0.3245	0.9327	0.2523	0.9978	0.0069
Protein	0.9212	0.5236	0.9145	0.3652	0.9928	0.0628
Ash	0.88815	0.6525	0.9121	0.3225	0.9920	0.0535

As can be seen from table 1. Compared with the traditional DBN prediction method and the traditional shallow learning BP neural network, the improved DBN based prediction method proposed in this paper has significantly improved the prediction results.

The wheat near infrared spectroscopy model using the improved DBN based analysis, the correlation coefficient of moisture, protein, ash content prediction can reach more than 0.99, the moisture, protein and ash content of the forecast result and the real value of fitting curve is shown in figure3, figure4 and figure5.

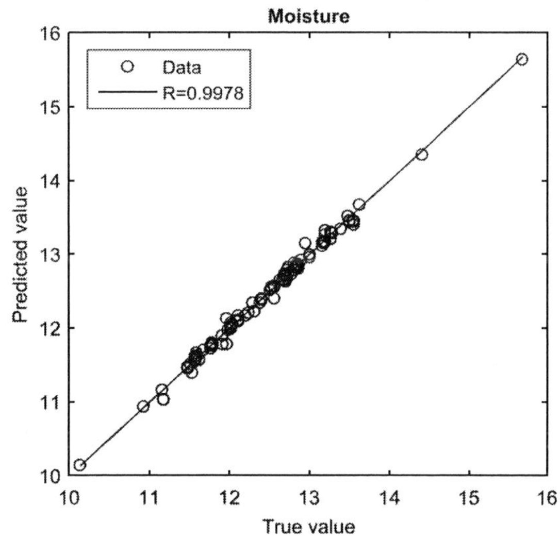

Figure 3. Fitting curves of predicted results and true value of moisture content in wheat

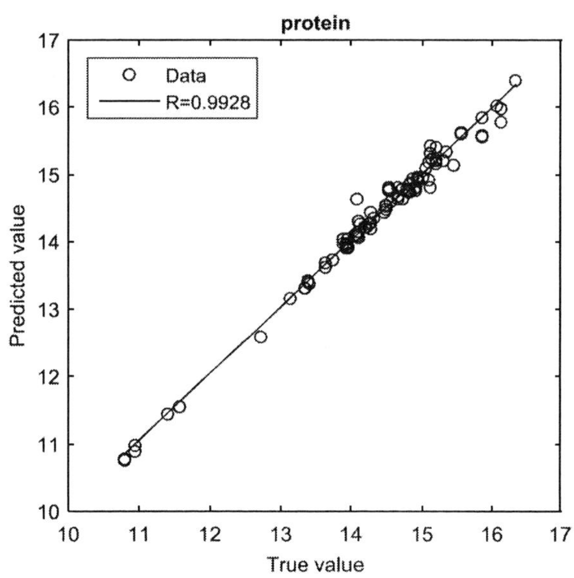

Figure 4. Fitting curves of predicted results and true value of protein content in wheat

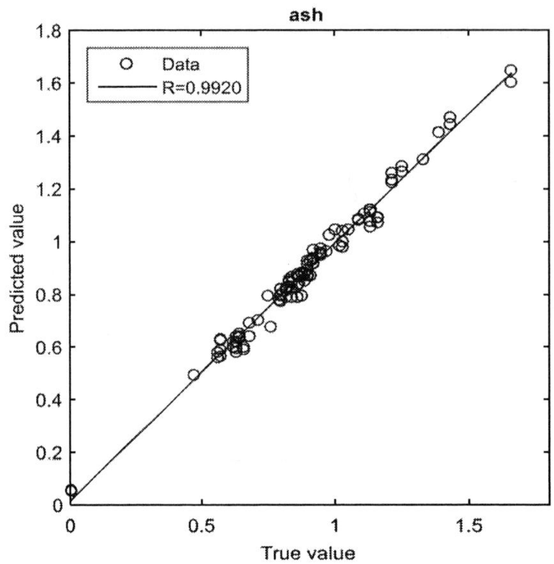

Figure 5. Fitting curves of predicted results and true value of ash content in wheat

5. Conclusion

The research work shows that it is feasible to use the near infrared spectroscopy and the improved DBN algorithm to detect the content of wheat components. Moreover, this method is also applicable to the detection of other agricultural products. An improved DBN network model is seen as a deep learning model. Firstly, the wavelet transform method is used to preprocess the original spectra of Wheat. Then, DBN is combined with random hidden algorithm, and the sparse structure of the network model is realized by changing the structure of the network, and the sparse network is obtained. Finally, using the improved DBN method to model and predict on the content of wheat, the correlation coefficient of prediction of the corresponding moisture, protein and ash content were 0.9978, 0.9928 and 0.9920, the predicted standard deviation were 0.0069,0.0628 and 0.0535. The experimental results prove that the model has good stability and accuracy, and enrich the modeling algorithm of near infrared spectroscopy technology, and lay the foundation for the transmission and sharing of the model.

6. Acknowledgment

This work is supported by National major scientific instruments and equipment development projects (2014YQ470377).The authors awfully thanks for anonymous reviewers who made constructive comments.

7. References

[1] Shiroma C, Rodriguez-Saona L. *Journal of Food Composition &Analysis*, 2009, 22(**6**): 596-605.

[2] Nicolai B M, Verlinden B E, Desment M, et al. *Postharvest Biology & Technology*, 2008, 47(**1**): 68-74.

[3] Ahmed Al-Harrasi et al. Application of NIRS coupled with PLS regression as a rapid, non-destructive alternative method for quantification of KBA in Boswellia sacra [J]. *Spectrochimica Acta Part A: Molecular & Biomolecular Spectroscopy*, 2017, 184:277.

[4] LI Bao Jian, CHENG Chun Tian. Monthly discharge forecasting using wavelet neural networks with extreme learning machine [J]. *Science China (Technological Sciences)*, 2014, (**12**):2441-2452.

[5] Nassim Laouti, Sami Othman, Mazen Alamir,Nida Sheibat-Othman. Combination of Model-based Observer and Support Vector Machines for Fault Detection of Wind Turbines [J]. *International Journal of Automation & amp; Computing*, 2014, (**03**): 274-287.

[6] Hinton G E. A practical guide to training restricted Boltzmann machines [J]. *Momentum*, 2012, 9(**1**): 599-619.

Research on key technologies of data processing in internet of things

Yangqing Zhu and Peiying Liang

Department of Internet of Things Engineering in Electronic and Information Engineering School, Foshan University, Teacher room 506, Foshan City, China.
Email: zhuyangqing_830516@163.com

Abstract. The data of Internet of things (IOT) has the characteristics of polymorphism, heterogeneous, large amount and processing real-time. The traditional structured and static batch processing method has not met the requirements of data processing of IOT. This paper studied a middleware that can integrate heterogeneous data of IOT, and integrated different data formats into a unified format. Designed a data processing model of IOT based on the Storm flow calculation architecture, integrated the existing Internet security technology to build the Internet security system of IOT data processing, which provided reference for the efficient transmission and processing of IOT data.

1. Introduction

Internet of things (IOT) can be taken as an important extension of Internet, are widely connected to the physical world by building the pervasive environment perception infrastructure, in the human world, the physical world and information world construction consists of a network, information and service ecosystem. In the traditional Internet era, information was shared among people. On the basis of the Internet, IOT further realize the information sharing of human, machine and things, to form a more close information flow ecosystem between the information world and physical world, and change the mode that people use the traditional Internet for information communication and resource sharing. [1]

IOT connects the digital world and physical world, the technological changes and challenges that it brings will be difficult to estimate. While relying on hardware devices, it needs more intelligent analysis and processing. Data processing is the most important part of the core technology of IOT. The large and real-time data into the system, how to transmit and process these massive, complex and real-time data to get the effective results, through the feedback allows users to perform intelligent management and control to the IOT objects, is the key technology of IOT. [2]

This paper analyzed the characteristics of IOT data, and researched some key technologies of IOT, such as data acquisition middleware, flow data processing mode and data security framework, provided a reference path for the construction of IOT data processing system.

2. Data Characteristics of IOT

2.1. Polymorphism and Heterogeneity of Data

There are a variety of sensors in wireless sensor networks (WSN), each of sensor has different function in different application systems. These sensors have different structures and different performances, and their collected data structures are different. In the Radio Frequency Identification (RFID) system, there are multiple RFID tags and multiple reader-writers. The micro computing devices in the machine to machine system are all different. Their data structures do not follow a unified pattern. The data in the IOT include text, images, audio, video and other multimedia data. There are also static data and dynamic data. The diversity of data and the heterogeneity of the sensing model lead to the heterogeneity of data. This heterogeneity exacerbates the difficulty of data processing and software development. [3]

2.2. The Magnanimity of Data
The IOT is often a dynamic network formed by the combination and connection of several wireless recognized objects. The quantity of goods in a medium-sized supermarket can be as many as millions or even tens of millions pieces. In a supermarket RFID system, assume that 10 million items need to be tracked, read 10 times a day, each time 100 bytes, the amount of data per day will reach 10GB, each year will reach 3650GB. In the field of real-time monitoring such as ecological monitoring, wireless sensor networks need to record multimedia information from multiple nodes, and the amount of data is amazing, reaching more than 1TB per day. In addition, in some emergency monitoring systems, the data are generated in real-time, high-speed and continuously in the form of stream, which aggravates the mass of data.

2.3. Timeliness of Data
The thing state of being perceived may change rapidly. Therefore, regardless of the WSN or RFID system, data acquisition of IOT is carried out at any time, every certain period of time to send data to the server , data update soon, historical data is only for recording the development process of transactions, although you can backup, but because of the massive not long-term preservation [4]. Only new data can reflect the existing state of the things perceived by the system, so the response speed or response time of the system is the key to the reliability and practicability of the system. This requires that the software data processing system of IOT must have sufficient speed of operation, otherwise it may lead to erroneous conclusions and even cause great losses.

2.4. Data Security and Transmission Reliability
When the IOT interacts with data, it may face malicious attacks from outside users, thus reducing the probability of success of network data transmission. With the increase of malicious behaviour, IOT interactive data outage probability will increase, which will also lead to increased data retransmission, thereby increasing the burden of communication network, and may even lead to paralysis of the network. At the same time, malicious eavesdropping may also exist, which will destroy the security of data transmission. In a word, the malicious attack and wiretapping behaviours of the external users not only reduce the reliability of the data transmission of IOT, but also deteriorate the security of IOT information. [5]

3. The Key Technologies of Data Processing in IOT

3.1. Middleware Technology
Service-Oriented Architecture (SOA) provides a better solution for heterogeneous information fusion in the IOT data processing. Using the SOA architecture to build IOT applications, can better integrate multiple sources and a wide range of services. [6]

3.1.1. Service-Oriented Application Architecture Description. SOA is a component model that uses different functional services of related applications and uses the interfaces and contracts defined between these services to build the application. The interface is defined in a neutral manner, which should be independent of the hardware platform, the operating system, and the programming language that implements the service. This allows services to interact in a unified manner in a wide variety of systems. Services are the foundation of SOA, so that components can interact directly and efficiently with application systems and software agents. In the SOA, a typical business operation involves a number of different components that typically reflect the needs of the underlying business process in an event driven or asynchronous way. In the context of IOT, traditional and emerging resources are open on the Internet in the form of services. Therefore, the research of data fusion application technology based on SOA has very important application value [7]. The SOA architecture usually consists of five main parts:

- Consumers. Obtain information from producers, at the same time, processing information.

- Application. Providing application interfaces or interoperable services that can be called from one another to different programs.
- Service. A task entity that implements certain specified actions.
- Service runtime environment support. The relevant service support functions required by the SOA application, such as the data interface of devices in the IOT, etc.
- Service provider. An interface entity that provides the specified function using related components.

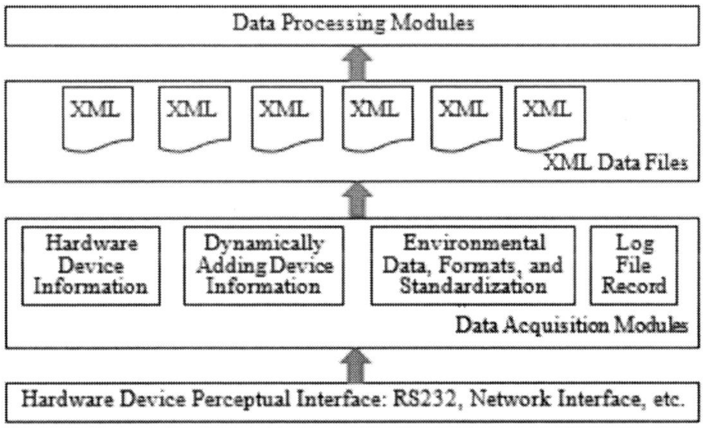

Figure 1. Middleware Architecture Diagram

Figure 1 mainly contains data acquisition unit and data format standard service unit. The middleware was between the perceived hardware layer and the service application layer, realized information exchange and management with front-end hardware sensing device, automatic acquisited the original information of hardware perception, refined the effective information, finished application of information exchange with the upper application and provided a standard format of data calls for the application software, completely got rid of the restrictions that reader-writer device of non-standardized protocol brought development, maintenance and extension.

3.1.2. Middleware Design of IOT Based on SOA. By comparing the characteristics of traditional Internet and data in IOT, the design, abstract and integration services of service oriented architecture middleware are used to be compatible with various types of data and protocols. Therefore, this paper presents the basic framework of IOT applications based on SOA, as shown in figure 1.

3.2. Flow Computing Technology

In view of the massive and real-time characteristics of the IOT data, the traditional query, statistics and model calculation based on relational database cannot meet the requirements. With the development of Web 2.0, large data processing technology developed rapidly, such as Google GFS, MapReduce, Hadoop platform, HDFS, MapReduce and so on, but these technologies are only suitable for dealing with batch static data. In recent years, stream computing has provided a very good model for data computing in IOT. [8]

Stream computing is a real-time calculation of stream data, and can be applied to a variety of scenarios, such as Web services, Machine Translation, IOT data processing, and so on. Take the sensor monitoring as an example, the atmospheric PM2.5 concentration can be real-timely monitored by placing the sensor PM2.5 in the atmosphere, the real-time monitoring data will be transmitted back to the data center, the monitoring system real-timely analyze the return data, to predict the air quality change trend, if the air quality to affect human health level in the next period of time, it start the emergency response mechanism.

Message processing is the foundation of real-time computing for real-time systems that need to process a large number of messages. The core of message processing is how to do not lose data in the process of message processing, and can make the whole processing system have good expansibility so as to be able to deal with the bigger message flow. [9]

At present, stream computing platform can be divided into three categories, the first category is commercial grade stream computing platform, such as IBM InfoSphere Streams, IBM StreamBase, the second is the open source stream computing framework, such as Twitter Storm, Yahoo S4, the third is the company occupied stream computing framework, such as Facebook Puma, Baidu DStream, etc.

In the IOT data processing system, the open source computing platform Storm is a very good solution. [10]

3.2.1. The Design Idea of Storm. Storm main includes the components design idea of Streams, Spouts, Bolts, Stream Groupings and Topology. Streams refers to an infinite sequence of Tuple. Spouts refers to reading stream data from the outside and continuing to send Tuple. Bolts refers to encapsulation of Tuple processing procedures, including filtering, aggregation, query and other operations. The function of Stream Groupings is to transfer Tuple between two components. Topology is a conversion diagram of stream data [11]. Figure 2 shows the flow diagram of the Topology of the Storm. In Figure 2, the stream data that Spout read is grouped by the Grouping component according to certain rules, such as Shuffle Grouping, Fields Grouping, All Grouping, Global Grouping, and Direct Grouping and so on. The tuple after grouped is send to the corresponding Bolt component for processing. The processed tuple data can be sent to the next Bolt to continue processing or to feed back to the user. Some processing results can be permanently stored on distributed disks for further data mining in the future.

3.2.2. Data Processing Flow of IOT Based on Storm. Figure 3 shows the data processing flow diagram of IOT based on Storm. In Figure 3, the actual IOT devices, such as a variety of sensors, cameras, RFID and other devices, they generated data can be processed into a unified data format specification by Fusion Middleware, and then transferred to the Storm data processing framework. The data after processing by stream data processing framework, a part directly in the text, table or the graphical form can be displayed to the user, some data can be saved to the HBase, ElephantDB or other non-structured database, to provide data source for data mining in the future.

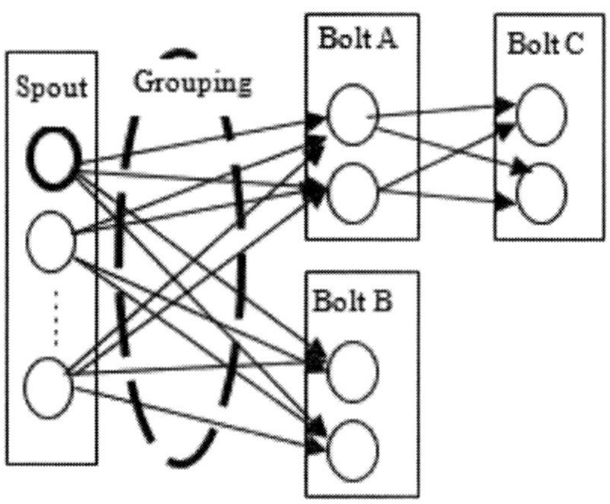

Figure 2. Storm Stream Grouping Flow

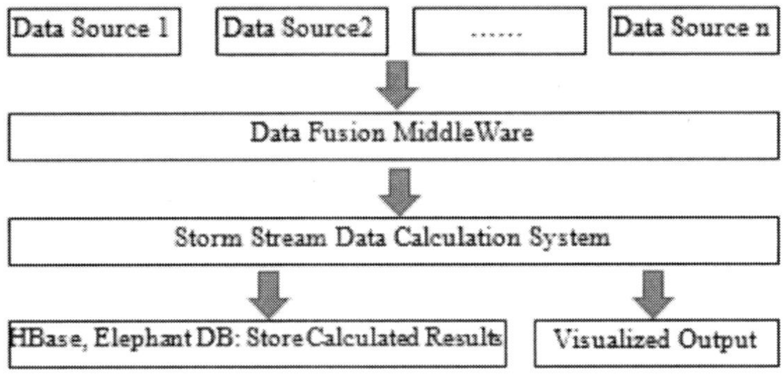

Figure 3. Internet of Things Data Processing Architecture Based on Storm

3.3. Data Security Technology

The current Internet security technology can be applied to the IOT. Figure 4 showed the data security transmission processing flow diagram of IOT. After a large number of sensors collect data, first obtain the corresponding digital summaries of these data, and then digitally sign them. The digital abstract and digital signature of the sensor data and the data itself constitute the original data packet of IOT. The original data packet of IOT can be transmitted by the network transmission environment (Internet, satellite communication network), the data need to be carried out data integrity verification and reliability verification in the data processing center. Their validation can be accomplished by digital abstraction techniques and digital signature techniques of the PKI security mechanism. The IOT sensor data that can be passed the security verification in the data center needs to be cleaned and filtered operation data, the valuable data can be preserved, and those not valuable can be filtered out. After cleaning and filtering, the data will be formed two types of data, one is the file type data, and the other is the data type of the database. The type of database data will be stored in the HBase database, and then protect data security using the security policy of database itself, the HBase database will eventually be stored in HDFS. All IOT data in the process of data migration, users can carry out certificate them with security authentication file type (PKI security) or the type of database security at any time. [12]

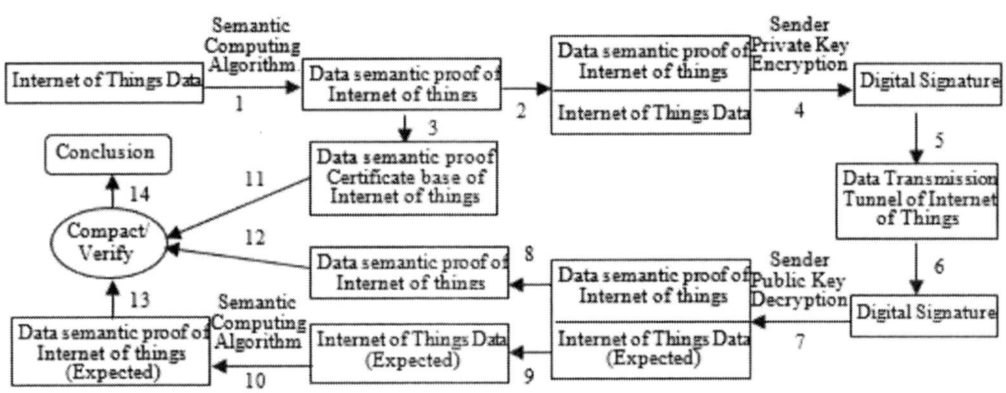

Figure 4. Data Security Scheme of Internet of Things Based on PKI Security System

4. Summary

In some extent, the research of this paper can eliminate the difference of heterogeneous data sources of IOT and provide users with standard and unified data processing interface. Storm streaming data architecture can effectively deal with real-time and large amounts of data in the IOT, and is an effective solution for data processing in the IOT system, such as sensor networks and RFID system.

Comprehensive use of the current Internet security measures to solve the IOT data transmission and processing security issues, can protect the IOT data efficient transmission and security.

In the future work, can combine block chain technology to further study the data transmission and data security system of IOT.

5. Acknowledgments

This study was supported by the innovation project of Guangdong Provincial Education Department (No.2014GXJK109) and Foshan University laboratory open innovation fund project.

6. References

[1] Zuquan Weng and Qi Zhang 2014 *China. Internet of Things Technology* 6 88
[2] Xiaofang Wu 2017 *China. Telecom World* 1 1
[3] Yingxin Hu, Xinna Ma and Lijuan Zheng 2014 *China. Internet of Things Technology* 4 79
[4] Ye Tian, Bo Yuan and Tingli Li 2016 *China. Acta Electronic Sinica* 2 247
[5] Chuankun Wu 2016 *China. Journal of Cryptologic Research* 2 40
[6] Jiuyun Xu, Huanxing Chi, Dan Jiang, Han Xiao, Hongxia Zhang and Jiashu Guo *2017 China. Journal of Chinese Computer Systems* 1 24
[7] Xingbang Lei and Jun Fang 2016 *China. Computer Science* 6 68
[8] Ziyu Lin 2015 *Principles and Application of Big Data Technology* (Beijing: Posts & Telecom Press) p 151
[9] Cheng Zhao and Jinlei Jiang 2016 *China. ZTE Technology Journal* 2 14
[10] Qianjin Zhang 2016 *China. Journal of Changzhou Institute of Technology* 6 30
[11] Minmin Chen, Xinchun Wang and Fengxian Huang 2015 *Storm technology insider and big data practices* (Beijing: Posts & Telecom Press) p 33
[12] Dan Wang, Wenbing Zhao and Zhiming Ding 2017 *China. Journal of Beijing University of Technology* 3 335

IOP Conf. Series: Journal of Physics: Conf. Series **887** (2017) 012048 doi:10.1088/1742-6596/887/1/012048

Human resource recommendation algorithm based on ensemble learning and Spark

Zihan Cong[1], Xingming Zhang[1], Haoxiang Wang[1] and Hongjie Xu[1]

[1] School of Computer Science and Engineering, South China University of Technology, Guangzhou, 510006, China.
Email: hxwang@scut.edu.cn

Abstract. Aiming at the problem of "information overload" in the human resources industry, this paper proposes a human resource recommendation algorithm based on Ensemble Learning. The algorithm considers the characteristics and behaviours of both job seeker and job features in the real business circumstance. Firstly, the algorithm uses two ensemble learning methods-Bagging and Boosting. The outputs from both learning methods are then merged to form user interest model. Based on user interest model, job recommendation can be extracted for users. The algorithm is implemented as a parallelized recommendation system on Spark. A set of experiments have been done and analysed. The proposed algorithm achieves significant improvement in accuracy, recall rate and coverage, compared with recommendation algorithms such as UserCF and ItemCF.

1. Introduction

With the rapid growth of data, human resources information management becomes complex and confusing. Job seekers are also easy to fall into the "information trail". The problem is mainly reflected as: for job seekers, the real interest in the job information only a small part of the information on the job seekers. Searching and screening process usually take a lot of time and effort. It is very challenging to accurately and quickly find their own interest in the recruitment of information.

In order to solve this problem, people has proposed a catalog and search engine two representative solutions [1]. These programs in the human resources industry has a more mature application. The current online recruitment service model can be divided into categories: Category recruitment model and Vertical recruitment model. Category recruitment model is mainly used in splitting catalogs. The recruitment of information is based on different topics for different categories. But the disadvantage is that it can only cover the limited popular information. Vertical recruitment mode does not produce recruitment data, but mainly by building a strong search engine, providing users with such as keyword search services, capturing the recruitment site to enrich the information to provide services for users. But the disadvantage is that in the real world, job seekers are often uncertain about their needs or difficult to describe their needs, hence the effect of search engines will greatly reduce [2].

The recommendation system is a kind of technology that can dig out the potential demand of the user. The recommendation system is considered to be the most effective way to solve the problem of "information overload"[3], compared with the split catalog and search engine. The system in the human resources research and application effect is unsatisfactory. Most human resource recommendation systems take into account only job seekers' application for jobs, and are not fully utilized for job seekers and job attributes (such as payroll, geography, industry) [4]; thus, based on Ensemble learning human resources recommendation algorithm, to help in-depth mining job seekers and business needs between the existing relationship between the realization of the precise recommendation of human resources.

IOP Conf. Series: Journal of Physics: Conf. Series **887** (2017) 012048 doi:10.1088/1742-6596/887/1/012048

2. Human Resource Recommendation Algorithm Based on Ensemble Learning

As the content, form and quantity of the original data are different. The recommendation algorithm is generally based on the specific business scenario. For the human resources circumstance, some common recommendation algorithm recommended effect is not obvious. In order to solve this problem, this paper proposes a human resource recommendation algorithm based on ensemble learning. It can be used to personalize the job seekers by combining the characteristics of human resources, including the following four aspects: data pre-processing, data sampling, ensemble learning, recommendation generation.

2.1. Data Pre-Processing

The application background of this algorithm is based on social security business data. In the system business scenarios, the data sources include user data, job data and behaviour data. These data are collectively referred to as raw data. There are problems of inconsistencies, incompleteness and incorrectness in the data format on the actual analysis, so the raw data must be processed which is known as the step of Extract-Transform-Load(ETL). Otherwise, it is futile and meaningless in the analysis and calculation of ambiguous, inaccurate numerical data [7]. So the need for pre-processing of the original data, the entire data pre-processing, includes the following sections. The flow chart is shown in figure 1:

1) Data extraction: The raw data is extracted from the original data in a streaming fashion.
2) Data cleansing: Clean dirty data from raw data.
3) Data transform: through the development of conversion rules, the original text information is transformed.
4) Data loading: the processed data is loaded into a well-designed data warehouse.

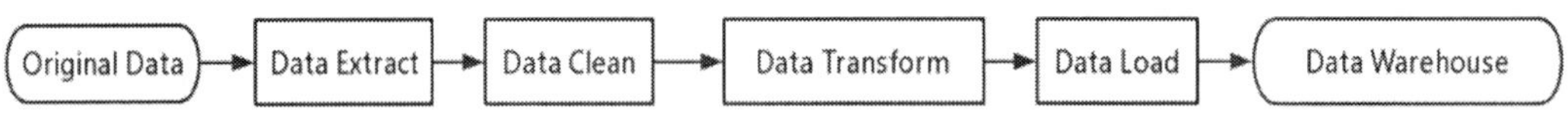

Figure 1. Data Pre-processing Flow Chart

2.2. Data Sampling

First, for the data in the data warehouse for statistical analysis, the recommended problem is a classification problem to produce recommended / not recommended two classifications. So 'not be interested behaviour' should be marked as a negative sample, and the behaviours of application, collection and browsing are marked as a positive sample. The total amount of behaviour data is 170,884, in which behaviour of 'not interested' is 23,516, and the behaviours of apply, collect, browse is 147,328. Hence, the sparse degree of the entire data set is:

$$\mathrm{p} = \frac{17084}{15000\times 5000} \cong 0.00227 \tag{1}$$

Due to the sparseness of the data, if these non-user behaviour data is not dealt with, they will have a great impact on the recommended quality. Taking into account the reality of human resources circumstance, if the user does not produce behaviour on the post, it may mean the user is not interested in the post. However, if all the data without user behaviour is marked as negative, the positive and negative sample ratio r for the entire data set is:

$$r = \frac{147328}{15000\times 5000-147328} \cong 0.00196 \tag{2}$$

In order to solve the problem of data sparseness and imbalance between positive and negative samples in the sampling process, the samples will be solved by sampling means. On one hand, sampling needs to ensure that the number of negative samples and the number of positive samples are equal; on the other hand, the implicit data and emphasis should be on those who are very popular. Based on this, the following sampling strategy is used to sample:

1) The positive samples were sampled using the SMOTE [8] algorithm, and the original 147328 positive feedback behaviour data was over-sampled for five times the original, i.e., 736640 positive samples.
2) The user is marked as "not interested" behaviour data using SMOTE algorithm, the original 23516 negative feedback behaviour data oversampling to 235160.
3) For each user, we take the highest view of the top 100 non-user behaviour of the job data as negative samples. Hence we sampled a total of 5000 * 100 = 500000 data.

2.3. Ensemble Learning

In the training phase, there are two ways to calculate: Bagging and Boosting [5]: The user interest model for each algorithm is output respectively, and then these two interest models are merged to get the fusion user interest model. The algorithm uses the decision tree as a basic classifier for integration [6]. The flow chart of generating user interest model is shown in figure 2.

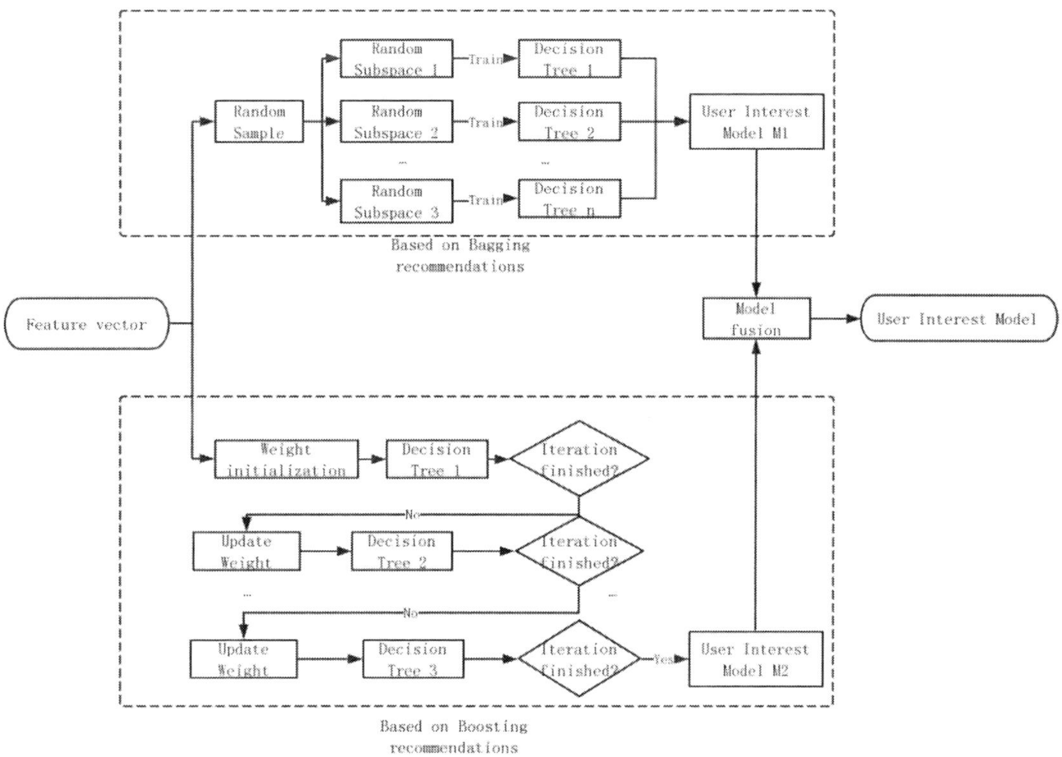

Figure 2. Generating a flow chart for the user interest model

Among them, the use of Bagging integration, random sampling to get n random sub-space and training for the n decision tree model, these decision tree model are combined, which let n models to each user U to predict job I of the classification, select the most frequent class p, and at the same time calculate the confidence. The Confidence formula is as follows:

$$Confidence(U,I,p) = \frac{\sum_{i=1}^{K} label(p)}{K} \quad (3)$$

Then the user's classification of the job results and classification confidence are added to M1, until the traversal of the entire user and post, the output interest model M1, it is a <user id, post id, predictive results, confidence> tuple.

Based on Boosting's user model interest generation, we first need to define a cost function, and the whole algorithm goal is to get the cost function to be optimized. In this process, the model is established for each tree, and the cost function of each tree is calculated. Then, the cost function is used to calculate the optimal solution in the descending direction of the gradient. After constant iterations, the output model predicts the effect gradually ideal.

The model of the user interest models M1 and M2 can be obtained. For each target user i and target post j, the following formula is used to define the user interest level of i in j: $l_{m1,i,j}$ represents the user's forecast category for the post. If the forecast is 1, the prediction is negative; $q_{m1,i,j}$ represents the user's confidence in the job prediction, the confidence interval range is [0, 1].

$$C_{m1,i,j} = l_{m1,i,j} * q_{m1,i,j} \quad (4)$$

The following formula defines the user interest degree of i in j, where $l_{m2,i,j}$ represents the user's prediction category for the job:

$$C_{m2,i,j} = l_{m2,i,j} \quad (5)$$

M2 has a degree of interest with value 1 or -1, and if the result is a direct result fusion, the contribution generated by M2will be greater than M1. In order to avoid this problem, we introduce the fusion factor in the fusion process. The following formula defines the degree of interest that occurs after fusion:

$$C_{i,j} = C_{m1,i,j} + r * C_{m2,i,j} \quad (6)$$

The calculated value of $C_{i,j}$ is positive if user i is interested in post j, and the higher the value, the higher the degree of interest. Conversely, the negative representative is not interested, the greater the absolute value stands for less interested. Calculate the corresponding position of each user and add to M, get the final integration of the user interest model, the model is a <user id, job id, interest degree> of the triple set, as the basis for the recommendation.

2.4. Recommended to Produce

The user interest model obtained in the previous section represents the user's classification results and the degree of interest for each post. In order to generate the final recommendation, a recommendation based on Top-N is used to generate the recommended results. First, the user interest model is grouped according to user id, the interest set for each user is selected to sort out the position of interest of the user, and the sorting operation is according to the degree of interest is obtained. The result set is from high to low, K posts. For online experiments, recommendations will be made for all users. For offline experiments, the behavioural records of the test set will be compared with the predicted behaviour and the corresponding recommended indicators will be calculated for the analysis and improvement of the algorithm. The flow chart is as follows:

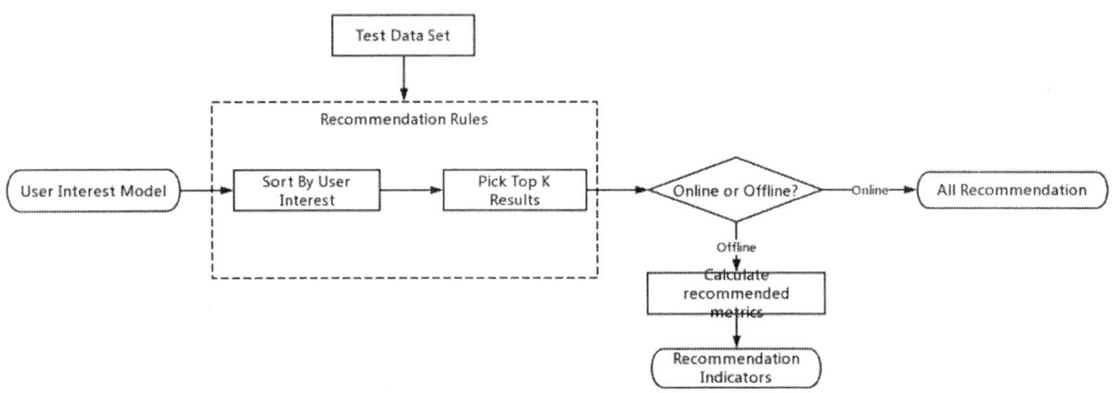

Figure 3. Recommends generating a flow chart

2.5. Algorithm Parallelization

In order to improve the efficiency of the implementation of the algorithm, here will be based on the Spark computational framework for the original algorithm parallelization. Spark first needs to interact with the data warehouse on the cloud platform to read and write raw data. Spark SQL is the interface through which Spark reads and writes to the data warehouse. It takes the records of the raw data from the data warehouse and converts it to RDD, providing a data foundation for the recommended calculation [9]. At the same time, the program main function builds Driver role through the Spark Context. Driver is responsible for generating computing tasks and directed acyclic graph of the scheduling. Tasks will be assigned to the Executor to perform. After executor completes parallel calculation of the training task, the data will be in the form of RDD as the output of the calculation results. Spark on the overall recommendation algorithm calculation process is shown as follow:

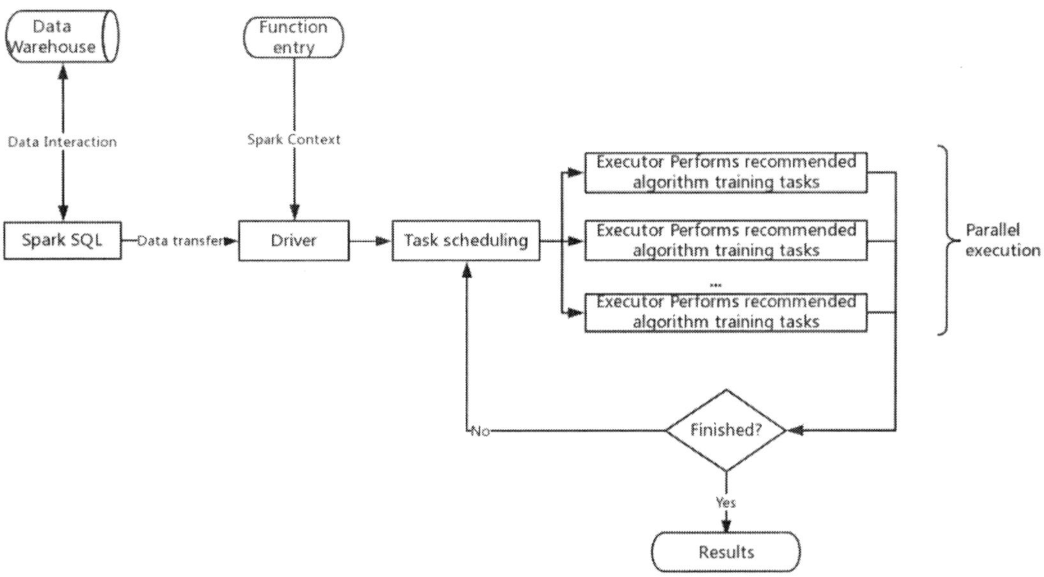

Figure 4. Algorithm parallelization flow chart

3. Experimental Results and Discussion

In order to verify the overall effect of the recommended system, the experiments are done in the three aspects of accuracy rate, recall rate and coverage [10]. For the algorithm, the following table describes the key parameter descriptions:

Table 1. Meaning of the meaning of the list

Parameter	Meaning	Source
D1	Maximum depth of Decision tree	Bagging
N	Number of decision trees integrated	Bagging
D2	Maximum depth of Decision tree	Boosting
T	Maximum number of iterations	Boosting
r	Fusion factor	Interest model fusion

In this paper, we use the parameter group of D1=12, N=25, D0=10, T=10, r=0.5 to generate the interest model. Meanwhile, in order to carry out the comparison experiment, this paper uses two kinds of most common algorithms based on user co-filtering (UserCF) and content-based collaborative filtering (ItemCF) on the same data set. The experimental results are as follows:

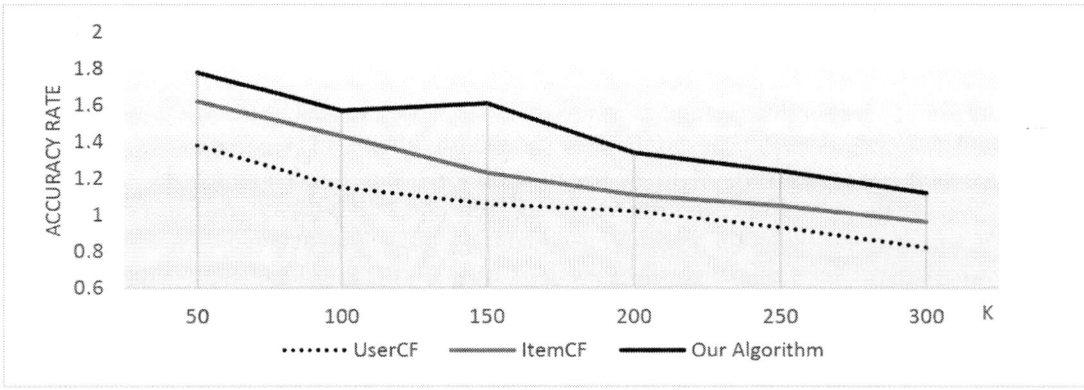

Figure 5. Accuracy curves for different recommended algorithms

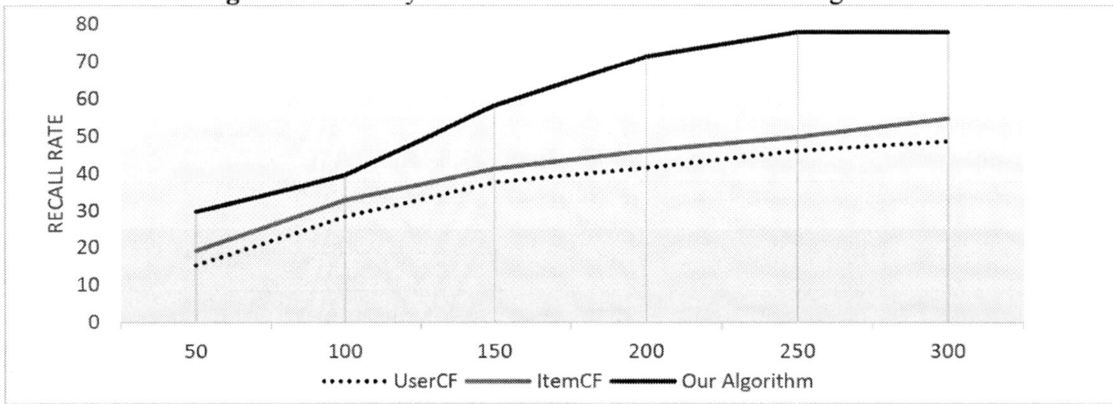

Figure 6. Recall curves **for** different recommended algorithms

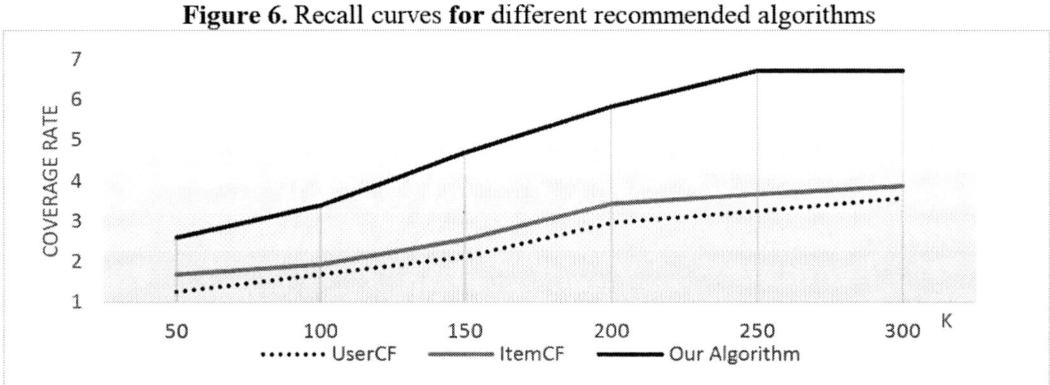

Figure 7. Coverage curves for different recommended algorithms

From the above graphs, it can be seen that the proposed algorithm has been more accurate than the traditional cooperative filtering algorithm with the increasing of recommended length. In the aspect of recall rate, since the recall rate calculation formula is based on the entire test data, so the recall rate curve always keeps increasing. At the particular points for K=250 and k=300, the recall rate of the algorithm is fixed, which shows that the proposed algorithm for the users predicted as "interested" entries has reached the bottleneck value. The recall rate of our algorithm can reach 77.9, which is higher than that of UserCF (48.6) and ItemCF (54.7). It shows that our algorithm has obvious advantages in recall rate. Similarly, for coverage, with the increase of K, the coverage of this algorithm has been significantly higher than the traditional co-filter algorithm coverage. Our algorithm's highest coverage rate can reach 6.71, which is about twice of the coverage of UserCF. Through the analysis of above experiments results, it can be approved that the proposed algorithm conducts a significant improvement compared with the traditional collaborative filtering algorithms.

The significant improvement is due to the reason that the traditional collaborative filtering recommendation algorithms only take into account the user's historical behaviour data, regardless of the user and project attribute information. In some recommendation scenarios, which have a large number of user interactions, traditional algorithms may have a good performance. Such as movie recommendation, user and film attributes are difficult to be facilitated, and the available data is mainly about user behaviours. For human resources recommendation, besides of user behaviour information, user and job themselves are abundantly informational, which reflects the characteristics of the user and the characteristics of the post. This article aims to use this information through the ensemble learning. In addition, in order to improve the recommended coverage, our algorithm is based on the post of the popularity of the post, focusing on the jobs which are popular but not interested by some users. This approach can improve the extraction of users' interests, solving the "long tail distribution" problem for recommendation.

4. Conclusion

This paper points out some shortcomings in solving the "information overload" problem for the current human resource recommendation. This paper proposes a recommendation algorithm based on ensemble learning, to generate user interest model. We use this algorithm to generate job recommendation for users. The implementation of the algorithm is parallelized on Spark platform. Finally, the performance of this algorithm is evaluated by a set of experiments. Compared with the traditional recommendation algorithms, such as UserCF and ItemCF, our algorithm shows significant advantages in the circumstance of human resource recommendation.

5. Acknowledgement

This work is financially supported by Guangdong Provincial Science and Technology Plan (No. 2016B030308002).

6. References

[1] Loeb S, Terry D. Information filtering [J]. Communications of the Acm, 1992, 35(12):26-28.
[2] Arayasirikul S, Chen Y H, Jin H, et al. A Web 2.0 and Epidemiology Mash-Up: Using Respondent-Driven Sampling in Combination with Social Network Site Recruitment to Reach Young Transwomen [J]. AIDS and Behavior, 2016, 20(6):1265-1274.
[3] Ricci F, Roach L, Shapira B, et al. Recommender Systems Handbook[M] Recommender systems handbook. Springer, 2011:1-35.
[4] White T J, Redner R, Skelly J M, et al. Examination of a Recommended Algorithm for Eliminating Nonsystematic Delay Discounting Response Sets[J]. Drug & Alcohol Dependence, 2015, 154:300.
[5] Freund Y. An improved boosting algorithm and its implications on learning complexity[C] ACM Conference on Computational Learning Theory, COLT 1992, Pittsburgh, Pa, Usa, July. DBLP, 1992:391-398.
[6] Podgorelec V, Zorman M. Decision Tree Learning [M] Machine Learning Models and Algorithms for Big Data Classification. 2016:1751 - 1754.
[7] Berro A, Megdiche I, Teste O. A Content-Driven ETL Processes for Open Data [M] New Trends in Database and Information Systems II. Springer International Publishing, 2015:29-40.
[8] Chawla N V, Bowyer K W, Hall L O, et al. SMOTE: synthetic minority over-sampling technique [J]. Journal of Artificial Intelligence Research, 2011, 16(1):321-357.
[9] Shen Z, Johnsson M, Zhao Z, et al. Spark Plasma Sintering of Alumina [J]. Journal of the American Ceramic Society, 2010, 85(8):1921-1927.
[10] Okita S C N C. EVALUATION SYSTEM AND EVALUATION METHOD: US, WO/2006/129711[P]. 2006.

IOP Conf. Series: Journal of Physics: Conf. Series **887** (2017) 012049 doi:10.1088/1742-6596/887/1/012049

Deep neural nets based power amplifier non-linear pre-distortion

Zhenyu Wang[1], Yanyun Wang[1], Chunfeng Song[1], Tao Chen[2] and Wei Cheng[1]

[1]School of Control and Computer Engineering, North China Electric Power University, Beijing 102206, P. R. China
[2]School of Business, Central South University, Changsha 410083, P. R. China
Email: zywang@ncepu.edu.cn

Abstract. This paper proposed a novel method based on deep neural networks (auto-encoder) model, to construct the pre-distortion model for non-linear feature of power amplifier. As auto-encoder nets are high non-linear function, with the optimization of object function to tune the weights, the nets can reach any non-linear model. For widely used power amplifier, this method can help setting the pre-distortion model. In this paper, deep (more layers) network structure have been adopted in the auto-encoder model. The experimental results show the effectiveness and efficiency of deep neural network based power amplifier non-linear pre-distortion technique.

1. Introduction

Wireless communication technology is attracting much research attention in recent years and is of importance to further information development. However, it has strict linearization requirements on power amplifier (PA) for suppressing adjacent channel interference and reducing bit-error rate. Digital pre-distortion is one of the most effective linearization techniques and widely used to compensate for the nonlinearity in the PA. The most commonly used pre-distortion methods [1] involve polynomial model, memory polynomial pre-distortion model [2-3], BP neural network pre-distorter [9], and so on. The deep neural network has also received a lot of attention in the literature and has been applied in various research fields successfully. This paper proposes a deep neural network [5-6] based power amplifier non-linear pre-distortion technique to construct the pre-distortion model and the results of this study show that the proposed method is effective.

2. Pre-distortion Technique

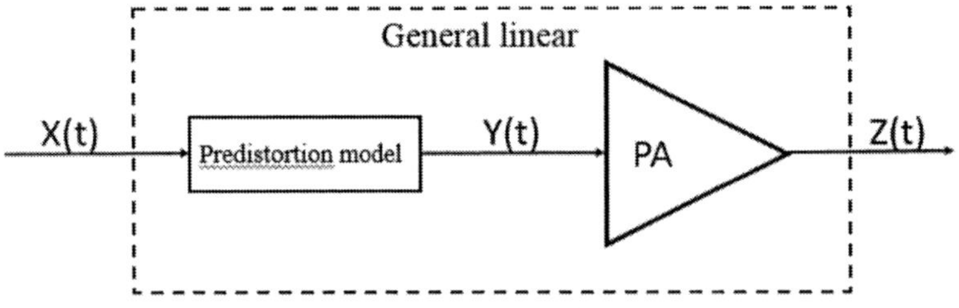

Figure 1. Pre-distortion technique

Pre-distortion is one of the main linearization techniques that have the inverse characteristic of the PA to nonlinearity.

2.1. Pre-Distortion Technology

Pre-distortion is based on the indirect learning architecture, as shown in Fig. 2. The benefit of the indirect learning architecture is that we should not assume a PA model and estimate the PA parameters, therefore pre-distortion is a common choice due to its effectiveness and ease of implementation.

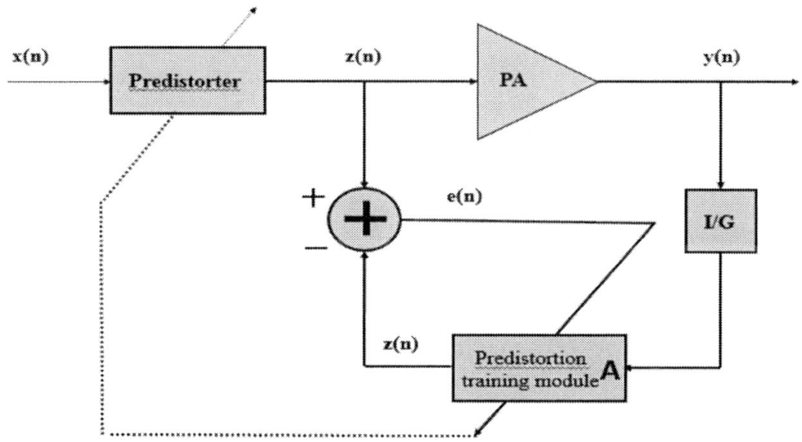

Figure 2. Indirect learning structure

2.2. Traditional Neural Network Pre-distortion Model

Among the various pre-distortion techniques, this paper focuses on black propagation (BP) neural network pre-distortion model [9].

BP neural network is a multi-layer feed-forward neural network, which has the function of non-linear reflection showing complex rule of cause and effect. BP neural network commonly having a single hidden layer consists of N input nodes, M hidden nodes and Q output nodes. The hidden layer function is defined as:

$$Y = f(W_1 X) \tag{1}$$

Where, f says activation function and usually uses purelin transfer function. Considering the nonlinear characteristics of pre-distortion model, it usually adopts the sigmoid function as the activation function in the pre-distortion model. W1 says the weights between the input layer nodes and the hidden layer nodes. Output layer function is as follows:

$$Z = g(W_2 Y) \tag{2}$$

Where, g adopts the sigmoid function as the activation function. W2 represents the weights between the hidden layer nodes and the output layer nodes. The error function of BP neural network is as follows:

$$E = \sum_{i=1}^{N} (Z_i - X_i) \tag{3}$$

Where, X_i says the ith input samples and Z_i says the ith output samples. The BP neural network model is of low computation complexity and easy hardware implementation and has certain practical value. But because of its low network layers with three layers and small network depth caused by the little nodes, the model can't learn into the exact model in a relatively short time. Aiming at the problem, this paper puts forward the pre-distortion model of deep neural network.

3. The Pre-distortion Model Based on Deep Neural Network

This section introduces the role of the deep neural network in the pre-distortion model and concrete implementation methods. In this paper, the Auto-encoder is adopted as the deep neural network.

3.1. The Deep Neural Network

The deep neural network is the result of artificial neural network and its motive is to establish and simulate human brain neural network for learning and analyzing. It imitates the human brain mechanism to explain the data and it is a kind of multilayer perceptron containing many hidden layers. The concept of depths [6] first put forward by Hinton et al in 2006. He proposed deep belief networks (DBN) and unsupervised greedy layer-by-layer training algorithm to solve the deep structure optimization problem. Then he put forward multi-layer automatic encoder (Auto-encoder) deep network. This paper uses the multi-layer Auto-encoder network structure. At present, deep learning has been widely applied to the image and text classification [10-12], clustering [13], object segmentation [14] and other fields.

3.2. Auto-encoder Network

Auto-encoder network composes of multiple single restrained Boltzmann machine network stack, and the network is divided into encoder and decoder network. It is as shown in figure 3. Single RBM network uses minimizing the energy function between the visible layer and hidden layer to optimize the weights W, but Auto-encoder network usually obtain initialized weights from multilayer RBM with the process called pre-training as shown in figure 3(a). Auto-encoder is got by do fine-tuning to the network based on the process. In fact it is also a kind of deep neural network called DNN, as shown in figure 3 (b).

3.3. The Pre-distortion Model Based on Deep Neural Network

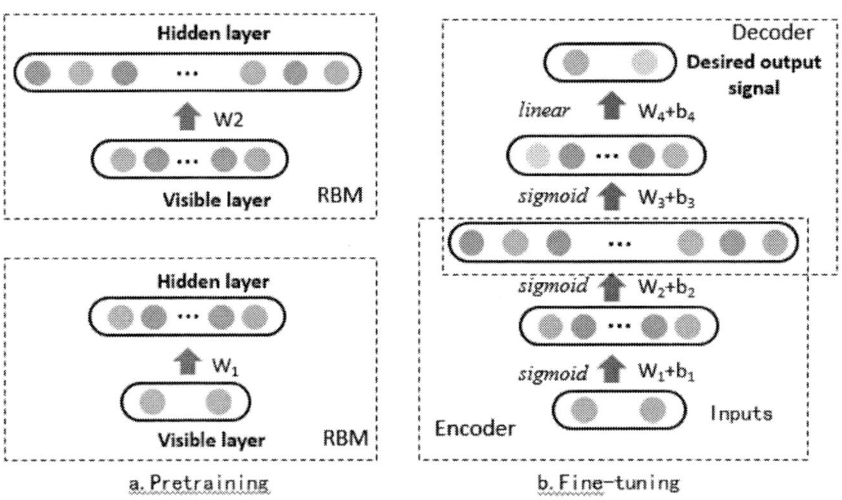

Figure 3. Auto-encoder nets pre-distortion structure

In view of the pre-distortion model, this paper adopts Auto-encoder network structure based on five-layer nonlinear mapping network. Among them the bottom layer uses the sigmoid function as the activation function. Considering the range of output is beyond the scope of (0, 1), on the top layer (output layer) the paper uses the linear function (linear) as the activation function. The specific network structure is 2-10-30-10-2. The structure of the neural network diagram is as shown in figure 3. Unlike typical Auto-encoder network, the input here is not the same as the output.

The objective function of the neural network is:

$$\min E = \frac{1}{N}\sum_{i=1}^{N}(Z_i - X_i) \tag{4}$$

Where, N says the number of signal sample, X_i says the ith input signal and Z_i says the ith desired output signal. In this article, the stochastic gradient descent method is used to optimize the objective function.

4. Experiments

In view of the algorithm of this paper, simulations are made through the matlab simulation platform and the power spectral density and ACPR got by calculating are used to evaluate the pre-distortion model of deep neural network.

4.1. Power Spectral Density

Using the pre-distortion model of deep neural network to train and iterate to 100 times, it can get ideal pre-distortion model. By the power spectral density diagram shown in figure 4 it can be observed visually the model has a better pre-distortion effect. In figure 4 there are as follows from left to right in turn: input signal, pre-distortion output signal, output signal after the pre-distortion processing based on the deep neural network.

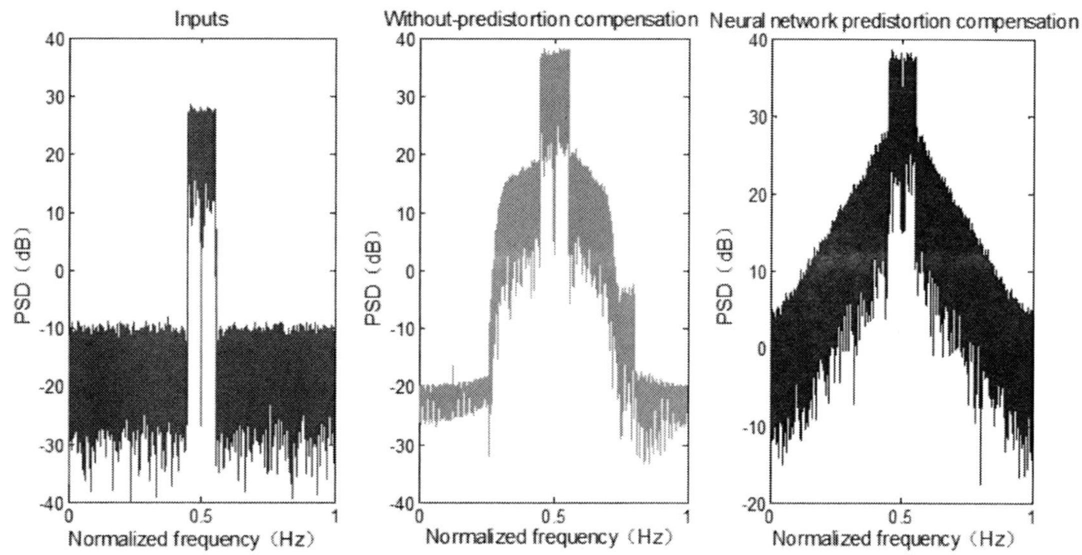

Figure 4. Curve of PSD

4.2. Quantitative Index Evaluation

The pre distortion model for power amplifier, usually use the adjacent channel power ratio ACPR (Adjacent Channel Power Ratio), Normalized mean square error NMSE (Normalized Mean Square Error) and Error vector magnitude EVM (Error Vector Magnitude) to evaluate. Among them, adjacent channel power ratio ACPR is defined as:

$$ACPR = 10\log_{10}\frac{\int_{f_2}^{f_3} s(f)df}{\int_{f_1}^{f_2} s(f)df} \tag{5}$$

S (f) is the function of power spectral density, [f1, and f2] is transmissionchannel, [f2, f3] is adjacent channel. The expression of NMSE is:

IOP Conf. Series: Journal of Physics: Conf. Series **887** (2017) 012049 doi:10.1088/1742-6596/887/1/012049

$$NMSE = 10\log_{10}\frac{\sum_{n=1}^{N}|z(n)-\hat{z}(n)|^2}{\sum_{n=1}^{N}|z(n)|^2} \quad (6)$$

Among them, where Z denotes actual signal value, $\hat{z}$ indicates the Model calculation signal value, NMSE reflects approaching degree between model and Physical actual module. The Expression of EVM is:

$$EVM = \sqrt{\frac{E[|e^2|]}{E[|X^2|]}} \times 100\% \quad (7)$$

In the formula, where X denotes the ideal signal output value, e indicates the errors between ideal signal output value and whole model output, the degree of distortion of global model to the extent of signal can be measured by EVM.

Evaluation parameters of the model are shown in table 1.the table shows that the degree of accuracy based on deep neural networks pre-distortion model is higher than the traditional pre-distortion model, linear input and output has been basically achieved after joining the pre-distortion, which Confirms the feasibility of the model. Figure 5 shows the training iterative NMSE graph of the deep neural networks pre-distortion model, it can be seen in the figure that its convergence speed is faster and achieves the ideal result in the 50th iteration.

Table 1. Deep neural network based non-linear pre-distortion

Model \evaluation index	ACPR(DB)	NMSE(DB)	EVM (%)
polynomial model	-36.02	-48.91	0.36
And memory polynomial model	-24.88	-43.71	0.70
Pre-distortion model based on DNN	-38.27	-50.12	0.33

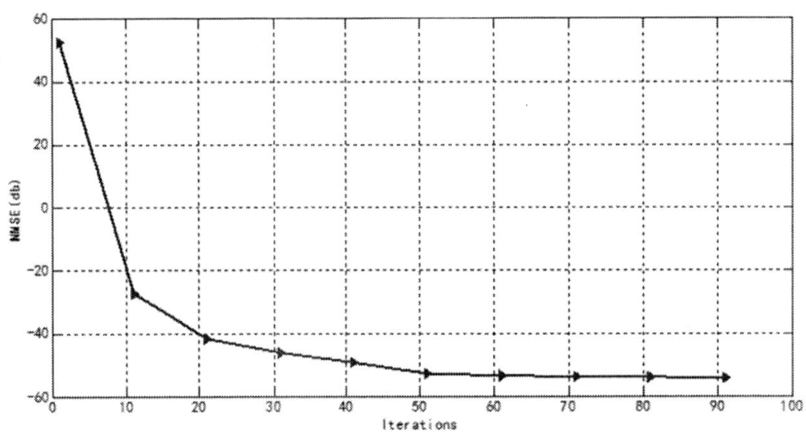

Figure 5. Iteration curve of NMSE

5. Conclusion

This paper presents a new pre-distortion model based on deep neural networks (auto-encoder), it can effectively realize the linearization of power amplifier (PA). Experimental results confirm the effectiveness of the proposed method. Deep neural network used in the pre-distortion model has two advantages, 1). The accuracy is higher, it can fit nonlinear model of the power amplifier effectively; 2). The iteration speed is faster, because of the deep neural network has deeper network structure than ordinary (monolayer) neural network, thus it has the stronger ability to learn, it can achieve faster speed of convergence in the practical application.

In future, we will focus on the number of nodes and layers of the deep neural networks to the further experiment. In addition, we will also consider joining the drop-out [7, 8] technology to improve the neural network training speed.

6. Acknowledgment

This work is supported by National Natural Science Foundation of China (61573139).

7. References

[1] Lefang Zhu, Research on Pre-distortion Technology of Power Amplifier, Nonlinear [D]. Shenyang Ligong University, 2012.

[2] Raich R, Qian H, Zhou G T. Orthogonal polynomials for power amplifier modeling and pre-distorter design [J]. Ve-hicular Technology, IEEE Transactions on, 2004, 53(5): 1468-1479.

[3] Morgan D R, Ma Z, Kim J, et al. A generalized memory polynomial model for digital pre-distortion of RF power am-plifiers[J]. Signal Processing, IEEE Transactions on, 2006, 54(10): 3852-3860.

[4] Bengio Y, Courville A, Vincent P. Representation learning: A review and new pe-rspectives [J]. 2013.

[5] Baldi P. Auto-encoders, Unsupervised Learning, and Deep Architectures [J]. Journal of Machine Learning Re-search-Proceedings Track, 2012, 27: 37-50.

[6] Hinton G E, Salakhutdinov R R. Reducing the dimensionality of data with neural networks [J]. Science, 2006, 313(5786): 504-507.

[7] Hinton G E, Srivastava N, Krizhevsky A, et al. Improving neural networks by preventing co-adaptation of feature de-tectors[J]. arXiv preprint arXiv:1207.0580, 2012.

[8] Wang S, Manning C. Fast dropout training[C]//Proceedings of the 30th International Conference on Machine Learning (ICML-13). 2013: 118-126.

[9] Hongmin Deng, Adaptive pre-distortion of the power ampli-fier based on BP neural network [J], JOURNAL OF CHINA INSTITUTE OF COMMUNICATIONS, 2003, 24(11): 141-145.

[10] A. Krizhevsky, I. Sutskever, G. E. Hinton. ImageNet Classification with Deep Convolutional Neural Networks [A]. Neural Information Processing Systems[C]. NIPS press, 2012: 1-9.

[11] C. Szegedy, W. Liu, Y. Jia, et al. Going Deeper with Convolutions [J]. arXiv preprint arXiv, 2014, 1409.4842

[12] K. Simonyan, A. Zisserman. Very deep convolutional networks for large-scale image recognition[C]. ICLR, 2015

[13] C. Song, Y. Huang, F. Liu, Z. Wang, L. Wang. Deep Auto-encoder based Clustering [J]. Intelligent Data Analysis, 2014, 6(18):65-76

[14] C. Song, Y. Huang, Z. Wang, L. Wang. 1000fps Human Segmentation with Deep Convolutional Neural Networks[C]. Asian Conference on Pattern Recognition (ACPR). 2015

IOP Conf. Series: Journal of Physics: Conf. Series **887** (2017) 012050 doi:10.1088/1742-6596/887/1/012050

Research on technology environment improvement of related industries based on internet governance

Jing Zhang[1] and Zhongliang Guan [1]

[1] Management Science and Engineering, Beijing Jiaotong University, School of Economics and Management, No.3 Shangyuan Village, Haidian District, Beijing, China.

Email: 447680873@qq.com

Abstract. The technology of Internet is an important factor of industry's development. Constructing a good technical environment is the foundation of the Internet and related industries' development. This paper demonstrates the necessity of the construction and improvement of the Internet and the related industries technology environment through comparing the current situation of the related industries. It also points out that China needs to improve the environment of the Internet technology urgently. The paper establishes the technology demand pattern of different related industries, and explores strategies of the different Internet technology environment's construction and perfection according to the different demand of the strong related Internet and the weak related Internet to the industries environment. This paper analyzes the factors that threaten the security of the Internet, and fully demonstrates the methods and tactics of establishing and improving the technology environment Internet hardware, the Internet and related industries in China under the basis of the framework of comprehensive management of Internet. This paper also studies the construction and improvement of the comprehensive management technology environment based on the Internet industry in China.

1. Introduction

The construction and perfection of the Internet technology environment is a kind of effective wayto control the Internet and related industries, which can effectively protect the policies and regulations and the effect of self-discipline in the industry. Technical supervision plays an important role in the construction of Internet technology environment. It can use information technology to protect the information system and information resources on the Internet, prevent the infringement, and provide effective network technical support for the development of Internet and related industries. Continuously development and innovation of Internet technology and the creation of a good technology environment is the basis of developing the Internet and related industries.

The Internet technology of US maintains a leadership in the world, which thanks to a good support of Internet technology environment. In the process of Internet application, the federal government and the state government fund building the industry development of the Internet application system and technical support environment. In the support of governmental financial, a unified Internet application technology support environment, which covers all industries, was formed, which achieves standardization and systematization. Its technical support and services areas related to various industries and business applications. Focus on Internet technology innovation, the next generation of Internet IPv6 protocol is a new Internet technology environment model, which makes e-commerce, online media, finance and other industries' Internet application achieve a smart link and dynamic support. At the same time, it also makes the construction and application of the Internet breakthrough its key technical bottlenecks.

German Internet technology environment and exchange capacity are at European leading level. According to the development of the Internet and the demand of domestic economic development, the German government has been increasing the Internet infrastructure and technology research and development investment, improving the network technology legal protection, paying attention to the development and management of Internet technology, carrying out the Internet dynamic label recognition technology environment with the United Kingdom, and the "Internet labelling Technology" interconnection of the two countries, which aims to strengthen the control of the Internet bad information.

The construction and application of Internet in China started later than developed countries, such as the United States. In the process of Internet operation, network security is often threatened and attacked, mainly manifested in: the information in the network transport process was intercepted, the transmission of documents were tampered, forged the e-mail, fake the identity of others, fake the identity of others, endangering the economic interests of others. These security problems seriously affect the Internet related industries to use the efficiency of the Internet, not only hinder the normal development of related industries, but also hinder the development of the Internet itself.

In order to create a good technical environment for the development of the Internet and related industries and complete the technology environment, China has taken some measures in the construction of Internet technology standards, the integration of Internet operation environment, Internet application, the construction of new generation Internet, the prevention of network hackers and the management technology of Internet. However, in the rapid development of the Internet, the challenge of implementing security management is further increased, and the difficulty of Internet content management is increasing. The development of the Internet technology environment and related industries in China faces some new security and social problems, mainly including: imbalance of Network technology security management and trade, imbalance of management and department linkage, network hardware technology and software application system development is slow, which hinders the healthy development of China's Internet and related industries. Therefore, the establishment of a green, sustainable network technology environment is the current problem that Internet and related industries need to solve.

2. The Demand Analysis of the Technology Environment

The construction and improvement of technology environment in Internet and related industries including network hardware technology environment and management technology environment construction are indispensable, as it shown in Figure 1. Therefore, the technical environment demand analysis of the Internet and related industries needs to be carried out from two aspects simultaneously.

Due to the different degree of use and dependence of the Internet in different industries, there is a certain difference in the demand for the technical environment. In this paper, through the cluster analysis, the Internet-related industries are divided into strong related industries and weak related industries. According to the classification of this article, e-commerce, online media, finance, information and communication industry and software industry belong to the strong correlation industry; equipment-manufacturing industry, mining industry, metallurgy industry, construction industry, logistics industry and other industries have low application of the Internet. These industries are weak related industries. This paper analyzes and compares the technical environmental requirements of strong and related industries.

2.1. The Technology Environment Demand of Strong Related Industries

In the field of network hardware technology, the development of strong related industries is technically more dependent on computer technology, communication technology and Internet technology. The construction of strong related industry technology environment focuses on the reform and innovation of information technology, and constantly develops the hardware, network topology and related technologies.

In terms of network management technology, the strong related industry are closely linked to the Internet, which face more security problems. In addition, its regulatory task is arduous. There is a strong demand for management technology. In the management process, managers or management departments must refine the management process, the implementation of the management of the responsibility to ensure that the Internet can be safe and efficient for the development of related industries.

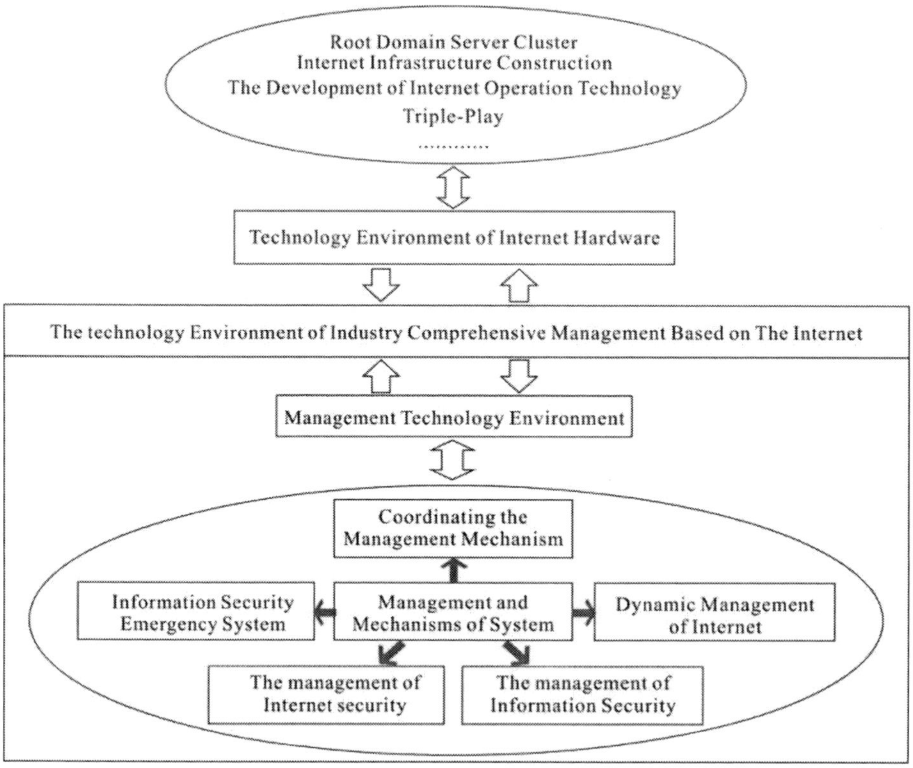

Figure 1. Internet and related industries comprehensive management technology environment to build the model map

2.2. Weak and Related Industries of the Technical Environment Needs

Weak related industries have less demand to the specialized information technology in the development process, but it rely on Internet network security environment very much. Most of the weak related industries are not related to the computer technology. Only the sales, supplier search, customer service and so on rely on the Internet for information dissemination, access and analysis. So ensuring the safety of information transmission process is the focus of the industry's technical environment needs.

Comparing with the Internet-related industries, weak related industries' demand of the management of technology is at a low level, the industry development faces less security problems. Regulatory task is relatively simple, the management process is clear, the management of the responsibility is relatively clear. It can ensure that the Internet for the development of the industry safe and efficient through the effective regulation.

2.3. Comparative Analysis of Technical Environment Demand

Due to the characteristics of the development of the industry itself and the depth of using the Internet, there is a different in technology environment between the strong related industryies and the weak related industries, which is mainly in the following areas. It is shown in Table 1:

Table 1. Comparison of technical environmental requirements for different industries.

	The basis of industry management	The requirements of technology environmental	The influence degree of technical environment on the effect of the industry
Strong related industries	Based on information technology	Innovation of Internet technology， Innovation of management technology	Medium
Weak related industries	Based on traditional production and operation	Improvement of network security, Innovation of management technology	General

2.3.1. The basis of industry management. The basis of industry management is the different industry background of the Internet. For the Internet strong related industry, the management and development of the industry are generally based on information technology, such as E-commerce industry, network media industry, information and communication industry, Internet technology is the effective driving force for its development. Therefore, the development of Internet Information technology will have a significant impact on this kind of industry.

The internet weak related industry is the traditional industry of production and management, such as manufacturing, metallurgy, construction and logistics industry, the improvement of competitiveness of these industries mainly depends on the production of technological innovation and production equipment. The role of the Internet technology environment in these industries is only as an application system network platform for ancillary production or marketing, the innovation of Internet technology has brought less improvement to the industry than the way of production and the change and innovation of production process equipment. So the innovation of Internet technology will not bring about great improvement of business condition for such enterprises.

2.3.2. Technical environmental requirements. Regardless of the strong related industry or the weak related industry. Because it is produced in the Internet environment, it has the technical environment demand. The demand of the strong related industry for the technical environment is mainly embodied in two aspects of the research and development of information technology and the innovation of management technology. The innovation of technology makes the production and operation of this kind of industry have changed remarkably, the competition ability of the industry has obviously improved. The demand of the weak related industry to the technology environment is mainly embodied in the need of the Internet as the security of the communication platform. The level of Internet security will directly affect this kind of industry in the Internet access to information, exchange information and improve the overall ability to operate and sell.

2.3.3. The influence of technology environment on the industry. The construction of the technology environment is often the technology improvement and innovation, which is based on the security of Internet network. The construction of the technology environment of the weak related industry should be based on the network security, and the influence of the technological environment changes on the weak related industry is quite remarkable. In addition to the construction of the technology environment, the guidance of policy and the management of the law are equally important, and the improvement of the pure technology environment has less influence on the strong related industries than the weak related industry.

By comparing and analyzing the difference of correlative degree industries' demand of Internet for technology environment, which can provide reference for our country's Internet and related industry to implement different technical strategy, and improve the technology environment of applying Internet in relevant industry.

3. The Improvement of Internet and Related Industries' Management of Technology Environment

This paper analyzes and studies the improvement of the governance technology environment of Internet and related industries from two aspects of network hardware technology environment and management technology environment.

3.1. Perfecting the Technology Environment of Internet Hardware

The improvement of network hardware technology environment should be carried out from the following aspects:

3.1.1. Establish our own root domain server. The domain Name System is a typical tree structure, the United States early occupy the "root" part. When other countries join the Internet, they can only join in the tree "branch" or "leaf" part. The safe operation of the Internet is very dependent on the root domain server, which is controlled by the United States alone, and it poses a great hidden danger to the safe operation of the Internet in our country.

If there is no domain root server, when the country's domain name server resolve a domain name at the first time, it need a foreign domain name root server to obtain a top-level index. Our country has introduced the domain name root server mirror server and.COM/.NET server's mirror server. Internet users ' network speed and security performance problems are expected to improve. After introducing the root domain server. How to manage it, ensure its orderly and efficient use and operation, will be an important issues that China's relevant departments will face. In addition, the country must accelerate the promotion of top-level domain names. The registration and use of CN. Registered country. CN domain name can improve the international competitiveness of national domain names, change the status that domestic users had registered foreign domain names too much, maintain network security and national interests, and increase our voice in Internet domain name supervision.

3.1.2. Environment Construction of Internet Infrastructure. The construction of Internet basic environment refers to the construction of basic network facilities and related network equipment. In view of the current development trend of our country, a variety of network integration has been or is becoming a reality, in recent years by the industry's popular "triple-play" (telecommunications network, Internet and radio and television network) mode has been widely regarded as the inevitable trend of China's Internet development.

On this basis, China's relevant industry institutions and associations put forward the establishment of a new Internet operation and management mechanism. That is the concept of "digital network competition". It is based on the integration of telecommunication network, Internet and broadcast television network. The country relaxes the policy of network integration, and allows the market of Internet information service to be further opened because of long-distance telephone, local telephone, wireless telephone, cable TV, data transmission network and open competition of service.

3.1.3. Urge Internet operators to establish high reliability Internet. The Internet operators must improve the Internet service environment and improve the credibility of the Internet business according to the real network security and business demand. Operators should enhance the function of network support system, improve Internet performance, standardize business quality test, optimize management system, realize the effective management of user terminal equipment, and resolve the perception and judgment of the whole Internet service quality. At the same time, the government and suppliers should jointly establish the Internet Information flow, the information protocol analysis and management system to implement the management and control of the business flow. In the construction of Internet support special network in China, the network management information is separated from the service bearer network, which gradually forms the three-layer integrated Internet management mode of basic network element layer management, IP service Internet layer traffic and performance management and IP service quality management, and realizes the intellectualization of three layer management Internet.

3.1.4. Promoting technological innovation, facilitating "triple-play", building a unified network technology platform. "Triple-play" refers to the integration of computer network, telecommunication network and CATV Network, that is, the traditional telecommunication network, computers and CATV network will tend to infiltrate and fuse with each other, the key technologies need include:

- Digital technology. So that all kinds of information content can be exchanged and transmitted between telecommunication network, computer network and cable TV net. [1]
- Access Network technology. This is the main difficulty of triple-play, responsible for the transmission of telecommunications business transparently to the user. [2]
- Broadband transmission problem. The broadband transmission of backbone network is the basis of triple-play. The optical communication network based on DWDM (dense wavelength division multiplexing) is bound to dominate the whole backbone network, and the three networks will be merged on this basis.
- Parallel computing and distributed systems. From the aspect of computer architecture, it has become the main means and technical way to improve computer's performance, and distributed system provides process communication and computer resource sharing as well as parallel computing ability, and provides users with transparent access to resource multi-element computing and service on other computers.
- Software technology. It provides a guarantee for the realization of "triple-play". The growth of large-scale service requests, the development of soft switching platform system with good scalability and fault-tolerant effect makes the three networks and their terminals can be transformed through software, ultimately supporting the characteristics, functions and services required by various users.

Telecommunications network, CATV Network, computer Internet in the course of the development of different shortcomings, cannot meet the user's various needs, the triple-play become an inevitable trend. Our country already basically has the technical condition that constructs triple-play fuses, network Foundation and market space, accelerate the integration of triple-play has entered a critical period, should be integrated planning, to determine the rational, advanced and applicable technical routes, to promote network construction, business applications, industrial development and supervision and management of the coordinated development of all work, to explore the establishment of a triple-play model that conform to our country's national conditions.

3.2. Improvement of the Management Technology Environment

With the development of the Internet hardware technology, the Internet has become more and more difficult to supervise the Internet in various industries and departments. We must build a good internet management technology environment to improve the Internet and related industry management system, implement the supervision mechanism, straighten each link management process, clearly the responsibility and promote the development of the Internet and related industries. The construction of Internet management technology environment can be started from the following aspects:

3.2.1. Improving the collaborative management and dynamic technology management. Strengthen the leadership of the organization, integrate the power of all relevant departments, clarify the power and responsibility of each functional department, unify the organization of the Internet and related industry governance action, form the mechanism of cooperative management and joint enforcement.

Continue to improve the site search system for the record and require the domestic access of all industries and enterprise sites are recorded compulsively to ensure the accuracy of the record information; Speed up the construction of website isolation technology, the quarantine of unregistered websites, illegal websites, etc., in order to combat network piracy, control harmful information dissemination and other Internet management measures to provide effective technical support; Accelerate the construction of network search technology, set up a website information search system to meet the information needs of Internet related industry, and provide the effective information of industry development in time.

3.2.2. Perfecting the Internet Information Security emergency system. Establishing the Internet information security emergency management system of unified two-layer structure and the "Unified" means that the state should set up the emergency coordination agency to be responsible for the emergency management of Internet information security; "Two-layer structure" means to exert the advantages of the industry and the provincial government and strengthen the emergency management. For the emergency of Internet Information security involving national security or economic development, the government must unify and coordinate the command to prevent the further deterioration of the situation and restore the normal operation and normal social order of the Internet as soon as possible. The government is in possession of a large number of network security information, should be at the critical time to start emergency plans to ensure the normal operation of the national infrastructure.

Internet information security emergency system should achieve five aspects to ensure the security of Internet information:

- Controlling the further actions of illegal intruders by removing the authority to safeguard the confidentiality of the system.
- Establishing the necessary mechanism to ensure the integrity of the information transmission.
- Ensuring the basic availability of information systems in the event of a catastrophic attack by establishing a minimal disaster backup system.
- Establishing the network blacklist to exclude the users of the information system from the legitimate use of the information system.
- Through using the blocking mode to ensure the controllability of the system in order to isolate the spread of the virus in time and avoid the network traffic anomaly caused by the further congestion.

3.2.3. Improve information security management technology environment. Because of the network communication security problem in the Internet, it is necessary to establish and perfect the technical environment of Internet Information security management in the light of the factors that threaten the security of Internet information transmission, including the following aspects:

- Establishing end-to-end secure communication protocol -- SSL protocol. SSL protocol has the characteristics of data confidentiality, identity authentication and data integrity. [3]
- Popularizing the use of secure transaction protocol--set in the process of electronic commerce. Set protocol is an internationally recognized standard for secure electronic transactions, which deals with the issue of electronic shopping and electronic payment through the network, which is the core of credit card holders, merchants and banks. Set technology can be used to further regulate the transaction process of electronic commerce in China.[4]
- Building virtual private network -- VPN technology. VPN technology is the Internet data transmission tunnel encryption security technology, which can guarantee the industry and the organization internal distributed (local area network) the special information in the interconnection on-line exchange, prevents and protects against the possible hacker attack and the network harm. VPN technology is mainly based on the security technology of IP layer, it is the combination of network encryption, key management, certificate and access control, and it is the ideal choice for the large-scale enterprise group in our country.[5]
- Security certification center--CA system. Digital certificates can solve the most critical identification problem in network security. The CA system consists of a series of certificate servers, which are generally divided into enterprise-level certificate server system and common certificate server system. CA system has played an active role in the authentication security of Internet application, electronic commerce and other management in our country.

3.2.4. Construction of network security management technology environment. In addition to the security technology above, network security management also needs the support of enterprise management. Building a good internal management of enterprises is the basis of network security management. The following aspects can be used to promote and supervise the management of the enterprise Internet security management:

- The network security management needs the leadership's high attention. Internet security management is a dynamic system engineering; the work of technical personnel cannot be completed. It need Internet managers to work together.
- According to the status of enterprises to determine the network security management strategy. The construction of an enterprise's network security Service cannot rely on the security services provided by the external Internet professional company, but also take appropriate protection and prominent management strategy according to the current situation of Internet application.
- Establishing a strict network security system file. The document of network security system mainly includes network construction plan, computer room management system, all kinds of personnel responsibility division system, security secrecy system, security policy document, etc.[6] Because the network security management system of the Internet is not sound enough, the file management of the respective Internet security system is difficult to unify, and it is suggested that the unified file management system and mode be popularized and popularized in order to perfect the file management system of Internet security system in China.

4. Conclusion

The construction and improvement of Internet technology environment is an effective way to govern the Internet and related industries, which can effectively guarantee the effect of policy and regulation and self-discipline. Continuous development and innovation of Internet technology, to create a good technical environment is the Internet and related Industry Development Foundation. Compared with the construction of Internet technology environment abroad, China's Internet construction and application start late and there are some problems in the process of Internet operation, which hinder the development of related industries and the internet itself. Learn from the successful experience of foreign Internet technology environment construction, combined with the actual situation of Internet and related industry in our country, this paper analyzes and contrasts the technical environmental requirements of different related industries, and puts forward the technical strategies suitable for the construction of technology environment in China's Internet and related industries.

The improvement of network hardware technology environment is mainly from four aspects: establishing our own root domain server, paying attention to network infrastructure construction, urging Internet operators to establish a high reliability of the Internet, promoting technological innovation, facilitating "three-network integration", building a unified network technology platform. The improvement of management technology environment is mainly from perfecting collaborative management, strengthening the construction of Internet dynamic management technology, perfecting the emergency system of Internet Information security, perfecting the technical environment of information security management, and constructing the technology environment of network security management.

5. References

[1] Hu Y X and Zheng Y 2008 Discussion on the development of triple-network fusion *Telecommunications Technology* (Sichuan: Electronic Technology Research Institute of Southwest China) 48(5):12-16

[2] Xue Y Y 2005 *Access network technology* (Beijing: China Machine Press)

[3] Yang F D and Ye Z L 2009 *Basic Computer Network* (Beijing: China Machine Press)

[4] Yu M 2006 Analysis and comparison of set protocol and SSL protocol for ecommerce secure transaction *Neijiang Technology* (Sichuan: Neijiang Research Institute of Science and Technology information) 27(7):111-111

[5] Wang J 2008 Discussing about the VPN technology and the development status in China *Network and Information* (Liaoning: Liaoning Information Center) (8):79-79

[6] http://www. Xici. net/b694738/d45811454. htm

IOP Conf. Series: Journal of Physics: Conf. Series **887** (2017) 012051 doi:10.1088/1742-6596/887/1/012051

Petri net modeling of encrypted information flow in federated cloud

Abdul Rauf Khushk[1] and Xiaozhong Li[2,]*

[1]College of Economics and Management, Tianjin University of Science and Technology, Tianjin 300222, China
[2]College of Computer Science and Information Engineering, Tianjin University of Science and Technology, Tianjin 300222, China, E-mail: lixz@tust.edu.cn
*Corresponding author

Abstract. Solutions proposed and developed for the cost-effective cloud systems suffer from a combination of secure private clouds and less secure public clouds. Need to locate applications within different clouds poses a security risk to the information flow of the entire system. This study addresses this by assigning security levels of a given lattice to the entities of a federated cloud system. A dynamic flow sensitive security model featuring Bell-LaPadula procedures is explored that tracks and authenticates the secure information flow in federated clouds. Additionally, a Petri net model is considered as a case study to represent the proposed system and further validate the performance of the said system.

1. Introduction

Recently, lots of combined systems are being introduced, proposed and developed to overcome the security issues of different organizations. For large organizations, data and information are very important to be protected so federated cloud system with combined Petri nets and other system integrated into it made a promising security system to protect data and information [1, 5]. Moreover, large organizations demands of keeping sensitive information on their more limited servers rather than in the public cloud which outlined the Federated Cloud Computing (FCC) in which both public and private cloud computing properties are used [2]. Likewise, the development of federated cloud system (FCS) with a different model of Petri nets modeling according to the needs of industries and several organizations for accessing the data and information and several applications are still being developed concerning the security policies. FCS has been more reliable as well as cost effective but the security risks of protecting data and information are still a big problem. Hence, minimizing and to overcome these security risk problems tracking and controlling of data and information flow in the system is essential [3]. So, a novel model system must be developed describing the information and tracking the flow of data and information [5].

Following the federated cloud systems, many models have been emerged with different methods of workflow security such as Petri nets especially to model the workflow integrating Bell-LaPadula model to workflow security which can be also called as the flow- sensitive analysis of programs[4]. But these kinds of models can only read and write policies having several drawbacks. Moreover, some of the Bell-LaPadula integrated models failed to get clearance for location level. Thus, realizing these several drawbacks and problems that occurred in different previous models developed which still lack protecting the securities policies. So, our research and dedication for developing a new combined model of federated intensively affect the information flow security of the entire system. In this paper, we present the user security levels for a distributive dynamic flow sensitive security model applying read with decryption and write with encryption in cloud system can be very promising and much more advantageous for overcoming the security risks.

Like industry standards such as UML activity diagrams, Business Process Model and Notation and EPCs, Petri nets offer a graphical notation for stepwise processes that include choice, iteration, and concurrent execution. Unlike these standards, Petri nets have an exact mathematical definition of their execution semantics, with a well-developed mathematical theory for process analysis [Carl Adam Petri].

In this study, we proposed a formal model of colored Petri Nets in which partition of workflows over a set of accessible clouds to meet maximum security requirements which are significantly based on a multi-level security model that extends Bell-LaPadula to encompass cloud computing [6]. We have used Colored Petri Nets (CPNs) to capture the flow of information flow in federated cloud computing systems, as well as the clouds and services, are fixed by applying decryption method in read and encryption method in write. Moreover, we focused on locating various kinds of data resources and users inside the formal model system. Several modifications and movements are not considered such as server-side components and dynamic movements, the execution of tasks in the system [7] etc. This formal model especially designed to capture the distributive dynamic information flow-sensitive security, where subjects (services) and objects (data) can migrate and change their security status dynamically.

2. Formal Model of Petri Nets Representing Distributive Dynamic Flow –Sensitive Security

Usually, a basic Petri Net (N) consists of two types of nodes, Pl and Tr, respectively called places and transition, a set $F \subseteq (Pl \times Tr) \cup (Tr \times Pl)$ of arcs that connect the nodes, and the initial marking $M0 : Pl \rightarrow N$ which is a mapping from the set of places to the set N of all non-negative integers [11].

As per the information lattice for security concerns, $L_{sec} = (L_{sec}, \leq_{sec})$ consists of a set L_{sec} and a partial order relation$\leq_{sec}$ such that, for all $l, l' \in L_{sec}$, there exists a least upper bound $l \cup l' \in L_{sec}$, and a greatest lower bound $l \cap l' \in L_{sec}$ [8,10].

Bell-LaPadula (1973) discovered a multi-level control security model which consist of five components such as (i) a set of possible access rights R, (ii) a complete lattice for security concern Lsec = (Lsec, ≤ sec), (iii) an access control matrix: B: $S \times O \rightarrow 2R$, (iv) a clearance map: c: $S \rightarrow$ Lsec and (v) security level map: l: $E \rightarrow$ Lsec [10] .We similarly adopt this model where services modeled as the subjects S and data as the objects O.

As progression with information lattices and Bell-LaPadula security requirements. Here we extend the Bell-LaPadula model to encompass cloud computing as a first step by including the extended security level map as follows - l: $E \cup P \rightarrow L_{sec}$. Moreover, for more cloud security concern we also introduce a new mapping loc to represent the location of each service and data item in the cloud system. - loc : $E \rightarrow X$

Furthermore, an additional rule is deployed with a security level which is greater or equal to the entity on the cloud system such as if an entity e is located in Cloud X, then we must have

$l(X) \geq_{sec} (e)$.

For an effective and efficient Petri net formal model, we have integrated all the basic requirements of components such as information lattices, Bell-LaPadula and cloud security described as below;

Definition (DDFSSM): A Distributive dynamic flow-sensitive security model for federated clouds is a tuple:

DDFSSM = $(S_{encry}, O_{decry}, L_{sec}, A, st_{init})$,

S is a finite non-empty set of encryption subjects/encryption services, O is a finite non-empty set of decryption objects/decryption data, L_{sec} is a complete security lattice, A is a finite set of actions partitioned into three sets, each action being a pair$\varphi = (\varphi^{in}{}_{encrypt}\ \varphi^{out}{}_{decrypt})$ Consisting of two finite multi-sets over the set of tuples $M = S_{encry} \times L_{sec} \times L_{sec} \times X \cup O_{decry} \times L_{sec} \times X$. where st_{init} is an initial state defined as a finite multi-set over the set of tuples M.

In general, a state of DDFSSM is a finite multi-set over the set of tuples M.

Considering the above concept we have proposed a formal Petri nets model as in figure (1). A tuple $(s, l, c, x) \in S_{encry} \times L_{sec} \times L_{sec} \times X$, later signified by (s, l, c) @ x, represents an encryption service (s) with the security level (l) and the clearance level (c) $(c \geq_{sec} l)$ residing on cloud p. Similarly, a tuple (o, l, p) $\in O_{decryption} \times Lsec \times X$, later signified by (o, l)@x, represents a decryption data item (o) with the security level l residing on cloud x. a set place followed by a set of transition such as places (P):(P1.P2.P3, P4,

P5) and transitions(T): (T1, T2,,,T11). Moreover, the initial marking as M_0 (0,1 ,0 ,0 ,0). The set of input and output in the system is done by the matrix for as in table (1). Set of Initial M_0 {P1=0,P2=1,P3=0,P4=0,P5=0} and the result is M_0{01000}.

Table 1. Representation of a set of input and output in matrix order.

Set of input											
	T1	T2	T3	T4	T5	T6	T7	T8	T9	T10	T11
P1	1	0	0	0	0	0	0	0	1	0	0
P2	1	0	1	0	1	0	0	0	1	0	0
P3	1	0	1	0	0	0	0	0	0	0	0
P4	0	0	0	0	0	1	1	0	0	1	1
P5	0	1	0	1	0	0	1	1	0	0	0

Set of output											
	T1	T2	T3	T4	T5	T6	T7	T8	T9	T10	T11
P1	0	1	0	0	0	0	0	0	0	1	0
P2	0	1	0	1	0	0	0	0	0	1	1
P3	0	1	0	1	0	0	0	0	0	0	0
P4	0	0	0	0	1	0	1	1	1	0	1
P5	1	0	1	0	0	1	0	0	0	0	1

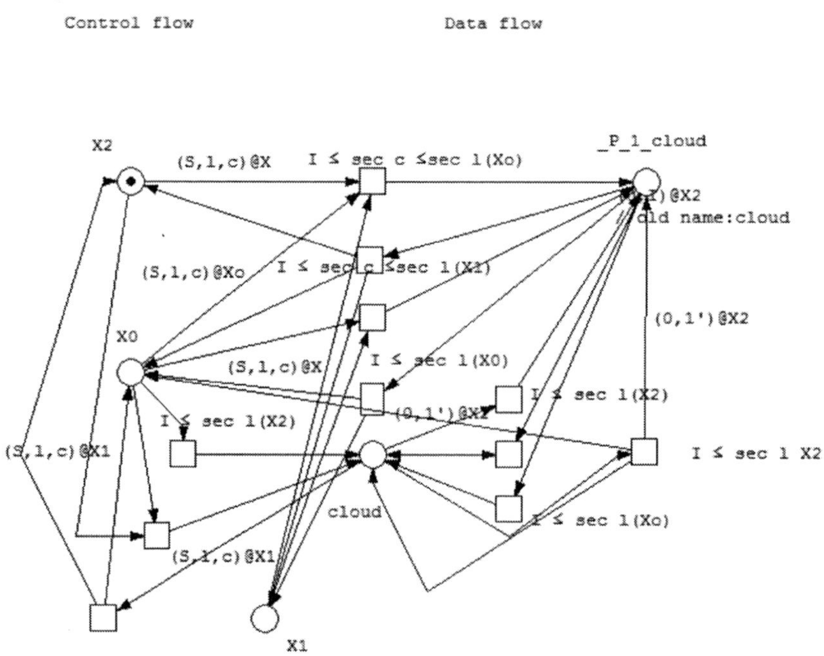

Figure 1. Shows the structure of the access control subnet (a central component of the Petri net model), representing the behavior of the subjects and objects residing on the same cloud (for simplicity, transitions relating to data destruction and creation have been omitted).circle denotes place from where token is fired and square box represents transition of fired token.

We take a persistent storage system with CRUD (create, retrieve, update, and delete) interface as an example, shown in figure 1, that highlights the importance of incorporating security infrastructure in clouds. We consider two public clouds, X0 and X1, and one private cloud, X2. Services and data can be deployed on different clouds, both services s0 and s1 can be deployed on X0, X1 and X2. However, the data item X0 can only be deployed on cloud X2. The left most sub-net (control flow subnet) shows how the services can migrate between different clouds under the security policy represented by the guards of the form $l \leq_{sec} c \leq_{sec} l$ X1 associated with the transitions. The data resources are represented by a single token, (d0, 1)@X2, inside another place labeled X2 (belonging to the data flow sub-net). It follows from the security levels of the data resources and clouds that such a resource can never enter X0 or X1. Note also that the security policy for data migration is represented by the guards of the form $l \leq_{sec}$ l(X0) associated with the transitions in the data flow sub-net.

3. Conclusions

In conclusion, we presented a Petri net distributive dynamic flow-sensitive security model which can be used to analyze the information flow in federated clouds with encryption and decryption method. The entities present in the cloud system can be assigned different security levels belonging to a given security lattice. Furthermore, each cloud is assigned a security level which captures the confidentiality level of the cloud. It is also potential to specify in a formal way different security strategies for the movement of entities between different clouds. The resulting formal model can then be represented by a suitable colored Petri net, and its dynamic properties analyzed using the existing verification methods and tools developed for Petri nets.

4. Acknowledgement

This work was supported by National Natural Science Foundation of China Grant Nos. 11301382 and 61603273, by Tianjin Application Foundation and Advanced Technology Research Project under Grant No. 14JCQNJC00300, and by Natural Science Foundation of Tianjin under Grant No. 16JCYBJC18500.

5. References

[1] W. Zeng, C. Mu, M. Koutny, and P. Watson, "Formal verification of secure information flow in cloud computing Workshop on information security applications, 2016: 103-116.
[2] P. Watson, "A multi-level security model for partitioning workflows over federated clouds," Journal of Cloud Computing, vol. 1, no. 1, pp.1–15, 2012.
[3] W. Zeng, C. Mu, M. Koutny, and P. Watson, "verifying secure information in federated Clouds 2014 IEEE 6th International Conference on Cloud Computing Technology and Science
[4] G. Smith, "A new type system for secure information flow," in In CSFW14. IEEE Computer Society Press, 2001, pp. 115–125.
[5] W. Zeng, C. Mu, M. Koutny, and P. Watson, "A flow-sensitive security model for cloud Computing systems," CoRR, vol. abs/1404.7760, 2014.
[6] K. Knorr, "Dynamic access control through Petri net workflows," in Proceedings of the 16th Annual Computer Security Applications Conference, ser. ACSAC '00, 2000, pp. 159–167.
[7] W. Zeng and M. Koutny, "Data resources in dynamic environments," in IEEE 8th International Symposium on Theoretical Aspects of Software Engineering (TASE), 2014.
[8] D. E. Denning, "A lattice model of secure information flow," Commun.ACM, vol. 19, pp. 236–243, May 1976.
[9] J. Landauer and T. Redmond, "A lattice of information," in Computer Security Foundations Workshop VI, 1993. Proceedings, Jun 1993, pp.65–70.
[10] N. A. Electronic, L. Lapadula, T. Original, D. E. Bell, and L. J. Lapadula, "Secure Computer Systems: Mathematical Foundations," 1973.
[11] W. M. P. V. der Aalst, "Petri-net-based workflow management software," in Proceedings of The NFS Workshop on Workflow and Process Automation in Information Systems, 1996, pp. 114–118.

Predicting protein amidation sites by orchestrating amino acid sequence features

Shuqiu Zhao[1, 2], Hua Yu[1, 2] and Xiujun Gong[1, 2,*]

[1]School of Computer Science and Technology, Tianjin University, Nankai, Tianjin, China, 30072

[2]Tianjin Key Laboratory of Cognitive Computing and Application, Nankai, Tianjin, China, 30072

* Corresponding author: gongxj@tju.edu.cn

Abstract. Amidation is the fourth major category of post-translational modifications, which plays an important role in physiological and pathological processes. Identifying amidation sites can help us understanding the amidation and recognizing the original reason of many kinds of diseases. But the traditional experimental methods for predicting amidation sites are often time-consuming and expensive. In this study, we propose a computational method for predicting amidation sites by orchestrating amino acid sequence features. Three kinds of feature extraction methods are used to build a feature vector enabling to capture not only the physicochemical properties but also position related information of the amino acids. An extremely randomized trees algorithm is applied to choose the optimal features to remove redundancy and dependence among components of the feature vector by a supervised fashion. Finally the support vector machine classifier is used to label the amidation sites. When tested on an independent data set, it shows that the proposed method performs better than all the previous ones with the prediction accuracy of 0.962 at the Matthew's correlation coefficient of 0.89 and area under curve of 0.964.

1. Introduction

Post translational modification (PTM) can affect the function and structure of a protein and play an important role in many physiological and pathological processes, such as immune response [1], cancer diseases [2, 3], neural diseases [4] and so on. Amidation is a type of PTMs, which can happen at the C-terminus of a peptide as known as C-terminal amidation. Amidation is crucial to the activation of many neuropeptides and hormones, such as receptor recognition and signal transduction [5, 6], and it also associates with a variety of diseases such as hypertension [7], cancer [8], neural dysfunction [9], and sleep apnea [10]. Therefore it is very necessary to identify amidation sites. Because traditional experimental methods are often time consuming and costly, identifying amidation using computational methods is a practical alternative.

In the past several years, a number of methods have emerged to predict PTMs sites, and some of them have achieved very good performance. For instance, the PseAAC algorithm proposed by Chou *et al.* [11] is widely used to represent sequence order information, such as in cysteine S-nitrosylation sites predictor [12]. Lv *et al.* [13] proposed a predictor named CarSPred using the high quality indices 8 (HQI8) feature to predict carbonylation sites of human proteins. The composition of k-spaced amino acid pairs (CKSAAP) method was used to predict Ubiquitination sites by Chen *et al.* [14]. Xu *et al.* [15] developed SUMO_LDA based on position-specific amino acid propensity (PSAAP) and modification of composition of k-space amino acid pair (MCKSAAP) feature extraction methods to predict sumoylation sites. Those methods are also applicable to the amidation sites prediction. Cui *et al.* [16] proposed an amidation sites predictor based on nearest neighbor algorithm. Wang *et al.* [17] developed PrAS, which combined four representative features to identify the amidation sites. Both of them achieved great performance.

In this paper, we aim at building a more accurate predictor by orchestrating amino acid sequence features making full use of position related information and physicochemical information. The position related features including PSAAP and PKSAAP can more correctly describe the distribution difference of the amino acids at each position between amidation samples and non-amidation samples, and HQI8 feature can capture the physicochemical properties of the amino acids. An extremely randomized trees algorithm is applied to choose the optimal features to remove redundancy and dependence among components of the feature vector by a supervised fashion. Finally the SVM classifier is used to label the amidation sites.

2. Materials and Methods

2.1. Dataset

Building an experimentally validated dataset is a critical step for the success construction of a classifier in identifying amidation sites computationally. In this paper, the dataset used to train the classifier is derived from Wang's dataset [17], which can be obtained from the UniProt database. There are 209 protein sequences in the dataset. 139 sequences are served as training set, and 70 are used as the independent test set. There are 332 amidation sites and 1202 non-amidation sites in the training set, 165 amidation sites and 632 non-amidation sites in the independent test set. A protein sequence fragment P that contains 21 amino acid residues can be expressed as:

$$P = R_1 R_2 \cdots R_{11} \cdots R_{20} R_{21} \tag{1}$$

where R_i denotes one of the 20 natural amino acids or the pseudo amino acid 'X'. R_{11} is the potential amidation site, and R_{12} is a Glycine residue.

2.2. Feature Extraction

In order to predict amidation sites using a computational method, we need to translate sequence fragments into feature vectors which are recognizable and learnable by a classifier, like Support Vector Machine (SVM). Three feature extraction methods are used to build the feature vector including HQI8, PSAAP and PKSAAP features for representing the sequence fragments.

2.2.1. HQI8 features. The AAindex database [18] contains a variety of physicochemical properties of amino acids. Because very large size of features in a classification problem may cause over-fitting [19], thus, Sahara *et al.* [20] use the fuzzy clustering method to cluster these properties, and each of the achieved clusters is represented by their central properties, namely, high-quality indices (HQI8). These amino acid indices' accession numbers in AAindex database are BLAM930101, BIOV880101, MAXF760101, TSAJ990101, NAKH920108, CEDJ970104, LIFS790101 and MIYS990104.

We use HQI8 to do feature extraction on the sequence fragments, and each sequence fragment with the length of L, will produce an $L\times 8$-dimensional feature vector. Without considering the amino acid R_{12}, each sequence fragment in our work will have a 160-dimensional feature vector, and when we encounter pseudo-amino acid 'X', all the property values are set to 0.

2.2.2. PSAAP features. According to the annotated situations of the center amino acid R_{11}, the training set can be divided into the amidated positive dataset and the non-amidated negative dataset. The basic idea of PSAAP is to obtain the frequency difference of 21 amino acids at each position between the positive training set and negative training set, and this method has been successfully applied to the prediction of many types of PTMs sites [21].

When the length of the fragment is L, the PSAAP generates a $21\times L$-dimensional position-specific amino acid propensity matrix $Z = Pos - Neg$, where Pos and Neg are defined as:

$$Pos=\begin{bmatrix} f^+_{1,1} & f^+_{1,2} & f^+_{1,3} & \cdots & f^+_{1,L} \\ f^+_{2,1} & f^+_{2,2} & f^+_{2,3} & \cdots & f^+_{2,L} \\ \vdots & \vdots & \vdots & \ddots & \vdots \\ f^+_{21,1} & f^+_{21,2} & f^+_{21,3} & \cdots & f^+_{21,L} \end{bmatrix} \qquad Neg=\begin{bmatrix} f^-_{1,1} & f^-_{1,2} & f^-_{1,3} & \cdots & f^-_{1,L} \\ f^-_{2,1} & f^-_{2,2} & f^-_{2,3} & \cdots & f^-_{2,L} \\ \vdots & \vdots & \vdots & \ddots & \vdots \\ f^-_{21,1} & f^-_{21,2} & f^-_{21,3} & \cdots & f^-_{21,L} \end{bmatrix} \quad (2)$$

The $21\times L$-dimensional matrix *Pos* is the occurrence frequency of 21 amino acids at each position of protein sequences contained in positive training dataset, and matrix *Neg* has the same size as *Pos*, but with the numbers that counted from fragments of negative training dataset.

By directly querying the matrix *Z*, we can obtain the feature vector *T* with the length of *L*, defined as equation (3). So the dimensions of PSAAP feature vector in our work will be 21.

$$T=(t_1,t_2,\cdots,t_l,\cdots,t_{L-1},t_L) \quad (3)$$

where t_l is defined as:

$$t_l=\begin{cases} z^{1,l} & when \quad R_l=A \\ z^{2,l} & when \quad R_l=C \\ \vdots & \quad\vdots \\ z^{20,l} & when \quad R_l=Y \\ z^{21,l} & when \quad R_l=X \end{cases} \qquad (l=1,2,\cdots,L) \quad (4)$$

2.2.3. PKSAAP features. The main idea of MCKSAAP [15] is to obtain the frequency difference of k-space amino acids pairs at different positions, and the method reduces the feature dimension relative to CKSAAP [14]. Inspired by MCKSAAP, we propose the PKSAAP (pertinence of k-space amino acid pair) method, which can get the pertinence difference of k-space amino acids pairs between the positive and negative training set. The pertinence value of events *x* and *y* is calculated as:

$$C_{x,y}=\frac{p(x,y)}{p(x)p(y)} \quad (5)$$

where $p(x)$ is the probability of occurrence of event *x*, $p(y)$ corresponding probability of event *y*, and $p(x,y)$ is the probability that event *x* and event *y* occur simultaneously.

We use $I_{l,k}$ to represent the pertinence difference of amino acid pair formed by *l*-th and $(l+k+1)$-th amino acids between the positive and negative training set, calculated as follows:

$$I_{l,k}=\frac{p^{b,(l,l+k+1)}}{f^+_{i,l}\times f^+_{j,l+k+1}}-\frac{n^{b,(l,l+k+1)}}{f^-_{i,l}\times f^-_{j,l+k+1}} \quad (6)$$

where $f^+_{x,y}$ and $f^-_{x,y}$ are defined in equation (2), and represent the frequency of corresponding amino acids. The $p^{b,(l,l+k+1)}$ and $n^{b,(l,l+k+1)}$ are defined in equation (7), and $p^{b,(l,l+k+1)}$ represents the frequency of the amino acid pair *b* formed by *l*-th and $(l+k+1)$-th amino acids in positive training set, $n^{b,(l,l+k+1)}$ represents corresponding frequency in negative training set.

We all know that 21 amino acids can form 441 amino acid pairs (AA, AC... XX). So, when the length of the sequence fragment is *L* and $k=0$, we can get $p^{b,(l,l+k+1)}$ and $n^{b,(l,l+k+1)}$ from the $441\times(L-1)$-dimensional matrices P_{pair} and N_{pair} respectively, which defined as follows:

$$P_{pair} = \begin{bmatrix} p^{1,(1,2)} & p^{1,(2,3)} & \cdots & p^{1,(L-1,L)} \\ p^{2,(1,2)} & p^{2,(2,3)} & \cdots & p^{2,(L-1,L)} \\ \vdots & \vdots & \ddots & \vdots \\ p^{441,(1,2)} & p^{441,(2,3)} & \cdots & p^{441,(L-1,L)} \end{bmatrix} \quad N_{pair} = \begin{bmatrix} n^{1,(1,2)} & n^{1,(2,3)} & \cdots & n^{1,(L-1,L)} \\ n^{2,(1,2)} & n^{2,(2,3)} & \cdots & n^{2,(L-1,L)} \\ \vdots & \vdots & \ddots & \vdots \\ n^{441,(1,2)} & n^{441,(2,3)} & \cdots & n^{441,(L-1,L)} \end{bmatrix} \tag{7}$$

If the length of the sequence fragment is L and k = 0, 1, 2... l, both P_{pair} and N_{pair} are 441× (L-1+L-2+...+L-l-1)-dimensional matrices. And the feature vector T is:

$$\underset{k=0,1,\cdots,l-1,l}{T} = (I_{1,0}, I_{2,0}, \cdots, I_{L-1,0}, I_{1,1}, I_{2,1}, \cdots, I_{L-2,1}, \cdots, I_{1,l}, I_{2,l}, \cdots, I_{L-l-1,l}) \tag{8}$$

In this work, we let k be 0, 1... 19, and the PKSAAP obtains 210-dimensional feature vectors.

2.3. Feature Selection

We union all the features and normalize their values in order to achieve better performance. Finally, we obtain a 391-dimensional (160+21+210) feature vector. The five-fold cross-validation and extremely randomized trees method [22] are used to select the optimal feature set.

The extremely randomized trees method is a tree-based ensemble algorithm. The basic idea of this method is splitting the samples in a dataset based on the information gain into a tree whose nodes denote the features, and calculate the importance of each feature using the samples routed to the corresponding node. For example, we have a sample set and a feature set. This algorithm randomly chooses k features from the feature set, and for each one of the k features, randomly chooses a feature value as the cut-point. Then the samples can be divided into two subsets, one with samples whose values on this feature are great than the cut-point and one with others. Then it can calculate the information gain of such a classification. Among those k features, the one that has the max information gain is chosen as the root node, and the same way is used to split each subset. The split results of the two subsets will be the left sub-tree and right sub-tree. Iteratively, this algorithm can construct a tree whose internal nodes are the features. In this tree, each feature's importance can be represented by the proportion of the samples routed to the corresponding node. Construction of multiple trees can increase the stability of feature selection, and the average value of each feature's importance is its importance score.

The information gain used in above procedure can be defined according to a previous research [22], whose value can be calculated using the following equation:

$$I_g = \frac{2M_{c,s}(P)}{H_c(P) + H_s(P)} \tag{9}$$

in which P is the sample set, c is a classification in P, s is a split of the feature, $M_{c,s}(P)$ is the mutual information of the split outcome and the classification, $H_c(P)$ is the entropy of the classification, and $H_s(P)$ is the split entropy.

2.4. Evaluation Criterion

In this paper, we use the five-fold cross-validation to select the best-performed prediction model on the training set, then test the model using the independent test to further evaluate the effectiveness. Five metrics are used to evaluate the predictor, which are Sensitivity (Sn), Specificity (Sp), Accuracy (Acc), Mathews Correlation Coefficient (MCC), and Area under ROC (AUC). The area under the curve (AUC) is the proportion of the space below the receiver operating characteristic curve (ROC) to the whole space, and this value can directly illustrate the stability of performance of a classifier.

We choose support vector machines (SVM) to train the prediction model, which has been successfully applied on the prediction of PTMs sites [23, 24]. With proper parameters, SVM has good performance in small sample sets and unbalanced datasets. Both in our training set and independent test set, the ratio of the positive samples and the negative samples is 1:4.

3. Results and Discussion

3.1. Performance Analysis

We use the above feature extraction methods and their combinations to train five classifiers separately. All of them are adjusted to their optimal status. And the results are list in table 1.

From table 1, among the three individual feature extraction methods, HQI8 is the best. Between the two combined methods, PSAAP+HQI8+PKSAAP method is better than PSAAP+HQI8. We can see that PSAAP and HQI8 features are useful for predicting amidation sites and PKSAAP features can slightly improve the performance by including the information of amino acid pairs. So we choose the best one, PSAAP+HQI8+PKSAAP for the final prediction.

Table 1. Performance of different methods in the independent test set.

	Sn	Sp	Acc	MCC	AUC
PKSAAP	68.5%	94.6%	87.7%	0.66	0.928
PSAAP	86.1%	96.4%	94.2%	0.82	0.955
HQI8	88.5%	96.7%	95.0%	0.85	0.961
PSAAP + HQI8	92.1%	97.0%	96.0%	0.88	0.961
PSAAP + HQI8 + PKSAAP	92.7%	97.2%	96.2%	0.89	0.964

We also compare our method with existing method PrAS [17]. The PrAS achieves 81.2% on Sn, 94.9% on Sp, 92.1% on Acc, 0.76 on MCC, 0.96 on AUC. We can see that our method is better than PrAs with great improvement while we can reach to 92.7% on Sn, 97.2% on Sp, 96.2% on Acc, 0.89 on MCC and 0.964 on AUC. Specifically, the values of Sn, Sp, Acc, MCC and AUC have an improvement by 11.5%, 2.3%, 4.1%, 0.13 and 0.004. The method proposed by Cui *et al.* [16] also can predict amidation sites, but we do not compare our method with it because the data set they upload is not complete.

Table 2. Performance of different classifiers in the independent test set.

	Sn	Sp	Acc	MCC	AUC
Naive Bayes	86.7%	94.5%	92.8%	0.79	0.945
KNN	89.1%	97.0%	95.4%	0.86	0.952
SVM	92.7%	97.2%	96.2%	0.89	0.964

Besides training the SVM classifier, KNN and Naive Bayes classifiers were also trained and tested on the independent test set with the 55-D optimal feature set. The performance of those classifiers are listed in table 2. We can see that SVM classifier is the best classifier on this prediction problem.

3.2. Feature Analysis

In the optimal feature set which is 55-dimensions, 21 dimensions are selected from PKSAAP, while HQI8 and PSAAP respectively contribute 22 and 12 dimensions. Although PSAAP features have the least size, more than half are selected when constructing the optimal feature set, indicating that these features are very suitable for the prediction of amidation sites.

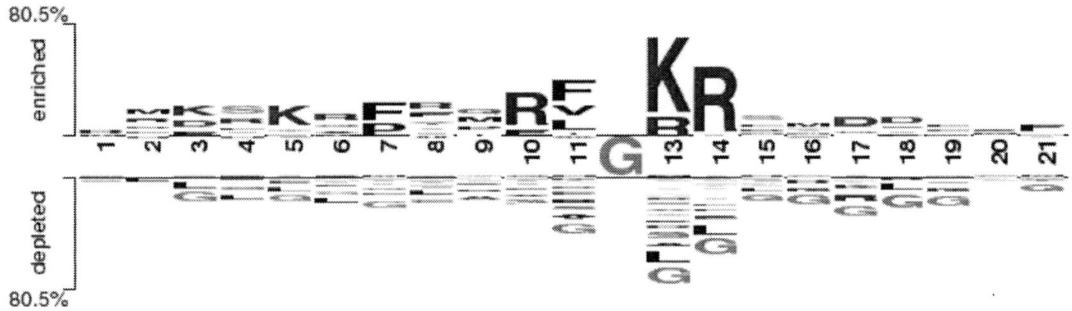

Figure 1. Two Sample Logo of the positive and negative training datasets. The enrichment difference between the two datasets are more significant at position 13, 14, 11 and 10.

We filter out the top 10 features, and list them in descending order as 13-PSAAP, 14-PSAAP, 12-13-PKSAAP, 13-MIYS990104, 13-NAKH920108, 13-BIOV880101, 10-PSAAP, 13-TSAJ990101, 14-TSAJ990101 and 11-PSAAP. Two Sample Logo [25] can show the difference between the two datasets about the tendency of amino acids at different positions. For example, in Figure 1, the same information about PSAAP features is illustrated that amino acids at position 10, 11, 13 and 14 of the fragment have more significant distribution. And we can find out that the 13-th position of amidated fragments is enriched with amino acid K and R, the 14-th position and 10-th position are enriched with R and 11-th position is enriched with F, V and L.

Our method can achieve very good performance which benefits from those advantages: 1) HQI8 is obtained by clustering a variety of physical properties. 2) Amino acids at some positions have specific distribution pattern, on which PSAAP is suitable. 3) PKSAAP method can obtain the information about amino acid pairs, which can slightly improve the performance.

4. Conclusion

In this paper, we propose a computational method to predict the amidation sites. Our method utilizes three feature extraction methods to represent the sequence fragments and uses extremely randomized trees algorithm to construct the optimal feature set. In independent test set, our method has great performance and can reach to 92.7% on Sn, 97.2% on Sp, 96.2% on Acc and 0.89 on MCC. The AUC score of our method is 0.964. Compared with other existing predictors, our method is better than them. We hope that our method can help researchers to predict and validate amidation sites. Most significantly, our approach provides a new perspective to the amidation sites prediction and can inspire more related research works on fragment's position specificity.

5. Acknowledgments

This work was supported by the National Natural Science Foundation of China(61930007), the National High Technology Research and Development Program of China (863 Program)(2015BA3005) and the National 973 Program (2013CB32930X).

6. Reference

[1] Rudd P M, Elliott T, Cresswell P, Wilson I A and Dwek R A 2001 *Science* **291** 2370-2376
[2] Moschos S J, Jukic D M, Athanassiou C, Bhargava R, Dacic S, Wang X, Kuan S F, Fayewicz S L, Galambos C, Acquafondata M et al. 2010 *Human pathology* **41** 1286-1298
[3] Driscoll J J, Pelluru D, Lefkimmiatis K, Fulciniti M, Prabhala R H, Greipp P R, Barlogie B, Tai Y T, Anderson K C, Shaughnessy J D et al. 2010 *Blood* **115** 2827-2834
[4] Bennett E J, Bence N F, Jayakumar R and Kopito R R 2005 *Molecular cell* **17** 351-365
[5] Eipper B A, Sto_ers D A and Mains R E 1992 *Annual review of neuroscience* **15** 57-85
[6] Mueller G P and Driscoll W J 2008 *Post-translational Modifications of Proteins: Tools for Functional Proteomics* **68(11)** 67-84
[7] Shimosawa T, Kanozawa K, Nagasawa R, Mitarai T, Isoda K, Takahashi K, Ando K, Tozawa Y, Nagase M, Sasaki N et al. 2000 *Hypertension Research* **23** 167-171
[8] Dennison S R, Harris F, Bhatt T, Singh J and Phoenix D A 2009 *Molecular and cellular biochemistry* **332** 43-50

IOP Conf. Series: Journal of Physics: Conf. Series **887** (2017) 012052 doi:10.1088/1742-6596/887/1/012052

[9] Bousquet-Moore D, Prohaska J R, Nillni E A, Czyzyk T, Wetsel W C, Mains R E and Eipper B A 2010 *Neurobiology of disease* **37** 130-140
[10] Kuyama H, Nakajima C, Nakazawa T, Nishimura O and Tsunasawa S 2009 *Proteomics* **9** 4063-4070
[11] Chou K C 2011 *Journal of theoretical biology* **273** 236-247
[12] Xu Y, Shao X J, Wu L Y, Deng N Y and Chou K C 2013 *PeerJ* **1** e171
[13] Lv H, Han J, Liu J, Zheng J, Liu R and Zhong D 2014 *PloS one* **9** e111478
[14] Chen Z, Chen Y Z, Wang X F, Wang C, Yan R X and Zhang Z 2011 *PloS one* **6** e22930
[15] Xu Y, Ding Y X, Deng N Y and Liu L M 2016 *Gene* **576** 99-104
[16] Cui W, Niu S, Zheng L, Hu L, Huang T, Gu L, Feng K, Zhang N, Cai Y and Li Y 2013 *Molecular genetics and genomics* **288** 391
[17] Wang T, Zheng W, Wuyun Q, Wu Z, Ruan J, Hu G and Gao J 2017 *Computational biology and chemistry* **66** 57-62
[18] Kawashima S, Pokarowski P, Pokarowska M, Kolinski A, Katayama T and Kanehisa M 2007 *Nucleic acids research* **36** D202-D205
[19] Trost B and Kusalik A 2013 *Bioinformatics* **29** 686-694
[20] Saha I, Maulik U, Bandyopadhyay S and Plewczynski D 2012 *Amino Acids* **43** 583-594
[21] Xu Y, Ding J, Wu L Y and Chou K C 2013 *PLoS One* **8** e55844
[22] Geurts P, Ernst D and Wehenkel L 2006 *Machine learning* **63** 3-42
[23] Zhou Y, Zhang N, Li B Q, Huang T, Cai Y D and Kong X Y 2015 *Journal of Biomolecular Structure and Dynamics* **33** 2479-2490
[24] Du Y, Zhai Z, Li Y, Lu M, Cai T, Zhou B, Huang L, Wei T and Li T 2016 *Journal of proteome research* **15** 4234-4244
[25] Vacic V, Iakoucheva L M and Radivojac P 2006 *Bioinformatics* **22** 1536-1537

IOP Conf. Series: Journal of Physics: Conf. Series **887** (2017) 012053 doi:10.1088/1742-6596/887/1/012053

Efficient detection of dangling pointer error for C/C++ programs

Wenzhe Zhang

Science and Technology on Parallel and Distributed Laboratory, State Key Laboratory of High Performance Computing, State Key Laboratory of High-end Server & Storage Technology,College of Computer, National University of Defense Technology, Changsha, PR China

Email: zhangwenzhe@nudt.edu.cn

Abstract. Dangling pointer error is pervasive in C/C++ programs and it is very hard to detect. This paper introduces an efficient detector to detect dangling pointer error in C/C++ programs. By selectively leave some memory accesses unmonitored, our method could reduce the memory monitoring overhead and thus achieves better performance over previous methods. Experiments show that our method could achieve an average speed up of 9% over previous compiler instrumentation based method and more than 50% over previous page protection based method.

1. Introduction

C/C++ programming requires programmers to manually organize dynamic memory [1]. Dynamic objects (memory spaces) are allocated via dynamic memory allocation API such as *malloc* and then freed via *free* [2]. This program paradigm has been used in many decades and it is criticized as being error-prone [3]. Often, programs may access a memory location that has been already freed, leading to unpredictableprogram results [4]. Moreover, it is very hard to notice this error during programming or debugging because reasoning all the *malloc* and *free* is complex [4].

This error, accessing already freed memory location, we call it dangling pointer error in this paper. Figure 1 shows an example of it. The last two lines in the sample code access the memory location pointed by *a*, which has been already freed. It is hard to notice this error during programming or debugging because that accessing the freed memory location (as shown in the last two lines in Figure 1) may not trigger any fault. Programs can run normally but give the wrong output, or programs may fault at a place that is very far from the error point, which makes debugging very complex.

Previous work [4] that aims to detect dangling pointer error mainly rely on dynamic instrumentation [5-6] or compiler instrumentation [7] to monitor every memory access. Normally, dynamic instrumentation is not acceptable because of its high overhead (more than 10X of the program's execution time). Compiler instrumentation method is more popular but it is also expensive for that it needs to insert a monitoring function call before every memory access. Recent work [4] proposed to give each object (allocated by *malloc*) a single page and rely on page protection to monitor all accesses to these objects. However, giving each object a single page will cause a huge waste of virtual memory space, leading to virtual memory explosion and much more TLB (Translation Lookaside Buffer) miss. For programs that make intensive memory allocation, its performance overhead is much worse than the method of compiler instrumentation [7].

```
int *a = malloc(sizeof(int));
*a = 1; //access *a
free(a);
b = *a; //error accessing freed memory location
*a = 3; //error accessing freed memory location
```

Figure 1. Sample code of dangling pointer error.

Above all, current mature method for monitoring memory access and detecting dangling pointer detection remains to be compiler instrumentation. In this paper we introduce an optimized compiler instrumentation method to detect dangling pointer error and at the same time instrumenting less memory accesses to improve performance. Our idea is straightforward: we only monitor the memory accesses that are after the program's first *free* operation. For all the memory accesses that are before the first *free* operation, there will be no dangling pointer error because at that time, no memory spaces are freed. To determine whether a memory access is before or after the first *free* operation, we developed a special pass based on LLVM [7-8], a popular complier framework, to do analysis at the intermediate code level. Experiments show that by monitoring less memory accesses, we can achieve 9% performance gain compared with the normal compiler instrumentation method and more than 50% compared with the page protection based method.

There are also a lot of methods proposed to detect dangling pointer error and these methods are mainly at higher levels such as overriding pointer dereference operations or using smart pointers. Moreover, the fast development of managed runtime system such as Java virtual machine offers new way to catch or eliminate dangling pointers. However, all these methods could not be used for C programs, which we believe is widely used in today, is commodity software development. In addition, those methods may be proved efficient for the higher-level languages, but the higher-level languages (such as Java) themselves are not as efficient as C. Thus in this paper we aim to raise a solution for the dangling pointer problem that can be adopted widely. The previous work would not replace our work in this paper.

The rest of this paper is organized as follows: We give the design and implementation of our detection method in Section 2. We show the experiments in Section 3 and conclude in Section 4.

2. Efficient Detection for Dangling Pointer Error

In theory, detecting dangling pointer error calls for two processes: (1) collecting dynamic memory allocation (*malloc*) or de-allocation (*free*) information (mainly the memory region of each object) during execution. (2) monitoring each memory access at runtime and check if the memory access refers to an already freed memory location. In this section, we will discuss these two processes respectively.

2.1. Dynamic Memory Allocation & De-Allocation

In order to get all the *malloc* and *free* information of the upper applications, we provide a dynamic linked library that overrides the *malloc* and *free* interface [1]. Thus, all the *malloc* and *free* operations called by upper applications will first go into our library. The code of *malloc* and *free* are shown in Figure 2. In our*malloc* and *free* we mainly first record the corresponding information and then call the original *malloc* and *free* that are provided by the standard glibc [9]. We keep all this information during the lifetime of a process thus at any time we can know if a memory access refers to a valid space or freed space.

```
void * malloc(size_t n){
    void * ret = ori_malloc(n);
    //record the malloc information
    record_malloc_info(ret, n);
    return ret;
}

void free(void *p){
    size_t n = get_size(p);
    //record the free information
    record_free_info(p, n);
    ori_free(p);
}
```

Figure 2. Code of malloc and free.

2.2. Memory Access Monitoring

We rely on the famous compiler framework LLVM [7-8] to achieve the goal of monitoring memory accesses in programs. LLVM first compiles the program's source code into a special intermediate code, at which level we can do a lot of analysis and code transformation such as loop unwinding [10]. Then the intermediate code will be complied into the final binary code for execution.

```
(a)
int main(){
    ...
    a = 1;
    foo(a);//the first fun in main
    b = 2;
    ...
}

void foo(int a){
    ...
    a = 1;
    free(...);//the first free in foo
    b = 1;
    ...
}
```

```
(b)
int main(){
    ...
    check(&a);
    a = 1;
    foo(a);
    check(&b);
    b = 2;
    ...
}

void foo(...){
    ...
    check(&c);
    c = 1;
    free(...);
    check(&d);
    d = 1;
    ...
}
```

```
(c)
int main(){
    ...
    //check(&a);
    a = 1;
    foo(a);
    check(&b);
    b = 2;
    ...
}

void foo(...){
    ...
    //check(&c);
    c = 1;
    free(...);
    check(&d);
    d = 1;
    ...
}
```

Figure 3. Example of complier instrumentation.

Here we developed an LLVM pass to insert checking function call before the memory accesses we want to monitor. The idea is shown in Figure 3. First, the original program code is show in Figure 3(a). Previous compiler instrumentation method is shown in Figure 3(b). We can see it just blindly inserts a check function before every memory access to check if the memory access will trigger a dangling pointer error. Thus all the memory accesses (the 4 accesses to *a, b, c, d*) shown in the figure are monitored.Our method is shown in Figure 3(c). We argue that there are some memory accesses that do not need to be monitored because at that time, no memory has been freed and thus it is not possible to trigger a dangling pointer error. As shown in Figure 3(c), the memory access to *a* in the *main* function and the memory access to *c* in the *foo* function can be left unmonitored. Thus we can get performance gain by eliminating the check function before them.

In order to determine if a memory access is before or after the first *free* operation in the program, we do a simple static analysis at the intermediate code level. We start from the *main* function and try to find the first function call of the main function. If this function call is *free*, we just mark all the memory accesses before the *free* can be left unmonitored. On the other hand, if this function is not *free*, we then start scan the code in the function (the case shown in Figure 3).Then if we find a free in the function, we can do the mark accordingly (the *foo* function shown in Figure 3). We can see in this simple example our method reduce half of the memory accesses that need to be monitored and thus we are likely to get performance gain at run time.

2.3. Dangling Pointer Error Detection

Above all, at execution time we got all the information of which piece of memory is valid and which piece of memory has been freed. We are also able to monitor all relative memory accesses. Then detecting dangling pointer error is simple and straightforward. At each memory access, we check if this memory location has been freed or not. If so, we report this to the programmer.

3. Experiments

This paper introduces an efficient method to detect dangling pointer error in C/C++ programs. In this section we mainly test and show our method (we call OPT in this section) in comparison with previous state-of-the-art work: (1) compiler instrumentation based work (we call Baseline in this section) and (2) page protection based work [4] (we call PPB in this section).The benchmarks we chose are mainly allocation intensive benchmarks from the olden benchmark suite [11]. They all make intense memory allocation and de-allocation, which is perfect to test and show our system.

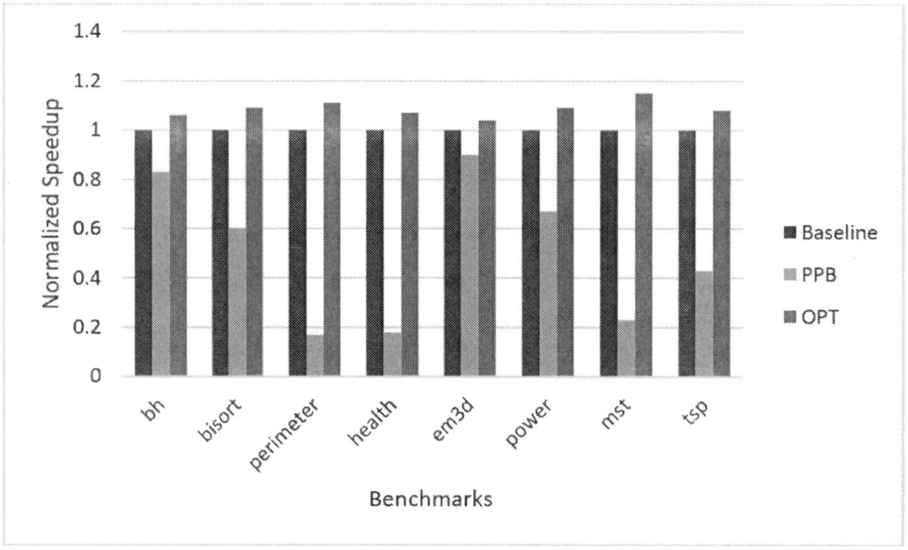

Figure 4. Performance Comparison.

Figure 4 shows the result. First, as we can see, the previous page protection method (PPB) achieves very poor performance. This is because it gives each object a single page. When the benchmarks make intense memory allocation, it will use much more virtual space than it will needs, causing severe memory bloating. This extensive use of virtual memory space will cause sever TLB miss, leading to poor performance.

Second, we can see compared with previous compiler instrumentation method, our method achieves better performance on all benchmarks. This is mainly because that we selectively leave some memory accesses unmonitored. Moreover, the later a program calls *free*, the more benefit we can get. The case is shown in the benchmark *perimeter* and *mst*.

Finally, our method averagely achieves 9% performance gain over baseline and more than 50% performance gain over PPB.

4. Conclusion

This paper introduces an efficient detector to detect dangling pointer error in C/C++ programs. By selectively leave some memory accesses unmonitored, our method could reduce the memory monitoring overhead and thus achieves better performance over previous methods. Experiments show that our method could achieve an average speed up of 9% over previous compiler instrumentation based method and 50% over previous page protection based method.

5. Acknowledgments

The work is partially supported by the The National Key Research and Development Program of China (No. 2016YFB0200401), by program for New Century Excellent Talents in University, by National Science Foundation (NSF) China 61402492, 61402486, 61379146, by the laboratory pre-research fund (9140C810106150C81001).

6. References

[1] Berger, E. D., McKinley, K. S., Blumofe, R. D., & Wilson, P. R. (2000). Hoard: A scalable memory allocator for multithreaded applications. ACM Sigplan Notices, 35(11), 117-128.
[2] Kanev, S., Xi, S. L., Wei, G. Y., & Brooks, D. (2017, April). Mallacc: Accelerating Memory Allocation. In Proceedings of the Twenty-Second International Conference on Architectural Support for Programming Languages and Operating Systems (pp. 33-45). ACM.
[3] Gay, D., Ennals, R., & Brewer, E. (2007, October). Safe manual memory management. In Proceedings of the 6th international symposium on Memory management (pp. 2-14). ACM.
[4] Dhurjati, D., &Adve, V. (2006, June). Efficiently detecting all dangling pointer uses in production servers. In Dependable Systems and Networks, 2006. DSN 2006. International Conference on (pp. 269-280). IEEE.
[5] Seward, J., Nethercote, N., &Fitzhardinge, J. (2004). Valgrind, an open-source memory debugger for x86-GNU/Linux. URL: http://www. ukuug. Org/events/linux2002/papers/html/valgrind.
[6] Luk, C. K., Cohn, R., Muth, R., Patil, H., Klauser, A., Lowney, G. ... & Hazelwood, K. (2005, June). Pin: building customized program analysis tools with dynamic instrumentation. In Acmsigplan notices (Vol. 40, No. 6, pp. 190-200). ACM.
[7] Lattner, C., &Adve, V. (2004, March). LLVM: A compilation framework for lifelong program analysis & transformation. In Proceedings of the international symposium on Code generation and optimization: feedback-directed and runtime optimization (p. 75). IEEE Computer Society.
[8] Lattner, C. (2008, May). LLVM and Clang: Next generation compiler technology. In The BSD Conference (pp. 1-2).
[9] Kegel, D. (2007). Building and testing gcc/glibc cross toolchains.
[10] Baskaran, M. M., Bondhugula, U., Krishnamoorthy, S., Ramanujam, J., Rountev, A., &Sadayappan, P. (2008, June). A compiler framework for optimization of affine loop nests for GPGPUs. In Proceedings of the 22nd annual international conference on Supercomputing (pp. 225-234). ACM.
[11] CARLISLE, M. Olden benchmark suite. 1995-01-03)[2007-06-30]. http://www. cs. princeton. edu/-mcc/olden, html.

IOP Conf. Series: Journal of Physics: Conf. Series **887** (2017) 012054 doi:10.1088/1742-6596/887/1/012054

Research and implementation of role-playing teaching mode supported by gamification

Xu Cui, Zhenglei Zhang and Lei Sun

Shandong Urban Construction Vocational College, Jinan, China, 250103
cuixu20021981@163.com

Abstract. The paper designs a Role-playing Teaching Mode Supported by Gamification to stimulate the interest of learners. In the process of creating the teaching mode, the factors of incentive factors, teaching mode and course selection are the most important factors gained by investigate and research. Then under the guidance of the three factors, a leaning framework of role-playing teaching mode which is called Gamification Learning Framework (GM1.0) is determined. In the design of GM1.0, First, collect problem cases which students interested in and select three courses which are Algorithm Design, Data Structure and Program Design. Then, extract the knowledge points of the three courses and merge into the problem cases to form game maps. Last, Learners gain a role-playing actor to join games with the support of game maps and finish selected tasks reaching a higher task level by upgrade checkpoints, experience promotions and award medals changing. After that, learners' enthusiasm for learning can be stimulated and the innovation abilities can also be improved gradually.

1. Introduction

One of the most authoritative reports "Horizon Report" [1, 2] said: Games and Gamification will be adopted in fields of higher education in two to three years, and the main relevant service fields are teaching, learning, and creative inquiry. Games and Gamification will be applied to education is not accidental. In recent years, Games and Gamification have already been used in the commercial [3], economic [4] and other industries [5, 6] including several education branches.

The "Horizon Report" results accord to the following facts: Almost everybody loves to play games and the game player's age distributes from several to dozens of years old. "Horizon Report 2013" [1] showed that the game players who are 18 to 35 years old accounted for 31% and "Horizon Report 2014" [2] showed that the game player number is 68% which was more than 18 years old. We know that 18 years old young people should learn a large number of knowledge, but in fact they waste a large amount of time playing various games. Students who are 18 years old in the University accepting higher education, are wasting more and more time playing games. They may do more meaningful things if they waste less time playing games.

This article will give an overall design method of "Research and Implementation of Role-playing Teaching Mode Supported by Gamification". The ultimate objective of design is to maximize the mobilization of the enthusiasm of learners. Section 2 is about the teaching mode of present learning framework and the teaching mode of Gamification Learning Framework. Section 3 gives an introduction of the technology related to the article. Section 4 describes how to design the framework of role-playing teaching mode supported by gamification. The application of the gamification learning framework has been described in section 5. Section 6 concludes the paper.

2. The Teaching Mode of Present Learning Framework and Gamification Learning Framework

The present learning framework adopts the advanced project teaching method which organize knowledge information according to knowledge points and knowledge units [7, 8]. Students learn knowledge information following the prescribed order without any incentives to guide them. So in the learning process, learners often do not have higher enthusiasms.

The main solution to the proposed problem in this paper is adding gamification design to the learning framework, which focus on several points as follows after referring to some existing results [10-17]:

- Adding incentive factors such as reward, score, ranking and experience points to the gamification learning framework, to improve participation motivation and learning enthusiasm.
- Designing a role-playing simulation game environment to motivate participants cooperating with one another, to improve students' learning, comment and sharing abilities.
- Designing comprehensive game maps as teaching mode, to guide players learning knowledge information with the help of the excitation mechanism (incentive factors).
- Introducing several courses, refining knowledge points and knowledge units according to the game maps, to reach the goal of solving practical problems and cultivating innovation abilities.

3. Related Works

In the process of gamification leaning framework construction, several parts which are incentive factors, teaching mode and course selection play an important role. Investigate and research are one of the most effective methods in gaining the first hand data. We obtain the materials by the follow ways: Adopting a number of methods including the questionnaire, BBS, QQ, Micro-blogs, Micro-letters, E-mail and Telephone to reach the excepted investigate and research results.

To gain the proper incentive factors, mainly through the questionnaire and Micro-letters. We send 2000 questionnaires to university and college students and 1000 questionnaires to fraternal institutions and 2467 questionnaires were recovered. After sorting the recovered questionnaires, the incentive factors mainly concentrated in reward, score, ranking and experience points.

To gain the proper game mode, mainly adopt the BBS, QQ and Telephone. We have listed fourteen teaching modes which are ACT, AVG, FPS, and RPG and so on. The result is nearly 50 percent students selected the mode of RPG (Role Play Games).

A BBS has been set up, in which nearly 100 teachers coming from more than 30 colleges and universities related to information technology (IT) are invited to discuss the proper courses of the project. Finally three courses which are Algorithm Design, Data Structure and Program Design are identified as a basis for the gamification learning framework construction.

The above investigate and research results are regarded as the first hand data to the gamification learning framework construction to stimulate the interest of learners.

4. The Gamification Learning Framework

The construction of Gamification Learning Framework (GM1.0) consists of four parts which are Problem Cases, Knowledge Points Set, Game Maps and Information Organization which are related each other. When well designed the components of the four implementation sections, the framework can guide learners to love the gamification learning process and can guide learners to solve problems, stimulate interests and cultivate abilities. Figure 1 shows the gamification learning framework which referring to the students' learning characteristics and some mature research results [10-17].

From Figure 1, we know that the final target of the project is gaining the Game Maps, which plays a leading role in the gamification learning framework construction.

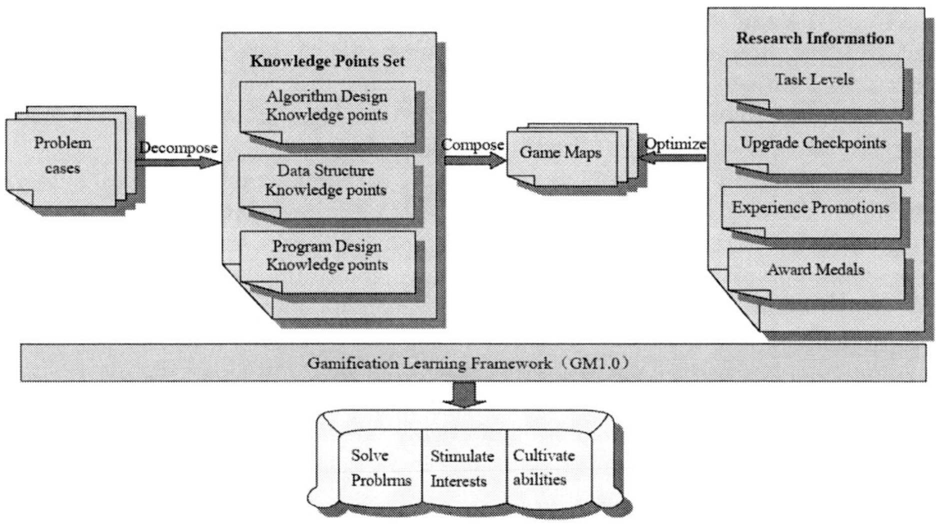

Figure 1. The gamification learning framework (GM1.0)

4.1. Gaining the Problem Cases

Problem Cases are questions needed to be solved. On one hand, the quality of case selection is directly related to the learners' learning interests, on the other hand, the final purpose of the case selection guides the students to learn and to master most of the knowledge points and knowledge units of selected courses. So we collect, arrange and extract cases interested in by students, mainly including entertainment, sports, science and technology information.

Then it need to classify the collected cases and gain the mathematical models which can do well in gaining the knowledge points set and making up all courses we selected. Figure 2 describes the problem cases working contents.

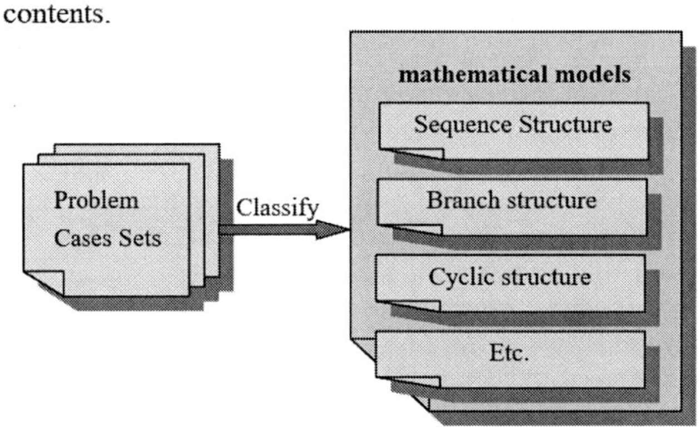

Figure 2. The problem cases sets classify

4.2. Gaining the Knowledge Points Set

In [8, 9], Research of Knowledge Points and knowledge Units and the Reuse Knowledge have been well done. A Personalized E-learning System Based on Portal Technology [7] had been constructed in 2011. In the E-learning System, the extraction of the knowledge points and knowledge units had been finished with the completion of the teaching research project "Research on learning resource management in E-learning environment" (we finished it in 2011-2013).

In this part, some adjustment has been done to the knowledge points and knowledge units description. The new description and organization method are as follows:

- The description and organization of knowledge points

Knowledge point name: briefly describe the knowledge point and cannot be repeated.

knowledge point attribution: detailed Describe knowledge point including text, graph, audio and video etc.

knowledge point relevance: describe one knowledge point's preamble, follow-up and relevant knowledge points information.

● The description and organization of knowledge units

knowledge unit name: briefly describe the knowledge unit and can not be repeated.

knowledge unit attribution: show the relevant knowledge points and can solve a complete problem.

Table 1 shows the organization of knowledge points and knowledge units and the relationship between knowledge points and knowledge units.

Table 1. The organization and relationship of knowledge points and knowledge units.

knowledge points				**knowledge units**		
name	***attribution***	***relevance***	***level***	***name***	***attribution***	***level***
can not be repeated	text, graph, Audio, Video. etc	preamble, follow-up, relevant knowledge points	difficulty level	can not be repeated	relevant knowledge points	difficulty level

4.3. The Research of Information Organization

In figure 1, we know that the research of information organization includes Task Level, Upgrade Checkpoints, Experience Promotions and Award Medals which can optimizing the game maps.

● Task Level: Task Level is the demarcation point of different knowledge points or knowledge units. The setting of task level should be able to meet the following requirements:

The task between every two task levels is a complete task which can described by a knowledge point or a knowledge unit.

The sequence of task levels can be adjusted if it do not affect the completion of the task.

There must be more than one task marked key task in a task level.

● Upgrade Checkpoints: Upgrade Checkpoints is related to the key task which determine if the task has been finished.

If all key tasks of a game map have been finished, then the learners can upgrade.

If all key tasks and all other tasks of a game map have been finished, then the learners can upgrade and gain extra points.

● Experience Promotions: Experience Promotions are the mark who have higher abilities, then more advanced tasks would be opened to them.

If learners finish only lower level tasks, the experience promotions improves slowly.

If learners finish higher level task, the experience promotions improves rapidly.

● Award Medals: Award Medals is the mark who can exempt from examination and gain more than 80 points.

If learners gain all award metals needed, then he can exempt from examination and gain 80 point at least.

If learners gain all award metals needed and take part in an examination, then he can gain 90 point at least.

4.4. Composing and Optimizing Game Maps

After gaining the problem cases and knowledge points set which coming from three courses (Algorithm Design, Data Structure and Program Design), composing game maps including as many questions as possible has come.

In the process of composing game maps, the more relevant organization of information is the field of relevance of knowledge points which including preamble, follow-up and relevant knowledge points. Then learners do not need to search knowledge points which they prefer to learning, and they only need to select the problem cases interested in to them, and they can obtain the knowledge points related to cases easily.

With the use of the mature cases, organizing relevant knowledge points into knowledge units, and optimizing all knowledge units to reach the purpose of no or less repeated knowledge points in different cases. Then the learners can obtain the highest learning efficiency.

4.5. The Gamification Learning Framework

A Gamification Learning Framework (GM1.0) is constructed with the support of problem cases set, knowledge points set, and incentive factors set etc. In GM1.0, every learner can register an account and can finish tasks by walking different level game maps and obtain upgrade checkpoints, experience promotions and award medals etc. Then learners can own different role-playing score, which is very close to playing games. Do not it attract the interest of learners and enhance the innovative abilities of learners? Yes, it can. Why? Please have a look at the application results in section 5.

5. Application

An experimental class which has 46 students and a regular class which has 42 students are selected to test the design result. In the process of teaching, the experimental class have privilege to use the GM1.0 while the regular class have no privilege and the rest teaching and learning conditions are the same. Table 2 shows the the comparison data of the two classes.

Table 2. The comparison data of two classes

class	*Learning interests*		*Innovation abilities (%)*
	Attendance (%)	*Failure rate (%)*	
experimental class	87.4[a]	2.2	82.7[b]/46.3[c]
regular class	82.3[a]	7.14	72.4[b]/23.1[c]

[a] The average attendance.
[b] The rate preferring to finish the new problems.
[c] The rate finishing the new problems.

We have found some differences in learning interest and innovation abilities of the two kinds of class. In learning interest aspect, the experimental class has higher attendance and lower failure rate. In innovation ability aspect, the experimental class has higher enthusiasm in solving unfamiliar problems. From the statistical data in figure 2, the difference is obvious.

6. Conclusions

The ultimate goal of higher education is enhancing the learners' learning interests and inspiring innovation abilities of every students. Then searching the most appropriate teaching mode plays an important role. The experimental data above shows differences in the use of different teaching mode. After using the Gamification Learning Framework (GM1.0), the learning interests and innovation abilities have been improved greatly. More and more students prefer to join the learning process and have higher learning efficiency through the use of Gamification learning framework. It is hoped that more and more students join in the GM, and the enthusiasm for learning is getting higher and higher.

7. Acknowledgement

Educational reform project: Teaching Reform Project of Vocational Education in Shandong, China (2017281).

8. References

[1] http://www.nmc.org/publication/nmc-horizon-report-2013-higher-education-edition/.
[2] http://www.nmc.org/publication/nmc-horizon-report-2014-higher-education-edition/
[3] Kai Huotari, Juho Hamari. Defining gamification: a service marketing perspective. Proceeding of the 16th International Academic MindTrek Conference. pp: 17 - 22.Oct. 2012.
[4] Juho Hamari. Transforming homo economicus into homo ludens: A field experiment on gamification in a utilitarian peer-to-peer trading service. Electronic commerce research and applications. 2013, 12(1/6)

[5] Luis de-Marcos, Adrian Dominguez, Joseba Saenz-de-Navarrete, Carmen Pages. An empirical study comparing gamification and social networking on e-learning. Computers & education. 2014, 75(Jun.)
[6] Adrian Dominguez, Joseba Saenz-de-Navarrete, Luis de-Marcos, Luis Fernandez-Sanz, Carmen Pages, Jose-Javier Martinez-Herraiz. Gamifying learning experiences: Practical implications and outcomes. Computers & education. 2013, 63(Apr.)
[7] Xu Cui, Shaotao Zhang, The Personalized E-learning System Based on Portal Technology, WISM 2011:433-440
[8] Xingwei Hao, Xiangxu Meng, Xu Cui, Knowledge Point Based Curriculum Developing andLearning Object Reusing, 6th International Conference on Web-based Learning, (2007LNCS Vol.4823):126-137
[9] Xingwei Hao, Xiangxu Meng, Xu Cui, a Network Course Content Authoring Method Based on Discipline Knowledge Ontology Base,International Conference on Machine Learning and Cybernetics, 2007, Volume: 7:4192-4197
[10] Ben W. Betts, Jay Bal, Alan W. Betts, Gamification as a tool for increasing the depth of student understanding using a collaborative e-learning environment, International journal of continuing engineering education and life-long learning, 2013, 23(3/4)
[11] Jorge Simoes, Rebeca Diaz Redondo, Ana Fernandez Vilas, A social gamification framework for a K-6 learning platform, Computers in human behavior, 2013, 29(2)
[12] Brock Dubbels, Gamification, Serious Games, Ludic Simulation, and other Contentious Categories, International journal of gaming and computer-mediated simulations, 2013, 5(2)
[13] Erik D. van der Spek, Towards Designing for Competence and Engagement in Serious Games, International Conference on Serious Games Development and Applications 3rd, Bremen, Germany.
[14] Diane Nahl, Leon James, Gamification in Instruction and the Management of Intersubjectivity in Online University Courses, International journal of web portals, 2013, 5(2)
[15] Maciej Laskowski, Aleksander Wojdyga, What Can Gamified University Classroom Teach Us? Advanced Science Letters, 2014, 20(2)
[16] Balraj Kumar, Gamification in Education - Learn Computer Programming with Fun, International Journal of Computer and Distributed System, 2013, 2(1)
[17] Baughman, A.K., Chuang, W., Dixon, K.R., Benz, Z., Basilico, J., DeepQA Jeopardy! Gamification: A Machine-Learning Perspective, IEEE transactions on computational intelligence and AI in games, 2014, 6(1)

IOP Conf. Series: Journal of Physics: Conf. Series **887** (2017) 012055 doi:10.1088/1742-6596/887/1/012055

Study on an effective algorithm of acquiring and tracking BD2 medium-frequency signals

Yanfei Liu, Yu Yang, Shiqiang Jiang, Ke Jiang and Jieling Wang

Xi'an Institute of High Technology, Xi'an, Shaanxi 710025, China

Abstract: According to the modulation characteristics of NH code in MEO/IGSO satellites, a simple and effective serial-parallel combined method of capturing the BD2 satellite signals was presented and a complete software receiver of the B1 frequency point was realized. Based on Kalman filter, a new kind of vector tracking loop structure of pseudo code and carrier wave to adapt to the high dynamic situation was put forward, and the half-vector deeply-coupled structure based on CKF pre-filter was further studied. Simulation and practical test results showed that the new acquisition algorithm had high sensitivity, stronger adaptability for weak signal, and less computation, and that nonlinear filtering methods could not only be applied to KFPLL, but also track high dynamic signal of 100g.

1. Introduction

Nowadays, as the construction of the Beidou Navigation Satellite System (BDS)[1] has been primarily accomplished and begun to serve China and its circumjacent regions[2], further and deeper studies on satellite signal acquisition and tracking algorithms of the 2nd generation of Beidou Navigation Satellite System (BD2) will be conducted.

BDS satellites are different from GPS/GEO satellites [3], as MEO/IGSO satellite signal modulation has Neumann-Hoffman Codes (NH codes). Therefore, only when the four-dimensional search of satellite serial number, pseudo-code phase, carrier frequency and NH code phase is executed, can the acquisition of such satellite signals be performed and to obtain complete navigation messages in the tracking process, the influence of NH codes must be removed. Many papers on BD2 Baseband algorithm failed to fully discuss the acquisition of NH codes. Due to the phase flip, the adopted coherent integral length in the NH code acquisition process cannot exceed 1ms [4]. Because the length is too short, the integrated energy amplitude of useful signals is not significant, and then it will be very difficult to acquire the weak signals. Xie F and Liu J [5] proposed the dichotomy to check the flipping position of data bits, which was not practical as it required tedious calculation.

Based on the analysis of original acquisition algorithmand combined with the features of second modulation, we propose a novel serial-parallel combined acquisition method of MEO/LGSO satellite signals, which realizes the complete BD2 software receiver based on traditional tracking loops. In the light of the nonlinearity of system model in high dynamic environment, combined with the requirements of the software receiver for PLL performance, Cubature Kalman Filter (CKF), Unscented Kalman Filter (UKF) and Extended Kalman Filter are imbedded into the PLL, then more accurate observation equation is determined and new vector tracking loop structure is designed. Through EKF, the combined tracking structure of carrier waves and code loops is realized, the realization means of the semi-vector deep composite structure based on CKF pre-filtering is given and algorithm design is verified through simulation.

2. Analysis and Verification of BD2 Acquisition Algorithm

After NH code modulation, the BD2 input signals can be expressed as:

$$x(n) = D(n)C(n)h(n)\cos\left(\frac{2\pi(f_i + f_d)n}{f_s} + \varphi\right) \quad (1)$$

In which, f_i denotes the intermediate frequency, f_d the Doppler frequency shift and f_s the sampling frequency. Through the traditional code phase acquisition process, obvious peak values can be obtained in the situation of code phase alignment. But because of the frequency spectrum shifting effects of NH codes, several peak values may appear, which makes it difficult to obtain correct Doppler frequency shift estimation.

3. Overall Design of BD2 Receiver

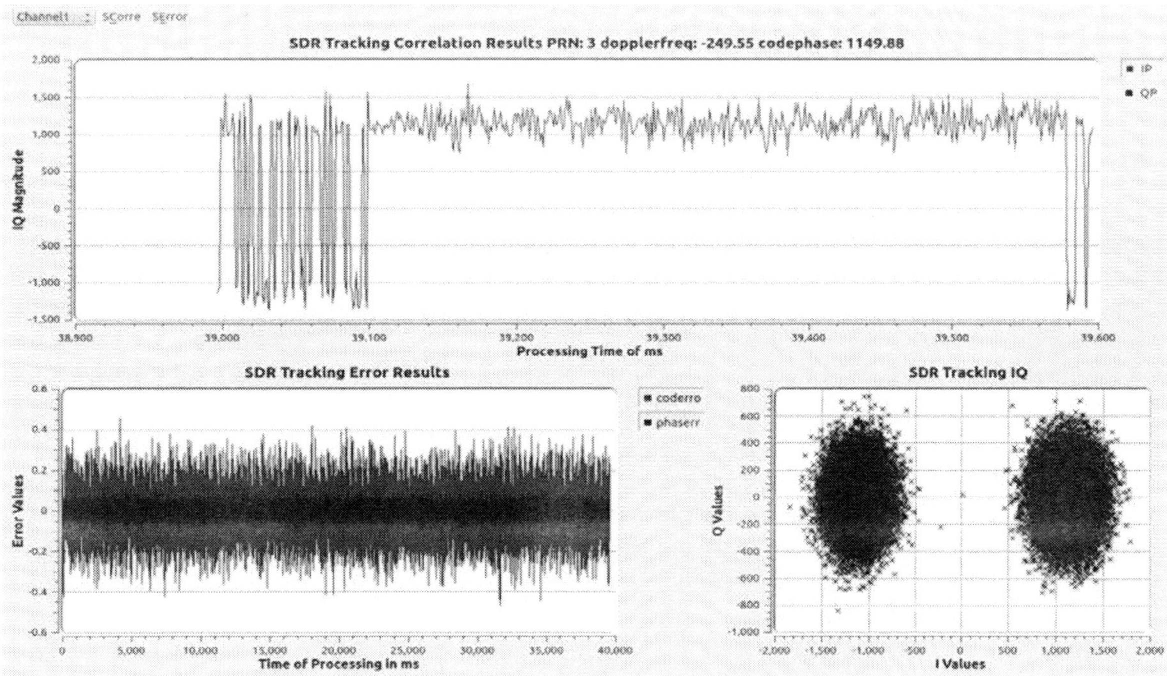

Figure 1. Tracking results of Satellite No.3

As to the BD2 software receiver designed in this paper, its corresponding frequency point is B1I frequency point, the intermediate frequency is 4.092MHz, the sampling frequency is 16.368MHz and the digitalizing bit is 2bit. The latest universal serial bus USB3.0 is adopted to apply the collection of intermediate-frequency data, and in Qt creator of Linux operating system, C/C++ is utilized to accomplish the functions of BD2 acquisition and tracking, as well as positioning resolution.

In the tracking phase [6], set the preliminary testing integral time length as 1ms, adopt third-order phase locked loop (PLL) in carrier wave cracking, utilize the arctangent phase discriminator with a noise bandwidth of 18Hz and set the loop gain 0.15. In code tracking, use a second-order delay locked loop (DLL), adopt a unitized early minus late power discriminator with a noise bandwidth of 5Hz and set the loop gain as 1. And then the tracking results are shown in Fig. 1. As for the navigation solution phase, refer to [7], and need not be repeated here.

4. Design and Simulation of High Dynamic Signal Tracking Loop

In practical application, high dynamic environment may result in strong nonlinear characteristics. In addition, as an indispensable part of the tracking loops of the receiver, code tracking loop will be relatively less influenced under specific conditions. Therefore, referring to [8] and [9], VDLL is designed and EKF is adopted, but the acceleration must be taken into consideration as a state quantity and the actual realization processes need to be repeated here.

4.1. Principle of Nonlinear Filtering

When either the dynamic process or the measurement process of the system presents nonlinear characteristics, it will be difficult to apply standard Kalman filter and it is necessary to adopt other nonlinear filters to solve the problem of nonlinear optimal filtering. Assume the following nonlinear state equation and measurement equation. In which, w_{k-1} and v_k are additive white Gaussian noises.

$$\begin{cases} x_k = f\left(x_{k-1}, k-1\right) + w_{k-1} \\ z_k = h\left(x_k, k\right) + v_k \end{cases} \tag{2}$$

Q_k and R_k are the system noise matrix and measurement noise matrix. In this paper, based on EKF, UKF and CKF, high dynamic signal tracking loops are designed. After the instruction of nonlinear filtering, the loop structure is shown in Fig. 2:

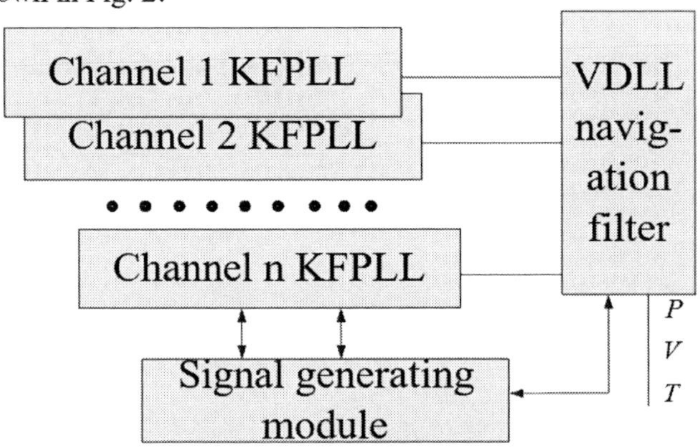

Figure 2. The overall structure of vector tracking loops

4.2. Modeling and Algorithm Design

In high dynamic environment, select the low order derivatives of the carrier wave phase error, frequency error and the low order derivatives of the frequency order as states, i.e. $X(k) = \left[\Delta\varphi, \Delta f, \Delta\dot{f}\right]^T$, select the in-phase accumulated value of the instant branch and the related accumulated value of the orthogonal branch as the observation value, and then the model of the PLL in a certain channel of the receiver may be expressed as:

$$\begin{cases} X(k) = FX(k-1) + \xi(k) \\ Z(k) = \begin{bmatrix} AR\left(\Delta\tau\right)\sin c\left(T\omega_e/2\right)\sin\left(LX(k)\right) \\ AR\left(\Delta\tau\right)\sin c\left(T\omega_e/2\right)\cos\left(LX(k)\right) \end{bmatrix} + \eta(k) \end{cases} \tag{3}$$

From the discussion above, it can be seen that the PLL model has strong characteristics of nonlinearity. In this paper, CKF and UKF are utilized to perform optimal estimation to the states [10].

4.3. KFPLL Simulation Verification

The high dynamic intermediate frequency data are obtained through simulation. From the 4th second, the receiver begins to execute the high dynamic process, the maximum acceleration on the X-axis can reach 100 times of g per second, and along other directions, the dynamic characteristics are not significant. The carrier-to-noise ratio of satellite signals is set as 40dB-Hz, and the refreshing cycle of the carrier wave loops is set as 1ms. Figure 3 and figure 4 are the Doppler frequency shift results of traditional carrier wave loops and KFPLL in tracking Satellite No. 23.

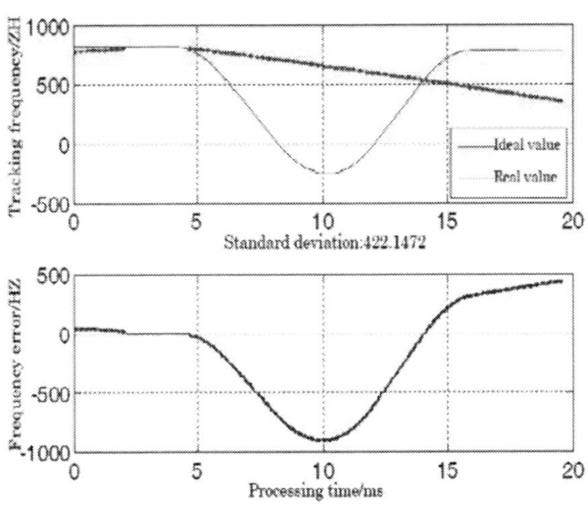

Figure 3. PLL frequency tracking results

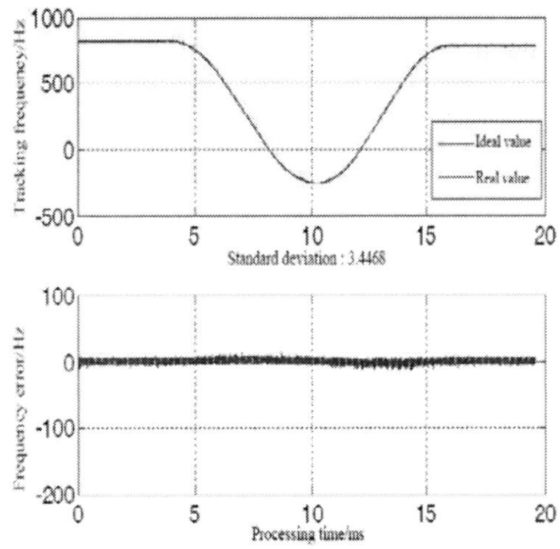

Figure 4. EKFPLL results

From the simulation results, it can be seen that when the high dynamic environment is serious, the traditional carrier wave loops may lose lock, but the carrier wave loops based on EKF, UKF and CKF nonlinear Kalman filtering don't lose lock. This indicates that the designed BD2 intermediate signal processing algorithm reaches the purpose to realize stable tracking of high dynamic signals.

5. The Design and Simulation of Semi-Vector Deep Composite Structure

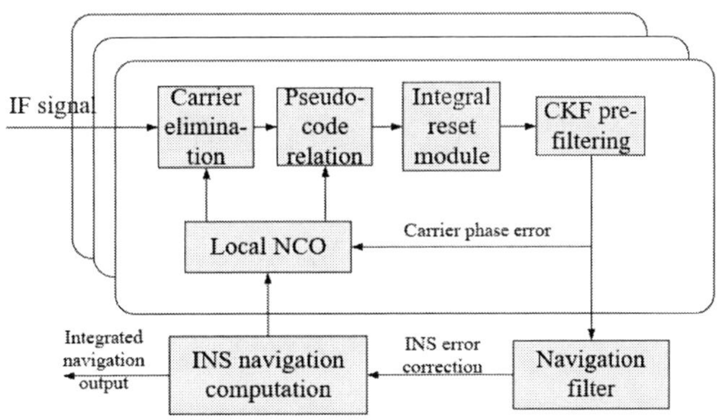

Figure 5. CKF semi-vector deep composite structure

As the full-vector deep composite structure [11] requires GNSS receiver to provide high-accuracy initial positioning results as the filtering initial value of the deep composite main filter in order to realize good carrier wave phase tracking, therefore, a semi-vector deep composite structure, i.e. all the channels tracking the carrier wave phase individually, can avoid the limitations of the full-vector deep composite structure.

The dramatic change of the receiver's motion state may result in the loss-of-lock of the traditional carrier tracking loops, which have fixed parameters. In order to adapt to the high-variability, CKF [12], a nonlinearity-adaptive filter, is introduced in the semi-vector deep composite to realize the stable tracking of carrier phases. The dynamic modeling of carrier tracking loops and the actual realization method of CKF-based carrier tracking are discussed above. The semi-vector deep composite structure of the CKF-based pre-filtering is shown in Fig. 5.

5.1. Establishment of Semi-Vector Deep Composite Model

The semi-vector deep composite structure of CKF-based pre-filtering composes of INS navigation computation, the computation of local NCO controlling quantity, pre-filter, navigation filter and the basic tracking loops. Navigation filter has the functions of estimating the error parameters of GNSS and INS, and the state quantity consists of user carrier position, errors of velocity and attitude angle, random drift of INS sensor, and the clock bias and clock drift of receiver.

$$\begin{cases} X_I = \left[\delta \upsilon_x \ \delta \upsilon_y \ \delta \upsilon_z \ \phi_x \ \phi_y \ \phi_z \ \delta L \ \delta \lambda \ \delta h \right. \\ \qquad \left. \varepsilon_{xb} \ \varepsilon_{yb} \ \varepsilon_{zb} \ \nabla_{xb} \ \nabla_{yb} \ \nabla_{zb} \right]^T \\ X_G = \left[\delta t_u \ \delta f_u \right]^T \end{cases} \tag{4}$$

Where, the state quantities in X_I respectively denote the velocity errors along three axial directions (East, North, Sky) of the navigation coordinate system (i.e. Geographic Coordinate System), 3-D attitude error, latitude error, longitude error, altitude error, and the random drift of gyroscopes and accelerometers along the three axial directions of carrier coordinate system. In X_G are the equivalent distance and distance rate, which are caused by the clock bias and clock drift of receiver.

The error state equation of GNSS/INS deep composite system is:

$$\dot{X}(t) = F(t)X(t) + W(t) \tag{5}$$

$$\begin{aligned} W(t) = \big[& w_{xa} \ w_{ya} \ w_{za} \ w_{xg} \ w_{yg} \ w_{zg} \ 0\,0\,0 \\ & w_{x\varepsilon} \ w_{y\varepsilon} \ w_{z\varepsilon} \ w_{x\nabla} \ w_{y\nabla} \ w_{z\nabla} \ w_{tu} \ w_{fu} \big]^T \end{aligned} \tag{6}$$

$$\begin{cases} X(t) = \left[X_I(t) \quad X_G(t) \right]^T \\ F(t) = \begin{bmatrix} F_I(t) & 0 \\ 0 & F_G(t) \end{bmatrix} \end{cases} \tag{7}$$

Select the estimated the Pseudorange and pseudorange rate error of the channels as the observation values of the navigation filter, and the observation equation value is:

$$Z(t) = H(t)X(t) + V(t) \tag{8}$$

Assume there are n channels and the observation value is denoted as:

$$Z = \left[\delta\rho_1 \ldots . \delta\rho_N \ \delta\dot{\rho}_1 \ldots . \delta\dot{\rho}_N \right]^T \tag{9}$$

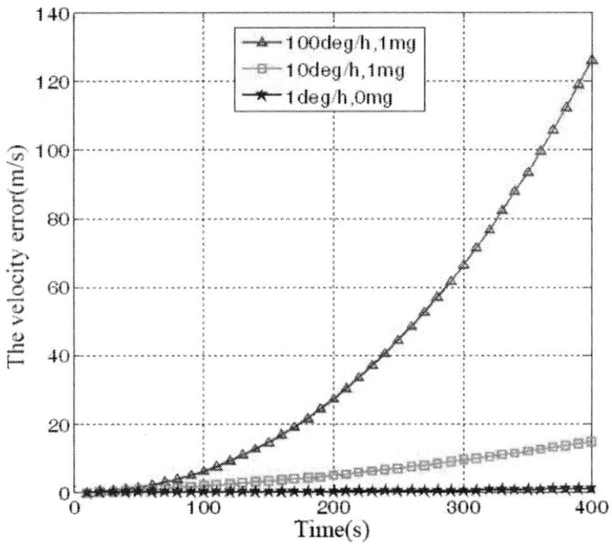

Figure 6. Velocity error of INS with different accuracy grades

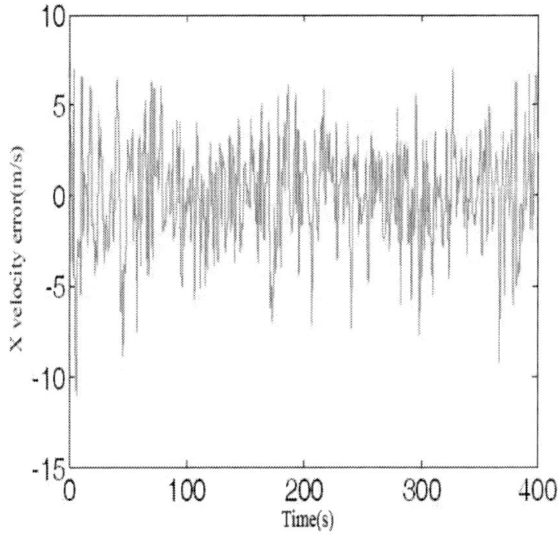

Figure 7. The velocity error of the East direction of deep composite navigation

5.2. Simulation Verification

Through setting such motor states as velocity, position and angular velocity of the carrier, calculate the motor tracks of the carrier first, and calculate the nominal outputs of the gyroscope and accelerometer. Finally, add different-grade noises as necessary to simulate IMU outputs with different accuracy grades. In the uniform motion state, as to the INS devices with different accuracy grades, the velocity errors by means of navigation solution are shown in Fig. 6.

High-dynamic medium-frequency signals still follow the high-dynamic scene mentioned above, and in this scene, add the output signal simulation module. As to the gyroscope, the random drift is 0.1°/h, and the constant drift is 10°/h; and as to the accelerometer, the constant drift and the random drift are 1mg and 0.01mg respectively. The simulation results are shown in Fig.7, which indicates that deep composite navigation can restrain the accumulated errors of INS satisfactorily.

6. Conclusion

In this paper, the influences of NH-code modulation on the acquisition of MEO/IGSO satellites are discussed in depth, and a simple but effective acquisition approach is proposed, which improves the ability to acquire weak signals [13-14]. Also, the processing algorithms of BD2 medium-frequency signals and baseband singles, and complete programs of BD2 software receiver are developed. In addition, based on EKF, UKF and CKF, the basic structure and the actual realization steps of KFPLL are developed. The simulation results verify that non-linear filtering method-based carrier tracking loops have higher adaptability to the high-dynamic conditions. The semi-vector deep composite structure, which is based on the CKF pre-filtering, can provide relatively high-accuracy navigation solution.

7. References

[1] Zeng T, Ao D, Hu C, et al. Multiangle BSAR Imaging Based on BeiDou-2 Navigation Satellite System: Experiments and Preliminary Results[J]. IEEE Transactions on Geoscience & Remote Sensing, 2015, 53(10):1-14.

[2] Zhao S, Yao Z, Zhuang X, et al. Analysis on coverage ability of BeiDou navigation satellite system for manned spacecraft [J]. Acta Astronautica, 2014, 105(2):487–494.

[3] Seo J, Walter T. Future Dual-Frequency GPS Navigation System for Intelligent Air Transportation under Strong Ionospheric Scintillation [J]. IEEE Transactions on Intelligent Transportation Systems, 2014, 15(15):2224-2236.

[4] Shi M, Peng A, Ou G. Analysis to the effects of NH code for Beidou MEO/IGSO satellite signal acquisition[C]// Industrial Electronics and Applications. IEEE, 2014:2075-2080.

[5] Xie F, Liu J, Li R, et al. A simultaneous multiple BeiDou signal acquisition algorithm for a software-based GNSS receiver [J]. Optik - International Journal for Light and Electron Optics, 2015, 127(4):1607-1614.

[6] Guo W, Yan K, Zhang H, et al. Double Stage NCO-Based Carrier Tracking Loop in GNSS Receivers for City Environmental Applications [J]. IEEE Communications Letters, 2014, 18(10):1747-1750.

[7] Kaplan E D, Hegarty C. Understanding GPS: principles and applications [M]. Artech House, 2006.

[8] Www.beidou.gov.cn. Beidou satellite navigation system space signal interface control documents. 2012.

[9] Hinedi S, Statman J I, Hinedi S, et al. High-dynamic GPS tracking [J]. NASA Publication, 1989.

[10] Lashley M, Bevly D M. Vector Delay/Frequency Lock Loop Implementation and Analysis [J]. Proceedings of the International Technical Meeting of the Institute of Navigation, 2009:1073-1086.

[11] Zhang Q, Li B. A low-cost GPS/INS integration based on UKF and BP neural network[C]// Fifth International Conference on Intelligent Control and Information Processing. IEEE, 2015.

[12] Babu R, Wang J. Ultra-tight GPS/INS/PL integration: a system concept and performance analysis [J]. Gps Solutions, 2004, 13(1):75-82.

[13] Luo DC,Liu ZG,Zhang S,Liu YF. Analysis on the Influence of the INS Precision in an Ultra-tightly Coupled INS/GPS Integration[C]// Proceedings of OTA 2016. IEEE, 2016.

[14] Luo DC,Liu Y,Liu YF,et al.Present Stauts and Development Trends of Inter-Statellite Link[J]. Telecommunication Enginnering, 2014, 54(7):1016-1024.

IOP Conf. Series: Journal of Physics: Conf. Series **887** (2017) 012056 doi:10.1088/1742-6596/887/1/012056

A new algorithm based on bipartite graph networks for improving aggregate recommendation diversity

Lulu Ma[1] and Jun Zhang[2]

[1]Department of Finance, Shandong Normal University, Jinan, Shandong, 250014, P. R. China.
[2]School of Information Science and Engineering, Shandong Normal University, Jinan, Shandong, 250014, P. R. China.
Email: malulu9311@163.com

Abstract. Most of the traditional recommendation algorithms focus on the accuracy of recommendation results; however, the diversity of recommendation results is also important, which can be used to avoid the long-tail phenomenon. In this paper, a new algorithm for improving aggregate recommendation diversity is proposed. Firstly, a candidate recommendation list based on predictive scores is constructed; and then a bipartite graph network model is constructed. Secondly, item capacity is set to limit the number of recommendations of popular items. Finally, the final recommendation result is generated by combining the recommendation augmenting path. Based on the real world movie rating datasets, experiment results show that the proposed algorithm can effectively guarantee the accuracy of the recommendation results as well as improved the aggregate diversity of the recommendation.

1. Introduction

Personalized recommendation technologies help users to obtain their preferred information from large amounts of data. It has become one of the most effective information filtering technology [1]. With the rapid development, personalized recommendation has been widely used in e-commerce recommendation, film and television recommendation, social recommendation, and so on [2, 3]. In recent years, more and more researchers have realized that the accuracy of recommendation should not be the only standard to evaluate the quality of recommendation. The diversity, novelty, and the coverage of the recommendation result are also important to the evaluation of recommendation systems [4]. The importance of diversity in recommendation results has been highlighted in several studies [5, 6]. Generally, the recommendation diversity includes personal diversity and aggregate diversity. Personal diversity is considered to be the view of a single user, which aims at recommending a series of items with low similarity to each other and satisfying the single user interests. Aggregate diversity is considered to be the view of the entire system, which aims at improving the capability of recommending those unpopular items in the system and weakening the long-tail phenomenon.

A good recommendation system should take into account both the accuracy and diversity of recommendation. However, accuracy and diversity are mutually exclusive. Some scholars adopt a heuristic strategy to improve the recommendation diversity at the expense of a certain degree of accuracy. Adomavicius Get et al. [7] model the recommendation to the maximum bipartite graph matching problem, and improve the coverage of the recommendation items. However, the long-tail phenomenon has not been effectively solved. Adomavicius Get et al. [8] also proposed a re-ranking algorithm, which improves the aggregate diversity by improving the unpopular item ranking. This method has low computational complexity, but it lacks the global awareness. From the global perspective, Huiting Liu et al. [9] proposed a diversity optimization algorithm based on the recommendation expectation. The author improved the aggregate diversity of the recommendation by controlling the recommended expectations of all items.

In this paper, aggregate diversity of recommendation systems is focused and an algorithm based on bipartite graph networks for recommendation aggregate diversity is proposed. Firstly, a threshold is set to filter the very unpopular item for each user. A candidate recommendation list based on predictive scores is constructed, and then the bipartite graph network model is constructed. Secondly, the item capacity is set to limit the number of recommendations of popular items. Finally, the final recommendation result is generated by combing the recommendation augmenting path. Based on the real world movie rating datasets, the proposed method is compared with the main algorithms, which focuses on improving recommendation aggregate diversity. This proposed algorithm has the following advantages: 1) it can optimize the aggregate diversity of most existing recommendation algorithms. 2) By setting the threshold, it can flexibly compromise between diversity and accuracy. Experiment results show that the proposed algorithm can effectively guarantee the accuracy of recommendation results as well as improving the aggregate diversity of the recommendation.

2. An Algorithm For Improving Recommendation Aggregate Diversity (Ad-Improved Algorithm)

The bipartite graph is a special model in graph theory, which is very important in the research field of complex network theory and practical application. An algorithm based on bipartite graph networks for improving aggregate recommendation diversity is proposed.

2.1. Related Definitions

Assume the set of user is defined as, $\mathrm{U} = \{\mathrm{u}_1, \mathrm{u}_2, \dots, \mathrm{u}_\mathrm{n}\}$and the set of item is defined as, $\mathrm{I} = \{\mathrm{i}_1, \mathrm{i}_2, \dots, \mathrm{i}_\mathrm{m}\}$. The user's rating of all items can be represented by a $\mathrm{n} \times \mathrm{m}$ dimensional matrix, R. $\mathrm{R(u, i)}$represents the true score of the user u for item i. The predicted score of the user for the unrated item is represented by another $\mathrm{n} \times \mathrm{m}$ dimensional matrix, $\mathrm{R'}$, where $\mathrm{R'(u, i)}$ represents the predicted score of the user u for item i. The predicted score can be calculated by user-based CF, item-based CF, matrix factorization CF and other existing algorithms. The user recommendation list is defined as Tu.The recommended list length is defined as N. The number of items that can be recommended for the item capacity is defined asIC. At the beginning of the recommendation, the item capacity of all items is the same.

Definition 1: Candidate recommendation list. Set the rating thresholdTr, and then the candidate recommendation list for user u is defined as,$\mathrm{Cu} = \{\mathrm{item} | \mathrm{item} \in \mathrm{I}, \mathrm{u} \in \mathrm{U} \text{ and } \mathrm{R'(u, item)} > Tr\}$. The items in the user candidate list can be regarded as items that the user most prefers.

Definition 2: Recommendation bipartite graphs.$\mathrm{G} =< U, I, E >$, U is the set of user nodes, I is the set of item nodes, and E is the set of relations between user u and itemi. The connection between user u and the edge of an item in Cu is called matching edge, and the edge of the connection user u and the edge of the item in Tu is called the unmatching edge.

Definition 3: Recommendation augment path. The recommendation augment path is a set of edges, which start from a user node and follow by a unmatching edge. The item capacity of final item node is not 0. In Figure 1, assume that the solid line represents the matching edge and the dotted line represents the unmatching edge. $\mathrm{u3} \rightarrow \mathrm{i1}$,$\mathrm{u1} \rightarrow \mathrm{i2}$ are un-matching edges, and $\mathrm{i1} \rightarrow \mathrm{u1}$is the matching edge. Assuming that the item capacity of item i2 is not 0, then $\mathrm{u3} \rightarrow \mathrm{i1} \rightarrow \mathrm{u1} \rightarrow \mathrm{i2}$ is a recommendation augment path.

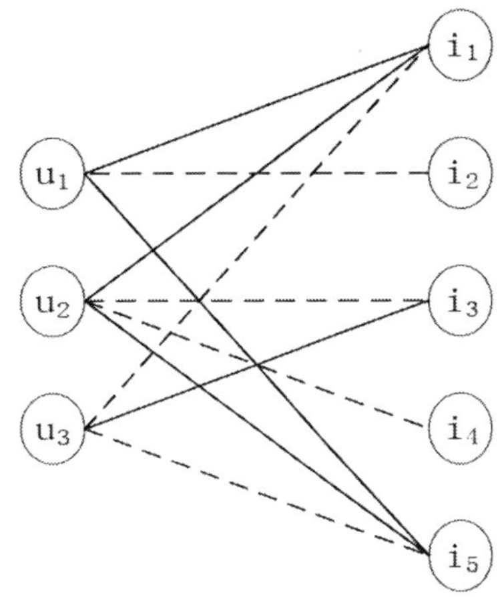

Figure 1. Recommendation system based on bipartite graph.

2.2. Main Idea of the Algorithm

The recommendation process is usually divided into two phases [10, 11]. One is to predict the score of the user's unrated item, and then is to generate the Top-N recommendation list. Therefore, there are often two ways to improve the recommendation diversity. One is to increase the diversity by increasing intervention in predicting scoring. The other is to optimize the recommendation generation phase, and try to find an optimal recommendation combination.

This paper adopts the second way. When the user recommendation list is not full, however, the items in the candidate recommendation list are up to the maximized recommendation times. If the item with the highest$R'(u,i)$ in Cu is added to the userTu, it is helpful to improve the recommendation accuracy rate and not helpful to the recommendation aggregate diversity. If the system recommends an unpopular item, whose IC is large to the user, will conducive to the recommendation aggregate diversity but not conducive to the recommendation accuracy. The proposed algorithm searches the recommendation augmentation path from the user node and replaces the matching and un-matching edges in the augment path.

In Figure 1, for example, suppose two items are the recommendation to each user, and the item capacity IC is 2. At this point, user u3 recommendation list is not full, and the candidate recommendation list item i1 and i5 item IC value are 0. Our algorithm will find augment pathu3 $\rightarrow$ i1 $\rightarrow$ u1 $\rightarrow$ i2. I1 will be recommended to u3, i2 will be recommended to u1. It achieves the u3 user's recommendation. It also increases the recommendation of unpopular item i2. However, since$R'(u1,i2) > Tr$, user u1 still have a high forecast score for item i2. Therefore, while increasing the aggregate diversity, our algorithm still has a good accuracy. In addition, the algorithm may be combined with any other recommendation algorithm with a score prediction step to improve its aggregate diversity recommendation.

The AD-Improved algorithm can be divided into the following processes:

1) Create a predictive scoring matrix; 2) Secondly, build a recommendation bipartite graph;3) Thirdly, generate recommendation results.

2.3. Create A Predictive Scoring Matrix

User-based CF algorithm, item-based CF algorithm, Matrix Factorization CF algorithm and other existing recommendation algorithm need predict the user's score on the unrated items. These algorithms select the items with highest score to build TOP-N recommendation list. Our algorithm uses the existing recommendation algorithm to generate the prediction score matrixR'.

2.4. Build Recommendation Bipartite Graph

The process of constructing bipartite graph is showed as follows. First, the score threshold Tr is set, and then the user candidate recommendation list Cu is obtained based on the predicted score matrixR'.

Second, connect the user u with its items in the un-matching edges. The recommendation bipartite graph is the initial state of the system.

2.5. Generate Recommendation Results

The step of generating recommendations to users includes three cases,

1) In this case, IC of a user is not 0.In order to ensure the recommendation accuracy, the item i with the highest predicted score is selected. Item i is added to Tu and it is removed from Cu. At the same time, since $item_i$ has made a recommendation, it is necessary to update the item capacity of item i, $IC_i - 1$. Finally, the edges between user u and item i in the bipartite graph are transformed from un-matching edges to matching edges.

2) When all the ICs in user Cu are 0, at least one recommendation augment path can be obtained from the starting point of the user node by using the breadth-first search strategy. It is recommended that the addition channel start at the user node and terminate at the item node. It is recommended to increase the $p = v_1 \rightarrow v_2 \rightarrow \cdots \rightarrow v_{2k-1} \rightarrow v_{2k}$. v_{2k-1} for the user node, v_{2k} for the item node. Therefore $v_{2k-1} \rightarrow v_{2k}$ is the non-matching edge, $v_{2k} \rightarrow v_{2k+1}$ is the matching edge. The proposed multiplication path$v_{2k-1} \rightarrow v_{2k}$ is transformed into a matching edge, and $v_{2k} \rightarrow v_{2k+1}$ is transformed into an unmatching edge. And the corresponding user's Tu and Cu are modified according to the edge, the item represented by v_{2k} is added to the user Tu of v_{2k+1}, and the item represented by v_{2k} is removed from the user Cu of v_{2k+1}. Finally, the item capacity of the item represented by the recommendation v2n is decreased by one operation, $IC_i - 1$.

3) The item capacity in the user candidate recommendation list is 0, and there is no recommendation addition channel from the user node. At this point, it cannot improve the aggregate diversity of the system, in order to ensure the accuracy of the algorithm, the candidate recommendation list of the highest score items added to the user recommendation list. The item capacity value of the item is not changed.

3. Experiments

In this paper, we use the Movielens1M data set as test data (http://www.grouplens.org), which is widely used in the recommendation field. The proposed algorithm, noted as AD Improved algorithm, is compared with the Expectation algorithm [7]. The pre-processed data set contains 2830 users with a score of 775,176 for 1919 items. The sparseness of the pre-processed data is 14.27% and uses the most popular collaborative filtering techniques (user-based CF, Item-based CF, matrix factorization CF) to predict the user's rating on the unrated items to obtain a list of candidate recommendations. User-based CF, item-based CF, select 50 neighbours with the highest similarity as nearest neighbours. Matrix factorization CF, the normalization constant was 0.043, the latent factor was 15, and the number of iterations was 30 times using the 5-fold cross-derivative method. The results showed that 70% of the items were rated less than 50 times. For this data set, we set the item capacity IC as 50.

3.1. Evaluation Indicators

In this paper, the prediction accuracy is used as evaluation indicator. The algorithm performs different recommendation algorithms and cross validation tests on the Netflix data set and the Movie lens dataset. The prediction accuracy and averaging the correlation of the predicted scores are 0.974 and 0.966. So the average forecast score can be used to estimate the accuracy of recommendation, the recommendation accuracy rate is:

$$\text{prediction} = \frac{\sum_{u \in U} \sum_{i \in Tu} r^*_{u,i}}{\sum_{u \in U} |Tu|} \tag{1}$$

In order to measure the recommendation aggregate diversity, we use coverage, Gini-Diversity and Entropy, which are showed as follows:

$$\text{coverage} = \left| \bigcup_{u \in U} Tu \right| \tag{2}$$

$$\text{Gini-Diversity} = -\sum_{i=1}^{N}(2i-1-n)\frac{rec(\mathrm{i})}{|U|*N} \tag{3}$$

$$\text{Entropy} = -\sum_{i=1}^{N}(\frac{rec(\mathrm{i})}{|U|*N})\ \ln(\frac{rec(i)}{|U|*N}) \tag{4}$$

The larger the coverage value, the more number of items covered by the system, and the aggregate diversity of the system is better. Gini-Diversity and Entropy measure the aggregate diversity from the quantitative view. The smaller the Gini-Diversity value, the better is the diversity. The greater the Entropy value, the better is the diversity. Where rec (i) stands for the number of times that item i is recommendation to different users.

3.2. Results and Analysis

Figure 2 shows different scoring threshold, Tr=(3.5,3.6, ...,5), Our algorithm(noted as AD-Improved) and the Expectation algorithm(Expection)[10] based on user-based CF, item-based CF, matrix factorization CF are compared. Top-5 Recommended results are used for comparison under accuracy and diversity. As can be seen from Figure 2a, the algorithm based on Item-based CF can flexibly control the compromise between coverage and accuracy. When the threshold Tr is 4.5, the accuracy is reduced by no more than 5%. However, compared with the Expectation algorithm, the algorithm has a small advantage in terms of accuracy and coverage. As can be seen from Figure 2b, compared with the Expectation algorithm, the Gini coefficient of the algorithm based on the item-based CF is relatively small and the diversity has a slight advantage in the case of the same threshold. According to the accuracy, our algorithm performs well and the recommended accuracy is about 3% when the threshold value is Tr ∈ [4, 4.6]. It can be seen from Fig. 2c that the algorithm based on matrix factorization CF is superior to Expectation algorithm in accuracy and diversity. In order to better compare with the Expectation algorithm, while showing the relationship between accuracy and diversity. Table 1 shows the performance of the algorithm diversity through the adjustment of the accuracy rate (achieved by the scoring threshold adjustment) during the Top-5 recommendation.

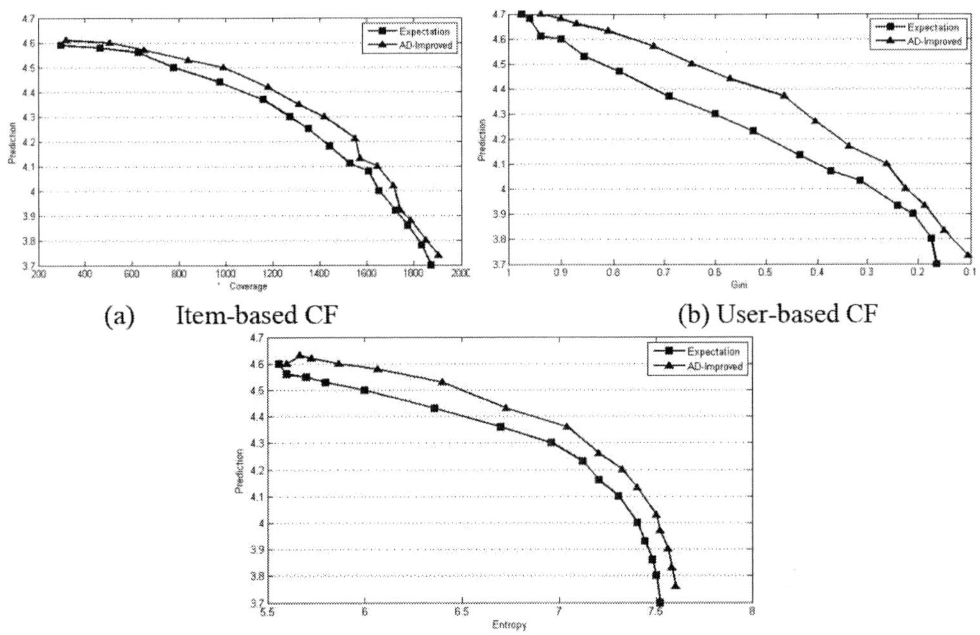

(a) Item-based CF (b) User-based CF

(c) Matrix factorization CF

Figure 2. Comparison of accuracy and diversity between AD-Improved algorithm and Expectation algorithm based on different existing recommendation algorithm

Table 1. Increased diversity under given accuracy conditions

	User-based CF	Item-based CF	MF
Standard	264 Accuracy =4.47	293 Accuracy =4.60	516 Accuracy =4.63
Accuracy level:Standard-0.01			
Expectation	480.8(82.1%)	520.2(77.5%)	657.9(27.5%)
AD-Improved	513.2(94.4%)	577.1(97.0%)	667.7(29.4%)
Accuracy level:Standard-0.05			
Expectation	734.2(178.1%)	752.6(156.9%)	816.8(58.3%)
AD-Improved	773.0(192.8%)	823.9(181.2%)	890.1(72.5%)
Accuracy level:Standard-0.1			
Expectation	890.5(237.3%)	894.2(205.1%)	1050.0(103.5%)
AD-Improved	998.9(278.4%)	1049.5(258.2%)	1150.2(122.9%)

In the table 1, the second line of the traditional user-based CF algorithm, item-based CF algorithm, matrix factorization CF algorithm recommended accuracy and diversity of the situation. From Table 1, it showed that, in the case of loss rate of 0.01, 0.05, 0.1, the algorithm of this paper has a specific improvement in diversity. Even if the accuracy rate is only 0.01, the diversity of the algorithm based on user-based CF and the algorithm based on item-based CF is still increased by more than 90%. In addition, compared with the Expectation algorithm, this algorithm performs better in terms of diversity improvement.

4. Conclusions

Recommendation diversity has become an important factor of the performance of the recommendation system. A recommendation diversity enhancement algorithm based on bipartite graph network is proposed in this paper. Compared with the existing algorithms, the results show that the proposed algorithm can effectively improve the aggregate diversity of the recommendation system and maintain high recommendation accuracy.

5. Acknowledgments

The project was supported by Shandong Province Higher Educational Science and Technology Program (No. J16LN10),

6. References

[1] C. Ziegler, S. M. McNee, J. A. Konstan, and G. Lausen. Improving recommendation lists through topic diversification. In WWW '05, 2005, pages 22-32.
[2] Resnick P, Varian H R. Recommender systems. Communications of the ACM, 1997, 40(3):56-58.
[3] Kim J E, Kim H G. Music recommendation method with respect to message service: US, US8410347. 2013.
[4] Hao Wu,Xiaohui Cui,Jun He,Bo Li, Yijian Pei, On improving aggregate recommendation diversity and novelty in folksonomy-based social systems, Personal and Ubiquitous Computing, 2014,18(8):1855-1869.
[5] M. Zhang and N. Hurley. Avoiding monotony: Improving the diversity of recommendation lists. In ACM RecSys '08, 2008, pages 123-130.
[6] Meyer, F., Fessant, F., Cl_erot, F. Toward a new protocol to evaluate recommender systems. ACM RecSys 2012 Workshop on Recommendation Utility Evaluation, pages: 9-14.
[7] Adomavicius G, Kwon Y O. Maximizing aggregate recommendation diversity: A graph-theoretic approach. The International Workshop on Novelty and Diversity in Recommender Systems. 2011: 3-10.
[8] Adomavicius G, Kwon Y O. Improving aggregate recommendation diversity using ranking-based techniques. IEEE Transactions on Knowledge & Data Engineering, 2012, 24(99):1-15.
[9] Huiting, liu. Kecheng, Yue. Expectation-based top-N recommendation approach for improving recommendations diversity. Science of Computer. 2014, 41(7):270-274.

[10] Katja Niemann, Martin Wolpers. A new collaborative filtering approach for increasing the aggregate diversity of recommender systems, Proceedings of the 19th ACM SIGKDD international conference on Knowledge discovery and data mining, 2013, pages 955-963.
[11] Kumar, Arun R; Buchibabu, B; Rao, T Venkat Narayana. Ranking based techniques using optimised recommendation diversity. International Journal of Science, Engineering and Computer Technology; 2016, pages: 387-391.

An ontology based trust verification of software license agreement

Wenhuan Lu[1, a], Xiaoqing Li[1, b], Zengqin Gan[1, c] and Jianguo Wei[1, d]

[1]School of Computer Software, Tianjin University, Tianjin 300350, China
Email: [a]wenhuan@tju.edu.cn, [b]xiaoqingli@tju.edu.cn, [c]zqg@tju.edu.cn, [d]jianguo@tju.edu.cn

Abstract. When we install software or download software, there will show up so big mass document to state the rights and obligations, for which lots of person are not patient to read it or understand it. That would may make users feel distrust for the software. In this paper, we propose an ontology based verification for Software License Agreement. First of all, this work proposed an ontology model for domain of Software License Agreement. The domain ontology is constructed by proposed methodology according to copyright laws and 30 software license agreements. The License Ontology can act as a part of generalized copyright law knowledge model, and also can work as visualization of software licenses. Based on this proposed ontology, a software license oriented text summarization approach is proposed which performances showing that it can improve the accuracy of software licenses summarizing. Based on the summarization, the underline purpose of the software license can be explicitly explored for trust verification.

1. Introduction

Once people install or download software, there is definitely showed up a Software License Agreement, which textually varies from license to license. Generally the agreement would be a long document with millions of words that is hard to be understand, even make users don't take time to read it before clicking agree button to install or download it. This situation makes users feel unsafe and uncomfortable when installing unfamiliar software.

Doesn't matter the different representations, forms and statements of the agreements, the intention of most software license agreements would be similar even be partially same. The semantic behind the literature of license agreements are same for lots of cases. As ontology is a leading edge technology for knowledge representation, we use conceptual modeling approach to build generalized software license agreement ontology, so as to figure out the common knowledge behind each license arguments [1, 2]. The ontology based knowledge model can be a shared representation for license agreement domain.

Software license agreement normally designed according to Copy-right Law [3]. The construction of software license agreement domain ontology is based on a large number of law articles and different software license agreements collected from internet. These sample data sets were taken from the official website of some popular software in computer science. It is important to analyze and summarize the concepts including their relations and conditions in this domain. Redundancy of concepts and properties can result the failure of the ontology construction.

The modeling of software license agreements is not a trivial issue and hardly to solve, especially concerning the diverse software and terminologies. Our research objective is to propose an appropriate methodology and a uniform conceptual model to provide a descriptive and strong expansibility semantic knowledge representation for the domain of software license agreement. Besides, the ontology model can also provide the visualization of the semantic of licenses in order to facilitating knowledge representation

A License Ontology plays an important role in knowledge acquisition of software license agreement. In this paper, we combine our License Ontology with existing text summarization technology [4]. The goal of our work is to take an information source, extract content from it and present the most important content to the user in a condensed form that is manner sensitive to the user's or applications' needs, and improve the algorithm of Classifier4J summarizer based on the features of the domain of software license agreement [5, 6]. With the help of License Ontology, we want to provide every software end users a utility recommendation system, as shown in figure 1, to acquire the most important part of software license agreement as soon as possible.

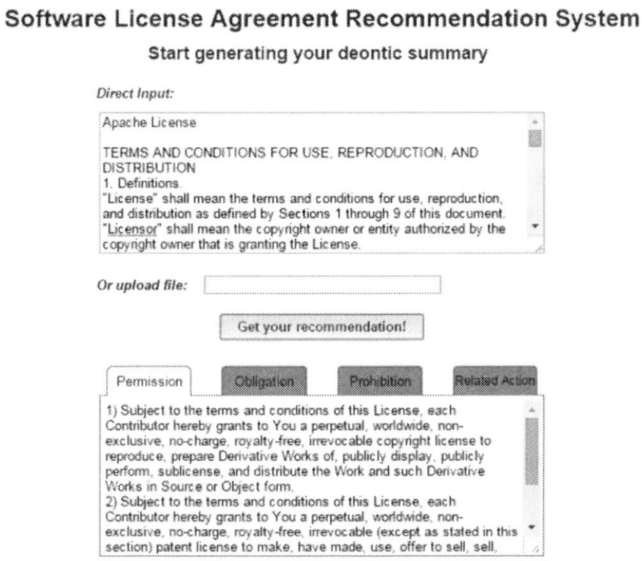

Figure 1. Prototype of Software License Agreement Recommendation System.

The interface is just a prototype so far and it is designed by Axure RP. All you need to do is upload a software license agreement and click the button and you will acquire the important part of the software license agreement as soon as possible.

The rest of this paper is organized as follows. First, Section 2 explores related work of the ontology construction. After that, methodology and model of domain ontology have been elaborated in Section 3. Then Section 0 presents our ontology for license case. Section 0 describes the experiment of software license agreement ontology. It combines ontology theory and text summarization and draws a conclusion by the result. Finally, the paper closes with concluding remarks and future perspectives.

2. Related Work

The methodology and model of domain ontology determine the describing and reasoning ability of ontology towards domain knowledge. A better model is the assurance of ontology construction.

As for the copyright domain which software license agreement belongs to, important efforts in this field have appeared in the last several decades overseas? However, related research is still at seed-time interiorly.

McCarthy introduced ontology into artificial intelligence for the first time in 1980. He proposed that we should construct an ontology to describe a world based on logical concepts [7]. The most frequently cited definition is that given by Gruber in 1993, that is, an ontology is defined as "an explicit specification of a conceptualization". In 1998, Studer et al. modified it stating that: "Ontology is a formal, explicit specification of a shared conceptualization. [8]" Valente's FOLaw distinguishes the various types of knowledge in legal reasoning, including normative knowledge, meta-legal knowledge, world knowledge, responsibility knowledge, reactive knowledge, and creative knowledge[9]. Breuker's LRI-Core, a core ontology for law, from the thought of a common sense foundational ontology, including five major categories in the top layer of LRI-Core: physical, mental, abstract, role, and occurrence [1[10]]. On the basis of LRI-Core, Rinke Hoekstra's LKIF core legal ontology, which means Legal Knowledge Interchange Format, is a library of ontologies relevant to the legal domain [11, 12]. It consists of 15 modules, each of which describes a set of closely related concepts from both legal and commonsense domains.

These works try to formalize and systematize legal theory or knowledge, and provide a uniform and standard cognition for legal domain understanding and concepts spreading. However, there is no ontology specially oriented to software license agreement documents, for which general legal article could not work properly. One reason is that the statements in software license differ with the legal articles. Secondly, the legal articles are more formal structured than software license agreements in normal. The model need refer to both copyright ontology and software license agreement domain knowledge.

3. An Ontology Model of Software License Agreement

The software license agreement domain is literature rich domain, and the same concept could have many different statements. We conceptualize this domain in three phases that allow facing this process in an incremental way. Starting from the Role Model, we combine entities in the domain of software license agreement with actions, and build a core framework of the domain ontology. Secondly, there is the model for the permissions, prohibitions, and obligations part, the Deontic Model. A model for the available actions, the Action Model, is built on top of the two previous models, and then we construct an ontology.

3.1. Role Model

Before the construction of ontology, we divided Software License Agreement Role into Legal Entity Role and Action Role. A role is not a character role in the usual sense, but rather an entity that is played by another entity in a context.

Each role has its context, shown in figure 2 by "context", we mean something as a whole, including a relation in which the former "entity" is defined. The context of a Legal Entity Role is software license agreement, which means all of the concepts involved in Legal Entity Role are from software license agreements and related legal regulations. By "Legal Entity Role", we mean role holders who are played by contributors or users. They have actions which are related actions in the legal field and impact on the productions. The context of "Action Role" is deontic model which will be introduced next. In other words, the actions involved are mainly from deontic model, and they are played by Legal Entity Role. That is the integrated framework of software license agreement domain ontology.

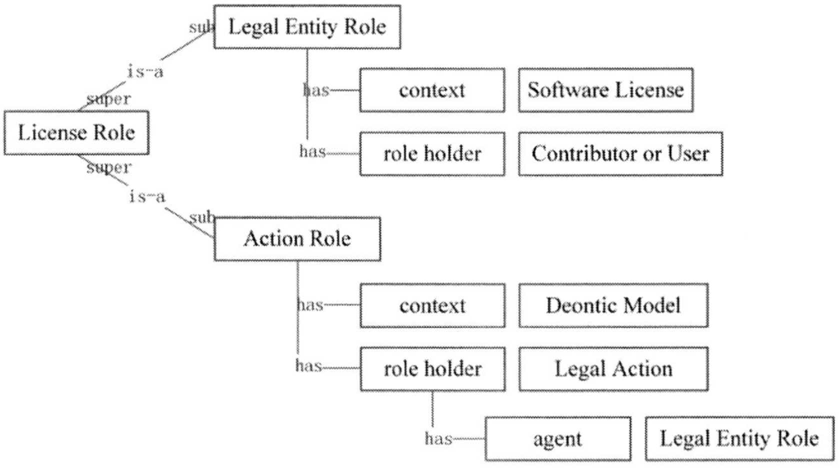

Figure 2. Role Model of software license agreement.

3.2. Deontic Model

After role modeling, we must build Deontic Model to integrate Role Model with software license agreement. It is important to note that legal regulations in this paper are from the World Intellectual Property Organization.

Legal regulations are similar in different software license agreement which can be divided into three parts. A permission allows an agent to perform the associated actions. A prohibition prevents an agent to perform the associated actions. An agent must perform an obligation which is the responsibility of the agent. The three parts contain most of the action in the domain of software license agreement.

A deontic act is performed by an agent under a certain condition and leads to a certain consequence. Sometimes the agent is obliged to do the deontic acts. We describe a deontic act as a tuple deontic (agent, action, condition, obligedTo, consequence).

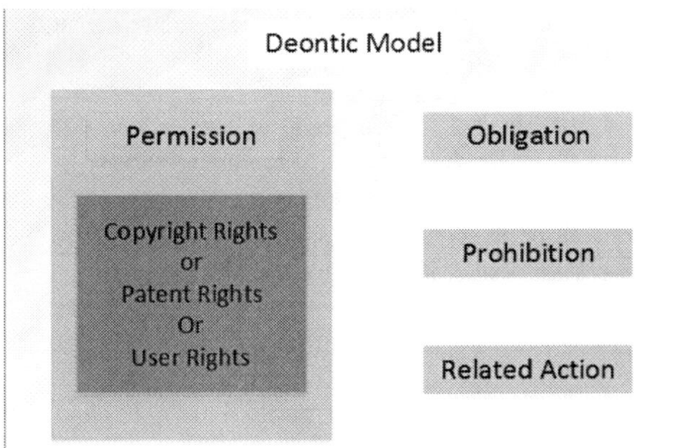

Figure 3. Deontic Model of software license agreement.

Figure 3 shows that deontic model contains permission, obligation, prohibition, and other related actions. Whether permission, prohibition, or obligation are describing the actions of the "agent". "Action" is the actions included in deontic model. "Conditions" signifies the precondition that an agent is allowed to execute the actions in usual. "ObligedTo" is the one who forces the agent to do the action. When you unfinished obligation or go against prohibition, "consequence" happen.

The most important part in deontic model is permissions of legal entity in the domain of software license agreement. Among all these grant licenses, related regulations of copyright rights and patent rights are from contributors, however, user rights are not from contributors, but from the law. In addition, related actions will be mentioned later.

3.3. Action Model

The last model, the Action Model, corresponds to the primitive actions that can be performed by the concepts defined in the Role Model and Deontic Model. These concepts are associated with a series of object properties, so that the ontology can illustrate what regulations expressing. Action model associate verbs primarily which is the main part of the model and the core of the software license agreement domain ontology.

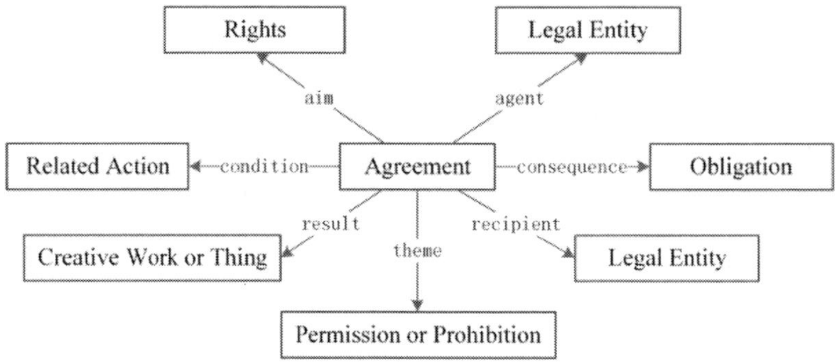

Figure 4. Action Model of software license agreement.

As it is shown in figure 4, there are main object properties related to actions:

- Agent: to execute, its domain is a legal entity.
- Aim: purpose of the action, its domain is rights.
- Consequence: result of the action, generally shows an action will certainly happen because of the other, its domain is obligation.
- Recipient: to accept, its domain is a legal entity.
- Result: product of the action, an inevitable product after legal entity is executed by the action.
- Condition: inverse of consequence, its domain is related action.
- Theme: object of the action, its domain is permission or prohibition.

The properties in our ontology are far more than these object properties shown in figure 4. Object properties mentioned above are basic properties which normalize the logical relationship of domain ontology, and there are some other properties, see Table 1, which play an important role during the construction of domain ontology using protégé 4.3.

Table 1. Object Properties of Action Model.

property	object	definition
agent	Legal Entity	voluntary initiator
aim	Rights	voluntary goal
consequence	Obligation	involuntary goal
recipient	Legal Entity	animate goal
result	Creative Work or Thing	inanimate goal
condition	Related Action	necessary circumstance
theme	Permission or Prohibition	essential participant
start	Date	dateminant temporal source
duration	Time	atemporal process resource
medium	Thing	physical resource for transmitting information
location	Place	spatial essential partipant

4. Instantiation of License Ontology

Ontology is the philosophical study of the nature of being, becoming, existence, or reality, as well as the basic categories of being and their relations. In 1993, Gruber originally defined the notion of ontology as an "explicit specification of conceptualization". In 1998, Studer et al. modified it stating that: "Ontology is a formal, explicit specification of a shared conceptualization." In recent years, after introducing into computer science, ontologies define domain concepts and the relationships between them, and thus providing a domain language is meaningful to both humans and machines. Ontologies have been shown to have benefits in a number of areas:

- knowledge sharing;
- knowledge reuse;

- verification and validation;
- domain theory development;
- Knowledge acquisition.

The previous pool of primitive legal entity and action roles, together with deontic properties that relate each action to the relevant roles, allows modeling the top-level of the software license agreement domain ontology, however, specific licenses should be associated with the top model to build the whole domain ontology. To reach the aforementioned research objective, we need to design and develop an ontology that conceptualizes different software license agreements from a uniform perspective. We use Protégé for ontology development of concepts and relationships. Protégé is a free, open source ontology editor developed at Stanford University. It provides a graphic user interface to define ontologies and deductive classifiers to validate the consistency of models and to infer new information based on the analysis of ontology. Recently, protégé has become the leading ontological engineering tool.

For instance, figure 5 shows how a software development product involves in the three top models between developing and using.

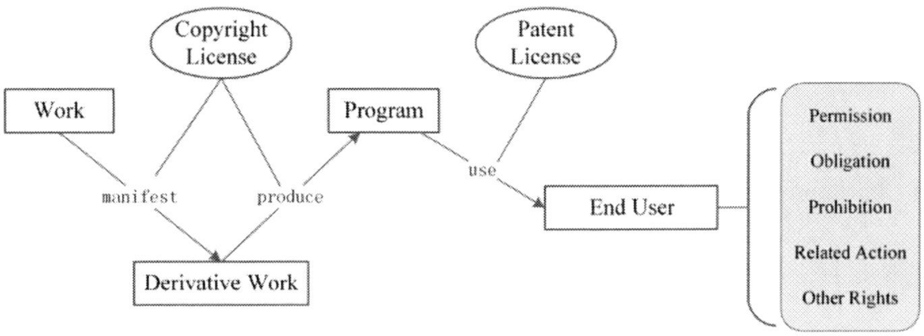

Figure 5. Software License Agreement Value Chain.

A work is realized as a manifestation after manufacture or other means that is used by end users under the software license agreement. In fact, every process can be regarded as an action model.

For example, Eclipse is open-source integrated development environment based on Java. By itself, it is simply a framework and a set of services for building a development environment from plug-in components. The version of Eclipse has been updated with the development and increasing demand, and after every update, there is a new derivative work. The compilation tool we use can be regarded as a product, and while the end user is using the product, it involves permission, prohibition, obligation, and other related actions stated in legal regulation as we mentioned before.

In the software license agreement field, there are numerous open-source software and business software, which have the very big different in regulation expression and the use of legal terms. Constructing domain ontology can formalize these software license agreements, and give an accurate description. Some classical legal ontologies that currently exist are wildly in legal field and clearly describe different object and the relationship between them. However, as an important branch of copyright, the domain of software license agreement needs a more applicable ontology. The ontology for software license agreement proposed in this paper is aimed at solving this problem.

In order to understand domain ontology better, event patterns should be introduced to state what is permitted, prohibited or obliged by a license and be naturally captured by the ontology terms described. The proposed actions and roles are used to model event patterns in the software license agreement.

4.1. Instantiation of Permission Pattern

For instance, figure 6 shows the sixth paragraph of article 10 of the Copyright Law of the People's Republic of China: The right of distribution, that is, the right to provide the public with original copies or reproduced copies of works by means of selling or donating. Licensor has the right to distribute software permitted by WIPO and charge for the copies from end user.

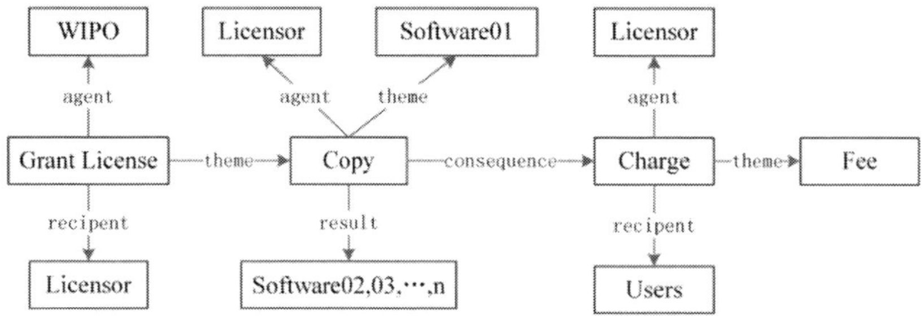

Figure 6. Instantiation of a Permission.

4.2. *Instantiation of Obligation Pattern*

Obligations are captured as event patterns that must be satisfied after the event pattern that triggers the obligation is exercised. For instance, figure 7 shows the legal regulation of Eclipse Public License v1.0: Under this section, the Commercial Contributor would have to defend claims against the other Contributors related to those performance claims and warranties, and if a court requires any other Contributor to pay damages as a result, the Commercial Contributor must pay those damages. That is, contributor has to compensate if there is any damage during use of the product, and meanwhile, called Commercial Contributor.

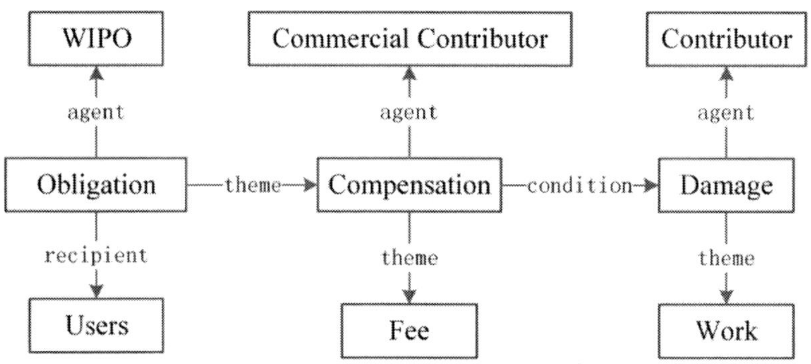

Figure 7. Instantiation of an Obligation.

4.3. *Instantiation of Prohibition Pattern*

As shown in figure 8, another legal regulation of Eclipse Public License v1.0, Contributors may not remove or alter any copyright notices contained within the program, means that you are not allowed to remove or alter any copyright notices if you are not a contributor.

Software license agreement domain ontology is based on the framework of three models above. Legal regulations in the software license agreement are represented by related actions and object properties, and merged into the ontology we built. However, the purpose of instantiation is to combine actions in regulations with deontic performance, which means permission, prohibition, and obligation, and present by the way of ontology that computers and humans can both understand.

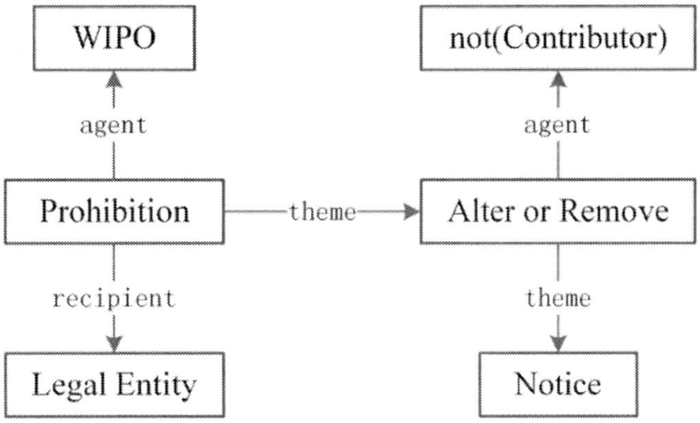

Figure 8. Instantiation of a Prohibition.

5. Evaluation of License Ontology

The previous two parts of this article series explained the modeling and instantiation of software license agreement domain ontology. On this basis, we improve the existing text summarizing technology combined with License Ontology to propose a more accurate summarizing method applying to a software license agreement.

The concepts appearing in License Ontology were weighted while calculating the frequency of each word. Because these words are more important in the domain of software license agreement, and we want to extract the summary that lets end users know which clause is permitted, obliged or prohibited. During reduplicative experiments, we adjust weight and modified our ontology continuously according to the experimental result. Consequently, the advanced text summarization tool is performing better than the original one.

To prove the accuracy and applicability of the ontology we built, classifier4J has been improved associated with software license agreement domain ontology.

As shown in figure 9, in the original algorithm, we put the first sentence where the most frequent word appears into summary. In a sense, that is right, however, after a series of experiments, we take a sentence as a unit and add the frequencies of 100 most frequent words to the sentences they belong to, and then we can get "the most frequent sentences".

```
while (it.hasNext()) {
    String word = (String) it.next();
    sflag=isSemanticWord(word);
    int workingSentencesweight = 0;
    for (int i = 0; i < workingSentences.length; i++) {
        if (workingSentences[i].indexOf(word) >= 0) {
                workingSentencesweight = workingSentencesweight+sflag;
                resultSentences.put(actualSentences[i],
                        String.valueOf(workingSentencesweight));
        }
    }
}
//order resultSentences
resultSentences=wsortMapByValue(resultSentences);
```

Figure 9. The core code of getting "the most frequent sentences".

Open-source summarizers apply to all domains, that is to say, it doesn't apply to a given domain perfectly. As for a software license agreement, we cannot get an accurate summary by traditional summarizers, for the reason that license agreement differs from other text. Therefore, in the field of software license agreement, it is important to know what we should do and what not.

The improved algorithm assigns weight to the concepts in License Ontology. As shown in figure 10, when the word appears in the text once, we add N times to its frequency. The weight is formulated based on License Ontology and it is improved continuously with the experimental result during reduplicative experiment. After that, License Ontology becomes the optimum ontology model of the domain of software license agreement. Consequently, we can find as showed below that the summary is better.

```
for (int i = 0; i < uniqueWords.length; i++) {
    if (stopWordsProvider == null) {
    // no stop word provider, so add all words
        result.put(uniqueWords[i], new
         Integer(Utilities.countWords(uniqueWords[i], words)));
    } else if (isWord(uniqueWords[i])
        && !stopWordsProvider.isStopWord(uniqueWords[i])
        &&(isSemanticWord(uniqueWords[i])==0)) {
        // add only words that are not stop words and not semanticWord
        result.put(uniqueWords[i], new
         Integer(Utilities.countWords(uniqueWords[i], words)));
    }else if (isWord(uniqueWords[i])
        && ((flag=isSemanticWord(uniqueWords[i]))!=0)){
        //semanticWord
       result.put(uniqueWords[i], new
         Integer(Utilities.countWords(uniqueWords[i], words))*flag);
    }
}
```

Figure 10. The core code of associating.

Sample dataset was taken from official website of common software in computer science that contained the software license agreements. The quality was measured by calculating the cosine similarity of the standard summary, which refers to several primary online summarizers based on Open Text Summarizer, MEAD, etc., with the summary produced by our program and subjective assessment, so we have quantification result to compare to the standard one. More formally, the cosine similarity between two summaries is given as follows:

$$cosine\ similarity(A,B) = \frac{AB}{|A||B|} = \frac{\sum_{i=1}^{n} A_i \times B_i}{\sqrt{\sum_{i=1}^{n}(A_i)^2} \times \sqrt{\sum_{i=1}^{n}(B_i)^2}} \quad (1)$$

Where A is sentence weight vector of the standard summary and B is of the summary produced by Classifier4J or advanced Classifier4J. We assign different weight to different sentences according to their importance. Our evaluation of license ontology is based on this method comparing differences of two sentences.

Assumed that there are 100 sentences in document and we want 10 of them to generate a summary. Then A and B are 100-dimensional vectors. If the n-th sentence appear in the summary, the n-th vector is not 0. We assign weighting factor 3 to the two principle vectors, which means the two most important sentences. Then we assign weighting factors 2 and 1 to the other 8 vectors according to their importance of the sentence. After that, the resulting cosine similarity ranges from 0 meaning totally different, to 1 meaning exactly the same and it is more precision avoiding that result is how many sentences are the same.

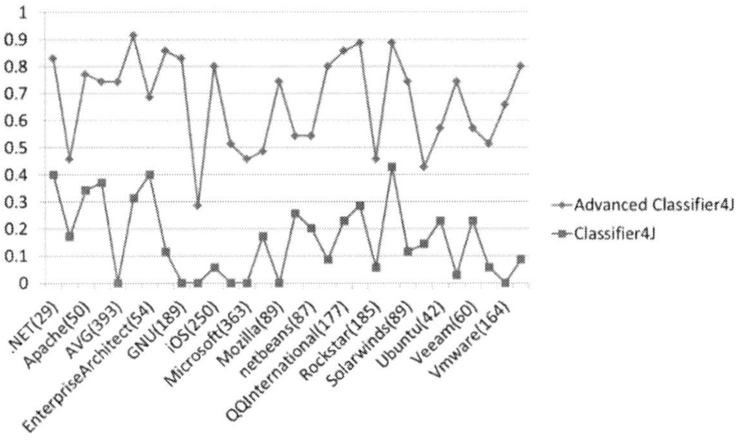

Figure 11. Comparison of Classifier4J and Method Combining with Domain Ontology.

30 documents which are presented by abscissa were tested and only 15 names of them are on the figure 11. The cosine similarity, which is presented by the vertical axis, between the summaries produced by Classifier4J and our program is given in Figure 11. There are altogether 3986 sentences in these documents. From the experimental data we can find that it enhances a lot after combining with domain ontology. Before the improvement of the algorithm, the summaries of Classifier4J are very different from the standard summaries. However, after combining with License Ontology, the average cosine similarity increases by 51 and is close to 70 percent. We can find the accuracy and applicability of the ontology we build. Unfortunately, the length of the license agreement really affects the accuracy.

However, it's worth mentioning that the main difficulty in evaluation comes from the impossibility of building a fair gold-standard. Furthermore, it is also very hard to determine what a correct summary is, because different users would choose different sentences as important information. Even a particular person may choose different sentences at different times. Hence, the quantitative evaluation might not be the only way to evaluate summaries. We also refer to subject assessment from researcher during the evaluating and analyzing of experimental results.

6. Conclusions and Future Work

The Software License Agreement Domain Ontology is an ontology based on the software license agreement framework that is capable of modeling complex value chains and legal regulations. During the construction of the domain ontology, this paper gives three top models associate with software license agreements to enrich domain ontology. It is pleased to see that the computer can understand what legal regulations say exactly.

Our approach to construct ontology is modular in that distinct sub-components of the ontology can be independently developed as we outlined in section 3. First of all, it provides a complete role model that takes into account the different roles of person and action. Second, there is the deontic model that captures concepts from the underlying legal regulations and gives shape to the third component, the action model. The latter includes the actions governed by the software license agreement that can be performed by the concepts defined in the role model and deontic model, which are associated with a series of object properties so that the ontology can illustrate what regulations express. However, for certain licenses, event patterns are essential to state what is permitted, prohibited or obliged by a license. The proposed actions and roles are used to model event patterns in the software license agreement. The purpose of instantiation is to combine actions in regulations with deontic performance and presented by the way of ontology.

Moreover, combined with text summarization, a more accurate summarizing method is creating applying to software license agreement. To prove the accuracy and applicability of the ontology we build, classifier4J has been improved associated with software license agreement domain ontology.

Protégé is used as our ontology editor and we build our domain ontology using Web Ontology Language. We use data from "Computer Software Protection Ordinance of the People's Republic of China", "Copyright Law of the People's Republic of China", "Patent Law of the People's Republic of China", and software license agreements like Eclipse, Apache, Microsoft, etc.

In a sense, the construction of domain ontology is endless. This paper aims to propose an informative, descriptive and scalability software license agreement ontology. The future work concentrates now on extending the domain ontology and offering a recommendation system that is easy to operate as shown before. This recommendation system will help end users understand licenses with less trouble, which is a sincere way to serve the users.

7. References

[1] Web Ontology Language (OWL) W3C recommendation, www.w3.org/2001/sw/wiki/OWL, last accessed 2017/3/11

[2] Chandrasekaran B, Josephson J R and Benjamins V R 1999 what are ontologies, and why do we need them? IEEE Educational Activities Department

[3] Lu W and Ikeda M 2005 An Intention-oriented model of copyright law for e-learning: international semantic mapping of copyright laws based on a copyright ontology. In: Conference on Towards Sustainable and Scalable Educational Innovations Informed by the Learning Sciences: Sharing Good Practices of Research, pp.757-760

[4] Dalal V and Malik L G 2013 A survey of extractive and abstractive text summarization techniques. In: Sixth International Conference on Emerging Trends in Engineering and Technology (ICETET). IEEE, pp:109-110

[5] Automatic summarization 2017 https://en.wikipedia.org/wiki/Automatic_summarization

[6] Jones, K. S 2007 automatic summarising: the state of the art. *Information Processing & Management An International Journal* **43(6)**, 1449-1481

[7] Mccarty L T 1989 a language for legal discourse I. basic features. In: International Conference on Artificial Intelligence and Law. pp.180-189

[8] Ontology, https://en.wikipedia.org/wiki/Ontology

[9] Valente A, Breuker J and Brouwer B 1999 Legal modeling and automated reasoning with ON-LINE. *International Journal of Human-Computer Studies* **51(6)**, 1079-1126

[10] Breuker J, Valente A and Winkels R 2004 Legal ontologies in knowledge engineering and information management. *Artificial Intelligence and Law* **12(4)**, 241-277

[11] Hoekstra R, Breuker J, Bello M D, et al 2007 The LKIF core ontology of basic legal concepts. In: The Workshop on Legal Ontologies & Artificial Intelligence Techniques June. DBLP. pp.43-63

[12] Wyner A and Hoekstra R 2012 A legal case OWL ontology with an instantiation of Popov v. Hayashi. *Artificial Intelligence and Law* **20(1)**,83-107

A pruning algorithm for Meta-blocking based on cumulative weight

Fulin Zhang[1], **Zhipeng Gao**[1] **and Kun Niu**[2]

[1] State Key Laboratory of Networking and Switching Technology, Beijing University of Posts and Telecommunications, Beijing 100876, China
[2] School of Software Engineering, Beijing University of Posts and Telecommunications, Beijing 100876, China
Email: zhangfulin@bupt.edu.cn

Abstract. Entity Resolution is an important process in data cleaning and data integration. It usually employs a blocking method to avoid the quadratic complexity work when scales to large data sets. Meta-blocking can perform better in the context of highly heterogeneous information spaces. Yet, its precision and efficiency still have room to improve. In this paper, we present a new pruning algorithm for Meta-Blocking. It can achieve a higher precision than the existing WEP algorithm at a small cost of recall. In addition, can reduce the runtime of the blocking process. We evaluate our proposed method over five real-world data sets.

1. Introduction

Entity Resolution(ER) first proposed by [1]. ER has a significant application in data cleaning and data integration. As the amount of the global data grows exponentially, it has been a hot topic. It is the task to determine whether two different records in one or more data sources describe the same real world object, sometimes referred to as entity matching, record linkage or object matching. ER needs to compute the similarity between records to detect the duplicates. Due to its characteristics (one record has to compared with all others), it is very expensive for large data sets especially the rapid development of the Internet causes a fast growth of the amount of entity. Assuming that the size of two data sets that need to be matched is m and n respectively, $m \times n$ comparisons required if we do it in a brute-force way. The consumption of computing resources is extremely large, and it will seriously affect the efficiency. It typically employs a filter process commonly called blocking before calculation which leaving the most similarity records in the same block, such that the comparisons can be reduced.

2. Related Work

Blocking is a crucial step for ER. It can improve the efficiency of the ER process. The earliest blocking method called Traditional Blocking [2], each record can only be placed into one block and records with the same blocking key values (BKV) [3] are placed in the same block. It can be easily implemented by inverted index. The Sorted Neighbourhood indexing [4] method sorts the records according to the BKV and then moves a sliding window of the size fixed to on the sorted data set, comparing the records in a window. The main disadvantage is that the size of the sliding window. Assuming that two similar records are not so close that they are not in the same window, and then they will be unmatched. [5] Proposed a method that can change the size of the sliding window dynamically. For the data sets that contains a large number of errors, the above two methods cannot perform well because the matching records are often placed into different blocks. Q-gram [6] based blocking inserts records into more than one block to make up for the above shortcomings. Its BKVs is the substring of lengths q. The Canopy clustering algorithm [7-8] uses a rough distance calculation method to allocate data into different overlapping subsets in the first stage, and then only calculates records in the same overlapping subset to reduce the number of records need for calculation. [10] Proposed a pay-as-you-go method for ER. It can efficiently process the ER with a limited resource and time (real-time system) by using hints. [12] Proposed an iterative blocking framework, which exploits the results of processed blocks to save many comparisons. It can achieve a higher accuracy. These methods above are not suitable for highly heterogeneous information spaces (HHIS) [11] where there have a lot of noise, no structural data, and the amount of data is large. In addition, they all need a predefined schema. Meta-Blocking [13], a schema-agnostic method, achieves a good balance between precision and recall. Its core idea is the notion of blocking graph, which can discard the redundant and superfluous comparisons by a pruning algorithm. It has four steps: (i) build the blocking graph from the input block collection. (ii) Then use weighting schemes to set edge weight. (iii) Pruning the blocking graph use edge-centric algorithms or node-centric algorithms. (vi) Transform the pruned blocking graph into a new block collection. It takes the entity resolution to the next level.

The contributions of our work are as follows: (1) A new proposed pruning algorithm for Meta-blocking that can raise precision at a small cost in recall and it is more efficient than the existing WEP algorithm. (2) A series of thorough experiments to verify our proposed method over five real-world data sets.

The remainder of this paper is organized as follows: Section 3 describes the main notions of the work. Section 4 propose our new pruning algorithm. Section 5 presents our experiments and the results of evaluation. Section 6 contains conclusions of the paper and some recommendations for the future work.

3. Preliminaries

In this section, we introduce some main notion of this work.

Entity Profile. An entity profile is something with distinct and real existence. It can be defined as follows: e_i = {id, $name_{i1}$, $value_{i1}$, $name_{i2}$, $value_{i2}$...}, id is a globally unique flag; $name_{ij}$ and $value_{ij}$ are the names and values of an attribute that the entity has.

Entity Resolution(ER). The ER process is illustrated in figure1.It commonly involves two steps.

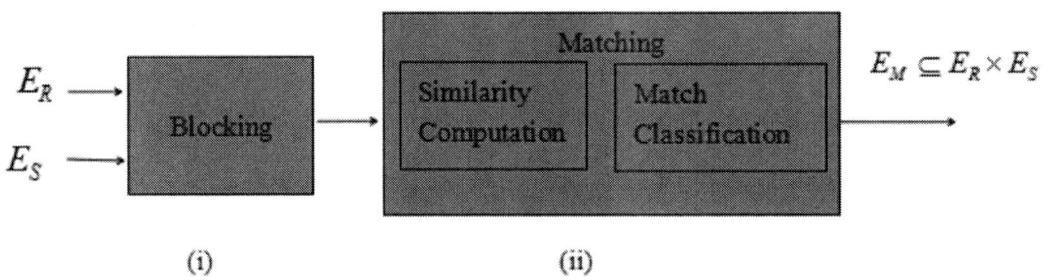

Figure 1. The traditional ER process

E_R, E_S is a set of entity profiles. (i) Blocking to reduce the total comparisons. Group similar records within blocks based on blocking key. (ii) Entities in the same block use a similarity function to Compared with each other. EM is the detected records that refer to the same real world object.

Clean-Clean ER. both E_R and E_S themselves do not contain duplicates. Dirty-Dirty ER: both E_R and E_S themselves contain duplicates.

Blocking. ER is an intrinsically quadratic task, the total number of record comparisons equals to $|M||N|$, with $|\cdot|$denoting the number of records in a data set. This is the bottleneck when ER scales to large entity collections. To reduce the large amount of comparisons, some blocking techniques are employed. Blocking can put the most similar entities together, and only candidate entity profiles in the same block are compared.

Meta-blocking. It is a schema-agnostic method that can reduce the redundant and superfluous comparisons by rebuild the block collection. It can improve the efficiency at a minor cost in effectiveness. The figure 2 is ER process with meta-blocking. We can note that it does not replace the existing blocking methods. In contrast, it can combine with the existing methods to make ER process better.

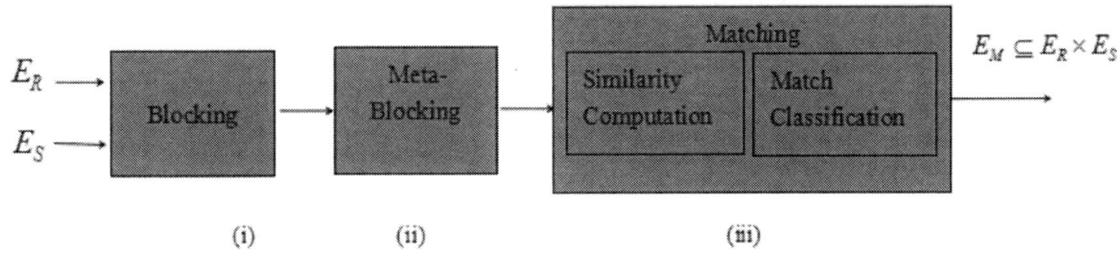

Figure 2. The ER process with Meta-blocking.

In our experiments, we use the following measures to evaluate our method. Pairs Quality(PQ): Estimate the proportion of non-redundant comparisons in the block. It is corresponding to precision used in the field of information retrieval. Its value between 0 to 1, A higher the value means a higher efficiency of the block. Pairs Completeness (PC): Estimate the proportion of the detected duplicates. It is corresponding to recall used in the field of information retrieval. Its value between 0 to 1, A higher value means a higher effectiveness of the block. Reduction Ratio (RR): Estimate the proportion of saved comparisons by blocking. Its value between 0 to 1, the higher value means the higher efficiency of the block.

4. Our Proposed Method

We introduce a new pruning algorithm Cumulative Weight Edge Pruning (CWEP) in this section. Differ from WEP's minimum edge weight and CEP's top K edges with the maximum weight [13], our method focuses on the cumulative weight. As is showed in the outline of Algorithm, the input of the algorithm is a blocking graph and a global thresholdθ. The blocking graph was built by block collections with edges weighted by different weighting schema. The output is an undirected pruned blocking graph. The algorithm first put graph's all edges into a sorted stack (line 1-3). The edges in the sorted stack are in descending order of weights. In addition, we can calculate the SumWeight of all edges. Then the algorithm removes edges from the sorted stack until the sum weight of edges in the sorted stack less than θ*SumWeight (line 4-5). Finally, the algorithm iterates again to remove edges that are not contained in the sorted stack (line 6-9). The remained edges in the sorted stack can form the new graph. The time complexity of our algorithm is O (|B|). |B| is the cardinality of the original block collections. In our work, we set θ to 0.5.

Algorithm :Cumulative Weight Edge Pruning(CWEP)

Input: (i) G_B^{in} the blocking graph (ii) θ the global cumulative weight pruning criterion
Output: G_B^{out} the undirected pruned blocking graph
1 SortedStack←{};
2 **foreach** $e_{i,j} \in E_B$ **do**
3 SortedStack.push();
4 **while** Cumulative Weight>θ*SumWeight **do**
5 SortedStack.pop();
6 **foreach** $e_{i,j} \in E_B$ **do**
7 **if** $e_{i,j} \notin$ SortedStack **then**
8 $E_B \leftarrow E_B - \{e_{i,j}\}$
9 **return** $G_G^{out} = \{V_B, E_B, WS\}$

5. Experiment

In this section, we do a series of experiments to demonstrate the performance of our method CWEP.

5.1. Setup

Our approach was implemented in Java 1.8. Our experiments were performed on a server with Intel(R) Core(TM) i7-4790 CPU 3.60GHz and 16.0GB of RAM. The operating system is Windows 7 Professional. The data sets we used in our experiments are showed in table 1.

Table 1. The datasets we used in our experiments.

	Entities	Name-Value pairs	Existing duplicates	Brute-force Comparisons
D1	1,076/1,076	2,568/2,308	1,076	1,157,776
D2	1,295	7,166	17,184	837,865
D3	4,910	19,626	2,224	12,051,595
D4	9,763	183,072	299	95,525,976
D5	1,354/3,039	1,354/3,039	1,104	4,114,806

D1 and D5 are clean-clean datasets, the others are dirty datasets. D1 is the "Abt_Buy" dataset containing 1076 product records of the Abt website and 1076 product records form the Buy site. D2 is the "Cora" dataset containing 1295 records of machine learning publications. D3 is the "ACM" dataset containing 4910 records from Association for Computing Machinery. D4 is the "Cddb" dataset containing 9763 records. It is a database containing audio CD information. D5 is the "Amazon-Google Product" dataset. It contains product records from amazon and google.

5.2. Discussion

The results are showed in the following figures. The x-axes are the different weighting schemas proposed by [13]. Figure 3(a) to 3(c) are the PC over different datasets, and figure 4(a) to 4(c) are the PQ (Due to the lack of space, we omit the other two data sets). We can observe that the interval between two lines in 3(a) to 3(c) is smaller than in 4(a) to 4(c) (except D4). It means that our CWEP algorithm can achieve a higher accuracy than WEP at a small cost of recall. It can increase precision by an average of 20 percent than WEP. In data set 2, it even reaches to 40 percent. Figure 5(a) to 5(c) are the RR over different datasets. The RR of CWEP is also higher than WEP (except D4). It means our algorithm have a higher efficiency than WEP.

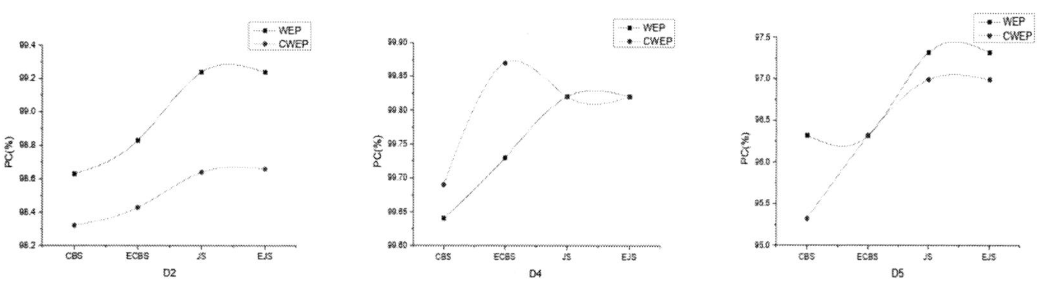

(a) PC over D2 (b) PC over D4 (c) PC over D5

Figure 3. Comparing effectiveness between CEP and CWEP across three datasets

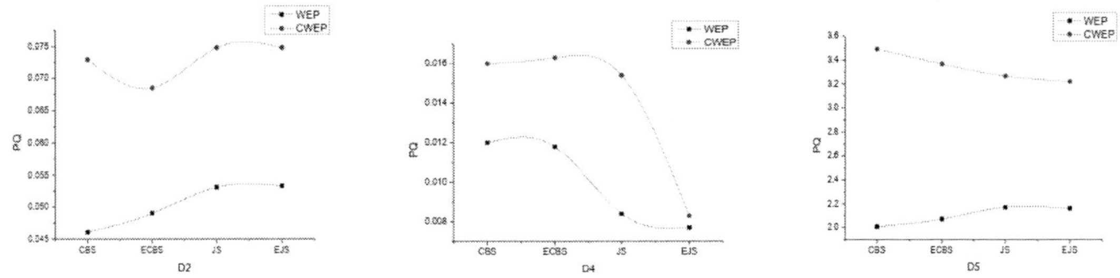

(a) PQ over D2 (b) PQ over D4 (c) PQ over D5

Figure 4. Comparing efficiency between CEP and CWEP across three datasets

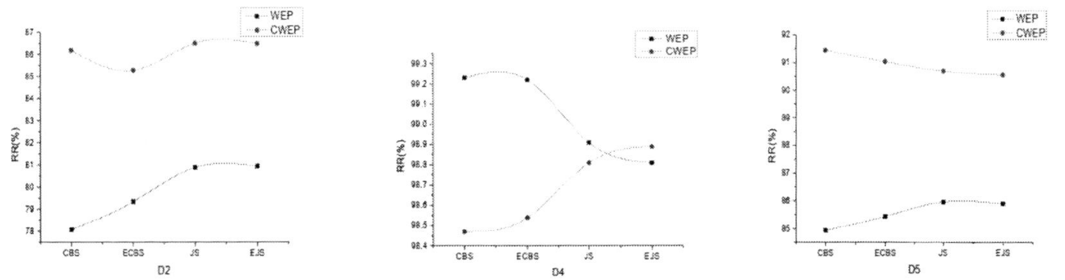

(a) RR over D2 (b) RR over D4 (c) RR over D5

Figure 5. Comparing RR between CEP and CWEP across three datasets

We also do the sensitivity analysis experiment. We set the pruning threshold θ from 0.1 to 0.9, we note that the higher the threshold, the lower the RR and the higher the PC in our experiment. It means our algorithm is robust. The total runtime of WEP and CWEP over different weighting schemas are showed in figure 6(a) to (c). As expected, our CWEP algorithm is faster than the WEP algorithm. Except for D4, the CWEP can save more than 17 percent runtime.

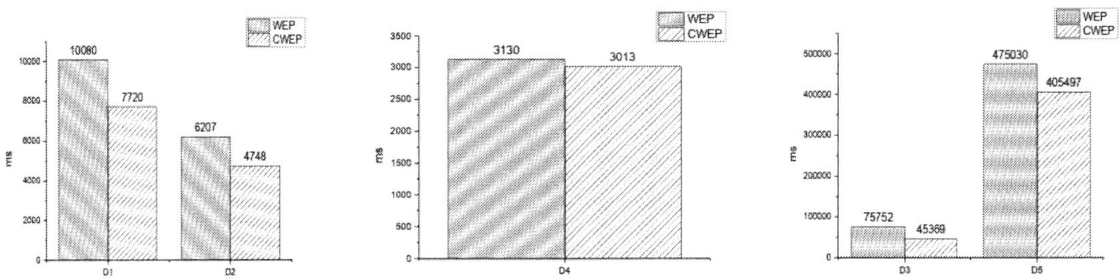

(a) (b) (c)

Figure 6. Comparing runtime of WEP and CWEP across all datasets

6. Conclusion

In this paper, we introduced a new pruning algorithm CWEP for Meta-blocking. We evaluated our algorithm over five real-world datasets. The result showed that our algorithm could achieve a higher accuracy than the existing WEP algorithm. In addition, it is more efficient. In the future, we plan to enhance its efficiency and apply our algorithm to the Parallel Meta-blocking.

7. Acknowledgments

This work is supported by the National Natural Science Foundation of China (61272515, 61372108), and National Science & Technology Pillar Program (2015BAH03F02).

8. References

[1] Newcombe HB, Kennedy JM, Axford SJ and James AP. "Automatic Linkage of Vital Records." 1959 Science 130.3381:954.

[2] P. Fellegi I and B. Sunter A. "A Theory for Record Linkage." 1969 Journal of the American Statistical Association 64.328:1183-210.

[3] Christen P. "A Survey of Indexing Techniques for Scalable Record Linkage and Deduplication." 2012 IEEE Transactions on Knowledge & Data Engineering 24.9:1537-55.

[4] Mauricio A. Hernández and Stolfo SJ. "The merge/purge problem for large databases." 1995 Acm Sigmod International Conference on Management of Data ACM: 127-38.

[5] Yan S, Lee D, Kan MY and Giles LC. "Adaptive sorted neighborhood methods for efficient record linkage." 2007 Acm/ieee-Cs Joint Conference on Digital Libraries ACM: 185-94.

[6] Baxter R, Christen P and Epidemiology CF. "A Comparison of Fast Blocking Methods for Record." 2003 Kdd Workshops: 25--seven.

[7] Richman J and Richman J. "Learning to match and cluster large high-dimensional data sets for data integration." 2002Eighth ACM SIGKDD International Conference on Knowledge Discovery and Data Mining ACM: 475-80.

[8] Mccallum A, Nigam K and Ungar LH. "Efficient clustering of high-dimensional data sets with application to reference matching." 2000 International Conference on Knowledge Discovery and Data Mining DBLP: 169-78.

[9] Faloutsos, C and Lin KI. "FastMap: a fast algorithm for indexing, data-mining and visualization of traditional and multimedia datasets." 1995 ACM SIGMOD Record 24.2:163-74.

[10] Whang SE, Marmaros D and Garcia-Molina H. "Pay-As-You-Go Entity Resolution." 2013 IEEE Transactions on Knowledge & Data Engineering 25.5:1111-24.

[11] Papadakis G, Ioannou E, Palpanas T, Niederee C and Nejdl W. "A blocking framework for entity resolution in highly heterogeneous information spaces." 2013 IEEE Transactions on Knowledge and Data Engineering 25.12: 2665-82.

[12] Whang SE, Menestrina D, Koutrika G, Theobald M and Garcia-Molina H. "Entity resolution with iterative blocking." ACM: 219-32.

[13] Papadakis G, Koutrika G, Palpanas T and Nejdl W. "Meta-Blocking: Taking Entity Resolution to the Next Level." 2014 IEEE Transactions on Knowledge & Data Engineering 26.8:1946-60.

IOP Conf. Series: Journal of Physics: Conf. Series **887** (2017) 012059 doi:10.1088/1742-6596/887/1/012059

Generation of Weibull distribution clutter based on correlated Gaussian sequence

Bin Wang* **and Fengming Xin**

School of Computer and Communication Engineering, Northeastern University at Qinhuangdao, Qinhuangdao, China

*Corresponding author: wangbinneu@qq.com

Abstract. With the continuous development of science and technology, the electromagnetic environment becomes more complex. Accurate clutter modeling is becoming increasingly difficult, which will have adverse effects in echo analysis. In this paper, in order to overcome electromagnetic interference, we use correlated Gaussian sequence to generate Weibull distribution clutter. Simulation results show that the estimated value of the proposed method is close to the theoretical value in the aspect of probability density and power spectral density. That demonstrates the validity of our method. Finally, the conclusions are given.

1. Introduction

As the increasing performance desire and new proposed system of radar, the demands for radar simulation system are increasing. Echo analysis is an important part in radar simulation, and the basis is clutter modeling. However, electromagnetic environment is more and more complex, and accurate clutter modeling is becoming increasingly difficult. So clutter modeling in complex environment is becoming hot issues of research.

Clutter simulation is the key link in the design of radar echo simulator, and many researchers have conducted a lot of work in related field. Simon Haykin proposes the concept of cognitive radar. Using all available knowledge, it can interrogate a propagation channel adaptively [1]. In [2], the authors add different clutter distributions into the framework of trackability, and significant differences are obtained. In [3], it is shown that when looking up or down wind local spectrum intensity is strongly linked to mean Doppler shift. The spectra will be given a temporal and spatial variability by related characteristics when combined with random fluctuations of spectrum width. In [4], different from traditional signal-dependent stochastic models, the authors propose a new approach of stochastic transfer function. In [5], the authors research on bistatic radar. The authors define bistatic clutter scenario and propose two land-clutter models. In [6], with the environment of Saudi Arabian urban, South African urban and Saudi Arabian date farms, the statistical analysis of three types of land clutter are presented. In [7], a new model is proposed, which is comprised by functions of radar frequency, polarization, sea state, and grazing angle. In [8], the authors describe X-band radar-based measurements of urban ground clutter, which is made from the CSIR campus in Pretoria, South-Africa. In [9], artificial neural network is used to classify four common radar clutter models. After simulating clutter models and extracting the important features, the authors make neural network with two hidden layers. In [10], the optimization of an information theoretic criterion for radar waveform design is described.

In this paper, after describing principles of monopulse radar and Zero Memory Nonlinear (ZMNL), we propose an improved mathematical modeling method for Weibull distribution clutter, which is based on correlated Gaussian sequence. In the simulation, we will compare the generated clutter with the theoretical value and verify the performance of our method.

2. Monopulse Radar

Since the invention of radar, single pulse technology has been a kind of important technology which has been greatly valued. In the early days, antenna microwave devices and other circuits do not get better treatment. Faced with such problems, cone scanning becomes an important method for automatic angle tracking. Since the advent of the sum and difference monopulse radar system, single pulse technology has more and more applications in radar system.

To determine location of the target in space, in many cases direction determination of the target is the primary task of radar, that is to say the angular coordinate of the target. Because of the advantages of strong anti-interference ability and high angular tracking precision of single pulse measuring angle system, it has been widely used.

Single pulse angle measurement is a simultaneous wave-lobe angle measurement method. In the same angular plane, if the same two beams overlap, then the equal signal axis is the overlap. Comparing the echo signals received by the two beams, the angular error signal of the target on this plane can be obtained. Then the transformed error voltage is added to the driving motor, the antenna can be controlled to move towards the direction of error reduction. As the two beams receive echo at the same time, monopulse angle measurement system will use shorter time in obtaining the target angular error information. Theoretically, the determination of angular error can be analyzed by only one echo pulse. The following advantages are not available in the conical scan radar:

(1) The time of obtaining angular error information is calculated in microseconds, with remarkable shortening;

(2) The fluctuation of echo amplitude has no effect on it;

(3) Angle measurement accuracy is improved significantly;

(4) Angle measurement branch has strong ability in anti- amplitude modulation jamming.

The methods of obtaining angular error signals are different, so there is a variety kind of monopulse radars. Sum and difference patterns of amplitude are most commonly used.

3. Generation of Weibull Distribution Clutter

The simulation of clutter data should meet a certain probability distribution in amplitude and the requirements in correlation properties simultaneously. That means the first-order and second-order characteristics of the data have to be produced to meet the requirements of the clutter.

The concrete block diagram of the clutter modeling with ZMNL method is as follows:

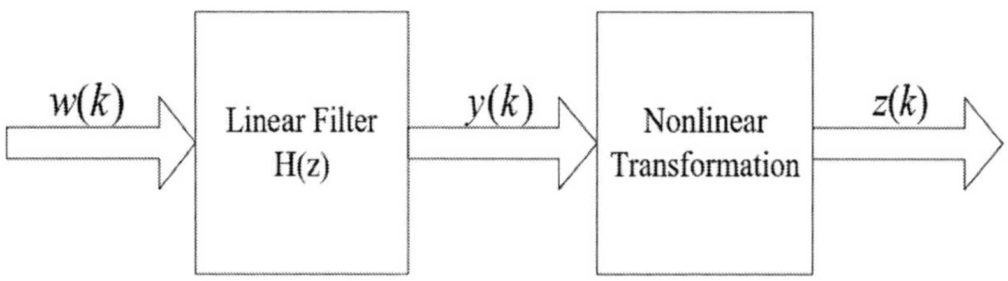

Figure 1. Block diagram of the clutter modeling with ZMNL method

As is shown in the figure 1, white Gaussian random process changes into correlated Gaussian process after getting through the filter $H(z)$, this process can obtain the required clutter sequence $z(k)$ after the nonlinear transformation. No breaking generality, assuming that $w(k)$ is a unit power white Gaussian random process with zero mean value, and the coefficient of the filter $H(z)$ is normalized.

Specific to monopulse radar, the key problem is how to generate correlated Gaussian sequence. Correlated Gaussian clutter can be viewed as the response to Gaussian white noise with zero mean value acting on the filter. It is supposed that Gaussian white noise's simple of a certain moment is $w(k)$ and $h(k)$ is the unit sequence response. So after getting through the filter, the response is

$$y(k) = w(k) * h(k) \tag{1}$$

where $y(k)$ is the simulation of the correlated Gaussian clutter sequence. On the basis of Fourier transformation, we can obtain

$$P_Y(w) = P_W(w) \mid H(w) \mid^2 \tag{2}$$

where $P_W(w)$ and $P_Y(w)$ are white noise and PSD function of correlated Gaussian noise. $H(w)$ is the frequency response function of $H(z)$. Assume that σ_W^2 denotes the variance of white noise, then

$$P_Y(w) = \sigma_W^2 \mid H(w) \mid^2 \tag{3}$$

Therefore, frequency characteristics of filter $H(z)$ are determined by the required correlated power spectrum characteristic of clutter. So we can use the method of filter design to achieve the requirements of correlation characteristics, and then the correlated clutter can be simulated. We can use the frequency domain method and time domain method to design filter. This paper uses the latter method to generate correlated Gaussian sequence Y. Principle diagram is shown in figure 2.

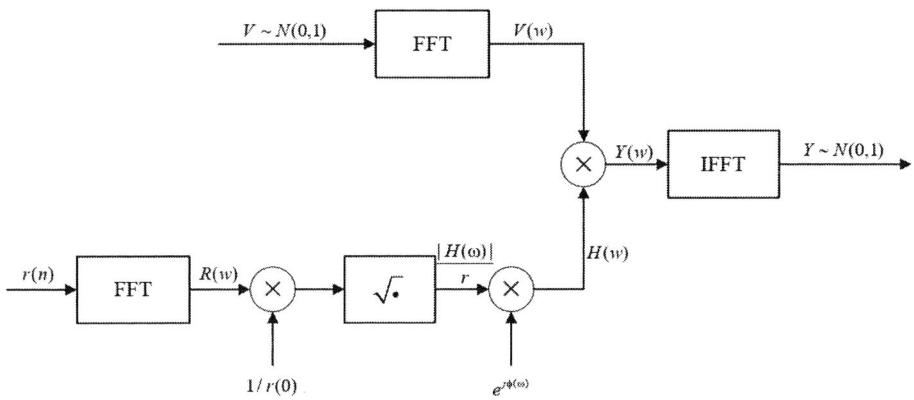

Figure 2. Generation principle diagram of correlated Gaussian sequence

As is shown above, to get the correlated Gaussian sequence two conditions should be known: Independent Gaussian sequence V and correlation coefficient sequence $r(n)$ of correlated Gaussian sequence. We can easily generate independent Gaussian sequence V and determine the correlated Gaussian sequence Y. Let

$$\mid H(w) \mid = \sqrt{FT(r(n))/r(0)} \tag{4}$$

An appropriate phase angle function should be selected to ensure physically feasibility of $H(w)$. Besides, we should make the independent and non-correlated standard Gaussian sequence V change into the frequency domain simultaneously.

So the PSD is

$$S_y(w) = S_v(w) * \mid H(w) \mid^2 \tag{5}$$

Change $Y(w)$ into time domain from frequency domain, then sequence Y denotes a correlated Gaussian random sequence.

Assume there are two independent random variables z_1 and z_2, and both of them obey the Gaussian distribution $N(0, \sigma^2)$. The nonlinear transformation is

$$x = (z_1 + z_2)^{1/p} \tag{6}$$

We can get the random variable x that obeys Weibull distribution with parameters p and q. The relationship between p and q is

$$q = (2\sigma^2)^{1/p} \tag{7}$$

The block diagram of generating incoherent Weibull distribution clutter is shown in figure 3.

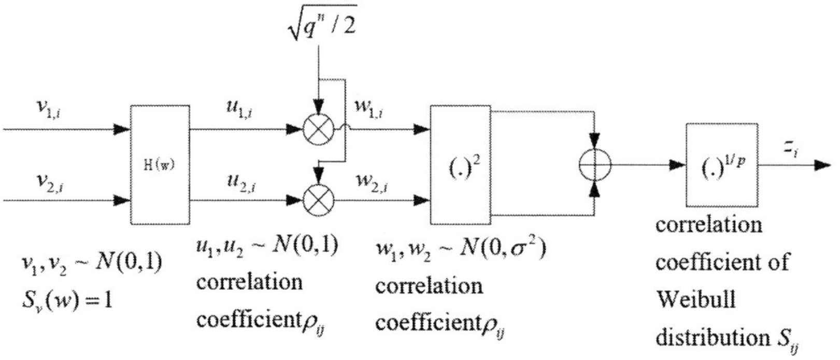

Figure 3. Generation principle diagram of Weibull distribution clutter

In figure 3, $n_{1,i}$ and $n_{2,i}$ $(i = 1, 2, ..., N)$ are independent Gaussian sequences which obey the distribution of $N(0,1)$. After getting through the filter, $y_{1,i}$ and $y_{2,i}$ are also correlated Gaussian sequences with correlation coefficient ρ_{ij} which obey the distribution of $N(0,1)$. $z_{1,i}$ and $z_{2,i}$ obey the distribution $N(0,\sigma^2)$ whose coefficient of correlation is also ρ_{ij}; The correlation coefficient of the generated Weibull distribution clutter x_i is S_{ij}, and the nonlinear relationship between ρ_{ij} and S_{ij} is

$$\begin{aligned} S_{ij} = &\{\Gamma^2(1+1/p)/[\Gamma(1+2/p) \\ &-\Gamma^2(1+1/p)]\} \\ &\times[{}_2F_1(-1/p,-1/p;1;\rho_{ij}^2)-1] \end{aligned} \tag{8}$$

where ${}_2F_1$ is Gaussian hypergeometric function and $\Gamma(\bullet)$ is Gamma function. We cannot directly solve ρ_{ij} with the formula above, so we will have a further discussion about how to get ρ_{ij} from S_{ij} in the implicit function.

According to the analysis above, the steps of generating Weibull distribution clutter are as follows:

(1) Select the required power spectral density $S(w)$ for clutter, and choose appropriate sampling rate Δw. After sampling from $S(w)$, we can get sequence $\{S_n\}, n = 1, 2, ..., N$.

(2) Make IFFT transformation to the sequence $\{S_n\}$, we can get the required correlation coefficient sequence S_{ij}, $i, j = 1, 2, ..., N$ of non-Gaussian random sequence.

(3) We can obtain correlation coefficient sequence $\rho_{ij}, i, j = 1, 2, ..., N$ of correlated Gaussian random sequence $\{y_{1,i}\}, \{y_{2,i}\}, i = 1, 2, ...N$.

(4) Through ρ_{ij} and normalized transfer function, we can generate correlated Gaussian sequence $\{y_{1,i}\}, \{y_{2,i}\}, i = 1, 2, ...N$.

(5) We can get σ^2 in $z_1, z_2 \sim N(0, \sigma^2)$, and through the nonlinear transform $x_i = (z_{1,i} + z_{2,i})^{1/p}$ we can generate noncoherent Weibull distribution random sequence by the Gaussian random sequence z_1, z_2.

The next section will make Weibull distribution clutter simulation in a certain condition through the steps we have discussed above, and analyze the results of simulation.

4. Simulations

In this section, we will compare the generated clutter with the theoretical value.

Simulation conditions are set up as follows. The length of the random sequence is 8000 points, power spectrum using Gaussian spectrum model. Sampling frequency $f_s = 1000Hz$, the center frequency $f_0 = 0Hz$, shape parameter $p = 1.5$, scale parameter $q = 0.5$. Simulation results are in figure 4 to figure 8.

Figure 4 is probability distribution of Weibull distribution clutter. The Rayleigh distribution model and the log normal distribution model are only applicable to the finite distribution, however, dynamic range of log normal distribution model is excessive exaggerated and Rayleigh distribution model is too conservative. Weibull distribution model lies halfway in between.

Figure 5 to figure 8 are the simulation results of Weibull distribution clutter with the proposed method. Figure 5 is independent non-correlated Gaussian random sequence. Figure 6 is Weibull distribution clutter obtained in simulation. Comparing figure 5 to figure 6, we can clearly observe the changes of sequences before and after the simulation.

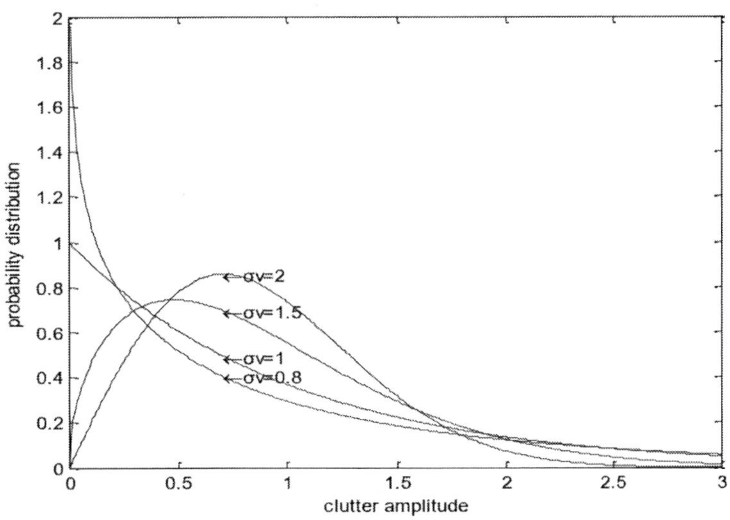

Figure 4. Probability distribution of Weibull distribution clutter

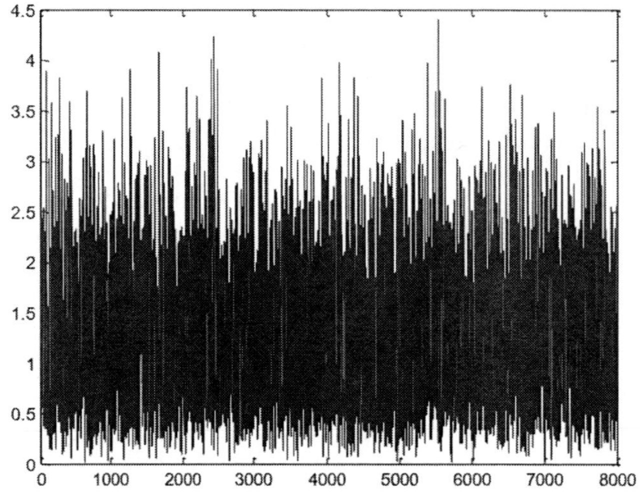

Figure 5. Independent non-correlated Gaussian random sequence

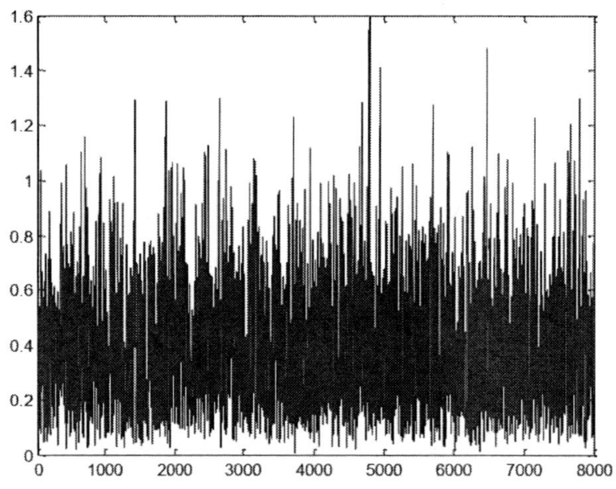

Figure 6. Weibull distribution clutter

Figure 7 shows the comparison between actual amplitude distribution and the theoretical amplitude distribution of clutter, which is also the first-order characteristic of clutter. It can be seen that the actual amplitude distribution and the theoretical amplitude distribution have the same changing trend. As PDF feature of modulation variable is display expression, so the difference in Figure 7 which peaks about at 0.5 at clutter amplitude of 0.2 is slightly obvious.

Figure 8 shows the comparison between the estimated power spectrum and the theoretical Gaussian power distribution, which is also the second-order characteristic of clutter. It can be seen that both of them have the same changing trend. The actual spectral width is larger than the desired width, while the error of the high frequency is large. It is mainly because that the required white Gaussian noise with zero mean cannot be accurately obtained, and the simulation can only use the pseudo-random number instead whose power spectral density is not constant in the strict sense.

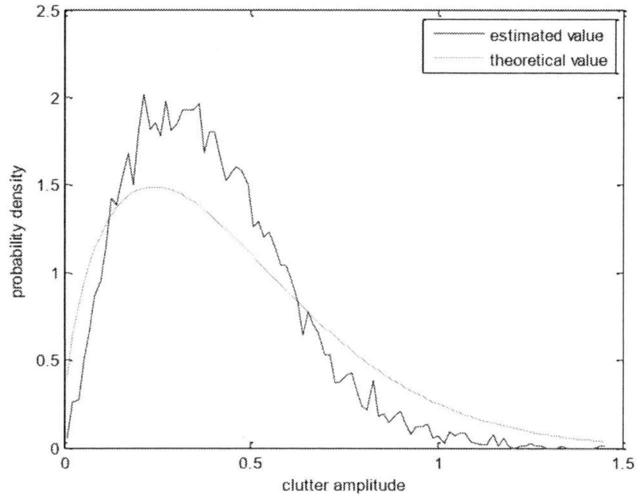

Figure 7. Distribution of clutter amplitude

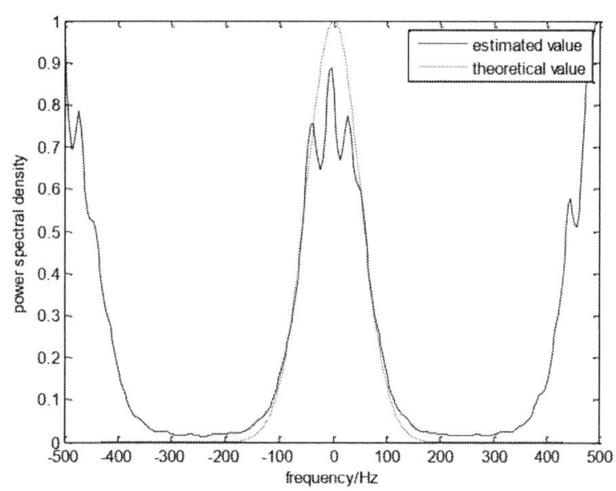

Figure 8. Clutter spectrum

5. Conclusions

In this paper, take monopulse radar for example, we give an explanation to the importance of radar clutter modeling and analysis. After describing principles of monopulse radar and ZMNL, we propose an improved mathematical modeling method for Weibull distribution clutter, which is based on correlated Gaussian sequence. Simulation results show that actual amplitude distribution of clutter and the theoretical amplitude distribution of clutter are basically consistent with each other, and they have the same changing trend. Actual Gaussian power distribution and theoretical Gaussian power distribution are consistent with each other, too. The actual spectral width is larger than the desired width, while the error of the high frequency is large. So we can conclude that the generated clutter approaches the theoretical value and our method is effective.

Echo analysis is an important part in radar simulation, and the basis is clutter modeling. With the electromagnetic environment becoming more and more complex, accurate clutter modeling is becoming increasingly difficult. It is very necessary to develop more accurate and faster clutter generation methods in the future research.

6. Acknowledgment
This work was supported by the National Natural Science Foundation of China (No. 61403067).

7. References

[1] S. Haykin, "Cognitive radar: a way of the future," IEEE Signal Processing Magazine, vol. 23, no. 1, pp. 30-40, January 2006.

[2] S. Schoenecker, P, Willett and Y. Bar-Shalom, "The Effect of K-Distributed Clutter on Trackability," IEEE Transactions on Signal Processing, vol. 64, no. 2, pp. 475-484, January 2016.

[3] S. Watts, L. Rosenberg, S. Bocquet and M. Ritchie, "Doppler spectra of medium grazing angle sea clutter; part 1: characterization," IET Radar, Sonar & Navigation, vol. 10, no. 1, pp. 24-31, January 2016.

[4] J. R. Guerci, J. S. Bergin, R. J. Guerci, M. Khanin and M. Rangaswamy, "A new MIMO clutter model for cognitive radar," Proc. IEEE Radar Conference, IEEE Press, 2016, pp. 1-6.

[5] M. Pola, P. Bezousek and Jan Pidanic, "Model comparison of bistatic radar clutter," Proc. Conference on Microwave Techniques, IEEE Press, 2013, pp. 182-185.

[6] Amer Melebari, M. Y. Abdul Gaffar and J. J. Strydom, "Analysis of high resolution land clutter using an X-band radar," Proc. IEEE Radar Conference, IEEE Press, 2015, pp. 139-144.

[7] V. Gregers-Hansen and R. Mital, "An Improved Empirical Model for Radar Sea Clutter Reflectivity," IEEE Transactions on Aerospace and Electronic Systems, vol. 48, no. 4, pp. 3512-3524, October 2012.

[8] J. J. Strydom, J. J. de Witt and J. E. Cilliers, "High range resolution X-band urban radar clutter model for a DRFM-based hardware in the loop radar environment simulator," Proc. International Radar Conference, IEEE Press, 2014, pp. 1-6.

[9] M. A. Darzikolaei, A. A. Ebrahimzade and E. Gholami, "Classification of radar clutters with Artificial Neural Network," Proc. 2nd International Conference on Knowledge-Based Engineering and Innovation, IEEE Press, 2015, pp. 577-581.

[10] A. Leshem, O. Naparstek, A. Nehorai, "Information theoretic adaptive radar waveform design for multiple extended targets," IEEE Journal of Selecte Topics in Signal Processing, vol. 1, no. 1, pp. 42-55, June 2007.

A multi-group firefly algorithm for numerical optimization

Nan Tong [1], Qiang Fu [1,2,*], Caiming Zhong [1] and Pengjun Wang[2]

[1] College of Science and Technology, Ningbo University, Ningbo, 315212, China
[2] Faculty of Information Science and Engineering, Ningbo University, Ningbo, 315211, China
*Corresponding author Email: fuqiang@nbu.edu.cn
Address: No.505 Yuxiu Rd. Zhuangshi St. Zhenghai Dist. Ningbo, China

Abstract. To solve the problem of premature convergence of firefly algorithm (FA), this paper analyzes the evolution mechanism of the algorithm, and proposes an improved Firefly algorithm based on modified evolution model and multi-group learning mechanism (IMGFA). A Firefly colony is divided into several subgroups with different model parameters. Within each subgroup, the optimal firefly is responsible for leading the others fireflies to implement the early global evolution, and establish the information mutual system among the fireflies. And then, each firefly achieves local search by following the brighter firefly in its neighbors. At the same time, learning mechanism among the best fireflies in various subgroups to exchange information can help the population to obtain global optimization goals more effectively. Experimental results verify the effectiveness of the proposed algorithm.

1. Introduction

Optimization problems in practical engineering are large-scale, nonlinear characteristics. The traditional optimization algorithms, such as hill climbing algorithm and chaotic search algorithm, are often difficult to solve this kind of problems. Therefore, swarm intelligent algorithm for the global optimization problems have been proposed in recent. Examples of notable swarm intelligent algorithms are particle swarm optimization algorithm (PSO) [1], artificial bee colony algorithm (ABC) [2], and ant colony optimization (ACO) [3].

Biological characteristics of firefly light is used to pass information for reference, by which Yang [4] proposed a new swarm intelligence algorithm in 2008, firefly algorithm (FA). The main idea of this algorithm is as follows. Multiple fireflies are randomly distributed in the whole search space, and all fireflies have its light intensity, corresponding to the fitness value of the optimization problem. Then each individual flies following the firefly with stronger light intensity in its visual range. After multi-iterations, all individuals gather around the best firefly, which represents the final optimization. As firefly algorithm has simple structure and can be implemented easily, it has been widely used in image processing [5], automatic control [6], economic emission dispatch [7], and other fields. However, the standard firefly algorithm also has problems of premature convergence and evolutionary stagnation.

To solve these problems, Gandomi [8] proposed a method combining chaos optimization algorithm to improve the fireflies' global optimization ability. Fu [9] put forward a multi-group firefly algorithm, in which the firefly population is divided into multiple sub-populations to enhance the diversity. Adil [10] developed an effective FA with partial random restarts and with an adaptive move procedure. Iztok Fister [11] used quaternion to represent individuals in firefly algorithm so as to enhance the performance of the firefly algorithm and to avoid stagnation. Although these methods improved the performance of optimization of firefly algorithm to some extent, "premature" phenomenon in solving high-dimensional or complex problem still exists.

In view of above, this paper analyzes the causes of premature convergence in the algorithm by studying the evolution mechanism of firefly algorithm, and puts forward an Improved Multi-Group Firefly Algorithm based on improved evolutionary mechanism (IMGFA). The simulation experimental results show that the modified firefly algorithm can avoid getting into local optimal value, and achieve better optimization performance than the standard firefly algorithm.

Section 2 presents and analyzes the standard firefly algorithm. Section 3 puts forward a multi-group firefly algorithm based on improved evolutionism. The experimental results and discussion on benchmark functions are presented in Section 4. Section 5 gives a conclusion to the work and suggests some directions for future studies.

2. Firefly Algorithm Description and Analysis

2.1. Algorithm Description

There are two important key factors arise in the standard firefly algorithm, namely, the light intensity and the attractiveness. Each firefly's fitness determines its light intensity. The better fitness means the stronger light intensity, and shows more intense attractiveness in the areas.

Assuming the total number of fireflies is m, which are randomly initialized in the search space. Suppose a N -dimensional space, where the ith firefly's location can be expressed as $x_i = (x_{i1}, x_{i2}, \cdots x_{iN})$.

Definition 1. Let x_i be the location of the ith firefly, then the light intensity of x_i is defined as:

$$I_i = I_{0i} e^{-\gamma r_{ij}} \tag{2.1}$$

Where, I_{0i} denotes the light intensity of the source, and equals to the fitness of x_i . γ is the light absorption coefficient which plays a important role in the convergence speed and equals to 1.The distance between two fireflies i and j at location x_i and x_j, can be obtained using the Euclidean distance as:

$$r_{ij} = \| x_i - x_j \| \tag{2.2}$$

Definition 2. The attractiveness of fireflies is defined as:

$$\beta = \beta_0 \cdot e^{-\gamma \cdot r_{ij}^{\omega}} \tag{2.3}$$

Where, $\omega = 2$. β_0 is the attractiveness at $r = 0$.

Definition 3. The movement of a firefly i to another more attractive firefly j can be expressed as:

$$x_i(t+1) = x_i(t) + \beta(x_j(t) - x_i(t)) + \alpha(randn - 1/2) \tag{2.4}$$

Where, α is a randomization parameter generated from interval [0, 1], randn is a random number drawn from Gaussian distribution.

2.2. Algorithm Optimization Mechanism

It can be seen from equation 2.4, the optimization process of firefly populations are realized by $\beta(x_j(t) - x_i(t))$ and $\alpha(rand - 1/2)$. $\beta(x_j(t) - x_i(t))$ reflects the learning behavior of firefly to the more attraction firefly in its visual range. $\alpha(rand - 1/2)$ can effectively expand the search range of the firefly populations, prevent premature local optimum. When a firefly cannot find better fireflies in the neighborhood, it is still able to have the ability to move in a random location.

Fireflies exchange information sufficiently in a small area so that they can influence each other, implement evolution quickly, and get the goal finally. But when in a larger search area or solving high-dimensional optimization problem, some fireflies maybe difficult to find and be attracted by the brighter one for the alienated distribution. As a result, they are unable to obtain the evolutionary information from other individuals, and become the "failure" vibrating randomly at their original position, which greatly affects the group's optimization ability, and cause premature convergence and evolutionary stagnation. If failed to produce a valid "Evolution Team" in the vicinity of the optimal target when random initialization, the firefly populations will be difficult to obtain the optimal target in the follow iteration. Therefore, only when it has close connection among fireflies, firefly algorithm can achieve the global optimization target quickly and efficiently.

3. A Multi-Group Firefly Algoritnm Based on Improved Evolutionism(Imgfa)

Each firefly seeks for the better optimization value by following the brighter firefly nearby, so there are multiple attractors to guide the population evolution, and to a certain extent, improve the robustness of the algorithm. From equation 2.4, we can see that, the attractiveness $\boldsymbol{\beta}$ exponentially decay with the increasing of $\boldsymbol{r_{ij}}$, which helps the fireflies to realize evolution when fully interaction among fireflies exists. But once r_{ij} increases exceeding a certain range, the interaction will interrupt. That means most of the fireflies will be isolated, and then the evolutionary stagnation occurs.

So this paper adjusts the evolutionary mechanism of the FA, and puts forward an improved firefly algorithm, which can appropriately enhance the interactive communication among fireflies.

3.1. Improved Evolutionary Computation Model

In firefly algorithm, each firefly with higher light intensity is able to attract its neighboring fireflies, but loses appeal when the distance between with others becomes far. In order to prevent the firefly populations from interrupting internal information transmission, and promote fireflies' activity, the best fitness value of the firefly population is emphasized in IMGFA. The improved evolutionary computation model is showed as:

$$x_i(t+1)=\begin{cases} x_i(t)+k*rand\ *(x_{best}(t)-x_i(t)) & \beta<ts \\ x_i(t)+\beta*(x_j(t)-x_i(t))+\alpha*(randn-1/2) & \beta\geq ts \end{cases} \tag{3.1}$$

Where, $x_{best}(t)$ is the location of the brightest firefly in the population at time t. $\omega*rand*(x_{best}(t)-x_i(t))$ indicates that the brightest firefly will attracts all others when information interruption arises. $\boldsymbol{rand}$ is a randomization parameter generated from interval [0,1]. $\boldsymbol{k}$ is the global attractiveness coefficient reducing gradually in iterations, which can avoid the growth of the disabled firefly efficiently . t_S is the global attractiveness threshold. When the attractiveness of a firefly is less than t_s, the firefly will be failure one for losing contact with the others. Then, the brightest firefly undertakes the mission of dragging the stray firefly to establish effective information interaction with the others.

Comparing with equation 2.4, the improved evolutionary computation model underlines the global attractor created from the brightest firefly in the population. From equation 3.1, each firefly can also receive evolution information even being out of touch with the neighbors. The global attractor also promotes the diversity of population evolution, locates the optimal solution quickly at the beginning of the evolution.

3.2. Simulated Annealing Rule

Simulated annealing rule works by simulating a stochastic process on heating and controlled cooling of a material to increase the size of its crystals and reduce their defects. Transitions between states are governed by a rule that favors states with lower cost or "energy". The temperature starts high and is gradually reduced with time according to a specified cooling schedule, in the belief that different features of the solution will "crystallize out" at different temperatures. At low-temperature equilibrium, low energy state occur much higher probability than high-energy ones.

Simulated annealing rule is used in this paper to avoid premature convergence. Assume the current status of fireflies as $x_i = (x_{i1}, x_{i2}, x_{i3}, \ldots, x_{iN})$, the energy in the state x_i is $E_i = (E_{j1}, E_{j2}, \ldots, E_{jN})$; and the status will change to $x_j = (x_{j1}, x_{j2}, x_{j3}, \ldots, x_{jN})$ after adding the disturbance, the energy in the state x_j is $E_j = (E_{j1}, E_{j2}, \ldots, E_{jN})$. The new state x_j will be accepted if $E_j > E_i$, otherwise determined by Probability $e^{-\frac{E_i - E_j}{k \cdot T}}$. Where, k is the Boltzmann constant? The three main steps of generating the new state x_j, calculating the energy E_j, and deciding to accept or not, are implemented in the iterations up to the lowest temperature.

3.3. Multi-Group Optimization Mechanism

This paper divides the firefly populations into n subgroups randomly disturbed in the feasible space. The parameters α, γ, and β_0, are set up in different model, by which each sub-group will implement evolution in diversified styles of "Individual cognition" and "Social cognition ", and achieve better global optimization results.

In each of sub-groups, a bulletin board records the best firefly's fitness and position in each subgroup per iteration, and recruits the best fireflies into elite team. Premature subgroup is found when the best fitness of it does not update in five successive iterations. And then, the best firefly in the premature sub-group can seek for help from the elite team, and jump out of the current local optimal area. Elite team constructs the information exchange channel between each subgroup for mutual learning, and provides more effective guidance to puberty subgroups in premature convergence.

If premature subgroup cannot get help from elite team lacking of a better guide, it will adopt Gauss mutation to healing itself, sort the fireflies by fitness value, choose the first firefly and the last firefly to exchange their position, and update the first one by Gauss mutation. The equation is given as:

$$x_i = x_i + x_i * N(\mu, \sigma) \tag{3.2}$$

Where, $N(\mu, \sigma)$ is a Gaussian distributed random vector, who's expectation is μ and variance is σ.

It increases the possibility of population diversity to adjust itself in the premature subgroup, which can take the subgroup out of the local area, and achieve the global optimization goal effectively.

3.4. Adaptive Distance Exponent Weight

The distance exponent weight ω is an extremely important parameter in firefly algorithm. When ω is low, a firefly owns wider vision space, and searches for more neighbor fireflies to information exchange extensively. And when ω is large, the fireflies' vision field will narrow for local search. As showed in equation 2.3, the ω equals to 2, which is not enough to coordinate the requirement of global search and local convergence. Bearing in mind these observations, we propose adaptive distance exponent weight showed in equation 3.3.

$$\omega(i) = \omega_{\min} + (\omega_{\max} - \omega_{\min})(\frac{i}{G})^2 \tag{3.3}$$

Where, $\omega_{\max}$ is the upper bound of ω, $\omega_{\min}$ is the lower bound of ω. G is the total number of iterations. i is the number of the current iteration.

Equation 3.3 indicates that ω will increase in each iteration. In the early period of the algorithm running time, the smaller ω is conducive to jumping out of local minimum area and good for accomplishing the task of global optimization, while in the later period of running time, the larger ω is conducive to rapidly reaching local convergence. Better adaptability is obtained for firefly algorithm by adjust ω according to the change of the iterations.

3.5. The Procedure of IMGFA

Step1: Parameter Initialization. Set the number of fireflies m, the number of subgroups n, and solution space's dimension N. Place different γ, β_0, and α for each subgroup. Setting convergence condition, and randomly generate the fireflies' initial position $x_{lj}(l=1,2,\cdots m/n \quad j=1,2,\cdots N)$ in the Solution space.

Step2: For all fireflies in the subgroup $i(i=1,2,\cdots n)$, carry out following operations:

1. Calculate the fitness $f(x_{ij})$ of each firefly as its light intensity I_{0ij}, sort $f(x_{ij})$, and record the best fitness f_{ibest} in the sub-group's bulletin board and the elite team;

2. Update the fireflies' positions according to equation 3.1, and judge the new positions' effectiveness by simulated annealing rule: accept the new position if better, otherwise selectively accept by probability;

3. Choose the best fitness of the new fireflies' group, and update f_{ibest} if better.

Step3: check each sub-group's bulletin boards, if the best fitness of any subgroup does not update in five successive iterations, turn to Step4, otherwise to step5;

Step4: Compare f_{ibest} with $f_{kbest}(k=1,2,\cdots n \quad k\neq i)$ in the elite team, if f_{kbest} is better than f_{ibest}, replace the firefly by f_{ibest} with the firefly by f_{kbest}; otherwise, in the subgroup i, choose the best firefly and the worst firefly to exchange their position, and update the worst one by Gauss mutation according to equation 3.2.

Step5: Adjust ω according to equation 3.3, and implement domain constraint for all fireflies;

Step6: Determine whether the convergence criterion has been met. If satisfied, go to step7; otherwise, return to step2;

Step7: Output results, end of running.

The time complexity of the algorithm is $O(m^2)$

4. Experimental Result and Discussion

In this section the IMGFA algorithm is benchmarked on six classical benchmark functions which have been extensively used. The benchmark functions include three fixed-dimension multimodal functions, and three varied-dimension functions showed in Table 1 and Table 2, in f_4 and f_5 are unimodal functions, and f_1, f_2, f_3, and f_6 are multimodal function.

Table 1. Fixed-dimension benchmark functions

Function	Range	f_{min}
$f_1(x)=-e^{-(x-4)^2-(y-4)^2}-e^{-(x+4)^2-(y-4)^2}-2e^{-x^2-(y+4)^2}-2e^{-x^2-y^2}$	[-10,10]	-2
$f_2(x)=0.5+\dfrac{\sin\sqrt{x^2+y^2}-0.5}{(1.0+0.001(x^2+y^2))^2}$	[-100,100]	0
$f_3(x)=20(1-e^{-0.2\sqrt{0.5(x^2+y^2)}})-e^{0.5(\cos(2\pi x)+\cos(2\pi y))}+e$	[-32,32]	0

Table 2. Varied-dimension benchmark functions

Function	Range	f_{min}
$f_4(x)=\sum_{i=1}^{N}x_i^2$	[-100,100]	0
$f_5(x)=\sum_{i=1}^{N}[100(x_{i+1}-x_i^2)^2+(x_i-1)^2]$	[-100,100]	0
$f_6(x)=\sum_{i=1}^{n}(x_i^2-10\cos(2\pi x_i)+10)$	[-10,10]	0

The total number of fireflies is 60 in each test function. There are 3 sugroups in the IMGFA algorithm, and 20 fireflies in each sub population. Other parameters are shown in Table 3.

Table 3. Algorithm parameters

	β_0	α	$\Delta\alpha$	γ
FA	1	0.3	0.96	1
	1	0.29	0.97	0.99
MFA	0.99	0.3	0.97	0.99
	1	0.31	0.98	1

In order to evaluate IMGFA algorithm intuitively and comprehensively, this paper also uses ABC and PSO to join the test. The population size in ABC and PSO is also set to 60. Each benchmark function runs 20 times independently. All of the varied-dimension benchmark functions are in 10, 20 and 30 dimensions. The statistical results (mean and standard deviation) are reported in Table 4~7.

Table 4. Result of fixed-dimension multimodal benchmark functions

FUNCTION	ALGORITHM	Ave	Std
f_1	IMGFA	-2	0
	IFA	-1.9793	0.2565
	ABC	-1.9990	0.0001
	PSO	-1.9996	0.0091
f_2	IMGFA	0	0
	FA	0.7163	0.1104
	ABC	1.1264E-12	9.3111E-12
	PSO	0.009716	0.0901
f_3	IMGFA	6.1391E-21	3.238E-20
	FA	1.7246	2.2781
	ABC	6.6538E-11	7.0944E-11
	PSO	6.5673E-10	1.0816E-09

Table 5. Result of varied-dimension benchmark functions in 10 dimensions

FUNCTION	ALGORITHM	Ave	Std
f_4	IMGFA	0	0
	FA	10.2693	2.1636
	ABC	7.3221E-10	5.3853E-10
	PSO	5.5215E-09	1.0416E-08
f_5	IMGFA	3.1936	0.2024
	FA	1.1369E+03	651.52
	ABC	12.1358	5.4593
	PSO	57.9097	27.0153
f_6	IMGFA	0	0
	FA	71.8171	10.3667
	ABC	1.0836E-06	3.6935E-06
	PSO	1.1762	0.7008

Table 6. Result of varied-dimension benchmark functions in 20 dimensions

FUNCTION	ALGORITHM	Ave	Std
f_4	IMGFA	0	0
	FA	328.0304	280.9303
	ABC	1.1042E-03	1.2729E-03
	PSO	0.63412	0.7982
f_5	IMGFA	11.2902	0.2501
	FA	6.9593E+05	1.3726E+05
	ABC	91.8931	55.8136
	PSO	163.6565	65.3945
f_6	IMGFA	0	0
	FA	331.2723	167.1311
	ABC	10.2825	3.8072
	PSO	38.3838	17.6047

Table 7. Result of varied-dimension benchmark functions in 30 dimensions

FUNCTION	ALGORITHM	Ave	Std
f_4	IMGFA	0	0
	FA	6.7378 E+03	1.1995E+03
	ABC	0.1907	0.2635
	PSO	5.6583	3.8593
f_5	IMGFA	17.7145	0.0056
	FA	1.8908E+06	1.5275E+05
	ABC	178.9736	143.5292
	PSO	189.3528	213.4426
f_6	IMGFA	0	0
	FA	872.3386	9.4243
	ABC	28.2192	4.7620
	PSO	132.2267	27.45305

Simulation results form Table 4 shows that, compared with FA, ABC, PSO, IMGFA has better evolution performance in accuracy, and better robustness with smaller variance. Especially for f_1 and f_2, IMGFA can obtain the minimum value in each test. According to the results of Table 5~7, IMGFA is able to provide very competitive results on the varied-dimension benchmark functions as well. IMGFA captures the global best value for f_4 and f_6 in 10, 20 and 30 dimensions. For f_5, IMGFA doesn't get the final goal, but it also shows better accuracy and stability of solution than all others.

5. Conclusions

Although having the advantages of simple structure, less adjustable parameters, and good optimization ability, the standard firefly algorithm is prone to premature stagnation when solving high dimensional optimization problems or in large solving area. This paper studies the evolution mechanism of the firefly algorithm, and proposals an improved firefly algorithm, IMGFA. IMGFA emphasizes the traction force of the best firefly to the "failure" firefly, sets the multi-subgroups net to enhance the diversity of firefly populations, and solves the problem of premature convergence by Simulated annealing rule and Gaussian mutation. Simulation results on benchmark functions show that IMGFA has good adaptability and robustness, and shows a perfect balance between exploration and exploitation.

6. Acknowledgment

The work was supported by Scientific Research Fund of Zhejiang Provincial Education Department (No. Y201326770), Ningbo Natural Science Foundation (No.2014A610069), China.

7. References

[1] Pratap P, Bhatia R S, Kumar B. Design and simulation of equilateral triangular microstrip antenna using particle swarm optimization (PSO) and advanced particle swarm optimization (APSO)[J]. Sādhanā, 2016, 41(7): 721-725.

[2] Doğan Aydin, Serdar Özyön, Celal Yaşar, Tianjun Liao.Artificial bee colony algorithm with dynamic population size to combined economic and emission dispatch problem. International Journal of Electrical Power & Energy Systems, 2014, 45(1), pp.144-153.

[3] Samà M, Pellegrini P, D'Ariano A, et al. Ant colony optimization for the real-time train routing selection problem[J]. Transportation Research Part B: Methodological, 2016, 85: 89-108.

[4] YANG Xin-she. Nature-inspired meta-heuristic algorithms.Luniver Press,pp.83-96, 2008.

[5] Mishra A, Agarwal C, Sharma A, et al. Optimized gray-scale image watermarking using DWT–SVD and Firefly Algorithm[J]. Expert Systems with Applications, 2014, 41(17): 7858-7867.

[6] Gorripotu T S, Sahu R K, Panda S. Firefly algorithm optimised PID controller for automatic generation control with redox flow battery[J]. International Journal of Computational Systems Engineering, 2017, 3(1-2): 48-57.

[7] Alham M H, Elshahed M, Ibrahim D K, et al. A dynamic economic emission dispatch considering wind power uncertainty incorporating energy storage system and demand side management[J]. Renewable Energy, 2016, 96: 800-811.

[8] A.H.Gandomi,X.-S.Yang,S.Talatahari,A.H.Alavi. Firefly algorithm with chaos. Communications in Nonlinear Science and Numerical Simulation, 18(1) , 2013 ,pp.89-98.

[9] FU Qiang,TONG Nan,ZHAO Yi-ming. Firefly algorithm based on multi-group learning mechanism, Application Research of Computers,30(12) , 2013 ,pp.3600-3617.

[10] Adil Baykasoğlu, Fehmi Burcin Ozsoydan. An improved firefly algorithm for solving dynamic multidimensional knapsack problems.Expert Systems with Applications,41(8), 2014,pp. 3712-3725.

[11] Iztok Fister ,Xin-She Yang, Janez Brest. Iztok Fister Jr.Iztok Fister. Modified firefly algorithm using quaternion representation.Expert Systems with Applications, 40(18) , 2013 ,pp.7220-7230.

A novel user-interest model based on mixed measure

Junkai Yi[1], Yacong Zhang[2], Mingyong Yin[3,*] and Xianghui Zhao[4]

1 College of Information Science and Technology, Beijing University of Chemical Technology, No. 15, Beisanhuan East Road, Beijing, China.
Email: yijk@mail.buct.edu.cn
2 College of Information Science and Technology, Beijing University of Chemical Technology, No. 15, Beisanhuan East Road, Beijing, China.
Email: zhangyacong84@163.com
3 Institute of Computer Application, China Academy of Engineering Physics, No. 64, Mianshan Road, Mianyang, China. Email: yinmingyong@126.com
4 China Information Technology Security Evaluation Center, No. 8, Shangdi West Road, Beijing, China. Email: zxhitsec@sina.com

Abstract. In this paper, we present a user-interest model based on mixed measure for finding users' interests, and in this model we can find the users' interests in webpages to improve the accuracy of users' interests and provide better personalized service for users. Information gain and mutual information are applied in this model for dimensional reduction, then find users' interests from the content of webpages by classifying-clustering. The results of experiments show that the model we propose has better performance, and the accuracy of the model is satisfactory.

1. Introduction

With the popularity of network, the Internet has become an indispensable information source for human. Meanwhile, our need to the information also increases gradually, but it is difficult to find the information we need exactly and quickly. Personalized service can meet users' demand by providing different services to the different users [1]. We focus on problems such as feature selection and user-interest exploration, and propose a user-interest model based on mixed measure in this paper. We divide the content of webpage into several words in the model, and apply information gain and mutual information for dimensional reduction. Then we use classification method for coarse grained partitioning to obtain several large groups, and apply clustering algorithm to each group for users' interest. The user-interest matrix is denoted by IR (interesting) based on calculating the degree of interest by the analysis of users' behaviours.

At present, many researchers study the user-interest modeling and achieve abundant research results. He L et al. proposed LDA-based theme mining model, and this model divided users' interest into initial interest and forwarded interest [2]. Jayarathna S et al. used four different user-interest theme model to generate fine-grained user-interest model, and made the degree of attention of all passages and fragments for visualization [3], which was in accordance with users' interests. Liu Z, Chen X et al. focused on the noisy of message and the complexity of words in users' microblogs, and proposed a keyword-mining approach to dig users' interests based on the combination of conversion and frequency [4]. Kumar S et al. extracted the users' interest from the history of webpages, recognized the importance of users' interest through computing the number of keyword transmission, and ranked the webpage according to the importance of interests [5]. Liu D et al. proposed an approach to analysis users' interests using fingerprint method. The fingerprint was binary strings obtained from the high-dimension feature vector based on Simhash algorithm, and used fingerprint sequence and distance between fingerprints to detect user-interest features [6]. Giri R et al. used an unsupervised theme model, deduced users' interests according to the files users browsed. [7]. LI C et al. proposed an election approach to extract high-dimensional keywords, and used the high-dimensional keywords to reflect users' interest. [8]. Li S et al. proposed a hierarchical clustering of Chinese words based on graph. The author created users' interest into a structure of hierarchical trees, and used maximum-matching mapping method to match the users' interests in the algorithm [9]. Ma Y et al. proposed a user-interest modeling approach of personal information integration and semantic reasoning based on multiple sources [10].

2. The User-interest Exploration

The main process of creating a user-interest model is to analyze the webpage content users browsed. We obtain the users' interests and compute the degree of users' interest. Moreover, we combine the users' interests and the degree of users' interest to denote users' interest. The procedure is shown as follows.

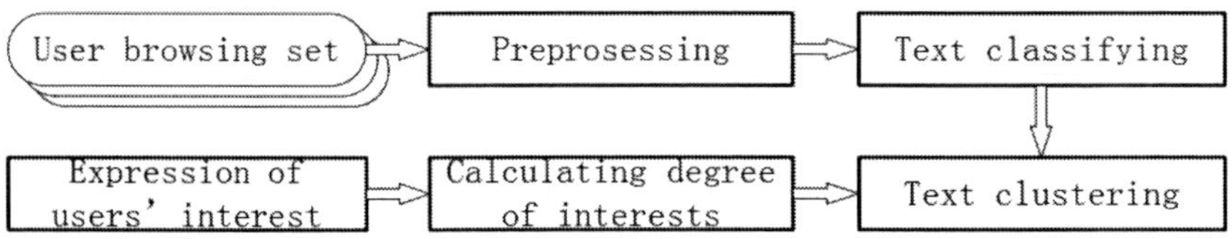

Figure 1. The procedure of user-interest model.

2.1. The Preprocessing of Webpage Texts

A better way of describing the webpage is vector space model (VSM). VSM uses feature words $(t_1, t_2, ..., t_n)$ to denote text d, and each feature words t_i has its weight w_i according to the importance in the text. So all texts could be denoted by feature vector $(w_{t1}, w_{t2}, ..., w_{tn})$. Before describing the webpage texts, we need to partition the words. We remove all tags, which are unrelated to the texts. We also extract webpage title and content as the object of partitioning words in this model.

To avoid the problem of curse of dimensionality, we combine information gain and mutual information to dimensionality reduction for the initial feature word sets. Information gain is used to describe the influence of feature to the whole classifying system. The formula is shown as follows.

$$IG(t)=-\sum_{j=1}^{k}P(C_j)\log P(C_j)+P(t)\sum_{j=1}^{k}P(C_j|t)\log P(C_j|t)+P(\bar{t})\sum_{j=1}^{k}P(C_j|\bar{t})\log P(C_j|\bar{t}) \quad (1)$$

Mutual information is used to evaluate the influence of feature to specific category. Mutual information is shown as follows.

$$MI(t,c_i)=\log\frac{p(t,c_i)}{p(t)\times p(c_i)}=\log\frac{p(t|c_i)}{p(t)} \quad (2)$$

We use the weighted sum for the information gain with the mutual information, and sort the weighted result. The weighted sum need to save k features to obtain the final feature word sets. We assume that k groups are $C_j(j=1,...,k)$, and t is the feature words for one group. $P(C_j|t)$ denotes the probability of feature words t belonging to the group C_j in training set. Let $p(t,c_i)$ denote the probability of feature words t belonging to group c_i in the training set. $p(t)$ denotes the feature t probability of text in training set. $p(c_i)$ is the probability of text belonging to group c_i. If feature t in group c_i has higher probability, feature t has lower probability in other groups. It means that feature t and group c_i has large correlation and obtain higher value of mutual information.

In the mixed measure approach of this paper, we use maximal value of mutual information and the value of information gain to the weighted sum for one feature words t. The formula is shown as follows, and the formula is the degree of feature importance to whole data set or one group.

$$Degree(t)=\alpha IG(t)+\beta MI(t) \quad (3)$$

where αandβ is 0.5. The words will be saved if the degree of value is 80% above in all training sets. We reduce dimension of words largely in this approach, and increase the efficiency of clustering. After obtaining feature word set, we use formula *tf-idf* to calculate the weight of feature words in this approach. In addition, the text can be described by vector space model with feature words and weight. Formula *tf-idf* is shown in formula (4).

$$\mathrm{w(t,d)}=\frac{(1+\log \mathrm{tf(t,d)})\times \log (\mathrm{N}/\mathrm{n_t})}{\sqrt{\sum_k [(1+\log \mathrm{tf(t,d)})\times \log (\mathrm{N}/\mathrm{n_t})]^2}} \tag{4}$$

tf(t,d) is the number of feature *t* in text *d*; *N* is the number of texts, which includes feature *t*. We use vector space model $d_i = (w_{i1}, w_{i2}, \ldots, w_{in})$ to denote each text d_i from all text set $D=\{d_1, d_2, \ldots, d_m\}$.

2.2. The Exploration of User-Interest Cluster

We obtain preliminary user-interest theme groups through sampling observing the history of webpages in this paper. User-interest theme groups mainly includes seven theme classes such as sport, shopping, video, game, military, technology, and education. We grab these theme webpages from some famous websites, and label these classes. Moreover, we divide the theme webpages into training set and testing set, and the ratio of two sets is 7:3. We use these webpages to train the classifying model through decision tree C4.5 algorithm. After classifying the webpages, we can obtain the user-interest distribution roughly.

We cluster each class next in this model. In the process of clustering, the common similarity function is Euclidean distance, Dice coefficient, Jaccard coefficient, cosine distance and so on. We use cosine distance to calculate the similarity of samples in this paper. The formula is shown as follows.

$$\mathrm{Sim(D,Q)}=\frac{\mathrm{D}\cdot\mathrm{Q}}{\|\mathrm{D}\|\times\|\mathrm{Q}\|}=\frac{\sum_i (d_i\times q_i)}{\sum_i d_i^2\times\sum_i q_i^2} \tag{5}$$

D and *Q* is the feature vector of different texts in the same theme, $D = (d_1, d_2, \ldots, d_n)$, $Q=(q_1, q_2, \ldots, q_n)$, and *Sim(D,Q)* means text *D* and *Q* has high similarity if the value of *Sim(D,Q)* is large. We apply K-means algorithm to process fine-grained clustering in this paper. K-means clustering algorithm has high efficiency, and deals with big text sets effectively. We determine *K* through the multiple experiments in this paper. In addition, we can obtain the user-interest clusters after finishing the clustering, and next step is to express users' interests directly.

3. The User-Interest Expression

The most direct way to describe the extent of users' interests is the time of users' staying and the content in the webpages. Generally, small amount of content and long-time of browsing has large value of interests. So we calculate the degree of users' degree through analyzing staying time and the content amount in this paper. Furthermore, we obtain the time of users staying in the webpages and the size of webpages from browsing information in this model, and calculate the degree of user' interests for each page.

T is the time of users staying in the webpages. Length is the size of webpages. The formula of calculating users' interests to the webpages is shown as follows.

$$\mathrm{P_d}=\frac{\mathrm{T}}{\mathrm{Length}} \tag{6}$$

If one user has *n* webpages in interest class I_i, and the degree of users' interests of each webpage is p_{ij}. The formula of the degree of interests in this class is shown as follows.

$$\mathrm{IR_i}=\frac{\sum_{j=1}^{n} \mathrm{p_{ij}}}{\mathrm{n}} \tag{7}$$

We need to normalize the result of formula (7), and obtain the degree of users' interests of one user-interest class. In addition, we construct matrix *IR* to describe the users' interests and the degree of users' interests in this model. *IR* is shown as follows.

$$IR=\begin{bmatrix} IR_1 & t_{1,1} & \cdots & t_{1,m} \\ IR_2 & t_{2,1} & \cdots & t_{2,m} \\ \vdots & \vdots & \ddots & \vdots \\ IR_k & t_{k,1} & \cdots & t_{k,m} \end{bmatrix} \quad (8)$$

We assume that we can obtain k user-interest webpage clusters after classifying and clustering in this model. IR_i is the degree of users' interests in each interest class. Moreover, we choose the feature words of m interest classes to denote the keywords of this class in this model. The interest matrix is denoted by IR.

Let $t_{i,j}$ denote *j-th* keywords of *i-th* interest subclass. The Figure 2 shows users' interests to the degree of interest in hierarchical structure according to the theme division of users' interests.

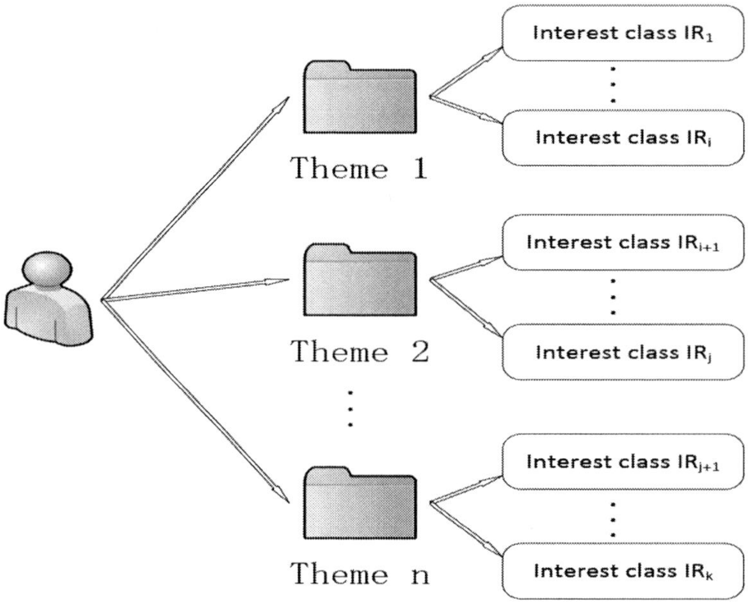

Figure 2. User-interest structure.

4. Experiment Result and Analysis

4.1. Data Set and Evaluation Standard

The experiment includes 13491 webpages. For evaluating the effectiveness of dimensionality reduction, this experiment divides the webpages into training set and test set in the ratio of 7 to 3. In addition, we use these webpages to train classifying model, and classify the webpages. After classifying the data, we finish the process of the user-interest exploration and the user-interest expression.

In the process of clustering, we apply *F-measure* to evaluate the result of clustering in this paper. *F-measure* is defined by accuracy rate and recall rate. For a given class, *TP* denotes the number of samples, which are classified correctly, *FP* denotes the number of samples, which are classified incorrectly, and *FN* denotes the number of the samples, which belong to this class but are classified to the other class. The accuracy rate *P* and recall rate *R* are defined as follows.

$$\text{Accuracy rate: } P=\frac{TP}{TP+FN} \quad (9)$$

$$\text{Recall rate: } R=\frac{TP}{TP+FP} \quad (10)$$

$$\text{F-measure: } F=\frac{2PR}{P+R} \quad (11)$$

4.2. Experiment Result

We preprocess the user browsing webpages in this paper, and obtain text feature. Furthermore, we use C4.5 algorithm to classify the texts according to the theme. The accuracy rate and recall rate of classifying is 82% and 85% respectively, and we use the model, which has been trained by theme class webpages, to classify the user browsing webpages. We need cluster the webpages in each class, and we obtain the user-interest cluster in this theme. Table 1 is the result of clustering, which MAC address is 00-22-15-7F-4A-D0, clustering in game theme.

Table 1. The Result Of Clustering

name	title	class	keywords
10700.htm	Secretly makeup	cluster0	Dress up, Makeup, Fashion, Operation, management
10953.htm	Bobbi doll's student outfit	cluster0	
11078.htm	Cosmetics shop	cluster0	
10736_3.htm	Ultimate Parkour	cluster1	Racing, driving, sports, off-road, riders
11675.htm	Altman motor racing	cluster1	
10896.htm	security police	cluster2	Adventure, challenge, fight, gun battle, shooting
10904.htm	Siege cannon	cluster2	
10519.htm	Escaping Chemistry Laboratory	cluster3	Leisure, puzzle, picture matching, the chamber of secrets, maze
10541.htm	Pet picture matching	cluster3	
30110.htm	Mermaid fault	cluster3	

Table 1 shows the name, title, cluster number of webpages users browsing. We select five feature words as the keywords of each class in this paper. User game theme class has four interest clusters in Table 1, and each interest class has several keywords. For evaluating the result, we calculate *F-measure* of clustering and classifying before clustering. The result is shown in Table 2.

Table 2. Clustering and Classifying-clustering

Clustering								Classifying-clustering							
class	N	TP	FP	FN	P	R	F	class	N	TP	FP	FN	P	R	F
1	331	201	142	130	0.61	0.59	0.60	1	331	258	74	73	0.78	0.78	0.78
2	452	324	134	128	0.72	0.71	0.71	2	452	394	52	58	0.87	0.88	0.88
3	401	296	98	105	0.74	0.75	0.74	3	401	329	78	72	0.82	0.81	0.81
4	437	301	143	136	0.69	0.68	0.68	4	437	362	74	75	0.83	0.83	0.83
5	302	198	137	104	0.66	0.59	0.62	5	302	236	70	66	0.78	0.77	0.78
6	420	294	104	126	0.70	0.74	0.72	6	420	343	81	77	0.82	0.81	0.81
7	436	325	132	111	0.75	0.71	0.73	7	436	354	79	82	0.81	0.82	0.81
8	320	213	122	107	0.67	0.64	0.65	8	320	263	59	57	0.82	0.82	0.82
9	468	365	104	103	0.78	0.78	0.78	9	468	421	79	47	0.90	0.84	0.87
10	231	182	51	49	0.79	0.78	0.78	10	231	207	53	24	0.90	0.80	0.84

The approach of classifying before clustering has higher accuracy rate and recall rate than clustering in Table 2. *F-measure* in classifying-clustering is almost 80% above. So we choose classifying-clustering to increase the accuracy rate. Figure 3 shows the result of classifying-clustering.

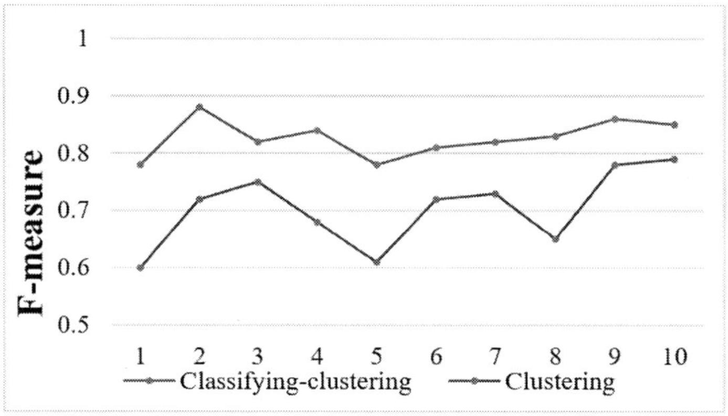

Figure 3. The comparison of *F-measure*.

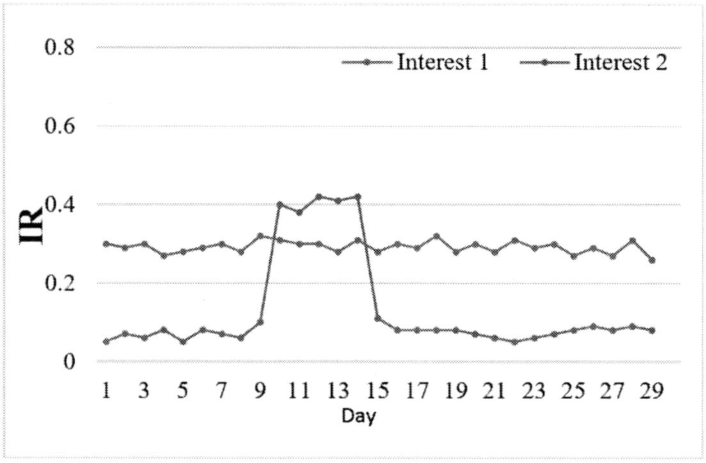

Figure 4. The change of users' interest.

Table 5 shows the interest distribution of the user *a* and *b*.

Table 3. User interest

user	Interest class
User *a*	{Shopping (makeup, facial mask, eye cream, 0.37), (floral skirt, shirt, 0.25), (computer, cell phone, charger 0.2)}, {video (romance, love, 0.4), (science fiction, exploration, space, 0.25)}
User *b*	{sports (rockets, basketball, home, 0.45), (swimming, diving, world championships 0.3)}, {video (detective, suspense, darkroom, 0.32), (terror, horror, crime, 0.28)}

From Table 3, we can know that user *a* like shopping and video. In these two interests, user *a* likes makeup best. In one interest video, the type of user *a* and *b* has obvious differences. User *a* likes romance, and user *b* likes suspense videos. In addition, the experiment studies the change of users' interests over time. Figure 4 shows the change of user *a* and user *b* over time. We can know that interest 1 has little changes, but the change of interest 2 is large in a certain period. So we can recognize interest 1 as stable interest and long-term interest, and interest 2 is the short-term interest. The division of long-term interest and short-term interest can offer better personalized service to the users.

5. Conclusion

The accuracy of user-interest model decides the quality of personalized service directly. We propose an information gain and mixed measure of mutual information approach for dimensionality reduction of webpages in this paper. In the process of user-interest exploration, we need classify the webpages into several classes. This approach narrows the range of clustering, and increases the accuracy of users' interests. Moreover, we calculate the degree of interest using user-staying time and the webpage content, and describe users' interests with the content of users' interests in this paper.

6. Acknowledgments

This work is supported by China Academy of Engineering Physics Project 2015A0403002, National Industrial Technology Development Program JCKY2016212C005， and National Natural Science Foundation of China Project U1636208 and U1536116.

7. References

[1] Organero M M, Kloos C D, Merino P M. Personalized service-oriented e-learning environments [J]. Internet Computing, IEEE, 2010, 14(2): 62-67.
[2] He L, Jia Y, Han W, et al. mining user interest in microblogs with a user-topic model [J]. Communications, China, 2014, 11(8): 131-144.
[3] Jayarathna S, Patra A, Shipman F. Mining user interest from search tasks and annotations[C]//Proceedings of the 22nd ACM international conference on Conference on information & knowledge management. ACM, 2013: 1849-1852.
[4] Liu Z, Chen X, Sun M. mining the interests of Chinese microbloggers via keyword extraction [J]. Frontiers of Computer Science, 2012, 6(1): 76-87.
[5] Kumar S. Mining User Interests from Web History[C]//Web Intelligence/IAT Workshops. 2013: 282-283.
[6] Liu D, Wu Q, Han W. Mining micro-blogging users' interest features via fingerprint generation[C]//Proceedings of the 2nd International Conference on Computer Science and Electronics Engineering. Atlantis Press, 2013.
[7] Giri R, Choi H, Hoo K S, et al. User Behavior Modeling in a Cellular Network Using Latent Dirichlet Allocation[M]//Intelligent Data Engineering and Automated Learning–IDEAL 2014. Springer International Publishing, 2014: 36-44.
[8] LI C, WANG J, GUO Z, et al. User Interest Modeling Based on Terminal User Behavior Analysis [J]. Journal of Computational Information Systems, 2015, 11(1): 349-356.
[9] Li S, Wu G, Hu X. Hierarchical user interest modeling for Chinese web pages[C]//Proceedings of the Third International Conference on Internet Multimedia Computing and Service. ACM, 2011: 164-169.
[10] Ma Y, Zeng Y, Ren X, et al. User interests modeling based on multi-source personal information fusion and semantic reasoning [M]//Active Media Technology. Springer Berlin Heidelberg, 2011: 195-205.

IOP Conf. Series: Journal of Physics: Conf. Series **887** (2017) 012062 doi:10.1088/1742-6596/887/1/012062

A risk evaluation model and its application in online retailing trustfulness

Ruyi Ye and Yingcheng Xu

China National Institute of Standardization, 100191, Beijing China PRC
Email: kyleyear@qq.com

Abstract. Building a general model for risks evaluation in advance could improve the convenience, normality and comparability of the results of repeating risks evaluation in the case that the repeating risks evaluating are in the same area and for a similar purpose. One of the most convenient and common risks evaluation models is an index system including of several index, according weights and crediting method. One method to build a risk evaluation index system that guarantees the proportional relationship between the resulting credit and the expected risk loss is proposed and an application example is provided in online retailing in this article.

1. Introduction

Risk evaluation is common in many areas, such as enterprise management planning, credit rating and information security. In many cases, the evaluation result is only a numeric score, it's important to guarantee the comparability of the result, and the reasonable way is to build an evaluation model in advance.

2. Model

2.1. Risk Evaluation

The expected loss, the product of the probability that the risk happens and the loss once it happens is used usually to measure the severity of risk:

$$S = P(R) \cdot L(R)$$

The methodology to assess the risk by the expected loss is not new. Early in 1947, Learned Hand put forward a famous "*B*>*PL*" rule [1] concerning civil liability, which later became the foundation of the study of "law and economics".

$P(R)$ refers to the probability that the risk R happens, and $L(R)$ the loss caused by the risk R. In information security, $P(R)$ is the product of the probability of being attacked and that of the system being breached under attacking, and the latter is the indicator of the vulnerability of the system [2]. $L(R)$ can be the amount of economic loss, or the utility changes if the type of the risk appetite is considered.

Usually multi-point risks exist in one system, and the total expected loss of all risks can be used to measure the severity of the risks, and the summation of expected losses of all risk points is:

$$\sum_{R_i} P(R_i)L(R_i) \tag{1}$$

2.2. Model of Risks Evaluation

For the convenience and comparability of repeating the risks evaluation, an index system is built in advance adaptable to a special area. Usually, the loss in money or the economic utility of the loss caused by a given risk happening do not change widely for a special risk bearer. For example, the loss caused by 10-day delay of materials delivery is almost the same whoever the seller is. Therefore theoretically, the loss can be evaluated in advance for all kinds of risk points once happening, and then the probabilities according to different sellers evaluated in order to make a better choice.

Assume that n risk points are treated$R_i(i = 1 \dots n)$, n indexes with n credits $C(R_i)$ are used to evaluated them, the according probabilities are$P(R_i)$, and losses $L(R_i)$ respectively.

To obtain a comprehensive score, it's required to appoint n weights $W(R_i)$ respectively, and the sum of them is 100 in centesimal scale without loss of generality:

$$\sum\nolimits_{R_i} W(R_i) = 100$$

According to Eq. 1, the expected loss caused by total risks is:

$$\sum\nolimits_{R_i} P(R_i)L(R_i) = \sum\nolimits_{R_i} \{1 - [1 - P(R_i)]\} \cdot L(R_i)$$

$$= \sum\nolimits_{R_i} L(R_i)/100 \cdot \sum\nolimits_{R_i} \{1 - [1 - P(R_i)]\} \cdot \frac{100 \cdot L(R_i)}{\sum_{R_i} L(R_i)}$$

Since $\sum_{R_i} L(R_i)/100$ is constant, if let the weights and the credits as following:

$$W(R_i) = \frac{100 \cdot L(R_i)}{\sum_{R_i} L(R_i)} \tag{2}$$

$$C(R_i) = [1 - P(R_i)] \tag{3}$$

Then the expected loss is:

$$K \cdot \sum\nolimits_{R_i} [1 - C(R_i)] \cdot W(R_i)$$

$$= K \cdot \sum\nolimits_{R_i} W(R_i) - K \cdot \sum\nolimits_{R_i} C(R_i) \cdot W(R_i)$$

$$= 100 \cdot K - K \cdot \sum\nolimits_{R_i} C(R_i) \cdot W(R_i) \tag{4}$$

Therefore, if let the every credit be the probability that the according risk does not happen, and the ratio of the loss caused if it happens to the total loss if all risks happen, the comprehensive score obtained is a multiple of the expected loss caused by all risks, and there are many advantages of such method to determine the weights and credits:

- The credit of single index is linear to the probability of the according risk happening, and the higher the probability is, the lower the score, and vice versa.
- The weight of an index is proportional to the according loss caused if happening. Then the weight will be small if the loss of the according risk is negligible though it can happen often.
-The comprehensive score is linear to the expected loss caused by all risks, and then the higher the score is, the lower the risk is. If the comprehensive score is close to 100, the expected loss is close to 0 or negligible, and if the score is close to 0, the expected loss is high and all risks have a high probability to happen. So the comprehensive score includes the total information of the expected loss caused by all

risks. In other word, the lost score is proportional to the severity level of the total risk under this ideal evaluation model.

3. Application Sample of Online Retailing

Online retailing is one kind of ecommerce, serving large number of consumers with large number of small orders. There are tens of thousands of online retailing website in China, and even millions of sellers on 3rd-part (such as taobao.com, jd.com) retailing platforms. Facing so many retailing sellers, a consumer may be confused, since he/she can't know their trustfulness and the according risk. In such situation, some independent organizations like BBB online provide the trustfulness evaluation service for the online retailers, and which help consumers to distinguish the retailers with lower according risk [3, 4].

Using Eq.3 and Eq.4, the risk evaluation index system for online retailing trustfulness is built as following 3 steps:

1) Sort out main online shopping risk

Consumers may face many kinds of risk when shopping online, among which only 4 usual risks are chosen to be analyzed here for illustration purpose:

- quality flaw
- discrepancies between the physical goods and descriptions
- delay on delivery
- bad after service

2) Evaluate the loss if the according risk happen and determine the weights

For simplicity, the monetary loss caused by risks is considered when the online shopping amount is 100 dollars. To calculate the loss according to a specific loss $L(R_i)$, several consumers can be inquired, and the average value can be used.

The data in table 1 are estimated according to some sampling survey:

Table 1. Average loss and weights according to 4 risk points

risk points	frequency	Average loss (ratio to the goods price)	weight
quality flaw	often	23%	28 [a]
discrepancies	often	36%	43
delay on delivery	often	5%	7
bad after service	relatively often	18%	22

[a] Notice, the weights are calculated using Eq. 2, for example the first weight of the quality flaw is obtained by 23% / (23%+36%+5%+18%) * 100 =28

3) Determine the standard of rating

According to Eq. 3, the score of every index is 1 minus the probability that the risk happens, in other word the probability that it doesn't happen. In reality, the frequency that a risk happens in a past period can be used to substitute the probability. Though the loss caused by a specific risk is almost the same to the consumer, the frequency is different from seller to seller, and so different seller can obtain different evaluation score of trustfulness.

4. Discussion and Conclusion

Theoretically, this index system with the loss caused by every risk point once happening as weight and the probability that the risk point doesn't happen as the rating standard is suitable to the risk evaluation case that many similar object is to be evaluated with the same risk points. The advantage and validity is that the comprehensive score is linear to the expected loss caused by all risks and the evaluation process is simple and convenient.

It's usual that some risks are correlated, and this doesn't affect the calculation of the expected of the total loss.

But problem comes since the Ls are assumed as constants in the Eq.2 if the danger L caused by a specific kind of risk changes widely, i.e. L cannot be looked as a constant for a specific risk, and for example a quality flaw can cause a big loss at this time or be trivial in another. There are two measures can be taken to solve this problem: the first is to split this risk into multiple risk point according to their danger once it happens, and the *L*s become constant for different risk point after splitting; and the second is to use the average loss to calculate the weight and consider the loss besides the probability when rating the according indicator.

Yet if the normal probabilities *P*s for every risk vary widely, it can become improper to deduce weights only from the according dangers when the risks happen. Some risks may cause serious danger but very seldom happen, for example, while deadly contaminated food and online economic crime can cause serious danger and loss, they are exceedingly rare on the familiar online retailing websites, and therefore if the weights are allocated too high, the discrimination of the whole index system will be reduced. Therefore, the probability of risk happening should be considered also to deduce a reasonable weight if the average probabilities are not comparable between different risks.

5. Acknowledgement

This research was financially supported by the 552017Y-5267 research fund.

6. References

[1] Learned Hand, U.S. v. Carroll Towing, 159 F.2d 169 (2d Cir. 1947), (http://en.wikipedia.org/wiki/Learned_Hand)
[2] Lawrence A. Gordon and Martin P. Loeb, The Economics of Information Security Investment, ACM Transactions on Information and System Security, Vol.5, No. 4, Nov. 2002, Pages 438-457
[3] Audun Jøsang, Roslan Ismail and Colin Boyd, "a survey of trust and reputation systems for
[4] online service provision", Decision Support Systems, 43(2) 2007, p.618-644
[5] Cheskin Research and Studio Archetype /Sapient, "eCommerce Trust Study", 2010

IOP Conf. Series: Journal of Physics: Conf. Series **887** (2017) 012063 doi:10.1088/1742-6596/887/1/012063

An ICP algorithm based on block path closest point search

Kuisheng Wang[1], Xing Li[1], Hongwei Lei[2] and Xiaorui Zhang[1]

[1]College of Computer Science and Technology, Xi'an Shiyou University, Xi'an Shanxi, 710065, China.
Email: kshwang@xsyu.edu.cn, 401164051@qq.com, 541750944@qq.com
[2] Guangdong Foshan Anda Technology company, Guangzhou, 510000, China.
Email: leihongwei@139.com

Abstract. At present, the traditional ICP algorithm has the problems of low efficiency and low precision. To solve these two problems, an ICP algorithm based on block path closest point search is proposed in this paper. The idea of the algorithm is as follows: firstly, the point cloud data is divided into blocks, and the nearest point block corresponding to the target point cloud is searched by the path method. Secondly, according to the global method, the nearest point can be determined only by finding the nearest point block, and complete all the closest match. The experimental results show that the improved ICP algorithm has faster speed and higher precision than the traditional ICP algorithm, for a large number of point cloud data advantage is more obvious.

1. Introduction

With the development of computer aided design technology, 3D reconstruction of real object has become the focus of computer vision and graphics and other related fields. Three dimensional reconstruction technology has a very wide range of applications in all fields of real life, including industrial automation, archeology, cultural heritage and computer-aided medicine.

At present, with the maturity of 3D scanning terminal, it is not a big problem to obtain accurate 3D model coordinates, but the acquisition of point cloud data on the surface of objects often needs to be completed by repeated measurements. In order to obtain the point cloud data of the whole surface of the object, it is necessary to integrate and match the local point cloud data. Data registration is one of the most important data processing objects in 3D object modelling. The precision of registration directly affects the accuracy of model building.

The iterative closest point algorithm keeps the point cloud closest to each other by iteration, at which point the cloud space is consistent [1-2].The traditional ICP algorithm is inefficient, so many researchers at home and abroad have made efforts to improve the ICP algorithm. In recent search, we propose to use point tangent planes to approximate point clouds, and finally to solve the least square distance from point to tangent plane, but this method is still relatively slow [3].There are two ways to improve the speed, which are reverse scaling and random search, but have some influence on the registration accuracy.

This paper focuses on the problem of long running time of the algorithm, the ICP algorithm was optimized ,through data block processing, reduce the number of calculations, improving the efficiency and accuracy of point cloud registration, it is of great significance for practical measurement and application.

2. The Principle of ICP Algorithm

ICP algorithm was first proposed by Besl, Mckay [4] and Chen in 1992. It is a set of point set (PSTPS) based on four variables.

The ICP algorithm is mainly used for 3D point cloud registration, the main process is: given two three-dimensional cloud data sets from different coordinate systems, we can find the spatial transformation of the two point clouds so that they can make spatial matching. Assuming {Pi, i=1,2,... } represents the first point cloud in the space, and the alignment, matching, transformation of the second point cloud Qi makes the objective function minimal[5].

$$f(\mathrm{R},\mathrm{T})=\sum_{i=1}^{N}\|Q_i-(\mathrm{R}\,P_i+T)\|^2=\min \quad (1)$$

The essence of the ICP algorithm is the optimal least squares matching algorithm, it repeats the process to determine the relation between point clouds computing optimal rigid transformation, until a convergence criterion is satisfied.

The purpose of the ICP algorithm is to find the rotation R and the translation T transform matrix between the target point cloud and the reference point cloud, so that the two cloud data can be optimally matched under some metric criteria. Specific steps of the ICP algorithm [6]: k represents the number of iterations.

1) Pick up clouds in the target cloud P, $P_i^k \in P$;

2) Calculate the corresponding points in the reference point cloud, making $\|Q_i^k - P_i^k\| = \min$;

3) Singular value decomposition, we obtain the rotation matrix R_K and the translation matrix T_K, making $\sum_{i=1}^{N}\|Q_i^k-(R^kP_i^k+T^k)\|^2=\min$;

4) Calculate $P^{k+1}=\{P_i^{k+1}=R^kP_i^k+T^k, P_i^k\in P\}$;

5) Calculate $d^{k+1}=\frac{1}{n}\sum_{i=1}^{N}\|P_i^{k+1}-Q_i^k\|^2$;

6) If the distance d^{k+1} does not reach a given minimum distance d, returns to (2) until the d^{k+1}<d or iteration number is greater than the default maximum iteration.

For each iteration of the ICP, the mean variance of the corresponding points is minimized so that point cloud P_i^k is closer to Q_i^k, and Q_i^k is the nearest point of P_i^k in Q_i. Thus, each iteration makes P_i closer to Q_i.Based on this idea, Besl et al demonstrate the convergence of the ICP algorithm [6].

The traditional ICP algorithm usually selects all points in the point cloud as the reference point, finding the nearest corresponding point.However, in actual measurement, the generated cloud data usually has hundreds of thousands or even millions of points. At this point, if point cloud registration using standard ICP algorithm, although it can basically meet the requirements of registration accuracy, it takes too long and the actual efficiency is not high.

3. ICP Algorithm Based on Block Path Closest Point Search

The main idea of ICP algorithm based on block path nearest point search is as following: firstly, the reference points data is partitioned into blocks, then find the nearest closest block according to the distance between the block and the registration point cloud, finally,calculating the distance between the registration point and all points in the nearest closest block,the point in block with minimum distance is the corresponding point.Similarly, all the corresponding points will be found. The algorithm can reduce the running time and improve the efficiency but didn't change the amount of data. Assuming that H represents the reference point cloud, P represents the point cloud to be matched.

The concrete steps of algorithm are as follows:

1) Get the reference point cloud H and the registration point cloud P;
2) The reference point cloud is partitioned into blocks;
3) Find the nearest block to the registration point;
4) Search the nearest point in the nearest block finding in the step 3 and get the closest point cloud B;
5) By the Singular value decomposition of matrix P and B, we obtain the rotation matrix R_K and the

translation matrix T_K;

6) Calculate the formula $P_{RT} = R * P + T$ and obtain the matrix P_{RT} after the coordinate transformation;

7) Calculate the distance between the reference point cloud H and P_{RT}, $d = \frac{1}{n}\sum_{i=1}^{N} \|P_i - H_i\|^2$;

8) To determine whether the d is convergent.If d is convergent,we can obtain the registration point cloud after coordinate transformation, or jumps to step 2 and calculate again.

4. Implementation of ICP Algorithm Based on Block Path Closest Point Search

According to the main idea of the ICP algorithm based on block path closest point search, the flow chart of its algorithm implementation is shown in Figure 1.

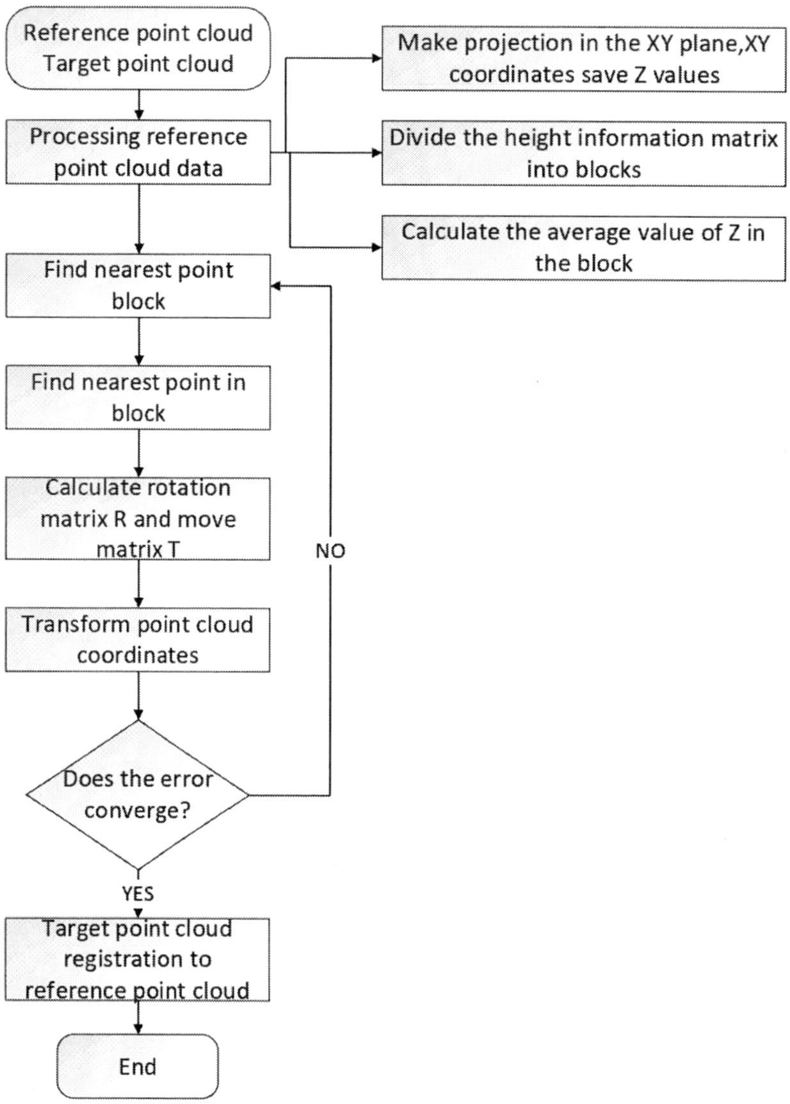

Figure 1. The flow chart of ICP algorithm based on block

The steps can be summarized as follows:

1) The reference point cloud H is projected on the XY plane and the coordinate values x and y of each point can be obtained. Then we create the height matrix which store the Z value of the reference point,the coordinate of height matrix is the value of x and y obtained by project.
2) The matrix Height is processed and divided into blocks of 4*4, 8*8, or 16*16 sizes;
3) Calculate the average of each block,which is the data stored by the new matrix ZMean after segmentation.
4) According to the path shown in figure 2 to find the nearest point block:
a) First, the initial block is determined according to the coordinates of the target points,making it a central point block;
b) Second, calculate the distance between the target point and the four point blocks adjacent to the central point block based on the path,finding the block which distance is the smallest as the new center point block.
c) Repeat the second step, until the nearest point block is found, which is the center point block itself;

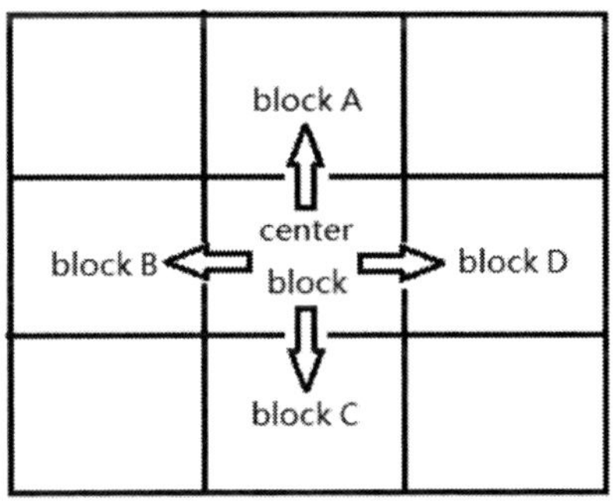

Figure 2. The calculate path

5) In the nearest point block, the nearest point is searched by the global method, and the point cloud B is obtained;
6) Computing the singular value decomposition (SVD) of the reference point cloud H and the nearest point cloud B, and the rotation and shift matrix R and T are obtained;
7) Calculate $P_{RT} = R * P + T$ (P is the registration point cloud), obtaining the registration point cloud after coordinate transformation;
8) Calculate the distance between the reference point cloud H and P_{RT}, $d = \frac{1}{n}\sum_{i=1}^{N} \| P_i - H_i \|^2$;
9) Determine whether the distance d is less than 0.1mm, less than is convergence, or jump to step 4);
10) If convergent,we can obtain the registration point cloud after coordinate transformation.

5. Experimental Results and Analysis

The experiment was conducted on Intel (R), Core (TM), i7-4790, CPU, 3.60GHz, MATLAB, 4G memory computer using simulation experiments.

Experiments and calculations were carried out on different block sizes of 4*4, 8*8 and 16*16.Balancing the running time and convergence error two factors,we could determine that when the block size is 16*16, the registration of point cloud data is more accurate and effective, as shown in table 1.

Table 1. Comparison of running time and convergence error of different block sizes

Block size (points)	The running time of the improved algorithm (s)	Convergence error (mm)
4*4	256.719	0.5992
8*8	402.639	0.0628
16*16	460.880	0.0326

Experiments were performed using points of the same amount of data,with the same time,the accuracy of the traditional ICP algorithm is lower than that of ICP algorithm based on block path closest point search,the registration was incomplete,but the algorithm based on block path closest point search has good matching effect, such as figure 3 and figure 4, we can see that the two point clouds has a little distance in figure 3.To the contrary, the two clouds have been coincide in the figure 4.

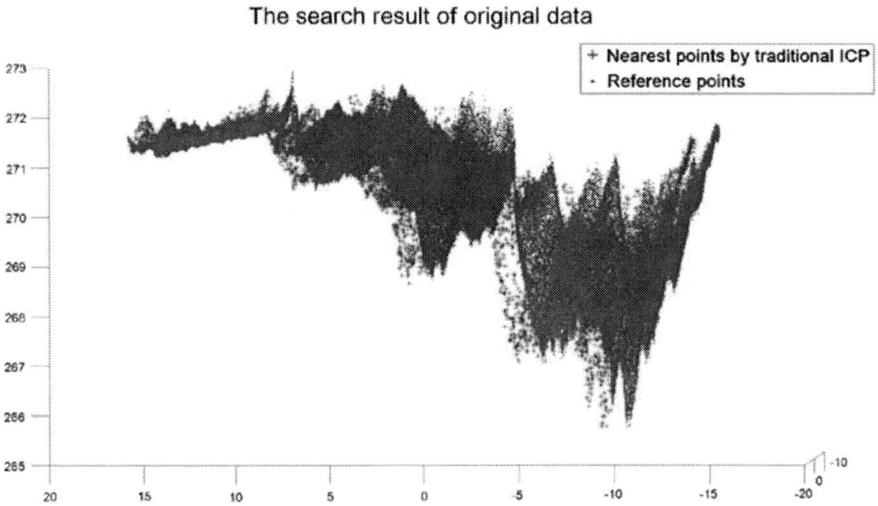

Figure 3. Traditional ICP algorithm

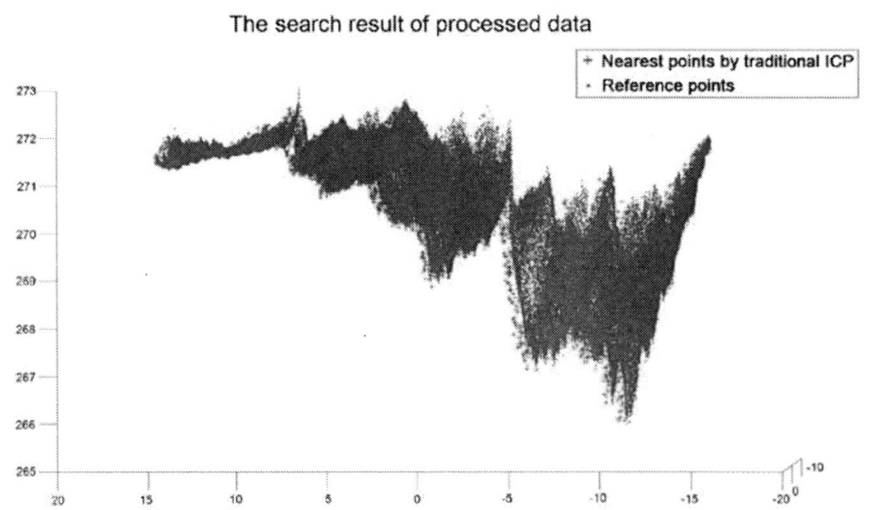

Figure 4. Improved algorithm

Comparing and analyzing the experimental data, we can see that:
1) Doing experiments with the same amount of data points when the registration accuracy is the same, although the number of iterations of the ICP algorithm proposed in this paper is more than that of the traditional ICP algorithm, its running time is much lower than the traditional ICP algorithm, and the efficiency is very high, as shown in table 2;
2) When the amount of point cloud data is larger, the advantage of ICP the algorithm based on block path closest point search is more obvious in registration efficiency and accuracy. It is important for practical measurement and application, as shown in table 2.

Table 2. Comparison of two algorithms for large amounts of data

The number of points	The running time (s)		Convergence error (mm)		Iteration number	
	Traditional algorithm	Improved algorithm	Traditional algorithm	Improved algorithm	Traditional algorithm	Improved algorithm
157058	199.917	188.936	0.0380	0.0321	15	30
200642	413.428	372.112	0.0304	0.0357	20	40
248255	872.783	460.880	0.0329	0.0326	20	40

6. Concluding Remarks

In this paper, an ICP algorithm based on block path nearest point lookup is proposed, the speed and accuracy of registration are optimized by looking for the appropriate block size. The experimental results show that it is more accurate and effective to use this algorithm for point cloud data registration when the block size is appropriate, especially when the amount of data is large.

The algorithm has higher registration accuracy and convergence speed, and has high practical value in the actual three-dimensional reconstruction.

7. References

[1] JIANG Chengcheng, HU Tongsen, ZHOU Wei, 2009, an improved Iterative closest point's algorithm.
[2] ZHOU Chun-yan, LI Yong,ZOU Zheng-rong,2011,Three -Dimensional Cloud ICP Algorithm Improvement.
[3] Blais G, Levine MD 1995 J. Registering multiview range data to create 3D computer graphics. IEEE Transactions on Pattern Analysis and Machine Intelligence, p 820-824.
[4] Besl PMakay 1992 J. A Method for Registration of 3D Shapes. IEEE Transactions on Pattern Analysis and Ma -chine Intelligence p 239-256.
[5] Ezra E,Sharir M,Efrat A 2008. On the performance of the ICP algorithm. Computational Geosciences.
[6] YANG Jun, ZHANG Yao, HUANG Liang 2016, Research on 3D model registration by improved ICP algorithm.
[7] ZHANG Xiaojuan, I Zhongke, ANG Xianze,etr al 2012.Research of 3D point could data registration algorithms based on feature points and improved ICP p 116-118.
[8] YANG Xaoqing 2015 vol 36, Point could registration based on improved ICP algorithm.Computer Engineering and Design.
[9] ZHU Xinyu,WAN Jianhua,LIU Shanwei,ZENG Zhe 2015, Improved ICP Algorithm for Point Cloud registration, Hydrographic Surveying and Charting,vol 35.
[10] ZHONG Ying, ZHANG Meng 2014, Automatic Registration Technology of Point Cloud Based on Improved ICP Algorithm, Control Engineering of China,vol 21.
[11] Wanxin XU, Hongli XU, 2012, Application of Improved ICP Algorithm in Point Cloud Registration, System Simulation Technology & Application (Volume 14).

IOP Conf. Series: Journal of Physics: Conf. Series **887** (2017) 012064 doi:10.1088/1742-6596/887/1/012064

A clustering algorithm based on maximum entropy principle

Yang Zhao[1,2] and Fangai Liu[1,2,3]

[1]College of Information Science and Engineering, Shandong Normal University, Jinan Shandong 250358, China;
[2]Shandong Provincial Key Laboratory for Distributed Computer Software Novel Technology, Jinan Shandong 250358, China;
[3]Corresponding author: E- mail: is.zhaoyang@hotmail.com

Abstract. Aiming at the shortcomings of clustering performance of many traditional text clustering methods, a clustering algorithm based on maximum entropy principle is proposed. The algorithm uses the cosine similarity measure cited in the traditional text clustering algorithm SP-Kmeans, and then introduces the maximal entropy theory to construct the maximal entropy objective function suitable for text clustering. The maximum entropy principle is introduced into the spherical K-mean text clustering Algorithm. The experimental results show that compared with DA-VMFS and SP-Kmeans algorithms, in addressing the large number of text clustering problem. The performance of CAMEP clustering algorithm is greatly improved, and has a good overall performance.

1. Introduction

With the growth of the World Wide Web and various text resources, people's desire for rapid, accurate and comprehensive access to information is increasing. Text clustering technology has received more and more attention and research as unsupervised clustering technology. In present, text clustering technology has become the key technology of automatic text categorization [1].

In a certain vector space model, the text can be expressed as a vector of high dimensional space by appropriate preprocessing, which has sparsity and Non-negative [1].In recent years, research shows that the text data also has the direction [2].This feature allows the text vector data to be normalized before clustering, and then the clustering analysis is performed. The SP-Kmeans [3] algorithm uses the cosine similarity to measure the correlation of the text vectors.

In recent years the great entropy principle has also been widely used in natural language processing [7] and text classification [8]. Surian D [9] in pointed out that the text clustering algorithm based on mixed vMF density model movMF in the text clustering process hidden variable entropy changes with self-annealing characteristics, Shi Zhong [10, 11] The deterministic annealing technique is used to improve the clustering performance of the movMF [14] algorithm, which provides the basis for introducing the maximum entropy principle in the traditional text clustering algorithm.

2. Algorithm for Maximal Entropy Clustering Algorithm and Spherical K-means Clustering Algorithm

Using the statistical physical degradation process, Yasuda proposed a deterministic annealing technique [4], which is an important branch of natural law. It is based on the annealing process, the optimal solution of the optimization problem into a series of temperature changes with the physical system of free energy function is minimal. Karayianni [5] Introduced deterministic annealing techniques into clustering. In this algorithm, a very large entropy clustering algorithm is proposed, and its essence is to use the deterministic annealing technique to find the objective function of clustering minimum. In a variety of versions of the maximum entropy clustering algorithm MEC [15], although the description is different, but only the formal differences. The MEC of the maximal entropy clustering algorithm is introduced only in the literature [6].

For the dataset $X=\{x_1, \cdots, x_N\}\subset R^d$, $V=\{v_1, \cdots, v_N\}$ is the K clustering center, $v_i\in R^d, 2\le K<N, U=\{u_{i,j}\}_{k\times n}$, $k>n$ is a membership matrix, $u_{i,j}$ for each sample belongs to the center of the probability.

$$u_{i,j}\in[0,1], 1\le j\le n$$
$$\sum_{i=1}^{K} u_{i,j}=1 \tag{1}$$

x_i is divided into K clusters $G_i\in(i=1,2,\cdots, K)$,and the clustering of each cluster is obtained by the maximum entropy fuzzy clustering algorithm MEC Center, the following objective function is minimized.

$$J_T(U,I)=\sum_{i=1}^{K}\sum_{j=1}^{N}u_{i,j}\|x_j-v_i\|^2+T\sum_{i=1}^{K}\sum_{j=1}^{N}u_{i,j}\ln u_{i,j} \tag{2}$$

Where $\|x_j-v_i\|^2=(x_j-v_i)^T\times(x_j-v_i)$,T is the Lagrange multiplier. The above equation can also be expressed as:

$$J_T=J_c(U,V)-TH(u) \tag{3}$$

Where $J_C(U,V)=\sum_{i=1}^{K}\sum_{j=1}^{N}u_{i,j}\|x_j-v_i\|^2$.

For a large T, the main attempt is to maximize the entropy H(u), the system is maintained at a higher temperature, with the decrease of T, entropy for the reduction of distortion, when T tends to zero, the minimum Jc (U, V) Directly obtain a non-random (hard) solution. Thus the Lagrangian multiplier there is equivalent to the temperature coefficient of the deterministic annealing technique, also known as the annealing coefficient.

The basic steps of the maximum entropy clustering algorithm MEC are as follows:

Initialization: Given the initial clustering center $V^0=\{v_1^0, v_2^0, v_3^0, \cdots, v_K^0\}$, and the fuzzy partitioning matrix $U=\{u_{i,j}\}$, l is the number of iterations, the maximum number of iterations M, set the annealing coefficient T, the minimum annealing coefficient MinT threshold ε .

Update $u_{i,j}^{l+1}$ with the following formula:

$$u_{i,j}=\frac{\exp(-\frac{\|x_j-v_i\|^2}{T})}{\sum_{h=1}^{K}\exp(-\frac{\|x_j-v_h\|^2}{T})} \tag{4}$$

Update v_i^{l+1} with the following formula:

$$v_i=\frac{\sum_{j=1}^{N}u_{i,j}x_j}{\sum_{j=1}^{N}u_{i,j}} \tag{5}$$

If T is minimized, stop; otherwise adjust the annealing factor T = T-! T to (2).

MEC algorithm can avoid the local minimum and get the global minimum, which has been widely used. However, one of the defects of the MEC algorithm is to use the European metric. For the high-dimensional vector data, the direction feature of the text vector is more important than the size feature, so the MEC is not suitable for clustering the text data.

3. The Clustering Algorithm Based on Maximum Entropy Principle

In this paper, the maximum entropy principle is introduced into the spherical mean clustering, and the clustering algorithm based on maximum entropy principle CAMEP is deduced for text clustering. $V=\{v_1,v_2,\cdots, v_K\}$ is the K clustering center for the sample set $X=\{x_1,\cdots, x_N\}\subset R^d$, $x_i^T x_i=1(1\le i\le N), V=\{v_1,v_2,\cdots, v_K\}$ and $K\times N$ is a membership matrix. $u_{i,j}$ is the degree of membership of the sample x_j belonging to the K center, the value of which is different from the hard division of the spherical K mean, but the fuzzy division between 0 and 1 truly reflects the data points and the center of the class Practical relationship, and meet $\sum_{i=1}^{K} u_{i,j} = 1$ .At this point the global maximum cost function can be considered:

$$J = \sum_{i=1}^{K}\sum_{j=1}^{N} u_{i,j} x_j^T v_i \tag{6}$$

In order to obtain the maximum value of the equation (6), we can get the maximum entropy principle by avoiding the local minimum and get the global minimum. In this case, we can define the minimized objective function:

$$J_T(U,V) = -\sum_{i=1}^{K}\sum_{j=1}^{N} u_{i,j} x_j^T v_i + \frac{1}{T}\sum_{i=1}^{K}\sum_{j=1}^{N} u_{i,j} \ln u_{i,j} \tag{7}$$

Note that the form of Eq. (7) is similar to Equation (2), and entropy terms are introduced. (2) Using the European measure, and (7) uses a cosine similarity measure. Equation (7) can also be expressed as:

$$J_r = J_c(U,V) - \frac{1}{T} H(u) \tag{8}$$

Where $J_c(U,V) = -\sum_{i=1}^{K}\sum_{j=1}^{N} u_{i,j} x_j v_i$,T is a Lagrange multiplier, which can be taken according to the need, and its value has some influence on the final clustering result. $H(u)$ is the entropy of the membership matrix. When the value of $\frac{1}{T}$ is large, minimizing $J_T(U,V)$ actually needs to maximize the entrop $H(u)$ .As the $\frac{1}{T}$ value decreases, the minimized $J_T(U,V)$ turns to the minimized $J_c(U,V)$, thus achieving global minima. The peak of the objective function under the condition of $\sum_{i=1}^{K} u_{i,j} = 1$, the minimum value of $J_T(U,V)$ is obtained. And define the following Lagrangian objective function $L(u,v,\lambda,\gamma)$:

$$L(u,v,\lambda,\gamma) = J_T(U,V) + \lambda\sum_{i=1}^{K}(v_i^T v_i - 1) + \gamma\sum_{j=1}^{N}(\sum_{i=1}^{N} u_{ij} - 1) \tag{9}$$

There is a partial derivative of each center vector v_i in $L(u,v,\lambda,\gamma)$:

$$\frac{\partial L(u,v,\lambda,\gamma)}{\partial v_i} = -\sum_{j=1}^{N} u_{i,j} x_j^T + 2\lambda v_i \tag{10}$$

Let equation (10) be equal to 0, then there is:

$$v_i = \frac{\sum_{j=1}^{N} u_{i,j} x_j}{2\lambda} \tag{11}$$

Since $v_i^T v_i = 1$, by (11) can be further introduced:

$$v_i = \frac{\sum_{j=1}^{N} u_{i,j} x_j}{\sqrt{(\sum_{j=1}^{N} u_{i,j} x_j)^T (\sum_{j=1}^{N} u_{i,j} x_j)}} \tag{12}$$

For $L(u,v,\lambda,\gamma)$ in each $u_{i,j}$ partial guide:

$$\frac{\partial L(u,v,\lambda,\gamma)}{\partial u_{i,j}} = -x_j^T v_i + \frac{1}{T}(\ln u_{i,j} + 1) + r \tag{13}$$

Let equation (13) be equal to 0, then there is:

$$\ln u_{i,j} = Tx_j^T v_i - (T_\gamma + 1) \tag{14}$$

Due to $\sum_i^N u_{i,j} = 1$, by (14) can be further introduced:

$$u_{i,j} = \frac{e^{Tx_j^T v_i}}{\sum_{i=1}^{K} e^{Tx_j^T v_i}} \tag{15}$$

The minimum process (9) is the clustering algorithm based on maximum entropy principle (CAMEP). It is noted that the Lagrange multiplier T is equivalent to the inverted annealing coefficient, When the T value is small, the system is maintained at a higher temperature, and the process of T increases is the process of system annealing, and the minimum point of the objective function is obtained by a series of changes with temperature T.

A complete description of the CAMEP algorithm is given below.

Step1:Giving $v_0 = \{v_1^0, v_2^0, \cdots, v_K^0\}$, and the fuzzy partitioning matrix $U = \{u_{i,j}\}_{k>n}=0$,the maximum number of iterations is M, the set annealing coefficient is T, the maximum annealing coefficient is MaxT, the threshold is ε, the number of iterations is r = 0;

Step 2: Find the update $u_{i,j}^{(l+1)}$ according to the formula (15);

Step 3: Update the center $v_i^{(l+1)}$ according to formula (12);

Step 4: If T is equal to the minimum, then stop; otherwise, if $\max_i \| v_i^{(l+1)} - v_i^{(l)} \| \le \varepsilon$ or l> M, adjust the annealing coefficient $T = T - \Delta T$, go to step 2.

4. Experimental Results and Analysis

First, the evaluation criteria of evaluating the performance of text clustering are described, then the various data sets and experimental setups of the experiment are described. Finally, the experimental results of each data set are given and analyzed.

4.1 Algorithm Performance Evaluation Criteria

The performance evaluation criteria of the algorithm based on the objective function are the internal evaluation standard and the external evaluation standard. If vi is the center of the normalized class Ki, the ACS is defined as follows:

$$ACS = \frac{1}{N}\sum_{i=1}^{K} \sum_{x_i \in K_i} x_j^T v_i \tag{16}$$

Where N is the number of samples, and the larger the ACS value, the higher the total tightness of the data and the center vectors. In addition, for text clustering experiments, the text of the class is often known, so the external evaluation criteria of the general use of mutual information (NMI).The NMI value is defined as follows: Assuming that X represents a known text class random variable, and Y represents the class random variable of the clustering result, then:

$$NMI = \frac{I(X:Y)}{\sqrt{H(X) \cdot H(Y)}} \tag{17}$$

Where X (Y) is the mutual information of variables X and Y, H (X) and H (Y) are the entropy of variables X and Y.Because clustering often does not know the number of clusters in advance, the NMI value can be used to evaluate the performance of the algorithm when the number of different clusters is better. The higher the NMI value, the more accurate the clustering result is. The NMI value is 1, marked exactly the same.

4.2 Description of The Experimental Datasets

The experiment uses 20 - Newsgroups data sets and some of the eight datasets from the CLUTO [12] text clustering toolbox. The data set contains the number of samples ranging from 690 to 19949, the smallest data dimension is 8 261 dimension, the largest is 43586 dimension, the actual number of clusters is 3, the largest is 20. From the above characteristics we can see that these data sets reflect the characteristics of the text datasets. Where the NG20 data is averaged from 20 different newsgroups, and the Bow toolkit [13] prepares the 20-

Newsgroups text with 19949 vector text data. NG17-19 is a subset of NG20 data, the actual number of categories for the three categories, each category includes nearly 1 000 from the political news of the text, and according to the characteristics of these news is divided into three categories, the previous clustering algorithm on the The clustering results of the data set show that the clustering of the data set is more difficult because of the overlap between classes and classes. The other data comes from the CLUTO toolbox [12], which has been pre-processed as vector text data. A detailed description of the data set is shown in Table 1.

It should be noted that the balance in Table 1 (n_d is the total number of document n_w is the total number of terms, k is the number of classes) is the balance of the data, that is, the ratio of the number of classes containing the minimum number of texts to the number of texts in the class containing the maximum number of texts, which reflects the balance between the class and the class in the data set. The NG20, NG17-19, and sports data sets used in the experiment are more balanced, ie, the number of samples is similar in each class, and the balance of other data sets is poor.

4.3 Experimental Results and Analysis

In order to test the maximum entropy sphere K-means algorithm proposed by the author, the clustering results of the above data sets are compared with each other when the number of clusters and the number of clusters are different. In the experiment, the algorithm is run 20 times for each case, and the NMI average of its clustering results is taken as the final evaluation value. At the same time, the specific NMI mean and the deviation table of the experimental results and the average cosine similarity of the clustering results degree. Experiments show that the maximum entropy spherical K-means algorithm has achieved satisfactory results for these data sets.

*4.3.1. Comparison of clustering effects of each algorithm when fixed clustering number.*From the clustering results NMI of the different data sets in Table 2 and Table 3, it can be seen that the SP-Kmeans algorithm has the lowest clustering NMI value for each data set, and its clustering effect is obviously lower than the other two clustering algorithms. The clustering effect of using CAMEP is better than that of DA-SPKM.

Table 1. Summary of text datasets

data	Source	n_d	n_w	K	balance
NG20	20 Newsgroups	19949	43586	20	0.991
NG17-19	3 overlapping subgroups from NG20	2998	15810	3	0.998
Reviews	San Jose Mercury (TREC)	4069	18483	5	0.998
Sports	San Jose Mercury (TREC)	8580	14870	7	0.636
Tr54	TREC	690	8261	10	0.088
La1	LA Times(TREC)	3204	31472	6	0.290
La12	LA Times(TREC)	6279	31472	6	0.282
La2	LA Times(TREC)	3705	31472	6	0.274

Table 2. NMI results on NG20, NG17-19, reviews, sports, and tr54 datasets

	NG20	NG17-19	reviews	sports	tr45
K	20	3	5	7	10
SP-KMeans	0.550±0.050	0.340±0.100	0.530±0.100	0.560±0.130	0.600±0.090
DA-VMFS	0.570±0.030	0.460±0.010	0.560±0.090	0.620±0.050	0.680±0.050
CAMEP	0.590±0.010	0.470±0.060	0.620±0.020	0.640±0.004	0.690±0.030

In addition, the authors find that the NMI deviation of the CAMEP clustering algorithm is much smaller than that of the SPK-Means and DA-VMF algorithms in most cases, which means that the algorithm of maximal entropy spherical clustering overcomes the sensitivity to initialization. In the clustering of the difficult data set NG17-19, the author finds that the NMI value of the algorithm CAMEP can reach 0.53, but the NMI value is very large, and its inner reason needs further study.

Table 3. NMI results on classic la1, la12, and la2 datasets

	classic	la1	la12	la2
K	4	6	6	6
SP-KMeans	0.540 ± 0.040	0.480 ± 0.100	0.480 ± 0.100	0.480 ± 0.070
DA-VMFS	0.510 ± 0.010	0.530 ± 0.030	0.520 ± 0.020	0.520 ± 0.040
CAMEP	0.560 ± 0.001	0.560 ± 0.006	0.560 ± 0.001	0.550 ± 0.006

Table 4. Average cosine similarity result on NG20, NG17-19, reviews, sports, tr45 datasets

	NG20	NG17-19	reviews	sports	tr45
K	20	3	5	7	10
SP-KMeans	0.1587	0.1531	0.2131	0.2450	0.3784
DA-VMFS	0.1583	0.1534	0.2140	0.2492	0.3798
CAMEP	0.1602	0.1535	0.2145	0.2476	0.3847

Table 5. Average cosine similarity results on classic la1, la12, and la2 datasets

	classic	la1	la12	la2
K	4	6	6	6
SP-KMeans	0.1505	0.1738	0.1747	0.1803
DA-VMFS	0.1523	0.1730	0.1755	0.1828
CAMEP	0.1527	0.1758	0.1762	0.1868

Table 4 and Table 5 show the average cosine similarity (ACS) of the clustering results of different clustering algorithms for different clustering algorithms. It can be seen from the table that the ACS values of CAMEP and DA-SPKM are greater than those of SP-Kmeans value.

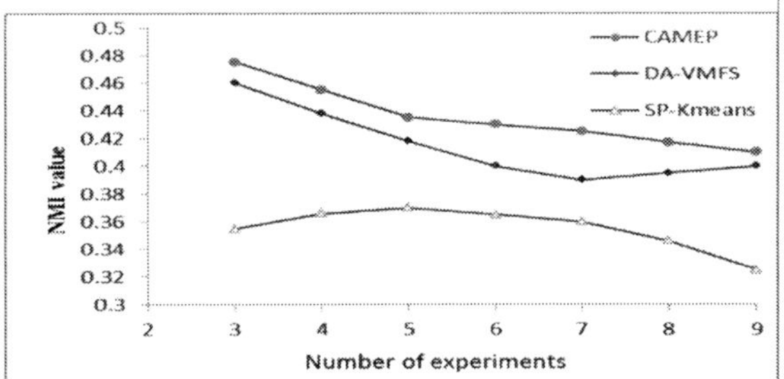

Figure 1. Results on NG17-19 dataset

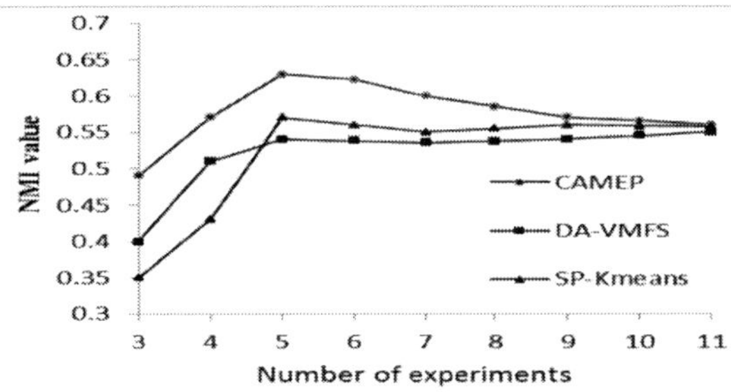

Figure 2. Results on reviews dataset

*4.3.2. Comparison of clustering results of different algorithms for different clusters.*The clustering algorithm often does not know the actual number of clusters in advance. Therefore, the authors compare the clustering performance of each algorithm in different clustering categories. In order to ensure the accuracy of the experiment, a clustering class running algorithm 20 times, and finally 20 times the average of NMI as the class of NMI value. Figure 1 and Figure 2 are the algorithm for some data sets in a variety of clusters under the NMI value comparison chart, we can see from the figure CAMEP due to the use of the maximum entropy strategy to avoid the local minimum point, so in different poly Class clustering performance is better than SP-Kmeans.

Table 6. Runtime results

	NG20	NG17-19	reviews	sports	classic
K	20	3			
SP-KMeans	84.5	6.4	12.2	25.8	5.1
DA-VMFS	1686.7	50.5	220.1	335.0	96.5
CAMEP	3177.9	55.2	326.3	449.2	125.5

*4.3.3. Comparison of clustering time of each algorithm.*Table 6 shows the clustering time comparison of the partial data sets in the actual clustering number. The clustering time is much smaller than the other two algorithms, Based on the analysis of the above sections, the author thinks that different clustering algorithms can be used for different data clustering tasks.

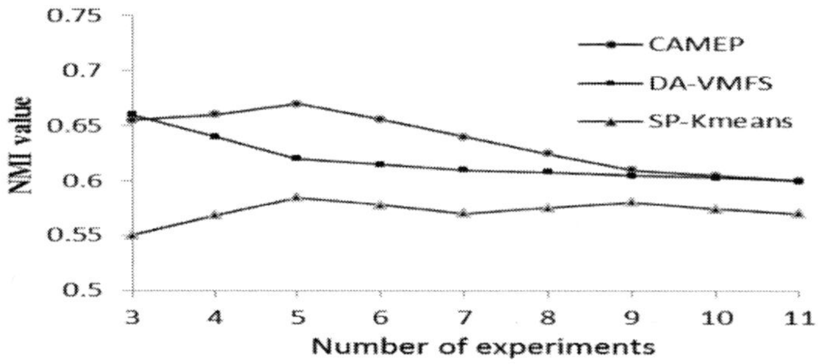

Figure 3. Result on sports dataset

5. Conclusion

In this paper, the maximum entropy principle is applied to the objective function of the spherical K-means algorithm, and the clustering algorithm based on maximum entropy principle is proposed. A large number of experiments show that the algorithm can effectively the clustering performance of text data set is better than that of traditional clustering algorithms. In addition, the author also found some problems: how to improve the clustering effect of CAMEP in the case of high difficulty data cluster, and how to further improve the CAMEP clustering effect reduce its clustering time. The above question is also the author's next research goal.

6. References

[1] Hashimi H, Hafez A, Mathkour H. *Selection criteria for text mining approaches* [M]. Elsevier Science Publishers B. V. 2015.

[2] Jammalamadaka S R, Sengupta A. *Directional statistics* / [J]. 2017.

[3] Dhillon I S, Fan J,Guan Yuqiang. *Efficient clustering of very large document collections*[C]//Data mining for Scientific and Engineering Applications Norwell. MA: Kluwer, 2011.

[4] Yasuda M. *Entropy Maximization and Deterministic Annealing Approach to Fuzzy C-means Clustering* [J]. Scis, 2012, 2010:1515-1520.

[5] Rose K, Gurewitiz E, Fox G A. *Deterministic annealing approach to clustering* [J]. Patter Recognition Letters, 1990, 11:589- 594.

[6] Javed K, Gouriveau R, Zerhouni N. *A New Multivariate Approach for Prognostics Based on Extreme Learning Machine and Fuzzy Clustering* [J]. IEEE Transactions on Cybernetics, 2015, 45(12):2626-2639.

[7] Li R, Tao X, Tang L, et al. *Using Maximum Entropy Model for Chinese Text Categorization*[C]// Asia-Pacific Web Conference. Springer Berlin Heidelberg, 2004:578-587.

[8] Li Ronglu, Wang Jianhui, Chen Xiaoyun, et al. *Using maximum entropy model for Chinese text categorization*[J].Journal of Computer Research and Development, 2005,42(1):94- 101.

[9] Surian D, Chawla S. *Mining Outlier Participants: Insights Using Directional Distributions in Latent Models*[C]// Joint European Conference on Machine Learning and Knowledge Discovery in Databases. Springer Berlin Heidelberg, 2013:337-352.

[10] Shi Zhong, Joydeep G J. *A unified framework for model- based clustering* [J]. Journal of Machine Learning Research, 2004, 4(6):1001- 1037.

[11] Shi Zhong, Ghosh J. *Generative model- based document clustering: a comparative study* [J]. Knowledge and Information Systems, 2005, 8(3):374- 384.

[12] ftp://www.cs.umn.edu/~karypis/CLUTO/flies/datasets.tar.gz.

[13] Mow: a toolkit for statistical language modeling, text retrieval, classification and clustering[EB/OL].http://www.cs.cmu.edu/mccallum/bow.

[14] Hornik K, Wu W W, Grün B, et al. *movMF: An R Package for Fitting Mixtures of von Mises-Fisher Distributions [J]*. Journal of Statistical Software, 2014, 058(10):1-31.

[15] Karayiannis N B. *MECA: maximum entropy clustering algorithm[C]//* Fuzzy Systems, 1994. IEEE World Congress on Computational Intelligence. Proceedings of the Third IEEE Conference on. IEEE, 2002:630-635 vol.1.
[16] ZHAO Yang, born in 1992, M. S. candidate. His research interests include data mining, personalized recommendation, etc.
[17] LIU Fangai, born in 1962, Ph. D., professor. His research interests include data mining, personalized recommendation, distributed computing, etc.
[18] This research was financially supported by the National Natural Science Foundation of China (No.61572301, No.90612003), the Natural Science Foundation of Shandong Province (No.ZR2013FM008, No.ZR2016FP07), and the Open Research Fund from Shandong provincial Key Laboratory of Computer Network (No. SDKLCN-2016-01).

A hybrid genetic algorithm for solving bi-objective traveling salesman problems

Mei Ma and Hecheng Li

School of Mathematics and Statistics, Qinghai Normal University, Xining, 810008, China.
Email: 1450670013@qq.com

Abstract. The traveling salesman problem (TSP) is a typical combinatorial optimization problem, in a traditional TSP only tour distance is taken as a unique objective to be minimized. When more than one optimization objective arises, the problem is known as a multi-objective TSP. In the present paper, a bi-objective traveling salesman problem (BOTSP) is taken into account, where both the distance and the cost are taken as optimization objectives. In order to efficiently solve the problem, a hybrid genetic algorithm is proposed. Firstly, two satisfaction degree indices are provided for each edge by considering the influences of the distance and the cost weight. The first satisfaction degree is used to select edges in a "rough" way, while the second satisfaction degree is executed for a more "refined" choice. Secondly, two satisfaction degrees are also applied to generate new individuals in the iteration process. Finally, based on genetic algorithm framework as well as 2-opt selection strategy, a hybrid genetic algorithm is proposed. The simulation illustrates the efficiency of the proposed algorithm.

1. Introduction

Traveling salesman problem is a typical combinatorial optimization problem. The problem can be described as follows: Given a list of n cities and the distances between each pair of cities, a salesman starts from some city, and then visit all other cities exactly once and returns to the origin city. How does the salesman plan his/her travelling rout such that the tour distance is the shortest among all possible routes? Despite the fact that the description is very simple, the problem is extremely difficult to solve, and has been proven to be a NP-hard problem [1]. In real life a lot of problems can be modelled as TSP problems, such as gene sequencing problems, the dartboard design problem, the hole punching problem, etc. [2, 3, 4]. In these TSP models, the single objective TSP, only minimizing tour distance, is the most common version.

In real world application, some problems always involve more than optimization objective, such as in a real trip, the tourists maybe take distance, cost and other comfort index into account at the same time. As a result, it is necessary to study some multi-objective TSP models and develop efficient solution methods for these problems. In recent 20 years, the multi-objective traveling salesman problems (MOTSPs) have been widely studied. One of the first papers considering the MOTSP is the study of Borges and Hansen [5], where the global convexity for a MOTSP was studied and the landscape of local optima was analysed.

In recently, evolutionary algorithm has been applied to solve MOTSP. Yan [6] proposed a new MOEA in 2003, in which an “inver-over” operator was defined to replace the traditional crossover and mutation operators, two kinds of selection operators were adopted, and niches are defined by using a sharing function to preserve the diversity of the population. The simulation results show that MOEA outperforms SPEA in some bi-objective instances. Elaoud [7] proposed a multi-objective genetic algorithm based on a dynamic selection of crossover and mutation operators. Experimental results show synergy effects among different operators. Samanlioglua [8] used a memetic random key genetic algorithm to find weakly Pareto optimal solutions for a symmetric multi-objective traveling salesman problem. The memetic algorithm adopted a "target-vector approach" and the local search. The random keys representation ensured that feasible tours are maintained during the application of genetic operators. The simulation results show the algorithm is better than that in [9]. Gao [10] developed a new approach, which combined MOEA/D with probabilistic model based methods for MOTSP. In this approach, an MOTSP is divided into a set of scalar objective sub-problems, and both priori and learned information are used to build a probabilistic model for each sub-problem. By the cooperation of neighbour sub-problems, MEDA/D could optimize all sub-problems simultaneously.

In [11], the authors considered the application of estimation of distribution algorithm (EDA) and genetic algorithm (GA) for bi-objective travelling salesman problem (BOTSP). In their paper, three existing EDAs are improved, and the refinement operator and a local search operator are developed. Subsequently, the EDAs are combined with GA to overcome the shortcomings of these two algorithms. The results show that the hybrid algorithms are capable of finding a set of good trade-off solutions. Psychås [12] dealt with multi-objective TSP using three hybrid evolutionary algorithms, and solved 2–5 objective TSP. Lust and Teghen [13] designed two phase Pareto local search (2PPLS) to find a good approximation of the efficient set of the BOTSP, in which PLS is a powerful tool to generate potentially efficient solutions.

Some variants of MOTSP are also studied. Wang et al. [14] described an uncertainty and multiple objectives in traveling salesman problem (UMTSP). They proposed a new approach to obtain Pareto efficient route in the uncertain multi-objective TSP, which converts the uncertain multi-objective TSP into an uncertain single objective TSP. A new ABC algorithm combined with genetic operators is designed to this problem. Li [15] introduced a parallel search system for dynamic multi-objective traveling salesman problem (DMO-TSP). He designed a multi-objective TSP in a stochastic dynamic environment. Yang et al. [16] proposed a multi-algorithm solver called multi-algorithm co-evolution strategy (MACS) for DMO-TSP, which can be used for solving small size DMO-TSP with low degree of conflict objectives in real time. In order to solve larger size DMO-TSP, Yang et al. [17] proposed a synchronized parallel multi-algorithm solver.

In the present manuscript, we consider a BOTSP in which both the distance and the cost are minimized, and proposed an efficient genetic algorithm based on two designed satisfaction degree indices of each edge. These two satisfaction degree indices are used to select edges and generate new individuals in the iteration process. Based on genetic algorithm framework as well as 2- opt selection strategy, a hybrid genetic algorithm is proposed.

This paper is organized as follows. The discussed problem is presented in Section 2. A hybrid genetic algorithm is given by applying 2-opt selection strategy in Section 3, and this is followed by the simulation in Section 4. We finally conclude our paper in Section 5.

2. The Bi-objective TSP Problem

BOTSP can be described as follows: give N cities and two costs c_{ij}^1 and c_{ij}^2, $i,j = 1,2,\dots,N$, on the edge from city i to city j. The purpose of solving BOTSP is to find a tour, i.e. a cyclic permutation ρ of the N cities, such that both the distance and the cost are minimized. The mathematical formulation can be written as

$$\begin{cases} \min \ f_1(\rho) = \sum_{i=1}^{N-1} c^1_{\rho(i),\rho(i+1)} + c^1_{\rho(N),\rho(1)} \\ \min \ f_2(\rho) = \sum_{i=1}^{N-1} c^2_{\rho(i),\rho(i+1)} + c^2_{\rho(N),\rho(1)} \end{cases} \quad (1)$$

In the problem as above, two objectives always conflict just as in other multi-objective optimization. When any one is minimized, the other maybe obtains the worse objective value. As a result, the optimal solution set of (1) is a set of trade-off solutions, i.e. Pareto optimal set.

3. Proposed Algorithm

In the section, we present a hybrid genetic algorithm (Algorithm 1) using two designed satisfaction degree indices.

3.1. The Two Satisfaction Degree Indices

In order to produce a good initial solution, we want the two weights to be as small as possible. We divide the two weights into three levels: high, middle and low, respectively. The distance weight classification is as follows:

Firstly, to sort the values in the distance matrix in descending order, and get the corresponding sequence S, the number of elements of S is $|S|$.

Secondly, to divide the sequence from left to right into three groups S_1, S_2, S_3, and the element numbers in these groups are $\left\lfloor \frac{|S|}{3} \right\rfloor, \left\lfloor \frac{|S|}{3} \right\rfloor$ and $|S| - 2\left\lfloor \frac{|S|}{3} \right\rfloor$, respectively. In these three groups, the elements in S_1 have largest distances while S_3 the smallest distances.

Similarly, the cost matrix is divided into T_1, T_2 and T_3.

Thus, according to the levels of distance and cost weights, all edges can be divided into nine categories: low-low, low-high, low-middle, middle-low, middle-middle, middle-high, high-low, high-middle, and high-high. Among them the former is the level of the distance weight, while the latter is the level of cost weight. In the paper, based on some numerical experiments, we suggest taking a value for each category as the first satisfaction degree, and these suggested values are shown in

Table 1. Values of the first satisfaction degree

Distance / Cost	high	medium	low
high	0.1	0.3	0.7
medium	0.3	0.5	0.8
low	0.7	0.8	0.9

From the Table 1, one can see that the lower the weight on the edge, the larger the first satisfaction degree indicator, and so these edges are more likely to be chosen. When two weights are located at the same level, the nine categories of the edges can be reduced to six categories (low- low, low- medium, low- high, medium - medium, high- medium, high- high). In the process of edge selection, we always choose the edge which has the larger value of the first satisfaction degree. But in the same class, in order to choose the edge as good as possible, the second satisfaction degree is executed for a more "refined" choice.

Given weight matrices $C = (c_{ij})_{N\times N}$ and $D = (d_{ij})_{N\times N}$. In order to define the second satisfaction degree effectively, a normalization procedure need to be executed as follows:

$$c'_{ij} = \frac{c_{ij}}{\sum_{k=1,k\neq i}^{N} c_{ik}}, \quad d'_{ij} = \frac{d_{ij}}{\sum_{k=1,k\neq i}^{N} d_{ik}} \tag{2}$$

The second satisfaction degree of any two cities is defined as:

$$F(i,j) = 2 - d'_{ij} - c'_{ij} \tag{3}$$

3.2. Population Initialization Based on Satisfaction Degree

In the execution process of genetic algorithms, the random generation of the population individuals can sometime affect the evolution of the populations. In this manuscript two satisfaction degrees as above are adopted to generate new individuals.

Set $L = (l_{ij})_{N\times N}$ is the first satisfaction degree matrix, l_{ij} is taken from a set composed of the different elements in Table 1, i.e. $l_{ij} \in \{0.9, 0.8, 0.7, 0.5, 0.3, 0.1\}$.The second satisfaction degree matrix is defined as $F(i,j)$. The details are given as follows:

1) Set $X = \{1,2,\dots,N\}, Y = \emptyset$. First, to randomly choose an integer t from 1 to N as the start city, then set $X = X\backslash\{t\},\ Y = Y \cup \{t\}$.

2) Determine city which follows city t. If there exists an unique $i \in X$ such that $l_{ti} = max\{l_{tk}\}, k \in X$. Then city t is followed by city i, set $X = X\backslash\{i\},\ Y = Y \cup \{i\}$. Otherwise, set $X_1 = \{i | l_{ti} = max\{l_{tj}, j \in X\}\}$, and use the roulette wheel selection to decide the city behind the city t:

i) Calculate the probability of the elements in the set X_1,

$$p(i) = \frac{F(t,i)}{\sum_{j=1}^{|X_1|} F(t,j)},\ i \in X_1 \tag{4}$$

ii) Calculate the cumulative probability of each element in the set X_1

$$q(i) = \sum_{j=1}^{i} P(j) \tag{5}$$

iii) To generate randomly numbers r in [0, 1];

iv) If $q(i-1) < r \le q(i), i = 2,\dots,n$, then set $X = X\backslash\{i\},\ Y = Y \cup \{i\}$;

v) Repeat step 2) until the set X becomes an empty set, and then an individual is generated;

3) Repeat step 1) and 2) N_p times, and then form the initial population with size N_p.

3.3. Crossover Operator

In this subsection, we propose a new crossover operator based on the 2-opt permutation algorithm in which the satisfaction degree information is used. The crossover process is given as follows:

1) Calculate the second satisfaction degree for a given path $\pi=(\pi_1, \pi_2, \dots, \pi_n)$;

2) The replacement probability of each edge (π_i, π_j) is determined according to the second satisfactory degree.

$$P(\pi_i, \pi_j) = \frac{\dfrac{1}{F(\pi_i, \pi_j)}}{\sum_{i=1}^{n-1} \dfrac{1}{F(\pi_i, \pi_{i+1})} + \dfrac{1}{F(\pi_n, \pi_1)}} \tag{6}$$

3) Delete two edges chosen by replacement probability, and add the new edges by 2-opt permutation algorithm.

3.4. Improved Genetic Algorithm

In the subsection, we propose an improved genetic algorithm based on new crossover operator and population initialization procedure.

Step1. (Initial population) using population initialization method to form the initial population $pop(0)$, we denote the crossover probability by p_c and the mutation probability by p_m. Let $t = 0$.

Step2. (Fitness) Evaluate each individual in $pop(t)$ by calculating objective values, and select non-dominated solutions according to the dominance and crowding distance just as NSGA-II does.

Step3. (Crossover and mutation) The crossover operator and the insertion mutation are adopted to generate genetic offspring.

Step4. (Next generation of population) The elitist selection scheme is applied to generate the next generation of population $pop(t + 1)$, and the non-dominated set is updated;

Step5. (Stopping criterion) If the termination condition is satisfied, then the algorithm is stopped, and output the dominated solution set; otherwise, let $t = t + 1$, go to Step 3.

4. Simulation

We select the several bi-objective TSP instances to illustrate the efficiency of the proposed algorithm. The instances (oliver30, eil51, rand100, rand300, kroab100, kroac100, krobc100, kroab150, kroab200) are selected from TSPLIB, and the first objective is to minimize the distance. For the instances oliver30, eil51, rand100 and rand300, the second cost between the edges are randomly generated. Three Euclidean instances (Euclab100, Euclab300, Euclab500) are considered, the costs on the edges correspond to the Euclidean distances between two points. We also use three mixed instances (mixed100, mixed300, mixed500), where the first cost comes from Euclidean instances while the second cost is randomly generated.

Two indicators are selected to measure the quality of the solutions.

1. The hyper volume indicator (to be maximized) [18]: the volume of the dominated space is defined by approximations sets, limited by a reference point. In Figure 1, $\{P_1, P_2, P_3\}$ is a set of the non-dominated objective vector, F_1, F_2 is the objective function. The hyper volume indicator of this set is the area of ABCDEFGH, where $Ref = (r_1, r_2)$ is a reference point, and $r_i = max(f_i(x)) + 0.1 \times (max(f_i(x)) - min(f_i(x))), i = 1,2$.

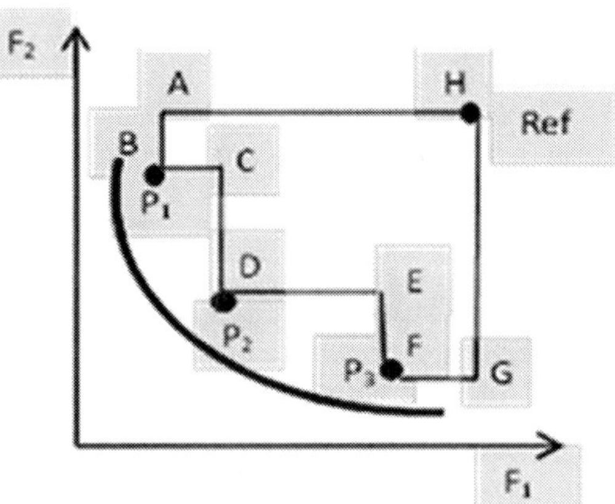

Figure 1. The hyper volume indicator in the two-objective case

2. Relative error: Let the optimal value of the single objective optimization problem be A (only when the first objective is considered in single objective TSPs) and the smallest value of the first objective function obtained by the proposed algorithm be B in bi-objective optimization, then the relative error between A and B can be obtained as follows:

$$e = \frac{A - B}{B} \tag{7}$$

Genetic parameters are taken as follow: the population size $N_p = 50$, the crossover probability $p_c = 0.85$, the mutation probability $p_m = 0.1$ and the maximum number of generations G=200 for instances oliver 30 and eil51, G=500 for other instances. The proposed algorithm (Algorithm 1) is executed 10 independent runs.

We first compare the results of the Algorithm 1, for the "kro" instances, with 2PPLS in [13]. The results are shown in Table 2, where H represent the average hyper volume of 10 runs, Time represent the average CPU time of 10 runs.

Table 2. Comparison of algorithm 1 with 2ppls on "kro" instances

Instances	Algorithm	H(10^8)	Time(s)	e
kroab100	Algorithm 1	228.52	34.83	-0.0394
	2PPLS	226.11	35.57	-
kroac100	Algorithm 1	223.90	28.71	-0.0323
	2PPLS	226.32	30.3	-
kroad100	Algorithm 1	228.01	25.63	0.0158
	2PPLS	227.41	26.44	-
krobc100	Algorithm 1	228.45	35.47	0.0271
	2PPLS	227.38	35.82	-
krobd100	Algorithm 1	228.61	35.29	0.0017
	2PPLS	226.12	37.18	-
krocd100	Algorithm 1	232.67	28.33	0.0186
	2PPLS	230.89	27.56	-
kroab150	Algorithm 1	601.49	100.37	0.0427
	2PPLS	592.51	91.21	-
kroab200	Algorithm 1	1012.19	259.41	0.0375
	2PPLS	1076.08	211.76	-

From Table 2, we can see that the hyper volume indicators on most of the instances are improved except for kroac100 and kroab200, and the relative error is small.

We compare the results of Algorithm 1 with 2PPLS on the other instances in Table 3. For the instances oliver30 and eil51, the relative error is 0. It follows that the proposed algorithm found the same solutions as those provided by the single objective optimization. We can see that the proposed algorithm finds better hyper volume indicators than 2PPLS for these the instances except for rand300, Euclab500 and mixed500. But the running time of 2PPLS is higher on the instances Euclab300, Euclab500 and mixed500.

Table 3. Comparison of algorithm 1 with 2PPLS on the other instances

Instances	Algorithm	H(10^8)	Time(s)	e
oliver30	Algorithm 1	24.83	7.92	0.0
	2PPLS	-	-	-
eil51	Algorithm 1	42.91	11.35	0.0
	2PPLS	-	-	-
rand100	Algorithm 1	204.67	21.46	-0.0051
	2PPLS	-	-	-
rand300	Algorithm 1	4780.29	380.52	0.0926
	2PPLS	4804.58	351.27	-
Euclab100	Algorithm 1	169.12	25.17	0.0078
	2PPLS	168.88	28.60	-
Euclab300	Algorithm 1	2309.91	459.72	0.0129
	2PPLS	2309.69	783.14	-
Euclab500	Algorithm 1	7163.26	5071.50	0.0412
	2PPLS	7165.39	6309.10	-
mixed100	Algorithm 1	334.57	25.06	0.0281
	2PPLS	331.99	28.28	-
mixed300	Algorithm 1	3411.99	440.27	0.0365
	2PPLS	3410.44	447.06	-
mixed500	Algorithm 1	10420.32	1946.81	0.0493
	2PPLS	10440.41	2682.45	-

5. Conclusions

In this manuscript two satisfactory degree indices are proposed, and used to generate individuals. Based on these procedures, a hybrid genetic algorithm is developed which is combined with 2-opt approach for solving bi-objective TSP. From the computational results, we can see the proposed algorithm is efficient and robust.

6. Acknowledgments

The work was supported by the National Natural Science Foundation of China (No. 61463045), the Natural Science Foundation of Qinghai Province (No. 2013-Z-937Q), and the Key Laboratory of the Internet of Things of Qinghai Province.

7. References

[1] Larranaga P, Lozano J A 2002 *Estimation of Distribution Algorithms: A New Tool for Evolutionary Computation* (Kluwer: Norwell).
[2] Blazewicz J, Kasprzak M and Kuroczycki W 2002 *J. Heuristics.* vol. **8**, pp 495-502.
[3] Eiselt H A, Laporte G 1991 *J. OR Society.* vol. **42**, pp 113–118.
[4] Reinelt G 1989 J. ORSA J Comput. vol. **4**, pp 206–217.
[5] Borges P C, Hansen M P 2000 Essays and surveys in metaheuristics (Operations *Research/Computer Science,* vol. **15**) (New York: Springer Science+Business Media), pp 129-150.
[6] Yan Z, Zhang L and Kang L 2003 *Proc. 2nd Int. Conf. on Evolutionary Multi-Criterion Optimization* (Berlin, Heidelberg: Springer-Verlag), pp 342-354.
[7] Elaoud S, Teghem J and Loukil T M 2010 *J. Electronic Notes in Discrete Mathematics.* vol. **36**, pp 939-946.
[8] Samanlioglu F, Ferrell Jr W G and Kurz M E 2008 *J. Comput. Ind. Eng.* vol. **55**, pp 439-449.
[9] Hansen M P 2000 *J. Heuristics.* vol. **6**, pp 419-430.
[10] Li Y, Zhou A M and Zhang G X 2013 *J. Comput. Math. Appl.* vol. **66**, pp 1857-1868.
[11] Ann S V, Kay T C and Yong C J 2011 *J. Flex Serv Manuf.* vol. **23**, pp 207–241.
[12] Iraklis-Dimitrios P, Eleni D and Yannis M. 2015, *J. Expert Syst. Appl.* vol. **42**, pp 8956–8970.
[13] Lust T, Teghem J 2010 *J. Heuristics.* vol. **16**, pp 475-510.
[14] Wang Z T, Guo J S and Zheng M F 2015 *J. Eur. J Oper. Res.* vol. **241**, pp 478–489.
[15] Li W Q 2014 *J. Mathematics and System Science.* vol. **4**, pp 295-314.
[16] Yang M, Kang L S and Guan J 2008 *Proc. Cong. on Evolutionary Computation (CEC08),* pp 466-471.
[17] Yang M, Kang L S and Guan J 2008 *Proc. 7th WSEAS Int. Conf. on Applied Computer & Applied Computational Science (ACACOS '08),* pp 288-293.
[18] Zitzler E 1999 *Evolutionary algorithms for multi-objective optimization: methods and applications (PHD thesis). (Zurich, Switzerland: Swiss Federal Institute of Technology (ETH)).*

IOP Conf. Series: Journal of Physics: Conf. Series **887** (2017) 012066 doi:10.1088/1742-6596/887/1/012066

Heterogeneous massive feature fusion on grassmannian manifold

Haichao Huang[1], Hongning Liu[1], Xiaoyun Kong[2], Xingdan Lou[3] and Zepeng Wang[4]

[1]State Grid Zhejiang Information &Telecommunication Company
[2]State Grid Zhejiang Electric Power Company
[3]Zhejiang Huayun Information Technology CO., LTD
[4]Department of CSIE, Hefei University of Technology

Abstract. Two issues remain unsolved on utilizing multimodal features for pattern recognition: the missing features and the curse of dimensionality. In this paper, we address the two issues by fusing the multimodal features on the Grassmann manifold. By defining grouping constrains on multimodal features, each multimodal feature vector is grouped into a set of subspaces, and is further represented as a point on the Grassmann manifold. To deal with missing features, L2-Hausdorff distance, a metric to compare multimodal feature vectors with different number of subspaces, is computed, and a kernel matrix can be obtained accordingly. Based on the kernel matrix, two feature selection criterions, one supervised and one unsupervised, are proposed to obtain a few representative features in the kernel space. Thus, the curse of dimensionality is alleviated. Experimental results on three multimodal dataset show the proposed feature fusion can outperforms the state-of –the-art by higher accuracy.

1. Introduction

When recognize an object, human usually adopts several types of features to identify the object more accurately and efficiently. For example, using color to discriminate oranges from bananas is difficult since both of them are yellow, similarly, it is difficult to discriminate apple from oranges using contour. If we fuse color and contour for classification, the three fruits will be recognized more easily and accurately. Motivated by this concept, researches in pattern recognition try to enhance the recognition accuracy by employing more than one single type of features, namely, multimodal features. In contrast with uni-modal feature, multimodal features contain richer information. Therefore, if we fuse multimodal features optimally, there is a great potential to achieve to a better recognition performance.

According to whether features in different modality will be reallocated, we classify the existing multimodal feature fusion methods into two categories, namely, multi-cue integration based fusion and modality identification based fusion, respectively.

Multi-cue based fusion integrates modalities by treating each type of feature as a modality, in the literature, a series of multi-cue based fusion methods have been proposed. In [1], each classifier's accuracy in local regions of feature space is estimated, and then the most accurate local classifier is output as the final decision. In [2], a theoretical framework of fusing decisions from multiple classifiers is provided. Specifically, each fusing scheme can be explained as a special case of the framework. Furthermore, the sensitivity to estimation errors of different schemes is analyzed. By allocating each modality with a sub-classifier, score level feature fusion[3] quantize the multimodal feature vector into a set of class labels predicted by the sub-classifiers, and the final label is obtained by fusing the sub-classifiers' labels through a combination strategy, such as the max rule or the majority vote rule. In

[4], each modality is represented as a graph and graphs from various modalities are integrated into a single graph by an adaptation scheme. A decision tree based modality integration method is proposed in [5]. To be specific, each modality is represented as a node in the decision tree and decisions at each node are taken using an accumulation scheme; decision from each node is multiplied by a coefficient that indicates the reliability of the corresponding modality. In [6], features from each modality is represented as one independent graph, and different learning tasks are formulated as inferring from the constrains in every graph as well as supervision information. The above multi-cue based fusion method can alleviate the dimensionality curse. However, by allocating different types of features into different modalities, the features in each modality are obtained in a heuristic manner.

In this paper, we propose an approach of multimodal features fusion which can handle missing features in recognition and, at the same time, alleviate the curse of dimensionality. We try to deal with the multimodal features on the Grassmann manifold and fuse the multimodal features into kernel matrix . Targeting this aim, we group the multimodal features into a set of subspaces either by minimizing inter-group feature correlation and maximizing intra-group feature correlation or, a greedy method. Next, L2-Hausdorff distance, a metric to compare multimodal feature vectors with different number of subspaces, is computed, and a kernel matrix can be obtained accordingly. To handle missing features, if one subspace contains missing features, it is abandoned in distance computation. Finally, two feature selection criterion in kernel space, one for classification and the other for clustering, is proposed to obtain a few representative features in the kernel space. Thus, the curse of dimensionality can be alleviated.

1.1. Minimize the Inter-Group Feature Correlation and Maximize the Intra-Group Feature Correlation

When employing the multimodal features for clustering, due to the absence of class labels, the grouping scheme is developed by making use of the features correlation.

To quantize the correlation between multimodal features, two measures, one based on classic linear correlation and the other based on information theory [7], can be adopted. For the linear correlation, the most well-known measure is linear correlation coefficient. Given a pair of features X and Y, the linear correlation coefficient r is given by the formula:

$$r = \frac{\sum_i (x_i - \overline{x_i})(y_i - \overline{y_i})}{\sqrt{\sum_i (x_i - \overline{x_i})^2}\sqrt{\sum_i (y_i - \overline{y_i})^2}} \tag{1}$$

where $\overline{x_i}$ is the means of X and $\overline{y_i}$ is the means of Y. It is noticeable that the value of r is restricts between -1 and 1. If X and Y are completely correlated, then r take the value of 1 or -1; if X and Y are totally independent, then r take the value of 0.

The linear correlation is an optimal choice to represent features' correlation, under the assumption that a pair of multimodal features are linear separable. However, in the real world, it is not safe to always assume linear correlation between features, since linear correlation measures may not be able to capture correlations that are not linear in nature. To overcome this shortcoming, an information-theoretical concept based correlation, which can be deemed as a measure of the uncertainty of a multimodal feature, is presented in our approach.

Given a multimodal feature X, its entropy is defined as:

$$EN(X) = -\sum_{n=1} p(x^n)\log_2(p(x^n)) \tag{2}$$

The conditional entropy of feature X is defined as:

$$EN(X|Y) = -\sum_{m=1} p(y^n)\sum_{n=1} p(x^n|y^n)\log_2(p(x^n|y^n)) \tag{3}$$

We can represent the feature correlation by computing the information gain [8]:

$$IG(X|Y) = EN(X) - EN(X|Y) \tag{4}$$

And symmetrical uncertainty [] can be applied to normalize IG:

$$SU(X,Y) = \frac{IG(X|Y)}{EN(X)+EN(Y)} \tag{5}$$

It is noticeable that these entropy-based measures require nominal features, but they can be applied to measure correlations between continuous features as well, if the values are discretized in advance. By defining symmetrical uncertainty, given D modalities $M_1, M_2, \ldots, M_D$, each containing a number of multimodal features, the inter-group feature correlation between a pair of modalities is defined as:

$$C(M_i, M_j) = \delta * \sum_{\substack{X_i \in M_i \\ X_j \in M_j}} SU(X_i, X_j) \tag{6}$$

where X_i and X_j are features belonging to modalities M_i and M_j respectively, and δ is a factor to normalize $C(M_i, M_j)$ between -1 to 1. The intra-group feature correlation within modality M_i is defined as:

$$C(M_i) = C(M_i, M_i) \tag{7}$$

Based on the definition of symmetrical uncertainty, a criteria to allocate the multimodal features into $|D|$ groups is proposed, to be specific, the inter-group feature correlation is minimized and the intra-group feature correlation is maximized. The criteria can be formulated into the following objective function:

$$\min \sum_{\substack{i=1 \\ j>i}}^{|D|} \left[\frac{C(M_i, M_j)}{C(M_i)} + \frac{C(M_i, M_j)}{C(M_j)}\right] \tag{8}$$

It is noticeable that the objective function yields $|D|$ modalities, with minimal inter-modality correlation and balanced features in each modality. Let n denotes the number of multimodal features, the computational complexity is$O(n^2)$, which is computational efficient.

2. Map Subspaces into Grassmann Manifold

2.1. Principle Angles on the Grassmann Manifold

As shown in Figure1, the Grassmann manifold $G(m, D)$[9] is the set of *m*-dimensional linear subspaces of theR^D. The m-dimensional linear subspaces correspond to a point on Grassmann manifold. The point can be seen as a matrix with size m*D.

Let X_1, X_2 be two matrices of size D by m, there are *m* principle angles. The *i*th principle angle can be defined as:

$$\cos\theta_i = \max_{u_k \in orth(X_1)} \max_{v_k \in orth(X_2)} u_k' v_k; \quad (i=1,\ldots,k) \tag{9}$$

where orth(X) represents theorthogonal basis of X.

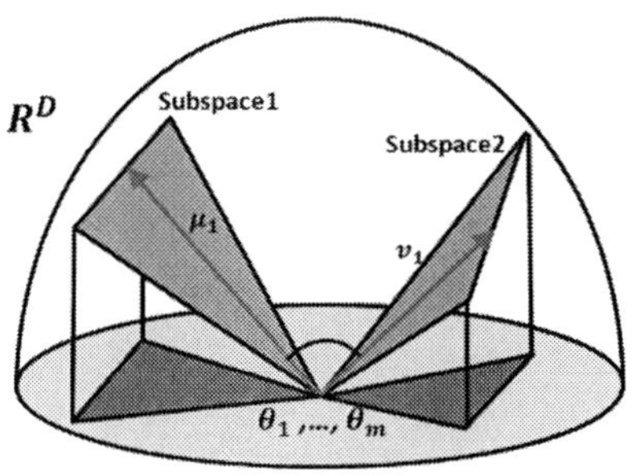

Figure 1. Principle angles in Grassmann manifold.

2.2. Distancein Grassmann Manifold

A real-valued function d: $X * X \rightarrow R$is called a metric if:

1. $d(x_1, x_2) \geq 0$,
2. $d(x_1, x_2) = 0$, if and only if $x_1 = x_2$
3. $d(x_1, x_2) = d(x_2, x_1)$,

4. $d(x_1, x_2) + d(x_2, x_3) \geq d(x_1, x_3)$,

for all $x_1, x_2, x_3 \in X$

Based on the definition of principle angle, a series of metrics are derived accordingly. In this paper, three distancesare used, they are Projection distance[10], Binet-Cauchy distance[11] and L2-Hausdorff distance[12], respectively.

If x_1 and x_2 contain the same number of subspaces, thenthe distance can be computed by:

1). The projection distance is defined as:

$$d_p(x_1, x_2) = (\sum_{i=1}^{m} \sin^2\theta_i)^{1/2} = (m - \sum_{i=1}^{m} \cos^2\theta_i)^{1/2} \tag{10}$$

2). The Binet-Cauchy distance is defined as:

$$d_{bc}(x_1, x_2) = (1 - \sum_{i=1}^{m} \cos^2\theta_i)^{1/2} \tag{11}$$

If x_1 and x_2 contain the different number of subspaces, thenthe distance can be computed by:

3). The L2-Hausdorff distance is defined as:

$$d_{l2}(x_1, x_2) = \sqrt{\max(m, n) - \sum_{i=1}^{m} \sum_{j=1}^{n} (u_i^T v_j)^2} \tag{12}$$

where $u_1, u_2, \ldots, u_m$ and $v_1, v_2, \ldots, v_n$ are orthonormal bases of U and V respectively.It is noticeable that the L2-Hausdorff distance is independent on the choice of basis.

By defining the distances on the Grassmann manifold, a kernel matrix can be constructed accordingly. To be specific, let n_{tr} and n_{te}be the number of training and testing samples respectively, the training kernel matrix is a $n_{tr} \times n_{tr}$ matrix M_{tr}; the testing kernel matrix is a $n_{te} \times n_{tr}$ matrix M_{te}. The kernel value of M_{tr} is computed as:

$$M_{tr}(i, j) = \exp(-\lambda * d(i, j)) \tag{13}$$

where λ is a parameter to be tuned, and $d(i, j)$ is the distance between the *i*th and the *j*th training sample.

Similarly, the kernel value of M_{te} is computed as:

$$M_{te}(k, l) = \exp(-\lambda * d(k, l)) \tag{14}$$

where $d(k, l)$ is the distance between the *k*th testing sample and the *l*th training sample.

3. Feature Selection in Kernel Space

By constructing kernel matrix on the grassmann manifold, we can select a few representative features in kernel space. In this section, two feature selection methods in kernel space, one supervised and one unsupervised are proposed.

4. Experimental Results and Analysis

To validate the performance of the proposed Grassmann manifold based feature fusion, we carry out three multimodal recognition tasks: one speech emotion recognition and two objects recognitions in images. All the three experiments run on a system equipped with Intel E8500 and 4GB main memory. In addition, thealgorithm of our approach is implemented on Matlab platform.

4.1. Multimodal Speech Emotion Classification

In this experiment, we collected 511 sentences speech. The dataset contains511 sentences emotional speech recorded by a female mandarin speaker. 41 acoustic features [11] from 6 types: pitch, log energy, the first three formant frequency, MFCC, PLCC and LSFs are extracted. We select half of the sentences for training and leave the rest for testing. The recorded speech data are labeled carefully with five basic emotions: Anger, Happiness, Neutral, Sadness and Surprise. Speech preprocessing including pre-emphasizing, blocking and Hamming windowing are performed on each speech frame of 25ms with 5ms frame shift. We extract the multimodal acoustic features on each frame. The C4.5 classifier [5] is used to evaluate the selected features' discriminative ability. We compare the kernel matrix of our algorithm with three representative kernel matrix (linear kernel, polynomial kernel and Gaussian kernel), which are learned from feature level fusions [2]. The hyper-parameters corresponding to the three kernels were learned by cross-validation. For simplicity, the number of subspaces is set to 5 in this experiment.

Table1. Comparison of recognition rate (percentage), Hap. Means Happiness, Neur. Neutral, Sad. Sadness, Sur. Surprise, Ave. Average.

	Anger	Hap.	Neur.	Sad	Sur.	Ave.
Full	0.901 ± 0.051	0.766 ± 0.013	0.301 ± 0.033	0.715 ± 0.025	0.502 ± 0.068	0.637 ± 0.038
Linear kernel	0.344 ± 0.012	0.232 ± 0.008	0.188 ± 0.016	0.311 ± 0.013	0.165 ± 0.021	0.248 ± 0.014
Poly. kernel	0.357 ± 0.024	0.246 ± 0.018	0.212 ± 0.015	0.501 ± 0.019	0.319 ± 0.034	0.327 ± 0.022
Gaus. kernel	0.404 ± 0.026	0.344 ± 0.022	0.237 ± 0.028	0.522 ± 0.021	0.433 ± 0.023	0.388 ± 0.024
Grass.	0.812 ± 0.009	0.701 ± 0.013	0.622 ± 0.011	0.922 ± 0.007	0.723 ± 0.015	0.756 ± 0.011

The experimental results in Table1 is measured in error rate by averaging five repeats on this dataset. As Table1 shows, the best average recognition accuracy was achieved by our approach. Recognition accuracy of our approach is competitive to the other methods on average. Furthermore, when comparing recognition accuracy of each emotion, our approach has partial advantages, such as neutral, sad and surprise .It is noticeable that features selected using the other three kernels are less discriminative that the features selected by our approach, since an obviously advantage in recognition accuracy is observed in our approach.

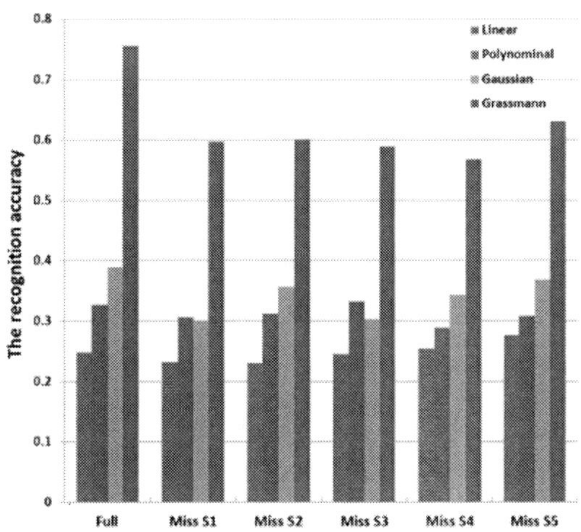

Figure 2. Comparison of the ability to handle missing features. Miss_i means the *i*th subspace is missing.

As shown in Figure2, we evaluate the testing sentences'recognitionaccuracy when features in one subspace are missing. Firstly, the recognition accuracy of the full feature set is the highest, since no feature is missing. Then, as we observed, if one subspace is removed, recognition accuracy decreases, andrecognitionaccuracy of our approach is the highest of all. That is to say, comparing with the the other three kernels, our approach can best reduce the decrease of recognition accuracy from missing features.

5. Conclusion

In this paper, we propose an approach to address two issues in multimodal feature fusion, the missing features and the curse of dimensionality. First, the multimodal feature vector is grouped into a set of subspaces, which is represented as a point on the Grassmann manifold. Then, to deal with missing features, L2-Hausdorff distance, is used to compare multimodal feature vectors with different number of subspaces. Next, to alleviate the curse of dimensionality, two feature selection algorithms, one supervised and one unsupervised, are proposed to obtain a few representative features. To validate our approach, experiments on three multimodal datasets is carried out. Experimental results show the proposed feature fusion can outperforms the state-of–the-art by higher accuracy.

6. References

[1] K. Woods, W. Kegelmeyer Jr, K. Bowyer, Combination of multiple classifiers using local accuracy estimates, newblock IEEE T-PAMI, 19(4), pages: 405–410, 1997.

[2] J. Kittler, M. Hatef, R. Duin, J. Matas, on combining classifiers, IEEE T-PAMI, 20(3), pages: 226–239, 1998.

[3] D. Greene, P. Cunningham, A matrix factorization approach for integrating multiple data views, Machine Learning and Knowledge Discovery in Databases, pages: 423–438, 2009.

[4] Z. Zhao, H. Liu, Multi-source feature selection via geometry dependent covariance analysis, JMRL, pages: 36–47, 2009.

[5] F. Bach, G. Lanckriet, M. Jordan, 'Multiple kernel learning, conic duality, and the smo algorithm, in Proc. of ICML, p. 6, ACM, 2004.

[6] P. Gehler, S. Nowozin, on feature combination for multiclass object classification, in Proc. of ICCV, pages: 221–228, 2009.

[7] T. Xia, D. Tao, T. Mei, Y. Zhang, Multiview spectral embedding, IEEE T-SMC-B, 40(6), pages 1438–1446, 2010.

[8] B. Xie, Y. Mu, D. Tao, K. Huang, m-sne: Multiview stochastic neighbor embedding, IEEE T-SMC-B, 41(4), pages: 1088–1096, 2011.

[9] X. Zhou, B. Bhanu, Feature fusion of side face and gait for videobased human identification, Pattern Recognition, 41(3), pages: 778–795, 2008.

[10] W. Zuo, H. Zhang, D. Zhang, K. Wang, and Post-processed lda for face and palmprint recognition: What is the rationale, Signal Processing, 90(8), pages: 2344–2352, 2010?

[11] Y. Wu, E. Chang, K. Chang, J. Smith, Optimal multimodal fusion for multimedia data analysis, ACM Multimedia, pages: 572–579, 2004.

[12] N. Cristianini, B. Scholkopf, Support vector machines and kernel methods: the new generation of learning machines, Ai Magazine, 23(3), pages 31–41, 2002.

A method for turbine blade temperature data segmentation

Chi Feng, Li Wang and Shan Gao

College of Information and Communication Engineering, Harbin Engineering University, Harbin, China.
Email: wangli19920609@foxmail.com

Abstract. Turbine blade, as one of the key components of the engine, operates in the badly working conditions. In order to better monitor the temperature status of turbine blades, research on temperature distribution of working blades is significant. The paper applies discrete Fourier transform to develop mathematical models, and the changes of period and peaks are summarized. The changing trends of temperature are reflected in each blade. The trends can be treated as one of the bases of the blade condition monitoring and fault diagnosis.

1. Introduction

In recent years, with the technical development and progress of the gas turbines, equipments and technical level have been observably enhanced and widely applied to various industries which include the electric power, the petroleum, the chemical metallurgy, the ship power and other fields [1].The turbine blades work as one of the most important parts of the gas turbine power, and due to long-term work in the harsh environment of high temperature and high pressure, gas turbines directly impact on the using lifetime of the gas turbine. Therefore, the current situation has received widespread attention. It is worth noting that the blade temperature which is an important assessment indicator of the blade's quality will become increasingly significant. After temperature data of turbine blades are obtained, the information of a single turbine blade can be acquired using segmentation methods. Ketui Daniel Kipngetich [2] took advantage of the fractal theory and the three fractal dimension features of blade temperature signals are extracted, with the weight of each feature being calculated by combining with K-means clustering analysis and ReliefF algorithm, and thereby established the temperature feature models of turbine blades. Jinyan Tong [3] used the density measurement data of the turbine blade with complex topological structures, and presented a CAD model directing method for density data of the turbine blade. Artur Kułaszka [4] extracted some characteristics of the blade signal under wavelet analysis, and he proposed waveform-based feature extraction, and creates the waveform-based eigenvectors.

The Continuous Fourier Transform (CFT) is a kind of integral transform, which is difficult to apply in the actual fields. So the Discrete Fourier Transform (DFT) plays an important role in digital signal processing and many other fields [5]. It is used in computer analysis for Fourier transform and estimating continuous time signal in the way of spectrum analysis. The paper is based on the DFT algorithm, and the Inverse Discrete Fourier Transform (IDFT) with better performance is proposed to extract the harmonic parameters [6].

2. Mentation Process of Turbine Blade Wave

2.1. Data Pre-Processing of Turbin Blades

In the process of turbine blades working in the condition of high temperature, on-line monitoring of the blades' temperature needs to be periodic segmentation of collected signal. Every point's temperature value of the blades needs to be measured before segmentation. The instrument using non-contact method measures temperature value of the surface and the sampling frequency is 500000. However, the measured data are voltage value in practice and need to be converted to temperature value.

The function relationship of output voltage and temperature is complex in system, and the fitting curves in different temperature measurement systems are different. It is connected with wavelength selection, phototube types, light attenuation, circuit parameters, chip types and many conditions. Generally, the expressions are not obtained directly, and fitting methods must be used.Through system calibration in standard black-body, the mathematical relationship between the temperature and voltage need be established. The fitting chart of voltage and temperature is presented in Figure.1. The value of the preceding 500 temperature points from turbine blades is presented in Figure.2, when the working condition parameter is 1.0 (the working condition parameter can reflect rotary velocity of turbine blades).

Various devices will have a disturbing effect in the process of measuring practically, and other noise and uncertain signals in the environment, and test signal will lead to huge disturbance. In order to better process and analyze some characteristics of blades, the voltage signal need to be filtered. Gaussian filter (where the window function is 8 and the standard deviationis is 1.7)adopted to remove noise jamming , and the filtering result becomes more smooth.

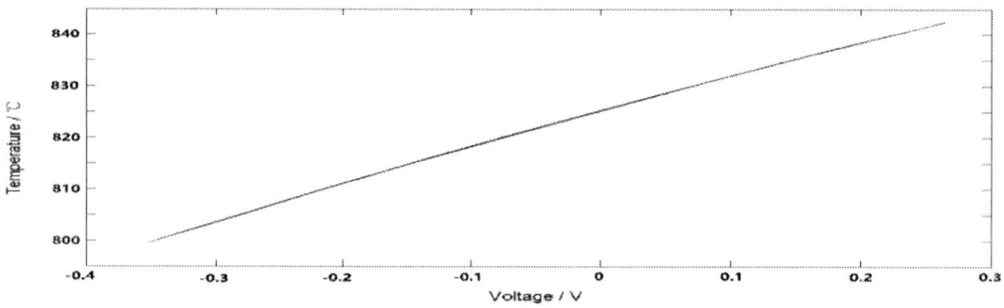

Figure 1. Fitting chart of voltage and temperature

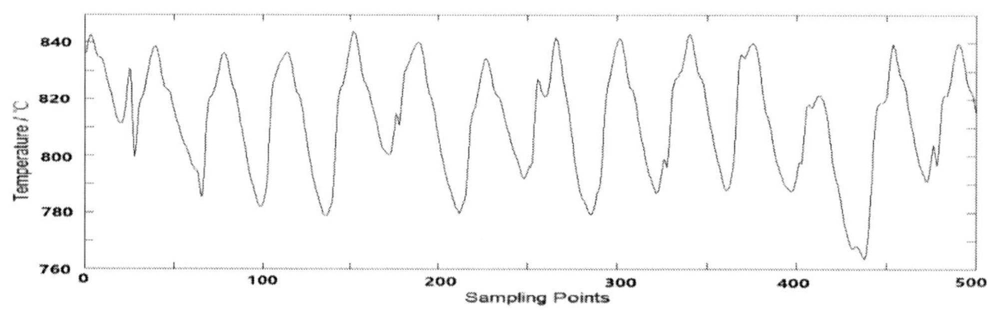

Figure 2. Temperature figure of voltage signal (1-500 points)

2.2. Creating Cosine Signal through Temperature Signal

In order to create cosine signal for segmenting turbine blades, the fundamental harmonic (first harmonic) of turbine blades' temperature data should be calculated by DFT. First, direct current component (subtract the mean value of temperature data) of the temperature data of blades is removed by subtracting the mean value of temperature data to acquire the new signal $LS(n)$, then the first harmonic frequency f of $LS(n)$is acquired by DFT. f is a numerical frequency and a integer, which needs be fine-tuned later. The cosine signal is showed in equation (1):

$$RS(n) = \mathrm{A}\cos(\frac{2\pi \mathrm{f}}{\mathrm{N}} n) \tag{1}$$

IOP Conf. Series: Journal of Physics: Conf. Series **887** (2017) 012067 doi:10.1088/1742-6596/887/1/012067

where A is an amplitude and arbitrary value. In the Figure.3, the red signal represents the cosine signal $RS(n)$ and the black signal represents $LS(n)$.

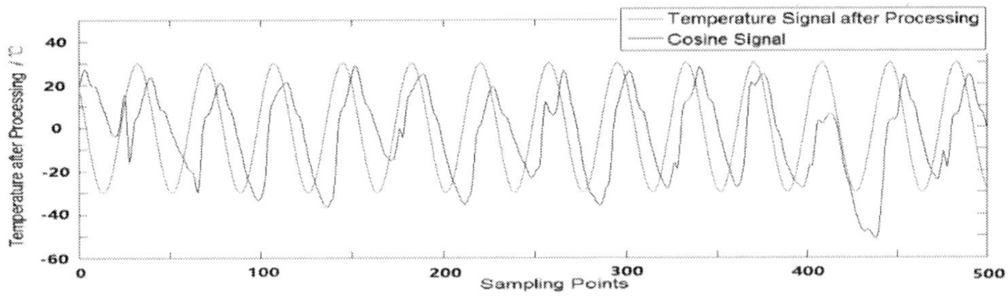

Figure 3. Waveform figure of temperature signal after processing and cosine signal (1-500 points)

2.3. Adjusting Phase Process of Cosine Signal

The phase of cosine signal and temperature is not matching through Figure.3. By adjusting the phase of cosine signal. The front 1000 points of temperature are got to adjust phase. In other words, the peaks of cosine signal $RS(n)$ align to the peaks of $LS(n)$ or the valleys of $RS(n)$ align to the valleys of $LS(n)$.

The concrete method:

1. The signal LS is a discrete sequence made up with 65536 points.

$$\mathrm{LS} = \{\mathrm{LS}[1], \mathrm{LS}[2], \ldots, \mathrm{LS}[65536]\} \tag{2}$$

2. Similarly, the cosine signal RS is also a discrete sequence made up with 65536 points.

$$\mathrm{RS} = \{\mathrm{RS}[1], \mathrm{RS}[2], \ldots, \mathrm{RS}[65536]\} \tag{3}$$

3. The peaks' and valleys' number of front 1000 points of $LS(n)$ is all r ($\mathrm{r} = INT(1000\mathrm{f}/\mathrm{N})$, $\mathrm{N} = 65536$), and f is the first harmonic frequency of $LS(n)$.

4. The point coordinates of front 1000 points of the peaks from RS are $[\mathrm{m}_1, \mathrm{m}_2, \ldots, \mathrm{m}_\mathrm{r}]$. The point coordinates of front 1000 points of the valleys from RS are $[\mathrm{n}_1, \mathrm{n}_2, \ldots, \mathrm{n}_\mathrm{r}]$.

5. When RS moves to right with i $(i = 0,1,2,\ldots, INT(\mathrm{N}/\mathrm{k}))$ points, m_i and n_i change following i.

$$m(i) = [\mathrm{m}_1 + i, \mathrm{m}_2 + i, \ldots, \mathrm{m}_\mathrm{r} + i] \tag{4}$$

$$n(i) = [\mathrm{n}_1 + i, \mathrm{n}_2 + i, \ldots, \mathrm{n}_\mathrm{r} + i] \tag{5}$$

6. The formula can be described as follows.

$$S_{peaks}(i) = LS[m(i)] = LS[\mathrm{m}_1 + i] + LS[\mathrm{m}_2 + i] + \cdots LS[\mathrm{m}_\mathrm{r} + i] \tag{6}$$

$$S_{valleys}(i) = LS[n(i)] = LS[\mathrm{n}_1 + i] + LS[\mathrm{n}_2 + i] + \cdots LS[\mathrm{n}_\mathrm{r} + i] \tag{7}$$

Where $i = 0,1,2,\ldots, INT(\mathrm{N}/\mathrm{k})$, on the basis of the formula calculating the corresponding location point i of maximal S_{peaks} and minimal $\mathrm{S}_{valleys}$, the cosine signal RS adjusts phase. The result is showed in Figure.4 and Figure.5.

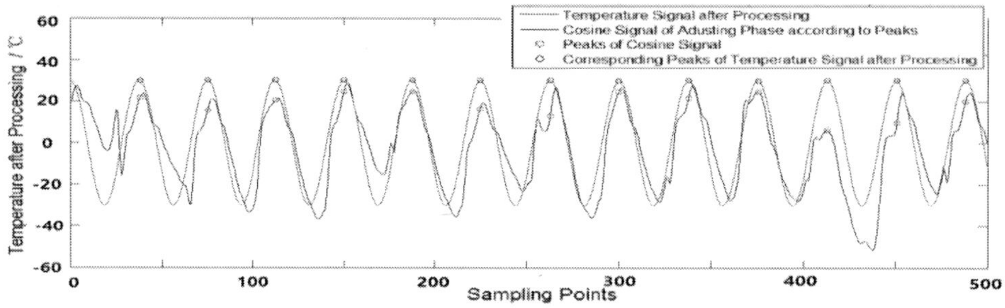

Figure 4. Waveform figure of temperature signal after processing and cosine signal of phasing control according to peaks (1-500 points)

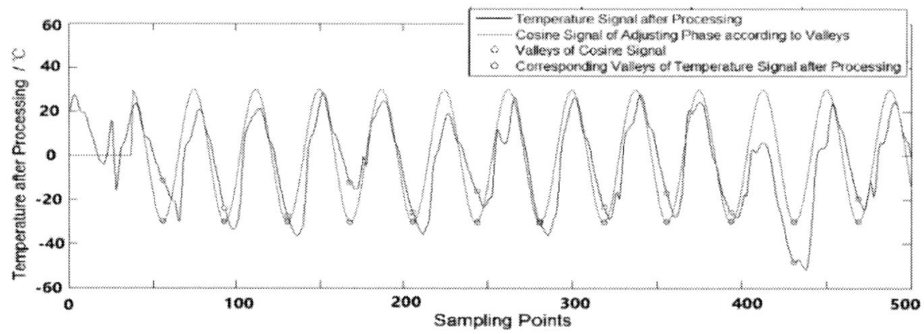

Figure 5. Waveform figure of temperature signal after processing and cosine signal of phasing control according to valleys (1-500 points)

2.4. Adjusting Frequency Process of Cosine Signal

The way of adjusting frequency f is similar to the one of adjusting phase.

The concrete method:

1. The peaks' and valleys' number of latter points of $LS(n)$ is all 50. The specific points where the cosine signal needs to be adjusted according to the frequency are different for different temperature waves. The best location of adjusting frequency is judged for different waves.
2. The point coordinates of the peaks from RS after 15000 points are $[\mathrm{m}_1, \mathrm{m}_2, \ldots, \mathrm{m}_{50}]$. The point coordinates of the valleys from RS after 15000 points are $[\mathrm{n}_1, \mathrm{n}_2, \ldots, \mathrm{n}_{50}]$.
3. The first harmonic frequency is F.

$$F = [\mathrm{f}-1, \mathrm{f}-0.95, \mathrm{f}-0.9, \ldots, \mathrm{f}+0.9, \mathrm{f}+0.95, \mathrm{f}+1] \tag{8}$$

The frequency of RS changes because of F. Meanwhile, both $\mathrm{m}(F)$ and $\mathrm{n}(F)$ change as well.

$$m(F) = [\mathrm{m}_1(F), \mathrm{m}_2(F) \ldots, \mathrm{m}_{50}(F)] \tag{9}$$

$$n(F) = [\mathrm{n}_1(F), \mathrm{n}_2(F) \ldots, \mathrm{n}_{50}(F)] \tag{10}$$

4. The formula can be described as follows.

$$\mathrm{S}_{peaks} = [\mathrm{S}_{peaks}(\mathrm{f}-1), \mathrm{S}_{peaks}(\mathrm{f}-0.95), \mathrm{S}_{peaks}(\mathrm{f}-0.9), \ldots, \mathrm{S}_{peaks}(\mathrm{f}+1)] \tag{11}$$

$$\mathrm{S}_{valleys} = [\mathrm{S}_{valleys}(\mathrm{f}-1), \mathrm{S}_{valleys}(\mathrm{f}-0.95), \mathrm{S}_{valleys}(\mathrm{f}-0.9), \ldots, \mathrm{S}_{valleys}(\mathrm{f}+1)] \tag{12}$$

On the basis of the formula calculating the first harmonic frequency f of maximal S_{peaks} and minimal $S_{valleys}$, the cosine signal RS is adjusted to frequency. The result is showed in Figure.6 and Figure.7.

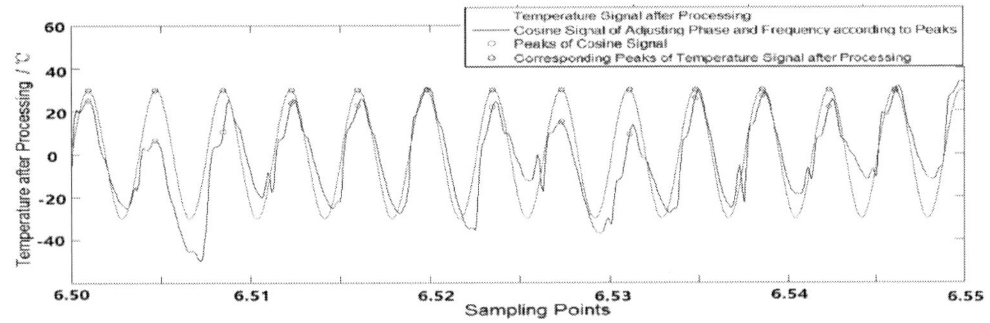

Figure 6. Waveform figure of temperature signal after processing and cosine signal of adjusting phase and frequency according to peaks (65000-65500 points)

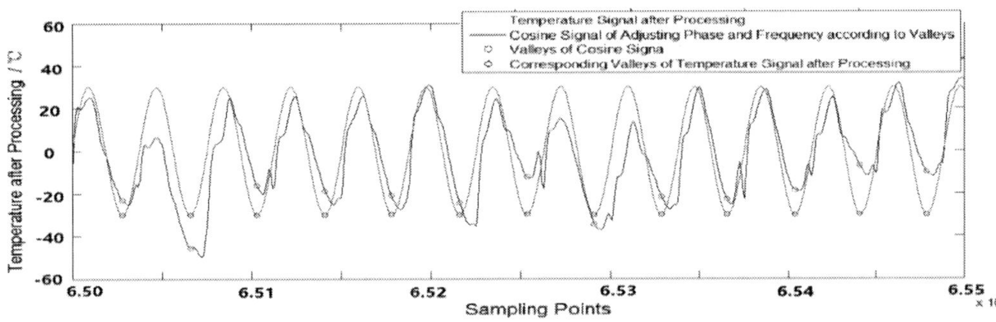

Figure 7. Waveform figure of temperature signal after processing and cosine signal of adjusting phase and frequency according to valleys (65000-65500 points)

2.5. Accurate Adjustment of Peaks And Valleys

The present peaks and valleys of temperature signal are found by the adjustive cosine signal, but the result is not so exact. In order to segment blade wave exactly in the whole wave, the exact peaks and valleys are searched in small scales through acquired them. That is, the local maximums and the local minimums are searched in small scales. The final result is showed in Figure.8.

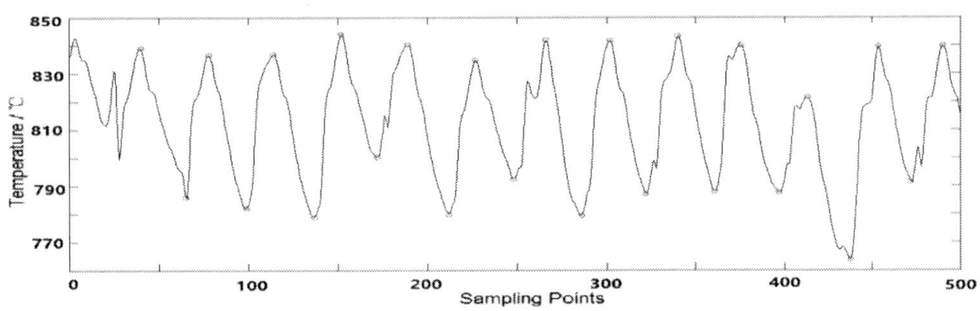

Figure 8. Waveform figure of temperature signal

3. Segmentation Result of Turbine Blade Wave

Between the peaks and valleys of the cosine signal RS adjusted phase and frequency are selected to segment tubine blade wave. Its aim is to reserve the key information of the peaks and valleys of turbine blades, so the value of the positional spots of segmentation is the half of the value of peaks and the value of valleys. The intersected result is showed in Figure.9. The accuracy rate of segmentation result is 100%.

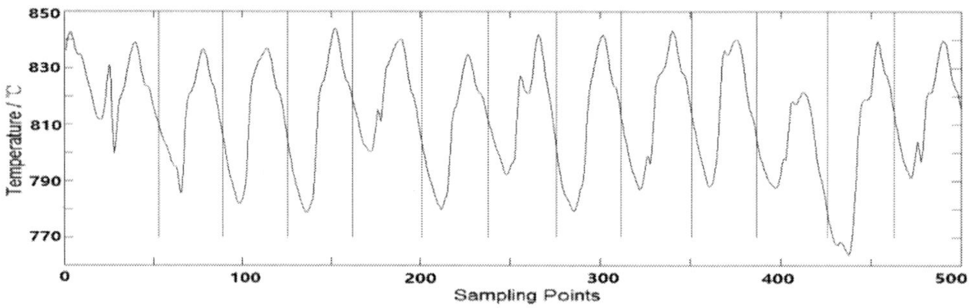

Figure 9. Waveform figure of temperature signal and segmentation

4. Conclusion

The paper mainly analyzes the characteristics of the turbine blades and introduces discrete fourier transform (DFT) to acquire the fundamental frequency of blades. By acquiring temperature data spots and applying discrete fourier transform, the first harmonic is calculated by summarizing the changes of period and peak. Thereby, a cosine signal is gained by the first harmonic. If the cosine signal matches the temperature signal of turbine blades totally, the phase and period must be accurately adjusted based on software to realize accurate segmentation of the data to the temperature of turbine blades and each blade data is obtained. Eventually, the experiment can reflect the changing trend of temperature from each blade. The result of segmenting turbine blade wave can be treated as one of the bases of the blade condition monitoring and fault diagnosis.

5. References

[1] Khan Z, Fida S, Nisar F, et al. Investigation of Intergranular Corrosion in 2nd stage gas turbine blades of an aircraft engine [J]. Engineering Failure Analysis, 2016, 68: 197-209.

[2] Kipngetich K D, Feng C, Gao S. Reflection error correction of gas turbine blade temperature[J]. Infrared Physics & Technology, 2016, 75: 153-159.

[3] Tong J, Ding X, Wang M, et al. Assessment of service induced degradation of microstructure and properties in turbine blades made of GH4037 alloy [J]. Journal of Alloys and Compounds, 2016, 657: 777-786.

[4] Artur Kułaszka,Józef Błachnio,Łukasz Kornas. Analysis of feasibility to assess microstructure of gas turbine blades by means of the thermographic method [J]. Journal of KONBIN, 2010, 13(1): 325-340.

[5] Mei L, Zhang Q Y, Sha X J, et al. Digital computation of the weighted-type fractional Fourier transform [J]. Science China Information Sciences, 2013, 56(7): 1-12.

[6] Gao H, Liang L, Chen X, et al. Feature extraction and recognition for rolling element bearing fault utilizing short-time Fourier transform and non-negative matrix factorization[J]. Chinese Journal of Mechanical Engineering, 2015, 28(1): 96-105.

Real-time vehicle detection and tracking in video based on faster R-CNN

Yongjie Zhang[1], Jian Wang[1, 3] and Xin Yang[2]

[1]Department of Electronic Engineering, Tsinghua University, Beijing 100084, China;
[2]Institute of Information Optics Engineering, Soochow University, Suzhou, Jiangsu 210056, China.
Email: jian-wang@tsinghua.edu.cn

Abstract. Vehicle detection and tracking is a significant part in auxiliary vehicle driving system. Using the traditional detection method based on image information has encountered enormous difficulties, especially in complex background. To solve this problem, a detection method based on deep learning, Faster R-CNN, which has very high detection accuracy and flexibility, is introduced. An algorithm of target tracking with the combination of Camshift and Kalman filter is proposed for vehicle tracking. The computation time of Faster R-CNN cannot achieve real-time detection. We use multi-thread technique to detect and track vehicle by parallel computation for real-time application.

1. Introduction

In auxiliary driving system, vehicle detection and tracking is a critical technology that plays an important role in vehicle active safety. The complex vehicle patterns in video frame from real scene make a challenge for real-time detection and tracking. The detection of vehicle using traditional image processing method is impossible in some complex conditions. Deep learning method is a promising alternative for natural scene target detection. Better results have been achieved using Faster R-CNN [1] proposed by Ren in some challenging datasets such as ImageNet and VOC, as well as professional datasets, such as KITTI [2].

In this paper, Faster R-CNN is applied to vehicle detection for auxiliary driving system. The neural network model is trained using partial KITTI dataset based on Faster R-CNN. An algorithm of target tracking with the combination of Camshift and Kalman filter is proposed for vehicle tracking. For real-time detection, multi-thread technique is used for vehicle detection and tracking. The experiment results validate our proposed method.

2. Overview of Faster R-CNN

Traditional methods for object detection based on deep learning employ sliding window or region proposal to generate object hypotheses, and then determine their classifications by a classifier. However, the process of region proposal computation is very time consuming, which is a bottleneck for real-time application.

To deal with this problem [3], Faster R-CNN proposes a Regional Proposal Network (RPN) [1], which shares the fully convolutional layers with a Fast R-CNN [4] object detection network. RPN is a fully convolutional network, which predicts both object proposals and objectness scores. The proposals and scores are fed into Fast R-CNN network for model training. The Faster R-CNN algorithm reduces time of computation and realizes good object detection performance with high mean Average Precision (mAP) at ILSVRC 2015 and the COCO 2015 competition.

IOP Conf. Series: Journal of Physics: Conf. Series **887** (2017) 012068 doi:10.1088/1742-6596/887/1/012068

3. The Tracking Algorithm

Camshift [5] algorithm is an extension of the meanshift [6] algorithm, based on color histogram. It can achieve the moving target tracking and adjust the size and position of the window in the simple background. However, in practice, camshift often fails in tracking when there are changes caused by illumination variation or partial occlusion, etc. A linear recursive Kalman filter using motion information including velocity and space direction to predict target position in the next frame.

In this paper, Camshift combined with Kalman filter is used for moving target tracking [7] [8]. Faster R-CNN is used for target detection. The detection results are used to initialize Kalman filter. The algorithm steps are as follows:

(1) To identify the target to be traced. For the first frame fetch the central coordinates of the target area (The coordinates of target areas coming from Faster R-CNN detection)., and then initialize the kalman filter with this coordinate;

(2) Based on the results of targets tracking in the first frame, use Kalman filter to predict the location in the current frame;

(3) With the coordinates acquired in the step(2) as the centre, set up the search window and calculate the back projection image, then use Camshift algorithm to do iteration, until it meets the conditions. Then the searched window coordinate is treated as the final coordinates, which are the real location of the tracking object in the current frame eventually;

(4) Correct kalman filter based on the coordinate from current frame, and set the current frame as first frame.

(5) If there is no new detection results from Faster R-CNN, making the program return to step (2), continue to track the target in the next frame. If the new detection results from Faster R-CNN are achieved, making the program return to step (2), continue to track the target in the next frame.

4. Experiment

4.1. Dataset

KITTI dataset is currently the largest computer vision dataset in the world, which is used to evaluate the performance of computer vision technology. Images were captured from real scenes such as urban, rural and highways. Figure 1 shows some images in KITTI dataset.

Figure 1. Some images from the KITTI dataset. Large variations in appearance and camera viewpoint, and severe occlusions are present in the dataset.

In this experiment, Subset from KITTI, including 7481 training images and 28742 labeled vehicle objects is used for model training. The car, van, truck, and tram are reclassified as part of the vehicle class. The test video was captured on the actual road using a usb camera on a car, with resolution 1280 * 720 and frame rate 30fts.

4.2. Training

A laptop with an Intel i7-6700 CPU, a GTX980M GPU with 8G memory, and operating system Ubuntu14.04 LTS is used for training and testing. The development language is python. Because of the limitation of computing power, a medium-sized VGG_CNN_M_1024 model is choosed for this experiment [9].

4.3. Detection and Tracking

The average computation time for target detection is 0.935s per image, and it is 0.0244s per image for tracking. Figure 2 and Figure 3 show the CDF (Cumulative Distribution Function) [10] of overhead time of detection and tracking. Figure 2 indicates that about 60% of the points fall within the range of [0.09, 0.11], while it's 75% of the points falling within the range of [0.003, 0.025] in Figure 3. The real-time vehicle detection and tracking system requires objects can be marked in the video frames without delay. Multi-thread technology can avoid blocking, while performing multiple tasks, maximizing multiprocessor performance. Multi-thread programming method is used, instead of the idea of "detection before tracking", to improve the real-time performance of the system and obtain good results.

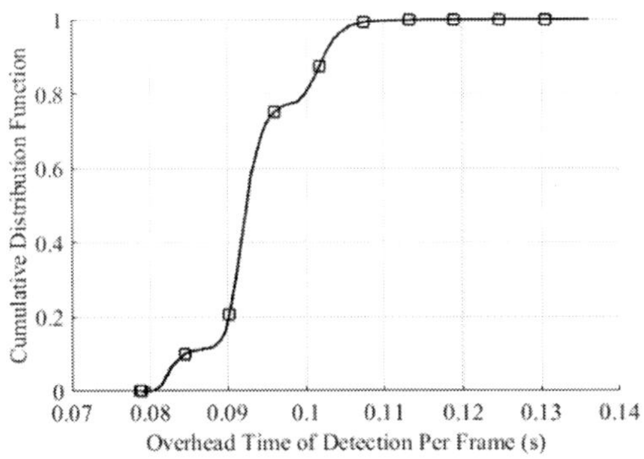

Figure 2. Overhead time of detection per frame.

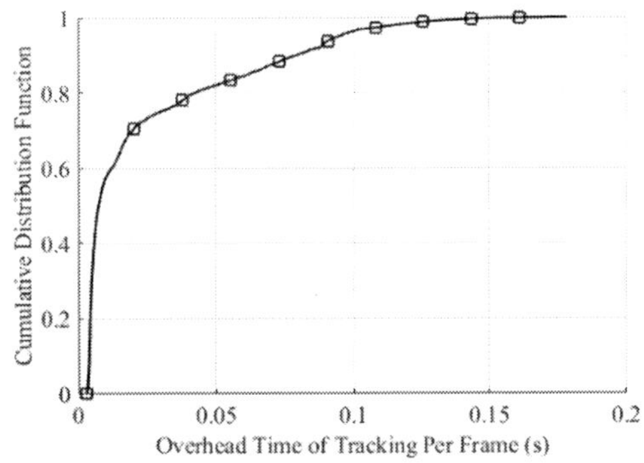

Figure 3. Overhead time of tracking per frame.

The flow chart of our algorithm for target detection and tracking is shown in Figure 4. Because the target detection computation is almost on GPU while the target tracking process uses CPU. We use two threads for real-time computation, detection thread and tracking thread, and define a global variable Flag to indicate whether new objects are detected or not. For the first frame, the detection thread detects targets using Faster R-CNN and the detected results are fed to tracking thread to start target tracking. The tracking thread uses new detection results for tracking if new detection results from detection thread are achieved. Two threads are running in parallel avoiding waiting for detection results.

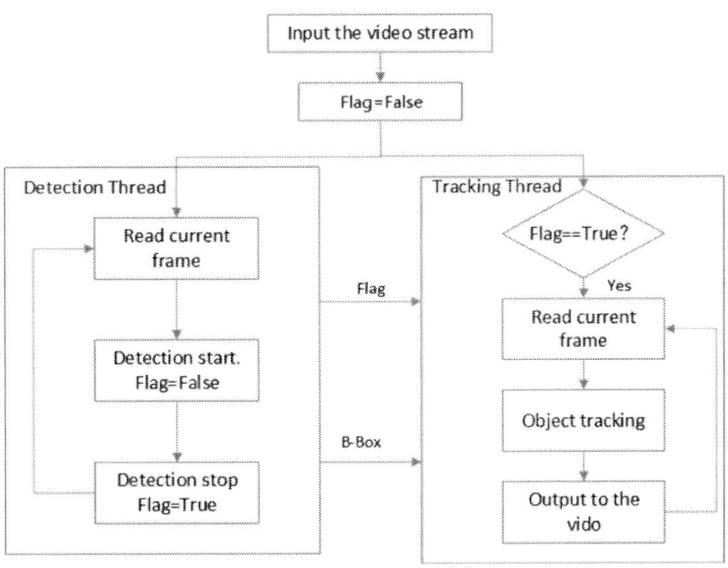

Figure 4. Flow chart of the multi-thread detection and tracking. Detection thread provides new object bounding-boxes and feed to the tracking thread.

Figure 5. Samples of vehicle detection and tracking.

Figure 5 shows some results of vehicle detection and tracking. Faster R-CNN demonstrates a strong performance on vehicle detection. Although the illumination changed and occlusion occur, Camshift combined with Kalman filter algorithm overcomes the interference better and achieves good tracking effect.

5. Conclusion

In this article, Faster R-CNN method is adopted to conduct the training of deep learning on the KITTI sub-dataset, which is effective for object detection in complex background. The combination of Camshift and Kalman filter method is used for vehicle tracking, by passing on unknown space parameters to initial search process, is effective to overcome the interference of the background and improves the reliability of object tracking. This method can be used in real-time application.

6. References

[1] Ren S, He K, Girshick R, and Sun J 2015 Faster R-CNN: Towards real-time object detectionwith region proposal networks *In Advances in neural information processing systems* 91-99

[2] KITTI benchmark, http://www.cvlibs.net/datasets/kitti/eval_object.php

[3] Wang X and Li X 2010 The study of moving target tracking based on kalman-camshift in thevideo *IEEE Int. Conf. on Information Science and Engineering* 1-4

[4] Girshick R 2015 Fast r-cnn *Proc. Int. Conf. on computer vision* pp 1440-8

[5] Exner D, Bruns E, Kurz D, Grundhöfer A and Bimber O 2010 Fast and robust camshift tracking*Computer Vision and Pattern Recognition Workshops. IEEE*, 9-16

[6] Leichter I, Lindenbaum M and Rivlin E 2010 Mean shift tracking with multiple reference color histograms *Computer Vision and Image Understanding* **114(3)** 400-408

[7] Zhang C, Qiao Y, Fallon E and Xu C 2009 An improved camshift algorithm for target trackingin video surveillance *9th IT&T Conference* 1-10
[8] Huang S and Hong J 2011 Moving object tracking system based on camshift and Kalman filter*IEEE Int. Conf. on Consumer Electronics, Communications and Networks* 1423-6
[9] Lee C, Kim H J and Oh K W 2016 Comparison of faster R-CNN models for object detection *IEEE Int. Conf. on Control, Automation and Systems* 107-110
[10] Xue J, Titterington M The p-folded cumulative distribution function and the mean absolute deviation from the p-quantile *J. Statistics & Probability Letters*, 2011, 81(8):1179–1182.

IOP Conf. Series: Journal of Physics: Conf. Series **887** (2017) 012069 doi:10.1088/1742-6596/887/1/012069

Vehicle positioning based on UWB technology

Siquan Hu[1, 2, a]*, Min Kang[1, b] and Chundong She[3, c]

[1]School of Computer and Communication Engineering, University of Science and Technology Beijing, Beijing, 100083, China
[2]Chengdu Visual Microcosm Technology Ltd. Chengdu, 610072, Sichuan, China
[3]School of Electronic Engineering, Beijing University of Posts and Telecommunications, Beijing, 100876, China
[a]husiquan@ustb.edu.cn, [b]kilometer4@sohu.com, [c]shurcd@vip.sina.com

Abstract. In recent years, with the rapid increase of the number of urban cars, the vehicle internet is becoming a trend of smart transportion. In such vehicle network, accurate location is very crucial in many new applications such as autopilot, semi-autopilot and Car-to-x communications. UWB technology has been used for indoor closed range positioning and ranging widely, while UWB outdoor positioning and ranging research is relatively less. This paper proposed UWB as the vehicle positioning technology and developed a method based on two-way-ranging (TWR) to solve the ranging problem between vehicles. At the same time, the improved TOA method was used to locate vehicles, which has higher precision compared with traditional GPS or LBS. A hardware module is introduced and the simulation results show that the modules are capable of precise positioning for vehicles in vehicle network.

1. Introduction

Ultra wideband(UWB) is a wireless technology that has many advantages such as high data rate transmission, the low power spectral density, low transmission power and large instantaneous bandwidth. These advantages make it attractive in different applications such as tracking and communication, radar and wireless body area network[1]. It is mainly used for short range communications.

In terms of positioning, current positioning systems, such as GPS, or future systems, like the European GALILEO, may not be available or do not have sufficient accuracy in specific areas where may not satisfying the need of future applications in vehicle internet. The poor performance of satellite systems in supporting those applications is mainly caused by attenuation due to signal blocking and/or multipath signals which cause distortion of the direct signal path. The primary reason of performance degradation is not the presence of multipath itself but the fact that multipath cannot be resolved due to the small signal bandwidth used by current positioning systems.

In terms of ranging, UWB is more powerful than others. The main advantages of UWB systems are the usage of ultra-short pulses, either for communication or for ranging systems[2]. Compared with UWB, the absolute location error of GPS may be very large, while the use of ultra-short pulse of UWB positioning can provide a higher accuracy [3] which may reach to centimeter level. In addition, the UWB wireless positioning cost is cheaper[4].

The aim of this paper is to propose a new method of inter vehicle ranging and vehicle location in outdoor areas. The main method used for ranging between vehicles is TWR, and the method of vehicle location in outdoor areas is the improved TOA. This research will be useful on the development of ranging and positioning for future vehicle internet.

2. Proposed Receiver and Transmitter Scheme

2.1. Description and Implementation

We used the DWM1000 module produced by Decawave company because it enables very accurate Time of Arrival measurements of RF signal. The DWM1000 module is based on Decawave's DW1000[5] UWB transceiver IC. It integrates antenna, all RF circuitry, power management and clock circuitry in one module. The DW1000 is a fully integrated single chip UWB low-power low-cost transceiver IC compliant to IEEE802.15.4-2011. It can be used in two-way ranging or TOA location systems to locate assets. It also supports data transfer at rates up to 6.8 Mbps.

DW1000 consists of an analog front end containing a receiver and a transmitter and a digital back end that interfaces to an off-chip host processor. A T_X/R_X switch is used to connect the receiver or transmitter to the antenna port. Temperature and voltage monitors are provided on-chip.

The receiver consists of an RF front end which amplifies the received signal in a low-noise amplifier before down-converting it directly to base-band. The receiver is optimized for wide bandwidth, linearity and noise figure. This allows each of the supported IEEE802.15.4-2011[6] UWB channels to be down converted with minimum additional noise and distortion. The base-band signal is demodulated and the resulting received data is made available to the host controller via SPI.

The transmit pulse train is generated by applying digitally encoded transmit data to the analog pulse generator. The pulse train is up-converted by a double balanced mixer to a carrier generated by the synthesizer and cantered on one of the permitted IEEE802.15.4-2011 UWB channels. The modulated RF waveform is amplified before transmission from the external antenna. The DWM1000 module is shown in figure 1.

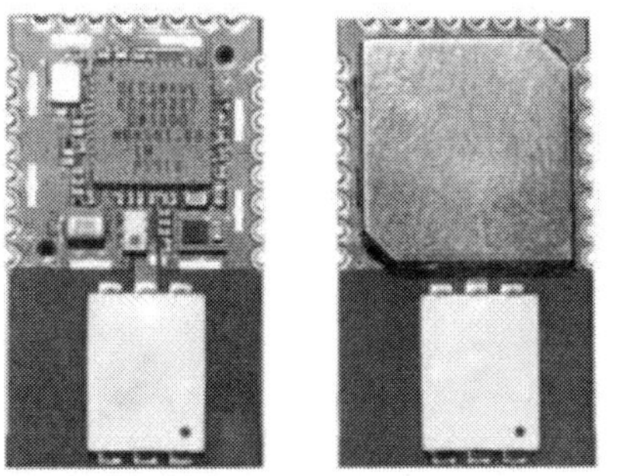

Figure 1. The DWM1000 Module

Figure 2. A node consists of a host MCU and a RF Module

A complete independent UWB node is showed in figure 2, where a host MCU and a RF Module are included. In our configuration, the host is a general ARM Cortex-M4 micro-controller. Each node can act as an anchor or tag. The anchor usually has a fixed location, whilst tag is a mobile node. Each vehicle is equipped with a node, and two nodes can be used for ranging between vehicles. The surrounding environment is configured with more nodes, the location of vehicle can be located. The results of distance measurement can be printed onto the serial port of a PC, and the location result needs several groups of distance. Then the relative position is calculated by the specific algorithm.

2.2. Theoretical Analysis of UWB Ranging

A classical time of arrival based method called two-way ranging (TWR) was originally proposed in [7]. The practical ranging demonstration is described in figure 3.

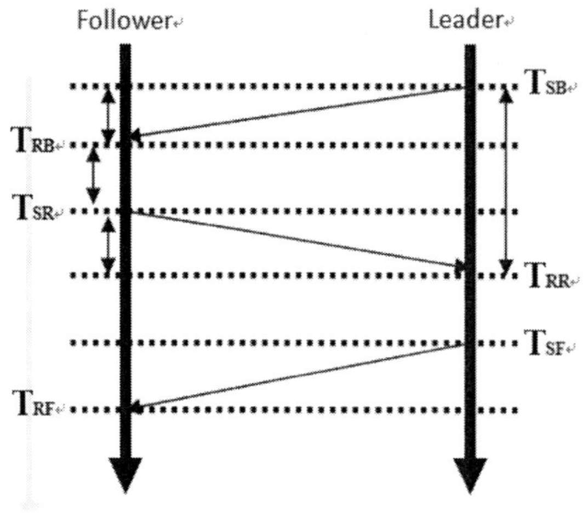

Figure 3. TW-Ranging Method

The leader observes a round trip time $L_{RT} = T_{RR} - T_{SB}$ and a turn around time $L_{TA} = T_{SF} - T_{RR}$, where T_{SB}, T_{RR} and T_{SF} are the leader send-time, receive-time and future send-time respectively. The follower observes a round trip time $F_{RT} = T_{RF} - T_{SR}$ and a turn around time $F_{TA} = T_{SR} - T_{RB}$, where T_{RB}, T_{SR} and T_{RF} are the follower receive-time, send-time and future receive-time respectively. The value of transmission time T is computed at both leader (T_l) and follower (T_f):

$$2T_l = (T_{RR} - T_{SB}) - (T_{SR} - T_{RB}) \tag{1}$$

$$2T_f = (T_{RF} - T_{SR}) - (T_{SF} - T_{RR}) \tag{2}$$

The follower or leader can then combine these two resultant round trip times (by averaging) to remove by effects of clock differences. The result is then divided by 2 to get one way trip time.

$$T = \frac{2T_l + 2T_f}{2 \times 2} \tag{3}$$

Thus the distance between the two prototypes is: (Here C is the speed of light)

$$d = T \times C \tag{4}$$

2.3. The TOA Proposed Approach

Here, it is assumed that input data of the localization phase are represented by the set of $TOA_{i,j}$, i= A,B, j= 1,2. Following the geometrical interpretation of the target localization problem, the four ellipses (E_i, i= 1, 2, 3, 4) with the foci Tx_i, $Rx_{i,j}$ and with the length of the semimajor axis $c.TOA_{i,j}/2$ can be created as figure 4. Then, the target position is obtained as the intersection of the particular ellipses. In the case of a single target scenario, the set of $TOA_{i,j}$ can contain 0-4 elements. Because of TOA estimation error, there are usually more than one intersections. Then, depending on the number of the estimated TOA, the localization of the target by TOA completing method can be described as follows:

Only one TOA or no TOA is estimated. The target position cannot be estimated. The only one $TOA_{A,i}$ for i=1 or i=2, and the only one $TOA_{B,i}$ for i=1 or i=2 have been estimated. The other TOAs are missing. The former TOA is given by RS_A, the latter TOA is given by RS_B. The target position in the discrete time instant n is given by the intersection of the ellipses E_1 or E_2 and E_3 or E_4 in figure 4.

Three TOAs have been estimated. Let us assume e.g. $TOA_{B,1}$, $TOA_{B,2}$, $TOA_{A,2}$ have been estimated and $TOA_{A,1}$ is missing. Then, the ellipses E_2, E_3, E_4 can be created. The target position estimate referred to as T_B is given by the intersection of E_3 and E_4 in figure 5. After that, $TOA_{A,1} = (\|Tx_AP\|+\|PRx_{A,1}\|)/c$. The final estimate of the target coordinates is obtained as the average of the points T_A and T_B in figure 4.

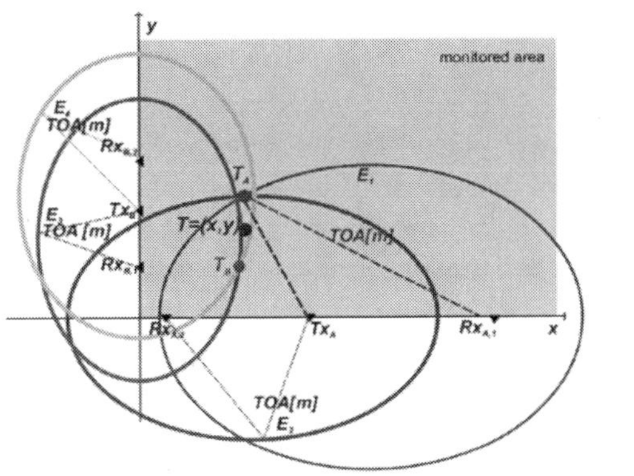

Figure 4. Geometrical interpretation of the target location

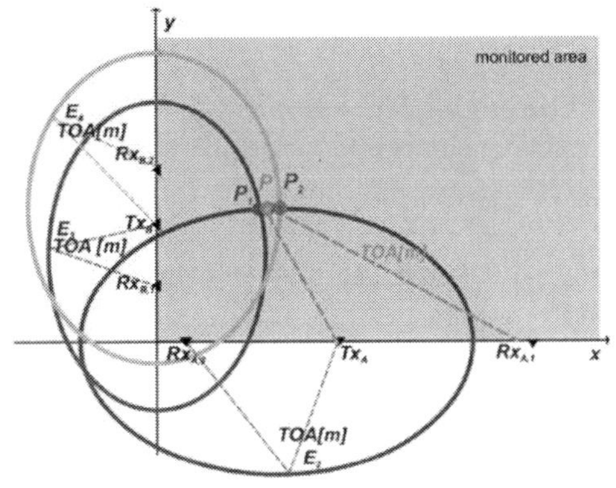

Figure 5. TOA completing method geometrical interpretation

3. Solution of the Problem

3.1. The ranging problem

To illustrate inter vehicle ranging problem based on UWB, we use the test scenario showed in figure 6. The red callout area is the reference location for the ranging module. The ranging algorithm is implemented on the host microcontroller which controls the DWM1000 module with a SPI interface. This method can be applied in a lot of scenes, such as two cars are moving at the same time, or one car is moving and the other is static. In range measurement, the two modules are identified as a tag and the other as anchor. It should be noted that the ranging module's configuration height should be consistent as far as possible, otherwise it will affect the ranging accuracy. The implementation is based on the TW-Ranging method mentioned in the previous section.

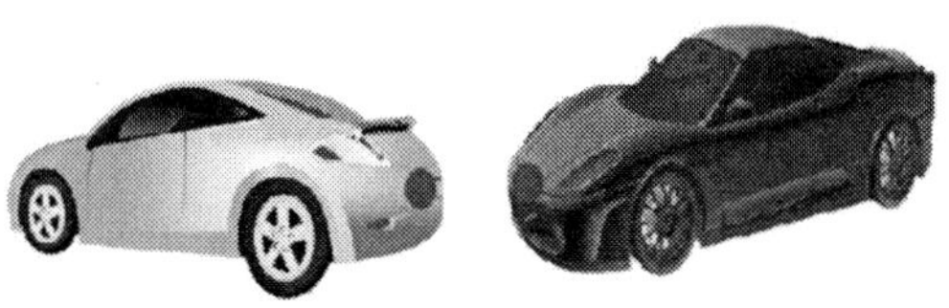

Figure 6. The ranging display

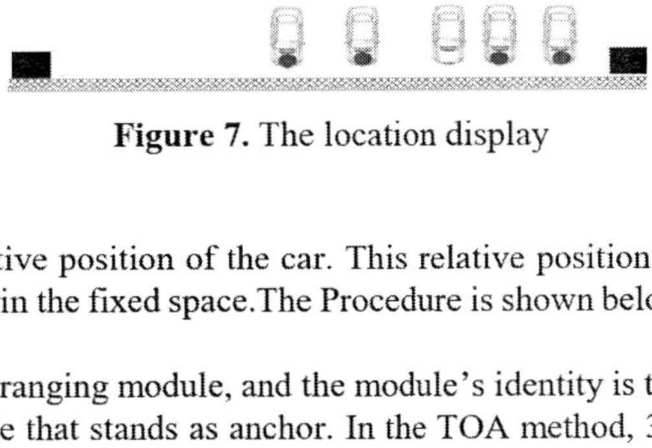

Figure 7. The location display

3.2. The location problem

The TOA method based on UWB can locate the relative position of the car. This relative position is defined as the location of the vehicle relative to a point in the fixed space.The Procedure is shown below with the scenario in figure 7.

The red solid circle represents the vehicle with the ranging module, and the module's identity is the tag. The black solid box stands for the ranging module that stands as anchor. In the TOA method, 3D positioning requires four points, that is, anchor points. Therefore, in outdoor space, you need to put four identity as anchor point ranging module. Through the method described in the preceding section, the location of outdoor vehicles is carried out. It should be noted that the use of the TOA method in positioning the vehicle, if the anchor and the tags' clock does not synchronize, there will be slight error. When the vehicle is stationary in a closed outdoor environment, the error does not affect the positioning effect.

4. Simulation Results and Discussion

4.1. The Ranging Simulation

The experiment was conducted outdoors. Transceivers were placed on a 1m high tripod. Between transceivers were LOS. Measurements were taken at known distances from 1 to 30 m. Measurements were taken at different operational modes of the module as in table 1. figure 8 and figure 9 show the distance measurements between transceivers and deviations from the pre–set distance. Denotations M1–M6 represent different types of operational modes.

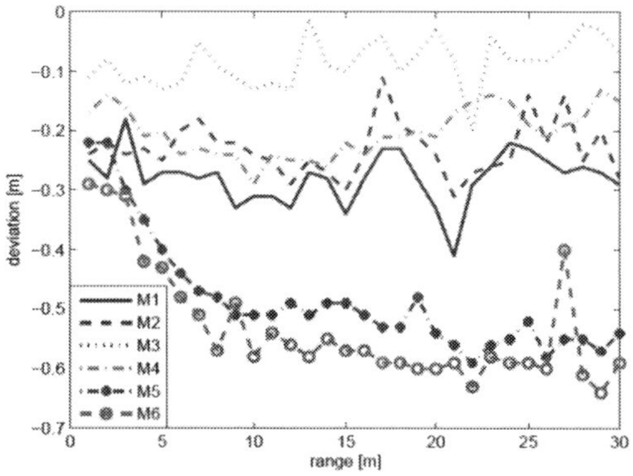

Figure 8. Outdoor distance measurements with non-calibrated module

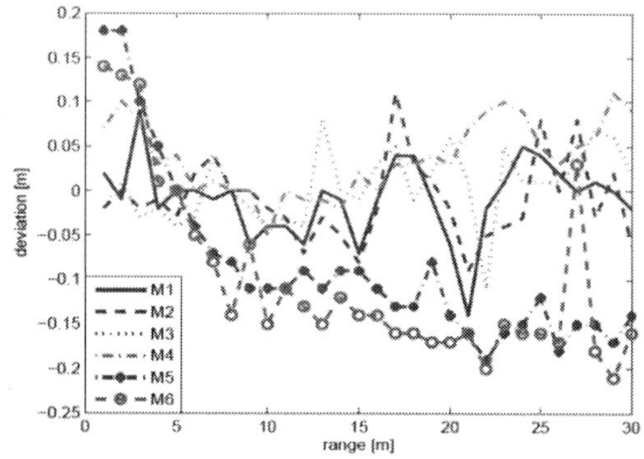

Figure 9. Outdoor distance measurements with calibrated module

Table 1. Operational modes.

Mode	Channel[GHz]	Data Rate [bps]	Bandwidth[MHz]	Preamble[bit]	PRF[MHz]
1	4.0	110k	500	1024	64
2	4.0	6.8M	500	128	64
3	6.5	110k	500	1024	16
4	6.5	6.8M	500	128	16
5	6.5	110k	900	1024	64
6	6.5	6.8M	900	128	64

Simulation results show that the modules are capable of precise measurements of distances in outdoor environments, where the measurements precision is less than 10 cm in line of sight. Modes from 1–4 returned almost the same precise results under 6 cm on average. Modes 5 and 6 have a channel bandwidth as 900 MHz, there accuracy of the measurements decreased but not exceed 20 cm.

4.2. The Location Simulation

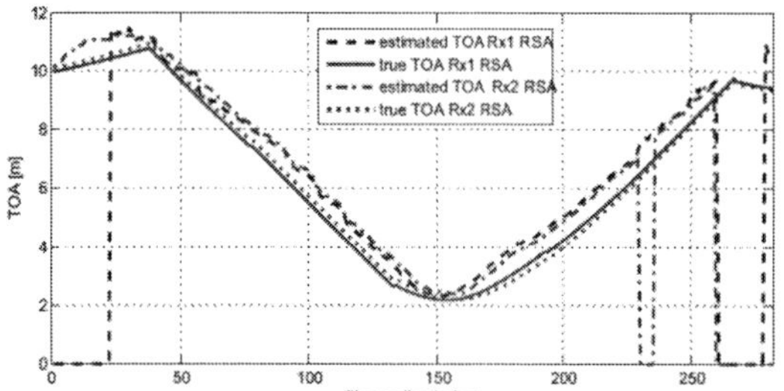

Figure 10. RS$_A$: the true and estimated TOAs

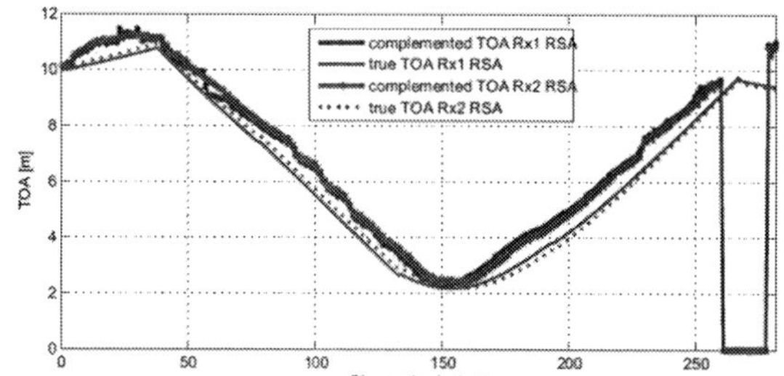

Figure 11. RS$_A$: the true and complemented TOAs

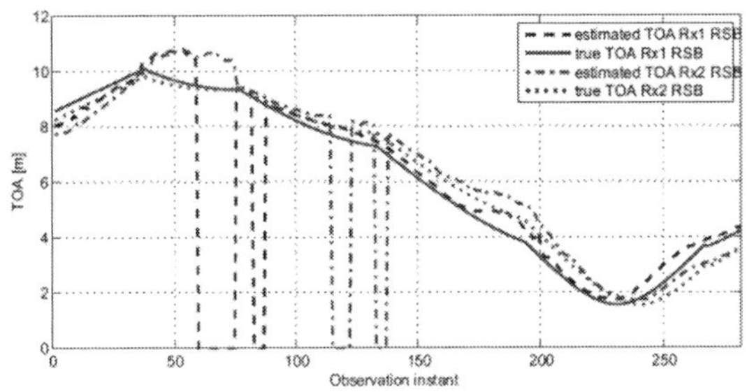

Figure 12. RS_B: the true and estimated TOAs

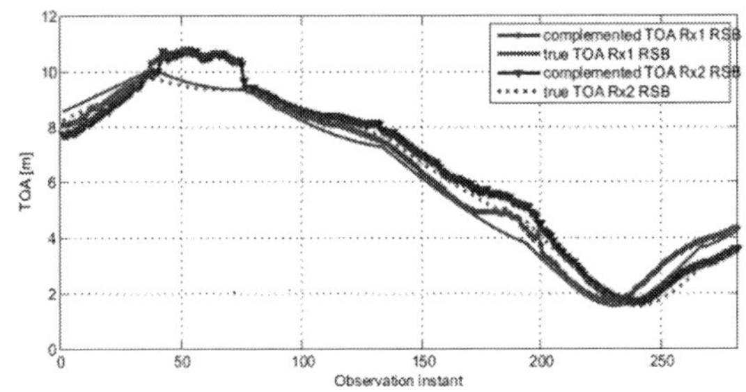

Figure 13. RS_B: the true and complemented TOAs

The true and estimated $TOA_{i,j}$ for i = A, B, j = 1, 2 are illustrated in figure 10 and figure 12. Then, the true, estimated and complemented $TOA_{i,j}$ for i = A, B, j = 1, 2 are illustrated in figure 11 and figure 13.

TOA's location method is based on the calculation algorithm whose input is from the ranging data. As shown in the simulation, the complemented TOAs method can restore the missing TOA fragments, more accurately smooth the RS_A and RS_B curve, so that the final positioning results are more reliable with higher accuracy.

5. Conclusion

In this paper, we propose a UWB based technique for inter vehicle ranging. Furthermore, the vehicle location of the TOA method is carried out with the ranging data obtained. With the help of UWB's ability of precise ranging to the distances below 10 cm, UWB modules could be very useful for outdoor vehicle positioning and may have significant potential in vehicle internet over traditional GPS or LBS positioning.

6. Acknowledgements

The study was supported by the National Natural Science Foundation of China (No. 91438120) and the Sichuan Science and Technology Support Program (No.2016GZ0339).

7. References

[1] Dokania R K, Wang X Y, Tallur S G, and Apsel A B 2011 *IEEE Trans. Circuits Syst. I* **58** pp 1458–1469
[2] Figueroa A 2015 *TU Dresden Technology Report*
[3] Gezici S, Tian Zhi, Giannakis G B, Kobayashi H, Molisch A F, Poor H V and Sahinoglu Z 2005 *IEEE Signal Processing Magazine* **22** pp 70–84
[4] Lanzisera S, Zats D, and Pister K S J 2011 *IEEE Sensors J.* **11** pp 837–845
[5] Decawave 2015 DW1000 User Manual 2.02 http://www.decawave.com/support
[6] IEEE802.15.4-2011(Revision of IEEE Std 802.15.4-2006) IEEE Standard for Local and metropolitan area networks – Part 15.4: Low-Rate Wireless Personal Area Networks (LR-WPANs). IEEE Computer Society Sponsored by the LAN/MAN Standards Committee. Available from http://standards.ieee.org/
[7] IEEE Std 802.15.4aTM-2007 IEEE Standard for information Technology

Fault diagnosis of wind bearing based on multi-scale wavelet kernel extreme learning machine

Siwen Zhu and Bin Jiao

Department of Electrical Engineering, Shanghai Dian Ji University, Electrical College, No. 300 blooms Road, Pudong New Area, Shanghai, China.
Email: 278630427@qq.com

Abstract. The principle of kernel Extreme Learning Machine (ELM) is demonstrated. On this basis, a multi - scale wavelet kernel extreme learning machine is proposed. The multi-scale wavelet kernel is used as the kernel function of the extreme learning machine. The test shows that it is an achievable extreme learning machine. Experiments show that, using the multi-scale wavelet kernel extreme learning machine in the wind turbine bearing fault diagnosis has higher classification accuracy and speed than the support vector machine classification algorithm, and has excellent application value.

1. Introduction

Wind turbines often work in harsh outdoor environments and have a high incidence of failures. As an important transmission parts of wind turbine, the bearing has the function of reducing the relative movement of the component and supporting the guide. Once the bearing broke out, it will cause the unit to stop running and bring great economic losses to the whole wind farm. Therefore, it is very important to quickly and effectively diagnose the failure of main bearing of wind turbine.

At present, many artificial intelligence algorithms are used in wind turbine bearing fault diagnosis. Vapnik and so on proposed support vector machine (SVM) on the basis of statistical theory [1], which effectively solved the problem of non-linear classification, but does not have a good classification for the limited sample and has long training time and low accuracy; On the basis of that he input layer connection weights of the feedforward neural network do not require iterative adjustment, Huang proposed a new neural network training architecture called Extreme Learning Machine (ELM) [2]. Compared with the traditional BP neural network based on gradient descent, this method can solve the least square solution of output weight at one time, and it is fast and generalized. It is widely used in classification problem. [3] Compared with the traditional classification method, ELM greatly reduces the training time on the basis of guaranteeing certain recognition accuracy [4]. ELM is faster and parameter-less, easier to deploy than SVM. [5]- [6] The method has good nonlinear mapping ability and is widely used in SVM. In Literature [7], the kernel method can also be applied to the limit learning machine algorithm. In Literature [9], the wavelet kernel function can be applied to the limit learning machine [8], and the performance is better. In this paper, the multiscale wavelet kernel function is applied to the ELM, and the fault diagnosis of the wind turbine bearing is carried out. The experiment shows that the method has high classification accuracy and speed.

2. ELM and its Improvements

2.1. ELM

N different training sample sets are $\{(x_i, t_i)|x_i \in R^n, y_i \in R^m\}$. The activation function of the feedforward network is g (x) and the number of hidden layer nodes is L. ELM training steps are as follows:

1) Determine the feedforward neural network structure;
2) Randomly set the connection weight of the feedforward neural network $(a_i, \cdots, b_i)$ and calculate the output matrix H of the hidden layer;

$$(a_i, \cdots, b_i), i = 1, \cdots, L \tag{1}$$

Let $h(x_i) = G(a_1 \cdot x_i + b_1) \cdots G(a_L \cdot x_i + b_L)$

$$\mathrm{H} = \begin{bmatrix} h(x_1) \\ \vdots \\ h(x_N) \end{bmatrix}_{N \times L} \tag{2}$$

H is an N × L matrix.

3) Solve the output weight of the least squares norm solution:

$$\beta = H^T(\frac{I}{C} + HH^T)^{-1}Y \tag{3}$$

Where I is the unit matrix and C is a constant. T is the expected matrix. The classification formula of the limit learning machine is shown below:

$$\mathrm{f(x)} = \mathrm{sign(h)}\left(H^T(\frac{I}{C} + HH^T)^{-1}Y\right) \tag{4}$$

2.2. Kernel Extreme Learning Machine (KELM)

Since the input weight of the ELM does not need to be adjusted during the training process, it can be given at once when the network is established. The implicit layer output of the ELM can be described as

$$\begin{matrix} \sum_{i=1}^{L} \beta_i G(a_i \cdot x_1 + b_i) = t_1 \\ \vdots \\ \sum_{i=1}^{L} \beta_i G(a_i \cdot x_N + b_i) = t_N \end{matrix} \quad , \mathrm{i} = 1,2, \cdots, \mathrm{L} \tag{5}$$

Where G (x) is the activation function. $a_i = [a_{i,1}, \cdots, a_{i,N}]^T$ is the input right. β_i is the output right. b_i is the offset of the i-th hidden layer unit. $a_i \cdot x_N$ is inner product operation. Equation (5) can be reduced to a matrix form expression of equation (6)

$$\mathrm{H}\beta = \mathrm{T} \tag{6}$$

Where H is the implicit output layer and T is the desired output.

$$\mathrm{H}(a_1, \cdots, a_L, b_1 \cdots b_L, x_1 \cdots x_N) = \begin{pmatrix} G(a_1 \cdot x_1 + b_1) \cdots G(a_L \cdot x_1 + b_L) \\ \vdots \\ G(a_i \cdot x_N + b_1) \cdots G(a_L \cdot x_L + b_L) \end{pmatrix}_{N \times L} \tag{7}$$

SVM determines that the optimal classification surface is based on the principle of structural risk minimization, but in the process of solving the ELM, it is necessary to consider not only the minimization of the empirical error, but also the structural risk minimization. Thus, a compromise is made between the minimized output weight and the minimized error, and the construction formula is as follows:

$$\min: \|\beta\|^2 \ and \ \sum_{i=1}^{N} \|\beta \cdot h(x_i) - t_i\| \tag{8}$$

Equation (8) is equivalent to equation (9):

$$\min: L = \frac{1}{2}\|\beta\|^2 + \frac{1}{2}C\sum_{i=1}^{N}\|\xi_i\|^2$$
$$s.t.: h(x_i)\beta = t_i^T - \xi_i^T, i = 1,2,\cdots,N \tag{9}$$

Where $\xi_i = [\xi_{i,1},\cdots,\xi_{i,m}]$ is the error between the network output value and the actual value which correspond to the training sample x_i.

According to the KKT condition, the Lagrange function is used to solve the most problem. That is, the problem of Equation (9) can be equivalent to

$$\min: L_{DELM} = \frac{1}{2}\|\beta\|^2 + \frac{1}{2}C\sum_{i=1}^{N}\|\xi_i\|^2 - \sum_{i=1}^{N}\sum_{j=1}^{M}(h(x_i)\beta_j) - t_{i,j} + \xi_{i,j} \tag{10}$$

Where β_j is the weight of the connection hidden layer and the j-th output node. The corresponding optimization constraints of $\beta = [\beta_i,\cdots,\beta_m]$ are

$$\frac{\partial L_{DELM}}{\partial \beta_j} = 0 \to \beta_j = \sum_{i=1}^{L}\alpha_i h(x_i)^T \to \beta = \boldsymbol{H}^T\alpha \quad \text{(a)}$$
$$\frac{\partial L_{DELM}}{\partial \xi_i} = 0 \to \alpha_i = C\xi_i, i = 1,2,\cdots,N \quad \text{(b)} \tag{11}$$
$$\frac{\partial L_{DELM}}{\partial \alpha_i} = 0 \to h(x_i)\beta_i - t_i^T + \xi_i^T = 0, i = 1,2,\cdots,N \quad \text{(c)}$$

The hidden layer output matrix H is only related to the number of samples and the number of nodes of the hidden layer but nothing to do with the number of output nodes of sample.

11 (a) and 11 (b) are incorporated into Formula 11 (c)

$$h(x_1)\boldsymbol{H}^T C\xi_1 - t_1^T + \xi_1^T = 0$$
$$\vdots \tag{12}$$
$$h(x_N)\boldsymbol{H}^T C\xi_N - t_N^T + \xi_N^T = 0$$

All the formula outputs are merged together, let

$$T = \begin{bmatrix} t_1^T \\ \vdots \\ t_N^T \end{bmatrix} = \begin{bmatrix} t_{11} & \cdots & t_{1m} \\ \vdots & \vdots & \vdots \\ t_{N1} & \cdots & t_{Nm} \end{bmatrix} \tag{13}$$

Then (13) can be combined into

$$\left(\frac{I}{C} + \boldsymbol{HH}^T\right)\alpha = T \tag{14}$$

In the end we can derive

$$\beta = \boldsymbol{H}^T(\boldsymbol{HH}^T + \frac{I}{C})^{-1}\boldsymbol{T} \tag{15}$$

ELM approximation function can be written as:

$$f(x_i) = h(x_i)\boldsymbol{H}^T(\boldsymbol{HH}^T + \frac{I}{C})^{-1}\boldsymbol{T} \tag{16}$$

where $H = \begin{bmatrix} h(x_1) \\ \vdots \\ h(x_N) \end{bmatrix}_{N\times L}$.

The hidden layer output $h(x_i)$ of each sample can be regarded as a non-linear mapping of the sample x_i, this mapping can be in the form of an added ax + b or RBF.

So,

$$\boldsymbol{HH}^T = \begin{bmatrix} h(x_1) \\ \vdots \\ h(x_N) \end{bmatrix}_{N\times L}^T \cdot \begin{bmatrix} h(x_1) \\ \vdots \\ h(x_N) \end{bmatrix}_{N\times L} = \begin{bmatrix} h(x_1)\cdot h(x_1) & \cdots & h(x_1)\cdot h(x_N) \\ \vdots & \vdots & \vdots \\ h(x_N)\cdot h(x_1) & \cdots & h(x_N)\cdot h(x_N) \end{bmatrix} \tag{17}$$

According to the theory of kernel function, implicit mapping can be constructed to replace the inner product of h(x) in equation (17), that is, we can construct kernel function instead of$\boldsymbol{HH}^T$, as shown below

$$\boldsymbol{HH}^T(\mathrm{i,j}) = \mathrm{K}(x_i, x_j)$$

$$\boldsymbol{HH}^T = \Omega_{ELM} = \begin{bmatrix} \mathrm{K}(x_1, x_1) & \cdots & \mathrm{K}(x_1, x_j) \\ \vdots & \vdots & \vdots \\ \mathrm{K}(x_N, x_1) & \cdots & \mathrm{K}(x_N, x_N) \end{bmatrix} \tag{18}$$

$$h(x)\boldsymbol{H}^T = \begin{bmatrix} \mathrm{K}(x, x_1) \\ \vdots \\ \mathrm{K}(x, x_N) \end{bmatrix}$$

Also,

$$f(x) = h(x)\boldsymbol{H}^T(\boldsymbol{HH}^T + \frac{I}{C})^{-1}\boldsymbol{T} \tag{19}$$

So the ELM solution formula can be written as

$$\mathrm{f(x)} = \begin{bmatrix} \mathrm{K(x, x_1)} \\ \vdots \\ \mathrm{K(x, x_N)} \end{bmatrix}^{\mathrm{T}} (\frac{\mathrm{I}}{\mathrm{C}} + \Omega_{\mathrm{ELM}})^{-1}\mathrm{T} \tag{20}$$

Through the above steps, a kernel extreme learning machine is constructed. The kernel extreme learning machine has a more powerful function approximation than the traditional extreme learning machine, and the ability to deal with sub-linear classification is also stronger.

2.3. Multi - scale Wavelet Kernel Function (MKELM)

Similar to the support vector machine, if a kernel function can satisfy the Mercer condition, then the kernel function can be used as the kernel function of the kernel extreme learning machine. To simplify the algorithm, the SVM kernel is applied to the ELM.

Known as a mother wavelet function h (x), its scalability and translation factors are a and b, then the wavelet basis function can be expressed as follows:

$$h_{a,b}(x) = \sqrt{|a|}h(\frac{x-b}{a}) \tag{21}$$

According to the theory of tensor product, a multidimensional wavelet function can be written as tensor product of multiple one-dimensional wavelet functions:

$$h(x) = \prod_{i=1}^{n} h(x_i) \tag{22}$$

According to (22), a translation of the kernel function can be constructed:

$$K(x, x') = K(x - x') = \sum_{i=1}^{n}(\frac{x_i - x_i'}{a}) \tag{23}$$

In this experiment, Morlet wavelet function is selected as $h(x) = \cos(1.75x)\cdot\exp(-x^2/2)$, then its corresponding wavelet kernel function can be expressed as follows

$$\mathrm{waveletkernel}(x, x') = \prod_{i=1}^{n}\left[\cos(1.75\times\left(\frac{x_i - x_i'}{a}\right))\exp(-\frac{(x_i - x_i')^2}{2a^2})\right] \tag{24}$$

Wavelet kernel function not only has the characteristics of strong nonlinear mapping, but also inherits the characteristics of wavelet analysis on non-stationary input parameters. Therefore, the wavelet kernel extreme learning machine with wavelet kernel function can approximate arbitrary function with high precision, which is not available in traditional kernel function.

Although the kernel function has a strong mapping ability and non-linear classification ability, but in some complex cases, the kernel extreme learning machine constructed by one wavelet kernel function does not meet practical application requirements such as data heterogeneous or irregular, large sample size, sample uneven distribution and so on. So combining multiple kernel functions to get better results is an inevitable choice.

Usually a mixed kernel function can be constructed by the way of the superposition of different kernel functions

$$K = a_1 \times K_1 + \cdots + a_n \times K_n \tag{25}$$

For the wavelet kernel function itself has the ability to expand the scale, different scale a can construct a different wavelet kernel function

The kernel ELM whose kernel function is wavelet kernel function is still the single-scale wavelet kernel ELM. In order to improve the function approximation ability with multi-scale characteristics, this paper proposes multi-scale wavelet kernel to improve the classification accuracy. Multiscale wavelet kernel function is constructed according to the properties of multi-scale wavelet kernel as follows

$$\begin{gathered} K = \prod_i^n \left[\cos(1.75 \times \left(\frac{x_i - x_i'}{a_1}\right))\exp(-\frac{(x_i - x_i')^2}{2{a_1}^2})\right] + \\ \prod_i^n \left[\cos(1.75 \times \left(\frac{x_i - x_i'}{a_2}\right))\exp(-\frac{(x_i - x_i')^2}{2{a_2}^2})\right] + \\ \cdots + \prod_i^n \left[\cos(1.75 \times \left(\frac{x_i - x_i'}{a_N}\right))\exp(-\frac{(x_i - x_i')^2}{2{a_N}^2})\right] \\ i = 1,2,\cdots,N \end{gathered} \tag{26}$$

Formula (26) can be abbreviated as

$$K = \sum_l^L \prod_i^n \left[\cos\left(1.75 \times \frac{x_i - x_i'}{a_l}\right) exp\left(-\frac{(x_i - x_i')^2}{2{a_l}^2}\right)\right] \tag{27}$$

Since the multiscale wavelet kernel function is a combination of wavelet kernel functions of multiple scales, the multi-scale wavelet kernel extreme learning machine(MKELM) can greatly relax the selection of the kernel parameters, and even can dilute the choice of kernel function, and its approximation ability higher, with the advantages of easily using, better recognition and so on.

3. Experimental Simulation

3.1. Simulation of Standard Data

In order to verify the validity of MKELM and the performance comparison with other kernel function ELM, this paper tests the standard classification test data of wine in UCI with Multi-scale wavelet kernel, wavelet kernel, gaussian kernel, poly kernel and no kernel. The test results are shown in Table 1; In order to verify the performance comparison between the MKELM and other algorithms, the standard data are classified by SVM, ELM, and MWKELM. Among them, SVM uses wavelet kernel function. The test results are shown in Table 2.

Table 1. Comparison of performance of different kernel functions

Kernel functions	Test accuracy	Test samples
Multi-scale wavelet kernel	0.991	200
Wavelet kernel	0.988	200
Gaussian kernel	0.965	200
Poly kernel	0.989	200
No kernel	0.967	200

As can be seen from Table 1, compared with other kernel functions, classification accuracy of the multi-scale wavelet kernel extreme learning machine is higher, the accuracy of no kernel and Gaussian is similar.

Table 2. Retrieval performance comparison of three classification algorithms

Classification algorithms	Test accuracy	Test samples
SVM	0.933	200
ELM	0.985	200
MKELM	0.991	200

As can be seen from Table 2, the ELM is more accurate than the SVM, and the MKELM has the best classification ability.

3.2. Simulation of Experiment

In order to further verify the effectiveness and superiority of MKELM in the fault diagnosis of wind turbine gearbox bearing, 25 sets of data were randomly selected as the training samples in 27 experimental data, and 21 sets of data were taken as test samples. The test results are shown in Table 3 Show.

Table 3. Comparison of experimental data test results

Classification algorithms	Test accuracy	Test samples
SVM	0.833	21
ELM	0.879	21
MKELM	0.905	21

As the Table 3 shows, the multi-scale wavelet kernel learning algorithm has the highest classification accuracy, the worst is the support vector machine SVM, the multi-scale kernel function method parameter selection complexity is much larger than the single kernel method parameter selection.[10] Therefore, the multi-scale wavelet kernel has better generalization ability. The accuracy of classification of multi-scale wavelet kernel learning machine is higher. However, the traditional ELM output weight is directly obtained by the least squares estimation method [11]. For the sake of learning robustness, by optimizing the selection parameters or solving the coherence of the training data and the gross error, such as the multi-hidden layer output matrix limit Learning machine algorithm [12], can further enhance the accuracy of learning machine

4. Conclusions

In this paper, a multi-scale kernel limit learning machine is proposed. The multi-scale wavelet kernel function is used as the kernel function of the extreme learning machine which satisfies the mercer condition and can be used as the allowable kernel of the extreme learning machine. In the application of the fault diagnosis of the gearbox bearing of the wind turbine, it is shown that the multi-scale wavelet kernel extreme learning machine has higher accuracy in fault classification, better performance than SVM and ELM, and has certain application value. However, the traditional ELM output weight is directly obtained by the least squares estimation method. For the sake of learning robustness, it is possible to further improve the accuracy of the learning machine by optimizing the selection parameters or solving the coherence of the training data and the gross error, such as the multi-hidden layer output matrix extreme learning machine. The selection of fault characteristics is also important and a more appropriate combination of features can be taken to get better retrieval results.

5. References

[1] Cortes C, Vapnik V. Support-Vector Networks [M]. Kluwer Academic Publishers, 1995.

[2] Huang G B, Zhu Q Y, Siew C K. Extreme learning machine: theory and applications J. Neurocomputing, 2006, **70(1/3)** 489-501.

[3] Wei W, Guang B Face recognition based on extreme learning machine *J. Neuromputing, 2011,***70(1/3)** 489-501.

[4] Nahvi B, Habibi J, Mohammadi K, et al Using self-adaptive evolutionary algorithm to improve the performance of an extreme learning machine for estimating soil temperature *J. Computers & Electronics in Agriculture*, 2016,**124** 150-160.
[5] Zhang S, Song Y, Wang Z, et al Structural analysis for the joint of the ITER ELM coil *J. Fusion Engineering & Design*, 2017,**114** 47-51.
[6] Colton A L, Buttery R J, Fielding S J, et al ELM studies on the COMPASS-D tokamak *J. Plasma Physics & Controlled Fusion*, 2017,**38(8)** 1359-1365(7).
[7] Lu Y, Zhang W, Shi J, et al. Unsteady aerodynamic modeling of biaxial coupled oscillation based on improved ELM *J. Aerospace Science & Technology, 2017,***60** 58-67.
[8] Fu H, Niu Z, Zhang C, et al. ASELM: Adaptive semi-supervised ELM with application in question subjectivity identification *J. Neurocomputing, 2016,* **207** 599-609.
[9] Gunn J P, Carpentier-Chouchana S, Dejarnac R, et al .Ion orbit modelling of ELM heat loads on ITER divertor vertical targets *J. Nuclear Materials & Energy*.2016.
[10] Huang G B, Ding X, Zhou H. Optimization method based extreme learning machine for classification *M. Elsevier Science Publishers B. V.* 2010.
[11] Yadav B, Ch S, Mathur S, et al. assessing the suitability of extreme learning machines (ELM) for groundwater level prediction *J. Journal of Water & Land Development*, 2017, **32(1)** 103-112.
[12] Kmiec′ K, Kot I Occurence of aphids from Eriosomatinae subfamily on elms in the gree area in Lublin. *J. Annales Umcs Horticultura,* 2010, **20(1)** 7-13.

IOP Conf. Series: Journal of Physics: Conf. Series **887** (2017) 012071 doi:10.1088/1742-6596/887/1/012071

Software detection method based on running state

Xiaolin Zhao[1,a], Quanbao Chen[1,b], Chun Shan[1,c], Ting Wang[1,d], Yiman Zhang[1,e]

[1]School of Software, Beijing Institute of Technology, Beijing 100081, China
[a]Email:zhaoxl@bit.edu.cn,[b]kenybens@sina.com,[c]sherryshan@bit.edu.cn,[d]mobao1005 0@163.com,[e]2164445931@qq.com

Abstract. It is extremely important for the study of software behavior modeling for software security research. This article determines whether the security of software operation by analyzing the credibility of software behavior, monitoring software running status. This paper focuses on a modeling algorithm, namely the GK-tail algorithm, which based on software behavior modeling method. At the same time, this paper improves the GK-tail algorithm, which focuses on data constraints and the interaction between software components. Restrictions on extending finite automaton can be obtained by using a combination of Daikon and ESC/JAVA tools. Restrictions can improve the accuracy of the generated model. So the generated behavior model can capture more accurate information. Finally, the paper designs and implements the software running state generator. It is feasible through the software state diagram to determine the feasibility of software security proved by the experiment.

1. Introduction

In recent years, with the rapid development of China's information industry, information technology has begun to be used in China's many other fields. It is an important issue how to protect the quality of the program.

It is an effective way to determine whether the software has been modified that analysis of software behaviour, monitoring software running status. The ultimate goal of behavioral analysis to obtain the semantic description and understanding of behavior by analyzing the behavioral feature data. Construction and detection is a key step to achieve the above goal. Dynamic analysis techniques are widely used to generate models that support test and verification software systems. The dynamic model of software behavior is used to dynamically detect anomalous behavior, static and dynamic checking of software components, validation protocols, generation of test cases, acquisition of exception event sequences, and validation of program attributes. The main technology of dynamic modeling of software behavior focuses on the constraints of data values or the interaction between software components. Some techniques that focus on data values are modeled as Boolean expressions, such as Daikon; some techniques that focus on component interaction are modeled as finite state machines [1].

This paper discovers it that software behavior modeling has an important application in software analysis, testing, and validation through the study of software security. The innovation of this research is to apply the software behavior modeling based on finite automata to the software security, and to improve the GK-tail algorithm so that the generated behavior model can capture more accurate information.

2. The Credibility of Software Behavior

At present, there are many views on the definition of credibility in academia and industry. The Trusted Computing Group (TCG) [2] defines credible with the expected nature of the entity's behaviour. If an entity is credible, then its behaviour is always in the expected way to achieve the desired goal. From this definition, trusted behaviour is a sufficient condition for a trusted entity. The key to a trusted behaviour is that the act should occur as expected and complete the intended function. IEEE Pacific Rim International Symposium on Dependable Computing (PDRC) and our scientists [3] have developed a credible computing perspective developed by fault-tolerant computing. They believe that a trusted computing system is a computer system that provides reliability, availability, information, and behavioural security which view on research from a reliable and secure perspective. Combining the above two perspectives, trusted computing is a measure of the credibility of the software from the perspective of the expected and behavioural security of software behaviour. Although the industries trusted computing platform is not enough to ensure the credibility of the system's life cycle. It is correct to determine the credibility of the software that the basic point of view according to the expected behaviour of the software. There is a lot of research is from this point of view [4-5].

2.1. Definition of Trusted Computing

The expectation of software behaviour is embodied in the behaviour feature, behaviour input and output, behaviour process, behaviour attribute and so on., It is about the software behaviour characteristics of knowledge [6], Whether the software behaviour in these areas conform to the compliance agreement, whether it is acceptable for the user.

First, we give a general definition of credible of software behaviour:

All acts or traces of conduct are expected behaviour for a given software. It is said that the behaviour of the software is credible, or that the software to meet the behaviour of the trusted requirements.

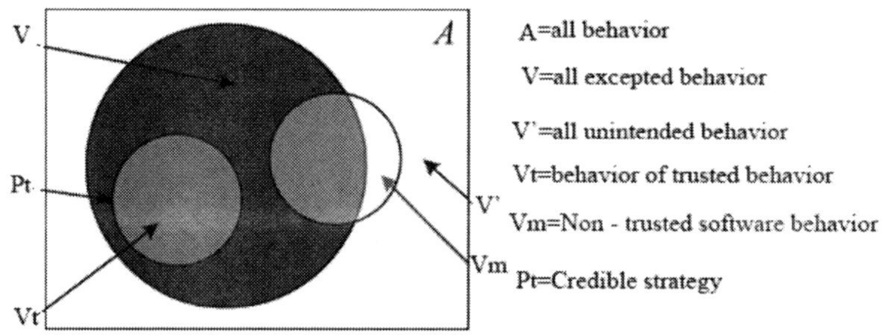

Figure 1. The credible behaviour definition can be visually represented. The behaviour of the trusted software V_t must conform to the trusted policy P_t; the verified untrusted software necessarily V_m has a behaviour that does not conform to the trusted policy.

2.2. The Importance of Trusted Computing

At present, trusted computing has become one of the research hotspots in information security field. Trusted computing technology and products continue to emerge [5]. People are increasingly aware of the importance and necessity of trusted computing technology. However, the development of trusted computing still has some problems. First, theoretical research lags behind technology development. There is no universally accepted credible computing theory model so far. And there is no software dynamic credibility of the measurement methods and theory; second, the current credibility of the measurement can only ensure that the static integrity of the system resources in the system boot. It does not ensure that the dynamic credibility of runtime system [7].

Wagner and Dean build a software behaviour model to create a Finite State Automaton (FSA) by statically analyzing the source code. But the FSA does not extract the system context information, leading to an impossible path problem. GoPalakrishna et al. [8] proposed a Push-down Automaton (PDA) model that uses additional stack save function call information. The algorithm reduces the uncertainty of the automaton by inserting an empty call before and after the program call and rewriting the executable file to solve impossible path problem. But the operating cost is too high. Feng et al [9] dynamically monitors the program's audit data, captures the control flow of the program by monitoring the system call sequence and call stack information when the program is executed. They establish VtPath between the two system calls for dynamic analysis to fix an impossible path problem to a certain extent. Liu et al. [10] extended the PDA model to combine the dynamic learning with the static analysis method. The constructed HPDA model has a stronger detection capability than the static method alone, which is lower than the dynamic method alone false alarm rate. Li Wen et al [11] proposed static analysis - dynamic binding of the hybrid model (HFA), which belongs to the combination of static analysis and dynamic learning method. But the model does not take into account the parameters of the system call. It can't effectively detect those who simulate the normal program behaviour for evading monitoring and detection of mimic attack behaviour. Frossi et al. [12] used the enhanced system call model for anomaly detection by combining the Finite State Automaton (FSA) with the Markov stochastic model.

Giffin et al. [13] pointed out that there would be a problem that the parameter values were insensitive to the context of the system call. The problem existed when the same set of parameters were obtained along the different executable paths, regardless of the context. This problem leads to the program loses the relationship between the system call parameter value and the executable path which uses the parameter value. An attacker can use the parameter values obtained from executable path A to perform the same system call on path B. If an executable path can't get the parameter value statically, all parameter values must be discarded from other executable paths for security. Therefore, it needs to be considered that the system called parameters in the context of the system call.

It is the easiest way to capture the current system call context that traversing the stack. But the cost of traversing the stack is too high. Spivey [14] proposed to dynamically create a call context tree (CCT), through the monitoring program in the CCT location to conduct a trusted evaluation. The shortcomings of the program runtime time are increased by 2 to 4 times the cost of most of the system It is unacceptable. Bond et al. [15] proposed a probabilistic call context model (PPC), which also reduces evaluation overhead. But its iterative method of calculating PCC values does not apply to multiple repetitions of the same function. Mutz et al. [16] proposed to create a set of normal parameter values for each system call. For each implementation path of the software, a set of parameters is specified for a system call. But the model is less accurate.

System invocation has its special status in software development. Most of the researches on software behaviour are focused on the system call sequence and its related information when the software is running. Forrest et al. proposed to use the system call short sequence to characterize the process. They monitor the privileged process of the system call sequence to establish the intrusion detection model. Yao Lihong et al [17] proposed CTBIDS detection model which using the system call feature tree description program behaviour characteristics. They discriminated program invasion through the abnormal finite accumulation or Hamming distance. Tan Xiaobin et al. [18] used the hidden Markov model to describe the regularity of local system calls during the normal operation of privileged processes.

Through the study of the trusted computing of software behaviour, it is found that there are many shortcomings in the feasible calculation of software behaviour, such as software behavior modeling ability, evaluation ability, evaluation accuracy, system practicality and so on. It still needs to be further improved.

The paper uses GK-tail algorithm to represent the software behaviour model. It is the key point to study of software behaviour. The paper will analysis and improvement of the GK-tail algorithm to get the better the software behaviour modeling.

3. Analysis and Improvement of GK - tail Algorithm

When describing the behavior of the software, it is found it that even if a very small software, it produces a huge amount of trace. It can be difficult to describe and analyze the problem with a large number of the trace. And trace description data will take up a lot of storage resources and be not conducive to long-term storage. In order to solve the above problems, the GK-tail algorithm uses the Extended Finite State Model (EFSM) to represent the software behavior model with data constraints.

3.1. Description of GK-Tail Algorithm

GK-tail is a dynamic model of software behavior based on extended finite state automata model (EFSM). It has four main steps:

1) Merge the software behavior trace which input equivalence;

2) Generate predicates;
3) Establish an initial extended finite state automaton;
4) Consolidate the equivalent state to obtain the final extended finite state automaton [19-20].

The paper calls the track to describe the sequence of the method call, the value of the label and the variable. The data is about the parameters, variables, or other information that invokes the method. They have their own values. Additional information is automatically added via the monitoring platform or manually by the tester. Figure 2 is an example of a call sequence. And Figure 3 is a corresponding set of interaction traces, as shown in the following figure.

In figure 2, the label under the label indicates the parameter values which online and the context variable which below the line. When a context variable exists, it is marked.

The GK-tail handles the interaction traces set, producing an EFA that handles all trajectories. The interaction traces represent the set of executions. Several interaction traces often trigger an identical behavior. This means that the different values of the parameters call the same method sequence. GK-tail uses this feature to simplify the input of the equivalent trace to the data set. Figure 4 shows the data set obtained from the interaction traces of Figure 2.

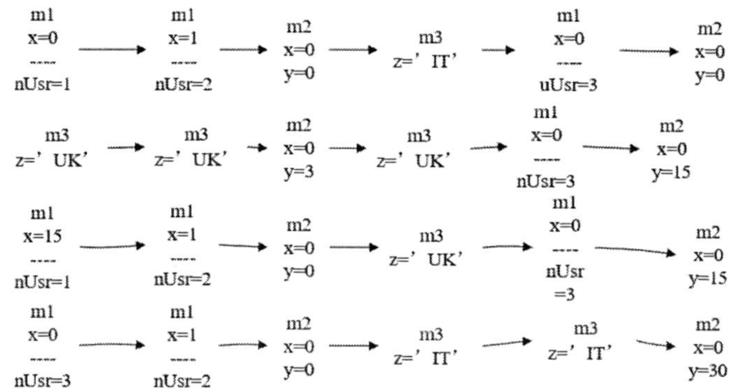

Figure 2. Examples of four call sequences

it1=(m1,0,1) (m1,1,2) (m2,(0,0),-) (m3,' IT' ,-) (m1,0,3) (m2,(0,0),-)

it2=(m3,' UK' ,-) (m3,' UK' ,-) (m2,(0,3),-) (m3,' UK' ,-) (m1,0,3) (m2,(0,15),-)

it3=(m1,15,1) (m1,1,2) (m2,(0,0),-) (m3,' UK' ,-),(m1,0,3) (m2,(0,15),-)

it4=(m1,0,3) (m1,1,2) (m2,(0,0),-) (m3,' IT' ,-) (m3,' IT' ,-) (m2,(0,30),-)

Figure 3. Interaction traces

In the second step, GK-tail generates predicates from the data set. Figure 5 shows an example of using Daikon to automatically generate predicates.

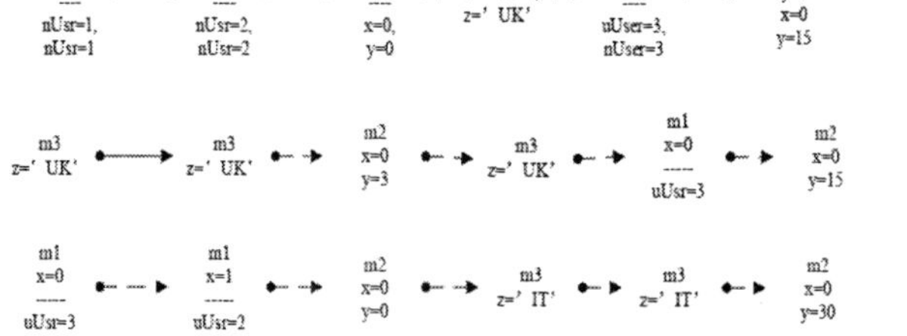

Figure 4. Enter the equivalent set of interaction traces

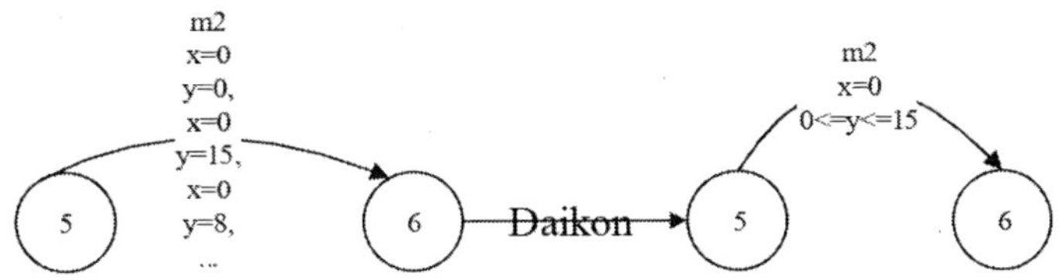

Figure 5. Automatically generates predicates. A predicate summarizes the conditions from the accepted corresponding method calls. GK-tail uses Daikon to generate predicates.

In the third step, GK-tail generates an initial EFA by adding a common initial state from the trace of the tagged predicate. Figure 6 shows an example of an initialized EFA.

In the fourth step, GK-tail simplifies the initial EFA by combining the equivalent state to obtain a compact model of the observed behavior.

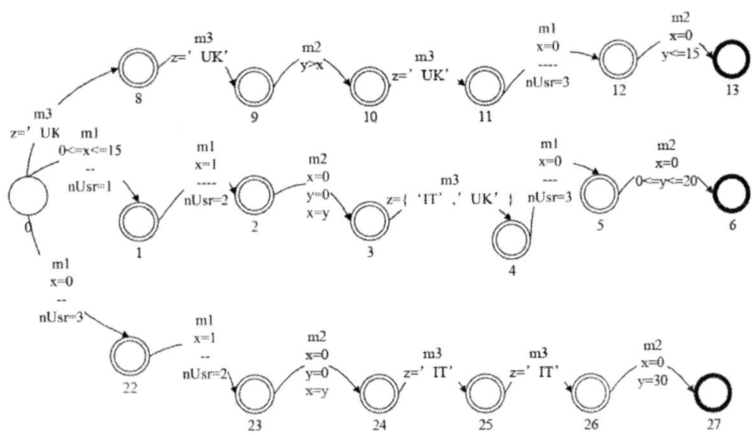

Figure 6. The initial EFA is obtained from the data set

Because the second step of the GK-tail algorithm uses the dynamic analysis tool Daikon, when using Daikon to obtain the predicate of the software behavior trace, the transition condition of the EFA state is obtained. This paper defines the improvement of the GK-tail algorithm by understanding the static and dynamic analysis tools. That means that the improvement of the analysis tools.

3.2. The Combination of ESC/Java and Daikon

In the second step of GK-tail algorithm, the resulting information is improved with higher accuracy because of the combination of the ESC/Java and Daikon tools.

The invariants of the program found by dynamic techniques are not necessarily correct, because of the limitation of times of running the tools. Static technology can verify that certain properties are always correct. But it is very difficult to choose a validation target for a static validator and add comments.

Integration of these two technologies, we can take their own excellent to make up for each other's shortcomings. Dynamic techniques can find invariants of programs which can be treated as annotations of programs or be used as static validation purposes. Static verifiers can verify invariants found by dynamic techniques [21].

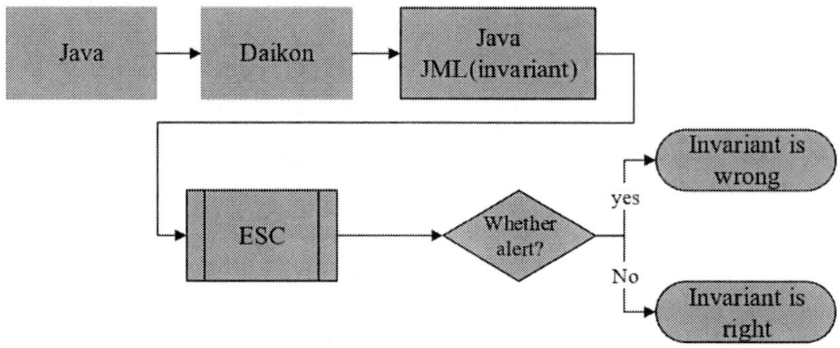

Figure 7. Daikon and ESC/Java tools combined workflow. First, run Daikon to get the program invariants. The program invariants can be inserted into the JML language; then the JML annotation with the Java code input to the ESC/Java. Finally, verify the invariants generated by Daikon.

The collection dynamic invariant discovery tool Daikon and static validator ESC/Java can filter the false invariants generated by Daikon.

This part describes two analysis tools, static analysis tools ESC/Java and dynamic analysis tools Daikon. Based on the analysis of GK-tail algorithm and tools, it is determined how to improve the GK-tail and design and implement the software behavior modeling system to verify the feasibility of the algorithm.

3.3. Improved GK-Tail Algorithm

Now we define the interaction traces. Given a finite set of a method X, a set of possible input parameters R and the corresponding domain D_{rx}, a possibly empty set of variables V and the corresponding domain D_V, a parameter trajectory is a triple (x, p_x, V), where $x \in X$, $p_x \in D_{Rx}$, $v \in D_V$. $it = (x_1, p_{x1}, v_1) \ldots (x_n, p_{xn}, v_n)$, where $x_i \in X$, $p_x \in D_{Rxi}$, $v_i \in D_v$.

Given a state of EFA and a parameter trace $P_t = (x, p_x, v)$, p_t is able to complete a transition $t = (s, x, P, s')$ when $P(p_x, v) =$ True. There is a transition sequence $it = (x_1, p_{x1}, v_1) \ldots (x_n, p_{xn}, v_n)$, when an EFSM has the initial state s0, the final state s_F accepts an interaction trace, (s_0, x_1, P_1, s_1) $(s_1, x_2, P_2, s_2) \ldots$ (s_{n-1}, x_n, P_n, s_n), where $P_i(p_{xi}, v_i) =$ True, $\forall\ i = 1..\ N$, $s_n \in s_F$.

The improved algorithm is divided into four steps: (1) monitor software to obtain the monitoring trace log, merge the input-equivalent trace, and record it; (2) use Daikon and ESC/Java to generate the predicate of the relevant trace (3) produce an initial EFA; (4) merge the equivalent state, optimize the initial EFA, get the final EFA. These four steps will be described in below.

A data set is defined as a sequence $(x_1, dp_{x1}, dv_1) \ldots (x_n, dp_{xn}, dv_n)$, where $x_i \in X$ is a method, $dp_{xi} \in \vartheta(D_{Rxi})$ is an input parameter evaluation set, and $dv_i \in \vartheta(DV)$ is a variable evaluation set. Given an input equivalent trajectory set $it_1 = (x_1, p^1{}_{x1}, v^1{}_1) \ldots it_n = (x_n, p^1{}_{xn}, v^1{}_n)$, ..., $it_m = (x_1, p^m{}_{x1}, v^m{}_1) \ldots it_m = (x_n, p^m{}_{xn}, v^m{}_n)$, GK-tail is merged into the data set $ds = (x_1, p^1{}_{x1} \cup \ldots\ldots \cup p^m{}_{x1}, v^1{}_1 \cup . \ldots\ldots \cup v^m{}_1) \ldots$ $(x_n, p^1{}_{xn} \cup \ldots \cup p^m{}_{xn}, v^1{}_n \cup \ldots \cup v^m{}_n)$.

In the second step, after the merging of the interactive trajectory, the paper uses Daikon and ESC/Java to generate the relevant trajectory predicate. Formal definition of Daikon and ESC/Java as a function daikonEsc: $\vartheta(D_{Rx}) \cup \vartheta(D_V) \rightarrow P$, where D_{Rx} is an input parameter evaluation set for any $x \in X$, D_V is an évaluation set of context variables, $P: D_{Rx} \times D_V \rightarrow \{\text{True, False}\}$ is a predicate based on parameters and context variables. Daikon and ESC/Java guarantee that when daikonEsc ($Data_x$, V_{ar}) = P, then $\forall\ P_x \in Data_x$, $d_v \in V_{ar}$, $P(p_x, d_v) =$ True. This step is the place where the GK-tail algorithm is improved.

The GK-tail algorithm uses the Daikon tool to get the predicate of the software behavior trace in this step. Before the merge process, the set of software behavior trajectories may contain incorrect data, and

Daikon will not check again. The improved GK-tail algorithm uses the combination of ESC/Java and Daikon to acquire the software. The improved GK-tail algorithm uses the combination of ESC/Java and Daikon to acquire the software the behavioral trace predicate will filter out the erroneous data information contained therein, thus ensuring that the resulting extended finite automaton's state and transition condition information are more correct and improved in accuracy.

The combination of Daikon and ESC/Java aims to avoid incorrect data in the input parameter set R and the corresponding domain D_{r_i}. It leads to bring the wrong results in the follow work. It is also the innovation of this paper. Through the previous section of the two analysis tools of the theoretical description, the algorithm GK-tail is proved that it is feasibility theoretically.

In the third step, an initial EFA is generated by adding a common initial state from the trajectory of the tagged predicate. This step corresponds to the system data analysis module in the system design. for $\forall$ i =1, ..., m, the initial EFSM (S, T, s_0, s_F) is the set of trajectories for a given predicate $\{seq_1, \dots seq_m\}$, where $seq_i = (x^i{}_1, P^i{}_1) \dots (x^i{}_{ni}, P^i{}_{ni})$, are defined as follows:

S= $\{s_0, s^1{}_1, \dots s^1{}_{n1}, s^2{}_1, \dots s^2{}_{n2}, \dots, s^m{}_1 \dots s^m{}_{nm}\}$; $s_F = \{s^i{}_{ni} \mid i = 1, \dots m\}$;$s_0$ is the initial state.

For each element in the input sequence set $(x^i{}_j, P^i{}_j)$, j>1, There is a transfer t= $(s^i{}_{j-1}, x^i{}_j, P^i{}_j, s^i{}_j) \in T$;

For each element in the input sequence collection $(x^i{}_1, P^i{}_1)$, There is a transfer t= $(s_0, x^i{}_1, P^i{}_1, s^i{}_1) \in T$;

In the fourth step, the compact EFA is reduced by combining the equivalent state. It obtains a compact model of the observed behavior. The merging of EFA is an extension of the main rules of the kTail algorithm, which proposed by Bierman and Feldmann. They thought that these states are equivalent when the states pass through the same path. It is a unique logical state of multiple expression and can be safely merged when different EFSM states generate an equivalent set of behaviors.

In general, the state of the path set may be infinite, so the contrast state may be very expensive. We compare the length of the sequence to k, similar to that proposed by Bierman, Feldmann. Given a state $s \in S$, define the k-future (s) of the sequence set $\{seq_1, \dots seq_r\}$, where $seq_i = \{(x_1, P_1) \dots (x_k, s_k)\}$, there exists a transfer sequence (S, x_1, P_1, s_1) (s_1, x_2, P_2, s_2) ... (s_{k-1}, x_k, P_k, s_k).

The above is an improvement to the GK-tail algorithm because Daikon's dynamically generated data information may be in error. ESC/Java verifies the data after using the Daikon to determine which data is incorrect, and then filter it out. The combination of Daikon and ESC/Java can the wrong information. It is shown that algorithm can improve the accuracy.

4. Implementation of Software State Monitoring System

Based on the above conclusions, the paper determines the Daikon as a monitoring program to run the tool. First, predicate is generated by Daikon from the data set, which is generated by the improved GK-tail algorithm. Finally, we can get the program running state diagram. This section will analyze the Daikon monitoring log with the related techniques such as program running trace, invariant detection method and program invariant extraction. In the GK-tail algorithm, the program invariant is found using dynamic detection technology, which is Daikon reasoning engine for invariant detection.

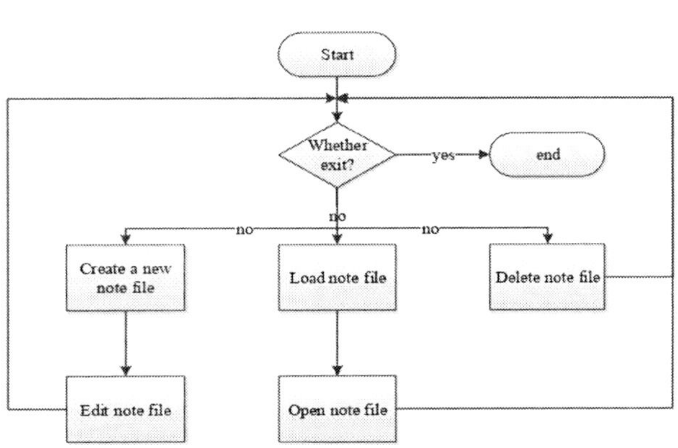

Figure 8. Notepad program flow chart

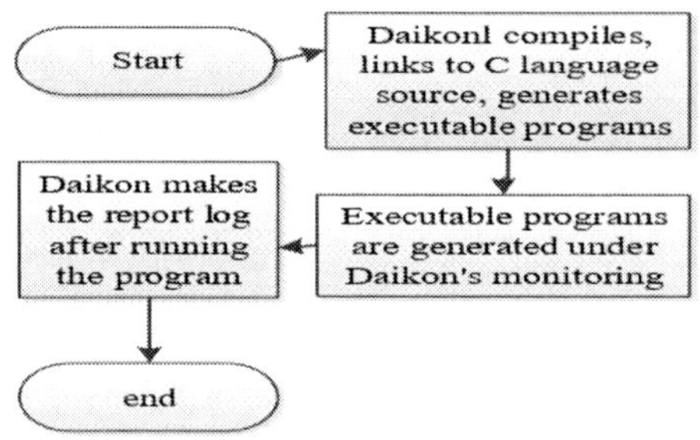

Figure 9. Flowchart of the Daikon monitor running

4.1. Daikon on Notepad Monitoring

Notepad is a simple program with create, loaded, and delete text functions. The Notepad can reload, modify, and delete the text file. The function flow chart shown in Figure 8.

Daikon monitors and generates a report file on the program. On the basis of the C source, Daikon integrates the compilation linker, which links the C code directly into an intermediate process file and generates an executable exe program. Daikon will execute this executable file with Purify's target code plug-in function and generate a log. The process shown in Figure 9.

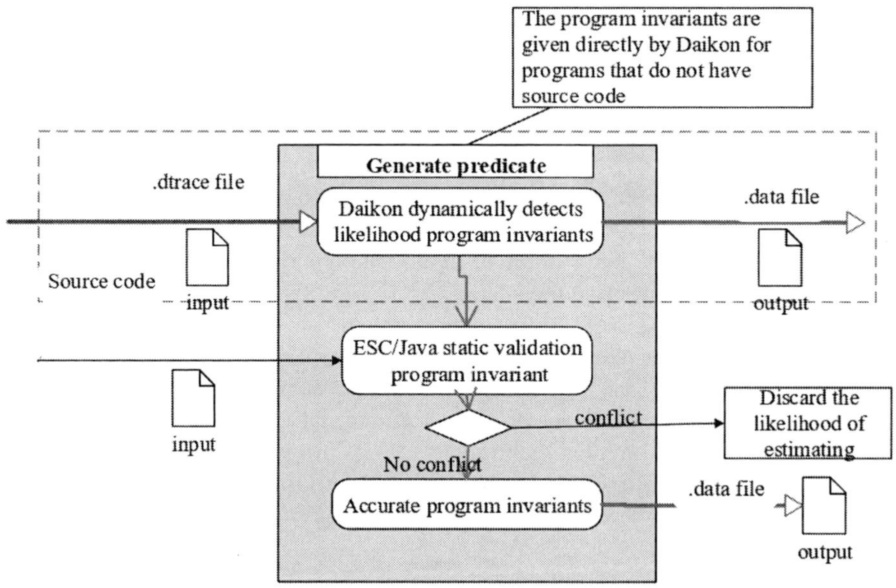

Figure 10. The figure 10 shows the predicate generation module implementation flow chart. First, we use the Daikon reasoning engine to invariably analysis the input .dtrace file to get the JML description form of the likelihood invariant. And then it is entered into ESC/Java that the JML description of the likelihood variable and the system source. The ESC/Java detection will discard conflicting program invariants and output more accurate program invariants, the predicates needed in the second step of GK-tail.

4.2. Design of State Chart Generator

The paper introduced how to monitor the operation of the program by the Daikon tool and how to generate predicates of the process previously by GK-tail algorithm. At the next step, it needs to generate software running state diagram from these results.

This procedure is running at java environment. The program draws graphics by using jGraph plug-in. The input text file is from the report log which logged by Daikon when the program is running. The output result is the program running state diagram corresponding to the log report. This procedure mainly draws the state graph by analysis the generated log report. First, the program read the target log file. And then the program records the function, call time and other information. Finally, the final state diagram is drawn. The process is shown in Figure 11.

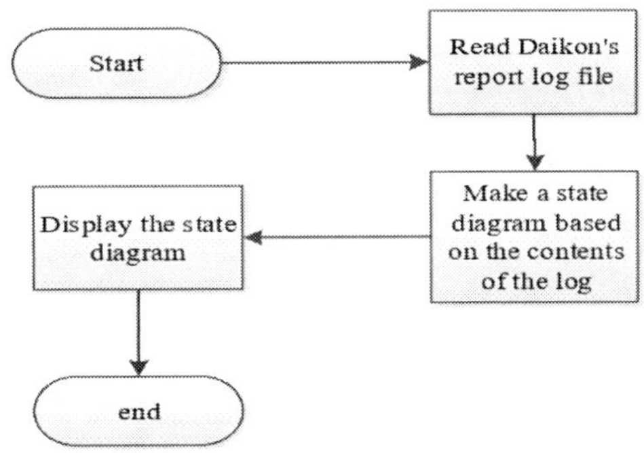

Figure 11. Generates the program flow diagram

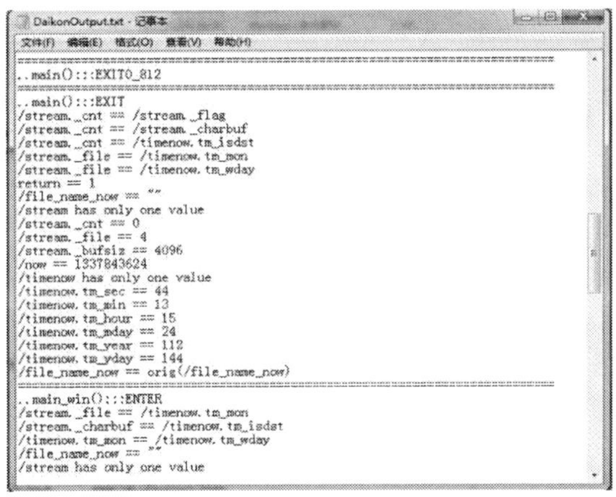

Figure 12. Daikon monitor part of the log files in Notepad

5. System Running Instance

The program finally achieved the above design. The paper completed the expected function. The paper will show it how to generate a log text file from Daikon monitor notepad program and how to draw the state diagram by jGraph.

First, the Notepad program is executed. Daikon generates the report as shown in Figure 12.

At this time, we run the drawing state diagram program. The program running state diagram is drawn as shown in Figure 13.

When the program is running, the operating status shown in Table 1.

Table 1. Shows the running status of the program consistent with the order of the actual operation of the user. Through the operation of this example, it is proved that state diagram which drawn by Daikon and the jGraph is achieve the desired demand and design.

Status unit contents	time	description
START	-	start
new_file()	19 : 46 : 38	Create new note
edit_file()	19 : 46 : 40	Edit note
load_file()	19 : 46 : 43	Load note
open_file()	19 : 46 : 45	Open note
del_file()	19 : 46 : 48	Delete note
END	19 : 46 : 52	end

It will change the process of the program by the injection tool to attack the program. It will get a different state diagram. It is only an examples which loaded without creating note of the state diagram, as shown in Figure 14.

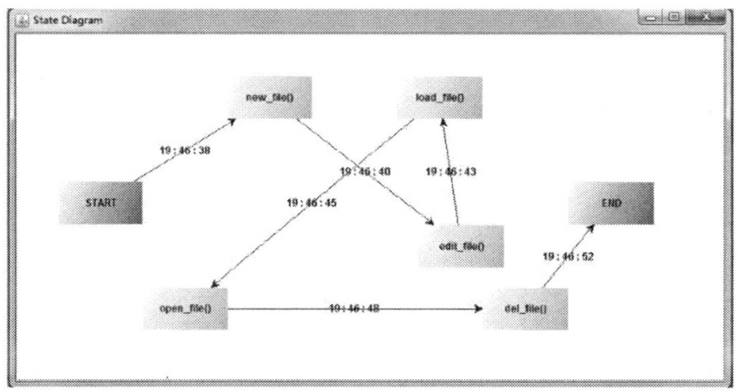

Figure 13. Normal calls to the software's state diagram

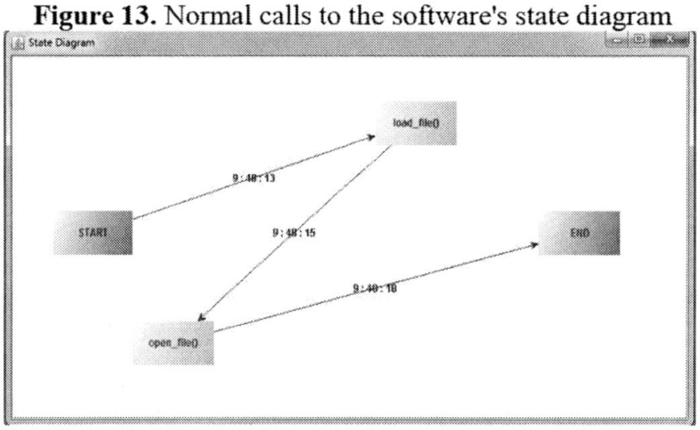

Figure 14. The software state diagram after being injected into the tool

6. Conclusion

Through the experiment, we get the following conclusions. First, the paper analyzes software behavior. The paper improves the software behavior dynamic modeling GK-tail algorithm based on extended finite state automata model. Second, the paper combines the GK-tail algorithm with Daikon, which is the invariant analysis software, to generate the invariant change. By summarizing their behavior, this paper gives a formal specification of the program. Third, on the basis of this result, this paper uses jGraph components to make software running state diagram. Through the analysis of behavioral feature, this paper obtains the semantic description of behavior. By comparing its normal running state, the software running behavior state is analyzed. Because the intrusion will change the normal implementation of the software or call the sequence, it can be found opportunely that the problem in runtime in the software. The GK-tail algorithm can judge whether software trusted behaviour is trusted.

7. Acknowledgment

This work was supported the National Key R&D Program of China under Grant No. 2016YFB0800700

8. References

[1] Davide Lorenzoli, Leonardo Mariani, Mauro Pezzè*.*Automatic Generation of Software Behavior Model* [J]. ICSE'08, May 10–18, 2008.

[2] Trusted Computing Group. TCG Specification Architecture Overview: Revision 1.4 [EB/OL].

[3] Zhang Huanguo, Luo Jie, Jin Gang, et al. *Development of Trusted Computing Research* [J]. Journal of Wu2 han University (Natural Science Edition), 2006, **52(5):**513-518.

[4] Li Xiaoyong, Gui Xiaolin, Mao Qian, et al.*Adaptive dynamic trust measurement and prediction model based on behaviour monitoring* [J]. Chinese Journal of Computers, 2009, 32**(4)**:664-674.

[5] Yang Xiao-hui. Research on Dynamic Trusted Theory Model of Software Behavior [D].Heifei: University of Science and Technology of China, 2010.

[6] Qu Yan-Wen.Software Bahavior [M].Beijing:Publishing House of Electronics Industry, 2004.

[7] YANG Xiao-hui, ZHOU Xue-hai, Tian Jun-feng, LI Zhen. *Novel Dynamic Trusted Evaluation Model of Software Behavior* [J]. Mini-Micro Systems, 2010, **31(11)**:2113-2120.

[8] Rajeev Gopalak rishna,EugeneH Spafford,Jan Vitek. Efficient intrusion detection using automata in lining[C]. Proceedings of the 2005 IEEE Symposium on Security and Privacy. Oakland, California, U SA, 2005, 18-31.

[9] Henry H Feng,Oleg M Kolesnikov,Prahlad Fogla,et al. Anomaly detection using call stack information[C]. Proceedings of the 2003 IEEE Symposium on Security and Privacy,Oakland, California,USA,2003,62-75.

[10] Zhen Liu,Susan M Bridges,Rayford B Vaughn. Combining static analysis and dynamic learning to build accurate intrusion detection models[C]. Proceedings of the 3rd IEEE International Workshop on Information Assurance College Park,Maryland, USA,2005,164-177.

[11] LI Wen, DAI Ying-Xia, LIAN Yi-Feng, FENG Ping-Hui. *Context Sensitive Host-Based IDS Using Hybrid Automaton* [J]. Journal of Software, 2009, **20(1)**:138-151.

[12] Alessandro Frossi,Federico Maggi,Gian Luigi Rizzo,et al. Selecting and improving system call models for anomaly detection[C]. Proceedings of the 6th International Conference on Detection of Intrusions and Malwareb,and Vulnerability Assessment,Como,Italy,2009,206-223.

[13] Jonathon T Gifhn,David Dagon,Somesh Jha,et al. Environment sensitive intrusion detection [C]. Proceedings of the 8th International Symposium on Recent Advances in Intrusion Detection,Seattle,W- ashington,USA,2005,185-206.

[14] Spivey JM. *Fast accurate call graph profiling* [J]. Software Practice &Experience, 2004, 34(3):249-264.

[15] Michael D Bond,Kathryn S Mckinley. Probabilistic call context[C]. Proceedings of the 22nd Annual ACM SIG PLAN Conference on Object-oriented Programming Systems and Applications, Montreal, Quebec, Canada, 2007,97-112.

[16] Akahashi T, Hayahi N, Hayashi S.Structure and Properties of Shape-Memory Polyurethane Block Copolymers [J]. Appl PolymSci, 1996, 60:1061-1069.

[17] YAO Li-hong, ZI Xiao-chao, HUANG Hao, MAO Bing, XIE Li. Research of System Call Based Intrusion Detection [J]. Acta Electroncica Sinica, **31(8)**:1134-1137, 2003.

[18] TAN Xiao-Bin, WANG Wei-Ping, XI Hong-Sheng, YIN Bao-Qun. *A Hidden Markov Model Used in Intrusion Detection* [J]. Journal of Computer Research and Development, **40(2):**245-250, 2003.

[19] LI Xin-Ming, LI Yi, XU Xiao-Mei. Software Vulnerability Analysis [J]. Computer Science, 30(8):163-165, 2003.

[20] Sekar R, Bendre M.Bollineni P,et al. A fast Automata-Based method for detecting anomalous program behaviors[C]//IEEE Symp.on Security and Privacy. Oakland: IEEE Press, 2001:144-155

[21] Jeremy W. Nimmer, Michael D. Ernst. *Static verification of dynamically detected program invariants: Integrating Daikon and ESC/Java* [J]. First Workshop on Runtime Verification, 2001.

IOP Conf. Series: Journal of Physics: Conf. Series **887** (2017) 012072 doi:10.1088/1742-6596/887/1/012072

Aided diagnosis methods of breast cancer based on machine learning

Yue Zhao, Nian Wang and Xiaoyu Cui

Sino-Dutch Biomedical and Information Engineering School, Northeastern University.
Email: zhaoyue@bmie.neu.edu.cn

Abstract. In the field of medicine, quickly and accurately determining whether the patient is malignant or benign is the key to treatment. In this paper, K-Nearest Neighbor, Linear Discriminant Analysis, Logistic Regression were applied to predict the classification of thyroid,Her-2,PR,ER,Ki67,metastasis and lymph nodes in breast cancer, in order to recognize the benign and malignant breast tumors and achieve the purpose of aided diagnosis of breast cancer. The results showed that the highest classification accuracy of LDA was 88.56%, while the classification effect of KNN and Logistic Regression were better than that of LDA, the best accuracy reached 96.30%.

1. Introduction

Breast cancer, ranking the first in the incidence of malignancy in women, is a major malignant tumors seriously influencing women's physical and mental health and even life-threatening. Breast cancer is the leading cause of death among women between 40 and 55 years of age and is the second overall cause of death among women exceeded only by lung cancer. Fortunately, the mortality rate from breast cancer has decreased in recent years with an increased emphasis on diagnostic techniques and more effective treatments. A key factor in this trend is the early detection and accurate diagnosis of this disease. So CAD systems are designed to support radiologist in the processing of visually screening mammograms to avoid miss-diagnosis. Classification systems can help minimizing possible errors that can be done because of inexperienced experts, and also provide medical data to be examined in shorter time and more detailed.

There have been a lot of research on medical aided diagnosis of breast cancer, and most of them have reported high classification accuracy. Quinlan reached 94.74% classification accuracy using 10-fold cross validation with C4.5 decision tree method [1]. The researcher, H. J. Hamiton, and N. Shan, reached 94.99% accuracy with RIAC technique [2]. While B. Ster and A. Dobnikar have got to 96.8% with linear discreet analysis method [3]. Using neuro-fuzzy techniques, the accuracy of method proposed by D. Nauck and R. Kruse was 95.06% [4]. In Abonyi and Szeifert, an accuracy of 95.57% was obtained with the application of supervised fuzzy clustering technique [5]. The fuzzy-GA method was introduced and a classification accuracy of 97.36% was achieved [6]. Three different methods, optimized learning vector quantization (LVQ), big LVQ, and artificial immune recognition system (AIRS) were applied and the obtained accuracies were 96.7%, 96.8%, and 97.2% respectively [7]. Multilayer perceptron neural network, four different methods, combined neural network, probabilistic neural network, recurrent neural network and SVM were used respectively by Ubeyli; and the highest classification accuracy of 97.36% was achieved by SVM [8]. Two different methods, Bayesian classifiers and artificial neural networks were applied and the obtained accuracies were 92.80% and 97.90%, respectively [9]. Murat Karabatak and M. Cevdet Ince, using the method combined with association rules and neural network, obtained the accuracy of 95.60% [10].

In the paper, K-Nearest Neighbor,Linear Discriminant Analysis,Logistic Regression combination with normalization, principal component analysis, 5-fold cross validation are applied to predict the classification of thyroid,Her-2,PR,ER,Ki67,metastasis and lymph nodes in breast cancer, in order to recognize the benign and malignant breast tumors and achieve the purpose of auxiliary diagnosis of breast cancer.

2. Basic Theories

2.1. K-Nearest Neighbor

K-Nearest Neighbor, abbreviated to KNN, is a simple and effective non-parametric classifier in the field of pattern recognition. The core idea of K-Nearest Neighbor is that there is a sample in the feature space, if the majority of the samples belong to a certain category, the sample also belongs to the same category. In the KNN algorithm, the similarity between objects is usually calculated by the distance between objects, generally using Euclidean distance or Manhattan distance, showed as in equation (1) and equation (2), respectively.

$$d(x,y)=\sqrt{\sum_{k=1}^{n}(x_k-y_k)^2} \tag{1}$$

$$d(x,y)=\sqrt{\sum_{k=1}^{n}|x_k-y_k|} \tag{2}$$

2.2. Linear Discriminant Analysis

Linear Discriminant Analysis, abbreviated to LDA, is one of the commonly used classification methods in statistics. The basic theory of the algorithm is to project a high-dimensional pattern of samples into the best discriminant vector space to guarantee model samples have the largest inter-class distance and the minimum inner-class distance.

2.3. Logistic Regression

Logistic Regression is actually a classification method for two classification problems. Its basic flow is to find a suitable predictor function firstly, generally expressed as h function, which is a classification function, used to predict the input data to determine the results, as in equation (3).

$$h(x)=Q_0+Q_1x_1+Q_2x_2+\cdots Q_mx_m=Q^TX \tag{3}$$

And constructing a Cost function which represents the deviation between the predicted output (h) and the training data class (y), taking the "loss" of all training data into consideration, and summing or averaging Cost as J (θ) function, which indicates the deviation of the predicted value of all training data from the actual category, as in equation (4). Obviously, the smaller the value of the J (θ) function is, the more accurate the predictor function is. So this step needs to find the minimum value of the J (θ).

$$J(Q)=\frac{1}{2}\sum_{i=1}^{n}(h_Q(x^{(i)}-y^{(i)}))^2 \tag{4}$$

3. Methodology and Experiment

3.1. Experimental Data

In this experiment, the data was taken from the First Affiliated Hospital of China Medical University, consisting of 77 samples. And every sample had 13 attributes and 7 classes corresponding to each record. These 7 classes represented 7 pathological indicators. Among them, Her-2, PR, ER, Ki67 these 4 indicators were important clinical data to guide doctors to make comprehensive treatment. They were also very important to decide the patient's treatment schedule. In the experiment, the 77 samples were all used to train to construct a classification model, and then randomly choose 20 samples for test.

3.2. Data Processing

Data normalization processing is a basic step of data mining. Different evaluation indicators often have different dimension units, which will affect the results of data analysis. In order to eliminate the impact of indicators, data must be standardized. After the raw data is normalized, the indicators are in the same order of magnitude, suitable for comprehensive evaluation. In this paper, the equation of data normalization processing is defined as in equation (5).

$$X_{norm} = \frac{2*(X - X_{\min})}{X_{\max} - X_{\min}} - 1 \qquad (5)$$

Principal component analysis is also an important issue in building classification systems. It refers to the export of a few principal components from the original variable so that they retain as much information of the original variable as possible and have as much good prediction as possible.

K-fold cross validation is shown to be efficient for small data sample and gives more reliable results than common test sample estimates. In the processing, the whole set is divided randomly into k nearly equally large subsets. The k-1 of subsets will be used to train, and the last will be used to test, which guarantee that each of subsets is involved exactly once in testing. K times results are obtained and then averaged as the accuracy of the algorithm.

3.3. Experimental Scheme

In the experiment, the 77 samples were all used to train to construct a classification model, and then randomly choose 20 samples for test. Normalization, principal component analysis, 5-fold cross validation were applied to processing the original data respectively. Based on the processed data and raw data, KNN, LDA,Logistic Regression three methods are applied to predict the classification of thyroid,Her-2,PR,ER,Ki67,metastasis and lymph nodes in breast cancer, respectively.

The performance of the classifier was evaluated using Sen,Spe and Accuracy. Accuracy indicated the percentage of all samples identified correctly. They could be written as in equations (6).

$$\begin{aligned} Sen(\%) &= \frac{TP}{TP+FN} \times 100 \\ Spe(\%) &= \frac{TN}{TN+FP} \times 100 \\ Accuracy(\%) &= \frac{TP+TN}{TP+TN+FP+FN} \times 100 \end{aligned} \qquad (6)$$

3.4. Results and Discussion

Based on the original features, KNN, LDA,Logistic Regression three methods were used to obtain the values of Sensitivity, Specificity and Classification accuracy in Table 1.

From the table 1, we could see clearly that in the method of logistic regression, metastasis, thyroid, ki-67 had relatively high accuracy, respectively 92.60%, 88.89%, 85.19%. And in LDA, the classification accuracy of these seven indicators were all relatively lower. However, in the algorithm of KNN, the classification accuracy of metastasis reached the highest, about 96.30%. Considering the results above, this showed that for the same data set, the classification effect varied from classifier to classifier. So in order to get a good classification effect, the choice of classifier was the key.

In addition, the sensitivity of thyroid,Her-2,metastasis in KNN, Her-2,metastasis in logistic regression and the specificity of ER,ki-67 in KNN and logistic regression were low, and even some was zero. After analyzing, that may be due to the reason that the sample data set was not large enough, and positive and negative samples were uneven distribution, so the learning and master of characteristics was not enough. Therefore, in the next time, the sample data would be collected further.

Table 1. The classification results of KNN, LDA,Logistic Regressionbased on raw features

Methods	indicators	TP	TN	FP	FN	Sen	Spe	Accuracy
KNN	thyroid	1	23	0	3	25.00	1	88.89
	Her-2	2	17	3	5	28.57	85.00	70.37
	PR	16	6	5	0	1	54.55	81.48
	ER	18	2	5	2	90.00	28.57	74.07
	ki-67	22	0	4	1	95.65	0	81.48
	metastasis	0	26	0	1	0	1	96.30
	lymph nodes	14	9	1	3	82.35	90.00	85.19
LDA	thyroid	3	20	3	1	75.00	86.96	71.24
	Her-2	5	16	4	2	71.43	80.00	70.83
	PR	12	8	3	4	75.00	72.73	79.08
	ER	15	6	1	5	75.00	85.71	71.84
	ki-67	12	3	1	11	52.17	75.00	65.48
	metastasis	1	23	3	0	1	88.46	87.54
	lymph nodes	11	8	2	6	64.71	80.00	74.23
Logistic Regression	thyroid	2	22	1	2	50.00	95.65	88.89
	Her-2	2	18	2	5	28.57	90.00	74.07
	PR	13	6	5	3	81.25	54.55	70.37
	ER	19	3	4	1	95.00	42.86	81.48
	ki-67	23	0	4	0	1	0	85.19
	metastasis	0	25	1	1	0	96.15	92.60
	lymph nodes	12	6	4	5	70.59	60.00	66.67

Table 2. The classification accuracy of the different processing methods about feature

Methods	indicators	Differently processed features				
		raw	normalization	PCA	5-fold cross validation	Normalization,PCA,5-fold cross validation
KNN	thyroid	88.89	96.30	88.89	79.17	81.92
	Her-2	70.37	70.37	70.37	68.92	69.08
	PR	81.48	85.19	81.48	62.42	66.50
	ER	74.07	81.48	74.07	68.92	78.08
	ki-67	81.48	85.19	81.48	77.75	79.33
	metastasis	96.30	92.59	96.30	83.08	83.17
	lymph nodes	85.19	81.48	85.19	68.67	74.17
LDA	thyroid	71.24	71.24	55.81	79.47	70.32
	Her-2	70.83	70.83	64.58	76.85	68.45
	PR	79.08	79.08	64.03	80.13	73.16
	ER	71.84	71.84	58.68	77.70	66.97
	ki-67	65.48	65.48	55.95	71.74	66.87
	metastasis	87.54	87.54	64.85	88.56	79.46
	lymph nodes	74.23	74.23	70.92	79.09	76.61
Logistic Regression	thyroid	88.89	88.89	85.19	76.67	79.33
	Her-2	74.07	74.07	74.07	76.67	75.59
	PR	70.37	70.37	66.67	72.83	70.17
	ER	81.48	81.48	74.07	71.67	75.42
	ki-67	85.19	85.19	85.19	77.67	80.75
	metastasis	92.60	92.60	96.30	84.58	88.25
	lymph nodes	66.67	66.67	62.96	69.00	79.17

After the data processing of normalization, principal component analysis, 5-fold cross validation, respectively, KNN,LDA,Logistic Regression three methods were used to obtain the values of classification accuracy again. We would compare the results with the previous results in Table 2.

As can be seen from Table 2, in the method of KNN, with the data normalization the results of thyroid,PR,ER were higher than the original results. When using the method of LDA, the classification accuracy of all indicators were slightly improved comparing 5-fold cross validation with original results. However, selecting the algorithm of logistic regression, the original features without any processing would achieve the highest accuracy. Experimental results showed that for the same data set, the classification effect was different with the input characteristics processed differently. So in order to get a better classification effect, the choice of way of data processing was the key.

Seen from Table 1 and Table 2, KNN classification efficiency was better, the highest accuracy rate of 96.30%, while in LDA and SVM, the highest accuracy rate was 92.60%, 85.19%. So for different classification samples, to choose a suitable classifier to get a higher accuracy was essential to achieve the purpose of aided diagnosis of breast cancer.

4. Conclusion

Diagnosis of breast cancer requires the comprehensive use of clinical, imaging, pathology, computer-aided diagnosis and other knowledge and technical means. Early detection and early diagnosis of breast cancer are the key to improve curative effect. In this paper, KNN, LDA and logistic regression three methods were used to the identification of breast cancer pathology, to help doctors comprehensive assessment to make specific treatment options and drug selection. Experiment results showed that KNN classification accuracy, better than LDA and logistic regression, was expected to become an effective and practical computer assisted diagnostic tools in breast cancer that could reduce the human subjective misdiagnosis rate and bring the gospel for the breast cancer patients.

Acknowledgments

The authors are grateful for the financial support from the National Natural Science Foundation of China (61501101, 61605025), the Natural Science Foundation of Liaoning Province (201601015).

5. References

[1] Quinlan J R Improved use of continuous attributes in C4.5 1996 *Journal of Artificial Intelligence Research* **4** pp 77–90

[2] Hamiton H J, Shan N, and Cercone N RIAC: A rule induction algorithm based on approximate classification 1996 *Computer Science Department*

[3] Ster B and Dobnikar A Neural networks in medical diagnosis: Comparison with other methods 1996 *In Proceedings of the international conference on engineering applications of neural networks* pp 427–30

[4] Nauck D and Kruse R Obtaining interpretable fuzzy classification rules from medical data 1999 *Artificial Intelligence in Medicine* **16** pp 149–69

[5] Abonyi J and Szeifert F Supervised fuzzy clustering for the identification of fuzzy classifiers 2003 *Pattern Recognition Letters* **24** pp 2195–2207

[6] Pena-Reyes C A and Sipper M A fuzzy-genetic approach to breast cancer diagnosis 1999 *Artificial intelligence in medicine* **17** pp 131–55

[7] Goodman D E, Boggess L and Watkins A Artificial immune system classification of multiple-class problems 2002 *In Proceedings of the Artificial Neural Networks in Engineering ANNIE* **2** pp 179–83

[8] Ubeyli E D Implementing automated diagnostic systems for breast cancer detection 2007 *Expert Systems with Applications* **33** pp 1054–62

[9] Maglogiannis I, Zafiropoulos E and Anagnostopoulos I An intelligent system for automated breast cancer diagnosis and prognosis using SVM based classifiers 2009 *Applied Intelligence* **30** pp 24–36

[10] Karabatak M and Cevdet Ince M An expert system for detection of breast cancer based on association rules and neural network 2009 *Expert Systems with Applications* **36** pp 3465–69

[11] Polat K and Gunes S Breat cancer diagnosis using least square support vector machine 2007 *Digital Signal Processig* **17** pp 694-701

[12] Salama G I, Abdelhalim M B and Zeid M A Breat Cancer diagnosis on three diffetent datasets using multi-classifiers 2012 *International Journal of Computer and Information Technology* **1** pp 36-43

Compact dual-band bandpass filter based on signal-interference techniques

Xingbing Ma[1, 2] and Ting Jiang[1]

[1]Key Laboratory of Universal Wireless Communications, Ministry of Education, Beijing University of Posts and Telecommunications, Beijing 100876, China
[2]Tianjin University of Technology and Education, Tianjin 300222, China
Email: maxingbing2008@sohu.com

Abstract. To realize good isolation between two signal passbands, a dual-band bandpass filter (BPF) in this article was presented using signal-interference techniques, in which five open loop resonators are adopted. The proposed filter topology is made up of two signal transmission paths in parallel, under signal-interference principles, overlap section of two original passbands, decided respectively by two different transmission paths, is selectively removed from the combined passband, as a result, two aim passbands are realized. In addition, good isolation between two aim passbands is established due to two new transmission zeros, produced by adopted signal-interference techniques. At last, good agreement can be observed between simulation and measurement.

1. Introduction

Dual-band bandpass filters (BPFs) have become attractive components in modern wireless communication systems. In recent years, many techniques have been carried out to implement dual-band BPFs. In [1-4], two filters, operating at different frequency bands respectively, are simply combined with common input and output ports. In [5, 6], stepped impedance resonators (SIRs) coupling with each other are used to achieve desirable dual-band passbands, in which resonators are directly connected to 50 Ω impedance matching line. In [7, 8], single shorted-circuited stub-loaded SIR and single side-coupled shorted lambda/4 SIR are respectively presented to generate two passbands, and several transmission zeros (TZs) are both created to improve selectivity. In [9], a dual-band filter with a mixture of bandpass and bandstop sub-filters is presented, which leads to a very compact size. In [10], to construct dual-band BPF, the input and output ports are connected to each other with a grounded feedline having two via holes located at the tapping point alignments of loading elements. In [11-16], signal interference techniques are adopted to design single-band, dual-band or multi-band BPFs and provide desirable TZs. Hereinto the transversal signal-interference filtering sections are formed by two connected transmission lines, directly connected to I/O feeling ports in [11, 12]. A structure with a short-ended coupled line coupler connected in parallel to a transmission line is introduced in [13]. To reduce size and improve stopband performance, artificial right- and left-handed transmission lines are used in [14]. In [15], a wide passband is constructed using transversal signal-interference technique, resulting in a pair of TZs, where the transversal resonator is used to couple with feedlines connected to I/O ports. In [16], a compact quad-band BPF using square ring loaded resonator is introduced, and multi-transmission zeros are created due to the different propagation paths based on signal interference technique.

In this article, based on conventional structure of miniature open-loop resonator filter [17], a new compact dual-band BPF with two signal transmission coupling paths in parallel has been reported based on the signal-interference techniques. Under signal-interference techniques, the overlap section of two original passbands, respectively decided by resonators at two paths is selectively removed. Good isolation between two designed passbands is achieved due to two new transmission zeros, produced by adopted signal-interference techniques. At last, the designed dual-band BPF has been validated by experiment.

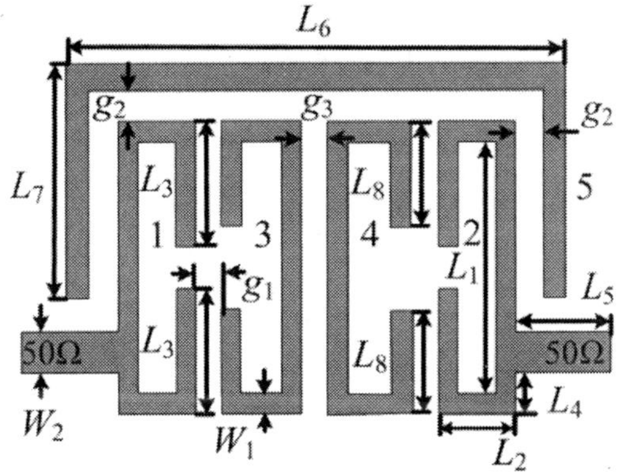

a

Figure 1 (a). The proposed dual-band BPF using signal-interference technique.

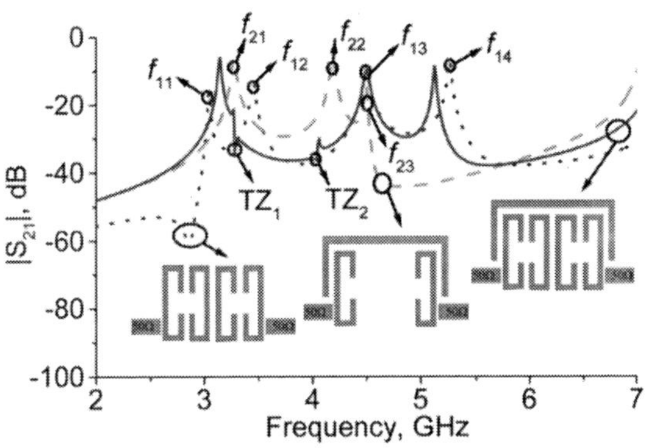

b

Figure 1 (b). Insertion loss characteristics of the proposed BPF under three different weak coupling structures.

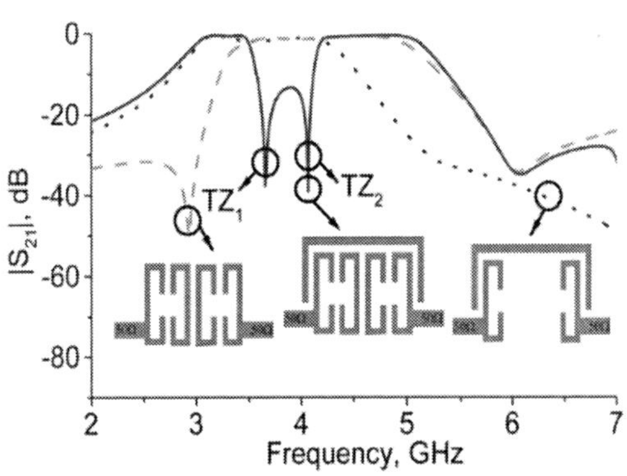

c

Figure 1 (c). Insertion loss characteristics of the proposed BPF under three different signal transmission patterns.

2. The Proposed Dual-band Bandpass Filter and Characteristic Analyses

Figure 1(a) illustrates structure of the proposed dual-band BPF, which is symmetrical about vertical axis. As illustrated, this proposed BPF consists of five open loop resonators with same line width. Tight gap coupling are established between these resonators. Hereinto resonators 3 and 4, 1 and 2 are with same sizes, which are directly connected to the I/O feedline ports. The only difference between resonator 3 and 1 is lengths (L_3 and L_8) of the open ends.

2.1. Weak Coupling Resonant Characteristics under Different Path Structures

Figure 1(b) shows insertion lossS_{21}characteristics of the proposed BPF under three different weak coupling structures, the adopted parameters in Figure 1(a) are as follow: W_1=0.5, W_2=2.72, L_1=10, L_2=2.5, L_3=5, L_4=1, L_5=8.25, L_6=13.4, L_7=7.8, L_8=5, g_1=0.4, g_2=0.55, g_3=0.5, and used substrate is TACONIC TLC-32 with thickness of 1.14 and dielectric constant of 3.2. (Unit: mm in this article)

It is noted for signal transmission path constructed of resonators 1, 2, 3 and 4, there are four resonant frequencies (f_{11}, f_{12}, f_{13} and f_{14}). Which represent four different current excitation modes between these four resonators, respectively named as O-S-O, O-O-O, S-S-S, and S-O-S. Hereinto, O and S respectively denote surface currents on symmetrical coupled arms of Figure 1(a) having the opposite direction and the same direction, described in [18]. Similarly, for path composed of resonators 1, 2 and 5, there are three resonant frequencies (f_{21}, f_{22} and f_{23}), representing surface current modes (S-S, S-S and O-O) between three resonators, respectively. As a result, based on signal interference techniques, two TZs are obtained, hereinto TZ_1 is located between f_{21} (S-S) and f_{12} (O-O-O), TZ_2 between f_{22} (S-S) and f_{12} (O-O-O), decided by two different transmission paths.

2.2. Resonant Characteristics under Three Different Signal Transmission Patterns

Based on same parameters above, Figure 1(c) showsS_{21} of the proposed BPF under three different signal transmission patterns, which verifies that overlap section of two original passbands, decided by two different transmission paths, is selectivity removed due to signal interference techniques. Two new passbands with two TZs are achieved. Due to coupling pattern differences, positions of TZ_1 and TZ_2 are shifted a little in Figure 1(c), different from ones in Figure 1(b). Note that L_3= L_8 in simulation about Figure 1(b) and Figure 1(c).

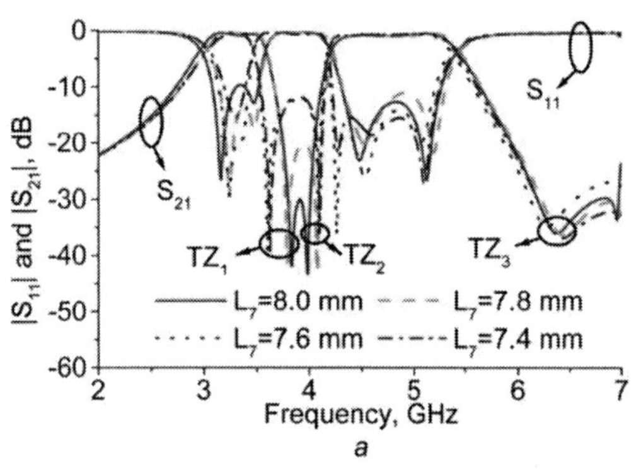

Figure 2 (a). The insertion loss and return loss characteristics with L_8=4.5 and varied L_7.

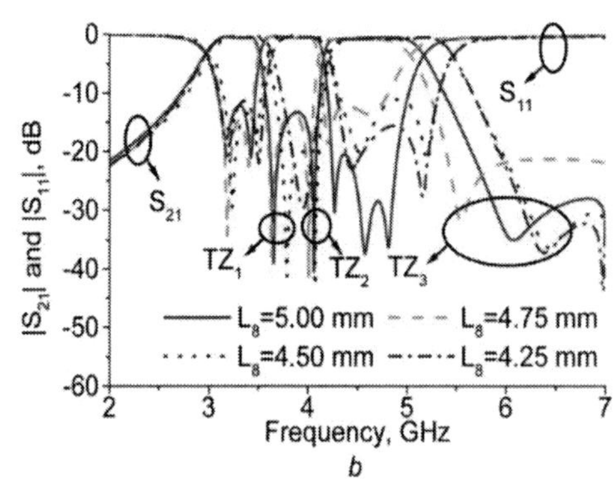

Figure 2 (b). Insertion loss and return loss characteristics with L_7=7.8 and varied L_8.

2.3. Dual-Band Characteristics with Varied L_7 Or L_8

To adjust dual-band bandwidths, parameters L_7 and L_8 are utilized to realize this aim. Figure 2(a) and Figure 2(b) show $S_{21}$and return loss S_{11} of the proposed BPF with varied L_7 or L_8, respectively. Other parameters in Figure 1(a) are given as follows: W_1=0.5, W_2=2.72, L_1=10, L_2=2.5, L_3=5, L_4=1, L_5=8.25, L_6=13.4, g_1=0.4, g_2=0.55, g_3=0.5. (Unit: mm)

Figure 2(a) is S_{21} and S_{11} characteristics with L_8= 4.5 and varied L_7.When L_7 decreases, the upper-side of the second passband and TZ_3 are almost fixed, and the second passband bandwidth with S_{11}= -10 dB only varies a little. Whereas the first passband bandwidth obviously decreases with L_7 decreased, which can be carefully adjusted to desirable bandwidth. In addition, TZ_1 and TZ_2 affected evidently, stopband bandwidth between two new passbands increases in a certain range when L_7 decreases.

Figure 2(b) provides S_{21} and S_{11} characteristics with L_7=7.8 and varied L_8, it can be observed that when L_8decreases, the lower-side of the first passband is almost fixed, and the first passband bandwidth below S_{11}= -10 dB is affected only a little, whereas the second passband bandwidth and TZ_3can been adjusted in a large range to desirable values.TZ_1 and $TZ_2$are affected evidently, the stopband bandwidth between two passbands decreases when L_8decreases.Especially, when L_8= 4.25 mm, TZ_1 and $TZ_2$are shifted and combined into one transmission zero.

2.4. Dual-Band Optimized Characteristics with Varied L_7 And L_8

Figure 3(a) and Figure 3(b) respectively show optimized S_{21} and S_{11} characteristics of the proposed dual-band BPF with varied L_7 and L_8, here L_3= L_8.In which the four resonators except of resonator 5 own same sizes. It can be observed that when L_7 and L_8 decrease, the center frequencies of two passbands both increase, and three TZs also move towards higher frequency region.

2.5. Summary

Based on the simulations and analyses above, it is verified that the proposed BPF can achieve a dual-band passbands due to signal interference technique. Hereinto, overlap part of original two passbands, decided by two different transmission paths respectively, is selectively removed, and two new passbands with two TZs, produced by signal-interference techniques, are obtained. In addition, two passband bandwidths can be carefully adjusted by changing lengths (L_7 or L_8), observed in Figure 2(a) and Figure 2(b). At last center frequencies of two passbands can be simultaneously adjusted and optimized, which is verified in Figure 3(a) and Figure 3(b).

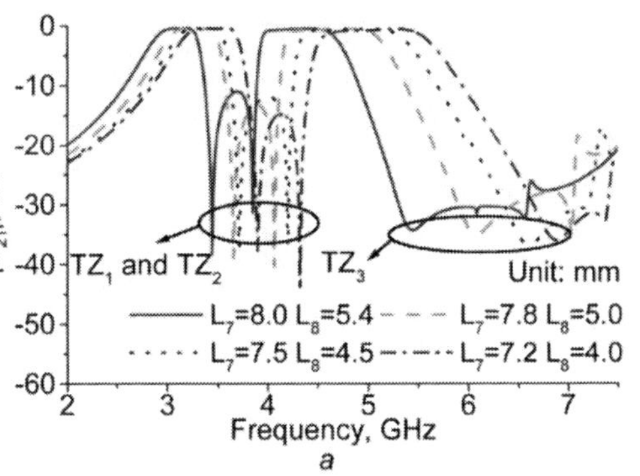

Figure 3 (a). Insertion loss characteristics with varied L_7 and L_8, and L_3= L_8.

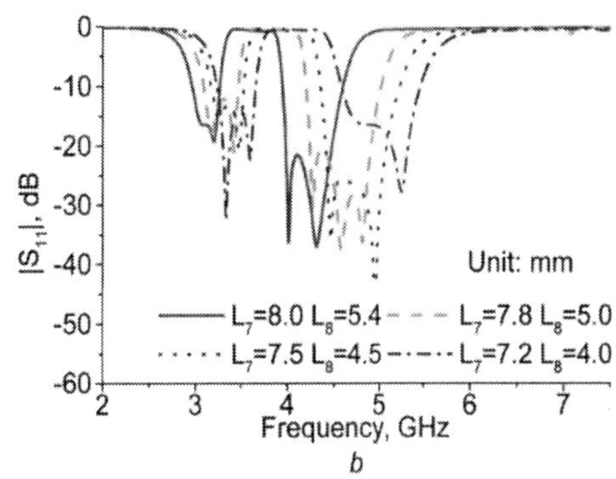

Figure 3 (b). Return loss characteristics with varied L_7 and L_8, and L_3= L_8.

3. Experimental Results

A microstrip dual-band BPF is designed, fabricated, and measured to validate the proposed concept. The substrate TACONIC TLC-32 used in article is with a relative dielectric constant of 3.2, a thickness of 1.14 mm, and loss tangent of 0.003. The BPF is designed and optimized using HFSS software, and Figure 4 is the photograph of the fabricated BPF. It has a size of 28×12 mm^2, Figure5 shows the simulated and measured results of the proposed dual-band BPF, the fabricated dual-band BPF in Figure 4 has two passbands below S_{11}= -10 dB from 3.18 to 3.58 GHz and from 4.17 to 4.90 GHz respectively, and one transmission zero at 3.87 GHz between two passbands is obtained. Difference between simulation and measurement is due to the fabricated error and unperfected ground, and it can be improved by more careful fabrication technology.

Figure 4. The photograph of the fabricated dual-band BPF.

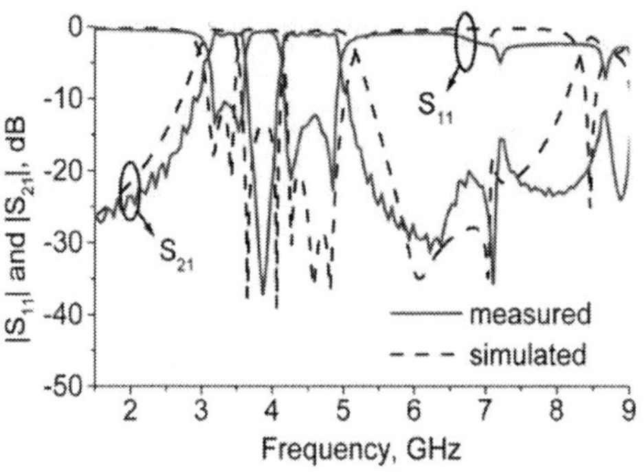

Figure 5. The simulated and measured results of the proposed dual-band BPF.

4. Conclusions

A compact dual-band BPF based on signal interference technique is proposed in this article, overlap part of two wideband passbands, decided respectively by two different signal transmission paths, is selectively removed from the combined passband, and two new passbands are achieved. Due to the used signal-interference techniques, good isolation between two aim passbands is realized with two introduced transmission zeros. Experimental results are in good agreement with predicted results. Because of design flexibility and good performances, the proposed dual-band BPF is attractive for wireless wideband bandpass systems.

5. Acknowledgments

This work was supported by National Nature Science Foundation of China under Grant 61671075 and BUPT Excellent Ph. D. Students Foundation under Grant CX2015203.

6. References

[1] Chen C and Hsu C2006A simple and effective method for microstrip dual-band filters design *IEEE Microw. Wireless Compon. Lett.***16** 246-8
[2] Wang J, Ge L, Wang K and Wu W 2011Compact microstrip dual-mode dual-band bandpass filter with wide stopband *Electron. Lett.***47** 263-5
[3] Lee J and Lim Y 2011 Compact dual-band bandpass filter with good frequency selectivity *Electron. Lett.***47** 1376-7
[4] Karpuz C, Gorur A2014Even-odd mode analysis for multiband bandpass filters having multiple resonators and its microstrip implementations *Microw. Opt. Technol. Lett.***56** 2664-7
[5] Yuan H and Fan Y2011Compact microstrip dual-band filter with stepped-impedance resonators*Electron. Lett.***47** 1328
[6] Zhang R and Zhu L2014Design of a compact dual-band bandpass filter using coupled stepped-impedance resonators*IEEE Microw. Wireless Compon. Lett.* **24**155-7
[7] Zhang X, Liu B, Zhao Y, Wang J and Chen W 2012Compact and high selectivity dual-band dual-mode microstrip BPF with five transmission zeros *Microw. Opt. Technol. Lett.***54** 79-81
[8] Li D, Zhang Y, Xu K, Song K and Li L 2015 Compact dual-wideband bandpass filter with good selectivity using side-coupled lambda/4 shorted SIR *J. Electromagn waves Appl.***29** 69-79
[9] Heng Y, Guo X, Cao B, Wei B, Zhang X, Zhang G and Song X 2013Compact superconducting dual-band bandpass filter by combining bandpass and bandstop filters*Electron. Lett.* **49** 1230-2
[10] Gorur A, Karpuz C, Ozek A and Emur M2014Metamaterial based dual-band bandpass filter design for WLAN/WiMAX applications *Microw. Opt. Technol. Lett.***56** 2211-4
[11] Roberto G and Alonso J 2005Design of sharp-rejection and low-loss wide-band planar filters using signal-interference techniques*IEEE Microw. Wireless Compon. Lett.***15** 530-2

[12] Roberto G, Manuel S, Bernard J, Julien L and Bruno B2009A class of microwave transversal signal-interference dual-passband planar filters*IEEE Microw. Wireless Compon. Lett.***19** 158-60

[13] Miguel A, Enrique B and German T2009Compact UWB bandpass filter based on signal interference techniques*IEEE Microw. Wireless Compon. Lett.***19** 692-4

[14] Ni J, Vendik I, Kholodnyak D and Hong J 2013a compact bandpass filter based on right- and left-handed transmission line sections*IEEE Microw. Wireless Compon. Lett.***23** 279-81

[15] Sun S, Zhu L and Tan H 2008a compact wideband bandpass filter using transversal resonator and asymmetrical interdigital coupled lines*IEEE Microw. Wireless Compon. Lett.* **18** 173-5

[16] Liu H, Ren B, Guan X, Wen P and Wang Y 2014Quad-band high-temperature superconducting bandpass filter using quadruple-mode square ring loaded resonator *IEEE Trans. Microw. Theory Tech.* **62**2931-41

[17] Zhang T, Yang K, Ning J, Bu S, Liu J and Luo Z 2008Development of miniature HTSC wide-band filter with open-loop resonators *Chinese Sci. Bull.* **53** 1300-3

[18] Hong J and Lancaster M1996 Couplings of microstrip square open-loop resonators for cross-coupled planar microwave filters *IEEE Trans. Microw. Theory Tech.*44 2099-109

IOP Conf. Series: Journal of Physics: Conf. Series **887** (2017) 012074 doi:10.1088/1742-6596/887/1/012074

Adaptive control for time-delay teleoperation systems with uncertain dynamics

Shan Liu, Xia Zhang, Wenfeng Zheng and Bo Yang

School of Automation Engineering, University of Electronic Science and Technology of China, No.2006, Xiyuan Ave, West Hi-Tech Zone, 611731, Chengdu, Sichuan, China

Email: shanliu@uestc.edu.cn

Abstract: In most teleoperation systems, the dynamics are uncertain and the communications exhibit time delays. In order to confront these problems, this paper reports a position error-based bilateral adaptive controller, in which the unknown operator dynamical parameters and environmental dynamical parameters are included in the unknown vector of the system to be evaluated, adaptive estimate laws are compensated by estimate errors and dissipation by time delays are compensated. By using Lyapunov-Krasovskii stability theorem, it is proved that both position errors and velocities of the teleoperation system asymptotically convergent to zero. Simulations are performed to compare the performance of the proposed controller with the traditional adaptive controller and to demonstrate the efficiency of the developed teleoperation control system.

1. Introduction

A typical teleoperation system is composed of the human operator, the master robot, the communication channel, the slave robot and the environment [1]. Since teleoperation systems allow people to perform complicated missions in a remote or inaccessible environment, the control of teleoperation system becomes a highly active field that is challenging due to the uncertain of their nonlinear dynamics and the time-delays in the signal transmission.

From 1989, passivity and scattering theory were used to design passive controllers of teleoperation system to guarantee stability independent of the time delay [2-3], which may give rise to position drift. To improve the tracking performance, H_∞ control method was applied to balance the robust and the performance [4]. Nevertheless, partial differential equation or inequation need be solved during designing H_∞ control method, which is hard to implement in real applications. To avoid these problems, some delay compensation methods were proposed based on model prediction theorem [5], which makes stability problem being implicit and needs a mass of criteria and constraints.

On the other hand, it is almost impossible to obtain exact dynamic models of the master robot and slave robot because of uncertainties from nonlinear frictions and flexibilities of robots' joints. Several methods have been proposed to deal with the tracking problems in the presence of parameters uncertainties, such as sliding mode control [6-8] and adaptive control [1, 7, 9-11]. When sliding mode controller is used in a teleoperation system, high frequency buffeting is easy to occur, which not only reduces performance of the control system, but also makes most of energy in the system be dissipated. Although current adaptive control methods can effectively deal with uncertainties of dynamic parameters in teleoperation systems to guarantee stability, uncertain external disturbance and nonlinear input uncertainties cannot be solved.

This paper proposes a position error-based bilateral adaptive controller (BAC). By using Lyapunov-Krasovskii stability theorem, we analyze the stability and tracking performance of the time-delay teleoperation system. Finally, the proposed controller is applied in a teleoperation system where the master robot and the slave robot are 2-degrees of freedom (DOF) robotic arm; simulations are performed to show the effectiveness of the controller.

The paper is organized as follows. The problem is formulated in Section 2. The design of the controller is presented and the stability is analyzed in Section 3. In Section 4, simulations are carried out to validate the proposed approach. The conclusion is drawn in Section 5.

2. Problem Formulation

The master and slave robots are modeled as a pair of n-DOF robotic arm. The nonlinear dynamics of teleoperation system that is robot dynamics with the human operator and environment in joint space are given by

$$M_m(q_m)\ddot{q}_m + C_m(q_m,\dot{q}_m)\dot{q}_m + G_m(q_m) = \tau_m,\ \ M_s(q_s)\ddot{q}_s + C_s(q_s,\dot{q}_s)\dot{q}_s + G_s(q_s) = \tau_s \quad (1)$$

where,

$$\begin{aligned} M_m(q_m) &= M_{q_m}(q_m) + J_m^{\ T} M_h J_m, & M_s(q_s) &= M_{q_s}(q_s) + J_s^{\ T} M_e J_s \\ C_m(q_m,\dot{q}_m) &= C_{q_m}(q_m,\dot{q}_m) + J_m^{\ T} B_h J_m + J_m^{\ T} M_h \dot{J}_m, & C_s(q_s,\dot{q}_s) &= C_{q_s}(q_s,\dot{q}_s) + J_s^{\ T} B_e J_s + J_s^{\ T} M_e \dot{J}_s \\ G_m(q_m) &= G_{q_m}(q_m) + J_m^{\ T} K_h h_m(q_m) - J_m^{\ T} f_h^*, & G_s(q_s) &= G_{q_s}(q_s) + J_s^{\ T} K_e h_s(q_s) + J_s^{\ T} f_e^* \end{aligned} \quad (2)$$

where the subscript $i = m,s$ denotes master and slave robot, $q_i, \dot{q}_i, \ddot{q}_i \in \Re^{n\times 1}$ are the joint positions, velocities and accelerations; $M_{q_i}(q_i) \in \Re^{n\times n}$ are the inertia matrices; $C_{q_i}(q_i,\dot{q}_i) \in \Re^{n\times n}$ are the Coriolis and centrifugal torque matrices; $G_{q_i}(q_i) \in \Re^{n\times 1}$ are the gravitational torques; $\tau_i \in \Re^{n\times 1}$ are the control signals, $J_i(q_i) \in \Re^{n\times n}$ are Jacobian matrices; $F_h, F_e \in \Re^{n\times 1}$ are human operator force and environmental force, respectively; $M_h, B_h, K_h, M_e, B_e, K_e \in \Re^{n\times n}$ are positive constant diagonal matrices and denotes mass, damping and elastic coefficient matrices of human operator and environment, respectively. Some important well-known properties of these robot dynamical modes are revisited here.

P1. The inertia matrix is positive definite, symmetric, lower and upper bounded, i.e., $0 < \lambda_{\min}\{M_i(q_i)\} I \le M_i(q_i) \le \lambda_{\max}\{M_i(q_i)\} I < \infty$.

P2. The matrix $\dot{M}_i(q_i) - 2C_i(q_i,\dot{q}_i)$ is skew-symmetric.

P3. These robot dynamic models are linearly parameterizable, i.e., $M_i(q_i)\ddot{q}_i + C_i(q_i,\dot{q}_i)\dot{q}_i + G_i(q_i) = \tau_i = Y_z(q_i,\dot{q}_i,\ddot{q}_i)\theta_z$ where $Y_z(q_i,\dot{q}_i,\ddot{q}_i) \in \Re^{n\times r}$ are matrices of known functions and $\theta_z = [\theta_{z1},\cdots,\theta_{zr}]^T$ are constant vectors of robotic physical parameters.

In this paper, the time delays are considered. It is assumed that the time delay from the master robot to the slave robot is T_m, and the delay of reverse channel is T_s. The definition of the position error is as follows,

$$e_m(t) = q_m(t) - q_s(t - T_s),\ \ e_s(t) = q_s(t) - q_m(t - T_m) \quad (3)$$

Then, the teleoperation dynamics in equation (1) can be transformed into

$$\begin{aligned} \hat{M}_m(q_m)\ddot{q}_m + \hat{C}_m(q_m,\dot{q}_m)\dot{q}_m + \hat{G}_m(q_m) = \hat{\tau}_m = Y_{zm}(q_m,\dot{q}_m,\ddot{q}_m)\hat{\theta}_{zm} \\ \hat{M}_s(q_s)\ddot{q}_s + \hat{C}_s(q_s,\dot{q}_s)\dot{q}_s + \hat{G}_s(q_s) = \hat{\tau}_s = Y_{zs}(q_s,\dot{q}_s,\ddot{q}_s)\hat{\theta}_{zs} \end{aligned} \quad (4)$$

where $\hat{M}_m(q_m), \hat{C}_m(q_m,\dot{q}_m), \hat{G}_m$ are estimates of $M_m(q_m), C_m(q_m,\dot{q}_m), G_m$, respectively; $\hat{M}_s(q_s), \hat{C}_s(q_s,\dot{q}_s), \hat{G}_s$ are estimates of $M_s(q_s), C_s(q_s,\dot{q}_s), G_s$, respectively; $\hat{\theta}_{zm}, \hat{\theta}_{zs}$ are estimates of the physical parameter vectors of the master robot and the slave robot, θ_{zm},θ_{zs}; $Y_z(q_i,\dot{q}_i,\ddot{q}_i)$, $i=m,s$ are matrices of dynamic regression function of known joint variables. Therefore, the human operator force f_h^* and environment force f_e^* are included in the dynamic uncertain vectors of the teleoperation system, θ_{zm},θ_{zs}.

3. Main Results

3.1. Control Design

Let us define new variables s_m and s_s as

$$s_m(t)=\dot{q}_m(t)+\lambda_m e_m(t),\quad s_s(t)=\dot{q}_s(t)+\lambda_s e_s(t) \tag{5}$$

where λ_i, $i=m,s$ are positive definite diagonal matrices.

The acceleration signals in teleoperation systems are usually unusable, thus we define a regression matrix of tracking error $\bar{Y}_i(q_i,\dot{q}_i,e_i,\dot{e}_i)$ as

$$M_i(q_i)\lambda_i\dot{e}_i+C_i(q_i,\dot{q}_i)\lambda_i e_i+G_i(q_i)=\bar{Y}_i(q_i,\dot{q}_i,e_i,\dot{e}_i)\theta_{zi} \tag{6}$$

The proposed controllers are given by

$$\tau_m=-\bar{Y}_m(q_m,\dot{q}_m,e_m,\dot{e}_m)\hat{\theta}_{zm}-k_m s_m-\beta_m\dot{e}_m,\quad \tau_s=-\bar{Y}_s(q_s,\dot{q}_s,e_s,\dot{e}_s)\hat{\theta}_{zs}-k_s s_s-\beta_s\dot{e}_s \tag{7}$$

where $\bar{Y}_i(q_i,\dot{q}_i,e_i,\dot{e}_i)\hat{\theta}_i$ is a item based on model to compensate errors, and $\hat{\theta}_i$ is estimate of θ_i; k_i, $i=m,s$ is positive diagonal matrix, denotes proportional gain and is used as additional item of error compensation; $\beta_i\dot{e}_i$ is used to eliminate extra energy dissipation due to time delay, β_i is positive diagonal matrix and denotes dissipation factor. Substituting the controllers in equation (7) on the teleoperation dynamics in equation (1), yields

$$\begin{aligned} M_m(q_m)\dot{s}_m+C_m(q_m,\dot{q}_m)s_m+2G_m(q_m)&=\bar{Y}_m\Delta\theta_{zm}-k_m s_m-\beta_m\dot{e}_m \\ M_s(q_s)\dot{s}_s+C_s(q_s,\dot{q}_s)s_s+2G_s(q_s)&=\bar{Y}_s\Delta\theta_{zs}-k_s s_s-\beta_s\dot{e}_s \end{aligned} \tag{8}$$

where $\Delta\theta_{zi}=\theta_{zi}-\hat{\theta}_{zi}, i=m,s$.

In former literatures, the adaptive law was usually given by $\dot{\hat{\theta}}_{zi}=-\Gamma_i^{-1}\bar{Y}_i^T s_i$, which is proposed by J Slotine in [9]. However, if reference signal does not satisfy endurable excitation condition, this adaptive law cannot guarantee accuracy of estimates. In this paper, in order to improve the accuracy and convergence rate of adaptive estimates, we define the adaptive law of uncertain parameter θ_{zi} as

$$\begin{cases} \dot{\hat{\theta}}_{zi}=-\Gamma_i^{-1}(t)\left(\bar{Y}_i^T s_i+(\xi_1+\xi_2)z_i+\xi_1 Y_i^T\varepsilon_i\right) \\ \dot{z}_i(t)=-\eta z_i(t)+\mu Y_{zi}^T\varepsilon_i-p_i(t)\dot{\hat{\theta}}_{zi}(t) \end{cases} \tag{9}$$

where $\Gamma_i(t)=diag(\gamma_1(t),\cdots,\gamma_p(t))$ is adaptive gain matrix and has the property $\Gamma_i(t)>0$, $\dot{\Gamma}_i(t)\le 0$, $\gamma_n(t)=a_n e^{-\int_0^t f_n(\upsilon)d\upsilon}+b_n$, $f_n(\upsilon)\ge 0, n=\{1,\cdots,p\}$, in which a_n and b_n are positive constants; $\varepsilon_i=Y_{zi}\theta_{zi}-Y_{zi}\hat{\theta}_{zi}=Y_{zi}\Delta\theta_{zi}, i=m,s$ denotes estimate errors of input torques; $p_i(t)$ is low-pass filtering signal of $Y_{zi}^T Y_{zi}$ and $p_i(t)=\mu\int_0^t e^{-\eta(t-\sigma)}Y_{zi}^T(\sigma)Y_{zi}(\sigma)d\sigma$, in which μ and η are positive constants; ξ_1

and ξ_2 are positive constants to decide the convergence rate of estimates and boundary of errors. Based on robust control method [12], ξ_1 is defined as

$$\xi_1 = \alpha \frac{\left\| \bar{Y}_i^T s_i \right\|}{\lambda_{\min}\left(Y_{zi}^T Y_{zi}\right) + \delta} \tag{10}$$

where α and δ are positive constants, $\lambda_{\min}\left(Y_{zi}^T Y_{zi}\right)$ is the minimum eigenvalue of $Y_{zi}^T Y_{zi}$.

3.2. Stability Analysis

Theorem 1: Consider the teleoperation system in equation (1) and the controller in equation (7) with adaptive laws in equation (9), the joint position tracking errors e_m, e_s and the estimate errors of the adaptive laws $\Delta\theta_{zm}, \Delta\theta_{zs}$ are bounded, and when $t \to \infty$, the joint position tracking errors converge to zero.

Proof: Let us define the following Lyapunov-Krasovskii function

$$V(t) = \frac{1}{2}\sum_{i=m,s} s_i^T M_i s_i + \frac{1}{2}\sum_{i=m,s} \Delta\theta_{zi}^T \Gamma_i(t) \Delta\theta_{zi} + \frac{1}{2}\sum_{i=m,s} e_i^T \lambda^T k_i \beta_i e_i + \frac{1}{2}\sum_{i=m,s} \int_{t-\Delta T}^{t} \dot{q}_i^T(\sigma) k_i \beta \dot{q}(\sigma) d\sigma \tag{11}$$

According to the property P1 of the robot, $M_m(q_m), M_s(q_s)$ are positive diagonal. $\Gamma_i(t)$ and k_i, β_i are positive definite matrices, thus $V(t)$ is positive definite.

The time derivative of equation (11) along equations (8)-(10), using P2, is given by

$$\begin{aligned} \dot{V}(t) &= \frac{1}{2}\sum_{i=m,s} s_i^T \dot{M}_i s_i + \sum_{i=m,s} s_i^T M_i \dot{s}_i + \frac{1}{2}\sum_{i=m,s} \Delta\theta_{zi}^T \dot{\Gamma}_i(t) \Delta\theta_{zi} + \sum_{i=m,s} \Delta\theta_{zi}^T \Gamma_i(t) \Delta\dot{\theta}_{zi} \\ &\quad + \sum_{i=m,s} e_i^T \lambda^T k_i \beta_i \dot{e}_i + \frac{1}{2}\sum_{i=m,s} \left(\dot{q}_i^T(t) k_i \beta \dot{q}(t) - \dot{q}_i^T(t-\Delta T) k_i \beta \dot{q}(t-\Delta T) \right) \\ &= -\sum_{i=m,s} \left(s_i^T (k_i + B_i) s_i + \frac{1}{2} \dot{e}_i^T k_i \beta_i \dot{e}_i + \Delta\theta_{zi}^T \left((\xi_1 + \xi_2) p_i + \xi_1 Y_{zi}^T Y_{zi} \right) \Delta\theta_{zi} - \frac{1}{2} \Delta\theta_{zi}^T \dot{\Gamma}_i \Delta\theta_{zi} \right) \\ &\le -\sum_{i=m,s} \left(s_i^T (k_i + B_i) s_i + \frac{1}{2} \dot{e}_i^T k_i \beta_i \dot{e}_i + \Delta\theta_{zi}^T \left((\xi_1 + \xi_2) p_i + \xi_1 Y_{zi}^T Y_{zi} \right) \Delta\theta_{zi} \right) \end{aligned} \tag{12}$$

B_i, k_i, β_i are positive constant, $(\xi_1 + \xi_2) p_i + \xi_1 Y_{zi}^T Y_{zi}$ is semi positive definite, thus $\dot{V}(t) \le 0$.

Due to $V(t) \ge 0$ and $\dot{V}(t) \le 0$, the variables s_i, the tracking errors e_m, e_s and estimate errors of adaptive laws, $\Delta\theta_{zm}, \Delta\theta_{zs}$ are bounded. By Barbalat Theorem, $\dot{V}(t)$ asymptotically converges to zero. Then, if $t \to \infty$, $s_i \to 0$ and $\dot{e}_i \to 0$. Thus, $\lim_{t\to\infty}(s_m + s_s) = 0$. Define $r(t) = x_m + x_s$, then $\lim_{t\to\infty}\left(\dot{r}(t) + \lambda\left(r(t) - r(t-\Delta T)\right)\right) = 0$ and its Laplace transform $\lim_{t\to\infty}\left(sr(s) - r(0) + \lambda\left(r(s) - e^{-s\Delta T} r(s)\right)\right) = 0$. Thus, it can be inferred that $\lim_{t\to\infty} s^2 r(s) = 0$, that is $\dot{x}_m + \dot{x}_s \to 0$ as $t \to \infty$.

In a similar way, we define $r(t) = x_m - x_s$, then $\dot{x}_m - \dot{x}_s \to 0$ as $t \to \infty$. Therefore, the velocities of the master robot terminal and the slave robot terminal asymptotically converge to zero, that is $\dot{x}_m, \dot{x}_s \to 0$ as $t \to \infty$. Due to $s_i = \dot{x}_i + \lambda_i e_i \to 0$, position tracking errors converge to zero.

4. Simulations

In this section, the performance of the proposed adaptive controller is evaluated by simulations and compared with the controller in [10]. The simulations are performed in Matlab/Simulink platform, with utilizing of the SimMechanics toolbox. The master and slave robot are modeled as a pair of 2-DOF robot arm, as shown in figure 1. Their nonlinear dynamics are modeled by (1) where for $i = m,s$,

$$M_{q_i}(\mathrm{q}_i) = \begin{bmatrix} p_{1i} + p_{2i} + 2p_{3i}\cos(q_{2i}) & p_{2i} + p_{3i}\cos(q_{2i}) \\ p_{2i} + p_{3i}\cos(q_{2i}) & p_{2i} \end{bmatrix},$$

$$C(\mathrm{q}_i,\dot{\mathrm{q}}_i) = \begin{bmatrix} -p_{3i}\dot{q}_{2i}\cos(q_{2i}) & -p_{3i}(\dot{q}_{1i} + \dot{q}_{2i})\sin(q_{2i}) \\ p_{3i}\dot{q}_{1i}\sin(q_{2i}) & 0 \end{bmatrix}, \quad G(\mathrm{q}_i) = \begin{bmatrix} p_{4i}g\cos(q_{1i}) + p_{5i}g\cos(q_{1i} + q_{2i}) \\ p_{5i}g\cos(q_{1i} + q_{2i}) \end{bmatrix},$$

$p_{1i} = m_{1i}l_{c1i}^2 + m_{2i}l_{1i}^2 + I_{1i}$, $p_{2i} = m_{2i}l_{c2i}^2 + I_{2i}$, $p_{3i} = m_{2i}l_{1i}l_{c2i}$, $p_{4i} = m_{1i}l_{c2i} + m_{2i}l_{1i}$, $p_{5i} = m_{2i}l_{c2i}$. The following kinematic Jacobian matrix is proposed for both robots,

$$J(\mathrm{q}) = \begin{bmatrix} -l_1\sin(q_1) - l_2\sin(q_1 + q_2) & -l_2\sin(q_1 + q_2) \\ l_1\cos(q_1) + l_2\cos(q_1 + q_2) & \cos(q_1 + q_2) \end{bmatrix}.$$

The coordinates of the end effector of both robots are represented as $[x,y]^T = h(q) = [l_1\cos(q_1) + l_2\cos(q_1 + q_2), l_1\sin(q_1) + l_2\sin(q_1 + q_2)]$.

The physical parameters are $l_{1i} = 1m$, $l_{2i} = 2m$, $l_{c1i} = 1/2m$, $l_{c2i} = 1m$, $m_{1i} = 1kg$, $m_{2i} = 3kg$, $I_{1i} = 1/12kg \times m^2$, $I_{2i} = 2/5kg \times m^2$, $i = m,s$, $g = 9.81$.

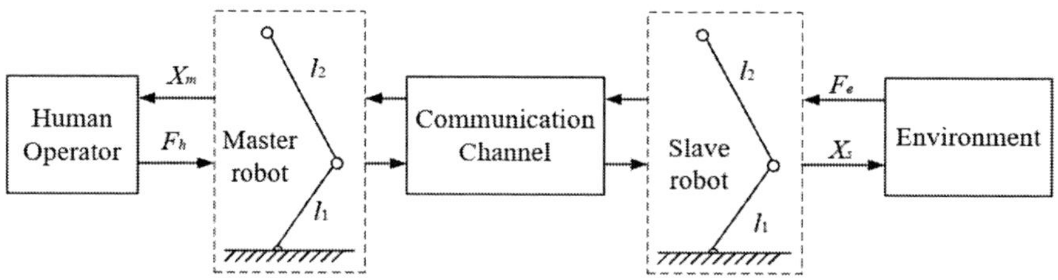

Figure 1. Teleoperation system considered in Section 4.

In the simulation, the environmental dynamic parameters are set as $M_e = m_eI$, $B_e = b_eI$, $K_e = k_eI$ where $m_e = 0.1kg$, $b_e = 20Ns/m$, $k_e = 100N/m$; the environment force is set as $f_e^* = [0 \quad 0]^T$ to simulate passive environment; the operation force is set as $f_h^* = [f_{h1}^* \quad 0]^T = [25\cos(\pi t) \quad 0]^T$, and the dynamic parameters of the human operator are set as $M_h = m_hI$, $B_h = b_hI$, $K_h = k_hI$, where $m_h = 0.2kg$, $b_h = 50Ns/m$, $k_h = 100N/m$. The dynamic parameters of the master robot and the slave robot are unknown. By the property P3, these dynamic models of both robots are linearly transformed, and then the real dynamic parameters are worked out to be $\theta_m = [3.8,3,0.4,100,200,200,3.933,50,25(-\cos(\pi t)),50(1-\cos(\pi t))]^T$, $\theta_s = [3.4,3,0.2,100,200,200,3.833,50]^T$.

The initial positions of both robots are $x_m(0) = [0.184,2.6389]$, $x_s(0) = [0.184,2.6389]$. The initial dynamical parameters are set as $\theta_m(0) = [1.9,1.7,0.5,90,190,180,2.0,40,20,45]^T$, and $\theta_s(0) = [1.9,1.7,0.1,90,190,190,1.9,40]^T$.

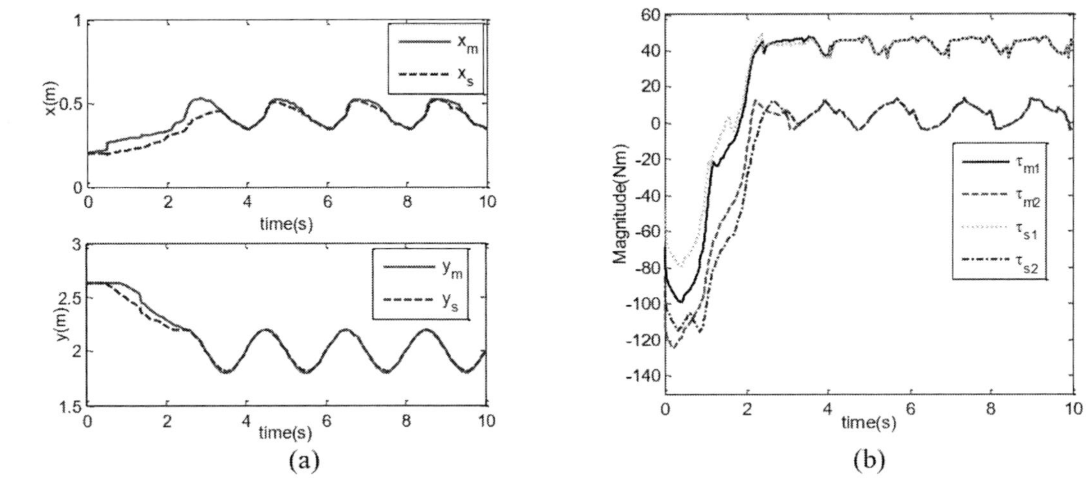

Figure 2. Simulation results. (a) Results for the Cartesian position when the teleoperation system interacts with the human operator and the environment; (b) Results for the torques of the master and slave controller when interacting with the human operator and the environment.

The controller gains are set as $k_m = diag(50,50)$, $k_s = diag(50,50)$, the dissipation factors are set as $\beta_m = \beta_s = diag(0.8,0.6)$, $\lambda_m = \lambda_s = diag(5,5)$. The adaptive gain matrices are set as $\Gamma_m(t) = (e^{-t}+2)I_{8\times8}$, $\Gamma_s(t) = (e^{-t}+2)I_{6\times6}$ where $I_{n\times n}$ expresses a $n\times n$ unit matrix; the constant parameters of the adaptive law are set as $\alpha_m = \alpha_s = 1$, $\delta_m = \delta_s = 1$, $\xi_{2m} = \xi_{2s} = 0.8$, $\mu_m = \mu_s = 1$, $\eta_m = \eta_s = 1$. The time delay of communication channel in the teleoperation system is $T_m = T_s = 0.5s$.

In order to analyze the tracking performance of the proposed controller with that of other controllers, the traditional adaptive controller (TAC) proposed in [10] is used in the same simulation. The control law of TAC is $\tau_i = Y_{id}\hat{\theta}_{id} - K_i(\dot{e}_i + \alpha_i e_i)$, its adaptive law is $\dot{\hat{\theta}}_{id} = -L_{id}Y_{id}^T r_i$, where $r_i = \dot{e}_i + \alpha_i e_i$, α_i, K_i, L_{id} are constant diagonal matrices. Its parameters are set as $\alpha_m = \alpha_s = 5$, $K_m = K_s = 200I$, $L_{md} = L_{sd} = 40I$.

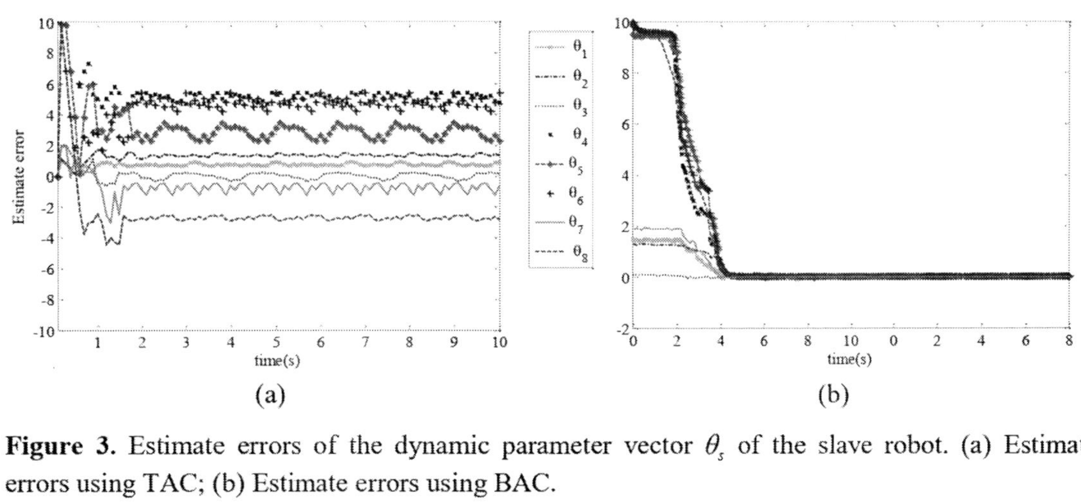

Figure 3. Estimate errors of the dynamic parameter vector θ_s of the slave robot. (a) Estimate errors using TAC; (b) Estimate errors using BAC.

The master and slave Cartesian positions are shown in figure 2, which denotes that the slave robot can follow the master robot. Figure 3 depicts the estimate errors of the BAC and the TAC. The estimate vector of uncertain dynamic parameters of the slave robot using BAC can converge to the real value $\theta_s = [3.4, 3, 0.2, 100, 200, 200, 3.833, 50]^T$ in 2 seconds, which is better than the performance of TAC.

5. Conclusions

In this paper, the problem of time-delay teleoperation system with uncertain dynamics was studied. The unknown operator dynamical parameters and environmental dynamical parameters were included in the unknown vector of the system to be evaluated, which avoids the influence of the uncertain parameters to the system. Estimate errors were used to compensate adaptive estimate laws to improve the accuracy and convergence rate of adaptive estimates. And dissipation by time delays were compensated to ensure the system stability and to improve the tracking performance of the control system. The stability of the controler was analyzed by Lyapunov-Krasovskii stability theorem. Simulations with two nonlinear robots evidence the performance of the proposed controller and its efficiency compared with the traditional adaptive controller. Future work includes the extension of this framework to the issue of variable time delay.

6. Acknowledgments

This paper is supported by the Fundamental Research Funds for the Central Universities (ZYGX2016J101).

7. References

[1] Liu Y C and Khong M H 2015 *IEEE/ASME Trans Mechatranics* **20** 2550-62
[2] Lee D and Spong M W 2006 *IEEE Trans Robot* **22** 269–81
[3] Li H, Kawashima K, Tadano K and Fofana M S 2011 *ASME Int Mech Eng Cong Exp* vol 7 (Colorado: ASME) pp 11–7
[4] Du H 2013 *the Institution of Engineering and Technology* 7 594-605
[5] Yu X and Man Z 2002 *IEEE Trans Circuits Sys* 49 261-264
[6] Bessa W, Dutra M and Kreuzer E 2010 *Robot Auto Sys*, 58 16-26
[7] Hace A and Franc M 2014 *IEEE/ASME Trans Mechatronics* 19 384-93
[8] Slotine J E and Li W 1991 *Applied Nonlinear Control* (New Jersey: Prentice hall)
[9] Nuno E, Ortega R and Basanez L 2010 *Automatica* 46 155-9
[10] Richert D, Macnab C J B and Pieper J K 2012 *IEEE Trans Sys Man Cyber Part A: Sys Hum* 42 558-70
[11] Sheng T 2000 *Robust control of robot* (Beijing: Tsinghua University Press)

IOP Conf. Series: Journal of Physics: Conf. Series **887** (2017) 012075 doi:10.1088/1742-6596/887/1/012075

Prediction of a service demand using combined forecasting approach

Ling Zhou

Department of Mechanical Engineering, Nanjing Institute of Industry Technology, 210014, Nanjing, China. Email: zhoul6@niit.edu.cn

Abstract. Forecasting facilitates cutting down operational and management costs while ensuring service level for a logistics service provider. Our case study here is to investigate how to forecast short-term logistic demand for a LTL carrier. Combined approach depends on several forecasting methods simultaneously, instead of a single method. It can offset the weakness of a forecasting method with the strength of another, which could improve the precision performance of prediction. Main issues of combined forecast modeling are how to select methods for combination, and how to find out weight coefficients among methods. The principles of method selection include that each method should apply to the problem of forecasting itself, also methods should differ in categorical feature as much as possible. Based on these principles, exponential smoothing, ARIMA and Neural Network are chosen to form the combined approach. Besides, least square technique is employed to settle the optimal weight coefficients among forecasting methods. Simulation results show the advantage of combined approach over the three single methods. The work done in the paper helps manager to select prediction method in practice.

1. Introduction

Short-term demand forecasting is significant for a logistic service provider. Not only can it facilitate effective resource allocation but also establish better marketing strategies. Practically service demand is estimated by schedulers based on their own experiences which often contain errors.

The objective of this paper is to exploit service demand forecasting for a LTL trucking company. In the paper, sample data are provided by company "Schmidt Transport" locating in Ruhr industrial area of Germany, and this company is one of the largest-size partners in a national logistics network "CargoLine" which is constituted of 43 members and operates over 2395 tractors and 3168 trailers.

2. Selection of Forecasting Method

Broadly, there are two categories of forecasting methods: combined approach versus single one. Inconsistency among different forecasts gives rise to a thought of combining forecasts of different technology [1]. Combined approach depends on several forecasting methods simultaneously, instead of a single method. It can offset the weakness of a forecasting method with the strength of another, which could improve the precision performance of prediction [2]. Therefore rather than to try to select a method which is the most appropriate, it may be better to try to combine the forecasts obtained by different methods. Up to now combined forecasting has been applied for various applications [3], [2]. Considering a relatively high demand on prediction precision in logistic industry, here combined approach will be considered.

3. Combined Forecasting Approach

During a combined prediction, each component of method provides different ways of useful information for the whole prediction. Main issues during combined forecast modeling are how to select methods for combination, and how to find out weight coefficients among methods.

IOP Conf. Series: Journal of Physics: Conf. Series **887** (2017) 012075 doi:10.1088/1742-6596/887/1/012075

3.1. Selection of Method

The principles of method selection include that each method should apply to the problem of forecasting itself, also methods should differ in categorical feature and underlying concept as much as possible [1].

Generally forecasting approaches can be grouped into quantitative approach vs. qualitative one, while quantitative approach further consists of time series method, regression[4], and intelligent methods such as Neural Network[5] [6] etc. Here only quantitative approach will be considered as qualitative one replies on subjective human judgment and doesn't be suitable for short-term forecasting as well [4].

Time series forecasting and regression analysis are two important short-term forecasting methods. In regression modeling, we must find explanatory variables which have theoretical relationships with forecasted output [7]. For national TL trucking industry, their service demand reflects national economic conditions to a large extent, thus some published national macroeconomic indexes can be selected as explanatory indicators [8]. However, LTL service providers mainly serve regional clients, and the regional demand and supply cannot represent national economic activities completely. So it may have no obvious correlation between the macroeconomic indexes and demand of LTL service providers. The viewpoint has been certified for the case investigated using correlation analysis [9]. Hence time series approaches instead of regression one will be taken into account.

Time series forecasts are those which are the result of extrapolating past data into the future [4]. There are two benchmark linear time series approaches --- single exponential smoothing and ARIMA [10]. Single exponential smoothing is a naive method in time series prediction, which involves weighting past data with weights that decrease exponentially with time [4]. It is often considered a simple and practical forecasting technique, since adaptations of the model can be made to account for changing conditions. Autoregressive Integrated Moving Average (ARIMA) is a flexible and efficient time series method with a solid basis of statistical theory. For all practical purposes, Box-Jenkins methodology can fit an ARIMA/ARMA model to all kinds of data patterns [11]. With automatic computer-based procedures, the usefulness of ARIMA method has been increased remarkably. Up to now, it has achieved a wide level of acceptance in many forecasting situations [12], [13].

Moreover intelligent approach is also a means to build short-term forecasting model. Among various intelligent tools, Neural Network excels at nonlinear approximation and self-adaptation in the process of prediction [14], [15].

Summarily, based on the two principles of suitability and differentiation during selecting methods, exponential smoothing, ARIMA and Neural Network are chosen to form a combined approach.

3.2. Determination of Weight Coefficients

The second issue of combined forecasting is to find out weight coefficients among methods. Generally, least square technique can be used to fix the optimal weight coefficients among forecasting methods, where the objective function is minimal of the sum of forecasting error at all relevant periods [1]. Then combined approach is implemented based on the optimal weight vector. In fact, optimal weight coefficients decide the extent of which each method influences the output of overall forecast.

4. Simulation and Results

The figure below shows the monthly transport demand of company Schmidt Transport from January 2006 to December 2010. To keep data confidentiality, here the unit of demand has been omitted. The plot reveals an upward trend from the 36th month to 60th month and a distinct seasonal pattern for which transport demand is relatively high in the third or fourth and tenth or eleventh month and relatively low in the eighth month.

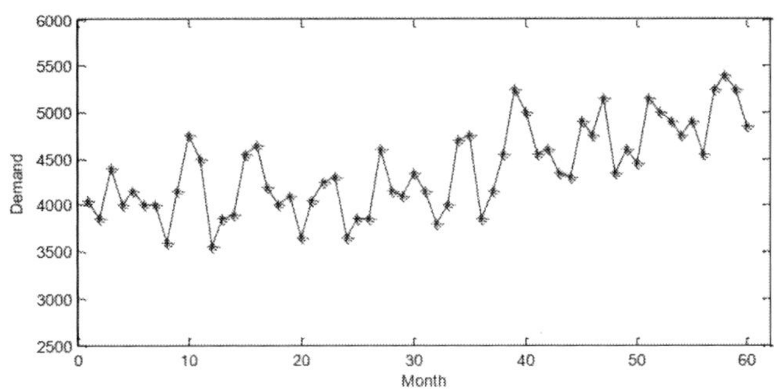

Figure 1. Original demand time series of five-year period

The total five-year data are divided into three parts: first four-year data for training, next six-month for validation and the last six-month data for testing. Validation sets is specifically used for the evaluation and selection of NN model; while ARIMA and exponential smoothing don't need a separate validation set, we use the first 54-month data for their identification and estimation; obviously the identification of combined model relates to the testing set of the original data .

In order to determine the value of adjustable parameter a [9] in the single exponential smoothing model, we use grid search which increases a with a step 0.1 in the range [0, 1] and then find the optimal value of a based on that the model with this a have the minimal error for training sets. Here the optimal value of a is 0.7.

Revised automatic model searching is used to identify ARIMA model; the ARIMA model (as in equation 1) has been found as the optimal one for the sample data.

$$y_t = y_{t-12} + 53.4364 + 0.2328(y_{t-1} - y_{t-13}) + 0.0063(y_{t-2} - y_{t-14}) + 0.0810(y_{t-3} - y_{t-15}) + 0.2261(y_{t-4} - y_{t-16}) + 0.1600(y_{t-5} - y_{t-17}) - 0.21\theta 2_{t-1} \quad (1)$$

Neural Network's architecture adopted is a three-layer MFNN. Here deseasonalization and detrending are adopted for the performance improvement [16]. The number of input nodes or the lagged observations used in the NN is often a more important factor than the number of hidden nodes for time series forecasting [17]. The hidden nodes in the experiment vary from 1 to 14. For the deseasonalized series, the input nodes vary from 1 to 6; for the original data, we use 8 lag numbers: 1-4, 12-14, and 24. Levenberg-Marquardt algorithm is employed for the training. The early stopping is implemented to avoid the overfitting in small-size sample modeling [14]. The maximum training epochs are 1000. Furthermore mean absolute percentage error (MAPE) is considered for the performance comparison. Table 1 records the optimal NN structure which has 6 lag numbers (lag1~lag6) of input and has 14 hidden number.

Table 1. The best NN model from deseasonalization+detrending datasets

Data type	Lag number	Hidden number
Deseasonalization + Detrending	6	14

Table 2 shows the performance of combined forecasting, NN, ARIMA and exponential smoothing under each sample sets. ARIMA model has 4,5% (MAPE) in training sets while 5.6% (MAPE) in testing set; correspondingly Exponential smoothing has an error of 7.5% in training sets while 6.0% in testing sets; NN model has a testing error of 4.5%, with training error 1,4% and validation error 1,8%. Comparatively, combined forecasting has the smallest testing error (4.1%).

Table 2. Model outputs using combinational forecasting, NN, ARIMA and exponential smoothing respectively.

Month	Actual	NN			ARIMA			Exponential Smoothing			Combinational Forecasting		
		Forecast	Error*	MAPE	Forecast	Error*	MAPE	Forecast	Error*	MAPE	Forecast	Error	MAPE
1.	4050.0	0	_		0	_		0	_				
2.	3850.0	0	_		0	_		2835.0	-26.4				
3.	4400.0	0	_		0	_		3545.5	-19.4				
4.	4000.0	0	_		0	_		4143.7	3.6				
5.	4150.0	0	_		0	_		4043.1	-2.6				
6.	4000.0	0	_		0	_		4117.9	3.0				
7.	4010.0	4074.2	1.6		0	_		4035.4	0.6				
8.	3600.0	3646.8	1.3		0	_		4017.6	11.6				
9.	4150.0	4116.8	-0.8		0	_		3725.3	-10.2				
10.	4750.0	4693.0	-1.2		0	_		4022.6	-15.3				
11.	4500.0	4540.5	0.9		0	_		4531.8	0.7				
12.	3550.0	3454.2	-2.7		0	_		4509.5	27.0				
13.	3850.0	3873.1	0.6		0	_		3837.9	-0.32				
14.	3890.0	3870.6	-0.5		0	_		3846.4	-1.1				
15.	4550.0	4645.6	2.1		0	_		3876.9	-14.8				
16.	4650.0	4561.7	-1.9		0	_		4348.1	-6.5				
17.	4200.0	4141.2	-1.4		0	_		4559.4	8.6				
18.	4000.0	3944.0	-1.4		4058.4	1.5		4307.8	7.7				
19.	4100.0	3972.9	-3.1		4144.4	1.1		4092.3	-0.2				
20.	3650.0	3558.8	-2.5		3840.1	5.2		4097.7	12.3				
21.	4050.0	4029.8	-0.5		4291.0	6.0		3784.3	-6.6				
22.	4250.0	4271.3	0.5		4745.1	11.6		3970.3	-6.6				
23.	4300.0	4424.7	2.9		4356.7	1.3		4166.1	-3.1				
24.	3650.0	3715.7	1.8		3559.4	-2.4		4259.8	16.7				
25.	3850.0	3788.4	-1.6		3889.4	1.0		3832.9	-0.4				
26.	3860.0	3871.6	0.3		3790.5	-1.8		3844.9	-0.4				
27.	4600.0	4498.8	-2.2		4493.9	-2.3		3855.5	-16.2				
28.	4150.0	4212.3	1.5		4727.8	13.9		4376.6	5.5				
29.	4100.0	4108.2	0.2		4029.5	-1.7		4.2180	2.9				
30.	4350.0	4332.6	-0.4		4039.1	-7.2		4135.4	-4.9				
31.	4150.0	4096.1	-1.3		4265.6	2.8		4285.6	3.3				
32.	3800.0	3841.8	1.1		3579.8	-5.8		4190.7	10.3				
33.	4000.0	4264.0	6.6		4110.7	2.7		3917.2	-2.1				
34.	4700.0	4681.2	-0.4		4336.7	-7.7		3975.2	-15.4				
35.	4750.0	4835.5	1.8		4613.7	-2.9		4482.5	-5.6				
36.	3850.0	3954.0	2.7		3877.6	1.0		4669.8	21.3				
37.	4150.0	4104.4	-1.1		3996.2	-3.7		4095.9	-1.3				
38.	4550.0	4531.8	-0.4		4147.1	-8.9		4133.8	-9.2				
39.	5250.0	5202.8	-0.9		5090.6	-3.0		4425.1	-15.7				
40.	5000.0	4965.0	-0.7		4534.2	-9.3		5002.5	0.05				
41.	4550.0	4531.8	-0.4		4609.1	1.3		5000.8	9.9				
42.	4600.0	4549.4	-1.1		4757.8	3.4		4685.2	1.9				
43.	4350.0	4480.5	3.0		4557.5	4.7		4625.6	6.3				
44.	4300.0	4287.1	-0.3		4190.6	-2.5		4432.7	3.1				
45.	4900.0	4914.7	0.3		4452.1	-9.1		4339.8	-11.4				
46.	4750.0	4731.0	-0.4	Training	5205.5	9.6		4731.9	-0.4				
47.	5150.0	5052.2	-1.9	set:	4850.9	-5.8		4744.6	-7.9				
48.	4350.0	4454.4	2.4	1.4	4277.7	-1.7		5028.4	15.6				
49.	4600.0	4678.2	1.7		4625.1	1.0		4553.5	-1.0				
50.	4450.0	4499.0	1.1		4893.8	10.0		4586.1	3.1				
51.	5150.0	5129.4	-0.4		5328.7	3.5		4490.8	-12.8				
52.	5000.0	4965.0	-0.7	Validati-	5205.5	4.1	Training	4952.2	-1.0	Training			
53.	4900.0	4777.5	-2.5	on set:	4733.3	-3.4	set:	4985.7	1.8	Set:			
54.	4750.0	4968.5	4.6	1.8	4811.2	1.3	4.5	4925.7	3.7	7.5			
55.	4900.0	4620.7	-5.7		4402.0	-10.2		4802.7	-2.0		4623.0	-5.65	
56.	4550.0	4668.3	2.6		4599.4	1.1		4870.8	7.1		4720.7	3.75	
57.	5250.0	5286.8	0.7		5096.0	-2.9		4646.2	-11.5		5004.2	-4.68	
58.	5400.0	4752.0	-12.0	Testing	5053.3	-6.4	Testing	5068.9	-6.1	Testing	4951.0	-8.32	Testing
59.	5250.0	5160.8	-1.7	set:	5598.5	6.6	set:	5300.7	1.0	Set:	5335.8	1.63	Set:
60.	4850.0	4641.5	-4.3	4.5	4530.5	-6.6	5.6	5265.2	8.6	6.0	4831.8	-0.37	4.1

* Error means relative error which equals to (forecast-actual)/actual*100.

It is concluded that:

- Combined modeling has the best forecasting performance in testing sets; while exponential smoothing/ARIMA the worst and NN somewhere between.

- Among NN, exponential smoothing and ARIMA, NN needs longest modeling time while it exhibits the relatively better forecast results; nevertheless exponential smoothing needs probably the shortest modeling time while shows the weakest precision; and ARIMA is somewhere between NN and exponential smoothing.
- Obviously combined modeling needs even a bit longer modeling time than NN, however it shows an outstanding performance in prediction.
- Forecasting facilitates the operation and management in logistics companies. Accurate forecasts can save operational and management costs which accounts for the large part of whole costs. Considering the importance of forecasting precision, combined forecasting is regarded as the best choice for the case studied.

5. Conclusion

The paper focuses on short term demand forecasting for a LTL carrier. Firstly multiply regression approach is not adopted considering not identifying suitable relevant influencing factors. Among time series forecasting methods, combined modeling is selected mainly on account of its feasibility and applicability. And the forecasting performance of combined model is further compared with three benchmark time series method --- ARIMA, exponential smoothing and NN. It is found that combined forecasting exhibit the most satisfactory performance in out-of-sample testing.

Obviously the results got will facilitate the final decision of forecasting approach selection for the logistic manager in practice.

6. References

[1] Kim J 2008 Combining forecasts using optimal combination weight and generalized autoregression Journal of Forecasting vol 27 pp419–432
[2] Chen K 2011 Combining linear and nonlinear model in forecasting tourism demand Expert Systems with Applications vol 38 pp10368–10376
[3] Song H 2013 Combining statistical and judgmental forecasts via a web-based tourism demand forecasting system International Journal of Forecasting vol 29 pp295–310
[4] Firth M 1977 Forecasting Methods in Business and Management (London: Edward Arnold)
[5] Nakamura E 2005 Inflation forecasting using a neural network Economics Letters vol 86 pp373-378
[6] Bansal K 1998 Neural network based forecasting techniques for inventory control applications Data Mining and Knowledge Discovery vol. 2 pp97-102
[7] Babcock W M 1999 Time series forecasting of quarterly railroad grain carloadings
[8] Transportation Research vol E35 pp43-57
[9] Fite T J 2002 Forecasting freight demand using economic indices International Journal of Physical Distribution & Logistics Management vol 32 pp299-308
[10] Zhou L 2005 Forecasting Demand in a Logistics Network based on ARIMA Proceeding of 35th International Conference on Computers and Industrial Engineering pp2257-2262
[11] Box G E P and Jenkins G M 1976 Time Series Analysis: Forecasting and Control (2nd edition) (Oakland: Holden-day)
[12] Makridakis S and Wheelwright S C 1989 Forecasting Methods for Management (New York: John Wiley&Son)
[13] Burger C J S C 1998 A practitioners guild to time series methods for tourism demand forecasting --- a case study of Durban, South Africa Tourism Management vol 22 pp403-409
[14] Cho V 2003 A comparison of three different approaches to tourist arrival forecasting Tourism management vol 24 pp323-330
[15] Hornik K 1989 Multilayer feedforward networks are universal approximators Neural Networks vol 2 pp359-366
[16] Wu R C 1997 Neural Network Models: Foundations and Applications to an Audit Decision Problem Annals of Operations Research vol 75 pp291-301
[17] Zhang G P and Qi M 2005 Neural network forecasting for seasonal and trend time series European Journal of Operational Research vol 160 pp501-514

IOP Conf. Series: Journal of Physics: Conf. Series **887** (2017) 012075 doi:10.1088/1742-6596/887/1/012075

[18] Zhang G et al. 1998 Forecasting with artificial neural networks: The state of the art International Journal of Forecasting vol 14 pp35-62

IOP Conf. Series: Journal of Physics: Conf. Series **887** (2017) 012076 doi:10.1088/1742-6596/887/1/012076

Research on regional intrusion prevention and control system based on target tracking

Yanfei Liu, Jieling Wang, Ke Jiang, Yanhui He and Zhilin Wu

Xi'an Institute of High Technology, Xi'an, Shaanxi 710025, China

Abstract: In view of the fact that China's border is very long and the border prevention and control measures are single, we designed a regional intrusion prevention and control system which based on target-tracking. The system consists of four parts: solar panel, radar, electro-optical equipment, unmanned aerial vehicle and intelligent tracking platform. The solar panel provides independent power for the entire system. The radar detects the target in real time and realizes the high precision positioning of suspicious targets, then through the linkage of electro-optical equipment, it can achieve full-time automatic precise tracking of targets. When the target appears within the range of detection, the drone will be launched to continue the tracking. The system is mainly to realize the full time, full coverage, whole process integration and active real-time control of the border area.

1. Introduction

Due to the long country's border line and complex terrain, the management and control is very difficult, the protection is more difficult, Current means of protection cannot effectively prevent the suspicious personnel from entering the country illegally. The system is established in this context. Through inspection and tracking, we can identify suspicious persons on the border line on the first time, and then take effective measures to prevent illegal immigration of suspected persons. This system saves us a lot of manpower, material and financial resources, and is conducive to the management and control of China's border lines, and also conducive to China's security and stability.

2. Overall Design Scheme of the System

The system intends to detect and track suspicious persons on the border, and transmit the acquired information to the command center in real time. The command center makes decisions on this basis. This system consists of three parts: fixed nodes, mobile nodes, data centers. As shown in Fig. 1.

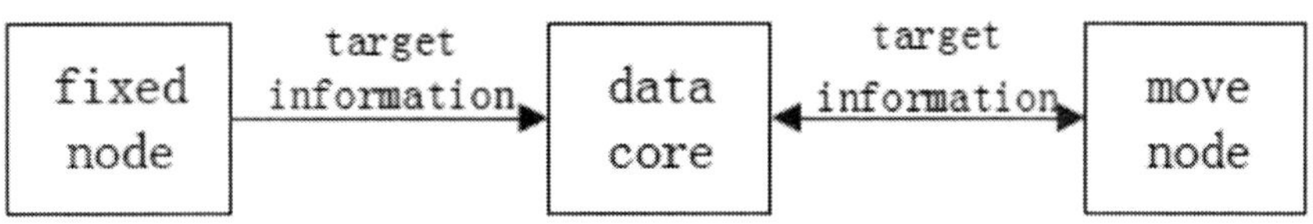

Figure 1. Overall structure drawing of the system

Fixed nodes are mainly composed of camera pan head, radar, wireless devices (data transmission, picture transmission) and solar panels. It mainly realizes the preliminary detection and recognition of suspicious personnel.

First, the radar equipment finds suspicious personnel and transmits the location signals of suspicious persons to the control panel. The control panel controls the camera pan head, which can realize the shooting and tracking of the suspicious personnel. The recorded video signals will be transmitted to the command center as the basis for the decision making

The mobile node is mainly composed of a foldable unmanned aerial vehicle, a four rotor unmanned aerial vehicle, a camera pan head and wireless devices. When suspicious personnel appear in the tracking range of the fixed nodes, the mobile node continues to track the target personnel, and transmits the location, image and other signals of suspicious persons to the command center in real time.

The data center mainly completes the acceptance and delivery of target data. Through large data analysis, the data center determines whether the target is suspicious, and then proceed to the next decision deployment.

3. System Hardware Composition

3.1. Fixed Node Part

The system is composed of the power supply system and target tracking system. The hardware structure of the power supply system points includes: solar panel module, 24V power module, voltage conversion module; target tracking system's hardware structure mainly includes: PTZ motor module, servo drive module, camera module, serial communication module and millimeter wave radar module. Their connection diagrams are shown in Fig. 2 and Fig. 3:

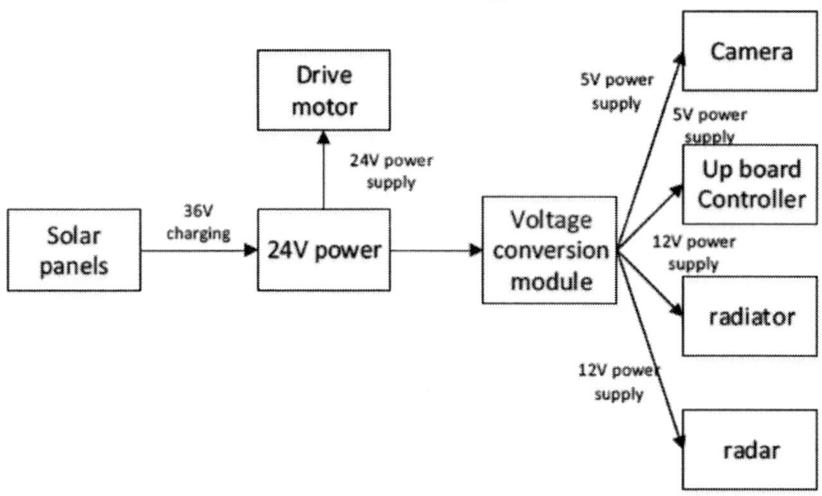

Figure 2. Power supply system hardware connection diagram

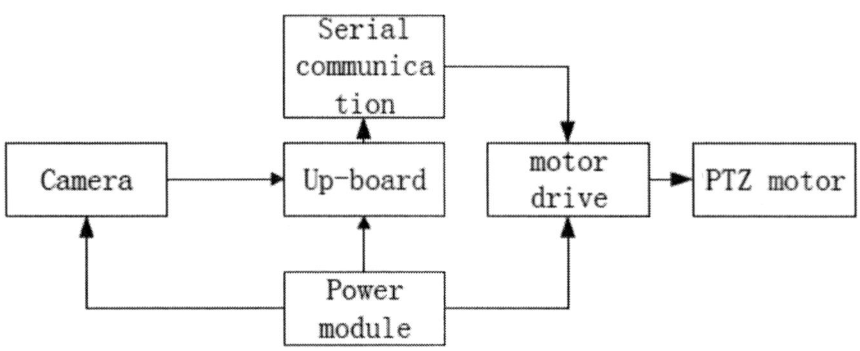

Figure 3. Hardware diagram of target tracking subsystem

3.2. Mobile Node Section

In this system, the mobile node is composed of folding wing UAV. The folding wing UAV is shown in Fig. 4.

Figure 4. Folding wing UAV

With the help of profili airfoil analysis software, we analyzed the resistance characteristics of open hundreds of airfoil, and choose the airfoil with the highest lift drag ratio, and then use CAD to design the fuselage and the folding mechanism. The structural assembly drawing of the folding wing is shown in Fig.5.

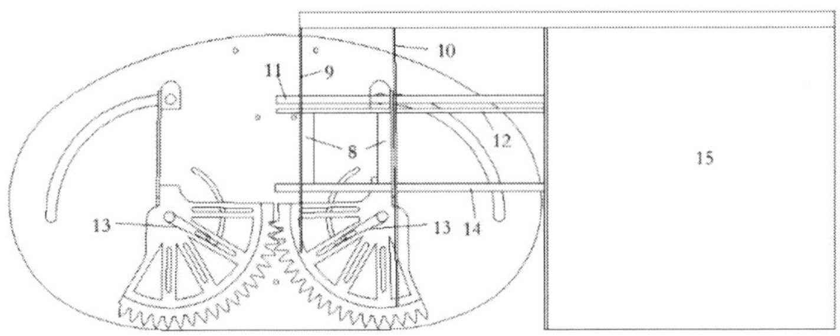

Figure 5. Folding wing structure assembly diagram

4. The Design of System'S Software Algorithm

4.1. The Design of Radar Ranging Precision Algorithm

The essence of increasing the number of FFT points is to increase the sampling points in frequency domain uniformly in the whole unit circle (that is, the whole distance spectrum), thus increasing the amount of computing. To improve the ranging accuracy of radar, it is only necessary to use FFT to determine the main lobe position of the echo distance spectrum, and then to increase the sampling points of the main lobe in the distance spectrum locally. The commonly methods used are ZOOM-FFT, direct selection method, several cascade FFT and so on, they are all the methods to increase the sampling points in frequency domain, this paper adopts a method which is more direct and fast, that is, use the Chirp-Z transform for equal interval sampling on a spiral line in the Z plane, The Chirp-Z transform of N point x (n) is

$$X(z_k) = \sum_{n=0}^{N-1} x(n) z_k^{-n}, z_k = AW^{-k} \tag{1}$$

z_k is the sampling point, N is the number of sampling point, A is the starting point position, and W determines the stretching rate and sampling interval of the helix. When $A = W = 1$, local refinement sampling can be realized on the unit circle, and the formula (1) is represented by the Chirp-Z transform

$$X(z_k) = W^{\frac{k^2}{2}} \sum_{n=0}^{N-1} g(n)h(n-k) = W^{\frac{k^2}{2}} \bullet g(k) * h(k) \tag{2}$$

In formula (2) $g(n) = x(n)A^{-n}W^{\frac{k^2}{2}}, h(n) = W^{\frac{k^2}{2}}$, and the FFT algorithm is used in the implementation of circular convolution in the frequency domain, the algorithm of Chirp-Z transform is divided into seven steps. The Chirp-Z algorithm calculation is less than $Nlog_2 N + 2(N+M) + (N+M-1)log_2 (N+M-1)$, because of the need to conduct a FFT operation to determine if the radar echo from the far main lobe when use the Chirp-Z transform, the computation calculation is $Nlog_2 N$, so the total amount of computation of the algorithm is $Nlog_2 N + 2(N+M) + (N+M)log_2 (N+M)$. If the method of increasing FFT number is used to improve the ranging accuracy of radar directly, its computation is $MNlog_2 NM$. When N=256, M=256, FFT calculation only using N algorithm for 2048 computation using FFT combining with Chirp-Z transform is 7680, using direct increase of computation of FFT points to 10485760 it can be seen that using FFT method combining with Chirp-Z transform, the computation uses only 3.75 times FFT however, the measurement accuracy is improved by 256 times. The method of increasing the number of FFT points directly can achieve the same precision, and the computation will be only 512 times of FFT.

The parameters of the radar are $T = 2.56ms, B = 500MHz$ (corresponding to the distance resolution of 0.30 m), $fs = 100KHz, A = 4.0V, N = M = 256$ (corresponding to the distance of the sampling interval is 0.0012m), the target from the radar distance is 12.0-12.3 m (step 5cm), the target is a single object, The range of local refinement is between the largest sampling point and the second largest sampling point in the echo intermediate frequency spectrum obtained by FFT. Its computer simulation results are shown in table 1.

Table 1. The simulation results of radar ranging accuracy improvement by using Chirp-Z transform

Actual distance	12.0000	12.0500	12.1000	12.1500	12.2000	12.2500	12.3000	maximum error
Maximum sampling point method	12.0000	12.0000	12.0000	12.3000	12.3000	12.3000	12.3000	0.1500
Chirp-Z transform	12.0012	12.0506	12.1000	12.1506	12.2012	12.2494	12.3000	0.0012

4.2. Target Tracking Algorithm

The system adopts the modified KCF algorithm, KCF tracking algorithm is one of the fastest and most stable tracking algorithms at the moment, but the traditional KCF algorithm does not have long-term tracking ability. Once the target is lost, it is difficult to find again. The lack of target detection mechanism is one of the major drawbacks of KCF tracking algorithms. Kalal et al. Proposed TLD tracking algorithm, which can make the tracking, learning and detection simultaneously, detect and track targets throughout the process, and constantly correct the template, but this method of testing target has seriously affected the speed of tracking. Therefore, in order to achieve long-term tracking, this paper adds the detection and localization mechanism based on the KCF tracking algorithm. Only when the target is lost can the restart detection mechanism be started, so that the tracking can be ensured, and the detection is taken as the second step, so as to ensure that the tracking speed is not affected greatly. The anew detection method used in this paper is to divide the whole image in the sampling window, respectively, calculated the sampling window peak response by KCF algorithm, the peak of the window, the peak is the largest window containing the target, this re positioning the target detection.

The overall process of the improved algorithm is shown in Fig. 6, when the peak does not appear abnormal, the algorithm will be in accordance with the way the KCF tracker to track the target frame;

when the peak is abnormal, stop immediately update the target template, re start the detection mechanism, upon detection of target positioning, reload the KCF tracker, continue to update the target template. To track the target according to the KCF tracker frame mode.

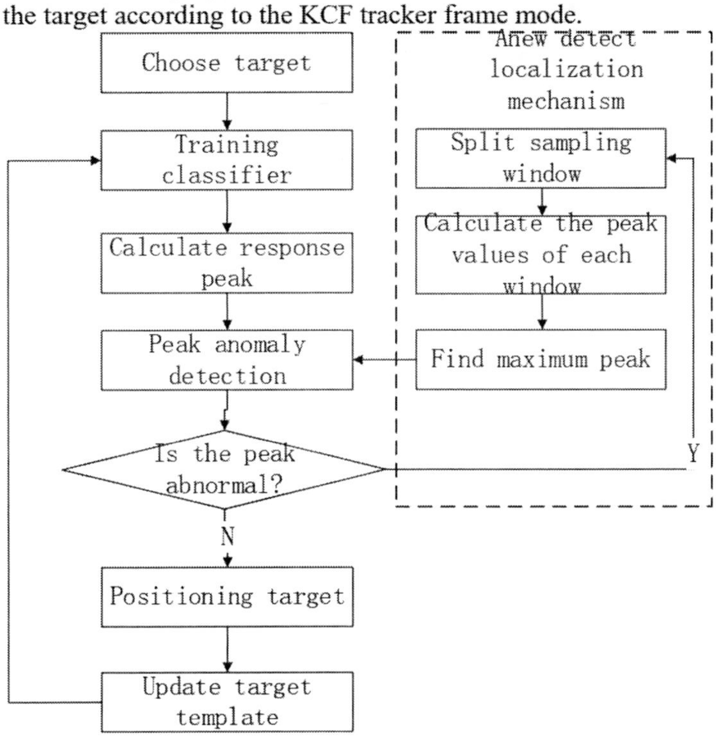

Figure 6. Overall flow chart of the improved algorithm

Fig.7 is the 4 properties: non rigid deformation, fast motion and illumination changes, occlusion success rate, it can be seen from the figure of this algorithm in 4 properties, the success rate of figure are ranked first, compared with the traditional KCF tracking algorithm performance were improved by 19.96%, 19.81%, 15.39%, 11.13%. It shows that the improved algorithm has good tracking effect in deformation, fast motion, illumination change, target occlusion and so on.

IOP Conf. Series: Journal of Physics: Conf. Series **887** (2017) 012076 doi:10.1088/1742-6596/887/1/012076

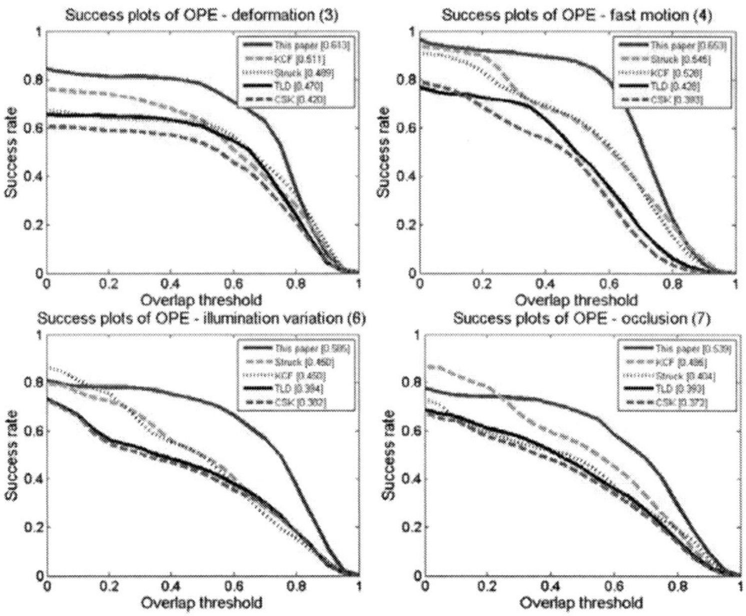

Figure 7. Success rate diagram of partial attributes

5. Conclusion

Aiming at some problems existing in border control, this paper designs a target tracking system based on Intrusion Prevention and control, the system can meet the real-time and all-weather requirements in radar target tracking, the two algorithms are improved, a good solution to the radar in the distance accuracy. The performance of target tracking algorithm is better than other algorithms.

6. References

[1] Jiang Haijiao, built here, Wang Chunyong, Li Zhenhua. Research on ranging characteristics and ranging accuracy of lidar [J]. Laser of China, 2011, (05): 234-240.

[2] Xiao Wei, Tu Qing, Liu Liangbing, Sultan, Zhang Lei. LFMCW radar ranging algorithm for frequency spectrum fusion of sinusoidal signals of the same frequency range [J]. Proceedings of the Chinese society of radio and instrument, 2013, (01): 94-103.

[3] Tan Xiaoheng, Zhou Shuai, Huang Zhenlin. 24GHz LFMCW radar ranging method based on wavelet packet [J]. Systems engineering and electronics, 2013, (03): 522-526.

[4] Wu Li, Peng Shusheng, Xiao Zelong, Hunan. Improve the millimeter wave short-range detecting method of radar range accuracy of [J]. Ordnance journal, 2009, (03): 301-305.

[5] Xin Zhekui, Fang Yongchun, Zhang Xuebo. Ground target tracking system for airborne PTZ adaptive tracking control [J]. Control theory and applications, 2010 small, (08): 1001-1006.

[6] Wu Anping, Hou Zhongxi, Wang Peng, Guo Tianhao. Control of steering wheel head based on neural network inverse model identification [J]. Journal of Huazhong University of Science and Technology (NATURAL SCIENCE EDITION), 2013, (S1): 13-16.

[7] Sun Pengfei, Guo Xiqing, Yang Jingxian. Machine vision control of two degrees of freedom pan machine vision, [J]. Opto Electronics Engineering, 2012, (10): 135-142.

[8] Guo Bingkun, Chen Shui, Wu Yundong, Li Wei. Fuzzy PID control of three axis stabilized tilt table of unmanned aerial vehicle [J]. Journal of Jimei University (NATURAL SCIENCE EDITION), 2013, (04): 308-313

[9] Sevillalara L. Distribution fields for tracking[C].IEEE Conference on Computer Vision and Pattern Recognition. IEEE Computer Society, 2012:1910-1917.

[10] Wang X, Hua G, Han T X. Discriminative Tracking by Metric Learning[C].Computer Vision - ECCV 2010, 11th European Conference on Computer Vision, Heraklion, Crete, Greece, September 5-11, 2010, Proceedings, PartIII. 2010:200-214.

IOP Conf. Series: Journal of Physics: Conf. Series **887** (2017) 012076 doi:10.1088/1742-6596/887/1/012076

[11] HENRIQUES J F, CASEIRO R, MARTINS P,et al..High-speed tracking with kernelized correlation filters [J].IEEE Transactions on Pattern Analysis and Machine Intelligence, 2015, 37(3):583-596.
[12] WU Y,LIM J,YANG M H.Online object tracking:A benchmark[C].Proceedings of the IEEE Computer Society Conference on Computer Vision and Pattern Recognition(CVPR),2013:2411-2418.
[13] SMEULDERS a W M,CHU D M,CUCCHIARAR,et al.Visual tracking; an experimental survey [J].IEEE Transactions on Pattern Analysis and Machine Intelligence, 2014, 36(7); 1442-1468.

IOP Conf. Series: Journal of Physics: Conf. Series **887** (2017) 012077 doi:10.1088/1742-6596/887/1/012077

Tweaked residual convolutional network for face alignment

Wenchao Du[1, a], **Ke Li** [1, b], **Qijun Zhao**[1, c], **Yi Zhang** [2, d] **and Hu Chen**[1,e*]

[1]National Key Laboratory of Fundamental Science on Synthetic Vision & School of Computer Science, Sichuan University, Chengdu, China

[2]School of Computer Science, Sichuan University, Chengdu, China

[a] E-mail: wenchao_du@stu.scu.edu.cn, [b] E-mail: keli@scu.edu.cn, [c] E-mail: qjzhao@scu.edu.cn

[d] E-mail: yzhang@scu.edu.cn, [e]E-mail: huchen@scu.edu.cn

Abstract. We propose a novel Tweaked Residual Convolutional Network approach for face alignment with two-level convolutional networks architecture. Specifically, the first-level Tweaked Convolutional Network (TCN) module predicts the landmark quickly but accurately enough as a preliminary, by taking low-resolution version of the detected face holistically as the input. The following Residual Convolutional Networks (RCN) module progressively refines the landmark by taking as input the local patch extracted around the predicted landmark, particularly, which allows the Convolutional Neural Network (CNN) to extract local shape-indexed features to fine tune landmark position. Extensive evaluations show that the proposed Tweaked Residual Convolutional Network approach outperforms existing methods.

1. Introduction

Face landmark detection plays a fundamental role in many computer vision tasks, such as face recognition, expression analysis, and 3D face modeling. Recently, many methods have been proposed to address this problem, with significant progress being made towards systems that work in the wild.

Regression-based approaches [4] [1] [20] have achieved impressive results by cascading discriminative regression functions that map facial appearance to landmark coordinates directly. In this framework, deep convolutional neural networks have proven effective as a choice for feature extraction and non-linear regression modeling [17] [23] [21]. Although these methods can achieve very reliable results in standard benchmark datasets, they still suffer from limited performance in challenging scenarios, e.g., involving large face pose variations and severe occlusions.

In this paper, we propose an architecture named Tweaked Residual Convolutional Networks, as illustrated in Fig. 1, which can further combines the advantages of traditional cascaded regression and convolutional neural networks, and get better performance than the other state-of-the-art methods.

The proposed TRCN method, a non-linear dynamical system, consists of TCN global feature extraction module and RCN refining module instead of single CNN. Specifically, the TCN feature extraction module predicts the face shape quickly by taking holistically a low-resolution version of the detected face as the input; the following RCN module refines the landmark location by taking local patches as the input, the joint local shape-indexed feature is extracted around the current landmarks (output of the previous TCN module) to fine-tune landmark position by the RCN. As illustrated in Fig. 1, our work is also motivated by the success of [21] and [17], however, there have a great difference from them. In summary, the contribution of our work is:

- We propose a non-linear cascaded framework to model non-linear mapping from the face image to the face shape with convolutional neural network.
- We introduce the concept of Residual Convolutional Net into the new deviation learning between the current location and the ground truth.
- TRCN can initialize the face shape better and extract shape-indexed feature with CNN, thus, it is more robust to challenging scenario than other methods.
-

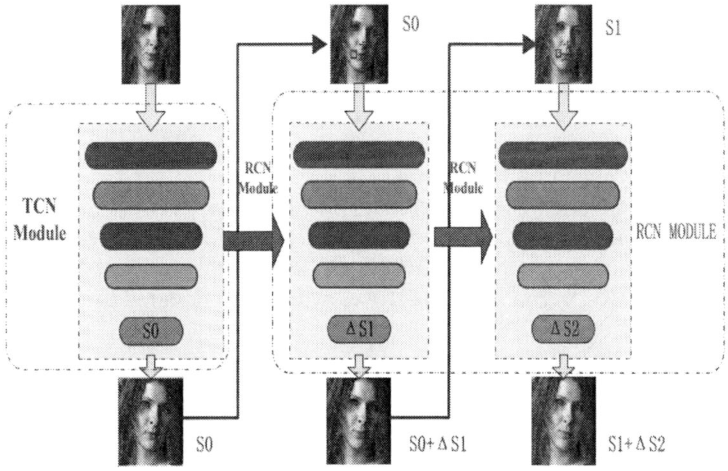

Figure 1. Overview of our Tweaked Residual Convolutional Network (TRCN).

2. Related Work

Significant progress on facial key point detection has been achieved in recent years. Remarkably, regression based methods[20] [1] [4] [4] [1] [4] [1] significantly boost the generalization performance of face landmark detection, compared to the algorithm based on statistical models such as Active shape models[13] and Active appearance models[7] [13]. A regression based approach regresses landmark locations where features extracted from face images serve as regressors directly. Landmark models are learned either in an independent manner, or in a joint fashion [21] [4].

Learning cascade-like regression models show superior performance on the face alignment task [20] [17] [21] [21]. Supervised descent method [20] learns cascades of regression models based on SIFT feature. Sun et al. [17] propose to use independent Convolutional Neural Networks (CNN) to perform coarse-to-fine shape searching. Zhang et al. [23] propose a novel approach that involves incorporating auxiliary information into the fitting process. Unlike other related methods, they do not incorporate a cascade of networks but instead frame the problem as a multi-task learning problem. Wu et al. [19] use a deep belief network to train a more flexible shape model, but do not learn any convolutional features. Finally, Baoguang Shi et al. [16] propose to jointly learn a cascade of linear regressors. Although the regressors are updated jointly via back-propagation, [16] uses linear regressors and employs hand-crafted HoG features rather than learning the features directly from the images. Zhang et al. [21] also utilise a coarse-to-fine shape search using first a global and then a set of local stacked autoencoders, However, each local autoencoder is trained with extracted shape-indexed feature (i.e,SIFT.), which seems sub-optical. As shown in the Fig. 2, which may be not accurate enough in some special conditions.

On the contrary, in this paper, we show that it is possible to obtain large improvements when, instead of utilising hand-crafted features, optimal features for the given problem are learnt with cascaded convolutional neural networks architecture.

Figure 2. Facial landmark detection under the partial occlusion, expression and pose scenario (from MTFL test set): Results of CFAN (top row) and our TRCN (bottom row).

3. Tweaked Residual Convolutional Networks

In this section, we present a novel Tweaked Residual Convolutional Network (TRCN) method for real-time facial landmark detection. Firstly, we introduce the overview of the proposed framework. Secondly, we describe the details about two components of TRCN, Tweaked Convolutional Network (TCN) module and Residual Convolutional Net (RCN) module. Finally we will give a detailed discussion about the difference from some existing works.

3.1. Method Overview

As shown in Fig. 1, the proposed TRCN designed the general regression framework by a two-level architecture, with the regression in each module modeled as a nonlinear deep network. Specifically, each module characterizes the nonlinear mapping from face image to face shape based on the shape predicted from the previous module. It is possible to obtain better improvements when optimal features for the given problem are learnt via deep convolutional network.

The first TCN module endeavors to approximate the facial landmark locations roughly, therefore, a low-resolution image is exploited for a large search step. The large step can alleviate the suffering from local minima and promise a fast model. Moreover, rather than local miniama feature from mean shape, the global image feature is employed as input to avoid the inaccuracy of mean shape. In order to improve the accuracy and robustness of network prediction, we redesign global feature extraction module.

General convolutional network use a large convolutional kernel (eg., 5*5, 7*7,.) to extract the feature of the image at the initial layer of the network, and capture higher abstract feature as deeper layer, their spatial concentration is expected to decrease suggesting that the ratio of 5×5 and 3×3 convolutional kernel should increase as we move to higher layers. However, it will lose more detail feature of image when using large convolutional kernel size because of a low resolution input image, and decrease the accuracy of predicted landmark. As the result, we redesign the convolutional network with smaller kernel size (e.g., 3*3, 2*2, etc) and adjust the deepth of the network, Tweaked Convolutional Networks approach ground truth of facial landmarks more accurately and robustly.

After getting an estimation face shape S0 from the TCN module, the RCN makes an effort to refine the shape by regressing the residual error ΔS between the current locations and the ground truth locations step by step. The nonlinear regression model RCN is exploited to model the nonlinearity between the current feature and the ground truth shape. Meanwhile, to characterize fine variations, generally speaking, the shape-indexed feature extracted from current local shape at different scale is exploited to enforce different search region. Furthermore, the feature-extracted of all facial points are concatenated together to enforce all facial points updated jointly so as to insure a reasonable solution, even under partial occlusion scenario or large pose.

3.2. Global TCN

The TCN module, the first module of the proposed two-level framework, directly estimates the face shape based on global raw features at a low-resolution image. Given a face image $I \in R^d$ of d pixels, $S_g \in R^p$ denotes the ground truth locations of p landmarks. The face landmark detection is to learn a mapping function T from the image to the face shape as follows:

$$T : S \leftarrow I \tag{1}$$

Generally, T is complex and nonlinear. To address this problem, k single convolutional layer is stacked as a deep neural network to map the image to the corresponding shape. Consequently, the face alignment task is formulated as minimizing the follow objective:

$$T^* = \arg\min_T \left\| S_g(I) - t_k\left(t_{k-1}\left(\ldots t_1(I)\right)\right) \right\|_2^2, \tag{2}$$

$$t_i(a_{i-1}) = \sigma\left(W_i a_{i-1} + b_i\right) \triangleq a_i, i = 1, \ldots, k-1, \tag{3}$$

$$t_k\left(a_{k-1}\right) = W_k a_{k-1} + b_k = S_0. \tag{4}$$

Where T = {t_1, t_2... t_k}, t_i is the mapping function of ith layer in the deep network, σ is a PReLU function and a_i is the feature representations of each layer. Nonlinear mapping in term of PReLU function is employed by the first k-1 layers to characterize the nonlinearity between the image feature and the face shape. Therefore, linear regression is exploited in the last layer t_k to get an accurate shape estimation S_0.

To prevent over-fitting, a regularization term $\sum_{i=1}^{k} \|W_i\|_F^2$ (a weights decay term) is added which tend to decrease the magnitude of the weights. The objective function is further re-formulated as bellow

$$T^* = \arg\min_T \left\| S_g(I) - t_k\left(t_{k-1}\left(\ldots t_1(I)\right)\right) \right\|_2^2 + a\sum_{i=1}^{k} \|W_i\|_F^2. \tag{5}$$

The function T contains lots of parameters and it is easy to fall into local minimum during optimization. To achieve a better optimization, firstly, we adopt the pre-train process to initialize the TCN module with part of train data, and then fine tune it with all data. As a result, the first few layers of convolutional network tend to capture the low-level feature such as texture patterns in an image, while the higher layers tend to capture abstract features containing context information of texture patterns.

After the optimization, the prediction of the facial landmarks is achieved as S_0, which is a rough but robust and fast approximation of the ground truth.

3.3. RCN Module

The global TCN described above gives a rough shape estimation S_0 of input image I, which is already close to the ground truth locations but not close enough due to the highly complicated variations in expression, pose, identity, etc. To achieve finer location, several RCNs are employed to iteratively predict the residual error ΔS_j between current shape S_{j-1} and ground truth S_g based on joint local shape-indexed features, referred as local RCNs.

Shape-indexed features extracted around the landmark points have been proven to be efficient for face alignment [4] [20] [4] [20] [20], which can only capture the information from itself, meanwhile, ignore the relevance with the other points. Most local feature is based on hand-crafted (e.g., SIFT, HOGS, LBF, etc.), as we know, it is data-driven and thus sub-optical for the face alignment task. Therefore, the facial points are modeled jointly with CNN in our RCN module, extracted-features are fused by RCN as output.

Single RCN maintains many of internal neural units that accumulates information extracted from the history of all past observations of the input space, this has the advantage that descent directions are naturally partitioned according to the previously calculated descent directions, so it can leverage this rich information and train an end-to-end face alignment method by learning a set of data driven features with Convolutional Neural Network directly from the images in a cascaded manner. Most importantly, RCN module uses residual error to impose a constraint on the descent direction.

Similarly as the global TCN, the successive RCN is also designed as a TCN to deal with the nonlinearity of predicting the face shape, but with the extracted local patches as input. With the estimated shape S_0 from global TCN, the CNN feature can be extracted around each facial point, denoted as Φ (S_0). The objective of the first RCN is to achieve a nonlinear regression R1 from the extracted feature Φ (S_0) to the residual error $\Delta S_1 = S_g - S_0$ as follow:

$$R_1^* = \arg\min_{R_1} \left\| \Delta S_1(I) - r_k^1\left(r_{k-1}^1\left(...r_1^1\left(\Phi\left(S_0\right)\right)\right)\right)\right\|_2^2 + a\sum_{i=1}^{k}\left\|W_i^1\right\|_R^2, \quad (6)$$

Where $R_1 = \{r_1^1, r_2^l, r_3^l, ..., r_k^l\}$, $\sum_{i=1}^{k}\left\|W_i^1\right\|_F^2$ is the weight decay term. Similar to the global TCN, the RCN is firstly initialized by using the part of train data, and then fine-tuned with all data.

After getting the face shape update (ΔS_1) by the first RCN, an updated face shape can be obtained $S_1 = S_0 + \Delta S_1$. then the successive RCN extracts local features around the new shape, and optimizes a deep network to minimize the new residual error between the current location and the ground truth. The objective of the jth RCN is shown as follow:

$$R_j^* = \arg\min_{R_j} \left\| \Delta S_j(I) - r_k^j\left(r_{k-1}^j\left(...r_1^j\left(\Phi\left(S_{j-1}\right)\right)\right)\right)\right\|_2^2 + a\sum_{i=1}^{k}\left\|W_i^j\right\|_F^2. \quad (7)$$

For each RCN, local features around the landmark points are extracted in a local patch of the same size but at different scale. Local patches of the different scale at low-resolution face image contain different context information and thus lead to different searching regions for each RCN. It is necessary for the anterior RCNs to approximate with a large search step when the current location is relatively far from the ground truth. On the other hand, the local patches of the same size but at smaller scale actually constrain the searching space within smaller region which means that the posterior RCN can refine the location with a tiny searching region leading to more accurate results.

4. Discussions

4.1. Differences with DCNN [17] and CFAN [21]

Both DCNN [17] and CFAN [21] follow the cascade framework and use a global nonlinear regression as the first stage to achieve a rough estimation of face shape. The differences lies on two aspects: in DCNN, after the global estimation, each facial point is refined with single CNN, which may be complicated and distort the whole shape without the constraint between facial points. While in CFAN, all facial points are refined jointly with SIFT feature, which seems suboptimal. On the contrary, the advantage of extracting image feature with CNN has been proved in the field of computer vision. Our TRCN extracts shape-indexed feature with CNN and refines jointly all facial points.

4.2. Differences with SDM [20]

A sequence of generic descent directions are learned by several successive RCN as well as SDM [20], but they differ in the following aspects: 1) SDM employs linear regression to model the mapping from shape-indexed features to a face shape, while our TRCN employs nonlinear regression, i.e., deep convolutional networks, to model the mapping from shape-indexed feature to face shape, which can achieve lower regression error. 2) SDM employs the mean shape as the initialization of the shape-indexed feature, which may be trapped when the initialization is far away from the ground truth, especially under the linear model. On the contrary, our TCN designs a deep convolutional network to directly predict a rough estimation of the face shape from the global image feature rather than shape-indexed feature, and it can obtain a more accurate initialization of the shape for the following RCNs.

5. Implemention Details

5.1. Data Augmentations

To train a robust global TCN module, we augment the training data by perturbing each training sample with random changes in translation, rotation and scaling. This can effectively prevent the deep models from over-fitting and achieve robustness to various changes in the wild data.

5.2. Parameters Setting

The global TCN has seven layers with four convolutional layers and three max-pooling layers after each convolutional layer, and then followed by two linear regression layer that is capable of learning non-linear mapping from a gray-scale full face with 48 * 48 pixels to a face shape. Numbers of convolutional output in each layer are respectively 32, 32, 64, and 128. For RCNs, have five layers with three convolutional layers and two max-pooling layers after them, numbers of convolutional output in each layer of RCN are respectively 28, 48, 64 with 24*24 pixels. All training samples are pre-processed by zero mean normalization. The weight decay parameter α controls the relative importance of the two modules, the average sum-of-squares error term and the weight decay term. Although α can be set different, the same value $\alpha = 0.0005$ is used for both global TCN and RCNs for simplicity.

6. Experiments

In this section, firstly we illustrate the experimental settings for the evaluations including the datasets and contrastive methods. Secondly, we investigate the alignment results of each module in our method. Finally, we compare the proposed TRCN with the other state-of-the-art methods.

6.1. Datasets and Methods for Comparison

To evaluate the effectiveness of the proposed TRCN architecture, four public datasets are used for our experiments, i.e., MTFL [22], BioID [9], COFW [3] and LFPW [2]. BioID dateset [9] is collected under laboratory conditions, while the images in MTFL [22], COFW [3] and LFPW [2] datesets are collected in the wild environment formulating a more challenging scenario. In all experience, our TRCN is only trained with the images from MTFL training set.

We compared our framework with a few state-of-the-art methods, i.e., DRMF [1], SDM [20], DCNN [17], CFAN [21] and TCDCN [22]. The available codes from [1] predict 66 facial points, and the released code of [20] only estimates 49 landmarks located in the inner regions of the face. Both of the released codes from [17] and [21] predict 5 landmarks, [22] predicts 68 facial landmarks. Since [22] does not detect eye centers, we only compare nose tip and mouth corners with them.

All these methods were recently proposed and reported state-of-the-art performance. The normalized root mean squared error (NRMSE) is employed to measure the between the estimated facial landmark locations and the ground truth. The NRMSE is normalized by the distance between centers of eyes. The cumulative distribution function (CDF) of NRMSE is applied for performance evaluation.

Table 1. Run time of each stage in terms of millisecond (ms)

	Global TCN	**RCN1**	**RCN2**	**total**
Run time (ms)	3.56	0.94	0.96	5.46

6.2. Investigation of Each Module in TRCN

As the proposed TRCN method consists of two module of TCN and RCN, we investigate how each module contributes to the performance improvement for facial alignment. The experiments are conducted on MTFL dataset [22] in terms of average detection accuracy of 5 facial points, i.e., NRMSE. The images from MTFL training set, are used for training and the test set are used for evaluation.

The evaluation results are shown in Fig. 3. As seen, the CDF of global TCN is 0.77 when NRMSE is 0.08. Therefore, if the initial estimated shape is directly used for estimating accurate facial shape, it is not accurate enough because of large pose, illustration, and so on. As the result, in the second stage, RCN module refine the landmarks accurately. In the first RCN, the detection accuracy is improved up to about 14% when NRMSE is 0.08 .Little improvement when NRMSE is 0.08, and it is only up to 0.05 in the second RCN. It demonstrates that the first TCN module mainly handle the large variations due to pose and expression and the latter ones precisely refine the landmark locations in a small search region.

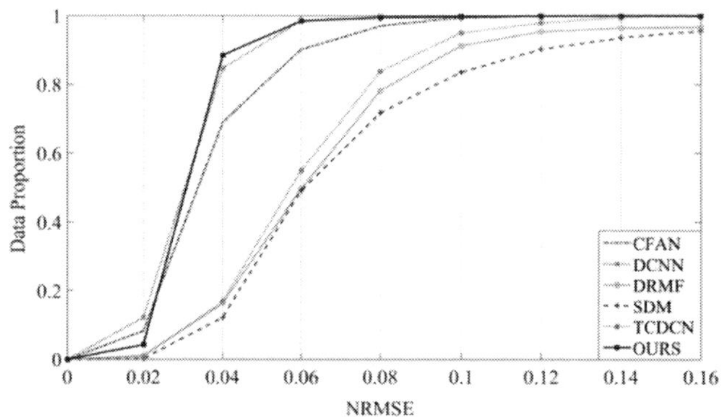

Figure 3. The cumulative error distribution curves of each module from MTFL

Besides the accuracy, another important factor of performance is the time complexity. We evaluate the run time of each module on MTFL testsets as shown in Table.1. The method is run on a desktop (Inter i7-4670 3.6 GHz CPU). To avoid the influence of random factors, the method is repeated several times, and the average of running time is reported.

6.3. Datasets

MTFL (Multi-Task Facial Landmark dataset) [22] has 13,466 face images, among which 5590 images are from LFW [11] and the remaining 7896 images are downloaded from the web. Each face is labeled with the positions of five key points. MTFL [22] consists of 10000 training images and 2935 test images with large variations in pose, expression, illumination, partial occlusion, etc, which makes the facial point detection quite challenging on this dataset. For our method, we only train our model with training set and evaluate each module on testset.

Bioid [9] has 1,521 images of 23 subjects, all the faces are frontal. The face detector detects faces of 1,477 images in Bioid. In this experiment, our TRCN is trained by using the images from MTFL training set. We compare with DRMF [1], SDM [20], DCNN [17] and CFAN [21] and TCDCN [22]. For a fair comparison, only the common images with face detected by all methods are employed for the testing.

The cumulative error distribution curves of these methods are shown in Fig. 4. As seen, compared to, our method reduces the detection errors significantly. Furthermore, we compare our TRCN with [21], [17] and [22] on this dataset in terms of five landmarks. As for [22], we only compare nose tip and mouth corners for a fair comparison since only the model of 68 landmarks is released. As seen, our TRCN outperforms [21] and achieve the same effect with [17].

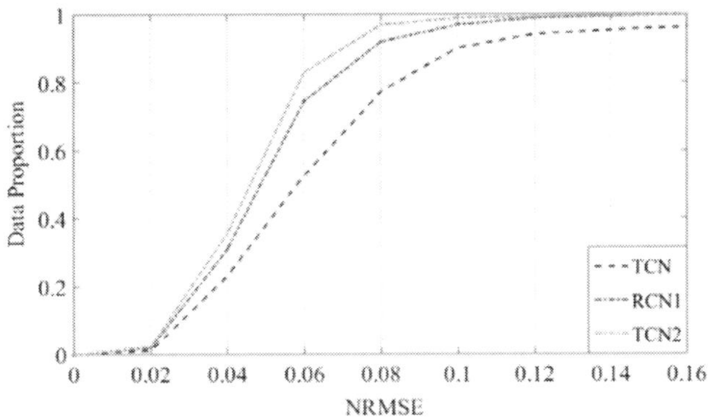

Figure 4. Comparison on BioID

LFPW [2] contains 1,432 face images from internet. Since some URLs of images are no longer valid, we adopt about 949 valid images for testing. All the testing images are out of our training sets. For DRMF [1] method, the tree-based face detector is used to achieve more accurate face detection. The performance of all methods are shown in Fig. 5. As seen, SDM [20] performs the best among the traditional methods. Similarly, we compare our TRCN with DCNN and CFAN in terms of five landmarks. Our approach again performs better than most methods.

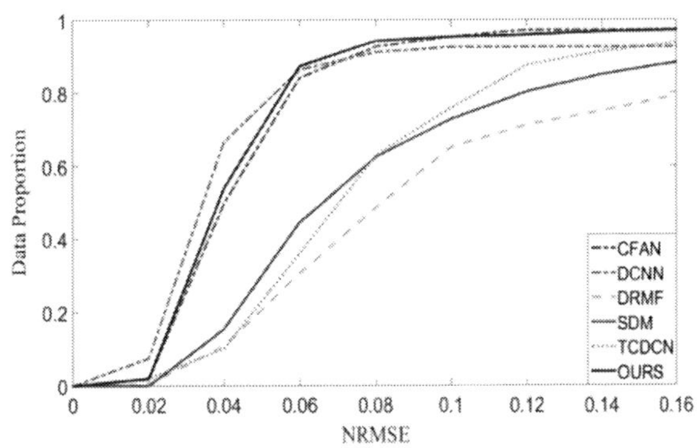

Figure 5. Comparison on LFPW

COFW (Caltech Occluded Faces in the Wild) database [3] has 1,007 images of faces obtained from a variety of sources. All images were hand annotated by with 29 landmarks. It is designed to present faces in severe occlusions due to pose, sunglasses, and interaction with objects (e.g., food, hands, and hat). The performance of all methods are shown in Fig. 6. As seen, our method shows better performance.

Fig. 7 shows the results of TRCN on some extremely challenging example faces from BioID, LFPW and COFW. It shows that our algorithm is robust to the variations from pose, expression, sunglasses and partial occlusion.

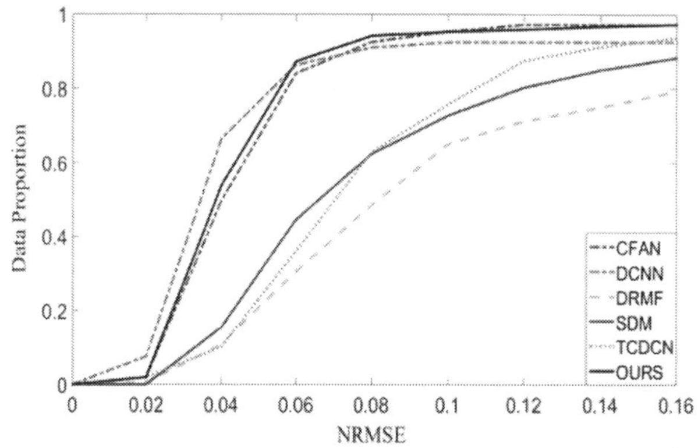

Figure 6. Comparison on COFW

Figure 7. Example results from BioID (top row), COFW (middle row) and LFPW (bottom row)

7. Conclusions and Future Work

To deal with the nonlinearity in inferring face shapes from face images, we proposed the Tweaked Residual Convolutional Net with a two-level architecture, each of which figures out part of the nonlinearity. The first TCN module takes a low-resolution version of the detected face as input directly, to estimate a roughly accurate shape globally. Then, the subsequent RCNs take the local patch at different scale as input to refine the shape better. Such a strategy achieves better results than the state-of-the-art methods, such as [17] and [17], on three databases with extensive variations. Furthermore, our method can work rather efficiently, with 70+ fps even with C++ codes on a common desktop with no parallel programming.

The proposed method provides a general framework, which can be further applied to other localization-sensitive tasks, such as pose estimation, object detection, etc. In the future, we plan to further exploit the proposed Tweaked Residual Convolutional Network for boarder impacts.

8. Acknowledgment

This work is supported by the National Natural Science Foundation of China (No.61202160) and the National Key Scientific Instrument and Equipment Development Projects of China (No. 2013YQ49087904).

9. References

[1] Asthana, A., Zafeiriou, S., Cheng, S., & Pantic, M. (2013). Robust Discriminative Response Map Fitting with Constrained Local Models. Computer Vision and Pattern Recognition (Vol.9, pp.3444-3451). IEEE.

[2] Belhumeur, P. N., Jacobs, D. W., Kriegman, D., & Kumar, N. (2011). Localizing parts of faces using a consensus of exemplars. IEEE Conference on Computer Vision and Pattern Recognition (Vol.35, pp.545-552). IEEE Computer Society.
[3] Burgos-Artizzu, X. P., Perona, P., & Dollar, P. (2013). Robust Face Landmark Estimation under Occlusion. IEEE International Conference on Computer Vision (pp.1513-1520). IEEE Computer Society.
[4] Cao, X., Wei, Y., Wen, F., & Sun, J. (2012). Face alignment by Explicit Shape Regression. IEEE Conference on Computer Vision and Pattern Recognition (Vol.107, pp.2887-2894). IEEE Computer Society.
[5] Cootes, T. F., Edwards, G. J., & Taylor, C. J. (2001). Active appearance models. IEEE Transactions on Pattern Analysis & Machine Intelligence, 23(6), 681-685.
[6] Cristinacce, D., & Cootes, T. F. (2006). Feature Detection and Tracking with Constrained Local Models. British Machine Vision Conference 2006, Edinburgh, Uk, September (Vol.41, pp.929-938). DBLP.
[7] Gao, X., Su, Y., Li, X., & Tao, D. (2010). A review of active appearance models. IEEE TRANSACTIONS ON SYSTEMS MAN AND CYBERNETICS PART C-APPLICATIONS AND REVIEWS, 40(2), 145-158.
[8] Gu, L., & Kanade, T. (2008). A Generative Shape Regularization Model for Robust Face Alignment. European Conference on Computer Vision (Vol.5302, pp.413-426). Springer-Verlag.
[9] Jesorsky, O., Kirchberg, K. J., & Frischholz, R. W. (2001). Robust Face Detection Using the Hausdorff Distance. International Conference on Audio- and Video-Based Biometric Person Authentication (Vol.2091, pp.90-95). Springer-Verlag.
[10] Le, V., Brandt, J., Lin, Z., Bourdev, L., & Huang, T. S. (2012). Interactive Facial Feature Localization. European Conference on Computer Vision (Vol.7574, pp.679-692). Springer-Verlag.
[11] Learned-Miller, G.B.H.E.: Labeled faces in the wild: Updates and new reporting procedures. Tech. Rep. UM-CS-2014-003, University of Massachusetts, Amherst (2014)
[12] Liu, X. (2008). Discriminative face alignment. IEEE Transactions on Pattern Analysis & Machine Intelligence, 31(11), 1941.
[13] Milborrow, S., & Nicolls, F. (2008). Locating Facial Features with an Extended Active Shape Model. Computer Vision – ECCV 2008. DBLP.
[14] Sagonas, C., Tzimiropoulos, G., Zafeiriou, S., & Pantic, M. (2013). A Semi-automatic Methodology for Facial Landmark Annotation. IEEE Conference on Computer Vision and Pattern Recognition Workshops (Vol.13, pp.896-903). IEEE.
[15] Saragih, J. M., Lucey, S., & Cohn, J. F. (2009). Face alignment through subspace constrained mean-shifts. IEEE, International Conference on Computer Vision (Vol.30, pp.1034-1041). IEEE.
[16] Shi, B., Bai, X., Liu, W., & Wang, J. (2014). Deep regression for face alignment. Computer Science.
[17] Sun, Y., Wang, X., & Tang, X. (2013). Deep Convolutional Network Cascade for Facial Point Detection. IEEE Conference on Computer Vision and Pattern Recognition (Vol.9, pp.3476-3483). IEEE Computer Society.
[18] Toshev, A., & Szegedy, C. (2014). DeepPose: Human Pose Estimation via Deep Neural Networks. Computer Vision and Pattern Recognition (pp.1653-1660). IEEE.
[19] Wu, Y., Wang, Z., & Ji, Q. (2013). Facial Feature Tracking Under Varying Facial Expressions and Face Poses Based on Restricted Boltzmann Machines. Computer Vision and Pattern Recognition (Vol.9, pp.3452-3459). IEEE.
[20] Xiong, X., & Torre, F. D. L. (2013). Supervised Descent Method and Its Applications to Face Alignment. Computer Vision and Pattern Recognition (Vol.9, pp.532-539). IEEE.
[21] Zhang, J., Shan, S., Kan, M., & Chen, X. (2014). Coarse-to-Fine Auto-Encoder Networks (CFAN) for Real-Time Face Alignment. Computer Vision – ECCV 2014. Springer International Publishing.

IOP Conf. Series: Journal of Physics: Conf. Series **887** (2017) 012077 doi:10.1088/1742-6596/887/1/012077

[22] Zhang, Z., Luo, P., Chen, C. L., & Tang, X. (2014). Facial Landmark Detection by Deep Multi-task Learning. European Conference on Computer Vision (Vol.8694, pp.94-108).
[23] Zhang, Z., Luo, P., Loy, C. C., & Tang, X. (2015). Learning deep representation for face alignment with auxiliary attributes. IEEE Transactions on Pattern Analysis & Machine Intelligence, 38(5), 918-930

IOP Conf. Series: Journal of Physics: Conf. Series **887** (2017) 012078 doi:10.1088/1742-6596/887/1/012078

Intelligent detection and removing of straight periodic streaks on digital images

Zehao Zhao[1], Hui Sheng[2] and Xianfeng Xu[3]

[1]Princeton International School of Mathematics and Science, Princeton NJ 08540 USA
[2]College of Geosciences, China University of Petroleum (East China), Qingdao, China
e-mail:hui_sh@163.com
[3]College of science, China University of Petroleum (East China), Qingdao, China

Abstract. Straight periodic steaks are common negative errors on images, an intelligent method is proposed to test and remove the periodic streaks on digital images. Because of many artificial errors or negative experiment factors, digital images are always contaminated by straight periodic streaks, which will correspondently incur error information on them. By Fast Two Dimensional Fourier Transform, the space spectra from these periodic streaks are discrete and can be test intelligently. After eliminate these frequency ingredient, the periodic streaks disappear after inverse Fourier Transform. The numerical simulation experiment shows that this method is robust in detesting and removing the periodic streaks on digital figures.

1. Introduction

During the recording process of optical pictures in fine optical lab, many environmental and man made mistakes always cause unwanted steaks on the result figures, especially for the laser source optical configuration with many optical devices. These fringes and steaks [1-3] are mostly the results of coherent interference between laser waves, which were reflected from different surfaces of various optical devices. These contaminated intensities can degrade the image quality and the measure precision in practice. Usually these additional figure intensities are removed by space filtering in 4f optical system for optical pictures or in computer for digital configures, where the space filtering method is more prosperous for digital figures because of the fast development of computing technology. Mostly, the space filtering is operated through two-dimensional Fourier Transform. Researcher ZHANG et al. [3] adjusted the noise effects on OCT images in Fourier domain. JIANPING LI AND YI XIAO [5] improved the fast Fourier Transform processing efficiency. MANDEEP SINGH AND KEDAR KHARE [6] used the Fourier Transform method to estimate the carrier parameters in digital holography.

However, it is a difficult task to test the steaks intelligently and accurately so that they can deleted completely. Here we provided an automatic a method to detect the target fringes on digital figures and remove it with high precision. The principle is discussed with the reasonable formula deduction first, then the numerical simulations is to show the working effect of our method, and at last the conclusion on the method is drawn.

2. Principles

In information optics research area, space frequency filtering methods are used to some negative information on images. To complete such work, Two Dimensional Fourier Transform (TDFT) is necessary for frequency component separation. For digital images, the TDFT process can be finished in computer automatically. Here the image with errors of straight steaks is expressed as IE, which is the summation of the original image intensity distribution Io and the streaks intensity distribution is. Then we have

$$I_E = I_O + I_S \tag{1}$$

After the TDFT operation of Eq. (1), the space frequency spectrum of I_E can be expressed as

$$\mathrm{FT}[I_E] = FT[I_O] + FT[I_S] \tag{2}$$

Where symbol 'FT [ ]'represents the operation of Fourier Transform. To express the frequency spectrum function, Eq. (2) is rewritten in the form of

$$F_E(f_x, f_y) = F_O(f_x, f_y) + F_S(f_x, f_y) \tag{3}$$

Here fx and fy are space frequency coordinates along the horizontal and perpendicular direction on space frequency plane respectively. The two items of the right side of Eq. (2) are space spectrum distributions of the original image and streaks respectively. For most conditions, they are located on the different areas on the frequency plane. On one hand, the distribution of the first term of the right side of Eq. (2) is almost random with small spectrum value expect the absolute value of the zero spectrum. On the other hand, the latter one are commonly characterized as one delta function distribution or the combination of a series of delta functions, which are one or many small bright points on one straight line on the TDFT plane. Obviously, it is easy to find these points by our observation. In fact, these points are located by human eyes traditionally and then these spectrums from straight streaks are removed from TDFT spectrum.

$$F_O(f_x, f_y) = F_E(f_x, f_y) + F_S(f_x, f_y) \tag{4}$$

After Inverse Two Dimensional Fourier Transform (ITDFT) of Eq. (5), the original image without streaks is achieved.

$$I_O = IFT[F_O(f_x, f_y)] \tag{5}$$

Where 'IFT [ ]' means ITDFT.

However, the current frequency filtering method needs human work to find the location of the corresponding spectrum of streaks on TDFT plane, which is inconvenient. Here we proposed an automatic detection method to found the streaks frequency for digital pictures in computer. Because the frequency spectrums of periodic streaks are delta functions, their absolute values are corresponding maximums in the space frequency data matrix. To search for these points, the method of maximum finding algorithm can be used automatically in computer. Unfortunately the absolute value of the zero frequency is commonly the biggest one on frequency plane. So before searching we should cross out the value of this point. The whole process of our method is illuminated step by step as the following.

Step 1: Read the target picture and carrying out the operation of TDFT on the picture with straight streaks by Eq. (2) and record the complex matrix of $F_E(f_x, f_y)$ in Eq. (3).

Step 2: Calculate the absolute value of the data matrix of abs $[F_E(f_x, f_y)]$ and keep it in computer memory.

Step 3: Replace the absolute value of zero frequency F_E (0, 0) with the value of zero. Then search the new data matrix for maximum value and keep down its coordinates(f_{x1}, f_{y1}), which is pixel number in data matrix. Replace this maximum value with zero again, find the next maximum value, and recorded the corresponding pixel number as(f_{x2}, f_{y2}),. Repeat this operation until all the maximum frequency points are found and their coordinates are recorded.

Step 4: Rewrite all the values of the recorded coordinates in complex data matrix FE (fx, fy) with zeros. The function $F_O(f_x, f_y)$ in Eq. (4) is achieved.

Step 5: ITDFT operation in Eq. (5) is carried out to recover the picture without the straight streaks. The five steps mentioned above can be realized by computer programs and this algorithm is convenient for us to complete the digital frequency filtering automatically, which is useful and welcome by engineers in real application practice.

3. Numerical Simulation

To verify the availability of the algorithm suggested here, many computer simulations are introduced. A typical example using this method is carried out and the corresponding results are discussed latterly.

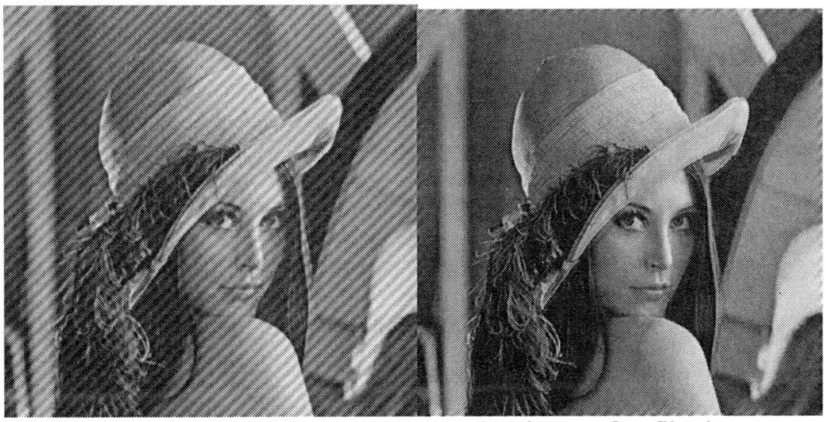

(a) Picture with harmonic streaks, (b) Picture after filtering

Figure 1. Pictures with and without harmonic streaks

Fig. 1(a) is a commonly used picture added with periodic straight streaks for testing. Pixel numbers of Fig. 1(a) on both horizontal and perpendicular directions are 512 pixels respectively. The intensity distribution of streaks in it is described as

$$I_S = 0.1\{\sin[0.4(x + y)] + 1\} \tag{6}$$

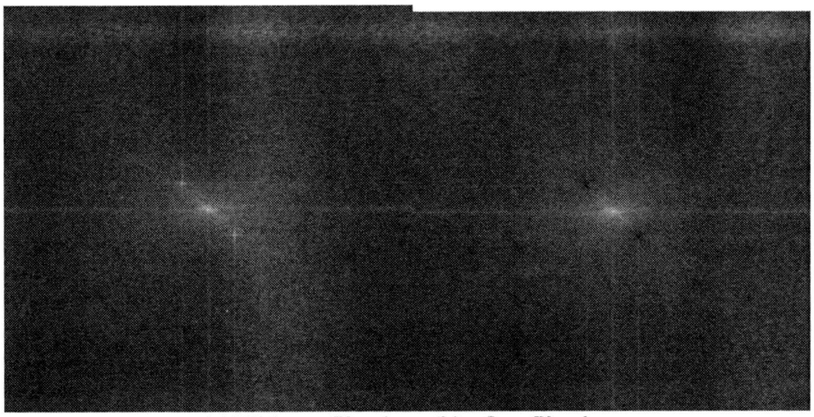

(a)Without filtering, (b) after filtering

Figure 2. Frequency Distribution on fourire transform plane

The distribution of the absolute value of streaks spectrum is shown in Fig. 2 (a), where it is easy to see the three bright points on spectrum plane. One point locates at (0, 0), the center of the spectrum plane and the other two points distribute symmetrically about the point (0, 0). By using the method suggested here, the coordinates of two streaks spectrum are found intelligently and they are (224,224) and (290,290) on spectrum plane respectively. The three absolute values are 2.6175×10^5, 7.4902×10^3, and 7.4902×10^3 respectively. Obviously, the point representing the zero frequency is much bigger than the absolute value of the streak frequency. In Fig. 2 (b), the two periodic streaks frequency are removed by zero replacement of step 4 in this method. By the operation of ITDFT in step 5, the picture without streaks is achieved in Fig 1. (b). obviously, there are very little difference between the original figure and Fig 1. (b). the results show that the method here is effective in detecting and removing the

IOP Conf. Series: Journal of Physics: Conf. Series **887** (2017) 012078 doi:10.1088/1742-6596/887/1/012078

figure streaks process.

4. Conclusion

On conclusion, an intelligent periodic streaks removing method based on the digital Fourier Transform filtering theory is proposed in this paper. This method can search and remove the streaks frequency spectrum on frequency plane automatically. The whole process of removing is completed by computer program. We hope that this technique has prosperous future in the application practice of image processing.

5. Reference

[1] J. W. Goodman, Introduction to Fourier Optics, 2nd edn, American Scientist, 1997(6):581-582

[2] Jian Li, Gang Dong Email author, Jianlei Zhang, Numerical study on evolution of subharmonic varicose low-speed streaks in turbulent channel flow, Applied Mathematics and Mechanics, 2016, 37,(3):325–340.

[3] Chernoray V. G,Kozlov V. V. , Löfdahl L. Chun H. H. Visualization of sinusoidal and varicose instabilities of streaks in a boundary layer, Journal of Visualization, 2006, 9(4): 437–444

[4] ANQI ZHANG, JIEFENG XI, JITAO SUN, AND XINGDE LI, Pixel-based speckle adjustment for noise reduction in Fourier-domain OCT images, BIOMEDICAL OPTICS EXPRESS, 2017, 8(3):1721-1730|.

[5] JIANPING LI AND YI XIAO, GPU accelerated parallel FFT processing for Fourier transform hyperspectral imaging, Applied Optics, 2015 , 54(13):D91-D98.

[6] MANDEEP SINGH AND KEDAR KHARE, Accurate efficient carrier estimation for single-shot digital holographic imaging, Optics Letters, 2016, 41(21): 4871-4874.

Multi-task convolution network for face alignment

Yang Sun[1], Xuan Zhang[2] and Chongrong Li[3]

[1]Institute for Network Sciences and Cyberspace, Tsinghua University, Beijing, China, Sunyang15@mails.tsinghua.edu.cn
[2]Institute for Network Sciences and Cyberspace, Tsinghua University, Beijing, China, zhangx@cernet.edu.cn
[3]Institute for Network Sciences and Cyberspace, Tsinghua University, Beijing, China, licr@cernet.edu.cn

Abstract. In order to solve face alignment more effectively, we proposed a multi-task deep convolution network for face alignment, which achieves good performance even in the case of large pose variations and severe occlusion. Instead of dealing with face alignment as a single task, we jointly trained the auxiliary task of pose estimation together with face alignment to guide the distribution of facial points. By doing so we are able to 1) avoid trapping in the local optimum due to the inaccurate face boxes, 2) improve the robustness in dealing with faces with pose variation and severe occlusion. Compared with the traditional methods, our method also improves the accuracy by providing better initialization instead of mean shape. Extensive experiments show that our method has great performance on various benchmarks.

1. Introduction

Face alignment [2,3,4,5,22] aims to locate facial key points automatically, which plays an important role in many facial tasks, such as face recognition [6], face attributes inference [7]. Although scholars have put forward many methods in the past decades, face alignment is still a challenging problem especially in the case of heavy occlusion and large pose variations.

However, both template fitting approaches [1] and regression-based methods [2] take face alignment as an independent problem and ignore the inherent relationship between face alignment and other auxiliary tasks. Researches have proved the effectiveness of multi-task learning. It's meaningful to explore the inherent correlation between face alignment and pose estimation (the angles of roll, yaw and pitch of head). Reference [20] takes the age estimation and gender classification as auxiliary tasks, but we should point out that the most helpful information is encoded in poses which related to the distribution facial points.

In this paper, we proposed a new framework to integrate these two tasks by using deep convolution network. We built a shared feature extraction model with two branches on the top to predict face landmarks and pose angles. Fig 1. Shows the results of our method in the case of large pose and occlusion. Thanks to the multi-task training, the deep multi-task convolution network gets the state-of-the-art performance.

Figure 1. Examples of landmark detection from challenging face images

Furthermore, we also evaluated the feasibility of using our network's output as the initial input in the traditional cascaded regression methods. As known to all, an obvious shortcoming of traditional cascaded regression method is the heavily dependence on initialization [8]. Particularly, it easily trapped into local optimum when the real shape is far from the mean shape. Besides this, the occlusion confuses the regression to estimate the next shape. To solve this problem, we used our network's output as the robust initialization. The landmark is closer to the ground truth shape by several iterates and can avoid being trapped into local optimum.

The following are our contributions: We proposed a multi-task convolution network for face alignment task which could works well even in challenging test datasets. Besides this, we took the deep convolution network's output as initialization of other methods to improve the regression results.

2. Related Work

Many face alignment methods have been proposed in recent years. The Regression-based methods and deep-learning-based methods are the most popular methods. The regression-based methods regress the face landmark explicitly by learning the features in the image, e.g. [18] use random forest [9] to achieve good performance with local binary features, [2] regresses features with random ferns. Convolution networks have also been used in face alignment task successfully, [19] detects face landmarks with three level networks.

Multi-task learning [23] has been used in many computer vision tasks successfully. Reference [20] proposes the method of TCDCN which first introduced multi-task learning into facial landmark detection. But TCDCN only works for five points with the auxiliary task of head pose estimation, gender classification and age estimation. Some tasks are not closely related to face alignment, in other words, the age doesn't have much relationship with where the facial points located in. Besides this, the most related task pose estimation only output right, left, or frontal but not the real angle. Focus on these problems, we removed the irrelevant tasks, and describe the pose by using three float numbers, which represent roll, yaw and pitch of a head to build our network.

In addition, many studies are concentrated on the initial problem. Reference [2] proposed to use different initializations and take the mean results as the final pose. Reference [17] solved the initialization problem by predicting the rough estimation from global image patch. These methods solve the problem by some assumptions which alleviate the problem but not solve the problem. We used our method to improve the robustness of initialization.

3. The Proposed Approach

3.1. Multi-Task CNN

In the field of machine learning, there have been some principles to guide the multi-task formulation. Now suppose we want to train T tasks together, and there are N samples to be trained, the training set can be represented as $\{x_i, y_i\}_{i=1}^{N}$, where x_i is the i-th image and y_i is the corresponding labels. The purpose of the multi-task network is to minimize:

$$\arg\min_{w} \sum_{t=1}^{T} \sum_{i=1}^{N} l(f(x_i^t, w), y_i^t) + \Phi(w) \tag{1}$$

where $f(.)$ is the function to compute the output using the parameter w. And the $l(.)$ is the loss function which represents the difference between the output and the real label. In this task we choice the Euclidean Loss for each sample:

$$l_i^t = \| y_i^t - f(x_i^t, w) \|_2^2 \tag{2}$$

But these two tasks have different levels of importance since our main goal is to optimize the face alignment with the help of auxiliary task. As a result, we use different coefficient to represent the level of importance. In our method, we share the same input for the task of landmark detection and pose estimation. We denoted the inputs as $\{x_i\}_{i=1}^{N}$ and their responding labels as $\{y_i^l, y_i^p\}_{i=1}^{N}$, where y_i^l is the real landmark and y_i^p is the pose angle. $y_i^l \in R^{136}$ represent 68 landmarks and $y_i^p \in R^3$ represent the yaw, roll and pitch angle of head. The problem is converted to following formula:

$$\arg\min_{w} \sum_{t=1}^{2} \sum_{i=1}^{N} \lambda^t l^t (f(x_i, w), y_i^t) + \Phi(w) \tag{3}$$

Our network use five convolution layers and four full-connected layers to predict the landmarks and pose. And the last full-connected layers output the results. Compared with the cascaded networks such as TCDCN, our model is deeper which could achieves robust performance.

3.2. Our Network Structure

In [20], multi-task CNN has been designed for face alignment and other relevant tasks. However, we find that maybe the results are limited by the following aspects :(1) There are so many irrelevant tasks such as gender estimate which may not help the accuracy improving. (2) The network is not deep enough so that couldn't capture sufficient information about the landmark location. (3) The input is only 40*40 gray-scale face image, and it is too small to get the state-of-the-art performance when dealing with 68 landmarks detection. What's more, the gray-scale image may loss some information, so we use the color face image as the network input.

Fig 2. Shows the structure of our network. Our network contains two part. The first part is a shared feature extraction model, which contains five convolutional layers, three polling layers, four fully connected layers and one dropout layer to avoid over-fit. The second part outputs the landmarks position and pose angles using the same feature vectors. In back propagation, the pose information is encoded into the network. The network's input is 224*224 BGR face image detected from the open library dlib. We resize the face image into 224*224 and then minus the mean value. We use the Rectified Linear Units (ReLU) as the activation function. The pooling layer we all use the max-pooling. The landmark location layer and pose direction layer are connected to one shared layer to predict the results, so they can share most of the computes to speed up. Compared with TCDCN, our network has more filters and deeper structures to capture more features to achieve better performance.

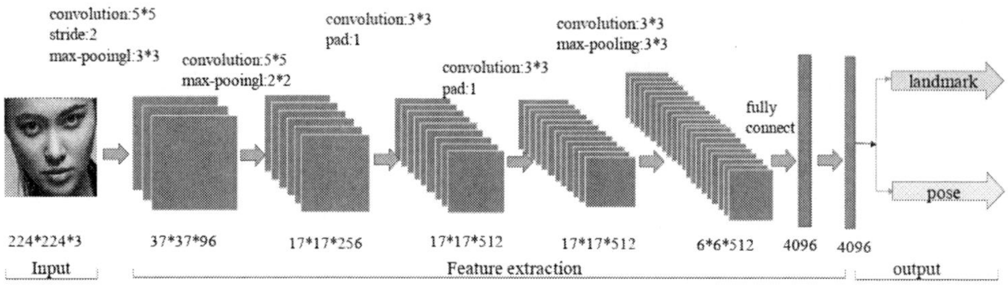

Figure 2. Structures of our Multi-Task Convolution Network

3.3. Training Matters

Our model is trained with the LFPW [12], Helen [13], AFW [24] and 300W [11] dataset. These datasets contain train and test images with 68 landmarks. In order to make the net more robust we enlarged the datasets to 22000 with data augmentation. In the training phase, we used the dlib to detect the face boxes, and transform the points to [-1 1]. Then in test phase, we need to translate the [-1 1] to the real point with the information of boxes. The pose angles are labeled with the software of intraface which can use the point's information to generate 3D head pose. In training phase, base learning rate is set to 0.001 and weight decay is 0.0005, the max iterations is 900000. The tasks of face alignment and pose direction share the same weights during the phase of feature extraction.

3.4. Experiment and Result

In our experiment, we test the performance of our method using the mean error which is measured by the distances between ground truth shape and estimated shape. Then the result was normalized by inter-ocular distance. The equation can be calculated as follows:

$$e_i = \frac{||y - S||}{n * D_i} \tag{4}$$

where D_i is the inter-ocular distance determined by the center of the two eyes, n is the number of landmarks, y is the output shape and S is the ground truth shape.

3.4.1. The Effectiveness of Related Task. Many works have proved that auxiliary tasks can help us improve the performance of main task. But there is no such theory to tell us which auxiliary task will help us or not. Because of this, we carefully choice the auxiliary task which is beneficial to landmark detection. Intuition tells us, compared with age or gender, the task of pose estimation is more relevant to face alignment. The reasons are as follows: Firstly, the pose angles can guide the landmarks' distribution, especially for the large pose variation. Secondly, it provides the constraint to avoid the landmark detection trapped into local optimum caused by the inaccurate of the face detection or large pose variation.

To evaluate the effectiveness of auxiliary task pose estimate, we tested the performance of two different nets (joint pose estimate learning and do not joint). The two networks have the similar constructs except the output layer of pose estimation. For simplicity, we call the two net as S-Net (Single net) and M-Net (Multi-task net) and evaluate them on the 300W dataset. The mean error of S-Net is 8.62 while M-Net is 7.39. This demonstrates that pose estimation task is beneficial to landmark detection.

3.4.2. Compare with CNN-Based Method. We use the CNN-based method TCDCN [20] as the baseline. Moreover, we also compared our method with the Cascade CNN which detect the landmark with three cascade networks. Thanks to the multi-task and deeper network, our method can achieve better performance than these above methods. These two works only detect five point (the center of eyes, nose and the corner of mouse) and the 68 landmarks don't conclude the points of center of eyes. We only compare the performance with the points of nose, left mouth corner and right mouth corner. TABLE I. shows the mean error calculated by (4), we can draw the conclusion that our method has better performance than the above two CNN-based method. Without much code optimization, our C++ implementation cost about 7ms for one image to pass the network on a NVDIA GTX970 GPU.

Table 1. Mean error of Different methods

Method	**Key Points**		
	Nose	***Left mouth corner***	***Right mouth corner***
Cascaded CNN	10.3	8.4	7.4
TCDCN	8.5	7.8	7.1
OURS	7.7	7.1	6.5

3.4.3. Evaluate With State-of-the-Art Methods. Besides the CNN-based methods, we also compared our method with some state-of-the-art methods on LFPW, Helen and 300W datasets. It's worth to point out that we use the dlib to detect the face boxes. The state-of-the-art methods includes DRMF [14], ESR [2], PCPR [15], SDM [16], CFAN [17], LBF [18] and CFSS [10]. TABLE II. Shows that our method has good qualities in different dataset.

Table 2. Evaluate with other Methods

Method	**TEST DATASET**		
	LFPW	***Helen***	***300W***
DRMF	6.57	6.70	9.22
ESR	--	--	7.58
PCPR	6.56	5.93	8.35
SDM	5.67	5.50	7.50
CFAN	5.44	5.53	--
LBF	--	--	6.32
CFSS	4.87	4.63	5.76
OURS	5.63	5.49	7.39

4. Good Initialation for Regresion

Besides the multi-task learning methods, regression-based methods are also very popular in the field of face alignment. Regression-based methods learn the regression function to approach the target shape.

4.1. Regression-Based Methods

Regression-based methods predict the landmark shape S from a given face image I, and we denote $[x_1, y_1 ... x_N, y_N]$ for the N landmarks, in this paper the N is 68. These methods consist of T iterations to update the initial shape to approach the ground truth shape:

$$2 \quad S^t = S^{t-1} + f(I, S^{t-1}) \tag{5}$$

where S^t is the shape after t iterations, and $f(.)$ is a function to calculate the Δs.

Although these methods work fast and accurate on frontal face, it usually fails to solve the faces with large pose well. The most important reason is these methods relied heavily on the initial shape. So faces with large pose are always trapped into local optimum when the initial shape is far away from the real landmarks. Many attempts were done to solve this problem and the most common method is use the mean shape as the initial shape. Burgos [25] proposed the smart initialization while the change between iterate is large then stop it and restart with another initial shape. Xiao [26] present progressive initialization with k-means of obvious points to handle challenging poses robustly. These methods solve the problem based on some assumptions which can mitigate the problem instead of solving it.

4.2. Initialize Using Our Method

To solve this problem, we used the result from our network as the initial shape of regression-based method. We implemented the ESR [2], and then compared the performance of these two initial methods using test images from 300W dataset.

Table 3. Compare of Two Networks

Table Head	**Initialization type**	
	Mean shape	***Our result shape***
Mean error	8.33	6.98

TABLE III. Demonstrates that the performance can be improved using our robust initialization. We find that this robust initialization has great influence especially on large pose faces and inaccurate detection boxes. For large pose faces, the mean shape usually far away from the ground truth shape, so it is easily to drop into local optimization during the regression. As for the inaccurate detection boxes, the mean shape is inaccurate too because it is calculated from the inaccurate boxes, so the performance is usually poor. In a word, using our method as initialization can give the regression a right start to obtain better result, which can avoid dropping into local optimize caused by the bad initialization. Fig 3 shows that our method is robust to the situations of inaccurate detection box, large pose variations and occlusion.

Figure 3. Compares the result with different initialization. The first row use the mean shape as initial shape, and the second row use our method to improve the robustness. And the green box is the detected face box.

5. Conclusion and Future Work

We proposed a multi-task convolution network that can works well for face alignment even in the case of challenging datasets. The experimental results proved the effectiveness of learning with pose estimate. Our method can also give good initialization instead of mean shape to improve the accuracy of traditional methods. In the future, we will focus on multi-task learning in other vision works.

6. References

[1] T. F. Cootes, G. J. Edwards, and C. J. Taylor. Active appearance models. TPAMI, 23(6):681–685, 2001.

[2] X. Cao, Y. Wei, F. Wen, and J. Sun. Face alignment by explicit shape regression. International Journal of Computer Vision, 107(2):177–190, 2014.

[3] D. Miller, I. Kemelmacher-Shlizerman, and S. M. Seitz. Megaface: A million faces for recognition at scale. arXiv preprint arXiv:1505.02108, 2015.

[4] A. Datta, R. Feris, and D. Vaquero. Hierarchical ranking of facial attributes. In AFGR, pp 36–42. IEEE, 2011.

[5] Cootes, T.F., Ionita, M.C., Lindner, C., Sauer, P.: Robust and accurate shape model fitting using random forest regression voting. In: ECCV, pp. 278–291 (2012)

[6] C. Chen, A. Dantcheva, and A. Ross. Automatic facial makeup detection with application in face recognition. In ICB, pp 1–8, 2013.

[7] A. Datta, R. Feris, and D. Vaquero. Hierarchical ranking of facial attributes. In AFGR, pp 36–42. IEEE, 2011
[8] B. M. Smith, J. Brandt, Z. Lin, and L. Zhang. Nonparametric context modeling of local appearance for pose-and expression-robust facial landmark localization. In CVPR, 2014.
[9] L. Breiman. Random forests. Machine learning, 45(1):5–32, 2001.
[10] S. Zhu, C. Li, C. C. Loy, and X. Tang. Face alignment by coarse-to-fine shape searching. In Proceedings of the IEEE Conference on Computer Vision and Pattern Recognition, pp 4998–5006, 2015.
[11] C. Sagonas, G. Tzimiropoulos, S. Zafeiriou, and M. Pantic. 300 faces in-the-wild challenge: The first facial landmark localization challenge. In ICCVW, pp 397–403, 2013.
[12] P. N. Belhumeur, D. W. Jacobs, D. Kriegman, and N. Kumar. Localizing parts of faces using a consensus of exemplars. In CVPR, pp 545–552, 2011.
[13] V. Le, J. Brandt, Z. Lin, L. Bourdev, and T. S. Huang. Interactive facial feature localization. In ECCV, pp 679–692, 2012
[14] A. Asthana, S. Zafeiriou, S. Cheng, and M. Pantic. Robust discriminative response map fitting with constrained local models. In CVPR, pp 3444–3451, 2013.
[15] X. P. Burgos-Artizzu, P. Perona, and P. Dollar. Robust face ′ landmark estimation under occlusion. In ICCV, pp 1513– 1520, 2013.
[16] X. Xiong and F. De la Torre. Supervised descent method and its applications to face alignment. In CVPR, pp 532–539, 2013.
[17] J. Zhang, S. Shan, M. Kan, and X. Chen. Coarse-to-fine autoencoder networks (CFAN) for real-time face alignment. In ECCV, pp 1–16. 2014
[18] S. Ren, X. Cao, Y. Wei, and J. Sun. Face alignment at 3000 fps via regressing local binary features. In CVPR, 2014
[19] Sun, Y., Wang, X., Tang, X.: Deep convolutional network cascade for facial point detection. In: CVPR. pp. 3476–3483 (2013)
[20] Z. Zhang, P. Luo, C. C. Loy, and X. Tang. Facial landmark detection by deep multi-task learning. In ECCV, pp 94– 108, 2014.
[21] K. Zhang, Z. Zhang, Z. Li, and Y. Qiao. Joint face detection and alignment using multi-task cascaded convolutional networks. arXiv preprint arXiv:1604.02878, 2016.
[22] Kostinger, M., Wohlhart, P., Roth, P.M., Bischof, H.: Annotated facial landmarks in the wild: A large-scale, real-world database for facial landmark localization. In: ICCV Workshops. pp. 2144–2151 (2011)
[23] Caruana, R.: Multitask learning. Machine learning 28(1), 41–75 (1997)
[24] X. Zhu and D. Ramanan. Face detection, pose estimation, and landmark localization in the wild. In CVPR, pages 2879– 2886, 2012.
[25] X. P. Burgos-Artizzu, P. Perona, and P. Dollar. Robust face ′ landmark estimation under occlusion. In ICCV, pp 1513– 1520, 2013
[26] S. Xiao, S. Yan, and A. Kassim. Facial landmark detection via progressive initialization. In Proceedings of the IEEE International Conference on Computer Vision Workshops, pp 33–40, 2015

Research status of operational environment partitioning and path planning for multi - robot systems

Xuefeng Dai, Qi Fan and Dahui Li

School of Computer and Control Engineering, Qiqihar University, Qiqihar, Heilongjiang 161006, China, E-mail: daixuefeng203@163.com

Abstract. There is a lot of applications for multi-robot systems instead of single robot systems. Therefore, coordination algorithms have become popular in the field of robotics for two decades. This paper analyzes and summarizes the current research status of coordinated algorithms for multi-robot systems, from perspectives of partitioning the environment, and path planning.

1. Introduction
Multi-robot systems have become popular in the field of robotics. However, the systems still face problems such as task allocations, cooperation, and resource competition. They are referred to as coordination of teammate robots. Therefore, this paper reviews the state of the art of coordinated algorithms for multi-robot systems from aspects of operational environment partitioning and path planning for robots.

2. Partitioning the Environment
Partitioned the environment can reduce the time consumed by the multi-robot system to search the task and avoid the repetitive search of the same regions by different robots. Factors that affect the partitioning are environment structure, target uncertainty, communication uncertainty, and mission urgency.

2.1. According to Target Moving Mode
In case of a static target search, the search map model can be cooperatively searched on the basis of the priori information of the environment and prediction of the target distribution, when the environmental is known in advance. Hu, Xie, Lum and Xu [1] studied the problem of multiple unmanned aerial vehicle (UAV) cooperative searches for multiple static targets on the ground using a search map. In the case of detection and communication constraints, the entire monitoring area is divided into units, each of which is associated with a probability of an individual target, which constitutes the probability map of the entire region. Each UAV preserves a single probability model for the existence of objects within each cell, and updates the probability model based on the distributed iteration Bayesian method. Non-linear transformation is used to linearism Bayesian analysis. A multi-protocol graph fusion scheme was proposed there and proved that all the individual probability maps converge to a same probability map, which reflects the real existence or non-existence of static objects in each unit. Peng, Shen and Zhu [2] proposed a distributed model predictive control for the multi-UAV cooperative search decision. The result is based on a Nash (optimality) and particle swarm optimization (PSO), and is used to solve the problem of small-scale distributed optimization of UAVs. Zigzag and the internal spiral patterns were used for unknown environment. Since the UAVs have minimal turning radius, they flied outside the area in order to ensure full coverage. This leads to a longer search path, resulting in a waste of search resources and reduced fleet search efficiency. The internal spiral patterns [3] require more turns near the center of the area. In side by side formation searching, because flight path length difference, a wide difference in the UAV speed is required. Therefore, the internal spiral patterns are not conducive to the

project. Therefore, Wu, Zhou, Yin and Liu [4] adjusted the turning time and turning position of each UAV in the formation, and successfully completed the area coverage search task.

If the targets are dynamic, then the above methods are not applicable. For example, the targets are randomly distributed evenly and dynamically in time. The moving target search algorithm employing a vertical line search pattern, with a slanted line search pattern or with Voronoi partitions and bionic algorithms, can be used. For the targets in a dynamic random environment with obstacles distributed uniformly, Xuan et al. [5] used a method based on Centroidal Voronoi Partitions (CVPs) for a cooperative distributed search decision. This research yielded the target assignment algorithm and proved the convergence of the algorithm. The proposed CVP strategy can be used to search the random target effectively and realize shorter target waiting times. The search space can be re-partitioned by remained UAVs when a UAV is out of order. Compared with the standard coverage search strategy, the proposed system has obvious advantages in case of the increasing number of multi-UAV, and the average waiting time of the target decreases as the number of multi-UAV increases and as the turning radius of the UAV decreases. At the same time, Wu, Zhou, Liu, and Yin [6] established improved centrally distributed multi-UAV cooperative search architecture. In this strategy, clustering analysis and Voronoi partitioning was employed. The collaborative search decision by controlling the improved model can effectively increase the UAV prediction range. Multi-UAV searches can be more effectively completed in the target area. Coordinated searching to solve the uneven distribution of resources at the same time. Based on the bionics algorithm, the problem of multi-UAV cooperatively track ground moving target is studied. Wang, Peng, Zhu and Shen [7] proposed a cooperative target tracking method for a multi-UAV system employing active perception. The distributed unscented information filter is used to estimate the state of the target, and the method of the combination of the receding horizon control and the genetic algorithm are used to realize the online cooperative trajectory planning of the UAV.

2.2. According to Specific Regions

To solve the multi-UAV target searching problem in an uncertain environment, priori information is employed to explore the environment division. Uncertain environment can be divided into unknown, known, and prohibited regions. Each of them will be processed individually. When making target searches, the search path should cover as many unknown areas as possible in areas of unknown attention and in areas of high attention with higher frequency of search; as far as possible, it should avoid known areas and completely avoid prohibited areas [8]. To guarantee complete avoidance of the prohibited areas, Wu, Huang, Song, Tang and Bai [9], in considering environmental information and data communication delay, introduced the search returns function, which generates prohibited areas to avoid decisions, identifies the unknown environment to search, and circumvents the known environment of a prohibited area completely avoid. This achieves better search results, but it is not the optimization on the spatial distribution of the UAV. In the approach proposed by Qiao and Wang [10] an unmanned combat air vehicle (UCAV) in collaborative route principle, will take both search efficiency and combat efficiency into account. They used search function returns and weighted average distance to guide of UCAV, strengthen patrols of the high attention areas, and improved the spatial distribution for UCAVs. On this basis, Li and Qiao [8] proposed the route planning model for the simulation of PSO. The results show that the proposed collaborative patrol route planning algorithm has better search efficiency than the results of document [10]. The particle swarm algorithm makes the UCAV completely avoid prohibited areas and reduces the search of the known area.

In searching for an uncertain environment, Fu, Wei and Gao [11] took multi-UAV kinematics and prohibited areas into account together when realizing cooperative searching. A search algorithm based on predictive control was proposed to consider current and future searching costs. It also analyzes the communication constraints for collaborative area searches and tests this with the Monte Carlo method.

2.3. According to Information Integrity

Communications among robots are suffered from limited distance, message delay, limited bandwidth, and so on. Communication among robots can be divided into three kinds: global communication, local communication and no communication. The factors that influence the inconsistency of environmental information are the initial prior information, the detection information, and the communication information. For the massage delay problem, the Kalman filtering method can effectively estimate the communication delay state. The information inconsistency compensation method, based on state estimation, can eliminate the environment information deviation caused by the global communication delay. Consequently, the auction algorithm and the consistency information filtering algorithm are proposed in local communications. Among them, Capitan, Spaan, Merino and Ollero [12] reduced the communication load through decentralized data fusion. Sun, Zhou, Zou and Ding [13] has improved the information consistency filtering (ICF) algorithm in order to realize the distributed information fusion of each UAV node in the communication and measurement range. In addition, the problem of UAV distributed filtering and control under conditions of limited communication and measurement is solved by using the communication connection robustness function. In no communication case, an extreme case in which each UAV cannot share information with other UAVs, leading to the environmental information of each UAV being completely inconsistent [14]. In addition, when the remote UAV is carrying out the continuous target tracking, Zhu and Zhou [15] used a nonlinear model predictive control (NMPC) to realize the UAV cooperative distributed online optimization to ensure the dynamic maintenance of the whole communication network. A variety of computational methods are inspired by biology. In addition to the aforementioned calculation method's usefulness in achieving a multi-robot system coordination strategy [16-17], it can also be used to solve the inherent problems of multi-robot system communication [18]. Feng and Swindlehurst [19] considered a collection of single-antenna ground nodes communicating with a multi-antenna UAV over a multiple- access ground-to-air communications link. They designed a device with multiple mobile ground-based terminals and developed an algorithm for dynamically adjusting the UAV heading.

In addition, Puig, Garcia and Wu [20] proposed a three-layer optimization coordination algorithm, which can reduce the waiting time of each robot in the local area. Lazarus, Tsourdos, White, Silson and Zbikowski [21] proposed a method of cooperative guidance and estimation. The computational efficient data information algorithm can lead to build and update the map which is based on geometrical tool known as two-dimensional splinegon. It also enables multi-UAV to efficiently search and explore unknown environments. Janchiv, Batsaikhan, Kim, Lee and Lee [22] proposed a region partitioning algorithm that combines exact cell decomposition and templates matching. But the algorithm only applies to cases where the coverage area and the barrier boundary are parallel to the rectangular edge.

3. Path Planning

The evaluation criteria of multi-robot path planning are real-time, reach ability, robustness, and optimization.

The ways to achieve path planning can be divided into traditional methods and intelligent optimization methods. The traditional methods of path planning include the visual graphical method, free space method, grid method, Voronoi diagram method, and artificial potential field method. The genetic algorithm, ant colony algorithm, immune algorithm, neural network, reinforcement learning, etc., are the main methods of using intelligent optimization for path planning [23]. Ma [24] used the method of constructing a Voronoi polygon graph to model. Dijkstra algorithm is introduced into the graph of the environment. The Voronoi diagram is searched, and the initial route is obtained. The algorithm improves real-time planning and can satisfy various constraints and obtain reasonable planning results. Based on a simultaneous consideration of the team cost and computation time, Savas and Ahmet [25] put forward a new method that combined with the Travelling Salesman Problem solution and Dijkstra shortest path solution for calculating bids. The approach has great advantages, including less computation, better real-time performance, a stronger ability to find the optimal result, etc. Compared with other algorithms, the genetic algorithm has strong universality and robustness. It is particularly suitable for solving complex problems that other methods cannot solve. The genetic algorithm is used to path planing of the multi-UAV.

Combining the traditional path planning method with the intelligent optimization path planning method can improve the robustness of multi-robot system path planning and improve the efficiency of motion planning.

Multi-robot path planning methods can be categorized also as centralized architecture, and distributed architecture, or hybrid architecture. Centralized architecture can make decisions on global information, with a high degree of coordination. But the real-time and dynamic of the centralized architecture is weak. Commonly used methods are ant colony algorithm, hybrid genetic algorithm and so on. Ebtehal and Hisham [26] researched the methods used to solve multi-autonomous vehicle path-planning for an application of heavy traffic control in cities. Considering the obstacles and other robots' paths, each robot has to reach its destination in the minimum time and number of movements. They used decoupled centralized approaches to solve the problem.

Distributed architecture make decisions on local information, and has better real-time performance and stronger fault-tolerant ability. But the global ability of the distributed architecture is poor. Commonly used methods are traffic rules, mutual exclusion method, the priority method [27], distributed colorless information filtering [7] and so on. Wang, Peng, Zhu and Shen [7] proposed a kind of multi-UAV cooperative target tracking method based on active perception. A distributed multi-UAV online coordinated flight path optimization algorithm is designed, based on distributed colorless information filtering, to achieve fusion target state estimation, a stochastic search, and rolling time domain control. It can get better tracking performance by optimizing the UAV track online with the predicted target states. Cichella [28] et al. introduced a time-coordination problem for multi-robots cooperating based on distributed control rules. It supports time-varying network to exchange information. Even if there are faulty in communication networks, temporary link losses, and switching topologies, it would ensure that UAVs meet the desired temporal assignments of the mission.

The hybrid architecture makes up for the poor global performance and poor real-time performance in a distributed environment. The hybrid architecture has better overall situation and real-time performance. Xiao and Yan [29] proposed an ICCEAA algorithm to improve the global coordination ability and the adaptive level of the robot effectively under the architecture of the fusion immune co-evolutionary algorithm and APF algorithm. The algorithm has better superiority in path planning in a dynamic environment. It can distribute tasks reasonably and accomplish tasks in a short time.

The path planning process requires a large amount of calculation. When the environment is complicated, the environment of the robot is too tight, especially when the number of robots is large; the time/space complexity of the algorithm is bound to be the key of the algorithm. In order to improve the efficiency of the path planning algorithm, the parallelism of path planning should be considered. Therefore, the parallelism of the method should be considered in the design and operation of the algorithm. As for the path planning of multiple robots, the distributed multi process and mixed multi process planning mechanism should be adopted as much as possible to realize the parallelism of each robot's path planning.

4. Conclusions

The multi-robot system has a good development prospect and is extremely challenging. In this paper, the environment partitioning and path planning of multi-robot system is reviewed and the corresponding solutions are proposed.

The research of path planning technology for multi robot systems has achieved fruitful results, but each method has its own advantages and disadvantages, and no one method can be applied to any occasion. Therefore, it is an open question to combine some methods and combine the advantages of various methods to achieve better results. In addition, in order to solve the uncertainty of environment partitioning and path planning for multi-robot systems, and improve the robustness of multi-robot system, this research is also the focus of future development. The commonly used methods to solve the problem of uncertainty are Markov Decision Process (MDP), fuzzy control method and rough set theory. Among them, the most classical method is based on the Markov decision process model in decision theory to study the coordination of multi-robot system, it is an effective tool to deal with the uncertainty of the robot system. In the future multi-robot system path planning, the behavior of each robot is modeled by Partially Observable MDP (POMDP), and then apply these models to the case of multi-robot system in environmental monitoring and target instances. The mobile robot's exploration process belongs to the continuous state of POMDP, and a large number of approximation methods have been given for the solution of POMDP. Besides the foregoing, the methods related to robot applications include the task shortening method, the evolutionary strategy method, etc.

With the development of science and technology, the multi-robot application area will continue to expand, and the area of multi-robot spatial environment division and path planning will also be further. With the practicality as the ultimate goal, constantly promote its development forward.

5. Acknowledgments
This article is supported by National Natural Science Fund under grant #61672304 and Natural Science Fund of Heilongjiang Province, China under grant #F201331 and #F201218.

6. References
[1] J. W. Hu, L. H. Xie, K. Y. Lum, J. Xu. Multiagent Information Fusion and Cooperative Control in Target Search. IEEE Transactions on Control Systems Technology, 2013, 21(4): 1223-1235.
[2] H. Peng, L. C. Shen, H. Y. Zhu. Multiple UAV Cooperative Area Search Based on Distributed Model Predictive Control. Acta Aeronautica et Astronautica Sinica, 2010, 31(3): 593-601.
[3] N. NIGAM.The multiple unmanned air vehicle persistent surveillance problem: a review [J]. Machines, 2014, 2(1):13-72.
[4] Q. P. Wu, S. L. Zhou, G. Y. Yin, W. Liu. Improvement of Multi-UAV Cooperative Coverage Searching Method. Electronics Optics & Control, 2016, 23(1): 80-84.
[5] Y. B. Xuan, C. Q. Huang, W. C. Wu, Y. Wang, X. W. Weng, W. X. Li. Cooperative Search Strategies of Multi-UAV for Random Targets. Control and Decision, 2013, 28(5): 711-715.
[6] Q. P. Wu, S. L. Zhou, W. Liu, G. Y. Yin. Multi-unmanned Aerial Vehicles Cooperative Search Based on Central-distributed Model Predictive Control. Control Theory & Applications, 2015, 32(10): 1414-1421.
[7] L. Wang, H. Peng, H. Y. Zhu, L. C. Shen. Cooperative Tracking of Ground Moving Target Using Unmanned Aerial Vehicles in Cluttered Environment. Control Theory & Applications, 2011, 28(3): 300-308.
[8] Y. S. Li, F. C. Qiao. Cooperative Patrol and Route Planning of Multi-UCAV under the Complex Environment. Journal of Naval Aeronautical and Astronautical, 2014, 29(2): 187-192.
[9] W. C. Wu, C. Q. Huang, L. Song, S. Q. Tang, R. C. Bai. Cooperative Search and Path Planning of Multi-unmanned Air Vehicles in Uncertain Environment. Acta Armamentar, 2011, 32(11): 1337-1342.
[10] F. C. Qiao, C. J. Wang. Multi-UAV Cooperative Patrol Route Planning Based on Degree of Attention[C]. // Marine Development and Command and Control Conference. Yantai, 2013: 144-147.
[11] X. W. Fu, G. W. Wei, X. G. Gao. Cooperative Area Search Algorithm for Multi-UAVs in Uncertainty Environment. Systems Engineering and Electronics, 2016, 38(4): 821-827.
[12] J. Capitan, M. T. J. Spaan, L. Merino, A. Ollero. Decentralized Multi-robot Cooperation with Auctioned POMDPs. The International Journal of Robotics Research, 2013, 32(6): 650-671.
[13] H. B. Sun, R. Zhou, L. Zou, Q. X. Ding. Distributed Cooperation Target Tracking for Heterogeneous Multi-UAV under Communication and Measurement Constraints [J]. Acta Aeronautica et Astronautica Sinica, 2011, 32(2): 299-310.
[14] H. Peng, Q. J. Zhang, Y. Li, L. C. Shen. Environment Information Coherence in Multi-UAVs Cooperative Area Searching. China conference on control and decision, 2009.
[15] Q. Zhu, R. Zhou. Using Multiple Unmanned Aerial Vehicles to Maintain Continuous Reliable Cooperative Communication in Target Tracking. Control Theory & Applications, 2015, 32(11): 1551-1560.
[16] E. G. Jones, M. B. Dias, A. Stentz. Time-extended Multi-robot Coordination for Domains with Intra-path Constraints. Autonomous Robots, 2011, 30(1): 41-56.
[17] W. C Wu, F. M. Zhang. Robust Cooperative Exploration with A Switching Strategy. IEEE Transactions on Robotics, 2012, 28(4): 828-839.
[18] M. O. F. Sarker, T. S. Dahl, E. Arcaute, K. Christensen. Local Interactions over Global Broadcasts for Improved Task Allocation in Self-organized Multi-robot Systems. Robotics and Autonomous Systems, 2014, 62(10): 1453-1462.
[19] J. Feng, A. L. Swindlehurst. Optimization of UAV Heading for the Ground-to-Air Uplink. IEEE Journal on Selected Areas in Communications, 2012, 30(5): 993-1005.
[20] D. Puig, M. A. Garcia, L. Wu. A New Global Optimization Strategy for Coordinated Multi-robot Exploration: Development and Comparative Evaluation. Robotics and Autonomous Systems, 2011, 59(9): 635-653.
[21] S. B. Lazarus, A. Tsourdos, B. A. White, P. Silson, R. Zbikowski. Cooperative unmanned aerial

vehicle searching and mapping of complex obstacles using two-dimensional splinegon. Proceedings of the Institution of Mechanical Engineers Part G Journal of Aerospace Engineering, 2010, 224 (2):149-170.

[22] Janchiv, D. Batsaikhan, B. S. Kim, W. G. Lee, S. G. Lee. Time-efficient and complete coverage path planning based on flow networks for multi-robots [J]. International Journal of Control, Automation, and Systems, 2013, 11(2):369-376.

[23] Y. M. Zhang, X. Y. Lei, S. Y. Yang, X. P. Fan, Z. H. Qu, Z. C. Jia. Path Planning Research for Multi-robot. Application Research of Computers, 2008, 25(9): 2566-2569.

[24] C. Y. Ma. Research on Auto Flight Path Planning Algorithm of Multiple Unmanned Air Vehicles [J]. Radio Engineering, 2015, 45(2): 5-7, 33.

[25] Savas Ozturk, Ahmet Emin Kuzucuoglu. Optimal bid valuation using path finding for multi-robot task allocation. J Intell Manuf 2015, (26):1049–1062.

[26] Ebtehal Turki Saho Alotaibi, Hisham Al-Rawi. Multi-Robot Path-Planning Problem for a Heavy Traffic Control Application: A Survey. (IJACSA) International Journal of Advanced Computer Science and Applications, 2016, 7(6): 179-188.

[27] W. Wang. Research of Multi-robot Path Planning Based on Multi-agent System. Chengdu: University of Electronic Science and Technology of China, 2015.

[28] V. Cichella, I. Kaminer, V. Dobrokhodov, E. Xargay, R. Choe, N. Hovakimyan, A. Pedro Aguiar, Antonio M. Pascoal. Cooperative Path Following of Multiple Multirotors over Time-Varying Networks. IEEE Transactions on Automation Science and Engineering, 2015, 12(3): 945-957.

[29] G. B. Xiao, X. H. Yan. New Cooperative Multi-robot Path Planning Algorithm. Computer Science, 2013, 40(4): 217-220, 248.

IOP Conf. Series: Journal of Physics: Conf. Series **887** (2017) 012081 doi:10.1088/1742-6596/887/1/012081

Research status of multi - robot systems task allocation and uncertainty treatment

Dahui Li*, Qi Fan and Xuefeng Dai

School of Computer and Control Engineering, Qiqihar University, Qiqihar, Heilongjiang 161000, China, E-mail:18645213579@163.com

Abstract. The multi-robot coordination algorithm has become a hot research topic in the field of robotics in recent years. It has a wide range of applications and good application prospects. This paper analyzes and summarizes the current research status of multi-robot coordination algorithms at home and abroad. From task allocation and dealing with uncertainty, this paper discusses the multi-robot coordination algorithm and presents the advantages and disadvantages of each method commonly used.

1. Introduction

Multi-robot systems have become a hot topic of research in the field of robotics. In addition, because of their high fault tolerance and robustness, they are widely used. The multi-robot system still faces problems such as task allocation and resource competition. They are referred to as coordination of teammate robots. Complex problems are divided into a set of sub-problems. The optimization problem is solved by competing and cooperation among robots. The operational capacity is improved. In the study of multi-robot system coordination algorithm, task allocation is an important guarantee for multi-robot to complete the task, and it is an important part of task planning. In the multi-robot system, there are many problems such as communication delay, dynamic environment change, sensor precision limit and system component failure. These uncertainties will affect the assignment of multi-robot system tasks, so the research on solving uncertainties is increasingly affected Attention. Therefore, this paper reviews the research of multi-robot systems in the process of exploration and summarizes the latest research trends of multi-robots in cooperation, task allocation and dealing with uncertainty.

2. Task Allocation

The goal of task allocation across a multi-robot system is to maximize the benefits of task execution and to minimize the cost for the entire system. The design goals for task allocation methods are real-time, robustness, reliability, optimization performance, communication requirements, computing requirements, and scale capabilities. There are two ways to distribute the tasks. The first is the way is emergent allocation and the second is intentional cooperative distribution. Emergent allocation is realized coordination by the interaction between each other and through impacts with the environment, which is more suitable for the tasks of large scale multi-robot systems. However, this method cannot predict the behavior of a robot accurately in the system and cannot guarantee the efficiency. Intentional cooperative distribution is achieved through robot display cooperation, easy-to-implement human-computer interaction, and has the potential to better utilize the capabilities of heterogeneous robotic teams. Intentional cooperative distribution can be divided into centralized distribution and distributed distribution. The centralized allocation has the characteristics of simple structure, strong calculation performance and easy implementation. But it is easy to cause the problem of communication congestion, and difficult to deal with real-time processing. The methods of centralized allocation are integer programming method, basic search algorithm and based on the satisfactory decision method, heuristic search algorithm, method based on graph theory, swarm intelligence method [1, 2]. Distributed allocation of communication load balancing is more suitable for a dynamic environment, involving frequent task changes and local autonomy [3]. The methods of distributed allocation are distributed cooperative structure ALLIANCE [4], vacancy chains method [1], contract net protocol and auction algorithm.

The market economy method [5-10] is the main method to realize the coordination of a multi-robot system. It solves the problem that the distributed system control is based on the decentralized subsystem. The market economy algorithm optimizes the control in a selfish way with the goal of maximizing the local index of the subsystem to get the optimal solution in the whole system. The advantage of the market economy approach is the simplicity of the calculation [1]. It uses a bidding-auction approach to achieve the task allocation, assigning the corresponding exploration task to the highest bidder. Auction types are usually divided into single auction, multiple auctions, and combined auctions. Dias, Zlot, Kalra and Stentz [7] summarized the early work and compared the time complexity and communication complexity of the three auction methods. From the auction to achieve the way, a centralized auction can achieve the optimal task allocation, where the decentralized auction not only makes each robot take a small amount of computation but also has robustness [8]. Guillaume, Lounis, Aurelie, Abdel-Illah and Philippe [11] proposed control architecture based on topological representation of the environment and a protocol based on sequential simultaneous auctions (SSA) which reduces the planning complexity to coordinate Robots' policies. The policies are individually computed using Markov Decision Processes oriented by several goal task positions to reach. When dynamic task allocation in multi-robot search and retrieval tasks, Wei, Koen and Catholijn [12] put forward a prediction method based on extension of the auction algorithm. The method uses a winner determination mechanism to allocate tasks to each robot. Robots implicitly coordinate their activities by team reasoning that leads to consensuses about task allocation. Based on the framework of a multi-agent system, the mathematical model of multi-UAV cooperative target assignment was developed by Liao, Chen and Zhou [13]. Considering the value of UAVs, fighting force, initial position, the value of targets, and battlefield threats on target allocation, he proposed a target allocation algorithm based on agents for a distributed cooperative auction, which can achieve the goal of multi-UAV system cooperative distribution. The target allocation algorithm has good optimization and time characteristics, which can meet the real-time requirements on the battlefield. However, the task allocation algorithm does not consider the problem of dynamic target assignments in the case of uncertain information. Through the analysis of the uncertainty of multi-UAV air combat situation information, Chen, Zhao and Xu [14] established the combat dominant function for a multi-UAV dynamic game, based on fuzzy information. A dynamic target assignment algorithm based on distributed cooperative auction is proposed. The dynamic expansion game was transformed into a static strategy game. The fuzzy structure element method and the particle swarm optimization algorithm are combined to give the Nash equilibrium solution of the mixed strategy under the fuzzy information. The proposed algorithm can achieve a optimal demand of project under the specified time or resource constraints. In order to achieve the goal of multi-UAV by the time of arrival of multiple targets in uncertain information, Zhao, Li and Wang [15] combined the auction algorithm with the interval consistency algorithm. Multi-target attack task allocation for multi-UAVs is proposed by using the auction algorithm. Based on the interval consistency, a control method is proposed for the simultaneous arrival of multi-UAV with uncertain information. In the combined auction, each robot will bid on all the subsets of the task; resulting in communication, evaluation of the tender, task clear, and other workload into an exponential growth. In recent years, there are two aspects of the combination auction method that need improvement. One is the hybrid structure of the form [5] of the auction and the other is the task after the grouping of the auction [6, 9-10]. The problem can be simplified by grouping the tasks before exploration, by clustering algorithms in cases where it is known to explore a priori information about a task or environment. Jones, Dias and Stentz [6] used two heuristic methods (clustering and opportunistic path planning) to simplify the problem and then, through the hierarchical auction, these methods will be coupled with each other. Two tasks were assigned to two different kinds of the robots. The patterns that are performed here after grouping the tasks can further balance the tasks between the robots. The method proposed by Nanjanath and Gini [9] is to re-auction the remaining tasks after a member of the multi-robot system has completed the task, so as to deal with the dynamic tasks and improve the robustness. Zhang, Collins and Shi [10] gave the results for heterogeneous multi-robot system coordination with different moving speeds and different task completion times. They proposed random clustering auction coordination based on the Markov chain search process and the simulated annealing algorithm, which can make the coordination performance index of the multi-robot system distribute between the results of global optimal and random allocation coordination strategies.

There are many research papers on the same information processing mode with the market economy algorithm. Cabrera-Mora and Xiao [16] proposed the exploration coordination based on the flood algorithm. Nieto-Granda, Rogers III and Christensen [17] proposed coordination algorithms of the redundant robot in the system. In addition, some scholars have researched the coordination of multi-robot systems at the logic level through the finite-state machine [18-19] or Petri nets [20-22]. This method can deal with different athletic abilities [18] or functions of robots [20], completely from the logical level of coordination achieved by the relatively limited behavior. However, it is possible to participate in the control loop of the multi-robot system by supervising the control [23].

3. Dealing with Uncertainty

Another trend in the research of coordination algorithms for multi-robot systems is that the uncertainties in the robot itself are receiving more and more attention. The uncertainties include problems with the communication delays, the dynamic change of the environment, the limitation of the precision of the sensor, the failure of the system components, and so on. In order to solve the problem of uncertainty, the common methods include the Markov decision process (MDP), fuzzy control method [14], and rough sets theory [24]. The typical method is based on the MDP model in decision-making theory to research the coordination of multi-robot systems. MDP is an effective tool [25, 26-28] to deal with uncertainties in robot systems.

Work in this area includes the studies of Huang and Luo [26] for multi-robot initiative targeting, seeking the difference between benefits and costs within the group to select a moving target maximum point for each robot. In this study, information entropy is used to represent the uncertainty of the robot positions, and the benefit is defined as the reduction of the position uncertainty of the robot after its movement to the target point. Capitan, Spaan, Merino and Ollero [25] modeled the behavior of UAVs with a partially observable Markov decision process (POMDP) and then applied these models to instances of multi-robot systems in environmental monitoring and target follow-ups. In order to achieve a number of UAVs to monitor the multi-target [29], POMDP was employed to infer the uncertainty in the application of dynamic model, so that more than UAV team to complete real-time tasks. To realize environment monitoring and a moving target tracking, an auction algorithm was used to solve the decision problem.

For the sake of promoting the cooperation between the UAVs, the different behaviors of UAVs are analyzed by using the factorization strategy to simulate the dynamic situation, and the decentralized data fusion system is used to obtain the belief of the joint factor state observed by all team members. The advantage of the proposed algorithm is that its computational complexity does not increase with the increase of UAVs or targets. Moreover, the complexity of the auction and decentralized data fusion depends only on the size of the local communication neighborhood of each UAV. Lozenguez, Adouane, Beynier, Mouaddib and Martinet [27] adopted a topological map representation instead of the grid method, which would lead to an explosive growth of the state space of the goal-oriented Markov decision process for each robot. Kim, Baik and Lee [30] studied the task allocation problem of a heterogeneous UAV team performing search and destruction tasks in a dynamic and uncertain environment. He put forward a distributed task allocation strategy based on resource benefits that provided the dynamic, unpredictable, events for robustness and a balanced way for the UAV team. The method achieved high resources and improved the welfare of aggressive responses. The scheme is superior to the real-time task allocation method. In the above work, the choice of the action strategy of the decision model is completed by the auction mechanism [25, 27, 29], the entropy is used as the measure of the uncertainty of the robot [25, 26], but the stratified social entropy [31] is not used in these literatures.

In the study of the multi-UAV close-formation problem [32], optimization of the control volume of the differential evolution strategy, and the Markov chain proves it is the optimal solution. The exploration process of a mobile robot belongs to a POMDP with continuous state [33]. A large number of approximate methods for the solution of POMDP have been given. Besides the foregoing, the methods related to robot applications include the task shortening method [34], the evolutionary strategy method [35], etc.

In addition, when the UAVs perform cooperative target positioning, the measurement data transmission will be delayed. Wang, Qin, Bai and Cui [36] proposed a nonlinear filter based on solving the Fokker-Planck equation to solve this problem. According to the measured arrival time, the proposed nonlinear filter can be divided into two parts. Non-delay measurements are fused in the first part, where the Fokker-Planck equation is used to propagate the conditional probability density function in the positive form. The time delay measurement is fused in the second part, where Fork-Planck is used in a backward form. The Bayesian formula is applied to the two parts during the measurement update process.

4. Conclusions

Multi-robot systems have many potential applications. From the point of view of task allocation, and dealing with uncertainty, this paper reviewed the state of the art of the multi-robot coordination algorithm and points out the advantages and disadvantages of each method.

5. Acknowledgments

This article is funded by National Natural Science Fund (No.61672304) and Natural Science Fund of Heilongjiang Province (No.F201218).

6. References

[1] S. Y. Tang, Y. F. Zhu, Q. Li, Y. L. Lei. Survey of Task Allocation in Multi Agent Systems. Systems Engineering and Electronics, 2010, 32(10): 2155-2161.

[2] L. L. Yu, J. L. Jiao, Z. X. Cai. Multi-robot Mission Planning Algorithm and Its System Implementation. Computer Science, 2010, 37(6): 252-255.

[3] L. Liu. Research on Multi-robot System Task Allocation and Formation Control. Changsha: National University of Defense Technology, 2006.

[4] L. E. Parker. ALLIANCE: An Architecture for Fault Tolerant Multi-robot Cooperation [J]. IEEE Transactions on Robotics and Automation, 1998, 14(2): 220-240.

[5] H. L. Choi, L. Brunet, J. P. How. Consensus-based Decentralized Auctions for Robust Task Allocation. IEEE Transactions on Robotics, 2009, 25(4): 912-926.

[6] E. G. Jones, M. B. Dias, A. Stentz. Time-extended Multi-robot Coordination for Domains with Intra-path Constraints. Autonomous Robots, 2011, 30(1): 41-56.

[7] M. B. Dias, R. Zlot, N. Kalra, A Stentz. Market-based Multirobot coordination: A Survey and Analysis. Proceedings of the IEEE, 2006, 94(7): 1257-1270.

[8] B. Di, R. Zhou, Q. X. Ding. Distributed Coordinated Heterogeneous Task Allocation for Unmanned Aerial Vehicles. Control and Decision, 2013, 28(2): 274-278.

[9] M. Nanjanath, M. Gini. Repeated Auctions for Robust Task Execution by A Robot Team. Robotics and Autonomous Systems, 2010, 58(7): 900-909.

[10] K. Zhang, E. G. Collins Jr, D. Shi. Centralized and Distributed Task Allocation in Multi-robot Teams via a Stochastic Clustering Auction. ACM Transactions on Autonomous and Adaptive Systems, 2012, 7(2), article21: 1-22.

[11] Guillaume Lozenguez, Lounis Adouane, Aurelie Beynier, Abdel-Illah Mouaddib, Philippe Martinet. Punctual versus continuous auction coordination for multi-robot and multi-task topological navigation. Auton Robot, 2016, 40: 599–613.

[12] Changyun Wei, Koen V. Hindriks , Catholijn M. Jonker. Dynamic task allocation for multi-robot search and retrieval tasks. Applied Intelligence, 2016, 45: 383–401.

[13] M. Liao, Z. J. Chen, R. Zhou. Design and Simulation of Multi-UAV Coordinated Task Allocation Based on MAS. Journal of System Simulation, 2007, 19(10): 2313-2317.

[14] X. Chen, M. M. Zhao, G. Y. Xu.Fuzzy Dynamic Game Based Operation Strategy for Multiple UAVs [J]. Electronics Optics & Control, 2014, 21(6): 19-23.

[15] M. M. Zhao, B. Li, M. L. Wang. Auction Algorithm Based Multi-UAV Arriving Simultaneously to Attack Multiple Targets with Uncertain Information. Electronics Optics & Control, 2015, 22(2):89-93.

[16] F. Cabrera-Mora, J. Xiao. A Flooding Algorithm for Multirobot Exploration. IEEE Transactions on Systems, Man, and Cybernetics-part B: Cybernetics, 2012, 42(3): 850-863.

[17] C. Nieto-Granda, J. G. Rogers III, H. I. Christensen. Coordination Strategies for Multi-robot Exploration and Mapping. The International Journal of Robotics Research, 2014, 33(4): 519-533.
[18] T. J. Koo, R. Li, M. M. Quottrup, C. A. Clifton, R. Lzadi-Zamanabadi, T. Bak. A Framework for Multi-robot Motion Planning from Temporal Logic Specifications. Science China (Information Sciences), 2012, 55(7): 1675-1692.
[19] A. Karimoddini, H. Lin, B. M. Chen, T. H. Lee. Hybrid Formation Control of The Unmanned Aerial Vehicles. Mechatronics, 2011, 21(5): 886-898.
[20] T. Ebadi, M. Purvis. A Framework for Facilitating Cooperation in Multi-agent Systems. The Journal of Supercomputing, 2010, 51(3): 393-417.
[21] N.Palomeras, P. Ridao, C. Silvestre, A. El-fakdi. Multiple Vehicles Misson Coordination Using Petri Nets. In: Proceedings of the 2010 IEEE International Conference on Robotics and Automation. Anchorage, Ak, USA: IEEE, 2010. 3531-3536.
[22] V. A. Ziparo, L. Iocchi, P. U. Lima, D. Nardi, P. F. Palamara. Petri Net Plans: A Framework for Collaboration and Coordination in Multi-robot Systems. Autonomous Agents and Multi-Agent Systems, 2011, 23(3): 344-383.
[23] F. Gao, M. L. Cummings, E. T. Solovey. Modeling Teamwork in Supervisory Control of Multiple Robots. IEEE Transactions on Human-Machine Systems, 2014, 44(4): 441-453.
[24] Q. J. Gao. Research on Autonomous Navigation of Mobile Robot Based on Rough Sets Theory [D]. Harbin: Harbin Institute of Technology, 2006.
[25] J. Capitan, M. T. J. Spaan, L. Merino, A. Ollero. Decentralized Multi-robot Cooperation with Auctioned POMDPs. The International Journal of Robotics Research, 2013, 32(6): 650-671.
[26] Q. C. Huang, R. H. Luo. Research on an Active Cooperative Location Method for Multiple Mobile Robots. Journal of Electronics, 2010, 38(11): 2644-2648.
[27] G. Lozenguez, L. Adouane, A. Beynier, A. I. Mouaddib, P. Martinet. Punctual Versus Continuous Auction Coordination for Multi-robot and Multi-task Topological Navigation. Autonomous Robots, 2016, 40(4): 599-613.
[28] T. Andre, C. Bettstetter. Collaboration in Multi-robot Exploration: To Meet or Not To Meet? Journal of Intelligent & Robotic Systems, 2016, 82(2): 325-337.
[29] J. Capitan, L. Merino, A. Ollero. Cooperative Decision-Making under Uncertainties for Multi-Target Surveillance with Multiples UAVs. Journal of Intelligent & Robotic Systems, 2016, 84(1): 371–386.
[30] M. H. Kim, H. Baik, S. Lee. Resource Welfare Based Task Allocation for UAV Team with Resource Constraints. Journal of Intelligent & Robotic Systems, 2015, 77: 611–627.
[31] T. Balch. Hierarchic Social Entropy: An Information Theoretic Measure of Robot Group Diversity. Autonomous Robots, 2000, 8(3): 209-237.
[32] X. Y. Zhang, H. B. Duan, Y. X. Yu. Time-domain Control of Multi-UAV Close Formation Based on Differential Evolution. Science China Press, 2010, 40(4): 569-582.
[33] M. Julia, A. Gil, O.Reinoso. A comparison of Path Planning Strategies for Autonomous Exploration and Mapping of Unknown Environments. Autonomous Robots, 2012, 33(4): 427-444.
[34] D. K. Grady, M. Moll, L. E. Kavraki. Extending the Applicability of POMDP Solutions to Robotic Tasks. IEEE Transactions on Robotics, 2015, 31(4): 948-961.
[35] B. Eker, H. L. Akin. Using Evolution Strategies to Solve DEC-POMDP Problems. Soft Computing, 2010, 14(1): 35-47.
[36] X. G. Wang, W. T. Qin, Y. L. Bai, N. G. Cui. Cooperative target localization using multiple UAVs with out-of-sequence measurements. Aircraft Engineering and Aerospace Technology, 2017, 89(1): 112-119

IOP Conf. Series: Journal of Physics: Conf. Series **887** (2017) 012082 doi:10.1088/1742-6596/887/1/012082

The analysis and detection of hypernasality based on a formant extraction algorithm

Jiahui Qian[1], Fanglin Fu[1], Xinyi Liu[1], Ling He[1], Heng Yin[2] and Han Zhang[1,*]

[1]School of Electrical and Information Sichuan University, Chengdu, Sichuan, 610065, China.

[2] West China Hospital of Stomatology, Sichuan University, Chengdu, Sichuan, 610041, China. Email: ling.he@scu.edu.cn

Abstract. In the clinical practice, the effective assessment of cleft palate speech disorders is important. For hypernasal speech, the resonance between nasal cavity and oral cavity causes an additional nasal formant. Thus, the formant frequency is a crucial cue for the judgment of hypernasality in cleft palate speech. Due to the existence of nasal formant, the peak merger occurs to the spectrum of nasal speech more often. However, the peak merger could not be solved by classical linear prediction coefficient root extraction method. In this paper, a method is proposed to detect the additional nasal formant in low-frequency region and obtain the formant frequency. The experiment results show that the proposed method could locate the nasal formant preferably. Moreover, the formants are regarded as the extraction features to proceed the detection of hypernasality. 436 phonemes, which are collected from Hospital of Stomatology, are used to carry out the experiment. The detection accuracy of hypernasality in cleft palate speech is 95.2%.

1. Introduction

Cleft palate is a birth defect with a gap in the roof of the mouth, which leaves a hole between nasal cavity and oral cavity. Velopharyngeal inadequacy is that the soft palate cannot separate the oral cavity and nasal cavity sufficiently, which can cause abnormal resonances due to the additional airflow through the oral cavity to nasal cavity during speech production. Thus, velopharyngeal inadequacy can result in hypernasal speech.

The cleft palate can be corrected with a series of surgeries. For many cleft palate patients, their treatments might last several years. If the opening of velopharyngeal port is not fully repaired after one-stage operation, the cleft palate patient will still have the resonance speech disorder. A further repair surgery is needed under this situation. Hypernasality is a symptom of resonance speech disorder. In the clinical environment, the grades of nasal speech is proportional to the opening size of velopharyngeal.

The hypernasality assessment in cleft palate speech can be classified into two categories: 1) invasive and 2) non-invasive techniques. In the clinical environment, the invasive techniques use invasive instruments which detect the movement of velopharyngeal. Non-invasive techniques can further be classified into two categories, a) clinical assessment and b) digital signal processing based techniques. The clinical assessment widely uses the nasometer, which is an expensive device. Another clinical assessment is the subjective judgment by experienced and well-trained speech-language pathologists. However, the result of subjective judgment can easily vary among the speech-language pathologists. Digital signal processing based techniques are fast and objective, thus making the detection process comfortable, simpler and more convenient.

Formant is an important cue to hypernsality detection. Three acoustic cues can be summarized from the literatures [1, 3]: 1) the existence of a new formant (around 250Hz for /a/, around 1000Hz for /u/ and /i/); 2) broadening of formant bandwidths; 3) presence of antiformants. Based on the above cues, in this paper, estimated formant is used to detect the hypernasality.

Formant extraction has been widely used for decades in the field of speech signal processing. The formant extraction consists of spectral peak picking and root extraction [4-7]. The two methods often suffer the problem of peak merger (two formants, which are closed to each other, are regarded as one formant). Dellar et.al and Kim et.al [8, 9] propose a method to resolve the peak merger using prediction-error polynomial obtained from LPC extraction. That method involves too much computation. Moreover, the Snell et.al put forward a method which uses the number of poles in a sector in z-plane to calculate the formant [4]. However, that method does not discuss the condition for peak merger. It may resolve the roots which peak merger does not occur to. Thus, the method in [4] may calculate the wrong formant from the LPC root. Peak merger is a serious problem of formant extraction. Thus most of current researches on the study of hypernasality use other acoustic parameters to detect hypernasality instead of formant. Dubey et.al [2] indicate that the changes in the vocal tract characteristics, and the extra nasal formant is a cue of hypernasality detection. They use the impulse-like ZTW (Zero Time Windowing) function, which is a high temporal resolution method, to resolve peak spectrum of hypernasal vowel sounds. Vjiayalakshmi et.al finds an appearance of a new resonant in the low frequency region. They use group delay function to resolve two closely spaced formants and define the Group Delay function-based Acoustic Measure (GDAM) to detect hypernasality [1]. Vijayalakshmi et.al [3] put forward a selective pole-defocusing based technique and use maximum of cross-correlation between the input and the resynthesized speech to detect hypernasality.

In this work, a method is proposed to calculate the formant using the LPC root extraction and the root classification algorithm. In the root classification algorithm, the condition for peak merger is discussed and the method mentioned in article [4] is used to classify the roots into three cases. Then, the estimated formants are used as acoustic features to detect hypernasality in cleft palate speech.

2. The Proposed Formant Extraction Method

In this work, a method is proposed to estimate the first, second and third formants in cleft palate speech. The flow chart is illustrated in figure 1. The LPC roots are extracted firstly. Then these LPC roots are classified into three cases. The final step is to calculate the formants.

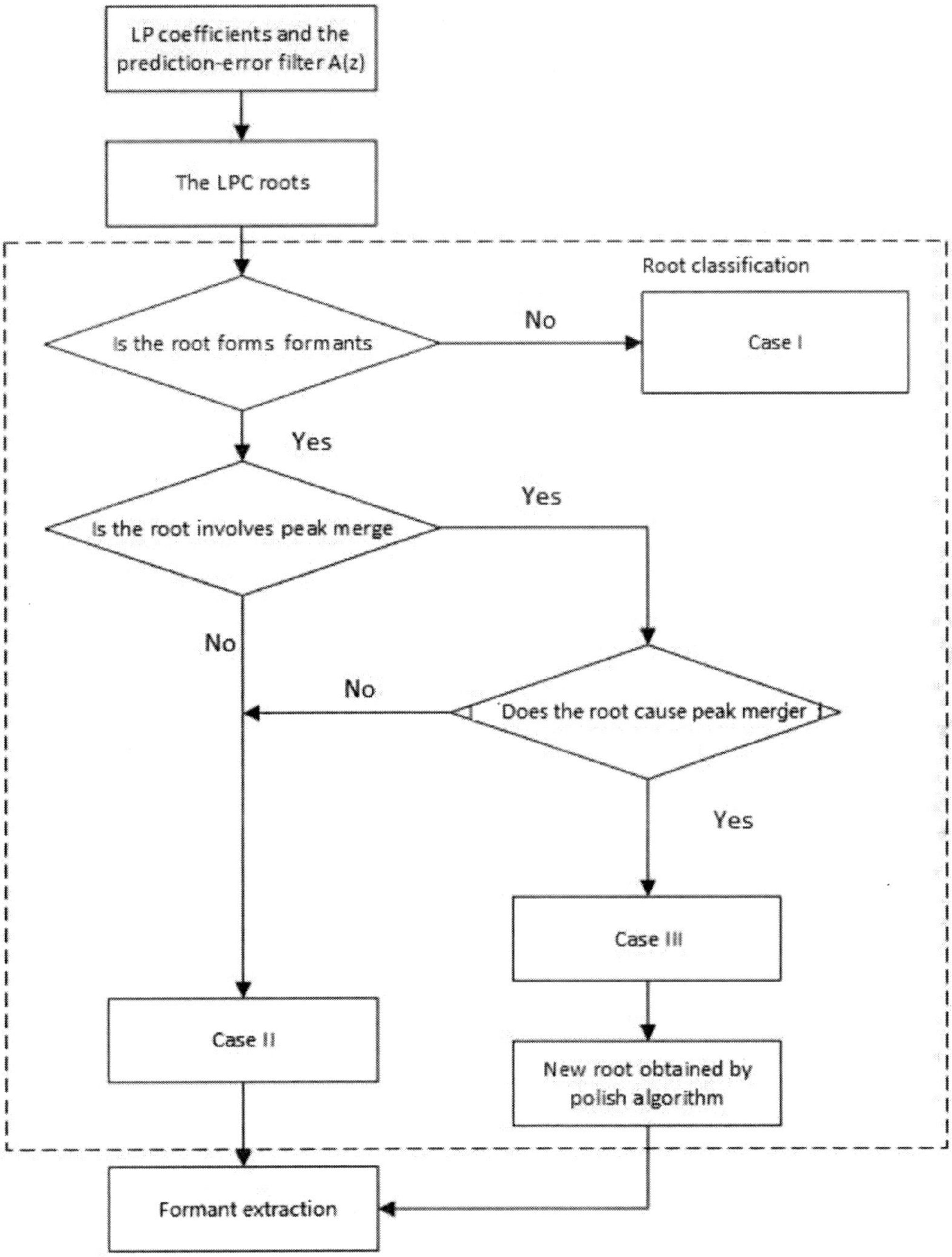

Figure 1. Flow chart of formant extraction

2.1. LPC Roots Extraction

The vocal tract system can be modelled as an all-pole system [4, 5]:

$$H_v(z) = \frac{G_v}{\sum_{k=0}^{I} a_k z^{-k}} \tag{1}$$

The method presented in article [5] is used to obtain linear prediction (LP) coefficients and the prediction-error filter A (z):

$$A(z) = \sum_{k=0}^{N_{LP}} a_k z^{-k} \tag{2}$$

Where G_v is the gain factor, N_{LP} is the prediction order, and a_k is the coefficients of Linear Prediction (LP).

The roots of A (z) correspond to the poles of the vocal tract system and relate to formants and formant bandwidths. The formant frequency F and formant bandwidth B can be calculated from the roots of A (z) by equations (3) and (4).

$$F = \frac{f_s}{2\pi}\phi_0 \tag{3}$$

$$B = -\frac{f_s}{\pi}\ln(r_0) \tag{4}$$

In equations (3) and (4), r_0 is the magnitude of the pole, ϕ_0 is the phase of the pole. And f_s is the sampling frequency, F is the formant frequency, and B is the 3-dB bandwidth.

This method could not solve the peak merger. As mentioned before, an additional nasal formant is an important cue in hypernasality speech. Therefore, resolving the peak merger problem and obtaining the accurate formants is crucial to detect the hypernasality.

2.2 Three Cases of LPC Roots

In order to solve the peak merger problem and separate the nasal formant, roots obtained from linear prediction coefficients are classified into three cases judged by their phase and magnitude. The root is defined in equation (5).

$$r_n = |r_n| e^{j\theta_n} \tag{5}$$

2.2.1. Case I. In case I, the roots do not form the formants. The range of formant bandwidths is given by Dunn [9].

Criterion i: $B > 700Hz$

The root obey to criterion i is the root which do not form the formant. If a LPC root forms a formant, its formant bandwidth should smaller than 700Hz.

2.2.2. Case II. In case II, the roots form the formants, and the peak merger does not exist. The roots obey to criterion ii and do not obey to criterion I.

Criterion ii: $|\theta_n - \theta_{nearest}| \geq 2\cos^{-1}(\frac{2r}{1+r^2})$ ($\theta_{nearest}$ is the phase of the nearest root to r_n. The r is decided by equation (4), when B is 700Hz)

To illustrate the condition for peak merger and the criterion ii, $|H_v(e^{jw})|$ can be considered as a two pole system as illustrated in equation (6).

$$|H_v(e^{jw})| \approx \frac{G_v^{'}}{|1 - r_1 e^{j\theta_1} e^{-jw}||1 - r_2 e^{j\theta_2} e^{-jw}|} \tag{6}$$

To obtain the largest phase difference, the two-pole system $|H_v(e^{jw})|$ can share the same magnitude r. Equation (6) can be substitute by equation (7).

$$|\mathrm{H}_v(\mathrm{e}^{jw})| \approx \frac{G_v^{'}}{\sqrt{(1 + r^2 - 2r\cos(w - \theta_1))(1 + r^2 - 2r\cos(w - \theta_2))}} \tag{7}$$

If the peak merge occurs, Equation (6) should have a single maximum value. The derivative of the squared value of equation (4) can be simplified into equation (8).

$$\frac{d}{dw}\left(\frac{G_v^{'2}}{|H_v(e^{jw})|^2}\right) = 8r^2\sin\left(w - \frac{\theta_1 + \theta_2}{2}\right)\left(\frac{1+r^2}{2r}\cos\left(\frac{\theta_1 - \theta_2}{2}\right) - \cos\left(w - \frac{\theta_1 + \theta_2}{2}\right)\right) \tag{8}$$

IOP Conf. Series: Journal of Physics: Conf. Series **887** (2017) 012082 doi:10.1088/1742-6596/887/1/012082

It is evident that only when the two poles satisfy equation (9), that the peak merger happens to the $[\theta_1, \theta_2]$

$$|\theta_1\text{-}\theta_2|<2\cos^{-1}(\frac{2r}{1+r^2}) \tag{9}$$

Because the root do not obey to criterion I. The bandwidth B of the roots in case II is smaller than 700Hz. Using equation (4) and (9) can calculate the criterion II easily.

2.2.3. Case III. In case III, the roots form the formants, and the peak merge may occur, if the roots do not obey to criterion1. To solve the problem of peak merge, a method is proposed to calculate the exact number of formants from LPC analysis in article [4]. Moreover, that work presents an algorithm to calculate the location of all formants. The algorithm which can obtain the number of the roots in a sector region in z-plane is discussed here. Return to the roots of LPC root extraction in z-plane. The change of θ is regarded as r moves along a curve τ. Moreover, the times of a closed curve circles the origin noted as $n(\tau)$. As the article [10] states, the number of roots between the θ_n and $\theta_{nearest}$ is equal to $n(A(\tau))$. τ is a closed curve, which is consisted of θ_n, $\theta_{nearest}$ and an arc as shown in Fig.2. In the figure.2, there is a sector which contain two roots. The figure.3 shows $A(\tau)$ in equation (2) where τ is a sector bounded by the θ_1, θ_2 and the arc as shown as figure.2. From the figure.3 the $A(\tau)$ circle the origin two times.

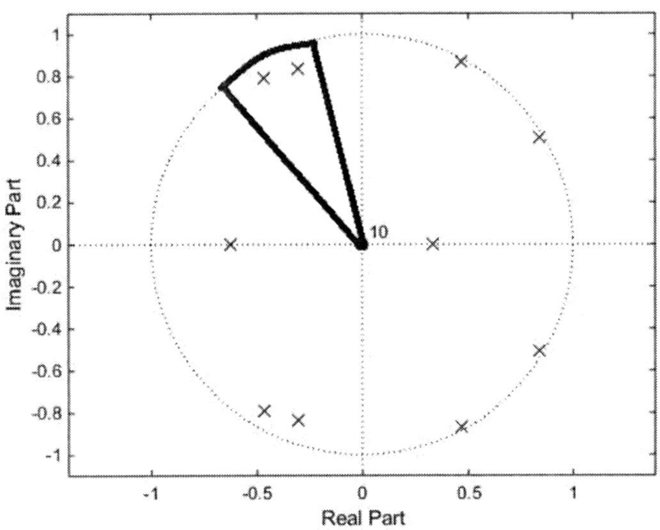

Figure 2. The root in the sector in z-plane

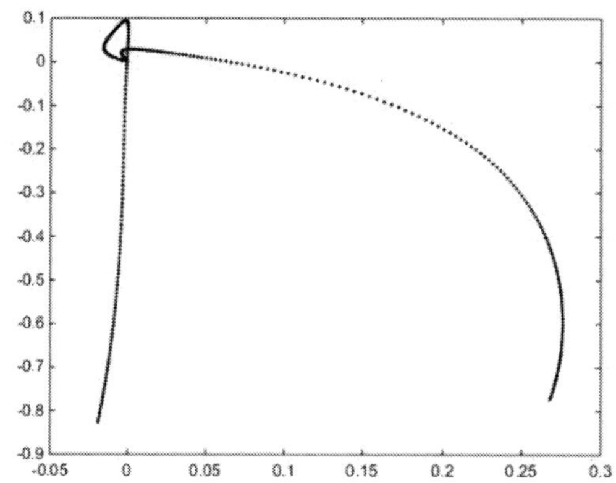

Figure 3. The $A(\tau)$ in the sector in z-plane

In order to avoid all polynomial evaluations on the arc, the R=1 is replaced by a large radius. As [4] notes, R=2 serve very well.

Separate the closed curve into two straight lines and an arc. Thus, n of the two straight lines can be calculated as follows:

(i) Substitute the two straight lines into a sequence r_n of 100 points with same distance. Then, take the sequence r_n into $A(z)$ as z.

(ii) Transform the discrete curve $A(z)$ into the polar quadrant and calculate the position in polar quadrant;

(iii) Count the times of the discrete curve $A(z)$ pass through the X negative axis N_- and positive axis N_+ ;

(iv) Calculate the number of formants:

$$n(\theta) = N_+ + N_- + [[\frac{N_{LP}\theta}{2\pi}]] \tag{10}$$

where N_{LP} is the number of LPC analysis.

Criterion iii: $n(P(\tau)) = 2$

If the roots in case III obey to the criterion iii, the peak merger actually occurs to the root. The algorithm which can obtain the number of a sector in z-plane is extended to determine the location of the two formants. The dichotomy in this sector is used here. Narrowing the area of the sector $[\theta_1, \theta_2]$ into two sub-sectors $[\theta_1^n, \theta_2^n]$ which are small enough and just obtain one root in one sub-sector. The way to divide the sub-sector is dichotomy. And the condition to stop the dichotomy is listed in equation (11).

$$|\theta_1^{'} - \theta_2^{'}| < \varepsilon \tag{11}$$

When the sub-sector $[\theta_1^{'}, \theta_2^{'}]$ meets the equation (11), the dichotomy is finished. Moreover, the $r = |r| e^{j\theta}$ is the final root, where the θ equal to($\theta_1^{'} + \theta_2^{'}$) /2. In this work, the ε equal to 0.001.

3. The Proposed Hypernasality Detection Method

The phonemes /a/, /i/ and /u/ are used to carry out hypernasal judgment in the papers [1] to detection the hypernasality. As shown in paper [1], the nasal formant of /a/ is around 250Hz, which is closed to F1. However, the nasal formant of the phonemes /i/ and /u/ is around 1000Hz. Merger peak may not occur to /i/ and /u/ as often as /a/ do. Thus, in this paper, the phonemes /a/ are used to carry out the experiment. In this paper, the 426 phonemes /a/ collected from Hospital of Stomatology are used to carry out the experiment. 216 speech utterances collected from normal speakers and 210 speech utterances collected from cleft palate patients are used in this work. Moreover, peak merger always occurs to the spectrum of nasal speech in low-frequency region. The nasal formant is close to the first formant F1. Thus, the peak merger in the second formant F2 and the thirst formant F3 is not discussed here.

In this work, a k-nearest neighbour's algorithm (k-NN) is used for classification. The classification of hypernasality applies the 10 times 10-folds cross validation method.

4. Experiments and Results

Table 1 shows the number of speech recordings with peak merger, for normal speech and cleft palate speech. It is observed that, 129 of 210 speech data from cleft palate speakers generate peak merger. For normal speakers, the 76 of 216 speech data form peak merger.

Table1. The number of speech data with peak merger

	The normal speech	The cleft palate speech
Total number of speech recordings	216	210
The number of speech recordings with peak merger	76	129
Percentage of the speech data with peak merger	35.1%	61.4%
$\frac{n}{N}\times 100\%$		

Figure 4 shows the frequencies of F1, F2 and F3. The frequencies of formants are calculated from the classical LPC root extraction method and the method presented in this paper. Take the ΔF=F1-F1', where F1 is obtained by LPC root extraction and F1' is obtained by our method to resolve the peak merger. The speech data of the cleft palate speakers have an additional formant which is around 250Hz. From figure.3, ΔF of the nasal speech data is obviously larger than the normal one. That is because that F1 which is substituted into formant in low-frequency region. It verifies the theory mentioned in the article [1] that there is a nasal formant in low-frequency region in the spectrum of nasal speech.

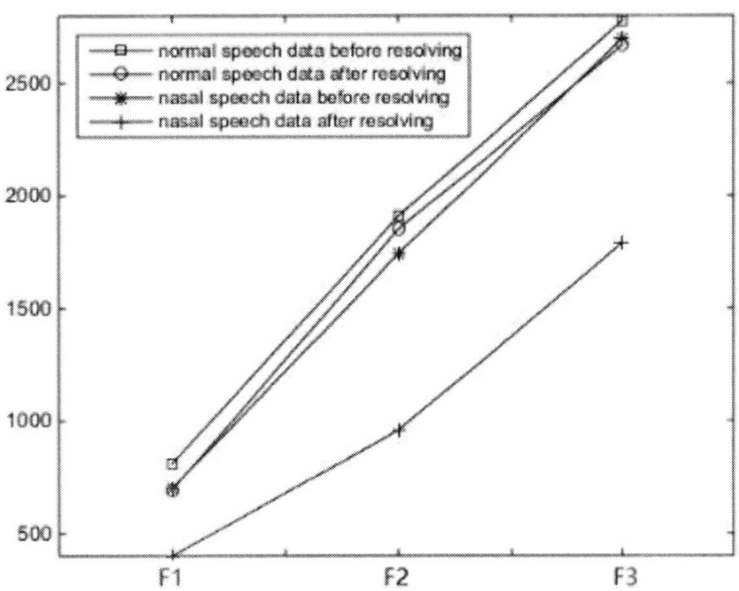

Figure 4. The comparison of the LPC root extraction method and our method

Formant frequency is regarded as an important feature to detect normal/hypernasal speech. The result of the detection of hypernasality is shown in Table.2. The accuracy is calculated by $n^{'}/n\times 100\%$. The $n^{'}$ is the number of the speech data which are classified into the right type and the n is the total number of this type.

The accuracy of hypernasality detection using the formant calculated by LPC root method is 57.4%. The average detection accuracy of this classification using the formant with our method is 88.3%.

Table2. The detection accuracy of each type and the average detection accuracy

	Normal speech	Hypernasal speech	Average accuracy
method proposed in this paper	81.5%	95.2%	88.3%
LPC root extraction method	53.2%	58.1%	57.4%

5. Discussions and Conclusions

This paper proposes a method to detect hypernsality in cleft palate. Formant is used as the acoustics feature. This method is based on the roots calculated from LPC root extraction method. These roots are classified into three cases, which are determined by the condition of peak merger, to test if peak merger occurs. In order to resolve the roots with peak merger, an algorithm which can determine the number of a sector in z-plane is proposed. Then, the formants can be calculated from the roots. This method can resolve the roots with peak merger to obtain an accurate formant frequency. As shown in Table 2, the detection accuracy of normal speech is slightly lower than that of hypernasal speech. As shown in table.1, peak merger occurs to 35.1% normal speech. If the peak merger happens, the formant for normal speech may be resolved into two formants, which are closed to the nasal formant. This normal speech is classified as nasal speech. Thus, the classification accuracy of normal speech is lower than that of hypernasal speech.

6. Conference

[1] P.Vijayalakshmi, M.Ramasubba Reddy and Douglas O'Shaughnessy,"Acoustic analysis and detection of hypernasality using a group delay function" IEEE Tras. *Biomedical Engineering,* **Vol. 53, No.4**, April 2007.

[2] Akhilesh Humar Dubey, S.R. mahadeve Prasanna and S.Dandapat, "" Communication (NCC), 10.1109/NCC.2016.7561188.

[3] Caplar R and Kulisic P 1973 *Proc. Int. Conf. on Nuclear Physics (Munich)* **vol 1** (Amsterdam: North-Holland/American Elsevier) p 517

[4] P.Vijayalakshmi, T. Nagarajan and Jayanthan Ra.V, "Select pole modification-based technique for the analysis and detection of hypernasality" TENCON 2009-2009 IEEE Region 10 Conference, pp. 1-5, 2009

[5] R. C. Snell and F. Milinazzo, "Formant location from LPC analysis data," IEEE Transactions on Speech and Audio Processing, **vol. 1,** no. 2, pp. 129–134, 1993.

[6] S. S. McCandless, "An algorithm for automatic formant extraction using linear prediction spectra," IEEE Transactions on Acoustics, Speech, and Signal Processing, **vol. 22,** no. 2, pp. 135–141, 1974.

[7] L.Welling and H. Ney, "Formant estimation for speech recognition," IEEE Transactions on Speech and Audio Processing, vol. 6, no. 1, pp. 36–48, 1998.

[8] Chanwoo, "A robust formant extraction algorithm combing spectral peak picking and root polishing" EURASIP Journal on Applied Signal Processing, vol.2006, pp1-16, 2006

[9] J.R Dellar Jr., J.G Proakis, and J.H.L.Hanson, "Discrete-Time processing of speech Signal," Macmillan, New York, NY, USA,Article in a conference proceedings:

[10] H. K. Dunn, "Methods of measuring vowel formant bandwidths," *Journal of the Acoustical Society of America*, **vol. 33**, no. 12, pp. 1737–1746, 1961.

[11] P. Henrici, "Applied and computational complex analysis", *Applied and ComputationalComplex Analysis*. Volume 1. New York: Wiley, 1974

Halftone information hiding technology based on phase feature of space filling curves

Jianhua Hu[1], Peng Cao[1], Zhihong Dong[1] and Xiaohe Cao[1]

[1]Beijing Institute of Graphic Communication, Beijing, 102600, China Email:315869117@qq.com

Abstract. To solve the problems of the production of interference fringes (namely moiré in printing) and improve the image quality in printing process of halftone screening for information hiding, a halftone screening security technique based on the phase feature of space filling curves is studied in this paper. This method effectively solves the problem of moire and optimizes the quality of the screening, so that the images presented after screening have achieved good visual effect. The pseudo-random scrambling encryption of the plaintext information and the halftone screening technique based on the phase feature of the space filling curves are carried out when screening,which not only eliminates the common moire in the screening but also improves the image quality and the security of information .

1. Introduction

Information hiding of halftone screening technique via using the non-perceivable ability of human eye on the tiny dots, screening the main map of the printing color separation, introducing hidden information and graphs in a color version or a few color plate and changing the dots location of the graphic hidden selection area, the purpose of information hiding has been achieved [1]. With the vigorous development of graphic information rasterization technology, information hiding in halftone screening has become the hotspot in today's research. Although there are many ways to solve the information hiding in the halftone screening, the general screening in the dots shape and its arrangement which has strong regularity make the screened image easily produce moire and affect image reproduction effect [2]. In this paper, the halftone screening and information hiding method based on the phase feature of space filling curves are presented. The hidden graphic information is characterized by different dot morphological templates and embedded in pseudo-random scrambling way, so that the problems of the prevalence of moire in the screening and the low image quality and so on were solved. This method has achieved ideal effect in information hiding and information anti-counterfeiting and so on.

2. Halftone Screening Based on Phase Feature of Space Filling Curves

In the traditional halftone screening, halftone screening is divided into amplitude modulation (AM) screening and frequency modulation (FM) screening. The amplitude modulation screening is the binarization of the image using different dots size, and the size of the dots corresponds to the corresponding gray value in the original image. Because of the regularity of its arrangement, the halftone image after screening has obvious moire, and the amplitude modulation algorithm is to control the gray scale of the image with the dots size, this leads to the fact that there is a large distortion in the bright and dark regions. To solve this problem, but also put forward the FM screening. FM dots size are fixed, which is through the control the intensity of the dots to show the tone, relative to the amplitude modulation can better prevent the occurrence of interference fringes, and the performance of the tone closer to the original continuous tone image. However, FM screening will cause the problem of dots expansion, which seriously affected the image reproduction effect. Moreover, it is difficult to popularize in the practical application because of the higher requirements of some materials and machines. Halftone screening based on phase feature of space filling curves is overcome the above problems, it not noly avoided the production of moiré, but also reduced the process requirements.

2.1. Space Filling Curves

Space filling curves[3] (SFC) were first discovered by the Italian mathematician Giuseppe Peano in 1890 for mapping from one-dimensional space to higher dimension, it is classified into fractal mathematical set because of its self-similarity, which is defined as follows:

Make a mapping c as

$$\{c: I \to R^2 \mid I = [0,1], R^2 = \{(x, y); x, y \in R\}\}$$

Then space filling curve is a mapping c covering the unit space S2 = [0, 1] × [0, 1]. For each point P in S there is a real number t, so that c (t) = P. When t is changed from 0 to 1, the map c provides the order of each point in the access space, that is, every point in the plane can be represented by a space filling curve on a continuous line (0,1)[4].

There are many forms of space filling curves. Following Giuseppe Peano, mathematicians Hilbert (1891) and Sierpinski (1912) also constructed different types of space filling curves. The typical space filling curves approximation are shown in Figure 1:

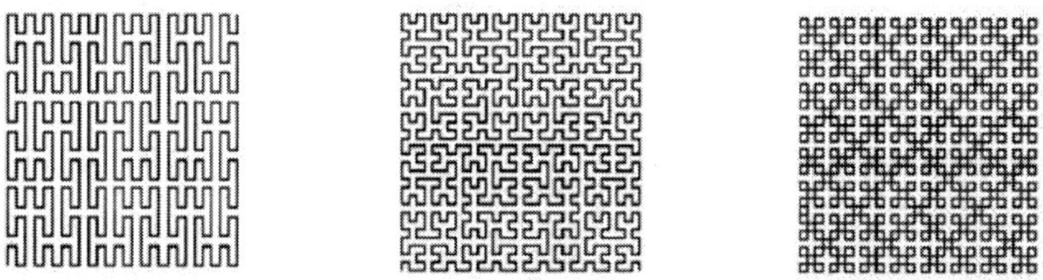

(a)Peano (b) Hilbert (c) Sierpinski

Figure 1. Typical space filling curves.

From the three space filling curves shown above, they all have the ability to access all the pixels, the only difference in traversing the path of the pixels, which are the value of the curves. Since the pixel arrangement of the digital image has a rectangular feature, the space filling curves are easier to traverse each pixel.

The most typical space filling curves are the Giuseppe Peano curve and the Hilbert curve. As the space filling curves have self-similarity, the Hilbert curve is taken as an example, as shown in Figure 2, (a) is the primitive of the Hilbert curve; (b) is the subunit, the four primitives are arranged in the order of four points in the primitive. In this way, the whole Hilbert curve is formed. The same is true of the Peano curve, which is easily achieved in recursive language. However, the space filling curves must occupy a square, such as the Hilbert curve occupied by the size of the area 2n * 2n, Peano curve of 3n * 3n. And the image is not necessarily just so big, especially when the image length and width difference is very large, the image must be divided into several small pieces, according to the order of the curve were dealt with separately [5-8].

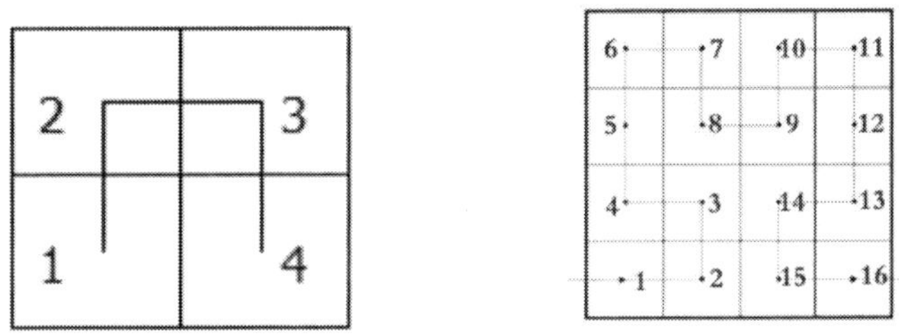

(a)Primitive (b) Subunit

Figure 2. Hilbert curve.

2.2. *Screening Process*

In this experiment, we first make a continuous image by multi-level quantization and error diffusion processing to make it a multi-level gray-scale image, and then call the halftone template to modulate the multi-level gray image and make it secret halftone image. Before modulation, we need to standardize the information to be standardized, spread spectrum, scrambling and other processing, the specific flow chart is shown in Figure 3

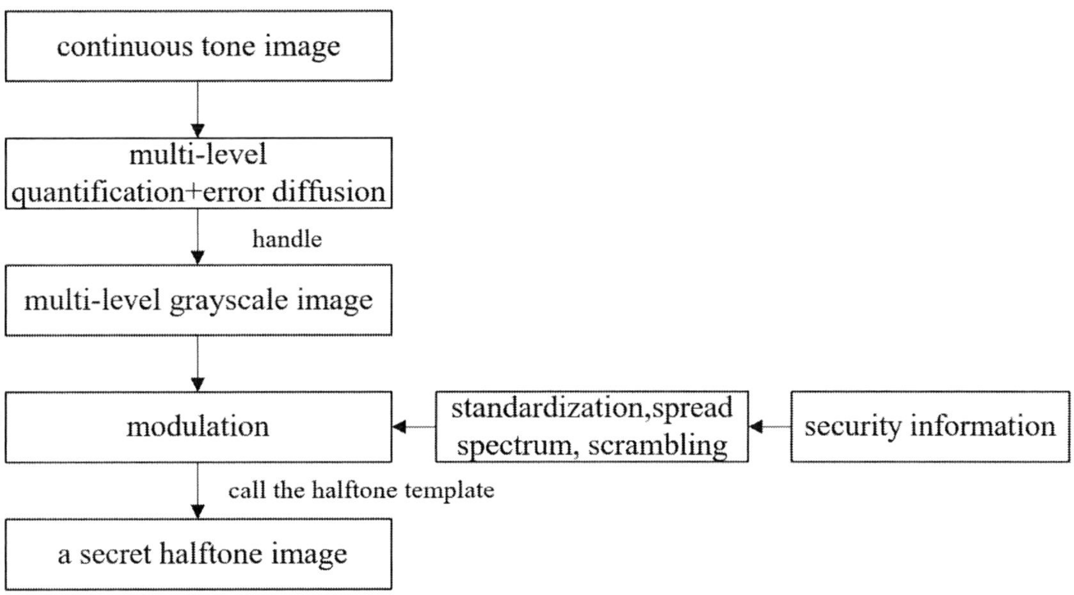

Figure 3. Screening process.

2.3. *Error diffusion*

In this experiment, the multi-value quantization is the original 256 levels of gray scale quantization to 16 grayscale, that is, the original image of the gray value becomes 0 to 15. From 256 gray to 16 grayscale, will naturally compress the details, resulting in errors. Therefore, we need to spread the original image error in order to better restore the original picture details.

In the colour conversion process of the image, due to the different color range, the conversion process may produce errors, the quantization error of the current pixel is spread to adjacent pixels in a certain proportion so the local quantization error is compensated on adjacent pixels [9, 10]. The basic processing consists of three steps, first, the pixel value of the original gray scale image at that point is added to the error value previously diffused there, and as a result of the current input; second, the current input value and the threshold is compared to get a binary output, the third, the input value and the output value of the error in accordance with a certain law spread to the untreated area. The block diagram of the error diffusion algorithm is shown in Figure 4:

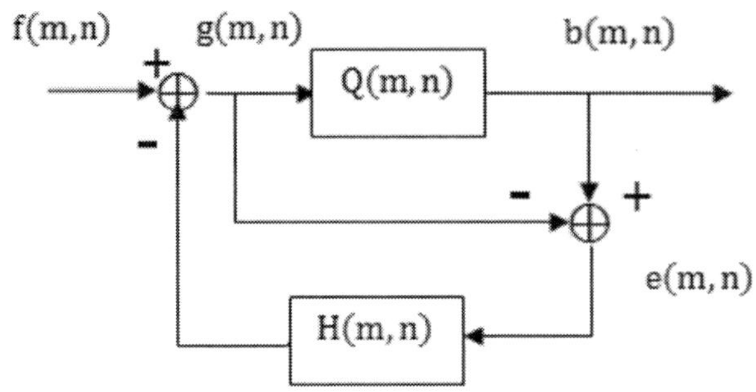

Figure 4. Principle block diagram of error diffusion algorithm.

The mathematical model corresponding to the error diffusion algorithm is:

$$g(m,n) = f(m,n) + \sum_{k,l} w(k,l) \cdot e(m-k, n-l)$$

$$b(m,n) = Q[g(m,n)] = Q[f(m,n) + \sum_{k,l} w(k,l) \cdot e(m-k, n-l)]$$

$$e(m,n) = b(m,n) - g(m,n)$$

H(m,n) is the error filter, $f(m,n)$ is the pixel value of the input image pixel, $e(m,n)$ is the quantization error value, $w(k,l)$ is the coefficient of the filter, $b(m,n)$ is the output pixel value.

2.4. Spread Spectrum Algorithm

Spread spectrum communication is an information transmission method, the spectrum is the frequency domain description of the electrical signal, and the spectrum carrying various information (such as voice, image, data, etc.) are generally represented in the time domain, that is, the information signal can be expressed as a function of time $f(t)$. The frequency domain representation $F(f)$ can be obtained by Fourier transform in its time domain representation $f(t)$ of the signal.The relationship between frequency domain and time domain is determined by equation (1-1):

$$F(f) = \int_{-\infty}^{\infty} f(t) e^{-j2\pi ft} dt$$

$$f(t) = \int_{-\infty}^{\infty} F(f) e^{j2\pi ft} df$$

The sufficient condition for the existence of the Fourier transform of the function $f(t)$ is that $f(t)$ satisfies the Dirichlet condition or is absolutely integrable within the interval (-∞, + ∞), that is, $\int_{-\infty}^{\infty} | f(t) dt$ must be a finite value.

The spread spectrum communication system refers to the spectrum of the information signal to be transmitted is extended into a wide band signal by a specific spreading function (independent of the information signal f (t) to be transmitted) and transmitted into the channel; and then the extension of the spectrum is compressed by using the corresponding technique or means on the receiving end, to restore the original information to be transmitted information signal bandwidth, and thus reach the purpose of transmission of information communication system. That is to say, the radio frequency bandwidth required to transmit the same information signal far exceeds the minimum bandwidth necessary for transmitting the information signal. Information is no longer an important factor in determining the bandwidth of the RF signal. The bandwidth of the RF signal is mainly determined by the spread spectrum function.

In this experiment, the purpose of spread spectrum is to widen the spectrum of the transmitted signal before transmitting the information, so that it can obtain stronger anti-interference ability and high transmission rate, and because the same frequency band different patterns can carry information from different users, so spread spectrum also increases the reuse rate of the band.. In this experiment, the stream 0 is expanded to 0101, and 1 is expanded to 1010.

2.5. Scrambling algorithm

In order to ensure the security of information, it is necessary to scramble the unencrypted information before modulating it. The common iterations are scrambled and the scrambled images have regular fringes, which are easy to be deciphered. Therefore, this paper uses a pseudo-random code chaotic scrambling method. The scrambling image obtained by this method has the advantages of high scrambling degree and good confidentiality, and can be encrypted and decrypted according to the parameter if the corresponding pseudo-random parameter value is set. The chaotic phenomenon is a deterministic, stochastic process that occurs in a nonlinear dynamical system. This process is neither periodic nor convergent,it is extremely sensitive to the initial value, and the long-term orbital behavior is unpredictable,the resulting sequence is called scrambling set by Li-York. The chaotic sequence scrambling algorithm of Logistic is adopted in this paper, and the mapping formula is [11, 12]:

$$x_{k+1} = \mu x_k (1 - x_k)$$

Where $0<4\leq \mu$ is called the branch parameter, $x_k \in (0, 1)$. The research work of chaotic dynamical systems indicates that when $3.5699457< \mu \leq 4$, the Logistic map operates in chaotic states. In other words, the sequence $\{x_k : k = 0, 1, 2, 3 \ldots\}$ generated by the initial condition x_0 under the action of the Logistic map is aperiodic, non-convergent and very sensitive to the initial value. The scrambling image obtained by this method has the advantages of high scrambling degree and good confidentiality, and can be encrypted and decrypted according to the parameter if the corresponding pseudo-random parameter value is set [13]. Figure 5 is a schematic diagram of the image before and after the modulated information is scrambled in this experiment. (a) An image that is scrambled before, (b) is a scrambled image:

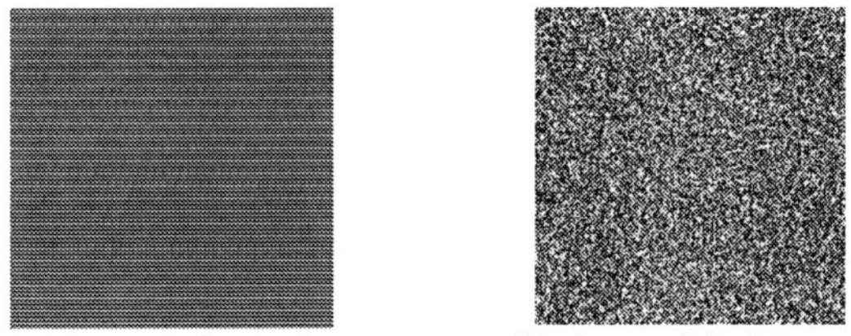

(a) The original picture (b) Scrambled image

Figure 5. Modulation information before and after scrambling.

As the amount of information embedded in this experiment is relatively small, the length of the spread spectrum discontent 128 digits part with 0 completion, making the image arrangement is very regular. But through Logistic chaotic sequence scrambling algorithm, completely disrupted the previous arrangement, the security margin has been improved.

3. Information Modulation

The information to be modulated in this experiment is a string of Baidu URLs, which is character information, but the modulation to the plaintext image is a string of 0/1 stream, so the string needs to be changed to 0 / 1 stream through certain rules. In this paper, the coding rule of the two-dimensional code is chosen to change the character information into 0/1 stream. Two-dimensional code character encoding table shown in Table 1:

Table 1. Two-dimensional code character encoding table.

Char	Value	Char	Value	Char	Value	Char	Value	Char	Value	Char	Value	Char	Value	Char	Value
0	0	6	6	C	12	I	18	O	24	U	30	SP	36	.	42
1	1	7	7	D	13	J	19	P	25	V	31	$	37	/	43
2	2	8	8	E	14	K	20	Q	26	W	32	%	38	:	44
3	3	9	9	F	15	L	21	R	27	X	33	*	39		
4	4	A	10	G	16	M	22	S	28	Y	34	+	40		
5	5	B	11	H	17	N	23	T	29	Z	35	-	41		

Through the two-dimensional code encoding rules Baidu Web site convert into a string of code stream, and then spread spectrum processing for this string of code streams. Since the mesh needs to be embedded in the same size as the original image, the array after the spread spectrum is placed in the same array as the original size, and the lack of part is filled zeros. At this point, a matrix of 0/1 equal to the original size is obtained, and then the matrix is scrambled, through a certain rule can be divided into different status classification, respectively, 00,01,10,11, by judging different status calls a different template for screening.

When the processed information is modulated, a screen template based on the phase characteristics of the space filling curves is selected. The space filling curve is used to fill the black regions of the map with a determined trajectory in the printing process, forming a modulation dot with different area rates. The filling route is shown in (b) above in Figure 2. According to this, we can get a 4 * 4 base template, through the basic template 90 °, 180 °, 270 ° rotation to get four different templates, according to the different forms of modulation information call different templates.

4. Screening Result

The test chart selected in this experiment is 128 * 128 color logo map, the modulated dark message is www.baidu.com, through the two-dimensional code encoding rules to convert the string to 0/1 code stream, the spread spectrum of its 0 to [0 1 0 1], 1 expanded to [1 0 1 0], the plaintext information is complemented by the tail zeros padding to 128 * 128 of the 0/1 matrix, and then the matrix is transformed into irregularly arranged matrix for information modulation. By dividing the bits in the matrix into groups with two bits in each group, and then call different templates according to different morphologies, the screening contrast diagram is shown in Figure 6, (a) the ordinary amplitude modulation screening, (b) the ordinary frequency modulation screening, (c) the space filling curves phase feature screening, (d) dots shape of space filling curves phase feature screening:

(a) AM screening (b) FM screening (c) space filling curves phase feature screening

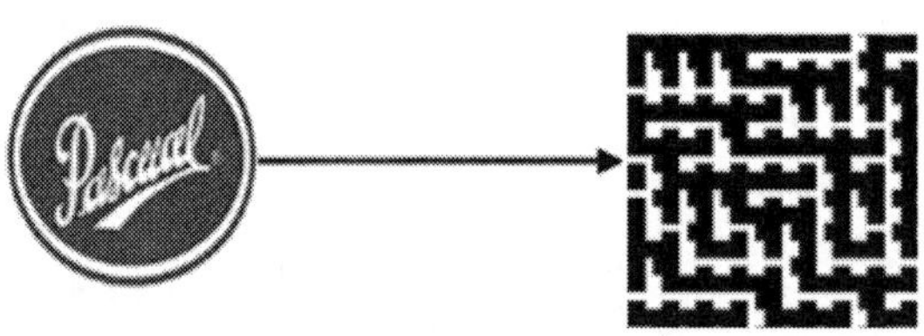

(d) dots shape of space filling curves phase feature screening

Figure 6. Screening effect comparison.

Comparing the figures, the results of the screening based on the phase feature of the space filling curve is more delicate, the reduction degree of original image is high and the reproduction effect of image is clear. In the amplitude modulation screening, we always directly can see the presence of moire , even more serious in the case of zooming . What's more, it has a great bad influence on the processing to the edge in the middle of the logo part for the image, causing a clearly fuzzy. In the FM screening, effectively solved the problem of moire, but lost the details of the image'edge .The same time, the recovery for the edge in the middle of the logo part of the image is low and the original picture reproduction is not good. The method that screening based on the phase feature of the space filling curves is proposed, saving the problems. And retain more details of the original image avoiding moire, the edge details are more delicately and smoothly handled making a better visual and significantly improving image reproduction effect. Make the security of encrypted information better by enlarging the dots which are 'back' type irregular distribution.

5. Conclusion

In view of the existence of the moire and image reproduction ineffective problem, this paper applies the idea of space filling curves to halftone image screening and information hiding, which can solve the above problems effectively. Experimental results show that the image after the screening can not only solve the problem of moire, but also a better reproduction of the original map, the algorithm of this paper has a broad development prospect in the field of security printing.

6. Acknowledgments

This paper is supported by the National Natural Science Foundation of China (No. 61370140) and Beijing science and technology special (Z161100005016007).

7. References

[1] GUO Ling-hua,XING Tie-dou,LIU Guo-dong,CHEN Yan. Appearing Effect of Information Hiding Technology Based on Halftone Screening [J]. Packaging Engineering, 2015, 36(15): 135-137+158.

[2] LIU Zhe-can,CAO Peng. Halftone screening and information hiding algorithm based on the location of dots [J]. Journal of Beijing Institute of Machinery, 2013, 28(06):25-28.

[3] Join J,Loren K P. Heuristics based on spacefill curve for combinatorial problems in euclidean space[J]. Man-agement Science, 1988, 34(3):290-305.

[4] HU Da-wei, LIU Gui-ying. Solving travelling salesman problem (TSP) based on space-filling curve and or-opt algorithm [J]. Journal of Chang'an University (NATURAL SCIENCE EDITION), 2007, 27(02):85-88.

[5] Lu Zhongkang,Shi Pengfei,Shan Ying. A Halftoning Algorithm Based on Space-Filling Curves [J]. Computer Applications and Software, 1998, (02):29-33+45.

[6] Yi Chuan Shao,Xing Jia Yao,Li Wei Tian. Peano Space Filling Curve Applied in Managing P2P Service Resources [J]. Applied Mechanics and Materials, 2012, 2025(220): 2508-2511.

[7] Yao Haigen. Digital halftone applications for space filling curves [J]. Publishing & Printing, 2013, 13(04):2-5.

[8] [8] Luiz Velho,Jonas de Miranda Gomes. Digital halftoning with space filling curves [J]. ACM SIGGRAPH Computer Graphics, 1991, 25(4): 81-90.

[9] Zhang Hanbing,Yuan Xin. Study on error diffusion algorithm of digital halftoning[J]. Journal of Computer Applications, 2010, 30(04): 925-928.

[10] Lee Je-ho,Allebach J P.Inkjet Printer Model-based Halftoning[J].IEEE Transactions on Image Processing, 2005, 14(5):674-688.

[11] Cao Guanghui,Hu Kayoing Wei. A Method of Binary Tree Traversal Scrambling Based on Logistic Chaotic Sequence [J]. Acta Physica Sinica | Acta Phys Sin, 2011, 60(11): 33-140.

[12] LIU Xiangdong,Zhang Junxing,Zhang Jinhai,He Xiqin.Image Scrambling Algorithm Based on Chaos Theory and Sorting Transformation[J].International Journal of Computer Science and Network Security, 2008,8(1):64-68.

[13] Niu Yongjie,Zhao Yaofeng. New Image Scrambling Hiding Algorithm Based on Chaotic Proliferation Encryption [J]. Computer and Modernization, 2013, (06):95-99.

IOP Conf. Series: Journal of Physics: Conf. Series **887** (2017) 012084 doi:10.1088/1742-6596/887/1/012084

Design of dual DC motor control system based on DSP

Peicheng Shi[1], Suo Wang[2], Zengwei Xu[3] and Ping Xiao[4]

Anhui Engineering Technology Research Center of Automotive New technique, Anhui Polytechnic University, Wuhu, Anhui, China 241000
Email: [1]shipeicheng@126.com, [2]1159461528@qq.com, [3]1421385950@qq.com, [4]tlxp95@163.com

Abstract. Multi-motor control systems are widely used in actual production and life, such as lifting stages, robots, printing systems. This paper through serial communication between PC and DSP, dual DC motor control system consisting of PC as the host computer, DSP as the lower computer with synchronous PWM speed regulation, commutation and selection functions is designed. It sends digital control instructions with host computer serial debugger to lower computer, to instruct the motor to complete corresponding actions. The hardware and software design of the control system are given, and feasibility and validity of the control system are verified by experiments. The expected design goal is achieved.

1. Introduction

DC motor features good linear speed regulation characteristics, high efficiency, simple control, good speed regulation performance and small size, which has been widely used in many areas [1]. DSP is the current mainstream microprocessor. With strong data processing capabilities and high running speed, it can be used for fields of automatic control, image processing, etc. To achieve motor motion control, the most widely used program is to take DSP as microprocessor of control system [2]. For example, in literature [3], PC and DSP master-slave structure is used to complete DSP control with PC as the host computer, and man-machine dialogue is achieved through serial communication, a large number of data processing is given to DSP, and finally motor movement is controlled by a motor driver. However, it only achieves DSP control of single motor. More exploration is needed to achieve synchronous control of multiple motors.

This paper designs a control system of dual DC motor with functions of synchronous PWM speed regulation, commutation and selection, which achieves communication between PC and DSP via RS232 serial port[4], sends digital control instructions to lower computer through the host computer serial debugger so that motor driver drives the motor to complete corresponding actions. Reliability and effectiveness of the control system are verified by actual system test. Compared with the traditional control scheme, this scheme can change the speed and turn of motor in real time, and can select the motor that is used.

2. Hardware Design in Control System

The overall control scheme of the system is shown in Figure 1, which mainly includes the following parts: PC, emulator, DSP, motor driver, lithium battery and DC motor. PC is mainly used to achieve communication with DSP and online debugging of system program through the emulator. DSP is used to receive the data sent by the host computer and generate PWM speed regulation signal, motor steering signal and select motor signal [5]. The three signals instruct two DC motors to complete the corresponding control instructions through motor driver, and achieve synchronous speed regulation, commutation and selection motor of the two DC motors. The system selects motor model: JS motor

5D90GN-24, parameters: 90W, DC24V, 7A, 1800r / min, reduction gear ratio of 5GN-180K. The two DC motors are powered by24V lithium battery.

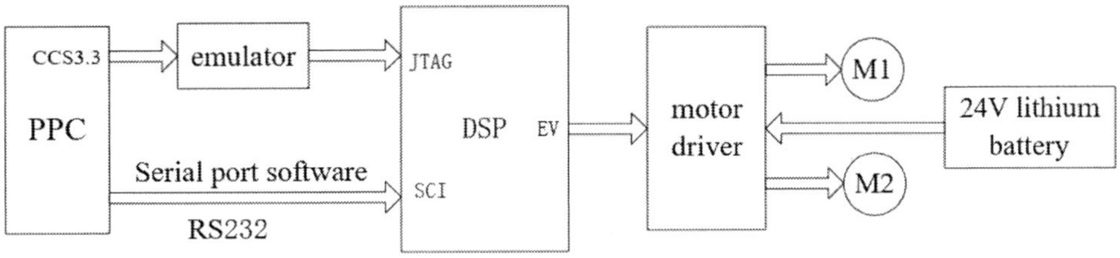

Figure 1. Overall scheme of the control system

2.1. Emulator Selection

The control system adopts Fourier electron XDS510V2 Emulator, which uses high-speed version USB2.0 interface supporting plug and play, with transmission speed up to 480MB/S; supports TI's C2000, C5000 full range of DSP and C6000 part model of DSP; adopts standard JTAG emulation interface, to avoid taking up user resources. The emulator is connected with the JTAG port on the DSP chip. The USB port of the emulator is connected with that on PC. By connecting PC and DSP with the simulator, on-line debugging of control system program can be done in CCS3.3 [6].

2.2. DSP Chip

The control system uses TMS320F2812 DSP as the main controller. TMS320F2812 is TI's 32-bit fixed-point DSP chip, which is currently one of the most cost-effective DSP chips. It not only has strong digital signal processing capabilities, but also has relatively complete event handling capabilities and embedded control capabilities [7]. DSP chip uses the advanced Harvard bus structure, in which the program and data are placed in different storage space to greatly improve DSP data processing rate [8]. CPU basic frequency is up to 150MHZ, the clock period is 6.67ns; the chip contains 18K RAM and 128K Flash. The chip also contains a wealth of peripheral resources, greatly reducing difficulty of hardware circuit design. In particular, two event manager modules EVA, EVB can produce 4 independent PWM waveforms and 6 pairs of 12 complementary PWM waveforms, which is very suitable for control of various types of motors [9]. Meanwhile, programming is compatible with C/C ++ language and assembly language.

2.3. Motor Driver

The motor driver is an important component that drives the motor to complete acceleration, deceleration, and braking, coasting plus positive and negative rotations. The system uses a dual-channel DC brushed motor driver which has a continuous output capacity of 12A current per channel, can drive DC motor of up to 290W electric power, has built-in overvoltage, undervoltage, overheat protection circuit to support full duty cycle input. In order to reduce and lower the impact of drive circuit on the entire control system and enhance security of the main control chip circuit, optocoupler isolation chip HCPL0630 is used for optical isolation of all the control signals input by DSP, and the optoelectronic device can significantly improve electromagnetic compatibility of the system; H bridge circuit consists of four N-channel enhanced AON6504 field effect tubes, drain source voltage of AON6504 field effect tube can reach 30V, gate-source voltage is 20V and the maximum current is up to 85A.

DSP -driver connection is shown in Figure 2, in which, T1PWM, T2PWM, PWM1, PWM2, PWM7, PWM8 are internal and external output pins of DSP chip, VCC1, VCC2 in the motor driver are optocoupler isolation power input of motors M1, M2, which are compatible with 3.3V, 5V power supply. EN1, EN2 are motor speed adjustment input; INA1, INB1 are control logic input of motor M1; INA2, INB2 are control logic input of motor M2; GND is grounded, this pin can be vacant.

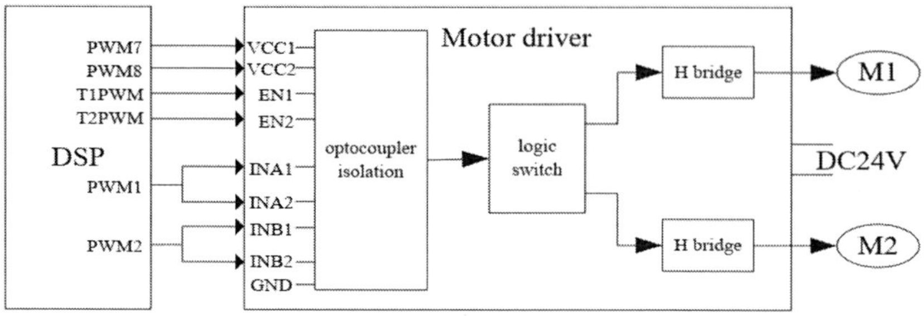

Figure 2. DSP is connected to the motor driver

The motor operation logic is shown in Table 1. Wherein, x can be 1 or 2, H is high level, L is low level, X means independence of level, and Z is high impedance state.

Table 1. Motor Operation Logic

Signal input			Power output
INAx	INBx	ENx	Motor status
L	L	X	Braking
L	H	TxPWM	Positive rotation
H	L	TxPWM	Negative rotation
H	H	Z	Coasting

2.4. Communication Circuit Design

In order to realize instruction digitization of dual DC motor speed regulation, steering and selection control system, this paper establishes communication module of PC and lower computer. The serial communication port is an asynchronous serial communication interface with two signal lines [10]. The serial communication driven by MAX3232 chip conforming to RS-232 standard has two receiving and transmitting channels [11]. Serial communication interface circuit is shown in Figure 3, JP3 corresponding to SCIA serial communication interface is used for DC motor speed regulation and commutation; JP4 corresponding to SCIB serial communication interface is used to select the DC motor used.

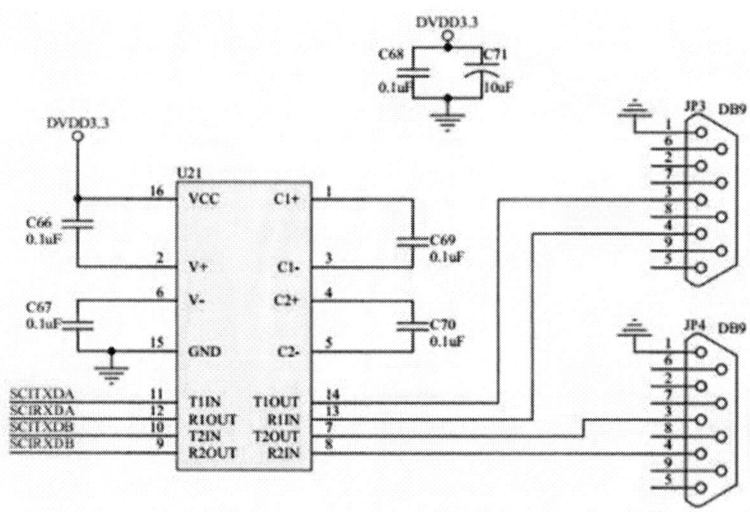

Figure 3. Serial communication interface circuit

3. Control System Software Design

Control system program includes initialization program, the main program, SCI serial communication program and motor control program. The initialization program is used to initialize the various function modules and global variables involved in the system. SCI serial communication program mainly consists of initialization function, receive interruption function of motor speed regulation and commutation data, receive interruption function of selection motor data. The motor control program mainly consists of motor PWM speed regulation program, motor steering adjustment program and selection motor program. The control system flow chart is shown in Figure 4. All the programs in the system are written in CCStudio v3.3 software, each module in CCS3.3 has its own fixed programming template, and each peripheral has its corresponding sub-function module. In programming, one only needs to modify or add corresponding code in the corresponding template.

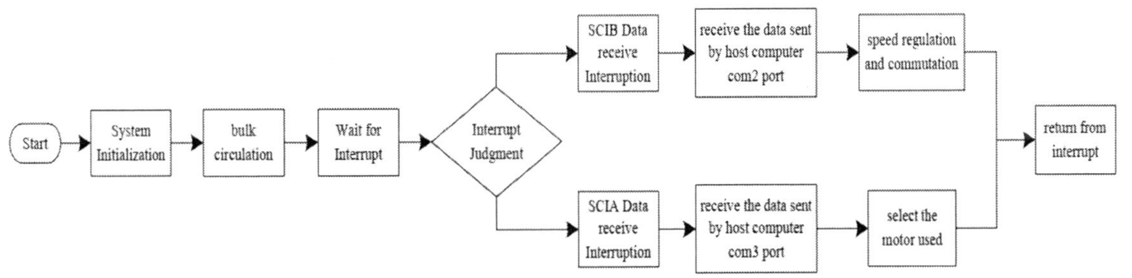

Figure 4. Program flow chart

3.1. Serial Communication Program Design

The serial communication program is used to receive the data sent by PC through serial port debugger and assign the received data to the corresponding global variable to realize speed regulation, commutation and selection of dual DC motor [12]. The system uses interrupt way to achieve data acceptance. When the data sent by host computer through the serial debugger reaches SCIRXDX pin, SCI starts to detect the start bit of the data and moves the data bit by bit into RXSHF register. When the RXSHF register receives a complete data and sends the data to receive data buffer SCIRXBUF, flag bit RXRDY is set, and generates a request interrupt to receive data. If the corresponding interrupt enable bit is enabled, it will generate SCI receive interruption.

Serial communication initialization function makes corresponding configuration of SCIA, SCIB, so that it can achieve communication and data acceptance between PC and DSP. SCIA serial port is used for motor speed regulation and commutation data acceptance, while SCIB serial port is used for selection motor data acceptance. SCIB communication data format is set to baud rate of 19200, data bit of 8, no polarity check and stop bit of 1; SCIB communication format is set to baud rate of 9600, data bit of 8, no polarity check and stop bit of 1. When configuring serial debugger parameters, they must be exactly the same as the corresponding SCI module settings. Speed regulation and commutation data receive interruption function receives the data sent by the host computer serial debugger through com2 serial port, and stores the received data in the array Scia_VarRx[i]. Scia_VarRx[0] to Scia_VarRx[3] data in the array are used for motor speed regulation, while Scia_VarRx[4] and Scia_VarRx[5] data are for motor commutation. Hence, the data sent by speed regulation and commutation serial port debugger each time is 6-bit decimal number. The first four digits of the array are combined and assigned to compare registers of the timers T1, T2 through intermediate variable CMPW_PWM. The last two digits of the array are combined and assigned to compare control register ACTR of EVA through intermediate variable PWM_DIRECT. Selection motor data interrupt function receives the 2-bit hexadecimal data sent by host computer serial port debugger through com3 serial port, stores the data in SCIRXBUF, and assigns it to compare control register ACTRB of EVB through intermediate variable choose.

IOP Conf. Series: Journal of Physics: Conf. Series **887** (2017) 012084 doi:10.1088/1742-6596/887/1/012084

3.2. Motor Control Program Design

The motor control program includes motor PWM speed regulation program, motor steering adjustment program and selection motor program. The system regulates motor speed by changing armature voltage at both ends of DC motor, that is, it regulates speed by changing DSP output PWM waveform duty cycle. This method can achieve inorganic speed regulation of motor and belongs to constant torque speed regulation. The motor commutation changes motor control logic input through forced level of PWM1, PWM2 output pin level, to achieve motor commutation. Selection motor is achieved by selecting conduction motor drive signal with PWM7, PWM8 output pin level.

The motor speed regulation program mainly configures the timer T1 and T2, and sets T1 and T2 to continuous up / down counting mode in the initialize EV function. T1PWM and T2PWM output polarity are set to active low; In this system, PWM waveform frequency is set to 18KHz to reduce current noise during motor driver operation; To change the value of TxCMPR in real time (i.e., to change duty cycle of PWM waveform), TxCMPR heavy load condition is set to load immediately. In order to achieve simultaneous change of two PWM waveform duty cycles, values of compare register T1CMPR, T2CMPR are determined by the value of variable CMPR_PWM. The motor commutation program is to configure output polarity of DSP output pins PWM1 and PWM2, of which, PWM1 and PWM2 are PWM function pins. PWM1, PWM2 output polarity is achieved by changing the value of comparison behavior control register ACTRA. Selection motor program makes appropriate configuration of timer T3, sets PWM7, PWM8 to PWM function pin, and output polarity is decided by the value of compare control register ACTRB.

4. System Testing

According to the above design, hardware connection diagram of the control system is built as shown in Figure 5. The system provides that data sent by motor speed regulation and commutation serial port debugger is 6-bit decimal number, while data sent by selection motor serial debugger is 2-bit hexadecimal number. The selection motor control word and motor running status are shown in Table 2. The system-related experiments and debugging results are shown in Figure 6, Figure 7, Figure 8, Figure 9, and Figure 10.

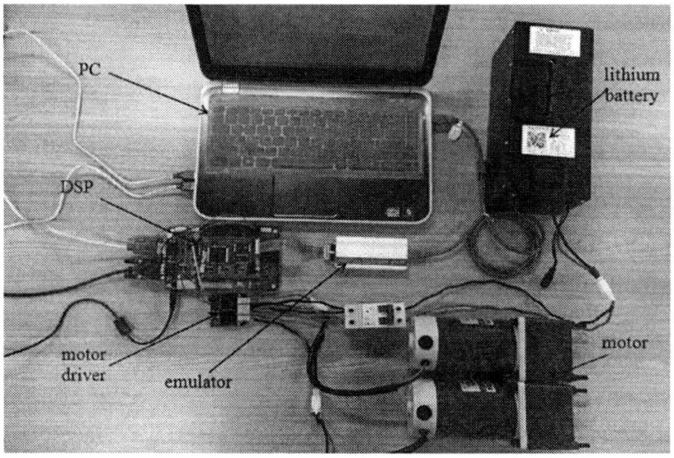

Figure 5. Hardware connection diagram of the control system

Table 1. Motor Operation Logic

Control word	motor running status
00	motors M1, M2 are not running
03	motor M1 is running
0C	motor M2 is running
0F	motors M1, M2 are running

Name	Value	Type	Radix
ScibRegs.SCIRX...	15	Uint16	unsigned
choose	15	Uint16	unsigned
EvbRegs.ACTRB.all	15	Uint16	unsigned

Figure 6. Global variable values at motor selection.

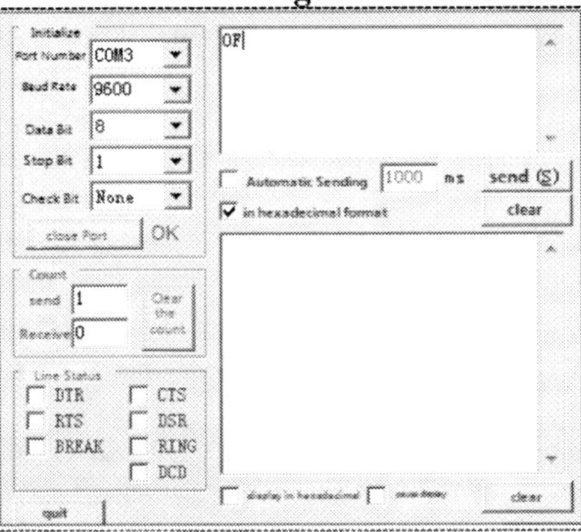

Figure 7. Motor selection window .

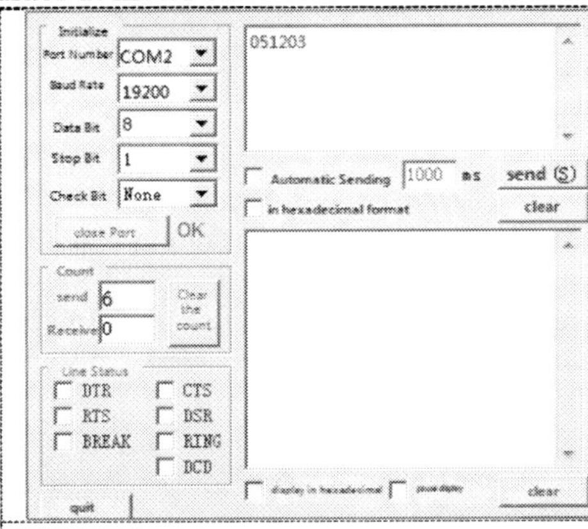

Figure 8. Motor speed regulation, commutation window.

First, "0F" is sent by selection motor data transmission window as shown in Fig.7, and the selected motor is M1, M2. After SCIB data receive interruption receives "0F", the global variable value related to the selection motor is shown in Figure. 6. Then, "051203" is sent by motor speed regulation and commutation data transmission window shown in Figure. 8. At this time, the two DC motors are in positive rotation and the rotation speed is 1/2 of that under rated voltage. After SCIA data receive interruption receives "051203", the global variable values of motor speed regulation and commutation are shown in Figure. 9. The PWM waveform output by T1PWM, T2PWM pin on DSP development board that is measured by oscilloscope is shown in Figure. 10, in which, yellow waveform is T1PWM wave, blue waveform is T2PWM wave. It can be seen from the figure that PWM wave duty cycle is 49.27%. And voltages at both ends of M1, M2 motor measured by multimeter are12.47V, 12.49V respectively at this time.

Name	Value	Type	Radix
Scia_VarRx	0x003F910D	unsigned int[6]	hex
[0]	0	unsigned int	unsigned
[1]	5	unsigned int	unsigned
[2]	1	unsigned int	unsigned
[3]	2	unsigned int	unsigned
[4]	0	unsigned int	unsigned
[5]	3	unsigned int	unsigned
CMPR_PWM	512	Uint16	unsigned
PWM_DIRECT	3	Uint16	unsigned
EvaRegs.T1CMPR	512	Uint16	unsigned
EvaRegs.T2CMPR	512	Uint16	unsigned
EvaRegs.ACTR.all	3	Uint16	unsigned
EvaRegs.T1PR	1042	Uint16	unsigned
EvaRegs.T2PR	1042	Uint16	unsigned

Figure 9. Global variable value chart of motor speed regulation, commutation.

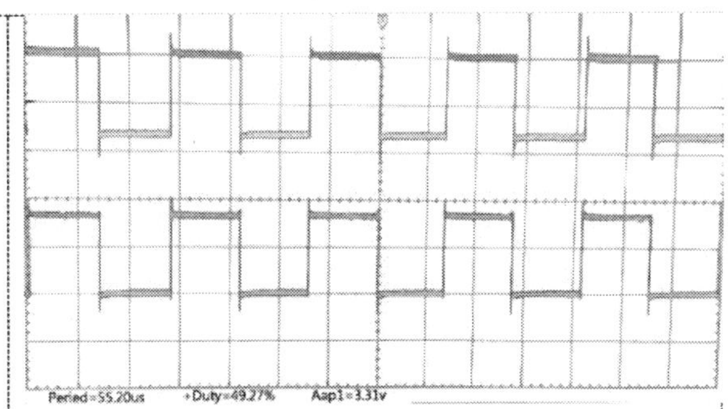

Figure 10. PWM waveform by DSP output.

5. Concluding Remarks

A dual DC motor control system with functions of dual DC motor speed regulation, steering and selection was constructed using TMS320F2812 DSP development board and motor driver, which not only simplifies the system structure, but also achieves digital control of the motor with host computer serial port debugger. Through the actual control system testing, it is verified that the designed control system can realize speed regulation, steering and selection of dual DC motor. The control system proposed in this paper can also be applied to control of more motors.

6. Acknowledgment

The authors would like to thank the financial supports of the National Natural Science Foundation of China (Grant No. 51575001), of Anhui province science and technology research key project (Grant No. 1604a0902158), of Anhui university scientific research platform innovation team building projects (2016-2018) and. of the Young and middle-aged top-notch talent project of Anhui Polytechnic University (2016BJRC010).

7. References

[1] Yanni Xiangli 2015 D. Design of Robot Motion Control System Based on DSP. *University of Chinese Academy of Sciences*

[2] Baofa Zhang,Hui Zhao and Youjun Yue 2010 J. Designing of stepper motor control system based on DSP. *Instrument Technique and Sensor*. **8** 63-65+81

[3] Yuansheng Xiong,Suxiang Qian and Jinfeng Gao 2004 J. Design of the control systems of stepper motor based on DSP and serial communication with host computer. *Mechanical & Electrical Engineering Magazine.* **9** 17-21

[4] Xun Wu,Yuesong Mei,Jianqiao Yu,Tianpeng Yu and Jingxu Li 2013 J. A Design of UART Serial Communication between the TMS320C6748 DSP and PC. *Applied Mechanics and Materials.* **380** 3657-60

[5] Wencheng Wang 2010 J. A Speed Regulation System of DC Motor Based on PWM Technology. *Applied Mechanics and Materials.* **29** 2194-99

[6] Guiyin Zhan 2014 J. Designed Controller of Permanent Magnet Brushless DC Motor Based on DSP. *Applied Mechanics and Materials*. **672** 1210-13

[7] Lv Yang,Jinghong Zhao 2012 J. The Design of DC Motor Excitation Regulation System Based on DSP. *Advanced Materials Research.* **383** 5123-29

[8] Jun Zhu,Wankui Li,Haixing Wang and Lili Han 2013 J. The Hardware System Design of PMSM Controller Based on DSP2812. *Applied Mechanics and Materials.* **273** 454-459

[9] Aicheng Zou,Qunying Wang and Yijie Cao 2014 J. Design of PWM DC Motor Control System Based on Single Chip. *Advanced Materials Research.* **1030** 1461-64

[10] Lina Duan,Jin Zhao 2014 J. Design of Control System of Brushless DC Motor Based on DSP. *Applied Mechanics and Materials.* **496** 1417-21

[11] Yun Yang,Yue Zhu 2013 J. Based on Serial Port Class Implements of SCM and PC Serial Communication Program Design. *Advanced Materials Research.* **629** 667-670

[12] Ľubica Miková,Ivan Virgala,Michal Kelemen 2015 J. Speed Control of a DC Motor Using PD and PWM Controllers. *Solid State Phenomena..* **220** 244-250

IOP Conf. Series: Journal of Physics: Conf. Series **887** (2017) 012085 doi:10.1088/1742-6596/887/1/012085

Dynamic formant extraction of wa language based on adaptive variational mode decomposition

Meijun Fu, Huazhen Dong and Wenlin Pan*

School of Mathematics & Computer Science, Yunnan Minzu University, Kunming, Yunnan, China, 650500.

Email: 779685578@qq.com, 1248571027@qq.com, panwenlin@sina.cn

Abstract. Wa language is one of Chinese minority languages spoken by the Wa nationality who lives in Yunnan Province, China. Until now, it has not been studied from the perspective of Engineering Phonetics. In this paper, for the above reason, by the adaptive variational mode decomposition (AVMD) we have investigated the dynamic formant characteristics of Wa language. Firstly, more precisely, use the synthetic dimension to split Wa language isolated words into voiceless and voiced segment, initials and finals. Secondly, use Linear Prediction Coding to estimate the first three formant frequencies and their bandwidths roughly. Thirdly, select the appropriate equilibrium constraint parameter and the number of decomposed layers so that Adaptive Variational Mode Decomposition (AVMD) can decompose the signal into some intrinsic mode functions (IMFs) without pattern aliasing. Fourthly, use the estimated formant frequencies and bandwidths to determine precisely the required IMFs. Fifthly, use the Hilbert transform to calculate the instantaneous frequency of the above determinate IMFs. Further, we implement the weight average operation on instantaneous frequencies to obtain the first three formant frequencies for each frame. Finally, comparing the first three formant frequencies obtained by the adaptive variance modal decomposition and by *Praat* software respectively, so we have drawn the conclusion that the relative correct rate of the former to the latter can reach 86% averagely in terms of the selected isolated words, which has shown that our method is effective on Wa language.

1. Introduction

The Wa nationality [1] is an important member of the Chinese nations. They live mainly in the southwest border of Yunnan Province, China and northern Myanmar. And the native language they use to communicate with each other is called Wa language. Wa language belongs to Va-De'ang branch, Mon-Khmer languages, and Austro-Asiatic family. Wa language can be divided into three kinds of dialects, which are "Barao dialect", "Lawa dialect" and "Wa dialect" respectively. Currently, Wa language is based on the Barao dialect and the voice of YanShuai is standard pronunciation. Until now, Wa language has not been studied by people from the perspective of Engineering Phonetics. For the above reason, we have decided to try some new things. So we try to investigate the dynamic formant frequencies on isolated words of Wa language. Here, the speech corpus used in this paper are all recorded in YanShuai.

It is well known that the glottal excitation signal at a quasi-periodic impulse into the sound channel will cause resonance characteristic, resulting in a group of resonant frequencies which are called formant frequencies or formants [2]. The formant includes two important parameters: the formant frequency and its width of band width (bandwidth), which are important parameters that distinguish different voiced sounds, of course, including vowels. The formant information is lied in the envelope of the speech spectrum. Therefore, the key to the extraction of formant parameters is to estimate the natural speech spectrum envelope, and thinking that the frequency corresponding to the maximum value of the spectral envelope is just the resonance peak. The resonant frequency can be arranged from low to high, respectively, the first formant ($F1$), the second formant ($F2$), the third formant ($F3$), and the like. In general, there are about five resonant peaks that can be told in a voiced, of which the first three have a critical effect on distinguishing different voiced sounds. While the voiceless short-term spectrum has no these two characteristics, it is very similar to a random noise spectrum.

Most of the traditional speech signal processing methods are based on deterministic linear system theory [3]. And Teager et al. [4] [5] in 1980 found that in speech signals there exists a nonlinear, random and chaotic mechanism. The chaotic dynamical system converges to a certain attractor, which can be fractal modelling in the phase space [6] [7] [8]. So the chaotic fractal theory can be used to study the speech signal. The study has found that in the voice signal, consonant (without periodicity) [2] chaos is much greater than the vowel (with periodicity) [2]. Multi-fractal [9] is an infinite set of singular measures defined by multiple scale exponents on fractal, and it can be fully presented probability measure distribution ratio and unevenness of speech signal in the overall fractal structure, it therefore, can be expressed as a sum of fractal subsets with different dimensions, which improves signal geometrical characteristic and local scale. Based on the above analysis, Multi-fractal can be used to split the voiceless and voiced segment, initials and finals.

Based on the nonlinearity of speech signals [4] [5], Maragos et al. [10] [11] [12] have proposed a nonlinear speech model----amplitude-modulated-frequency-modulated model (AM-FM model), which can be used to describe the amplitude and frequency of real-time speech formant [10]. At present, based on the AM-FM model, the main adaptive decomposition methods are: Empirical mode decomposition (EMD) [13], ensemble empirical mode decomposition (EEMD) [14], local mean decomposition (LMD) [15] and variational mode decomposition (VMD) [16]. When the signal is decomposed by EMD, there will be mode mixing and end effect [17] [18] [19] in the intrinsic mode functions (IMFs). In order to solve mode mixing, Huang proposed intermittent criterion [17] and ensemble empirical mode decomposition (EEMD) [14]. Due to that the EEMD has to add Gaussian white noise continuously, thereby increasing the amount of calculation. LMD, only to a certain extent weakens mode mixing and end effect. VMD [16], as a new signal processing method, it is assumed that the signal is composed of some IMFs which possess finite bandwidths with different center frequencies. In order to minimize the sum of the estimated bandwidth of each IMF, through the conversion to solve the variational problem, the IMFs are demodulated to the corresponding base band, and finally extract the IMFs and the corresponding center frequencies.

This paper is organized as follows. In Section 2, introduce the synthetic dimension that can better split Wa-language isolated words into voiceless and voiced segment, initials and finals. In Section 3, introduce the variational mode decomposition (VMD) that can eliminate the mode mixing, which helps us to decompose the formant frequency into the responding intrinsic mode function (IMF). In Section 4, introduce the adaptive variational mode decomposition (AVMD) on the basis of VMD. Section 5 illustrates the formant extraction algorithm based on AVMD; Section 6 includes experiment and results.

2. Synthetic Dimension

2.1. Definition of Generalized Fractal Dimension

Let δ be the side length of a box and $N(\delta)$ the least box number covering the set Γ, where $\Gamma = \{x(i)\}_{i=1}^{N}$ be a time series. And let P_i be the probability of falling into the i th box, Then, for any $q \in \mathbb{R}$, define Renyi's information entropy as follows.

$$K_q(\delta)=\begin{cases}\left(\ln\sum_{i=1}^{N(\delta)}(P_i)^q\right)\Big/(1-q), q\neq 1;\\ -\sum_{i=1}^{N(\delta)}P_i\ln P_i, \qquad q=1.\end{cases} \tag{1}$$

So we can define the following Renyi generalized dimension:

$$D_q=-\lim_{\delta\to 0}\left(K_q(\delta)/\ln\delta\right) \tag{2}$$

Especially, if $q=0,1,2$ respectively, D_0, D_1, D_2 is called Box, Information and Correlation dimensions respectively. Concretely, they are:

(1) Box dimension

$$D_0=-\lim_{\delta\to 0}\left(\ln N(\delta)/\ln\delta\right) \tag{3}$$

(2) Information dimension

$$D_1=-\lim_{\delta\to 0}\left(\ln\sum_{i=1}^{N(\delta)}P_i\ln P_i\Big/\ln\delta\right) \tag{4}$$

(3) Correlation dimension

$$D_2=\lim_{\delta\to 0}\left(\ln\sum_{i=1}^{N(\delta)}P_i^2\Big/\ln\delta\right) \tag{5}$$

2.2. Calculation of Generalized Dimension

Let Δt be the sampling interval, and $\delta_j=2^j\Delta t$ the side length of box, where $j=1,2,\cdots,J$. The steps of calculation on generalized dimension are the following:

1) According to the different value δ_j, compute the total number $N(\delta_j)\equiv N_j$ of boxes covering a frame of speech signal and the total number d_{ji} which fall into the ji th box, where $i=1,2,\cdots,N_j, j=1,2,\cdots,J$. As a result, $p_{ji}=d_{ji}/L$ represents the probability of a point falling into ji-th box. Here, L is the length of a frame of speech signal, i.e., the total number of points in the set Γ.

2) Substitute p_{ji} for P_i, then we are able to obtain $K_q(\delta_j), j=1,2,\cdots,J$. Construct the following Equation (6), such that it attains the minimum.

$$\min=\sum_{j=1}^{J}\left(y_j+D_qx_j+b\right)^2 \tag{6}$$

Here $x_j=\ln\delta_j, y_j=K_q(\delta_j), j=1,2,\cdots,J$.

3) We get easily the computational formula of D_q as follows.

$$D_q=\left(J\sum_{j=1}^{J}x_jy_j-\sum_{j=1}^{J}x_j\sum_{j=1}^{J}y_j\right)\Big/\left(J\sum_{j=1}^{J}x_j^2-\left(\sum_{j=1}^{J}x_j\right)^2\right) \tag{7}$$

4) For any definite value q, by the above formula we can obtain different character dimension D_q.

2.3. The Definition and the Calculation of Synthetic Dimension

Box dimension D_0 describes the geometric scale of a set, information dimension D_1 gives the point distribution in a set, and correlation dimension D_2 reflects the relational degree amongst points in a set. Therefore, we can define synthetic dimension as follows.

For $0 \le \alpha, \beta, \gamma \le 1; \alpha + \beta + \gamma = 1$, define

$$S_D = \alpha D_0 + \beta D_1 + \gamma D_2 \tag{8}$$

The above synthetic dimension possesses are better ability to identify voiceless and voiced segment, initials and finals than single fractal dimension D_0, D_1 and D_2. We finally obtained a set of optimal solutions by a large number of experiments, $\alpha = 0.8$, $\beta = 0.1$ and $\gamma = 0.1$.

3. Variational Mode Decomposition [16]

3.1. The Definition of IMF [16]

We define an intrinsic mode function (IMF) as a frequency-modulated-amplitude-modulated signal (FM-AM signal), written as:

$$u_k(t) = A_k(t)\cos(\varphi_k(t)) \tag{9}$$

Here, $A_k(t)$ and $\varphi_k(t)$ represent respectively the instantaneous amplitude and phase of $u_k(t)$, and satisfying

$$A_k(t) \ge 0;\ \omega_k(t) = d\varphi_k(t)/dt \ge 0 \tag{10}$$

Further, $A_k(t)$ and $\omega_k(t)$ vary much slower than the phase $\varphi_k(t)$, that is to say, in the long interval $[t-\delta, t+\delta]$ (where $\delta = 2\pi/\omega_k(t)$), $u_k(t)$ can be seen as a pure harmonic signal with amplitude $A_k(t)$ and instantaneous frequency $\omega_k(t)$.

3.2. The Construction of VMD [16]

The variational mode decomposition (VMD) algorithm can be described as follows: in terms of constraints, the sum of the intrinsic mode functions (IMFs) is equal to the input signal $x(t)$, seeking k IMFs $u_k(t)$ such that the sum of the estimated bandwidth of each IMF $u_k(t)$ is minimized. Then the corresponding constraint variational mode function expression is as follows.

$$\min_{\{u_k,\omega_k\}}\left\{\sum_k \left\|\partial_t\left[\left(\delta(t)+j/(\pi t)\right)*u_k(t)\right]e^{-j\omega_k t}\right\|_2^2\right\},\quad s.t.\ \sum_k u_k = x(t) \tag{11}$$

where, $\{u_k\} = \{u_1, \ldots, u_k\}$ represents the k IMF components; $\{\omega_k\} = \{\omega_1, \ldots, \omega_k\}$ represents the center frequency of each IMF component $u_k(t)$ correspondingly.

3.3. The Solution of VMD [16]

In order to obtain the optimal solution of the constrained IMFs, by the VMD transform, we convert the constraint variational problem to the unconstrained variational problem by introducing the penalty factor a and the Lagrange operator $\lambda(t)$, in which a can be guaranteed the reconstruction precision of the signal and $\lambda(t)$ can be strengthened constraints. The extended Lagrange expression is as follows.

$$L(\{u_k\},\{\omega_k\},\lambda)=a\sum_k\left\|\partial_t\left[\left(\delta(t)+j/(\pi t)\right)*u_k(t)\right]e^{-j\omega_k t}\right\|_2^2 \\ +\left\|x(t)-\sum_k u_k(t)\right\|_2^2+\left\langle\lambda(t),x(t)-\sum_k u_k(t)\right\rangle \tag{12}$$

where, a is the penalty factor (also called the equilibrium constraint parameter) and λ is the Lagrange multiplier. In the VMD, Alternate Direction Method of Multipliers (ADMM) is used to solve the above variational problem, and by alternately updating the parameters u_k^{n+1}, ω_k^{n+1}, and λ^{n+1} to seek to extend the "saddle point" of the Lagrange expression. The VMD calculation step can be taken as follows.

1) Initialize the parameters $\{u_k^1\}$, $\{\omega_k^1\}$, and $\{\lambda_k^1\}$, and set $n=0$.

2) Set $n=n+1$, and execute the entire loop.

3) Update the parameters u_k and ω_k, according to Formulas (13) and (14).

$$\hat{u}_k^{n+1}=\arg\min_{\omega_k}\left(\int_0^\infty(\omega-\omega_k)^2\left|\hat{u}_k(\omega)\right|^2 d\omega\right) \tag{13}$$

$$\omega_k^{n+1}=\int_0^\infty\omega\left|\hat{u}_k(\omega)\right|^2 d\omega\Big/\int_0^\infty\left|\hat{u}_k(\omega)\right|^2 d\omega \tag{14}$$

4) Set $k=k+1$, repeat Step 3 until $k=K$.

5) According to Formula $\lambda^{n+1}=\lambda^n+\tau\left(f-\sum u_k^{n+1}\right)$, update λ, where τ is a noise margin parameter and here take $\tau=0$.

6) Give any accuracy $\varepsilon>0$, if the iteration stop condition that

$$\sum\left\|u_k^{n+1}-u_k^n\right\|_2^2\Big/\sum\left\|u_k^n\right\|_2^2<\varepsilon \tag{15}$$

is satisfied, then stop the entire cycle, output the results and get K narrowband IMF components u_k. Otherwise, repeat Step2~Step5.

3.4. Simulation Example

Let the simulation signal $x(t)=x_1(t)+x_2(t)+x_3(t)$, where $x_1(t)=\cos(4*\pi*t)$, $x_2(t)=1/4*\cos(48*\pi*t)$ and $x_3(t)=1/16*\cos(576*\pi*t)$. $x(t)$ is decomposed by VMD, 3 IMFs with $u1$, $u2$ and $u3$, which respectively correspond to the input signals $x1$, $x2$ and $x3$. The simulation input signals and IMFs are shown in Figure 1.

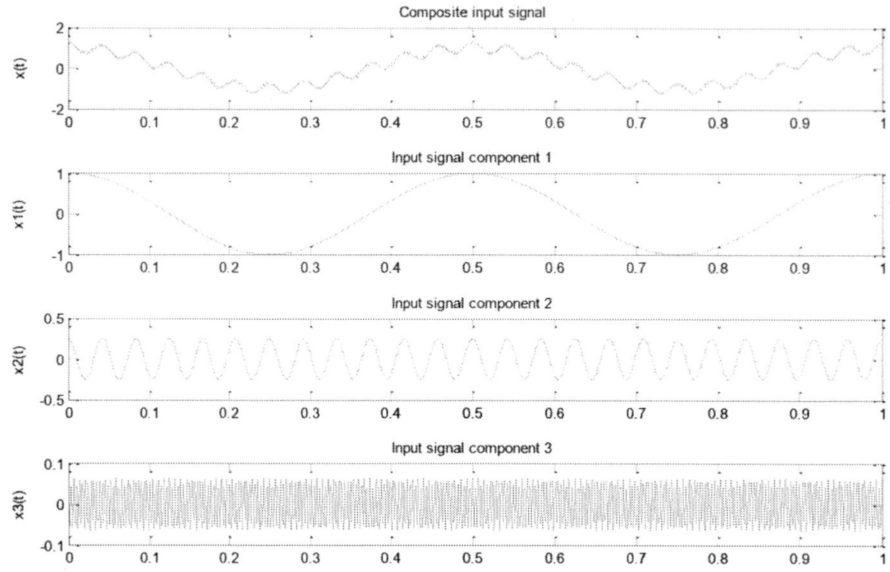

Figure 1. Simulation Signal and VMD Decomposition.

Comparing $u1$, $u2$, $u3$ and u ($u = u1 + u2 + u3$) with $x1$, $x2$, $x3$ and x respectively, we can find that they are not much difference in whether the amplitude or the frequency, which are shown in Figure 2.

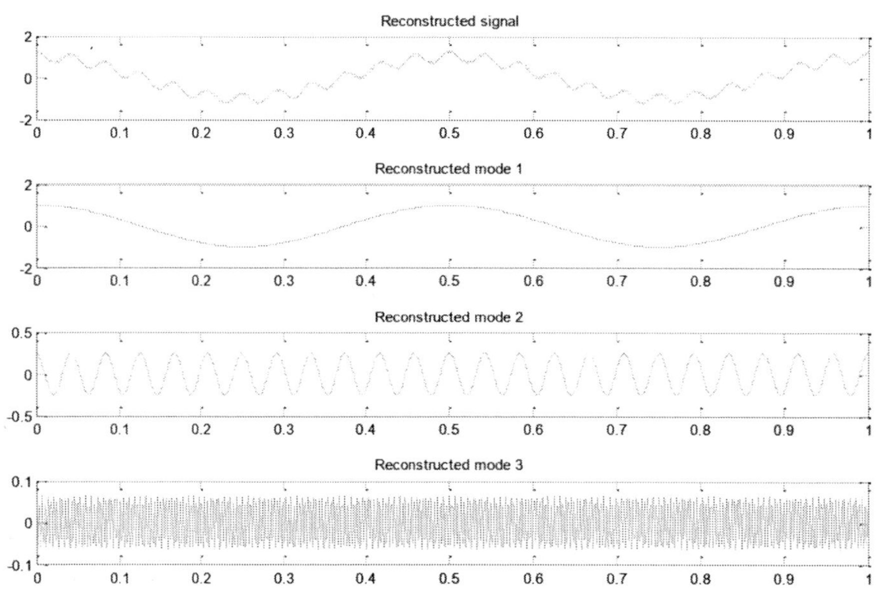

Figure 2. The mode contrast figure of the simulation signal and signals decomposed by VMD.

It can also be seen from the spectrum distribution of input signal x (See Figure 3) and 3 IMFs $u1$, $u2$, $u3$ (See Figure 4), which can be well matched with the original signal in the frequency domain, and their center frequencies are shown in Figure 5.

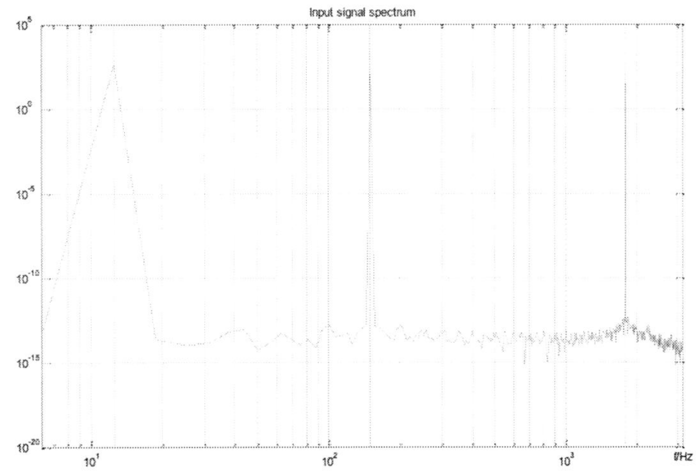

Figure 3. The frequency spectrum distribution of the input signal.

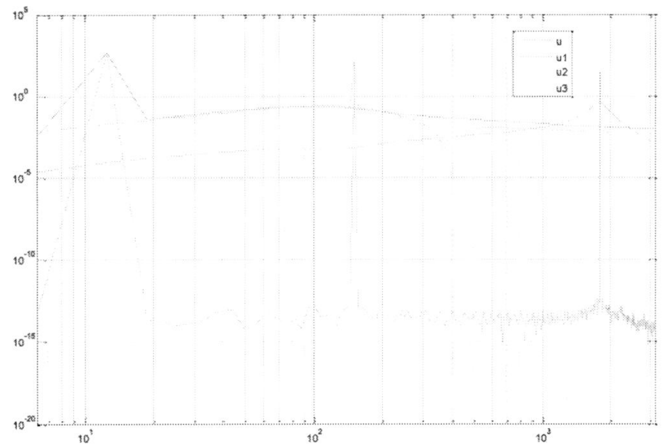

Figure 4. The Mode Signal Spectrum by VMD.

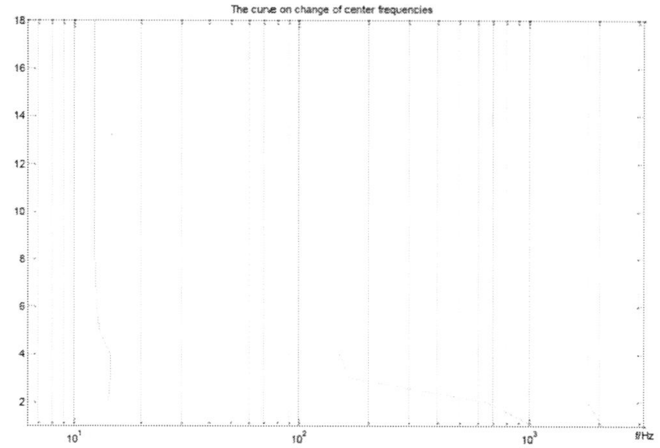

Figure 5. The Change of Center Frequency.

4. Adaptive Variational Mode Decomposition

Based on VMD, we decompose the simulation signal into different layers ($K = 2,3,4,5$). As a result, we have found that the decomposition accuracy is the highest if $K = 3$, that the signal is under-decomposed if $K < 3$, and that the signal is over-decomposed if $K > 3$. Under different decomposition layers K, the center frequencies of the above simulation signal are shown in Figure 6 respectively.

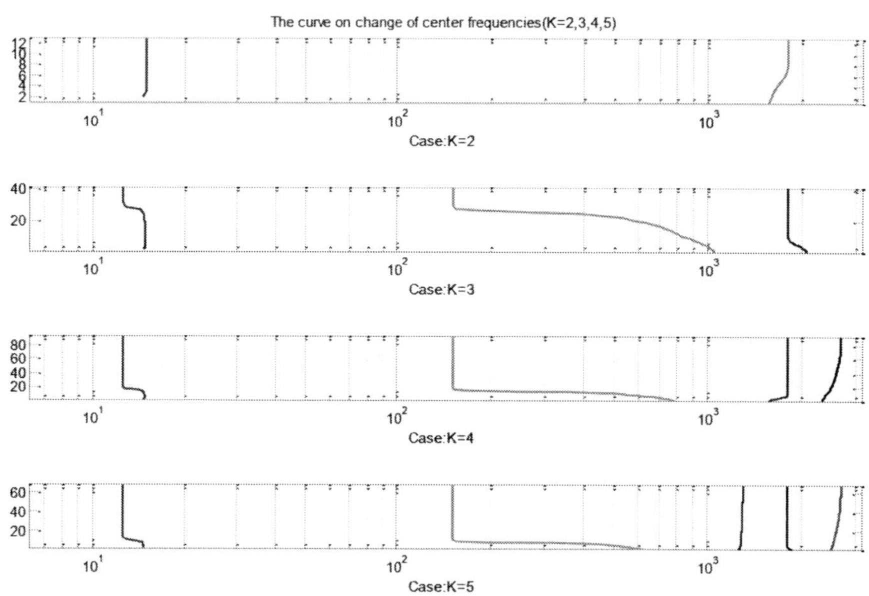

Figure 6. The change of center frequencies under different decomposition layers K

In addition, the equilibrium constraint parameter a also has an influence on the signal decomposition result. The smaller the value of a, the larger the bandwidth of each IMF, and the result is prone to center frequency overlap and mode aliasing phenomenon; the larger the value of a, the smaller the bandwidth of each IMF, and the result is that the center frequency overlap and the aliasing phenomenon disappears. Based on the above analysis, we proposed an Adaptive Variational Mode Decomposition (AVMD) method. Firstly, concretely speaking, preset value a according to the signal sampling frequency. Secondly, decompose the signal by VMD, and calculate the correlation coefficients ρ between each IMF and the original signal and set the threshold. Thirdly, if the minimum value of ρ is less than the threshold, stop the decomposition. Otherwise, increase the number K of decomposition layer, and continue to break down until the stop condition is met. The flow chart of AVMD is shown in Figure 7.

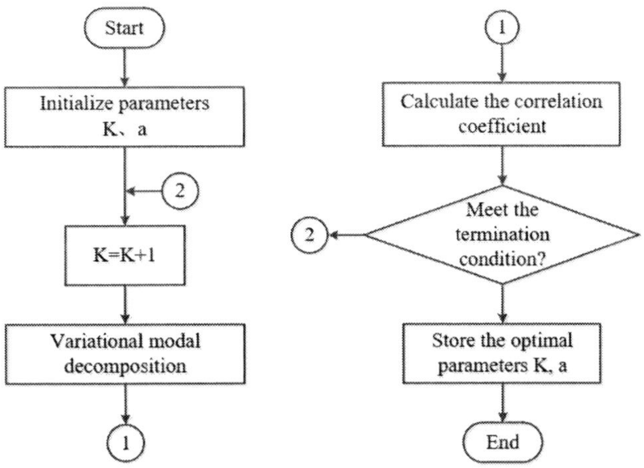

Figure 7. The Flow chart of AVMD

5. Formant Extraction Algorithm Based on AVMD

In this paper, we propose a formant detection algorithm based on the above AVMD. The algorithm uses the AVMD to decompose the formants into different IMFs, and then uses the Linear Predictive Coding (LPC) to extract the first three formants and its center frequencies of each frame. The flow chart of formant detection algorithm based on AVMD is shown in Figure 8.

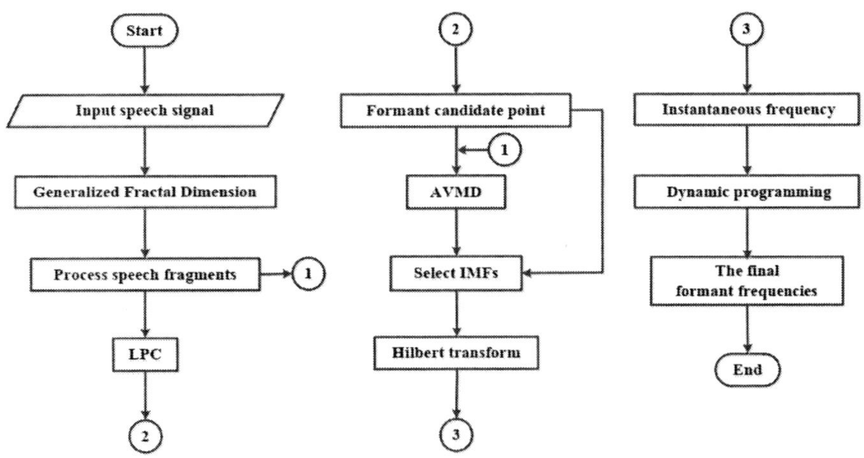

Figure 8. The Flow chart of Formant Extraction Algorithm Based on AVMD.

6. Experiment and Results

In order to verify the validity to extract the formant by AVMD, we select 100 isolated words from Wa language, which were recorded in YanShuai, as mentioned before. Now we illustrate the main steps of our experiment as follow.

1) Using Synthetic Dimension to split the 100 selected Wa language isolated words into voiceless and voiced segment, initials and finals. The division result of the Wa language word "romma", meaning "Heaven" in Chinese, is shown in Figure 9.

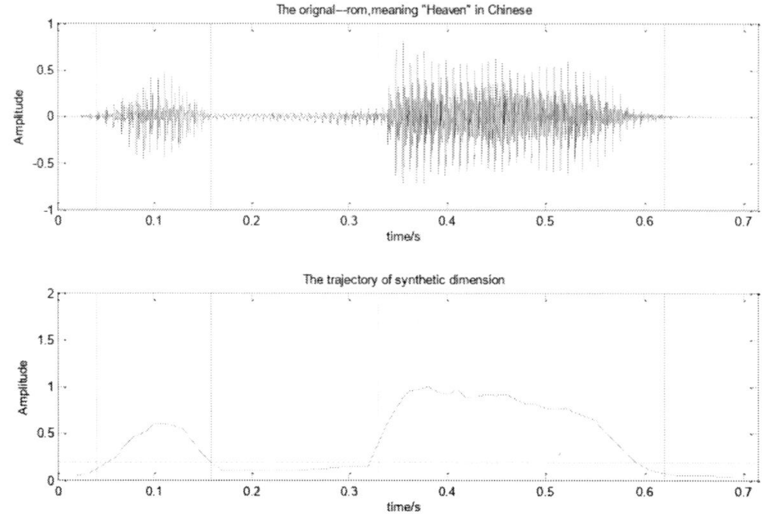

Figure 9. The Segmentation Result by Synthetic Dimension.

2) For every isolated word, using LPC to estimate the first three formant frequencies and their bandwidths roughly.

3) Selecting the appropriate equilibrium constraint parameter a (generally taking the half of sampling frequency as the value a) and the threshold of the correlation coefficient ρ such that AVMD can decompose the signal into some IMFs without pattern aliasing.

4) For every isolated word, using the estimated formant frequencies and bandwidths to determine precisely the required IMFs.

5) For every isolated word, using the Hilbert transform to calculate the instantaneous frequency of the determinate IMFs. Weighted average operations on instantaneous frequencies are carried on to obtain the first three formant frequencies for each frame. The result of the Wa language word "romma", meaning "Heaven" in Chinese, is shown in Figure 10.

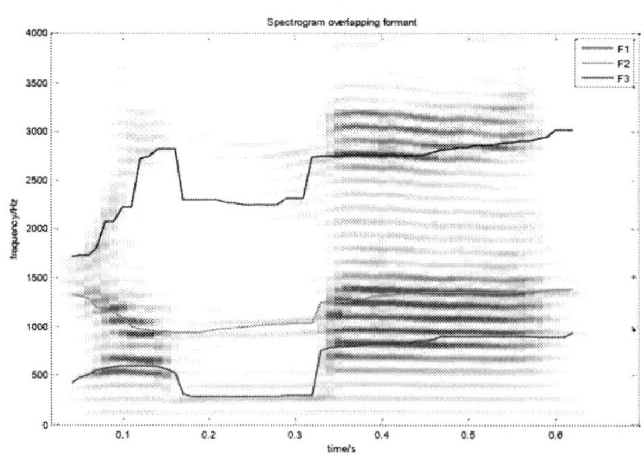

Figure 10. The First Three Formant Frequencies by AVMD.

6) Comparing the first three formant frequencies depending on the AVMD method and extracted by *Praat* software respectively, we can draw the conclusion that the relative correct rate of the former to the latter can reach 86% averagely in terms of the 100 selected isolated words, which shows that our AVMD method is effective on Wa language. The first three formant frequencies extracted by *Praat* software, belonging to the Wa language word "romma", meaning "Heaven" in Chinese, are shown in Figure 11.

7)

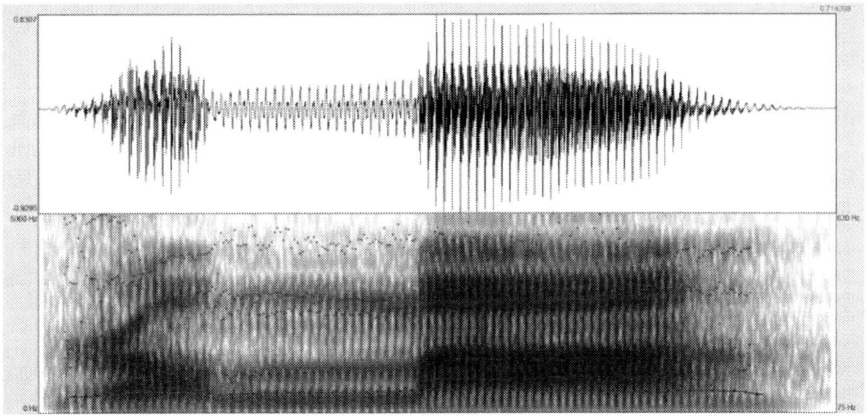

Figure 11. The First Three Formant Frequencies by PRAAT software.

7. Acknowledgments

This work is financially supported by Yunnan Provincial Department of Education Science Research Fund Project (2016YJS078), Yunnan Minzu University Overseas Master's Program (3019901).

8. References

[1] Zhou Zhizhi, Yan Qixiang, Chen Guoqing. The wa dialect study [M]. Yunnan: Nationalities Publishing House, 2004.
[2] Kong Jiangping. Experimental phonetics basic course [M]. Peking University Press, 2015.
[3] Ritsema C J. Acoustic theory of speech production [M]. Mouton, 1970.
[4] Teager H. Some observations on oral air flow during phonation [J]. Acoustics Speech \& Signal Processing IEEE Transactions on, 1980, 28(5):599-601.
[5] Teager H M, Teager S M. Evidence for Nonlinear Sound Production Mechanisms in the Vocal Tract [M]. Speech Production and Speech Modelling. Springer Netherlands, 1990:241-261.
[6] Karonen J O, Ostergaard L, Vainio P, et al. Practical numerical algorithms for chaotic systems [M]. Practical numerical algorithms for chaotic systems. Springer Verlag, 1989:125-8.
[7] Thompson C, Mulpur A, Mehta V, et al. Transition to chaos in acoustically driven flows [J]. Journal of the Acoustical Society of America, 1991, 90(4):2097-2108.
[8] Peitgen H O, Jürgens H, Saupe D. Chaos and fractals: new frontiers of science [J]. Mathematical Gazette, 2004, 79(484):241–255.
[9] Grassberger P. Generalized dimensions of strange attractors [J]. Physics Letters A, 1983, 97(6):227-230.
[10] Maragos P, Quatieri T F, Kaiser J F. Speech nonlinearities, modulations, and energy operators[C]. International Conference on Acoustics, Speech, and Signal Processing. IEEE, 1991:421-424 vol.1.
[11] Maragos P, Kaiser J F, Quatieri T F. On amplitude and frequency demodulation using energy operators [M]. IEEE Press, 1993.
[12] Maragos P, Kaiser J F, Quatieri T F. Energy separation in signal modulations with application to speech analysis [J]. IEEE Transactions on Signal Processing, 1993, 41(10):3024-3051.
[13] N. E. Huang, Zheng Shen, Steven R. Long … et. al. The empirical mode decomposition and the Hilbert spectrum for nonlinear and non-stationary time seri [M]. The Royal Society, 1998.

[14] Zhaohua W U, Huang N. E. ENSEMBLE EMPIRICAL MODE DECOMPOSITION: A NOISE-ASSISTED DATA ANALYSIS METHOD [J]. Advances in Adaptive Data Analysis, 2011, 01(01):0900004-.

[15] Smith J S. The local mean decomposition and its application to EEG perception data [J]. Journal of the Royal Society Interface, 2005, 2(5):443.

[16] Dragomiretskiy K, Zosso D. Variational Mode Decomposition [J]. IEEE Transactions on Signal Processing, 2014, 62(3):531-544.

[17] Huang N E, Shen Z, Long S R. A NEW VIEW OF NONLINEAR WATER WAVES: The Hilbert Spectrum1 [J]. Annual Review of Fluid Mechanics, 2003, 31(1):417-457.

[18] Datig M, Schlurmann T. Performance and limitations of the Hilbert–Huang transformation (HHT) with an application to irregular water waves [J]. Ocean Engineering, 2004, 31(14–15):1783-1834.

[19] Rilling G, Flandrin P. One or Two Frequencies? The Empirical Mode Decomposition Answers [J]. IEEE Transactions on Signal Processing, 2008, 56(1):85-95.

IOP Conf. Series: Journal of Physics: Conf. Series **887** (2017) 012086 doi:10.1088/1742-6596/887/1/012086

Vibration factors impact analysis on aerial film camera imaging quality

Jun Xie, Wei Han, Zhonglin Xu, Haifeng Tan and Mingquan Yang

Department of Aerospace Intelligence, University of Air Force Aviation, Changchun, 130022, China.
Email: xiejun1981@gmail.com

Abstract. Aerial film camera can acquire ground target image information advantageous, but meanwhile the change of aircraft attitude, the film features and the work of camera inside system could result in a vibration which could depress the image quality greatly. This paper presented a design basis of vibration mitigation stabilized platform based on the vibration characteristic of the aerial film camera and indicated the application analysis that stabilized platform could support aerial camera to realize the shoot demand of multi-angle and large scale. According to the technique characteristics of stabilized platform, the development direction are high precision, more agility, miniaturization and low power.

1. Introduction

Imaging is an important means of military reconnaissance. It is the main method for obtaining battlefield intelligence in modern war. The aerial camera is an important implement tools in imaging reconnaissance activities, electromagnetic wave radiations or reflections of the ground targets were collected through the optical system, and the lights were focusing on the focal plane and it was provide energy for recording medium. For the air film cameras, the recording medium is film. Digital imaging and digital image material has been developed rapidly since the 1990s, it has the characteristics of imaging convenient storage and real-time transmission, but the sensitive material with the characteristics of low cost, high resolution, high cost performance, high sensitivity, so it is still the most mature technology. While obtaining the target image, the vibration of the vehicle will be passed on to the aerial camera through the connection platform, so as to affect the work of aerial camera optical system, also has a great influence on the quality of imaging [1]. In the paper, by analyzing the characteristic of air film and camera vibration on the camera image quality, the domestic and foreign application situation of stabilized platform were summarized, the necessary of effective vibration reduction of the camera stable platform method was put forward, it can provide reference for the design of the vibration and stability of the aerial camera platform and the selection of film.

2. The Film Features and its Imaging Principle

The film is a kind of photographic material which can accept lights information, then after the chemical or physical process, a fixed image can be obtained. The nitric acid fiber, acetate fiber or polyester was taken as the base material, on the base coated by gelatin and suspension of photosensitive material made of silver halide particles in emulsion. When taking the photo, the lights through the camera lens illuminate on the film emulsion layer, the light reaches the silver halide crystal mass formed by the coalescence and silver halide crystals, and different intensity of illumination can make the film emulsion layer micro field have different amount of crystal structure changes and mutual coal escence, film exposure, latent image, developing the film for operation, the structure change of silver halide crystals into black metallic silver particles get nexus of contracts, produce images; The without photo sensitive part on film, the crystals without structure change is washing out by fixative, present light grey or transparent; The result is negative as the dark parts are more exposed, and the bright part is less exposure, the transparent part is not under the light irradiation, it is the black and white film process. There are three emulsion layers in the color film, respectively in the emulsion layer can produce dye containing different color coupler. The emulsion layers are colourless, when it was developing, the color development developer oxide coupling to be colored dyes. Compared to the common film, the air film have more excellent physical and mechanical properties such as excellent resistance to impact, high tensile strength, bending strength, compression strength, small creep, size stable, and the relationship between imaging quality and its characteristic parameters depends on the material structure and properties of the film itself, so the air film imaging quality influence factors analysis has important practical significance [1-3].

3. Film Imaging Quality Influence Factors

There are many influencing factors to film imaging quality, we only focus on the photographic properties factors [1].

3.1. Contrast

Contrast is the brightness of the image contrast imaging after the degree of difference. For film it is the difference of maximum density and minimum density, it can affect the deepest and the most shallow tone differences of the image, so it is one of the factors that influencing the quality of film imaging, it is usually called contrast coefficient. The contrast coefficient of air film is larger than normal film, usually between 1.5 to 2.6.

3.2. Tolerance

Tolerance refers to the film according to the proportional relationship between the scopes of the right record scene brightness difference portion of straight line on both ends of the photosensitive characteristic curve corresponding to the exposure of the logarithmic difference. In characteristic curve, the linear part is recorded according to the proportional relationship between the image luminance differences. The bigger of tolerance, the bigger scope of brightness, the range of tolerance is larger than the brightness of the scenery, the greater the range of correct exposure scalable.

3.3. Identification Rate

The identification rate is the resolution, it is the film's ability to identify and record the scene details, also called as force. It is showed to distinguish the number of lines every mm. The major influencing factors for film materials is the silver halide emulsion particle size. Generally speaking, the higher of sensitivity of emulsion particles, the lower identification rate, and low sensitivity of the emulsion particles is small, with high identification rate.

3.4. Fog Density

The density that did not photosensitive part was called fog density. Any film has a certain fog density, mainly produced by the instability of emulsion itself, and it also had relation to the exposure in air too long and friction factor and so on.

3.5. Maximum Density

Photosensitive characteristic curve can achieve the maximum density value is called the maximum density, is also the most black that film can be achieved. In reconnaissance plan, the exposure time should require the brightest part of the density value not exceed this value.

4. Effect of Vibration to Image Quality

Camera vibration can be divided into the external and internal vibration. External vibration is generally characterized by frequent vibration, such as the change of the flight, flying in the air with a plane and the plane engine, weapons and equipment of the fire control system, can cause the vibration of plane emergency or frequent periodic vibration. In the vibration, especially when the landing impact effect on camera parts is bigger, and frequent vibration caused by engine working on image quality is bigger also, the influence of the system must be eliminated [1,4,5].

Internal vibration is generally characterized by sudden, random vibration, such as camera internal parts, the system of work, such as the rotation of the motor, the shutter open, the rise and fall of platen, the vibration of the swing seat frame and so on, also can cause camera flash or frequent vibration. It can be found that the vibration of the camera is irregular, that is to say, the camera can be moving along at any direction or around any axis reciprocating vibration, and the vibration can produce very big effect on the imaging quality of the camera.

In order to study the effect of vibration to image quality, the vibration of the camera can be decomposed into respectively along the perpendicular of x, y, z axis vibration and around the triaxial vibration, as shown in figure 1.

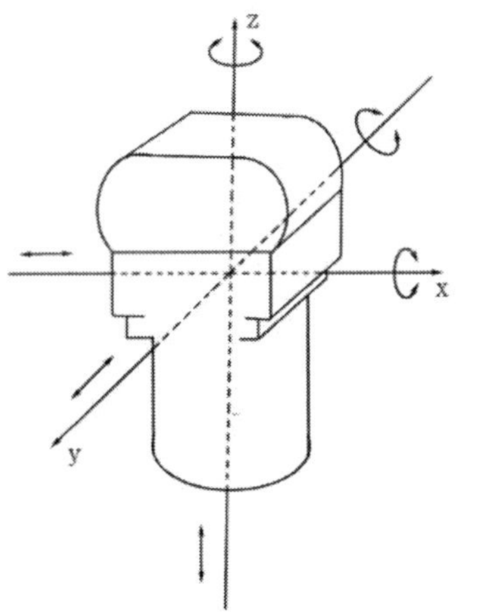

Figure 1. The directions of camera vibration

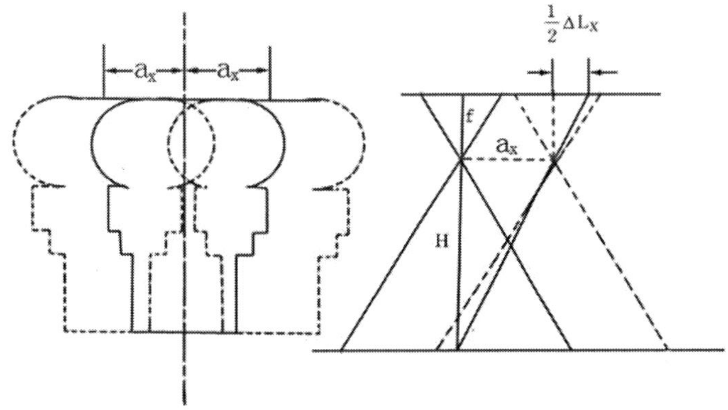

Figure 2. The camera image point vibration along the X axis

The biggest influence on image quality is the regular vibration of the camera, the vibration is a harmonic oscillator with sine law generally, and the six points of vibration can be written as the following form:

$$x = a_X \cos(\omega t + \alpha_X) \tag{1}$$

$$y = a_Y \cos(\omega t + \alpha_Y) \tag{2}$$

$$z = a_Z \cos(\omega t + \alpha_Z) \tag{3}$$

$$\varphi_X = \psi_X \cos(\omega t + \beta_X) \tag{4}$$

$$\varphi_Y = \psi_Y \cos(\omega t + \beta_Y) \tag{5}$$

$$\varphi_Z = \psi_Z \cos(\omega t + \beta_Z) \tag{6}$$

a_X、a_Y、a_Z、ψ_X、ψ_Y、ψ_Z are each vibration amplitude and angle amplitude, α_X、α_Y、α_Z、β_X、β_Y、β_Z are all the points in the early phase of the vibration, ω is the angular frequency of vibration, general aircraft engine vibration angular frequency. The following concrete analysis the influence of various vibration on image quality.

4.1. The Vibration Along the x, y, z Direction

In the process of exposure, the camera along the horizontal axis x or y axis vibration, vibration will cause the image accordingly. Figure 2 shows the image move caused by camera along the x axis vibration.

When the exposure time is greater than the vibration period T ($T=2\pi/\omega$), the image point in completing a vibration plate, the length of the image point vibration is set to Δl_x:

$$\frac{1}{2}\Delta l_X (a_X)^{-1} = \frac{f}{H} \tag{7}$$

$$\Delta l_X = 2a_X \frac{f}{H} \tag{8}$$

When the exposure time is less than the period of vibration, the exposure of the image point has not yet completed a vibration, the image point move less than $2\alpha_x f/H$, its length depends on the exposure time and its beginning and ending time in the process of vibration. Along the y axis vibration on image point and the influence of the above situation is the same.

When making imaging reconnaissance aircraft, α_x and α_y are small, reconnaissance height H generally greater than the camera focal length, so the absolute value of $2\alpha_x f/H$ is small; The vibration of the camera along the vertical axis z axis is equivalent to the change of the camera height, As a result, the vibration of the camera along the x, y, z axis had no real effect on image quality.

4.2. The Vibration Around the x, y Axis

In the process of exposure, the camera around the horizontal axis x or y axis vibration, will cause the image point make corresponding reciprocating movement on the plate.

Figure 3 represents the camera around y axis vibration, both sides of the fulcrum of vibration respectively. Set the same amplitude a_z, phase are 180 DHS, camera around y axis vibration amplitude (the angle amplitude) ψ_y is largest:

$$tg\psi_y = \frac{2a_z}{l} \tag{9}$$

Because the amplitude a_z in general is far less than mount fulcrum spacing l, also ψ_y is very small:

$$\psi_y = \frac{2a_z}{l} \tag{10}$$

In a vibration period, the optical axis of camera lens angle of vibration is $2\psi_y$. If the exposure time is greater than the vibration period, the main axis of image point to complete a reciprocating movement on the plate, so the length $\Delta l\varphi_y$ of the image point was as follows:

$$\Delta l\varphi_y = 2\psi_y \cdot f = \frac{4a_Z}{l} f \tag{11}$$

Generally the camera pivot spacing and the camera focal length roughly equal, so $\Delta l\varphi_y = 4a_Z$, the longer the camera focal length, the more serious the image motion. The influence of the x axis vibration is also the same.

4.3. The Vibration Around the z Axis

In the process of exposure, the camera around the vertical axis z axis vibration will cause as click to the reciprocating movement of the main focus for the center of the circle arc, as shown in figure 4.

Amplitude for each fulcrum is a, and it goes around the z axis direction, around the radian of the angle of the z axis vibration amplitude value is:

$$\psi_Z = \frac{a}{l'} \tag{12}$$

In (12), l' is the distance protection from z axis and mount.

Within a vibration period T, vibration Angle is $2\psi_z$, when exposure time is more than T, the image point in completing a vibration plate, arc length as follows:

$$\Delta l\varphi_Z = 2\psi_Z \cdot r = 2\frac{ra}{l'} \tag{13}$$

In (13), r is the distance between the point and the main focus.

On that, as the size and image point position and image point outside edge, the closer the negatives as move. Usually l' is greater than r. If exposure time is less than T, the image motion will be smaller. So kind of vibration caused by image motion is usually close to or smaller than the vibration amplitude of the camera.

Known from the analysis of the above that in the six kinds of vibration, the vibration around the horizontal axis x, y, impact on the quality of image is the largest, its influence degree and saddle point up and down is directly proportional to the amplitude of the vibration. Minimize the amplitude values, therefore, is stabilized platform vibration device to complete one of the main tasks.

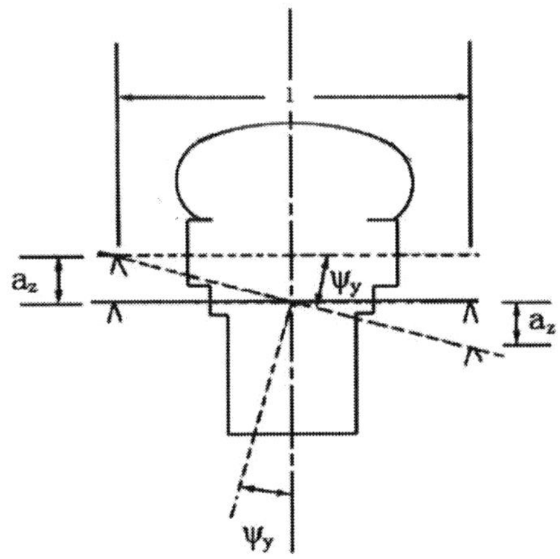

Figure 3. The camera vibration around Y axis

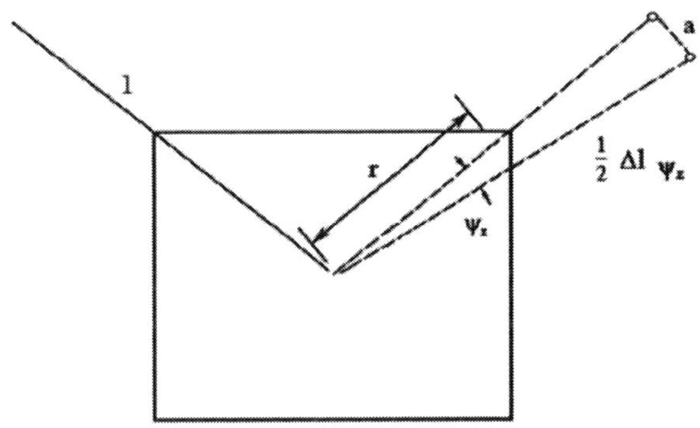

Figure 4. The camera vibration around the Z axis

5. Conclusion

Because the aerial reconnaissance is given priority to with photoelectric devices, not just the visible and infrared aerial camera, and synthetic aperture radar, laser range finder, etc., so in recent years abroad has stabilized platform to develop into a dedicated opto-electronic reconnaissance platforms, and its related technology developing rapidly, such as payload towards platform lighter, sensors, more smaller and higher stable precision in the direction of development. In recent years, the domestic aviation photoelectric reconnaissance platform also has great development, whether in the quality of platform, and precision of payload, stability, greater progress, yet there is a large gap compared with abroad, especially in terms of volume and quality, its reason mainly lies in the framework form and material, and the miniaturization of optical structure design and manufacturing technology level. According to the characteristics of modern warfare and the developing trend of the development of the aviation surveillance will be towards the three-dimensional space, intelligence information in real time, the means of diversification, reconnaissance and combat integration direction, thus determines the reconnaissance platform will be toward more accurate, more flexible, small volume, small energy consumption in the direction of development.

6. Acknowledgments

This work was financially supported by the Jilin Province Education Department "twelve-five" Science and Technology Research Project ([2015] no. 449).

7. References

[1] Jinhe B 2008 *Optical imaging reconnaissance equipment system* (Changchun: The blue sky press) p 62-81

[2] Ping J and Bao Z J 2003 *Critical technologies and their development for airborne opto-electronic trconnaissance platforms* (*Optics and Precision Engineering* vol 11) p 82-85
[3] Yongxin M, Kun Z and Jin G J 2008 *Design of the Absorber System of the Airborne Camera (Laser and Infrared* vol 38) p 1118-21
[4] Wujun M, Leyong W and Wenpu W and Fangyi D J 2009 *Design and Simulation of Aerial Camera Stabilized Platform(Aero Weaponry* vol 4) p 48-52
[5] FANG T, XUE P. J 2000 *Vibration Theory and Applications* (Xi'an: North West Industry University press) pp 32-46

Design of verification platform for wireless vision sensor networks

Juanjuan Ye, Fei Shang and Chuang Yu

College of Mechanical Engineering, Nanjing University of Science and Technology, Nanjing 210094, China.
Email: yjj15250982072@163.com

Abstract. At present, the majority of research for wireless vision sensor networks (WVSNs) still remains in the software simulation stage, and the verification platforms of WVSNs that available for use are very few. This situation seriously restricts the transformation from theory research of WVSNs to practical application. Therefore, it is necessary to study the construction of verification platform of WVSNs. This paper combines wireless transceiver module, visual information acquisition module and power acquisition module, designs a high-performance wireless vision sensor node whose core is ARM11 microprocessor and selects AODV as the routing protocol to set up a verification platform called AdvanWorks for WVSNs. Experiments show that the AdvanWorks can successfully achieve functions of image acquisition, coding, wireless transmission, and obtain the effective distance parameters between nodes, which lays a good foundation for the follow-up application of WVSNs.

1. Introduction

In recent years, wireless sensor networks (WSNs) have attracted widespread attention in academia. However, the traditional WSNs generally measure scalar data, such as temperature, light intensity, humidity, etc. The information collected and processed is limited. It is urgent to bring image, video and other visual information into the sensor networks to achieve more refined measurement, thus wireless vision sensor networks (WVSNs) came into being[1]. As most of the research of WVSNs only stay in the physical simulation stage, and simulation results have difference from the actual running status of WVSNs. It is urgent to set up a verification platform for WVSNs, which provides the physical support for the basic theory of WVSNs. The verification platform is composed of wireless vision sensor nodes through a specific way, it can reflect the large-scale network application environment. It supports the basic functions required for applications of WVSNs and provides physical support for specific protocols and algorithms.

At present, the typical verification platforms designed for WSNs at home and abroad mainly include Motelab of Harvard University[2], Kansei designed by Ohio State University[3], MoteWorks of Crossbow, MSRLab6 of Beijing Jiaotong University[4] and ComWSN of Central South University[5]. However the existing verification platforms for WVSNs are very few. Therefore, this paper puts forward the idea of constructing the verification platform for WVSNs. This paper designs a node for WVSNs at first, and then build a verification platform on this basis and runs it.

This paper is organized as follows: The first part briefly introduces the design scheme of the verification platform AdvanWorks designed in this paper. The second part introduces the routing protocol of AdvanWorks. The third part introduces the sensor node design of AdvanWorks, including hardware design and software design. In the fourth part, we test the AdvanWorks platform and analyze the experimental results. Finally, the last part summarizes the verification platform AdvanWorks.

2. Overall Program of Verification Platform

In order to realize the acquisition, processing, transmission of visual information, WVSNs have high requirements on performance of node processor and wireless communication capability of the network. Therefore, this paper designs a high-performance wireless vision sensor node to set up verification platform without additional configuration control system. In addition, only a communication protocol is used to simultaneously realize the data communication and transmission of PC configuration and monitoring information when the network is running. The basic framework of the verification platform AdvanWorks is shown in Figure 1, it consists of the source node, the intermediate forwarding node and the destination node. The number of nodes is less than that of actual network, and the nodes are distributed randomly. These nodes automatically set up the network in Ad-Hoc mode, in which the source node collects, processes and transmits the image data, the intermediate nodes forward the data packets, the destination node receives the data packets, and the PC acts as the network control center.

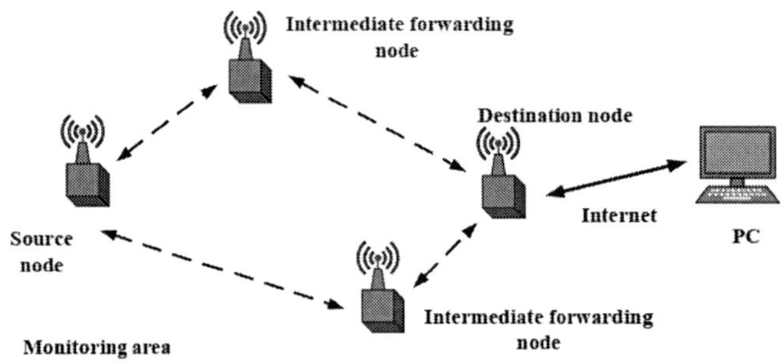

Figure 1. Basic framework of AdvanWorks.

3. Selection of Routing Protocol

The design of AdvanWorks only considers providing high-performance nodes with multimedia processing capability. For selection of routing route, it is possible to select the best transmission path by avoiding the communication between unrelated nodes outside the routing path and integrate coding and energy consumption during transmission, to reduce energy consumption during transmission and balance energy consumption of network. Since most nodes in the WSNs cannot communicate with the destination nodes directly, they need to communicate through the intermediate nodes, thus selecting an appropriate routing path to save energy has become an important goal of the network layer protocol. For the WVSNs with large amounts of data transmission, it is necessary to take energy efficiency into consideration when participating in the design of routing protocol.

Ad-Hoc On-Demand Distance Vector Routing (AODV) is an on-demand routing protocol that is based on the need of data transmission of the source node. The consideration of the distance is determined by routing hops between the source and the destination node [6].

The process during establishing AODV protocol routing involves three key packets: routing request packet (RREQ), routing reply packets (RREP) and routing acknowledgment package (RREP -ACK) [7]. During the sending of RREQ and RREP, each entry in the AODV routing table uses the destination sequence number created by the destination node in order to avoid the occurrence of routing loops [8]. In the process of route establishment, when the routing path has the same sequence number of the destination node, the smallest number of hops will be selected to shorten the distance of the data transmission and reduce the energy consumption. Therefore AODV can be selected as the routing protocol of WVSNs authentication platform.

4. Design of Sensor Node

The WVSNs are composed of multiple nodes. The functional requirements of the verification platform will eventually be transformed into the requirements of the design of nodes. The wireless vision sensor node in this paper has the function of the source node and the intermediate forwarding node at the same time. The node system is divided into hardware system and software system.

4.1. Design of Node Hardware

The node hardware system includes a microprocessor module, a wireless transceiver module, a visual information acquisition module and a power management module. The node hardware system block diagram is shown in Figure 2. The microprocessor module acts as the core module of the visual sensor node, exchanges information with other modules through the interface circuits and coordinates the node operation, and it also has the function of video processing. The wireless transceiver module controls the wireless communication between nodes and realizes exchange of information between the nodes. The visual information acquisition module collects the video information of the monitoring area and performs the preliminary processing on the data. The power management module uses the independent power to supply the function modules separately.

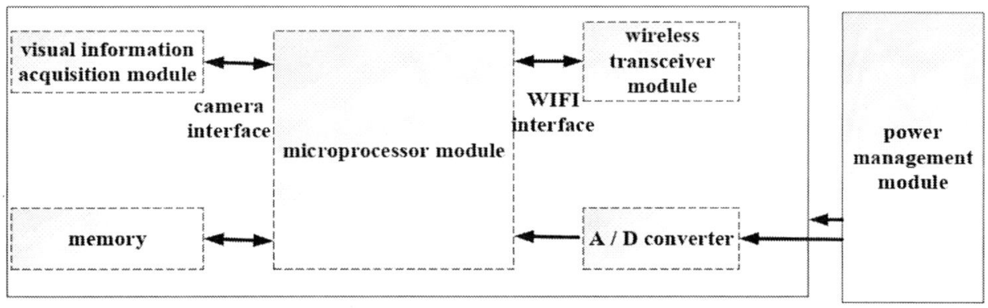

Figure 2. Block diagram of node hardware system.

4.1.1. Microprocessor Module. In the microprocessor module, RISC (reduced instruction set) architecture of the ARM11 microprocessor is used, and it is the hardware core of the wireless vision sensor node.

4.1.2. Wireless *Transceiver Module.* On account of the WVSNs transmit large amounts of data, and the transmission rate must reach the MB level, this paper uses WIFI technology to achieve data transmission[9].The wireless transceiver module is composed of RTL8189ES WIFI chip, which supports IEEE802.11n / g / f standard and protocol, forms a wireless transceiver module with a maximum speed up to 11 Mbps. It supports the Ad-Hoc operating mode required for WVSNs, when it works, the distance of transmission outdoors is up to 150 m.

4.1.3. Visual Information Acquisition Module. The core of visual information acquisition module is the image sensor, considering wireless vision sensor node is one kind of device with large-scale layout, the cost of control is stringent and the node itself has limited energy. Therefore, it is more appropriate to use CMOS sensor which has small size and low power consumption to build the node system [10]. In this paper, OV9650 camera with 1.3 megapixel and low-power video decoder chip with highly integrated TVP5150 are used to achieve visual information collection.

4.1.4. Power Management Module. Power supply module's function is to provide the required power supply for each modules. This system uses 3.7 V lithium battery-powered, battery capacity of 3.5 Ah, but the output voltage fluctuations with the remaining power, so it is easy to cause abnormal system work. In this paper, nodes use the DC boost circuit to convert Lithium battery output voltage to a stable 5V output voltage to supply node system, its maximum output current is 0.65 A.

4.1.5. Peripheral Interface Circuit. Peripheral interface of the microprocessor including UART0 debugging serial port, CMOS camera interface and WIFI interface are necessary to be designed to implement the exchange of information between the microprocessor with the visual information acquisition module and the wireless transceiver module.

4.2. Design of Node Software

The node software system is composed of the application layer and the Linux system layer. The Linux system layer contains the driver corresponding to the hardware system and provides the programming interface for the application layer through the system call. Application layer write the application program through the interface file to achieve the function of collecting and processing visual information, transporting wireless data, calculating and achieving the required physical parameters of routing protocol.

4.2.1. Building of Software development platform. This paper chooses the Linux operating system based on ARM as the node software platform. The software development platform is to install the Linux system on the host computer, establish the cross-compiling environment and the file sharing mechanism between the host Linux system and the node, and transplant the Linux system to nodes [11]. After the successful transplant, PC can be connected to node through the RS232 serial port, and can login node Linux system through the Secure CRT software to do the simulation of the terminal.

4.2.2. Development of Application software. The node application software adopts multi-threads scheme instead of single thread which is divided into visual information collection and processing thread, wireless communication between nodes thread, and parameter acquisition thread. Each thread occupies different resources at different times and improves system resource utilization rate. The visual information collection and processing thread sets the frame rate and the output format according to the user's request received by the wireless communication thread between the nodes. Wireless transceiver is in Ad-Hoc mode, the wireless communication thread between nodes needs to monitor information acquisition request sent by base station, collect image or video information, and realize the function of forwarding data collected by other nodes and transmitting data collected by itself. Based on the wireless transmission energy model, it can be seen that the data forwarding energy consumption is positively related to the distance between nodes. Therefore, it is necessary for the parameter acquisition thread to obtain the distance between nodes in the whole WVSNs.

5. Experiment and Analysis of Verification Platform

5.1. Test of Platform Basic Function

Firstly we set up the test scene, both the sending end and the receiving end have successfully transplanted the Linux system and are equipped with the user layer wireless network configuration tool (Wireless Tools.29). After the Linux system is started, a wireless self-organizing communication network must be established between the nodes. PC is connected to node through the RS 232 debug serial port, then uses the Secure CRT software to operate Linux system terminal of the node. Then the application software is executed at the sending end and the receiving end respectively, the image data are collected, processed and transmitted successfully, and the collected video information is decoded and displayed on the receiving end.

Tests are operated in 56 m long corridor indoors and square outdoors respectively. The distance between the transmitter and the receiver is changed in the long corridor, the average acceptance rate of 200 frames at 50 m is 98.55%, Although the image loss frame is occasionally present at 20 m, the frame loss rate is less than 50 m. The reason is that there are so many indoor corridor obstacles that wireless signal is unstable. In the case changing the distance between the sender and the receiver in the square, no obstacles in the range of 50 meters, the average acceptance rate of the 500 frames is 98.55%, When the range is more than 50 m, it begins to appear more obstacles so that acceptance rate decreases faster and the acceptance rate even drops to 15% in the range of 70 m.

In the receiving end decoding display with H.264, in the circumstances of long corridor, when the loss rate of frame is low, the displayed image at the receiving end is clearer as shown in Figure 3 (a). In the opposite, when loss rate is high, the displayed image at the receiving end is clearly blurred as shown in Figure 3 (b). In the open space environment, when the frame loss rate is low, the decoded image is clearer at the receiving end as shown in Fig 4 (a). When the frame loss rate is high, the decoded display of image is blurred as shown in Fig 4 (b).

Figure 3 (a). Display of low loss rate indoors. **Figure 3 (b).** Display of high loss rate indoors.

Figure 4 (a). Display of low loss rate outdoors. **Figure 4 (b).** Display of high loss rate outdoors.

5.2. Distance Parameter Acquisition between Nodes

The devices of the sending node and the receiving node are placed on the campus for testing, establishing a wireless Ad-Hoc network between two nodes and executing the receiving and receiving application program respectively. In the receiving program, there are some code about getting RSSI (Received Signal Strength Information). Put the two nodes at different distances, to obtain the received signal strength value from the receiving end. In order to simplify the process of solving the model's parameters, the Equation 1 is used as the RSSI ranging model.

$$RSSI = -(10n \log 10r - k) \tag{1}$$

In the formula, parameters n and k are related to the specific environment, k (dBm) is the value of RSSI (dBm) when the distance between nodes is 1m. It can be seen that k = -37dBm according to the actual test results. Using matlab to deal with the measured data to get the fitting curve. By fitting the matlab parameters, the signal attenuation factor n = 3.042 is obtained. In the signal transmission environment, the more obstacles, the greater the signal attenuation factor, we can see that there is less obstacles in our test environment. The results are shown in Figure 5. The measured data fluctuating around the theoretical curve, the closer the wireless transmission distance, the higher the degree of data fitting. In the fitting results, the negative correlation coefficient R-square is equal to 0.9811 that is close to 1, we can see that the overall fitting results are better. It is feasible that the method of node distance measurement is used in wireless vision sensor network. Thus, Equation 1 can be converted to 2. And according to the received signal strength, that can be converted to Equation 3.

$$r = a * 10^{\frac{RSSI}{b}} \tag{2}$$

$$r = 0.045 * 10^{\frac{RSSI}{27.47}} \tag{3}$$

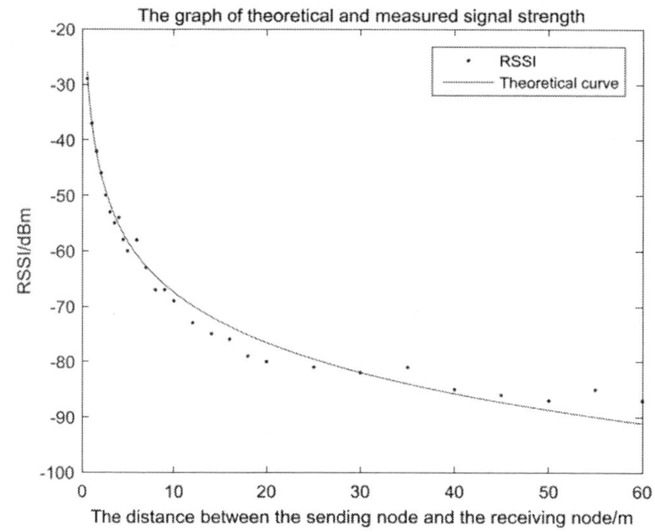

Figure 5. The fitting curve of distance between the nodes and the received signal strength and the measured data.

6. Conclusion

This paper designs a high performance wireless vision sensor node based on ARM 11 microprocessor, and builds an AdvanWorks verification platform for WVSNs. The results of different environmental experiments show that AdvanWorks platform successfully achieves functions of image acquisition, coding and wireless transmission, and obtains the effective distance parameters between nodes, which lays a good foundation for the follow-up application of WVSNs.

7. Acknowledgment

This work is supported by Natural Science Foundation of China (Grant No. 61401211)

8. References

[1] Redondi A, Baroffio L and Bianchi L, et al. 2016 Compress-then-Analyze versus Analyze-then-Compress: What Is Best in Visual SensorNetworks? America. J. IEEE Transactions on Mobile Computing. 15(12) 3000-13.

[2] Werner-Allen G, Swieskowski P and Welsh M 2005 Int. Symp. On Information Processingin Sensor Networks (Los Angeles) vol 1 (M Vetterli and K Yao:IEEE Press) p 68.

[3] Ertin E, Arora A and Ramnath R, et al. 2006 Int. Conf. on Information Processing in Sensor Networks(Nashville) (John S: ACM New York, NY,USA) pp 399-406.

[4] Hongwei H, Hongke Z and Yanchao N, et al. 2006 Int. Conf. On Signal Processing (Guilin) vol 4 (B Z Yuan and X F Tang et al.: EEE Press) pp 2950-56.

[5] Jianxin W, Pengwei Z and Weiping W 2010 Com-WSN: A component-based testbed for wireless sensor network. China. J. Computer Engineering and Applications. 46(31) 111-114.

[6] Yuli L, Xilong M and Baokang Z 2016 Research on Transplantation Technology of AODV Protocol on ARM Platform. China. J. Computer and Telecommunication. (06) 78-80.

[7] Mianlu H, Wei Z and Huizhou L 2015 Research and Improvement of AODV Routing Protocol. China. J. Computer Engineering.41 (01) 110-114.

[8] Cuiping W and Ming C 2012 Improvement of AODV routing protocol. China. J. Computer Engineering and Applications. 48(24) 91-94.

[9] Libo Y 2014 The design and implementation of wireless sensor networks node based on Wi-Fi. (Shanghai: Shanghai Jiaotong University).

[10] Xudong W and Yutang Y 2010 Comparative research and future tendency between CMOS andCCD image sensor. China. J. Electronic Design Engineering. 18(11) 178-181.

[11] Huangqing Z, Li G and ChengXiang S 2012. China. J. Computer and Digital Engineering. 40(02) 151-153.

The modeling and solution of course of action generation based on IN-DEA

Lujun Wan[1], Wen Li[2], Jiangbin Dai[1] and Aqian Huang[3]

[1]Air Traffic Control and Navigation College, Air Force Engineering University, No.1# Chang-le East Road Xi'an City, Shanxi Province, China.
[2]Equipment Management and Safety Engineering College, Air Force Engineering University, No.1# Chang-le East Road Xi'an City, Shanxi Province, China.
[3]Air and Missile Defense College, Air Force Engineering University, No.1# Chang-le East Road Xi'an City, Shanxi Province, China.
Email: L J.Wan pandawlj@126.com

Abstract. The course of action (COA) generation is the course of designing the task planning flow, which is one of the key supporting techniques while the military organization executing aerial operation. Due to the outstanding advantage of influence nets (IN), describing the causal logic between random variable, this proposal will introduce IN into the fast building mechanism of COA model. The COA generation model utilize influence nets is established. The probability propagation method is designed, which is based on influence intensity. An improved differential evolution algorithm (IDEA) is designed to solve the action policy optimized selected model. At last, the simulated result of aerial arrack campaign case show the action policy optimized selected in variety influence constants can improve the capability of cause-effect modeling, and the improved differential evolution algorithm has fine convergent and optimized capability.

1. Introduction

The Course of Action (COA) is a type of overall orderly behavior that the joint commanding organization of arms of the services conducts rational planning of the combat troops' action, allocation of the task sequence, scheduling of military resources, to achieve its operational objectives according to the analysis results of the battle-field-related information; it is a crucial component in the joint operational plan. Because the action plan is essentially an optimal process for COA to achieve the overall mission, it can be seen as a directed evolution process with the goal of combat mission, meaning that according to the combat process, select a series of COA so that the strength of the organization or cluster can achieve the desired evolutionary results. Probabilistic network can represent the interrelationships between variables in a simple and effective way, and becoming a powerful tool with broad application prospects in the uncertainty reasoning system [1]. In the application of COA generation, the probability network has prominent advantages in describing the association relations of COA elements, so it has gradually become the main tool for COA modeling and production in recent years. Probably the most representative of the probabilistic network is markov networks (MN), bayesian networks (BN) and influence networks (IN). In the actual practice of COA problem modeling, BN and MN has two limitations; one is probabilistic reasoning difficulties and the other one is the need to specify a large number of conditional probabilities. The former limitation leads to low time efficiency of variable probability calculation and the latter limitation leads to the complex modeling under a small amount of information [2]. IN simplifies the process of enlightening knowledge, and is ideal for conditional probability modelling between node variables that require subjective estimation. Standard IN can only describe the static causal relationship between random variables, but fail to give a valid description of the condition that the impact of variables has certain time-dependent relationship. This paper focuses on the generation of uncertain COA of air operations, uses the influence network model to establish its mathematical model, and uses the Differential Evolution Algorithm (DEA) to optimize the model of COA.

2. Modeling of COA Based on Influence Networks

2.1. Course of Action and Model Elements

Enemy action (E): it is the action taken by the enemy according to the changes in the battlefield, associated with the action process; it is the uncontrollable resistance factor in the generation of COA. Suppose that the action set that the enemy may adopt is $\mathrm{E}=\{E_1,E_2,\cdots,E_{NE}\}$ and $NE=|\mathrm{E}|$ is the action number of the enemy. Under normal circumstances, the enemy action is unpredictable and often randomly occurs; the commander needs to work with staff officers to give its occurrence probability in a certain stage of combat according to their own experience. Suppose that the combat process can be divided into T stages in the implementation of joint air operations and the occurrence probability of enemy action E_i in the battle t_k stage is $P_k\{E_i=1\}=P_k(E_i)$, $1\le i\le NE$ and $P_k(E_i)$ is an interval value. To meet the computing needs, set $P_k(E_i)$ to follow the approximate normal distribution.

Basic Action (A): targeted at the mission of joint air operations, in the operational conception stage, the decision-makers should determine the set of basic action $\mathrm{A}=\{A_1,A_2,\cdots,A_{NA}\}$ and $NA=|\mathrm{A}|$ is the number of basic action. If $P_k\{A_i=1\}=P_k(A_i)=1$ and $1\le i\le NA$, it shows that the basic action A_i is carried out during the combat t_k phase. In the unconstrained conditions, there will be the basic action A_i of 2^T types of combinations. Set the basic action A_i as $\pi(A_i)=\{a_i^j\,|\,1\le j\le|\pi(A_i)|\}$, $|\pi(A_i)|$ is the number of A_i, COA is the set of all the basic actions A_i combinations $G=\{a_i\,|\,a_i\in\pi(A_i),1\le i\le NA\}$ and the combat COA is $\Pi_G=\pi(A_1)\times\pi(A_2)\times\cdots\times\pi(A_{NA})$.

Desired combat goal: the ultimate operational goal of joint air operations and the combat effect is to make the greatest operational goals achieve joint probability. Set the expected goal set of combat in a certain battlefield situation as $\mathrm{D}=\{D_1,D_2,\cdots,D_{ND}\}$ and $ND=|\mathrm{D}|$ is the number of combat goals.

Stage combat goal: the stage goal of the combat organization or entity in the process of achieving a combat goal. Suppose that in a certain battlefield situation, the stage goal set is $\mathrm{S}=\{S_1,S_2,\cdots,S_{NS}\}$ and $NS=|\mathrm{S}|$ is the number of stage goal.

2.2. Generation Model of COA Based on the Influence Network

In order to effectively describe the influencing model based on influence intensity, the enemy-us action in joint air operations is deemed as the action in a unified way. Take the example of the influence of the action set $\{A_i\}_{1\le i\le n}$ on Action B to analyse the influence of the implementation of $\{A_i\}_{1\le i\le n}$ (Event True/False) on the implementation probability of Action B. Suppose X^n is a random vector of 0 or 1 value of n dimension and X_j is the j component. If the action A_j is implemented, $X_j=1$. If the action A_j is implemented, $X_j=0$. Suppose x^n is the real value of the random vector X^n, representing the specific state vector of the action set $\{A_i\}_{1\le i\le n}$. In order to conduct a quantitative description of the influence of the state vector x^n on Action B, the following expression is used to describe the influence intensity $h_n(x^n)$ of the state vector x^n on Action B:

$$h_n(x^n)=\begin{cases}1 & \text{Set } x^n\text{, Action B is bound to occur}\\ -1 & \text{Set } x^n\text{, Action B is bound not to occur}\\ 0 & \text{The occurrence probability of Action B is unrelated to } x^n\end{cases} \tag{1}$$

For a given state vector x^n, set the occurrence probability of Action B is $P(B|x^n)$, then the quantification of the influence intensity is described as 1, $P(B)$ or 0 correspondingly. $P(B)$ is the probability of Action B. If $h_n(x^n) \in [-1,1]$, use the influence intensity $h_n(x^n)$ and $P(B)$ to deduce the conditional probability of the occurrence of Action B as:

$$P(B|x^n) = \begin{cases} P(B) + h_n(x^n) \cdot [1 - P(B)] & h_n(x^n) \in [0,1] \\ P(B) + h_n(x^n) \cdot P(B) & h_n(x^n) \in [-1,0) \end{cases} \tag{2}$$

Therefore, the influence network model can be formally described as $(\mathrm{V,E,C,A,B})$, of which V is the node set of the model, and the single node is a random variable of 0-1 value. E is the directed edge set of the model and is the description of the cause-effect influence between nodes. C is the set of cause-effect influencing intensity between nodes, $E \to \{[h^{(i)}(x_i = 1), h^{(i)}(x_i = 0)], h's \in [-1,1]\}$. A, as the subset of V, represents the state of the influence action set $\{A_i\}_{1 \le i \le n}$ and its vector X^n. B is the probability distribution that affects the state vector X^n of the action set $\{A_i\}_{1 \le i \le n}$; the child node is the reference probability, and the root node is the prior probability. If $x = 1$, then $h(x) = h$; if $x = 0$, then $h(x) = g$. h and g represents the influencing intensity of Action A on Action B when it is implemented or not (true or false) respectively.

In this way, the cause-effect relationship model of various variables in the course of setting up COA through the influence network is to conduct quantization and modeling of the cause-effect relationship among the basic action, the enemy action, the stage goal and the expected operational target and to use the influence intensity value to quantify and describe the degree of cause-effect influence. In the model, the combat target node and the stage target node will influence each other in the course of the combat, and the current state of the node is associated with the state of the last moment. The parameter $P_t(\mathrm{V}_i^0)(t = 1,2,\cdots,T)$, $\mathrm{V}_i \in \{\mathbf{S} \cup \mathbf{D}\}$ carried on the dashed line indicate the state value passed at the previous stage to the current stage, namely $P_t(\mathrm{V}_i^0) = P_{t-1}(\mathrm{V}_i)$. Under the constraints of the entity resource and the time resource, the best combat goal, which is described as follows:

$$\Omega^* = \arg\max_{\Omega \in \Omega_G} \left\{ \prod_{i=1}^{ND} P\{D_i(t_T) | TC(t_0), G\} \right\} \qquad s.t. \quad R(\Omega) \le R_{TC} \ \& \ T(\Omega) \le T_{TC} \tag{3}$$

According to the combat under the implementation of the best COA Ω^*, the initial state $TC(t_0)$ is finally converted to the goal state $TC(t_T)$; if the action program Ω is followed to implement the combat, the consumption of combat resources should be within the limitations of the entire resource amount R_{TC}. If the action program Ω is followed to implement the combat, the completion time needs to be within the set combat time limit T_{TC}.

2.3. Probability Propagation Calculation Influencing the Network

Probabilistic propagation calculation is essentially a probabilistic reasoning mechanism [3]. The use of this reasoning mechanism needs to satisfy the requirement that the nodes of the probability-conduction association are independent. The probability of the root node changes with the process of the combat. In the new stage, the new probability will influence other nodes. Therefore, it is necessary to update the root node probability, calculate and update the probability of other nodes layer by layer.

Step 1: For the constructed influence network G = (V, E), select a node V_r; use the Dijkstra method to calculate the minimum distance $W(V_r, V_d)$ to the target sub-node V_d $(d = 1,2,\cdots,D)$. Divide it into $L = \max_{d \in D}[W(V_r, V_d)]$ layers;

Step 2: if $t_k \le t_T$, update the prior probability of the layer number L and root node of the influence network, update the influencing intensity value between the root node and sub-node and turn to Step 3.

Otherwise, the calculation is ended, output the independent probability value and joint probability value of the sub-node;

Step 3: Calculate the conditional probability distribution of the sub-nodes;

Given the influence value $\{h_i^{(i)}(x_i)\}_{1\le i\le n}$, the influence value of a two-dimensional random state vector X'' on the target sub-node B is $h_n(x'')$ and the conditional probability $P(B|x'')$ of the sub-node B is calculated as follows:

(1) Polymerized negative influence intensity value $I_{\text{negative}} = \prod_{i:h_1(x_i)<0} \left(1-\left|h_1^{(i)}(x_i)\right|\right)$;

(2) Polymerized positive influence intensity value $I_{\text{positive}} = \prod_{i:h_1(x_i)>0} \left(1-\left|h_1^{(i)}(x_i)\right|\right)$;

(3) Merge the positive and negative influencing intensity values and get the calculation formula of the overall influence value $h_n(x'') = \left[I_{\text{negative}} - I_{\text{positive}}\right]\cdot\left[\max\left(I_{\text{negative}}, I_{\text{positive}}\right)\right]^{-1}$;

(4) Calculation of the conditional probability $P(B|x'') = \begin{cases} P(B)+h_n(x'')\cdot[1-P(B)]; & h_n(x'')\ge 0 \\ P(B)+h_n(x'')\cdot P(B); & h_n(x'')<0 \end{cases}$;

Step 4: Calculate the occurrence probability of a sub-node

Given the conditional probability $P(B|x'')$ of the sub-node B and the prior probability $P(x'')$ of the root node, according to the mutually independent assumption of the occurrence probability of the general formula and the root node, it can get $P(B) = \sum_{x''\in X''}\left[P(B|x'')\cdot P(x'')\right]$;

Step 5: Determine whether the occurrence probability of each layer of nodes is calculated; if it is not completed, $L = L+1$ and return to Step 2; if completed, update the time period $[t_{k-1}, t_k]$, and return to Step 2.

3. Model Optimization and Solution Based on IDEA

3.1. Coding Form of the Individual

The feasible strategy set composed of the basic action A_d is $\pi(A_d) = \{a_d^1, a_d^2, \cdots, a_d^{r_d}\}$, r_d is the number of feasible strategies and the combination of the entire action is shown as $G = \{a_i \,|\, a_i \in \pi(A_i), 1\le i\le NA\}$. The experience knowledge of decision-makers should be considered in the single basic action and the action number tends to be integer, so the real number is used to encode the individual: the individual includes NA real sequences $\omega = (\omega_1, \omega_2, \cdots, \omega_d, \cdots, \omega_{NA-1}, \omega_{NA})$, $\omega_d \in \{1, 2, \cdots, r_d\}$, $d = 1, 2, \cdots, NA$. Set $\omega_d = k$, represent the k basic strategy of the strategy set $\pi(A_d)$, set $p_l \in [0,1]$ as individual learning probability.

3.2. Fitness Function

For the evaluation of the preferred COA, the ultimate goal of the joint air operations is to achieve the desired target intention. Therefore, the program to make the target effect reach the largest probability is the best COA, $f = E_{\max}[P(D_i(t_T))]$. In the uncertain environment, in order to accurately estimate the natural state of the enemy action E, the new random test observation can be used to gain the new signal and the simulation data is M. The joint probability value f can be evaluated by calculating the mean μ and variance σ^2 of M f_1. For the joint air operations, it is desirable to maximize the joint probability of the combat target, that is, the signal-to-noise ratio (SNR) can be defined to integrate the mean μ and variance σ^2, replacing the single joint probability f, the fitness function as follows.

$$\text{SNR} = -10\cdot\log_{10}\left[\frac{1}{\mu^2}\cdot\left(1+\frac{3\sigma^2}{\mu^2}\right)\right] \tag{4}$$

3.3. Differential Variance

Regardless of whether the individual is feasible before and after the mutation, the individual $\omega = \{\omega_1, \cdots, \omega_d, \cdots, \omega_{NA}\}$ generates a new variant individual according to the linear combination of the parent individuals, and the basic variation component is the difference variable of the parent individual. The following rules are used to generate new individuals:

$$v_i = \omega_{r_1} + F \cdot (\omega_{r_2} - \omega_{r_3}) \quad i = 1,2,\cdots,N_d \tag{5}$$

v_i is the variant of the individual ω_i; ω_{r_1}, ω_{r_2} and ω_{r_3} are three different individuals randomly selected in the parent population, and ω_i is not included. F is a real constant valued between the interval of $[0,2]$, which can control the influence of the difference vector $(\omega_{r_2} - \omega_{r_3})$. The size of the disturbance is positively correlated with the absolute value of the difference vector. When the population approaches the optimal value, the disturbance will also become smaller.

3.4. Cross Operation

For the time being, regardless of whether the individual can be decoded before and after the cross operation, this paper encodes the individual in integer form and performs the individual cross-operation with the following mechanism: the individuals that perform the cross operation are the mutation vector $v_i = \{v_{i1},\cdots,v_{id},\cdots,v_{iNA}\}$ and the target vector $\omega_i = \{\omega_{i1},\cdots,\omega_{id},\cdots,\omega_{iNA}\}$ respectively. Through the cross rules in the formula (6), generate new individual $\omega_j = \{\omega_{j1},\cdots,\omega_{jd},\cdots,\omega_{jNA}\}$.

$$\omega_j = \begin{cases} v_i & rand(0,1) \le CR \\ \omega_i & rand(0,1) > CR \end{cases} \tag{6}$$

In formula, $rand(0,1)$ is the random value between 0 and 1; CR is a cross constant, the value of which is the real number between $[0,1]$. The probability of cross increases as the CR value increases. The cross operation ensures that at least one phase element is exchanged between individuals, thus creating new individuals and preventing the evolution of the population from interruption.

3.5. Selection Operation with Learning

The selection operation in IDEA is a preference model based on the "greedy mechanism", and the size of the adaptive function of the individual is the only reference standard. If and only if the fitness value of the new vector individual v_i is better than that of the target vector individual ω_i, the new individual will be selected. Otherwise, the new generation of population ω_j is still reserved for iterative calculation, and in the process, it continues to conduct variation and cross-operation of target vector. IDEA's selection strategy is relatively harsh; parent individual and offspring candidate individual compete against each other and the preference is selected, making the offspring individuals always inferior to the parent individuals and promoting the population to seek for preference towards the best orientation in the shorted path. With the ranking result, seeking common gene of the whole chromosomes to get the fine individual and marked the former-rank chromosome gene bit. While individual learning probability $p_l \ge rand(0,1)$, utilize the superiority common gene instead of the latter-rank gene bit, to enhance the algorithm convergence rate.

4. Simulation and Calculation Analysis

4.1. Case Scenario

In order to organize and apply all kinds of air power in a unified way, the scenario destroys most of the important goals on an island reef, forces them to abandon the island and thus gets rid of the barrier for the friendly forces to carry out landing operations. In order to ensure that the operational intent is achieved, the air force will carry out the following actions: A_1 is air cover on the air attack; A_2 is air cover on marine attack; A_3 is air ground assault of high value fixed island goal; A_4 is air-ground destruction of the ground's fixed air defense weapons target; A_5 is air-found destruction of high maneuver air defense target; A_6 is air-ground assault of high-value island target; A_7 is air command and control operations; A_8 is regional cover; A_9 is electronic interference to the enemy; A_{10} is organization of air refueling. The enemy counterattack may include: E_1 is the enemy aircraft carries out air interception to us; E_2 is the enemy ships organizes the attack of air plane; E_3 is the island air defense weapon system conducts attack of air plane; E_4 is the enemy organizes electronic anti-interference; E_5 is air interception of the enemy's second echelon. The expected battle effect under the joint function of basic action and enemy action: S_1 is air-to-air combat effect; S_2 is air-to-sea combat effect; S_3 is attack effect of air-to-ground air defense firepower; S_4 is gathering of air formation; S_5 is destruction degree of island objective; S_6 is return and meet. Focus on the operational intent, the final target effect is summarized into: D_1 is task completion; D_2 is battle loss. Combined with staff experience and knowledge of experts in the field, the influence network is used to set up the COA generation model, as shown in Figure 1. According to the operational guidance, under the common analysis of the planning staff and experts in the field [4], divide the combat into six stages for organization and implementation: $[t_0,t_1]$, $[t_1,t_2]$, $[t_2,t_3]$, $[t_3,t_4]$, $[t_4,t_5]$ and $[t_5,t_6]$.

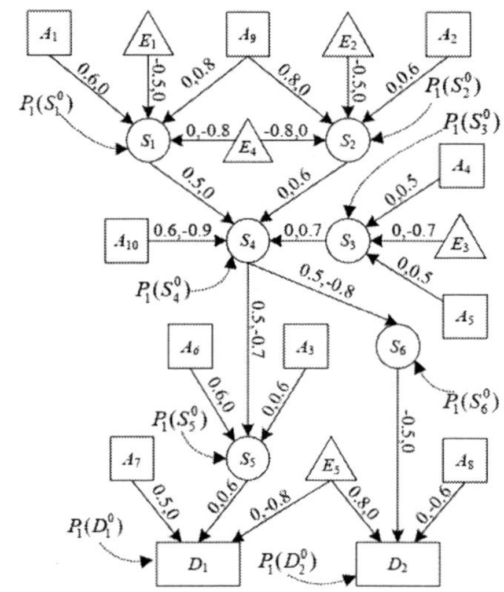

Figure 1. Influence network Model Generated by COA

4.2. Simulation and Calculation Analysis

The parameters of IDEA and standard DEA are set as follows: population size $N_d = 30$, calculation iteration number $n_{\text{iterative}} = 50$, scaling factor $F = 0.5$, cross constant $CR = 0.5$ and individual characteristic learning probability $p_l = 0.75$. The GA parameters: population size $N_d = 30$, the number of calculation iterations $n_{\text{iterative}} = 50$, the cross probability $p_c = 0.8$ and the individual mutation probability $p_m = 0.05$. The electronic interference action in Action A_9 has the best effect on S_1 and S_2 in the stage of $[t_0,t_1]$ and $[t_1,t_2]$. Subsequently, the enemy conducts electronic anti-interference and the interference effect gets worse in the stage of $[t_1,t_2]$ and $[t_2,t_3]$. This action will not be organized in the stage of $[t_3,t_4]$, leading to the same influencing intensity as before. The joint probability of the target functions D_1 and D_2 is the largest.

The objective function $f_1(s)$ is the independent occurrence probability of the node D_1; $f_2(s)$ is the independent occurrence probability of D_2, and the objective function $f_3(s)$ is the joint probability that the node D_1 occurs and D_2 does not occur, as shown in Fig.2 and Fig.3. The final target function value 0.8897 is better than the target function value 0.7875 of Ω_1. Focusing on the generation of COA Ω_2, in the case of operating for 30 times, as shown in Fig 4 and Fig 5, the three algorithms can find the optimal under certain parameter setting conditions, although IDEA and DEA have more running time than GA, and when approaching the global optimal solution, it is necessary to go through several iterations to achieve the optimal, which is decided by DEA's own mechanism[5], but compared to GA, the global search ability of IDEA and DEA is stronger and they can seek for the global optimal solution more quickly in the feasible solution space. Compared with the standard DEA, IDEA has better average convergence efficiency through the feature learning of the dominant individual points.

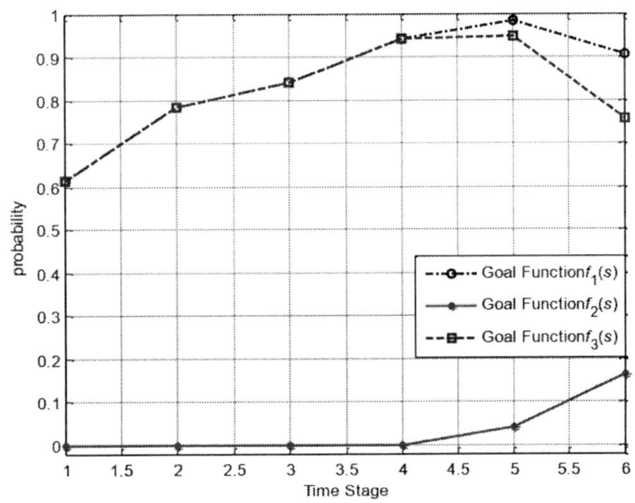

Figure 2. The target node probability of COA Ω_1

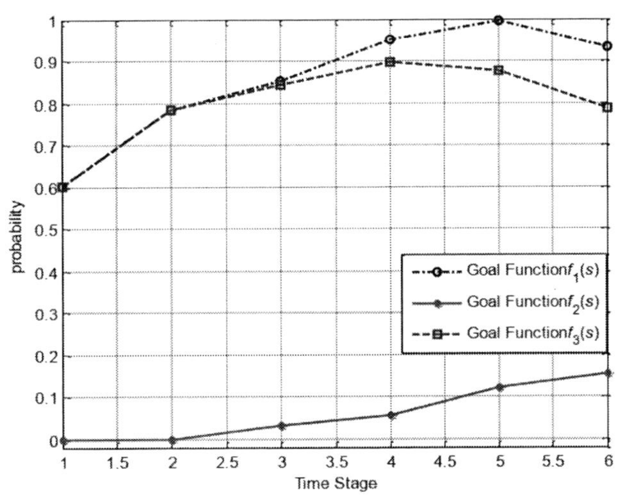

Figure 3. The target node probability of COA Ω_2

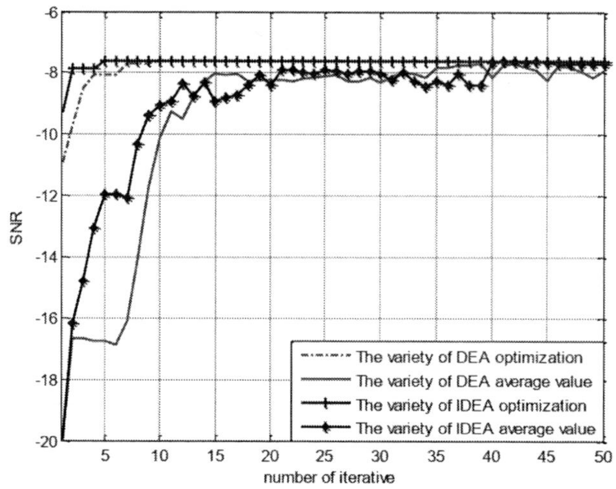

Figure 4. Results of IDEA and standard DEA algorithm

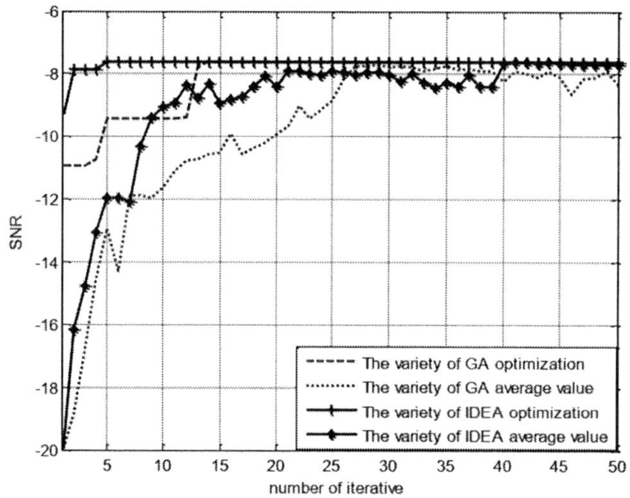

Figure 5. Results of IDEA and standard GA algorithm

5. Conclusion

Aiming at the characteristics of the generation of air operation COA, such as a large number of influencing factors and complex correlations, in order to solve the generation problem of uncertainty COA based on cause-effect relationship and obtain the best action strategy under different battle periods, an influence network based on influence intensity is used to set up the generation model of COA and the calculation step of corresponding probability propagation is given. Combining with the generation model features of COA, the coding, mutation and crossover strategy are designed based on IDEA. The feasibility and validity of IDEA are verified by simulating the case of air attack.

6. Acknowledgments

This work was financially supported by the National Natural Science Fund (61472443, 61573017), Program of ShanXi Natural Science Foundation (2016JQ6062).

7. References

[1] Liu J W, Li H E, Luo X L Study on Probabilistic Graphic Learning Technology Journal of Software 40 1025-44

[2] Lei T, Zhu C, Zhang W M Dynamic Bayes network based target-attacking course of action generation 2013 Proc. Int. Conf. on Intelligence and Security Informatics(Changsha) pp 1-5

[3] Zaidi A K, Papantoni K P 2010 Theory of influence networks. Journal of Intelligent & Robotic Systems, 60 457-91
[4] Yang D S, Zhang W M, Liu Z 2006 Mathematical Description and Solution Algorithm Research of Battle Task Plan Journal of Systems Engineering and Theory 6 26-34
[5] Gong D W Qin N N, Sun X Y 2010 Study on large-scale evolution of large population based on optimization problem of hybrid performance Control Theory & Applications 27 769-74

IOP Conf. Series: Journal of Physics: Conf. Series **887** (2017) 012089 doi:10.1088/1742-6596/887/1/012089

Rock images classification by using deep convolution neural network

Guojian Cheng and Wenhui Guo

School of Computer Science, Xi'an Shiyou University, Xi'an Shanxi, 710065, China.
Email: gjcheng@xsyu.edu.cn, 846522705@qq.com

Abstract. Granularity analysis is one of the most essential issues in authenticate under microscope. To improve the efficiency and accuracy of traditional manual work, an convolutional neural network based method is proposed for granularity analysis from thin section image, which chooses and extracts features from image samples while build classifier to recognize granularity of input image samples. 4800 samples from Ordos basin are used for experiments under colour spaces of HSV, YCbCr and RGB respectively. On the test dataset, the correct rate in RGB colour space is 98.5%, and it is believable in HSV and YCbCr colour space. The results show that the convolution neural network can classify the rock images with high reliability.

1. Introduction

For the effective development of reservoirs, it is necessary to provide a comprehensive reservoir description and characterization to determine the underground gas content. Granularity analysis is an important work of it [1]. The traditional method for rock classification is a manual work with many problems such as time-consuming and low accuracy. With the development of science and technology, Artificial intelligence is successfully applied in all walks of life. Many domestic and foreign scholars have done researches in the automatic classification of rock images, such as, Cheng Guojian and Liu Ye [2-3] used shallow neural network and SVM to classify rock images. Mariusz Młynarczuk et al.[4] performed the Classification of thin rock images respectively in RGB, CIELab, YIQ and HSV colour spaces using the nearest neighbour algorithm, K-nearest neighbour, the nearest pattern algorithm, and the optimized spherical neighbourhood; Hossein Izadi et al.[5] established a neural network to identify the rock mineral, whose accuracy was 93.81%. The above methods show that the application of machine learning in rock classification can improve its efficiency and accuracy.

However, using machine learning to classify rock images still has the following shortcomings. Firstly, to classify rock images by machine learning is based on the premise of artificial extraction of image features. Secondly, if the images are large, training a shallow neural network is almost impossible.

Convolution neural network (CNN) is an important deep learning architecture. It can extract the image features automatically and has a high classify accuracy. CNN has achieved a wide range of applications such as plant classification, face recognition, handwritten Chinese character recognition and so on [6-8]. In this paper, we construct a new convolution neural network for rock classification, rock images respectively in RGB, HSV, YCbCr colour spaces are used to train it, then contrasted the results and choose the best one.

2. The Rock Images

It is usually determined by professional geological researchers for types and structural parameters of rocks after identifying rock thin section under polarized light microscopy. The rock images used in this paper are all feldspar sandstones collected from an oil field in Ordos, they are divided into three types: (1) Coarse feldspar sandstones, as is shown in figure 1(a). (2) Medium granular feldspar sandstone, as is shown in figure 1(b). (3) Fine feldspar sandstone, as is shown in figure 1(c).

(a) Course feldspar sandstone (b) Medium granular feldspar sandstone (c) Fine feldspar sandstone

Figure 1. Example of three types of rock images

3. Convolution Neural Network

CNN was proposed by LeCun et al. [9] inspired by natural mechanisms in the 1990s.CNN are composed of several convolutional layers alternately connected with a number of pooling layers and can effectively characterize the essential features of the original image. It needs very little pre-process with original image. Convolution operations include continuous convolutions and discrete convolutions. The formulas are shown as Formula (1) and Formula (2):

Continuous convolution:

$$y(t)=\int_{-\infty}^{+\infty} x(p)h(t-p)\,dp = x(t)*h(t) \quad (1)$$

Discrete convolution:

$$y(n)=\sum_{i=-\infty}^{+\infty} x(i)h(n-i) = x(n)*h(n) \quad (2)$$

There are some challenges in using CNN such as how many convolutional layers, how large the size of convolutional kernel and the learning rate of the network should be set are all worth to study carefully.

4. Experiment

4.1. Data Sources

In this paper, 4800 rock images in RGB colour space are selected as experimental samples, normalize the size of them to 224*224. There are 1600 coarse-grained rock images, 1600 medium-grained rock images and 1600 fine-grained rock images. And 1200 rock images of each type are used as the training set, the remaining 400 are used as the test set. Then converting the 4800 rock images to YCbCr colour space and HSV colour space respectively, whose training set and test set distribution remains unchanged.

4.2. The Architecture

The convolution neural network structure designed in this paper is shown in figure 2, which is a 6-layer structure, 4 layers are convoluted and 2 layers are fully connected. The convolution layers use ReLU [10] as the activation function and the fully connected layers do classification by the Softmax classifier.

IOP Conf. Series: Journal of Physics: Conf. Series **887** (2017) 012089 doi:10.1088/1742-6596/887/1/012089

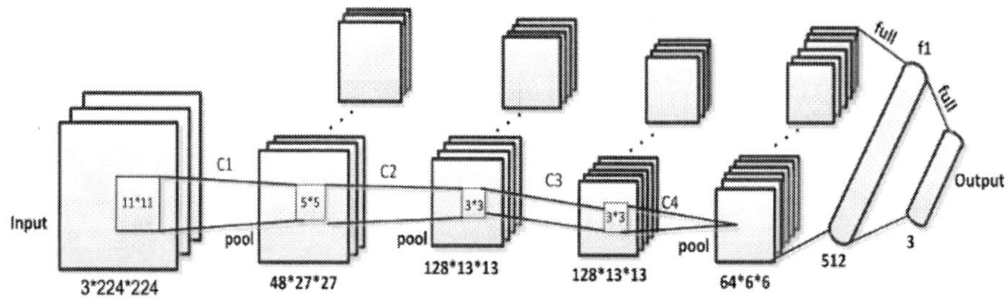

Figure 2. CNN structure

The original image is in RGB colour space, convert it to an RGB three-channel grayscale image as input. As is shown in figure 2 (The marks of the pooling layers S1, S2 and S3 are omitted in the figure), the size of the input image is 224*224. The first convolution layer C1 uses 48 convolution kernels of 11*11 (convolution step is 4 pixels) to obtain 48 feature maps of 56*56 by convolution operation, and then let the 48 feature maps go through a pooling layer with a window whose size is 3*3 (moving step is 2 pixels) obtaining 48 feature maps of 27*27(S1). The second convolution layer C2 takes S1 as input and convolves them with 64 convolution kernels of 5*5 (convolution step is 1 pixel) to obtain 64 slices of 27*27 feature map, then let the 64 feature maps go through a pooling layer with a window whose size is 3*3 (moving step size of 2 pixels) to get 64 feature maps S2 whose size are 13*13(S2).The third convolution layer C3 takes S2 as input and convolves them with 128 convolution kernels of 3*3 (convolution step size of 1 pixel) to obtain 128 slices of 13*13 feature map. The fourth convolution layer C4 takes the 128 slices feature map as input and convolves them with 64 convolution cores of 3*3 (convolution step size of 1 pixel) to obtain 64 feature maps whose size are 13*13. Then let these 64 feature maps go through a pooling layer with a window size of 3*3 (moving in steps of 2 pixels) to obtain 64 feature maps whose size are 6*6(S3). Finally, transforming the result into a one-dimensional vector and use it as the input of the fully connected layer, which is connected with the fully connected layer f1 which has 512 neurons and classifies the rock images according to its granularity size by the Softmax classifier.

4.3. Architecture Adjustment

The experimental data is loaded into memory by batch whose size is 100. An iteration is performed while a batch of data is loaded, where the dropout is set to 0.5.

4.3.1. Number of layers. Setting an appropriate number of convolution layers is one of the key steps in constructing a CNN. If the convolution layers are too less, the network will be not able to learn the essential features of the original image, if there are too many convolution layers, it may lead the network over-fitting. In this paper, the number of convolution layers of 5, 4 and 3 are studied respectively. Compared the convergence and error rate while each network has been subjected to 400 iterations, the results are shown in table 1 and figure 3.

Table 1. Error rate of different convolution layers

Number of convolution layers	5	4	3
Network structure	11-7-5-3-3	11-7-5-3	11-7-5
Error rate	16.4%	3.1%	5.1%

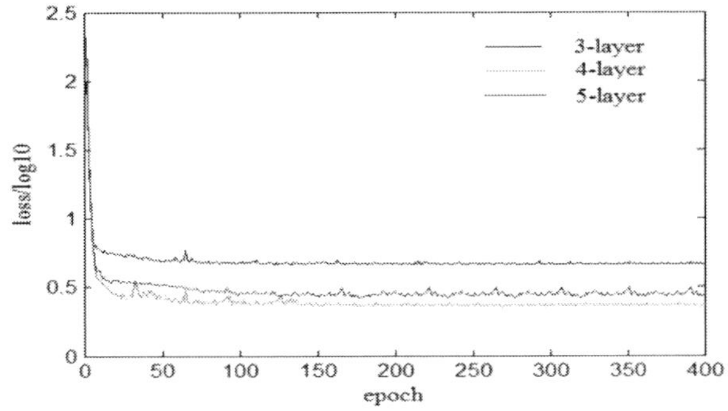

Figure 3. Network loss function for different convolution layers

It can be seen from figure.3 and table 1 that whenever the number of convolution layers is 3, 4 or 5, the network can converge quickly. However, when the 4-layer convolution structure is used, the loss function and the error rate are the lowest.

4.3.2. Convolution kernel. Convolution kernel size is another important factor affecting the efficiency of CNN. When using a 4-layer CNN, the convolution kernel size of it is set to 11-7-5-3, 11-5-3-3 and 11-3-3-3 respectively for experiment in this paper. After 400 iterations, the error rate is shown in Table 2 and the loss function curve is shown in Figure.4.

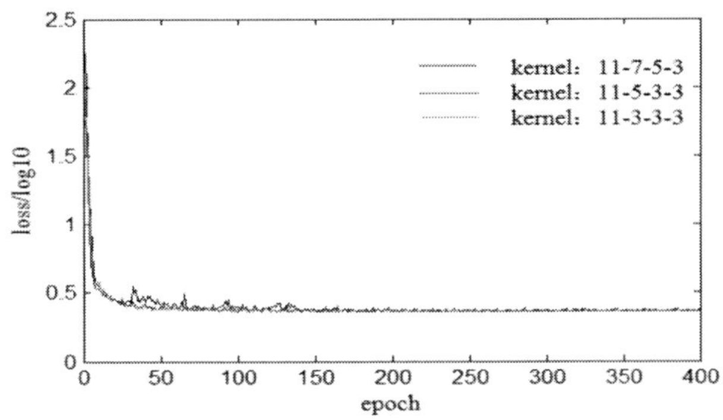

Figure 4. Loss function of different convolution kernel network

Table 2. Error rate of different convolution kernel network

Network structure	11-7-5-3	11-5-3-3	11-3-3-3
Error rate	3.1%	1.5%	2.4%

From table 2 and figure.4, we can see that setting the convolution kernel size to 11-7-5-3, 11-5-3-3, 11-3-3-3 respectively, the network all can get convergence fast. However, when it is 11-7-5-3, the network convergence fluctuates greatly than the other two. But, the network with convolution kernel of 11-5-3-3 has the lowest error rate. Therefore, a 4-layer CNN whose convolution kernel is 11-5-3-3 is used in this paper.

4.3.3. Learning rate. Learning rate is a key factor influencing the convergence rate and classification accuracy. When a 4-layer CNN whose convolution kernel size is 11-5-3-3 by setting the initial learning rate as 0.01, 0.001, 0.0005, 0.0001 respectively, the error rate and iterations times are shown in Table 3.

Table 3. Relationship between convergence and learning rate

Learning rate	0.01	0.001	0.0005	0.0001
Convergence iterations	Doesn't converge	26	21	39
Error rate of test set	56.8%	3.6%	1.5%	7.5%

It can be seen from table 3 that when the initial learning rate is 0.01 the network does not converge because of the gradient step is too large and deviates from the optimal point. When the initial learning rate is 0.001, the gradient step is larger and the convergence speed of the network is slower and the error classification rate is lower. When the initial learning rate is 0.0005, the network convergences fast and the error classification rate is the lowest. When the initial learning rate is 0.0001, the network can convergence and finds the best point, but needs to iterate more times and the classification accuracy is lower. In summary, setting the initial learning rate as 0.0005 is the most appropriate.

5. Validation

In order to prevent the network from over-fitting, this paper use cross-validation. Each type of rock image data set is randomly divided into ten subsets, which is 160 per subset for cross-validation. The production of data set and experimental process in YCbCr colour space and HSV colour space are consistent with that in RGB colour space. Table 4 shows the cross-validation results for RGB, YCbCr and HSV colour spaces.

Table 4. Cross-validation error rates

Time	1	2	3	4	5	6	7	8	9	10	Average
HSV	0.21	0	0.21	0.21	0	0.21	0	0	0	0.21	0.117
YCbCr	0.41	0.21	0	0.21	0.41	0.21	0	0.41	0	0.21	0.253
RGB	2.1	1.04	1.2	2.3	1.0	0.83	1.04	0.83	2.3	2.5	1.66

It can be seen from table 4 that the error rates of cross-validation in HSV, YCbCr and HSV colour spaces are all small. So the network does not appear the over-fitting phenomenon in the three colour spaces.

Granularity analysis can provide authenticate researchers essential basis for determining types and structural parameters of rocks, but it is a manual work with low reliability in traditional. However, the experiment results above shows that do the rock granularity analysis using CNN is with high accuracy which could improve the reliability and efficiency.

6. Conclusion

In this paper, CNN is used to identify the rock granularity. The experiments show that it has high reliability whether in HSV, YCbCr or RGB colour space. In RGB colour space, the classification accuracy achieves 98.5% with high efficiency. However, the experimental results still have deviation may be caused by the use of single polarized images. The next step of study can be committed to multi-polarized light in the rock image. In view of the high reliability of the application of CNN in rock image classification based on rock granularity, it can be considered to apply to the classification of rock components.

7. Acknowledgments

The main project is supported by the National Science and Technology (No.2011ZX05044) and Shaanxi Province industrial science and technology (No.2015GY104).

8. References

[1] Wang Pujun, Jiao Yangyang, Uang Kaikai, Zengbao, Bian Weihua, Classification ofVolcanogenic Successions and Its Application to Volcanic Reservoir Exploration in the Junggar Basin, NW China, Journal of Jilin University(Earth Science Edition), Volume 46,No.4,July 2016, Pages1056-1070

[2] Cheng Guojian,Ma Wei,Wei Xinshan, Rong Chunlong,Nan Junxiang, Research of rock texture identification based on image processing and neural network, Journal of Xi'an Shiyou University(Natural Science Edition), Volume 28, September 2013, Pages 105-109

[3] Guo Chao, Liu Ye, Recognition of Rock Images Based on Multiple Colour Spaces, Science Technology and Engineering, Volume 14, June 2014, Pages 247-251+255

[4] Mariusz Młynarczuk, Andrzej Górszczyk, Bartłomiej Ślipek, The application of pattern recognition in the automatic classification of microscopic rock images, Computers & Geosciences, Volume 60, October 2013, Pages 126-133

[5] Hossein Izadi, Javad Sadri, Mahdokht Bayati, An intelligent system for mineral identification in thin sections based on a cascade approach, Computers & Geosciences, Volume 99, February 2017, Pages 37-49, ISSN 0098-3004

[6] Yuma Mikia, Chisako Muramatsua, Tatsuro Hayashib, Xiangrong Zhoua, Takeshi Haraa,Akitoshi Katsumatac, Hiroshi Fujitaa, Classification of teeth in cone-beam CT using deep convolutional neural network[J], Computers in Biology and Medicine 80 (2017) 24–29

[7] Mads Dyrmann, Henrik Karstoft, Henrik Skov Midtiby, Plant species classification using deep convolutional neural network[J], biosystemsengineering151(2016)72-80

[8] André Teixeira Lopes a, Edilson de Aguiar b, Alberto F. De Souza a, Thiago Oliveira-Santos, Facial expression recognition with Convolutional Neural Networks:Coping with few data and the training sample order[J], Pattern Recognition 61 (2017) 610–628

[9] Hinton G E, Salakhutdinov R. Reducing the dimensionality of data with neural networks [J].Science, 2006, 313(5786):504-507.

[10] Krizhevsky A, Sutskever I, Hinton G E. Imagenet classification with deep convolutional neural networks[C] Advances in neural information processing systems (NIPS 2012), 2012: 1097-1105.

Intelligent acoustic data fusion technique for information security analysis

Ying Jiang, Yize Tang, Wenda Lu, Zhongfeng Wang, Zepeng Wang and Luming Zhang

[1]State Grid Zhejiang Information &Telecommunication Company
[2]Zhejiang Huayun Information Technology CO., LTD
[3]Department of CSIE, Hefei University of Technology

Abstract. Tone is an essential component of word formation in all tonal languages, and it plays an important role in the transmission of information in speech communication. Therefore, tones characteristics study can be applied into security analysis of acoustic signal by the means of language identification, etc. In speech processing, fundamental frequency (F0) is often viewed as representing tones by researchers of speech synthesis. However, regular F0 values may lead to low naturalness in synthesized speech. Moreover, F0 and tone are not equivalent linguistically; F0 is just a representation of a tone. Therefore, the Electroglottography (EGG) signal is collected for deeper tones characteristics study. In this paper, focusing on the Northern Kam language, which has nine tonal contours and five level tone types, we first collected EGG and speech signals from six natural male speakers of the Northern Kam language, and then achieved the clustering distributions of the tone curves. After summarizing the main characteristics of tones of Northern Kam, we analyzed the relationship between EGG and speech signal parameters, and laid the foundation for further security analysis of acoustic signal.

1. Introduction

Among all human language, tone is an essential component of tone languages, and is used to build words much as consonants and vowels do [1]. For instance, in Chinese mandarin, the syllable /ti/, when pronounced with a level pitch pattern, means "embankment"; with a rising pattern, the meaning is "whistle"; with a falling-rising pattern, the meaning is "bottom"; and with a falling pattern, the meaning is "land". Thus, semantic distinction in tonal languages depends not only on articulatory composition, but also on tone patterns. Thus, more and more researchers pay attention to research the distinguishing characteristics of different tones. Here, in continuation of this work, we use the physiological and acoustic signals information to describe the tones in a tone language.

From a linguistic or physics point of view, the voice generation of larynx towards the lip involves three components: the voice source (phonation), the acoustic resonance (articulation), and the lip radiation (loudspeaker) (Fig. 1) [2]. The vibration of vocal cords generates speech signal, and the vowels and consonants are highly connected to the vibration difference of vocal cords, namely the phonation types. The vibration of vocal cords is a synergistic effect of different parts of a throat. During the process of generating voice, F0 is decided by the vibration frequency of vocal cords [3]. After tuning by soundtrack, mouth cavity and nasal cavity, the source signals from vibration can be made into different sounds. This is called resonance, which can be seen as a modulating process to the source signals. Finally, the sounds are transferred into the air through the radiation effect of mouth to form the speech signals. Thus, the generation of speech is a process of stimulation and modulation. In the research of traditional linguistics and speech processing, what we collect is the speech signals, and many in-depth studies have been done as well as numerous acoustic models are raised. Meanwhile, the generation of tones is also fully analyzed. However, tone is much more than just a physical quantity. As a linguistic term, tone is a kind of rational cognition of pitch by linguist, belonging to the categorical perception.

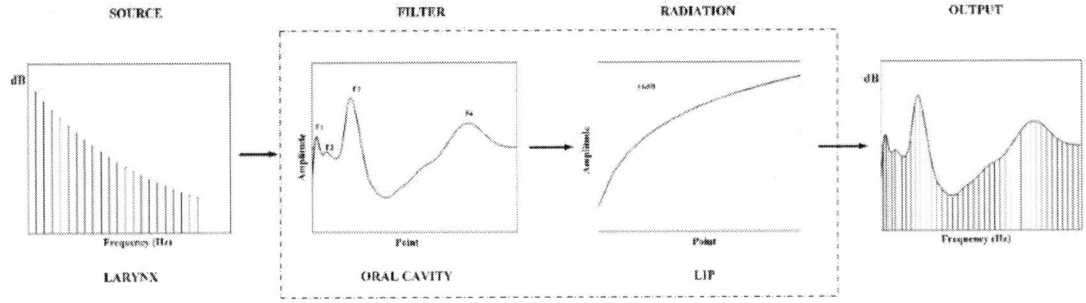

Figure 1. Voice generation of an Sine curve.[4]

Though F0 is the frequently used physical quantity in the present speech research, F0 cannot indeed represent the original vibrational state of vocal cord with the influence of phonation. An electroglottograph (EGG) signal, for example in Fig. 2, can be used to measure the vocal information directly and index the degree of contact between the vocal folds, so it has been a useful tool in measuring and describing phonations [5-11]. Childers and Lee used EGG measures to extract source related features for modal, vocal fry, falsetto and breathy voice, which are produced by subjects with and without vocal disorders [12]. Four factors were found to be the most important in distinguishing the above four phonation types: glottal pulse width, pulse skewing, abruptness of glottal closure, and turbulent noise. Results for glottal pulse width showed that modal phonation was produced with medium glottal width, vocal fry with a short glottal width, falsetto and breathy voice with a long glottal width. For pulse skewing, modal phonation was characterized by medium skewing, vocal fry with high skewing, falsetto and breathy voice with low skewing. In terms of abruptness of closure, modal voice and vocal fry were characterized by abrupt closure, while falsetto and breathy phonations were characterized by progressive closure. Finally, breathy phonation was the only phonation produced with high turbulent noise.

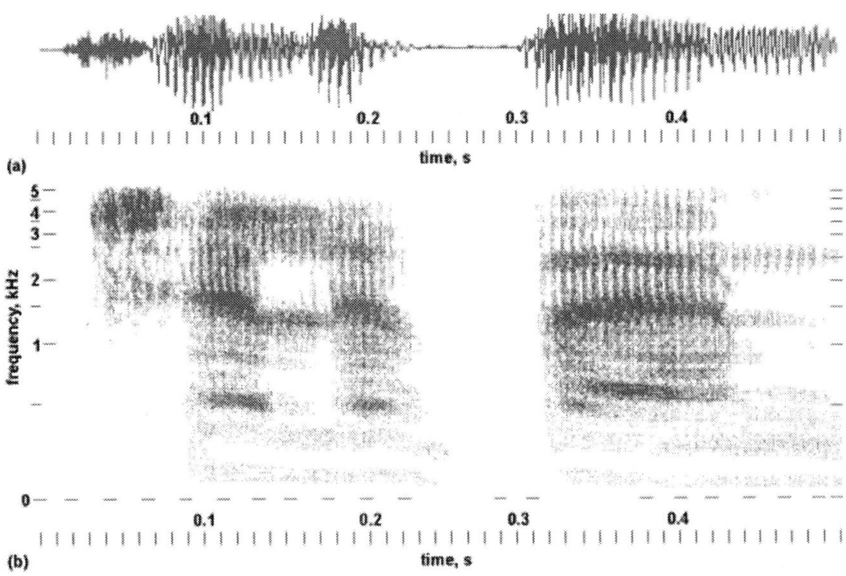

Figure 2. The voice signal and Electroglottograph in Northern Kam Language / **ʔa**$^{\mathbf{5}\ \mathbf{5}}$ / **and** /la^{22}/.

2. Methods

Corpus: In this paper, a corpus of synchronous EGG and speech signals is established. It is acquired under clean environment from a 53-year-old man who is familiar with Kam language and mandarin. Speech signals are recorded by FieldPhon software with sampling frequency 44100Hz, and the strict processing of listening, identifying and tuning is implemented after recording. The corpus includes 770 monosyllable words, covering all tones on 9 unchecked syllables and 4 checked syllables in Northern Kam Language. Moreover, the universality and proportionality of the samples of tones, consonants and vowels are ensured as far as possible. EGG signals are synchronously acquired with the recording of speech signals.

Data processing: Seven kinds of F0 parameters are acquired based on the basic acoustic parameters extracted by STRAIGHT [22]: 1) the amplitude difference between starting point and end point of fundamental frequency (pitch variation), 2) the length of fundamental frequency curve (duration), 3) the maximum of fundamental frequency, 4) the minimum of fundamental frequency, 5) the extreme points of fundamental frequency curve, 6) the slope of fundamental frequency curve, 7) the slope variation of fundamental frequency curve. Meanwhile, two parameters of EGG, CQ and DECPA, are obtained based on the parameters including CQ, Peak Vel and Min Vel extracted by EggWorks and processed with Matlab.

Analysis Methods: The rule of minimum opposition is adopted for analyzing the experimental data so as to accurately attain the influence of EGG parameters and acoustic parameters on the acoustic characteristics. There is only one kind of characteristic difference (tone, consonant or vowel difference) for data comparison. For example, when the opposition characteristic is consonant, the rule of minimum opposition ensures the consistency of vowel and tone. The opposite characteristic is tone, and the consonant and vowel are consistent. During analysis, the pathological voice is ignored while the outliers and interferences of erroneous recording are eliminated.

The method of minimum opposition used in this paper for one person and one language system can also be applied to multi-person and multi-language system. However, the person and the language type should be respectively taken as the condition of opposition for analysis and discussion.

3. Results

3.1. Analysis of Tone Identification by EGG Signal

At present, the tone definition is mainly confined to the value and modes of pitch. The value of pitch in physiology field corresponds to the vibration speed of vocal fold, and in acoustics or information field corresponds to the value of the fundamental frequency. On the other hand, tone is a linguistic term. We can think that the tone is the combination of phonation type and vibration frequency. Phonation type is a suprasegmental feature related to timbre, while vibration frequency determines the value of fundamental frequency. As shown in Fig. 3, with the influence of resonance, labial radiation and consonant, it is often difficult to accurately acquire the value of F0 through the speech signal. Meanwhile, EGG can obtain the glottal information of larynx directly without the influence of the above disturbances. So the value of F0 extracted by EGG signal is more precise than by the speech signal. As shown in Table 1, F0 extracted by EGG signal has smaller variation range and is smoother compared to F0 extracted by the speech signal.

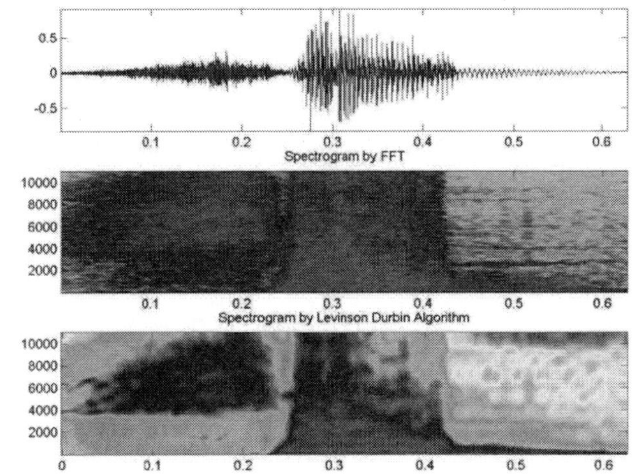

Figure 3. F0 extracted by EGG and speech signals. The red line indicates F0 extracted by speech signal, and the blue line indicates F0 extracted by EGG. F0 extracted by EGG has smaller variation range and is smoother.

Table 1. Analysis of F0 of /wən[33]/ extracted by EGG and speech signals. F0 slope is calculated after fitting for F0 values. F0 range is the variation range of F0 values. DRF is the variation range after five points smoothing for F0 values.

	F0 slope	F0 range	DRF
F0 extracted by EGG	**0.027**	**11.27Hz**	**6.17Hz**
F0 extracted by speech	0.167	31.92Hz	21.64Hz

3.2. Analysis of Acoustic and EGG Parameters of Checked and Unchecked Syllables

In the Northern Kam Language, like most Eastern Asian languages, the same tone value can correspond to the checked and unchecked syllables. For example, as shown in Fig. 3, tone7 is on unchecked syllable and tone10 checked syllable. They have the same tone value 55, but with the different durations, the ends of the vowel and the variation ranges of F0. In Tai-Kadai languages, the tones on checked and unchecked syllables differ in duration, and the checked syllables end in the consonants, leading to big difference in the phonation type.

Figure 4 and Table 2 show the parameters of the checked and unchecked syllables with the same tone value. We can find that DECPA of the checked syllable is higher than that of the unchecked syllable with the same tone value, but the trends of DECPA of the checked and unchecked syllables are similar. Furthermore, from Table 2, the difference between the values of DECPA of the checked and unchecked syllables is beyond 100. The F0 slope of the checked syllable is also higher, which is related to the higher DECPA of the checked syllable. Meanwhile, the values of CQ of the checked and unchecked syllables with the same tone value are approximately equal. From Fig. 4, the vowel is /a/ in syllables with different tone values, and the values of CQ are around 0.55.

			Egg weight (g)		Haugh unit		Breaking strength (kg cm⁻²)	
Zeolite (%)	Ca (%)	NPP (%)	30-36 weeks	36-42 weeks	30-36 weeks	36-42 weeks	30-36 weeks	36-42 week
0.0	3.26	0.25	56.34	56.63	81.54	74.46	3.93	3.33
0.0	3.26	0.19	57.26	56.38	81.29	71.30	3.95	3.61
0.0	2.45	0.25	56.72	57.42	75.92	73.69	3.75	3.38
0.0	2.45	0.19	56.93	57.61	76.61	77.90	3.50	3.06
1.5	3.26	0.25	57.85	57.90	73.14	78.10	3.73	3.64
1.5	3.26	0.19	57.43	58.24	72.21	70.13	4.24	3.84
1.5	2.45	0.25	56.89	57.36	73.10	72.60	3.86	3.75
1.5	2.45	0.19	55.92	56.79	78.14	74.77	3.56	3.45
3.0	3.26	0.25	56.95	56.53	73.36	73.82	3.54	3.45
3.0	3.26	0.19	57.86	59.28	73.79	73.58	3.60	3.86
3.0	2.45	0.25	56.38	57.24	73.29	73.73	3.91	3.50
3.0	2.45	0.19	57.87	58.07	75.76	74.43	3.51	3.79
4.5	3.26	0.25	57.65	58.03	75.00	72.92	4.03	3.93
4.5	3.26	0.19	56.81	57.51	75.89	74.79	3.46	3.30
4.5	2.45	0.25	58.54	58.64	77.69	78.38	3.77	3.72
4.5	2.45	0.19	58.31	58.73	76.25	69.37	4.01	3.53
SEM			0.463	0.632	2.089	3.113	0.241	0.219
			Probability					
Zeolite			*	0.066	**	NS	NS	NS
Calcium			NS	NS	NS	NS	NS	NS
Phosphorus			NS	NS	NS	NS	NS	NS
Zeolite×Calcium			**	0.086	*	NS	NS	NS
Zeolite×Phosphorus			*	0.089	NS	NS	NS	NS

Figure 4. EGG parameters of the checked and unchecked syllables with the same tone. The red line indicates the checked syllables, and the blue line indicates the unchecked syllables.

Table 2. Acoustic and EGG parameters of uncheckd and checked syllables.

tone value	**55**		**31**		**13**		**25**	
syllable	unchecked	checked	unchecked	checked	unchecked	checked	unchecked	checked
	Sa	sap	Sa	sap	so	sot	sa	səp
CQ	0.588	0.585	0.564	0.587	0.552	0.589	0.571	0.524
DECPA	536.77	**682.84**	431.03	**639.78**	275.91	**391.65**	365.78	**432.30**
F0 range	5.5819	8.9108	44.1751	40.0741	12.5073	13.0353	46.5851	69.5686
F0 slope	0.1925	**0.3171**	-1.2271	**-1.2523**	0.4467	**0.5925**	1.2591	**3.8649**

3.3. Analysis of Similar Tones

The similar tones, for example, the tone4 and tone9 shown in Fig 4 have much resemblance in tone contour. The curves of them are nearly parallel, and the durations of them are approximately equal, with only the values of F0 are different. For these similar tones with little difference in tone characteristics, it is difficult for the current tone recognition algorithms to distinguish them.

The parameters DECPA and CQ of similar tones (tone4, tone9) for /tu/ and /ta/ respectively. DECPA can distinguish the similar tones with the same vowel and consonant. For /tu/ DECPA of tone4 is larger than that of tone9, and the values of DECPA of the two tones are clearly distributed in two intervals. For /ta/ DECPA of tone9 is larger than that of tone4. At the same time, the distributions of DECPA for /tu/ and /ta/ are also different as the result that the articulation positions of /a/ and /u/ are different. So DECPA is related to not only the energy change of pronunciation but also the articulation positions.

Meanwhile, the similar tones cannot be well discriminated by CQ in /tu/ and /ta/. In addition, the curves of CQ of different tones in the same vowel are approximately similar. But the values of CQ of different vowels in the same tone are inconsistent, which may imply that CQ to some extent has distinction of the vowels.

4. Conclusion

Although EGG is a widely used instrument in physiology and medicine, it is still a relatively new means of measurement in the fields of linguistics and speech. EGG signals were initially used in linguistics to examine phonation types; the EEG parameters CQ and DECPA have been mainly used to measure the opening and closing glottis states, as well as velocity variations. Because Northern Kam is a relatively rich tonal language, it is very meaningful to study its tones. In this paper, we collected EGG signals of all the monosyllables in the language, and then we analyzed the relationship between the EGG signals and tones.

5. References

[1] Gang P., William S.-Y. Wang. Tone recognition of continuous Cantonese speech based on support vector machines. *Speech Communication* ***45***.1, 49-62 (2005).
[2] Kong J.P. On Language Phonation. *Central Nationalities University Press*, Beijing, 274-291 (2001).
[3] Yip M. Tone. *Cambridisge University Press*, London (2002)
[4] Hardcastle, William J., John Laver, and Fiona E. Gibbon, Eds. The handbook of phonetic sciences. Vol. **79**. *John Wiley & Sons,* (2012).
[5] Guion, Susan G., Mark W. Post, and Doris L. Payne. Phonetic correlates of tongue root vowel contrasts in Maa. *Journal of Phonetics* ***32***.4, 517-542 (2004).
[6] Esposito, Christina M. An acoustic and electroglottographic study of White Hmong tone and phonation. *Journal of Phonetics* ***40***.3, 466-476 (2012).
[7] Childers, Donald G., and Ashok K. Krishnamurthy. A critical review of electroglottography. *Critical reviews in biomedical engineering* ***12***.2, 131-161 (1984).
[8] Michaud, Alexis. Final consonants and glottalization: new perspectives from Hanoi Vietnamese. *Phonetica* ***61***.2-3, 119-146 (2005).
[9] Esposito, Christina M. An acoustic and electroglottographic study of phonation in Santa Ana del Valle Zapotec. *Poster presented at the 79th meeting of the Linguistic Society of America*, San Francisco, CA. (2005).
[10] Michaud, Alexis, and Martine Mazaudon. Pitch and voice quality characteristics of the lexical word-tones of Tamang, as compared with level tones (Naxi data) and pitch-plus-voice-quality tones (Vietnamese data). *Proceedings of Speech Prosody 2006* (2006).
[11] DiCanio, Christian T. The phonetics of register in Takhian Thong Chong. *Journal of the International Phonetic Association* ***39***.02, 162-188 (2009).
[12] Childers, Donald G., and C. K. Lee. Vocal quality factors: Analysis, synthesis, and perception. *The Journal of the Acoustical Society of America* ***90***.5, 2394-2410 (1991).
[13] Vu-Ngoc, Tuân, Christophe d'Alessandro, and Alexis Michaud. Using open quotient for the characterisation of Vietnamese glottalised tones. *Proceedings of Eurospeech-Interspeech 2005: the 9th Annual Conference on Speech Communication and Technology*. 2885-2889 (2005).
[14] Michaud, Alexis. A Measurement from Electroglottography: DECPA, and its Application in Prosody. *Speech Prosody 2004, International Conference*. 633-636 (2004).
[15] Long Y. Research of Kam Language Dialect. *Shanghai Normal University, P R China*. (2012).
[16] Long Y. and Zheng G. The Dong language in Guizhou Province, China. *Arlington: SIL and University of Texas at Arlington* (1998).
[17] Wang CC, Ding QL, Tao H, Li H. Comment on "Phonemic diversity supports a serial founder effect model of language expansion from Africa". *Science.* ***335***.6069,657 (2012,).
[18] Chao, Y. R. A system of tone letters. *Le maître phonétique* ***45***, 81-83 (1980).
[19] SY, William. Phonological features of tone. *International Journal of American Linguistics*, 93-105 (1967).
[20] Shi, Lin. An experimental analysis of the five level tones of Gaoba Dong language. *Journal of Chinese Linguistics* ***15***.2, 335-361 (1987).
[21] Xiaonong Z. Phonetics. *The Commercial Press.* Beijing (2010).
[22] Han X., Li L.,Pan WY. Computer based field investigation and processing system for language. *Journal of Tsinghua University (Science and Technology).* ***53***.6, 888-892 (2013).

Efficient large-scale graph data optimization for intelligent video surveillance

Quanhong Shang[1], Shujun Zhang[1], Yanbo Wang[2], Chen Sun[3], Zepeng Wang[4] and Luming Zhang[4]

[1]State Grid Zhejiang Electric Power Company
[2]State Grid Zhejiang Electric Power Company Information & Telecommunication Branch
[3]Zhejiang Huayun Information Technology CO., LTD
[4]Department of CSIE, Hefei University of Technology

Abstract. Society is rapidly accepting the use of a wide variety of cameras Location and applications: site traffic monitoring, parking Lot surveillance, car and smart space. These ones here the camera provides data every day in an analysis Effective way. Recent advances in sensor technology Manufacturing, communications and computing are stimulating.The development of new applications that can change the traditional Vision system incorporating universal smart camera network. This Analysis of visual cues in multi camera networks makes wide Applications ranging from smart home and office automation to large area surveillance and traffic surveillance. In addition, dense Camera networks, most of which have large overlapping areas of cameras. In the view of good research, we focus on sparse camera networks. One Sparse camera network using large area surveillance. As few cameras as possible, most cameras do not overlap Each other's field of vision. This task is challenging Lack of knowledge of topology Network, the specific changes in appearance and movement Track different opinions of the target, as well as difficulties Understanding complex events in a network. In this review in this paper, we present a comprehensive survey of recent studies Results to solve the problem of topology learning, Object appearance modeling and global activity understanding sparse camera network. In addition, some of the current open Research issues are discussed.

1. Introduction

2011 terrorist attacks in the United States in September Lead to review of security measures to see the use of a wide range of surveillance tasks for camera networks become one important research topic. All kinds of valuable science and engineering applications based on the development of effective the use of multi camera network. Under the assumption of a dense With the large angle crossing field camera network (FOV) Camera, early researchers use geometric information Calibrate the camera and reconstruct the shape and trajectory Objects in 3D space 1,2,3,4,5. Despite intensive cameras In recent years, the network has been well studied The network still has a challenging problem as it is It is necessary to locate, track and analyze the target in a wide range Use as many cameras as possible. Camera in a sparse camera the network does not necessarily have overlapping FOVs. An ideal sparsity Camera network monitoring system should generate tracking A sequence of independent moving objects The amount or extent of a camera in a network Overlap between different FOVs [6].

To successfully carry out automatic video surveillance in a sparse camera network, several challenges must be tackled. The first key challenge is track correspondence modeling to meet the requirement of accurate maintenance of person identity across different cameras. The goal of track correspondence modeling is to estimate which tracks result from the same object even though the tracks are captured by different cameras and at different times. Ideally, by combining the local trajectories of individuals obtained from different cameras, the activities of the individuals can be understood in a global view (Fig. 1). The identification of corresponding tracks is made more difficult by the changes in the appearance of an individual from the FOV of one camera view to that of another. Causes of these changes include variations in illumination, pose and camera parameters.

The second key challenge for a sparse camera networkbased video surveillance system is how to learn the relationship between the cameras. Spatial topology refers to distribution and linkage and is utilized to predict the trajectories of targets in a sparse camera network. For example, if a target disappears from the FOV of camera 1, then knowledge of the position of all cameras may allow the inference that the exit from camera 1 is linked to the entrance of the target into the FOV of camera 2 and camera 3. This inference may be supported by similarities in the appearance in the different images. Such inferences depend on the topological relationships between the cameras in the network. As shown in Figure 2, these relationships may be of several different types [6]. If the camera network is dense, there are various methods for multi-camera tracking with overlapping FOVs and, as discussed in the literature, there are various geometric matching methods for manually calibrating the cameras [1], [2], [7], [8], [9], [10]. The methods used in these papers can compute the transformation between 2D images coordinates and 3D spatial coordinates for a ground plane. However, it is unrealistic to expect that the FOVs of all the cameras always overlap one another, and the need for all cameras to be calibrated to the same ground plane usually cannot be satisfied. For instance, a target may appear in the FOV of two different cameras initially and may disappear from one of the FOVs, reappearing after some time in the FOV of a third camera. Several approaches have recently been presented to estimate the spatial topology and linkage in a camera network [11], [12], [13], [14], instead of directly calculating the 3D position in the ground plane, as in [1], [2] and [7], [8], [9], [10].

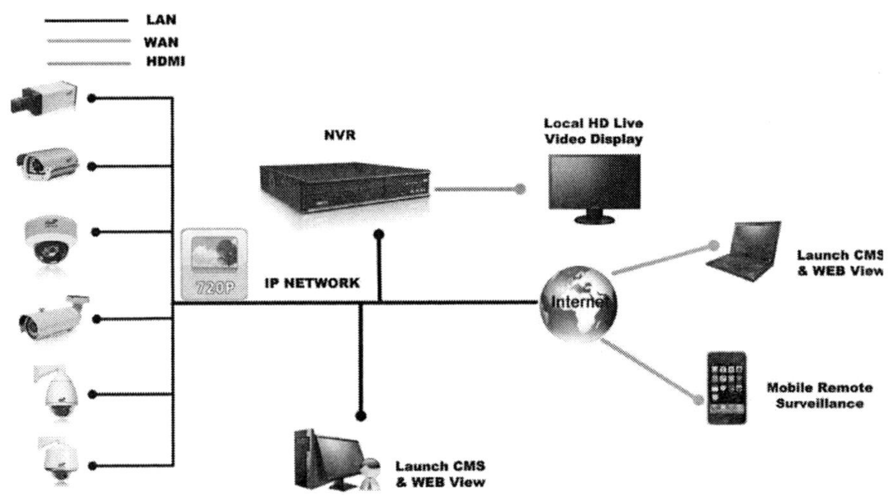

Figure 1. Tracking targets in a camera network by combining the local trajectories of each individual.

The third challenge is global activity understanding. For an intelligent video surveillance system, it is not enough to only track the targets without further analysis. Through the global automatic and comprehensive analysis of targets in a sparse camera network, the activities of the targets are understood and anomalous events are detected. Several previous surveys, *e.g.*, [15], have discussed activity understanding. A sparse camera network-based video surveillance system usually covers a much larger area, and hence it usually provides more information about the targets trajectory and activities than a single camera or a conventional dense camera network monitoring a small area. However, because the camera network is sparse, it is necessary to carry out global activity understanding from a new point of view.

2. Inter-Camera Tracking Correspondence

Sparse camera networks, the relationship between the two Camera, system, topology and transition period Camera, is based on the camera identification. In the case of sparse camera network vision monitoring, therefore, it is desirable to determine whether a target the interest of a camera has been observed elsewhere Network. This problem can be easily solved by conventional dense camera network by simply matching the 3D position Target location for each candidate, as camera the network is well calibrated. However, a sparse camera Network based visual surveillance systems typically contain a number. There is no overlap of views because of camera requirements. It is impractical to maintain calibration. The most basic idea Existing methods for non-overlapping camera configurations Target matching between cameras is formulated as recognition. The problem of the system, the target is described Visual appearance cues compared to candidates in video Captured by other cameras in a sparse network. This kind of the recognition problem is often referred to as the"object" Recognize ',' object re identification 'or' appearance Modeling". When modeling a person's goals in an A general assumption for sparse camera networks is Personal clothes remain unchanged on camera. However, even if to accept this assumption, the camera-tracking task remains Challenges due to lighting, pose and camera changes.Parameters in different cameras on the network. * The general camera-tracking framework can be Decomposition into three layers:

- Low level features for visual appearance description: extraction of concise features.
- Integration of low-level features.
- Inter-camera target identification.

2.1. Low Level Features for Visual Appearance Description

The low-level features for visual appearance description can be divided into two categories: global visual features and local visual features. The global visual features encode the target as a whole. The local visual features describe the target as a collection of independent local descriptors, *e.g.*, local patches and Gabor wavelets. A global feature usually contains comprehensive and rich information, so it is very powerful if it is accurate. On the other hand, it is very sensitive to noise, partial occlusions, viewpoint changes and illumination changes. In contrast with global features, local features are less sensitive to these factors; however, as the features are extracted locally, some information, especially spatial information, may be lost.

Global visual features: A global feature encodes the tracking target using a single multidimensional descriptor. For instance, Huang et al. used the size, velocity and mean value of each channel in HSV color space to describe the appearance of vehicles. In many applications, the target is more complex than a vehicle, and thus stronger descriptors are needed. General global features used to represent a target's appearance include the global histogram GMM and newly defined global descriptors for human reidentification, e.g., panoramic appearance map (PAM) and color position.

a) Global histogram: Histogram-based methods can represent large amounts of information. They are efficient, easy to implement, and have a relatively high tolerance to noise.

One of the earliest attempts to represent targets by histograms was carried out by Kettnaker and Zabih. Orefej *et al.* also extracted histograms of HSV color values of pixels in the target area. PCA is applied to 6,000 histograms of HSV color values and HOG descriptors, and the eigenvectors corresponding to the top 30 eigenvalues are extracted as the final representation. Morioka *et al.* calculated the covariance matrix of color histograms and projected the set of color histograms onto the eigenspace spanned by the eigenvectors corresponding to the first d largest eigenvalues. Lin and Davis proposed a normalized rgs1 color to deal with illumination changes since the independence of chromaticity from brightness in this color space allows using a multivariate Gaussian density function to cope with the differences in brightness. Nakajima *et al.* described a system that learns from examples to recognize persons in images taken indoors. Images of full-body persons are represented by the RGB color histograms (Fig. 2(a)), the normalized color histograms and the shape histograms.

b) Newly defined global descriptors for human reidentification: Gandhi *et al.* developed the concept of a PAM (Fig. 2(c)) for performing person re-identification in a multicamera setup. A PAM centering on the person's location is created with the horizontal axis representing the azimuth angle and the vertical axis representing the height. PAM models the appearance of a person's body as a convex generalized cylinder. Each point in the map is parameterized by the azimuth angle and height, and the radius of the cylinder is treated as constant. This allows PAM extracts and combines information from all the cameras that view the object features to form a single signature. The horizontal axis of the map represents the azimuth angle with respect to the world coordinate system, and the vertical axis represents the object height above the ground plane. Cong *et al.* proposed a color-position histogram (Fig. 2(b)), in which the silhouette of the target is vertically divided into equal bins and the mean color of each bin is computed to characterize that bin. Compared to the classical color histogram, it consists of spatial information and uses less memory. Moreover, this new feature is a simpler and more reliable measurement for comparing two silhouettes for person re-identification.

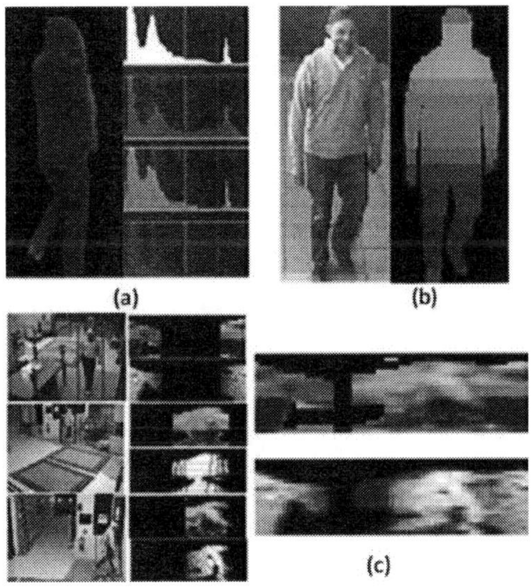

Figure 2. Examples of global representation features: (a) global histogram of RGB color, (b) color position, (c) panoramic appearance map.

Local visual features*: a) Local visual feature detection:* Examples of local visual features include corners, contour intersections and pixels with unusually high or low gray levels. Contour intersections often take the form of bi-directional signal changes and the points detected at different scales do not move along a straight bisector line; hence, corners are often detected using local high curvature maxima. Lowe proposed localizing points at local scale-space maxima of the difference of Gaussians. Doll'ar applied Gabor filters separately in the spatial and temporal domains. By changing the spatial and temporal size of the neighborhood in which local minima are selected, the number of interest points can be adjusted. Laptev extended the Harris-Stephens corner detector to space-time interest points where the local neighborhood has significant variation in both the spatial and the temporal domains.

b) Local visual feature descriptor: An image can be described using a collection of local descriptors or patches that can be sampled densely or at points of interest. Compared to extracting local descriptors at the interest point, dense sampling retains more information, but at a higher computational cost. A performance evaluation of different local visual feature descriptors has been given by Mikolajczyk and Schmid. We discuss the different types of descriptors as follows.

(1) Distribution descriptors: Distributions of intensity, gradient, and shape are widely used as local descriptors. Zheng *et al.* extracted SIFT features for each RGB channel at each pixel to associate a group of people over space and time in different cameras. These features are clustered and quantized by K-means to build a codebook. The original image is then transformed to a labeled image by assigning a visual word index to the corresponding feature at each pixel of the original image. Wang *et al.* extracted Histogram of Gradients (HOG) features in the Log-RGB space. They argued that taking the gradient of the Log-RGB space had an effect similar to homomorphic filtering, and made the descriptor robust to illumination changes. Hamdoun *et al.* utilized a method in known as SURF to detect interest points; it was Hessian-based and used an integral image for efficient computation. The descriptors SURF are 64 dimension vectors which coarsely describe the distribution of Haar-wavelet responses in sub-regions around the corresponding pixels of interest. Laptev and Lindeberg presented a local spatio-temporal

descriptor that included position dependent histograms and the PCA-based dimensionality reduced spatial-temporal gradients around the spatio-temporal interest points.

(2) Frequency descriptors: The spatial frequencies of an image carry important texture information and can be obtained by the Fourier transform and the wavelet transforms, such as Haar transform and Gabor transform. In contrast to the Fourier transform and the Haar transform, in which the basis functions are infinite, the Gabor transform uses an exponential weighting function to localize the decomposition of an image region into a sum of basis functions. The region to analyze is first smoothed using a Gaussian filter, and the resulting region is then transformed with a Fourier transform to obtain the time-frequency analysis. Gray *et al.* used two families of texture filters, taken from Schmid and Gabor respectively, on eight color channels corresponding to the three separate channels of the RGB, YCbCr, and HSV color space, in which the used eight color channels contain only one of the luminance (Y and V) channels.

3. Conclusions

This paper formulated a novel aggregation-based camera network for representing GSP, based on which multiple aesthetic communities can be online discovered according to human visual perception. More specifically, by constructing the GSP of each photo using active learning, salient regions discovery and gaze shifting sequence prediction are incorporated into an aggregation-based deep CNN. Based on this, a so-called deep perception graph is built which effectively measures the aesthetic discrepancy between users, and mutliple aesthetic communities can be identified using graph shift. Thereby, an online GMM describes the aesthetic distribution toward each community, based on which a series of media applications (*e.g.*, retargeting and aesthetic assessment) can be tackled in a unified framework.

4. References

[1] S. N. Sinha, M. Pollefeys, and L. McMillan, Camera network calibration from dynamic silhouettes, *in Proc. of CVPR*, pages: 195–202, 2004.

[2] S. N. Sinha and M. Pollefeys, Multi-view Reconstruction using Photoconsistency and Exact Silhouette Constraints: A maximum-Flow Formulation,*in Proc. of ICCV*, pages: 349–356, 2005.

[3] R. Szeliski, Rapid Octree Construction from Image Sequences, *CVGIP: Image Understanding*, 58(1), pages 23–32, 1993.

[4] J. Starck and A. Hilton, Surface Capture for Performance-Based Animation, *IEEE Computer Graphics and Applications*, 27(3), pages: 21–31, 2007.

[5] P. R. S. Medonca, K.-Y. K. Wong, and R. Cipolla, Epipolar Geometry from Profiles Under Circular Motion, *IEEE T-PAMI*, 27(3), pages: 21–31,2007.

[6] C. Stauffer and K. Tieu, Automated Multi-camera Planar Tracking Correspondence Modeling, *in Proc. of CVPR*, pages: 259–266, 2001.

[7] D. C. Brown, Close-Range Camera Calibration, *Photogrammetric Engineering*, 37(8), pages: 855–866, 1971.

[8] G. Champleoux, S. Lavallee, P. Sautot, and P. Cinquin, Accurate Calibration of Cameras and Range Imaging Sensor: The Npbs Method, *in Proc.of ICRA*, pages 1552–1557, 1992.

[9] Q. Cai and J. K. Aggarwal, Tracking Human Motion in Structured Environments Using a Distributed-camera System, *IEEE T-PAMI*, 2(11), pages:1241–1247, 1999.

[10] R. T. Collins, A. J. Lipton, H. Fujiyoshi, and T. Kanade, Algorithms for Cooperative Multisensory Surveillance, *Proc. of IEEE*, 89(10), pages:1456–1477, 2001.

[11] R. Farrell, D. Doermann, and L. S. Davis, Learning Higher-Order Transition Models in Medium-Scale Camera Networks, *in Proc. of ICCV*,pages: 1–8, 2007.

[12] Y. A. Sheikh and M. Shah, Trajectory Association across Multiple Airborne Cameras, *IEEE T-PAMI*, 32(2), pages: 361–367, 2008.

[13] D. Makris, T. Elis, and J. Black, Bridging the Gaps between Cameras, *in Proc. of CVPR*, pages: 205–210, 2004.

[14] O. Javed, Z. Rasheed, K. Shafique, and M. Shah, Tracking across Multiple Cameras with Disjoint Views, *in Proc. of ICCV*, pages: 952–957, 2003.

[15] O. Williams, A. Blake, and R. Cipolla, Sparse Bayesian Learning for Efficient Visual Tracking, *IEEE T-PAMI*, 27(8), pages: 1292–1304, 2005.

Research on control law accelerator of digital signal process chip TMS320F28035 for real-time data acquisition and processing

Shuangle Zhao, Xueyi Zhang, Shengli Sun and Xudong Wang

Applied Liberal and College, Tianjin University of Science and Technology, Tianjin economic and Technology Development Zone No. 13, 29 Avenue.
Email: zhsl347new@qq.com

Abstract. TI C2000 series digital signal process (DSP) chip has been widely used in electrical engineering, measurement and control, communications and other professional fields, DSP TMS320F28035 is one of the most representative of a kind. When using the DSP program, need data acquisition and data processing, and if the use of common mode C or assembly language programming, the program sequence, analogue-to-digital (AD) converter cannot be real-time acquisition, often missing a lot of data. The control low accelerator (CLA) processor can run in parallel with the main central processing unit (CPU), and the frequency is consistent with the main CPU, and has the function of floating point operations. Therefore, the CLA coprocessor is used in the program, and the CLA kernel is responsible for data processing. The main CPU is responsible for the AD conversion. The advantage of this method is to reduce the time of data processing and realize the real-time performance of data acquisition.

1. Introduction

In real-time data acquisition, if the data analysis program running time is longer than the data acquisition time, it will cause the ADC not work for a period of time, resulting in the omission of the detection results [1]. The core is characterized by the parallel operation between the master kernel and the CLA kernel, and the real-time performance of the data acquisition is achieved. DSP TMS320F28035 introduces its inner structure as shown in figure 1. The CLA core frequency is the same as the main CPU, which is 60MHz, which guarantees the speed required for high-speed data operation.

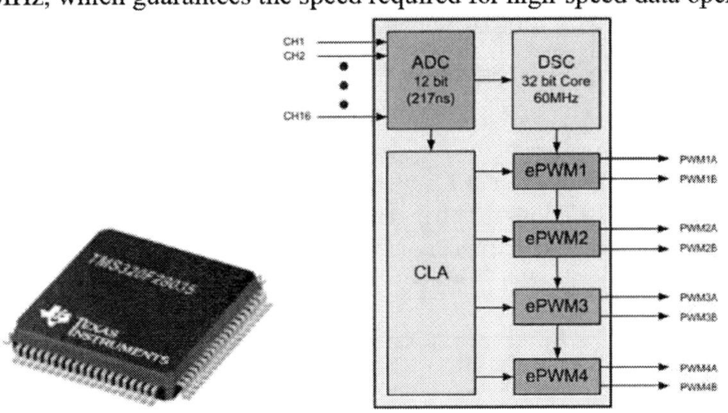

Figure 1. Chip 28035 (left) and Internal CLA Architecture (right)

IOP Conf. Series: Journal of Physics: Conf. Series **887** (2017) 012092 doi:10.1088/1742-6596/887/1/012092

CLA allows direct access to the ePWM, the comparator, and the ADC result register. CLA can be programmed using the C language and assembly language, and the CLA support settings need to be prepared in the pre-processing section before compiling [2].

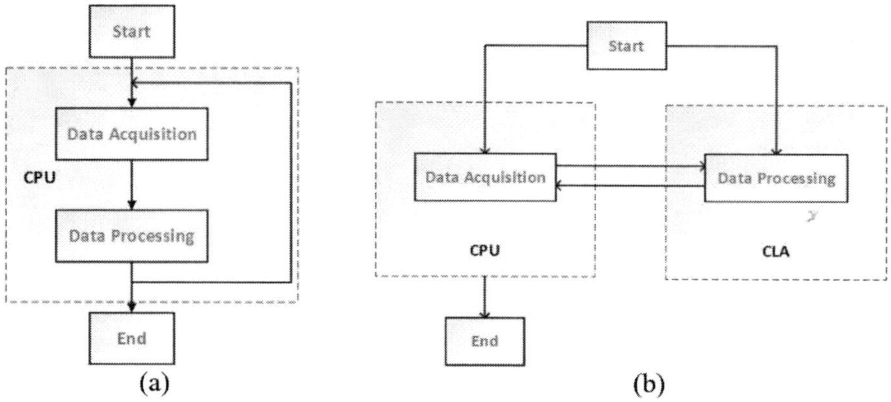

Figure 2. Normal DSP operation model (a) and operation mode with CLA (b)

The DSP with CLA kernel can synchronize the main processor and the CLA kernel seen as figure 2. This parallel operation reduces the running time and improves the operation efficiency. It is very important for time sensitive programs.

2. Steps for the Use of the CLA Core

CLA kernel programming, you can use the C language or assembly language, here focuses on the use of C language [3].

2.1. CLA Interrupt Settings

The PCLKCR3 registers enable the CLA clock, i.e., SysCtrlRegs.PCLKCR3.bit.CLA1ENCLK = 1;

2.2. Initialize the Peripheral interrupt (PIE) Controller, and then Copy the PIE Vector Table into the DSP Memory by Programming;

2.3. CLA ISR Set

Distribute the user-defined interrupt service program (ISR) to the peripheral interrupt vector table (PIE), copy the CLA program code into memory, and get the initial address of the CLA program memory. Then, the offset address of the CLA task vector is computed based on the initial address, and the 8 CLA task vectors (MVECT) are filled.

2.4. Mapping the Task Interrupt Source and Opening the CLA Interrupt

The interrupt source for 8 tasks in the CLA program, using the Cla1Regs.MPISRCSEL1.bit.PERINT1SEL statement setting, 0 for the ADCINT interrupt trigger, 1 for none, and 2 for the EPWM1INT interrupt trigger. You do not need to set up this interrupt when using software to run tasks. Interrupt enable register MIER.

2.5. Enable CLA Interrupts

This section includes peripheral interrupt extensions PIEIER, IER, EINT, ERTM which will be enabled for 1.

2.6. IACK Software Triggers and CLA Data Space and Program Space Mapping Registers

IACK=1 IACK is the software triggers.
MMEMCFG.bit.RAM0E=1 CLA RAM0 data space.
MMEMCFG.bit.RAM1E=1 CLA RAM1 data space.
MMEMCFG.bit.PROGE=1 CLA program space.

When allocating memory to the CLA program and data, the CPU cannot access the program and data of the corresponding address.

2.7. CMD File Configuration

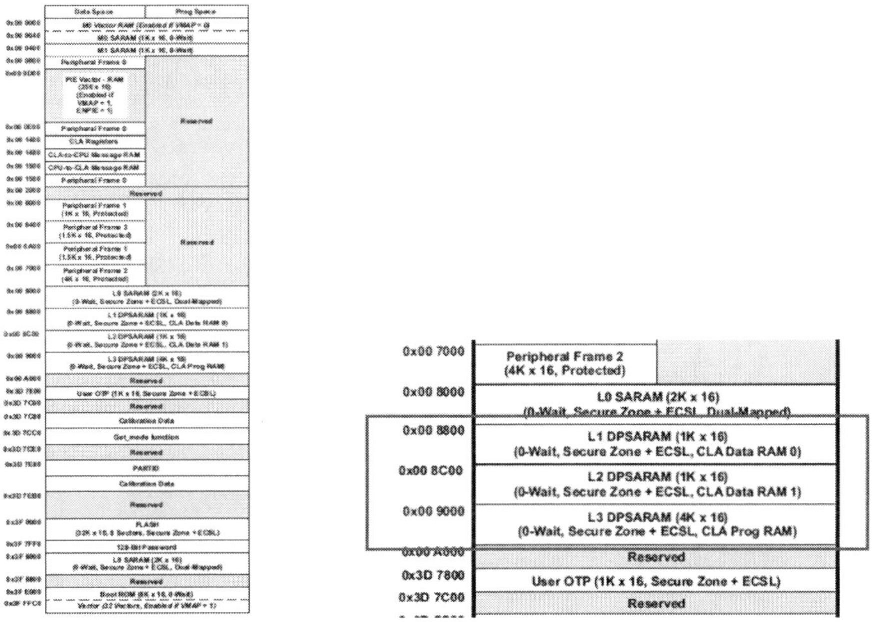

(a) (b)

Figure 3. DSP 28035 RAM mapping (a) all the memory allocation (b) CLA using memory

The CMD file allocates memory space to each module of the DSP [4]. The memory map for the 28035 chip is shown in figure. Figure 3 shows the DSP 28035 the memory allocation, the CLA program must be kept in the range of 0x009000-0x00A000 addresses, CLA data can be stored in the 0x008800-0x008C00 address range is within the range of L1DPSARAM, or the 0x008C00-0x009000 address range that is within the range of L2DPSARAM, when CLA use these programs or data storage space, and CPU will not be able to use this space to write the program, so CPU and CLA core part based on chip memory capacity.

3. CLA Core Debugging Platform

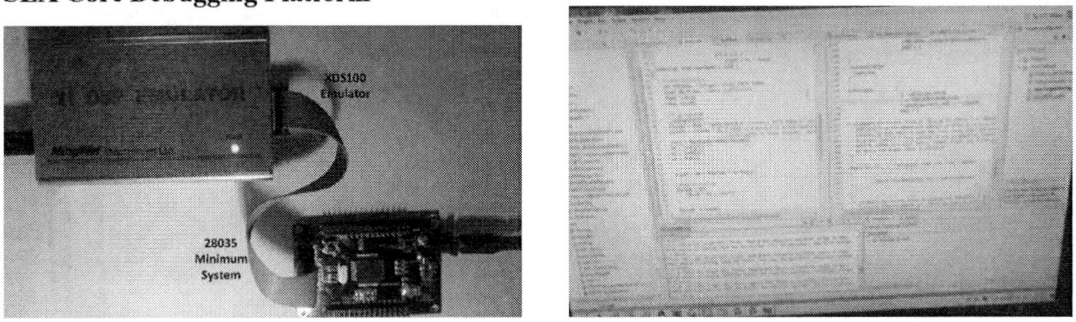

Figure 4. CCS 5.5 Developing Environment. Hardware (Left) and Software (Right).

Using TI's TMS320F28035 minimum system, the DSP integrates a CLA core, using CCS5.5 as a program debugging platform. In debugging, you can observe the value of the control register in the CLA kernel.

4. Experimental Result

For testing the CLA work, there are two same projects written using CLA core and not-using CLA. Figure 5 shows the two experimental test codes. When AUX1 pin is high-voltage, the DoArcDetect function executes twice, and then the AUX1 is low-voltage. The oscillometer DL850E saves the AUX1 signal. The DoArcDetect function can use CLA core or without it. The cost time of sub-function is described in table 1.

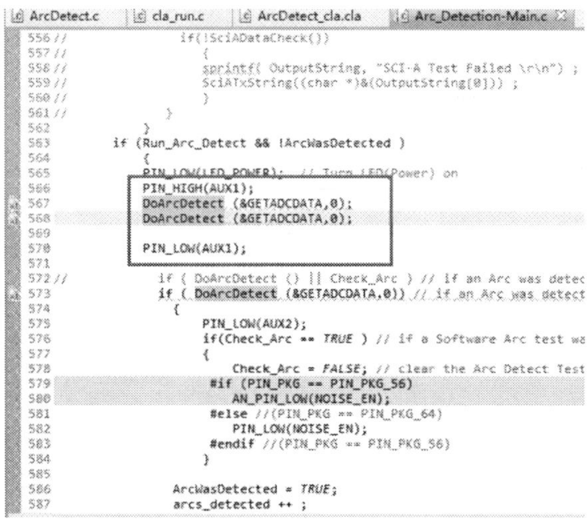

Figure 5. The test project code.

Table 1. The function time

Sub Function	Time(ms)
Data Acquisition	10
Data Processing	19

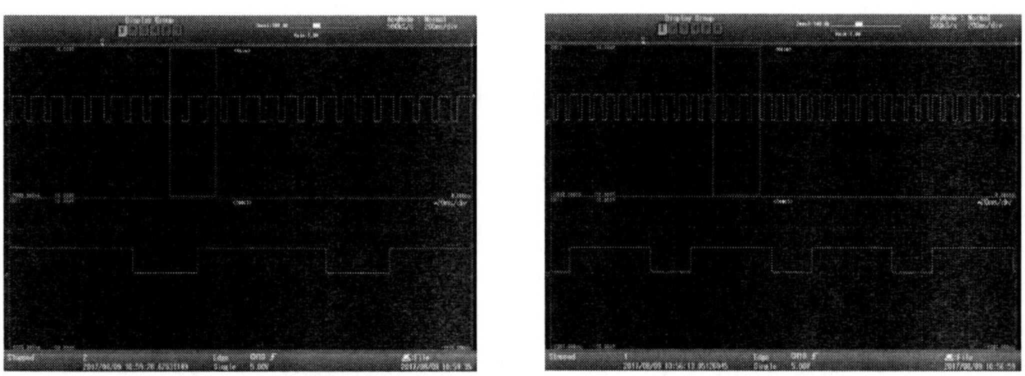

(a) (b)

Figure 6. The experimental result. (a) not-using the CLA core. (b) Using the CLA core.

The experimental result is shown as in figure 6. The function execution time of the non-using CLA core is 29ms, and the other execution time is only 18ms when using the CLA core. So it is losing 9ms data when the software project acquisition data, whereas using the CLA core the data processing section execution time is 8ms, shorten then the data acquisition section 10ms, so this is not losing the data.

5. Conclusion

In conclusion, the CLA core help the main CPU core in parallel program execution. It is import when the program includes the AD data acquisition section and data processing section. When data processing time is larger than the data acquisition time, and software is also sequence run, the data must be lost. So the CLA core support an effective methods to solve the problem.

The appearance of CLA kernel gives more choices to designers and programmers, reduces the workload of main CPU and improves the real-time performance of data analysis. But CLA also has its shortcomings, the use of many constraints, cannot be nested interrupts, cannot use more than 64 data, using the judgment statement of lower efficiency in CLA, in addition to the ADC result register, a comparator and a ePWM, CLA cannot access other peripherals. If CLA takes up too much data or programs, CPU will not be able to access the remaining memory space, resulting in a waste of memory space.

6. Acknowledgments

Project supported by Tianjin University of Science and Technology Youth Innovation Fund (2016LG31)

7. References

[1] Yang Chuanhai, Xiao Bing. Design of Infinite Impulse Response Low Pass Filter Based on TMS320F28035 Control Law Accelerator [J]. Electric Drive, 2012, (05): 26-29.
[2] DSP application development utility subroutine [M]. People's post and Telecommunications Press. Wang Anmin, Cheng Yu edited, 2005.
[3] TMS320x2803xPiccolo Control Law Accelerator (CLA) Reference Guide. Texas Instruments Incorporated. 2009.
[4] C2000 CLA C Compiler. http://processors.wiki.ti.com/index.php/C2000_CLA_C_Compiler?keyMatch=cla%20c&tisearch=Search-EN [EB/OL]. Texas Instruments Incorporated.

IOP Conf. Series: Journal of Physics: Conf. Series **887** (2017) 012093 doi:10.1088/1742-6596/887/1/012093

The vision guidance and image processing of AGV

Tongqing Feng and Bin Jiao

Electrical Engineering College, Shanghai Dian Ji University, No. 300 blooms Road, Pudong New Area, Shanghai, China.
Email: 780862201@qq.com

Abstract. Firstly, the principle of AGV vision guidance is introduced and the deviation and deflection angle are measured by image coordinate system. The visual guidance image processing platform is introduced. In view of the fact that the AGV guidance image contains more noise, the image has already been smoothed by a statistical sorting. By using AGV sampling way to obtain image guidance, because the image has the best and different threshold segmentation points. In view of this situation, the method of two-dimensional maximum entropy image segmentation is used to solve the problem. We extract the foreground image in the target band by calculating the contour area method and obtain the centre line with the least square fitting algorithm. With the help of image and physical coordinates, we can obtain the guidance information.

1. Introduction

Visual navigation technology is a kind of advanced guidance mode, it has the advantages of intelligence, high reliability and flexibility. It can capture around more complete information, so it has wide application prospect in AGV navigation system. At present, the more common application is to install the camera on the AGV and use the partial vision navigation method. Through the on-board computer, the high-level decision-making tasks such as path identification and tracking control are completed. The production workshop of vision navigation AGV has less illumination change and better flatness of ground. In this condition, stability and real-time are two main performance indexes of vision system. Therefore, the faster the image processing speed is, the shorter the AGV will obtain the path information. Therefore, to improve the performance of AGV, it is necessary to establish an efficient image processing system.

2. Principle of Visual Navigation Technology

AGV vision guidance uptake the pavement images by the dynamic guiding camera. The pavement image collection is three-dimensional scene in a 2D space perspective [1]. It needs the inverse process to identify the information of the path in the process of machine vision and realize the measurement of partial difference and deflection angle. At present, the AGV vision system usually uses the corresponding point calibration method to obtain the path information of the image. The image coordinate system is one level of coordinates of the vision system [2]: It is divided into image pixel coordinate system and image physical coordinate system. The image pixel coordinate system is fixed on the image in pixels of the Cartesian coordinates, the origin is located at the upper left corner of the image. The unit of the image physical coordinates is millimetre and its origin is the intersection point and the image plane of the camera axis.

Firstly, we establish the image pixel coordinate system of μ-o-V and image physics coordinate system of x-c-y image pixel coordinate system and the reference coordinate system is as shown in Figure 1.

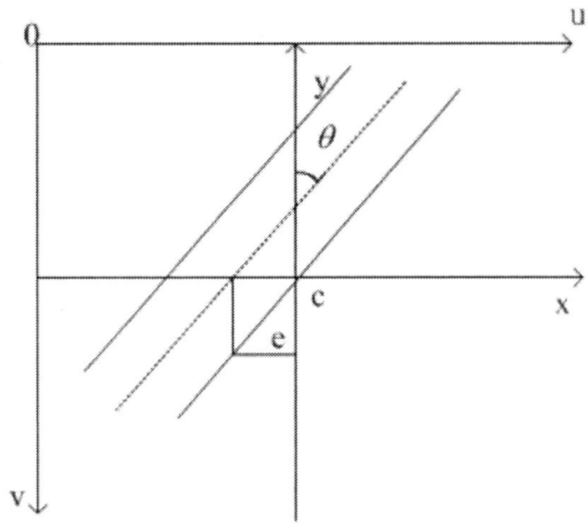

Figure 1. Reference coordinate system

AGV travels along the Y axis negative direction, its trajectory is divided into straight line and arc segments. It is because that the ribbon width is 6cm, the camera view height is about 15cm and the arc track radius is larger than 1m, so the camera view range is relatively small. In order to simplify the problem, we can take the arc in the field of view as a straight line approximately. The deviation e is the abscissa value of the intersection point between the fitted line and the X axis and the angle of departure θ is the angle between the fitted line and the positive axis of the Y axis.

3. The Information Processing of Guided Image

3.1. Two-dimensional Maximum Threshold Segmentation.

The two-dimensional maximum entropy threshold [3] method introduces the concept of entropy of the information theory. The image can be regarded as a two-dimensional gray function, which contains $N \times N$ pixels and the gray value of pixel is divided into L grades. Firstly, the region gray of the original image is obtained and the target pixel and its adjacent pixels are used as the template. In the original image, the corresponding coordinate of the gray value of the pixel and its region gray mean constitute the data pair (i, j).We set $n_{i,}j$ as the number of pixels of the point gray i and the region gray j and $p_{i,j}$ is the probability density. Therefore, we can have the Formula (1) as follow:

$$p_{i,j} = \frac{n_{i,j}}{N \times N} \tag{1}$$

As shown in Figure 2, we take the point gray value and the region gray mean value as the coordinate axis to establish the two dimensional image of gray distribution [4].

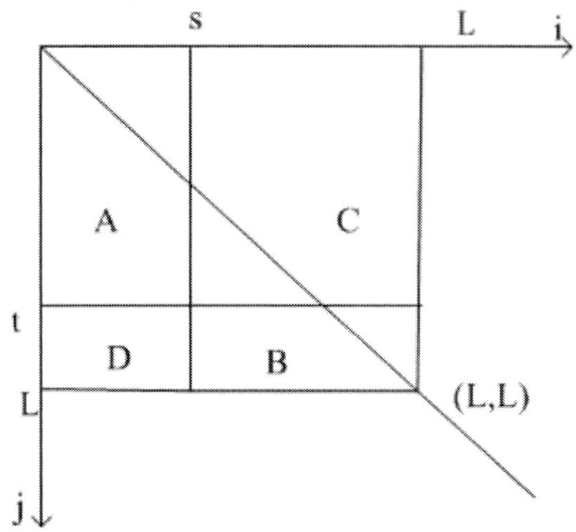

Figure 2. The two-dimensional segmentation of image

If the segmentation threshold is located (*s, t*), the two-dimensional image can be divided into four regions, in which the target pixels are distributed in the A region, the background pixels are distributed in the B region. The C region contains the boundary pixel points and the D region contains the noise signals. Therefore, in order to make the image target and background have good segmentation effect, the best entropy threshold should be determined by two-dimensional maximum entropy method in A and B regions. In order to make the partition entropy more additive, the posterior probabilities of A and B regions are normalized respectively:

$$p_A = \sum_i \sum_j p_{i,j} \quad i = 1,2,\cdots,s; j = 1,2,\cdots,t \tag{2}$$

$$p_B = \sum_i \sum_j p_{i,j} \quad i = s+1, s+2,\cdots,L; j = t+1, t+2,\cdots,L \tag{3}$$

The discrete two-dimensional entropy is defined as the Formula (4) as follow:

$$H(A) = -\sum_i \sum_j p_{i,j} \lg p_{i,j} \tag{4}$$

Then the two-dimensional entropy of the A region and the B region are respectively shown in the Formula (5).

$$\begin{aligned} H(A) &= -\sum_i \sum_j (p_{i,j}/P_A) \lg (p_{i,j}/P_A) \\ &= -(1/P_A) \sum_i \sum_j (p_{i,j} \lg p_{i,j} - p_{i,j} \lg p_A) \\ &= \lg P_A + H_A/P_A \end{aligned} \tag{5}$$

H_A is shown in Equation (6):

$$H_A = -\sum_i \sum_j p_{i,j} \lg p_{i,j} \quad i = 1,2,\cdots,s; j = 1,2,\cdots,t \tag{6}$$

$$H(B) = -\sum_i \sum_j (p_{i,j}/P_B) \lg (p_{i,j}/P_B) = -(1/P_B) \sum_i \sum_j (p_{i,j} \lg p_{i,j} - p_{i,j} \lg P_B) \tag{7}$$

The final H_B is shown in Equation (8):

$$H_B = -\sum_i \sum_j p_{i,j} \lg p_{i,j} \quad i = s+1, s+2, ..., L; j = t+1, t+2, ..., L \tag{8}$$

$$P_B = 1 - P_A \quad , \quad H_B = H_L - H_A \tag{9}$$

Form Equation (9), we have:

$$H_B = \lg(1 - P_A) + (H_L - H_A) / (1 - P_A) \tag{10}$$

The discriminant function of entropy is defined as:

$$\phi(s,t) = H(A) + H(B) \tag{11}$$

The best threshold is shown in Equation (12):

$$\theta(\overset{*}{s},\overset{*}{t}) = \max\{\theta(s,t)\} \tag{12}$$

According to the above algorithm, the two-dimensional maximum entropy segmentation results are shown in Figure 3.

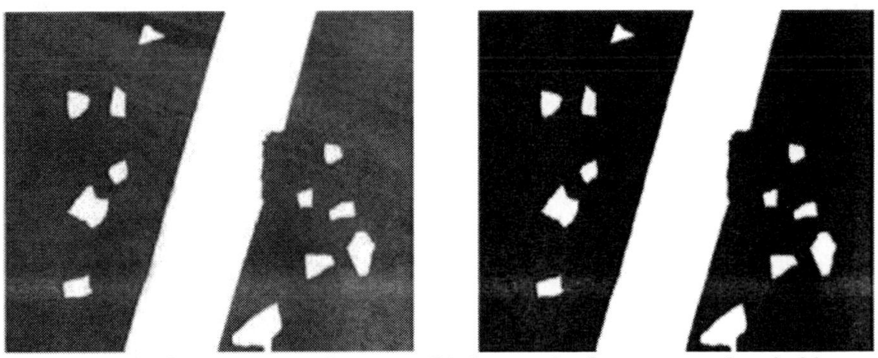

(a)The image before segmentation (b) the result after entropy threshold segmentation

Figure 3. Threshold segmentation image

It can be seen that the two-dimensional maximum entropy threshold segmentation algorithm is more accurate for the calculation of the optimal threshold and it is suitable for processing the image in this paper. So, the foreground and the background of the image are separated completely and clearly.

3.2. Edge Detection Based on Canny Algorithm

After threshold segmentation, the guiding image hope that through an edge detection algorithm [5] to extract the image edge features of the guiding ribbon and sundries in the foreground object. The edge of the image corresponds to the local maximum of the first derivative and the detection of the weak edge of the image is easily disturbed by the noise signal. Therefore, we introduce canny edge detection [6] which uses double thresholds to detect strong edges and weak edges respectively.

The original input image is *f (x, y)*. Firstly, the quasi Gauss function is used as a smoothing operation and the gradient of the smoothed g (x, y) is shown in Equation (13).

$$\nabla g(x,y) = \begin{bmatrix} \dot{g}_x \\ \dot{g}_y \end{bmatrix} = \begin{bmatrix} \frac{\partial g}{\partial x} \\ \frac{\partial g}{\partial y} \end{bmatrix} \tag{13}$$

By convolution operations, we have:

$$\nabla g(x,y)=\nabla f(x,y)\otimes G(x,y)=f(x,y)\otimes\nabla G(x,y) \tag{14}$$

Edge features are smoothed by the Gauss function and paid a fuzzy price. The width of the detected edge is increased and the fuzzy edge of the image is sharpened by NMS (Non-Maxima Suppression, NMS) [7].The NMS method compares the gradient amplitudes of adjacent pixels to each other in the direction perpendicular to the edge and removes the gradient magnitude, which is smaller than the neighbourhood. According to this operation, the non-maxima of the gradient amplitude are removed and the edges become thin.

The NMS method suppresses the amplitude image and still contains many false edges caused by noise and fine texture, which can be removed by double threshold algorithm. Double threshold algorithm uses dual threshold *T1* and *T2* and *T2 = 2T1* on non-maximum suppression amplitude image. Two double threshold edge images $G_1[i,j]$ and $G_2[i,j]$ are obtained. The algorithm will constantly collect edges in $G_1[i,j]$ until all the gaps in the $G_2[i,j]$ are joined. Based on the above algorithm, canny edge detection results are obtained as shown in Figure 4.

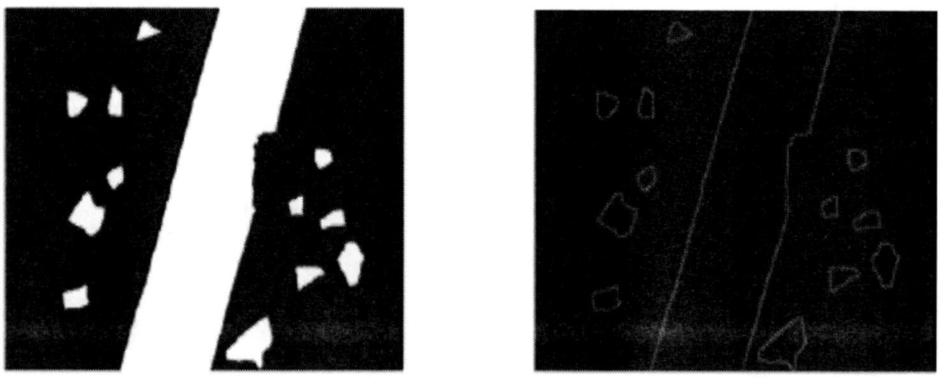

(a) Input image (b) Test results

Figure 4. Edge detection of Canny

4. Information Extraction of Path Guidance

According to the definition of the guidance parameters e and θ and the fitting line L of the guidance path can reflect the path information more accurately. Therefore, the extraction of navigation information is realized by the aid of the image physics coordinate system and the equation of the fitting line in the image physical coordinate system is obtained. The least square method is used in the line fitting algorithm and the linear equation of the fitting line L in the image physical coordinate system is assumed to be shown [8]:

$$y=ax+b \tag{15}$$

The formula contains two undetermined parameters, a represents the intercept and b represents the slope. In this paper, 320 sets of data (x_i, y_i), i=1,2,3……,320 are obtained by averaging the coordinates of each row of pixels in the outline of the guided ribbon and the least squares method requires the weighted sum of squares of the deviation of the observed value y_i to be minimum:

$$\sum_{i=1}^{320}\left[y_i-(a+bx_i)\right]^2 \tag{16}$$

The Equation (17) and Equation (18) is for the partial derivative of a, b respectively of above formula:

$$\frac{\partial}{\partial a}\sum_{i=1}^{320}\left[y_i-\left(a+bx_i\right)\right]^2=-2i=\sum_{i=1}^{320}(y_i-a-bx_i)=0 \tag{17}$$

$$\frac{\partial}{\partial b}\sum_{i=1}^{320}\left[y_i-\left(a+bx_i\right)\right]^2=-2i=\sum_{i=1}^{320}\left(y_i-a-bx_i\right)\cdot x_i=0 \tag{18}$$

The Equation (19) can be obtained after sorting:

$$\begin{cases} 320a+b\sum x_i=\sum y_i \\ a\sum x_i+b\sum x_i^{\;2}=\sum x_i y_i \end{cases} \tag{19}$$

By solving the equation set, the best estimate of the parameters a and b of the fitted line can be obtained:

$$\hat{a}=\frac{(\sum x_i^{\;2})(\sum y_i)-(\sum x_i)(\sum x_i y_i)}{320(\sum x_i^{\;2})-(\sum x_i)^2} \tag{20}$$

$$\hat{b}=\frac{320(\sum x_i y_i)-(\sum x_i)(\sum y_i)}{320(\sum x_i^{\;2})-(\sum x_i)^2} \tag{21}$$

According to the geometric relation $\theta=arctan\ (1/a)$, the e is shown in the Formula (22) by making y=0.

$$e=\frac{-b}{a} \tag{22}$$

The results of path guidance information extraction are shown in Figure 5.

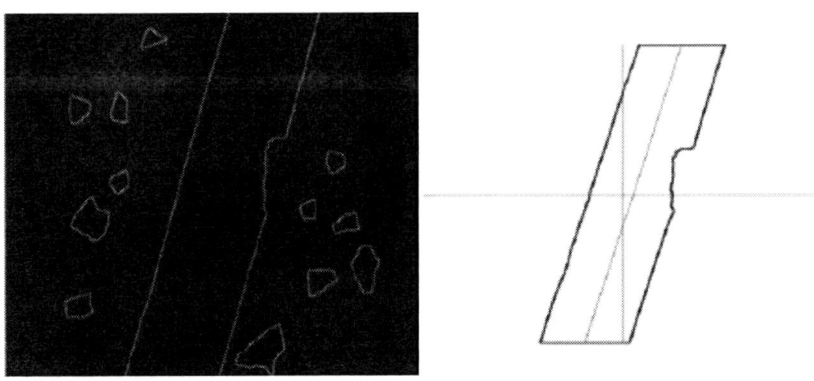

(a) Input image (b) The image of linear fitting

Figure 5. The result of linear fitting

Therefore, we can have the final results that we want: e=17.763 (unit: pixel), θ=18.336° (unit: angel).

5. Conclusion

Through the establishment of vision guided image processing platform, the camera provides original visual information. In the process of image processing, two-dimensional maximum threshold segmentation method, edge detection method and Canny least squares method make the original trajectory guidance information can be more clear and the deviation and deviation angle of the processing results more effective. It can effectively solve the problem of unstable operation and AGV track caused by other guidance. In general, we can determine the location of AGV, eliminate the error and improve the system stability and can effectively solve the actual operation of AGV. It can solve practical problems of the AGV.

6. References

[1] Peng W, Sun C and Zhang Z 2011 Linear pose estimation with a monocular vision system *J. Chinese Journal of Scientific Instrument.* **32 (5)** 1126-31
[2] Mei Y C, Wang C L and Liao Q W 2009 Applied research of OpenCv-based camera calibration *J. Computer Engineering & Design.* **30(16)** 3856-55
[3] Gao L, Yang S Y and Li H Q 2007 New unsupervised image segmentation via marker-based watershed *J. Journal of Image & Graphics.* **12(6)** 1025-32.
[4] Luo Z, Liu H P, Hu X F and Xu W 2016 Research of Vision-Guided AGV Deviation-rectifying Algorithm *J. Computer Simulation.* **33 (1)** 373-377
[5] Villagra J and Herrero P D 2012 a comparison of control techniques for robust docking maneuvers of an AGV *J. IEEE Transactions on Control Systems Technology.* **20(4)** 1116-23.
[6] Jia Z, Balasuriya A and Challa S 2008 Sensor fusion-based visual target tracking for autonomous vehicles *J. Artificial Life and Robotics.* **12** 317-328.
[7] Zhang J P, Lou P H, Qian X M and Wu X 2016 Research on precise positioning technology by multi-window and real-time measurement for visual navigation AGV *J.Chinese Journal of Scientific Instrument.* **37** 1356-63
[8] Zhu L, Lin X and Wu W J 2016 Work piece positioning of industrial robot based on machine vision *J. Journal of Chinese Computer Systems.* **37(8)** 1873-77

IOP Conf. Series: Journal of Physics: Conf. Series **887** (2017) 012094 doi:10.1088/1742-6596/887/1/012094

(26)Research on quasi-dynamic calibration model of plastic sensitive element based on neural networks

Fang Wang[1], Deren Kong[1], Lixia Yang[2] and Zouzou Zhang[2]

[1]School of Mechanical Engineering, Nanjing University of Science & Technology, Nanjing 210094, China;
Email: wangfang@mail.njust.edu.cn
[2]Xi'an Modern Chemistry Research Institute, Xi'an 710065, China

Abstract. Quasi-dynamic calibration accuracy of the plastic sensitive element depends on the accuracy of the fitting model between pressure and deformation. By using the excellent nonlinear mapping ability of RBF (Radial Basis Function) neural network, a calibration model is established which use the peak pressure as the input and use the deformation of the plastic sensitive element as the output in this paper. The calibration experiments of a batch of copper cylinders are carried out on the quasi-dynamic pressure calibration device, which pressure range is within the range of 200MPa to 700MPa. The experiment data are acquired according to the standard pressure monitoring system. The network train and study are done to quasi dynamic calibration model based on neural network by using MATLAB neural network toolbox. Taking the testing samples as the research object, the prediction accuracy of neural network model is compared with the exponential fitting model and the second-order polynomial fitting model. The results show that prediction of the neural network model is most close to the testing samples, and the accuracy of prediction model based on neural network is better than 0.5%, respectively one order higher than the second-order polynomial fitting model and two orders higher than the exponential fitting model. The quasi-dynamic calibration model between pressure peak and deformation of plastic sensitive element, which is based on neural network, provides important basis for creating higher accuracy quasi-dynamic calibration table.

1. Introduction

Copper cylinders as a sensitive element of crusher gauges are widely used for measuring the peak pressure in weapon chamber. Since the 1860s Nobel invented the copper cylinders for pressure measuring so far, with its convenience, simple operation, without drilling of the gun barrel as well as the consistency, copper crusher method has always been the main technical means of chamber pressure measuring in "practical ballistics" area and has been widely used in the world by far [1-4].

In order to ensure the accuracy of the chamber pressure test, copper cylinders need to be calibrated before use. If a static calibration method is used, an obvious dynamic error (up to 5% to 30% of the measured value) will be produced when measuring rapidly changing chamber pressures [4-6]. In the late 1970s, the quasi-dynamic calibration technology was first used by the United States to eliminate this dynamic error. This method was subsequently adopted by European countries, and became one of the most important methods for the calibration of plastic sensitive element. The so-called quasi-dynamic calibration refers to use a half-sine pressure pulse, which is similar to the chamber pressure waveform and has known peak value and pulse width, to calibrate sensitive elements of crusher gauges (copper cylinders or copper sphere), and obtain corresponding relationship between the pressure and deformation of the sensitive elements. In order to accurately calibrate a copper cylinders or a copper sphere with a drop-weight quasi-dynamic calibration device, it is necessary to know the relationship between the operating parameters of drop-weight device (such as weight, drop height, piston area, initial volume of oil cylinder, the diameter of the piston, the starting diameter of the copper sphere, etc.) and the pressure peak and pulse width by the device. And in order to ensure the accuracy of quasi-dynamic calibration, the relationship between the deformation of the plastic sensitive elements and the peak of the pressure pulse must meet certain precision requirements. Zhu Mingwu carried out a quasi-dynamic calibration practice aimed at the $\phi 4mm \times 8mm$ copper cylinders [9] and analyzed the characteristics of quasi-dynamic calibration of copper cylinders and the problems which should be paid attention to. In Reference [11], a corresponding mechanics-mathematical model is established to guide the research work by using the method of theoretical analysis and according to the laws of physics which play a dominant role in the working process of the device. In Reference [12], based on the dimensional analysis theory, the dimensional analysis of copper sphere's dynamic calibration process is carried out, and a dimensionless model between drop height, peak pressure and pulse width is established. The maximum relative error of the model for peak pressure is less than 3%.

Based on the excellent nonlinear mapping ability of RBF neural network [13-15], a quasi-dynamic calibration test on copper cylinders is carried out based on drop-weight device. The relationship between the deformation of copper cylinders and the peak of pressure pulse is established, and the accuracy of the neural network fitting model is compared with the accuracy of the traditional fitting model.

2. The Principle of Quasi - Dynamic Calibration for Plastic Sensitive Element

The quasi-dynamic calibration device mainly consists of the pressure pulse generator and the reference pressure measurement system. The principle of the half-period sine pressure pulse generator [7] is shown in Figure 1. The pressure pulse generator consists of a piston and a mass that can be dropped onto the piston from various heights. When the mass is released from a certain height, it transmits its Kinetics energy through the piston to the fluid inside the oil-filled cylindrical chamber. The drop weight and the piston move downward with the same velocity. At the same time, the piston skins rapidly into the oil-filled chamber into which the crusher gauges and the reference transducers are mounted. Its energy is then transformed to the fluid as compression energy which causes a pressure increase. After releasing the total energy, the maximum pressure is reached and the reverse motion of the piston and drop weight starts. They are pushed upwards with the same velocity until the piston is stopped and the falling mass rebounds, and is generally caught. During this process, a pressure pulse is generated, which is similar to a single half cycle of a sine wave. This pressure curve pattern is quite similar to real gas pressure variation inside fired ammunition; both in its duration and in its shape. Hence, the pressure pulse generator is still used for a comparison calibration based on a reference sensor which provides the reference pressure measurements. The aim of this method is to check that the dynamic behavior of the crusher gauge can mimic the real gas pressure which certainly improves the accuracy of the measurement. The plastic sensitive element is calibrated by this method, which can eliminate or reduce the static and dynamic difference produced by the past static calibration and the dynamic use. The peak value P_m and pulse width τ of the pressure pulse can be changed by adjusting the falling mass m, the drop height h, the piston area S, the initial volume of the oil-filled chamber V_0, etc. [8, 10].

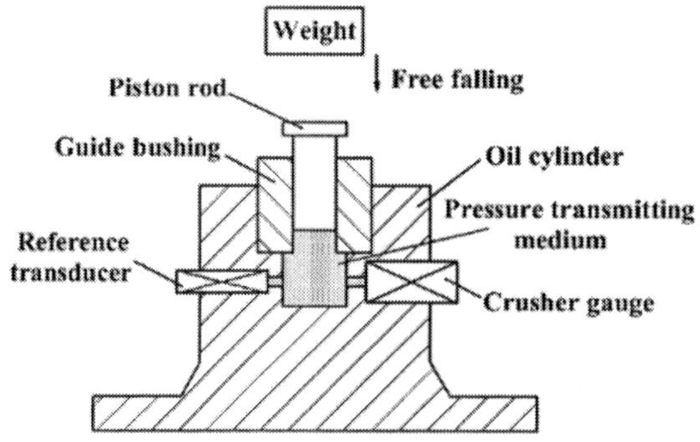

Figure 1. The principle of the hydraulic pressure pulse generator.

Four crusher gauge mounting holes (A, B, C, D) and four reference sensor mounting holes (1, 2, 3, 4) are evenly distributed around the cylinder chamber as shown in Figure 2. To perform a quasi-dynamic calibration, the reference transducers should be installed into the sensor mounting hole on the cylinder. The plastic sensitive elements should be placed into the crusher gauge, and the crusher gauges should be tighten in its mounting holes. In the crusher gauge measuring range, 7 to 9 pressure points are set evenly for calibration. The average pressure value of four reference pressure sensor which is measured by the reference pressure measurement system is regarded as the standard pressure. The piezoelectric pressure transducers Kistler type 6213BK (RS) delivers low amplitude and high impedance charge signal, generally expressed in pico-Coulomb (pC). Thus, a charge amplifier Kistler type 5018A1000 with scale factor of 100 MPa/V is used. The data acquisition (DAQ) board consists in a multi-channel device of four high speed digitizers. Each digitizer has 4 channels of parallel with a resolution of 14 bits and a maximum sampling rate of 10 MHz. The major conversion occurring in the DAQ board is an analog to digital conversion. In addition, signal processing was performed with programs developed with the LabVIEW software [16]. A Butterworth phaseless filter was used. For all signals, the frequency domain transform was computed by DFT for optimizing the filter cut off to the signal. After the experiment, the relationship between the standard pressure and the mean value of the deformation of the plastic sensitive element is established. Using the polynomial regression technique, the regression equation between the peak pressure and deformation of copper cylinders (height after compression for copper sphere) can be obtained. According to this, the dynamic pressure conversion table of the plastic sensitive element is compile. [8-10].

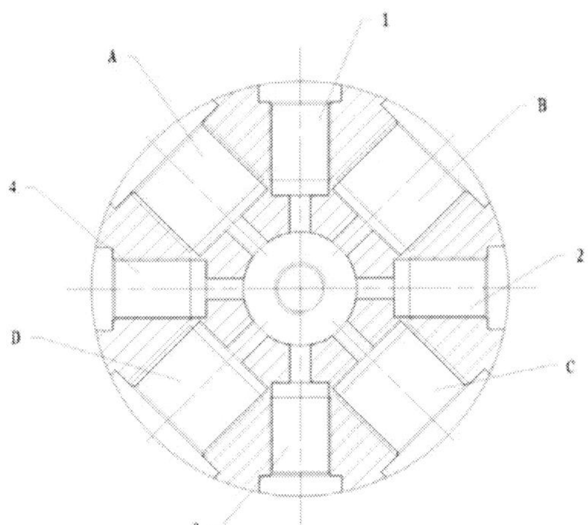

Figure 2. The position of the mounting hole of crusher gauges and reference sensors.

3. The Neural Network Model of Crusher Gauges Based on Quasi - Dynamic Calibration

3.1. The Basic Principles of RBF Neural Network

(1) The structure of RBF neural network

RBF neural network is a kind of multi-layer forward network structure with single hidden layer. RBF neural network includes the input layer, the hidden layer and the output layer [17]. The structure of typical RBF neural network is shown in Figure 3.

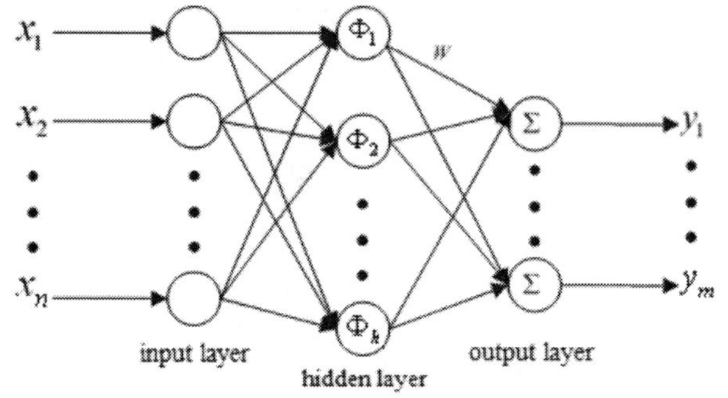

Figure 3. Structure of radial-based function (RBF) neural network.

The network has n inputs, h hidden nodes, m outputs, in which $x=(x_1,x_2,...,x_n,)^T \in R^n$ is the input vector of the neural network, $W \in R^{h\times m}$ is the weight matrix, $y=[y_1,y_2,...,y_m]^T$ is the output of the RBF neural network, $\Phi_i(*)$ is the activation function of the i-th hidden layer node, and the radial-based function is used as the activation function of the hidden layer neuron. Generally, the Gauss function is chosen as the radial-based function.

The output of hidden layer neurons is:

$$\Phi_i(x,c_i)=\exp\left(-\frac{1}{2\sigma_i^2}\|x-c_i\|^2\right) \tag{1}$$

Where c_i the center of Gaussian basis is function and σ_i is the variance of the Gaussian basis function.

The output layer neuron nodes use the Purelin function to weight the output generated by the neurons of the hidden layer, which can be expressed as:

$$y_k=\sum_{i=1}^{h} w_i\Phi_i(x,c_i) \tag{2}$$

Where w_i represents the weight between the i-th neurons in the hidden layer to the one in the output layer, and h represents the number of neurons in the hidden layer.

(2) The training process of RBF network

There are two types of parameters to be determined in the network: center c_i and width σ_i of the radial basis function, and the weights between the hidden layer and output layer. The training process of the network is composed of two steps: the determination of the center and width of the radial basis function, and the training of the weights W between the output layer and the hidden layer. The K-means clustering algorithm is usually adopted to determine the center and the width of the radial basis function, and improved LMS algorithm is applied to tune the weight.

3.2. Establishment of RBF Neural Network Fitting Model

According to the basic principle of the RBF neural network, the peak pressure P_m is selected as the input parameter of the RBF neural network based on quasi-dynamic calibration model. The output parameter is the deformation of the plastic sensitive elements y. So that it can be determined that the number of the nodes in input layer is 1, and that in output layer is the same. The structure of RBF neural network model for the plastic sensitive element based on quasi-dynamic calibration is shown in Figure 4.

IOP Conf. Series: Journal of Physics: Conf. Series **887** (2017) 012094 doi:10.1088/1742-6596/887/1/012094

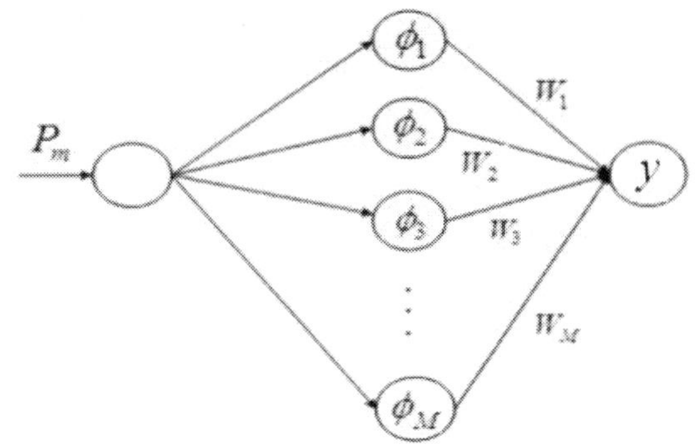

Figure 4. Structure of RBF neural network for quasi-dynamic calibration model.

The quasi-dynamic calibration experiments will be carried out for a batch of copper cylinder in order to test the prediction effect of the neural network model for quasi-dynamic calibration.

4. Quasi - Dynamic Calibration Experiments

4.1. Quasi-Dynamic Calibration Experiment and Data Processing

The quasi-dynamic calibration device which was developed by our research group is used as experimental equipment. The four Kistler 6213BK quartz high-pressure sensor are used as the "standard pressure sensor", and its calibrated range is 0 to 800MPa. The objects to be calibrated are $\phi 3.5mm \times 8.75mm$ copper cylinder. The experimental temperature is 20 ± 2°C. The pulse width of the half-sine wave pressure pulse is 6 ± 0.6ms. 9 pressure points are set approximately in the range of copper cylinders, in which the peak pressure and deformation of plastic sensitive elements are measured.

The experimental data of the copper cylinders in the pressure range of 200MPa to 700MPa (that is, the training sample of neural network) are listed in Table 1. Among them, P_m is the average peak pressure of four "standard pressure sensors", and y is the average of deformation of four copper cylinders.

Table 1. Experimental data for quasi-dynamic calibration of 200MPa~700MPa pressure range

Calibration point	1	2	3	4	5	6	7	8	9
P_m/(MPa)	214.78	269.24	379.95	429.85	486.25	528.49	563.22	655.96	689.31
y/(mm)	0.073	0.391	1.063	1.378	1.746	2.011	2.219	2.750	2.962

In order to further test the accuracy of the neural network model, the test samples shown in Table 2 are given.

Table 2. The testing samples

P_m/(MPa)	y/(mm)
331.51	0.767
612.73	2.530

4.2. Training and Learning of Neural Network Model

In order to improve the training efficiency of the network, the training samples are normalized, and the data in Table 1 and Table 2 are mapped to [0,1] by using formula (3).

$$\hat{P} = \frac{P - P_{\min}}{P_{\max} - P_{\min}} \tag{3}$$

After training, we also need to use the formula (4) to restore the data.

$$P = \hat{P} \cdot (P_{\max} - P_{\min}) + P_{\min} \quad (4)$$

where P is the original data; $P_{\min}$ and $P_{\max}$ are the minimum and maximum of the sample; $\hat{P}$ is the data after normalized pretreatment.

Under the environment of MATLAB R2013a, the RBF neural network prediction model is designed, trained and tested for the network after training. Set the training accuracy to 0.00001. Use the training sample data to train the model. During the training process, adjust the expansion factor to tend to the correct value and minimize the error. K-means clustering algorithm is used to select the center and width of the neural network, and the iterative LMS algorithm is used to adjust the connection weights between the output layer and the hidden layer. The RBF neural network model is trained by the newrb function provided by MATLAB neural network toolbox. The training results for the RBF neural network model of copper cylinders based quasi dynamic calibration are shown in Figure 5. It can be seen that the training goal can be achieved with a smaller number of iterations.

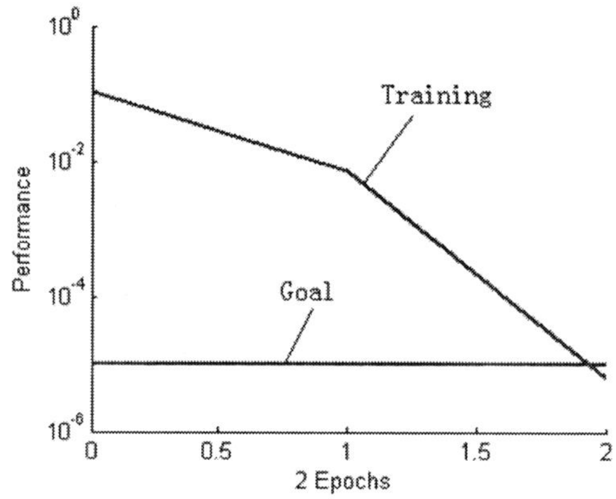

Figure 5. Training results of RBF network.

Finally, the parameters of RBF neural network prediction model with the best prediction effect are listed in Table 3.

Table 3. The parameter setting of RBF neural network

Parameters	Parameter Value
Neural Network Structure	1-2-1
Training function	newrb function
Radial basis function	Gaussian function
Test Function	sim function
The mapping function between hidden layer and output layer	Purelin function
Maximum number of neurons in hidden layer	100
Target error(Goal)	1×10^{-5}
Spread factor	1.9

4.3. Comparison of the Fitting Accuracy between the Neural Network Model and the Traditional Empirical Model

According to the polynomial regression model in references [10], the relationship between the peak pressure P_m and the deformation y of the copper cylinders can be obtained by polynomial fitting to the data in Table 1 as shown in formula (5).

$$y = -1.2939 + 0.0064 \times P_m - 2.5855 \times 10^{-7} \times P_m^2 \tag{5}$$

The residual standard deviation of the model is 0.0145mm and the correlation coefficient is 0.9999.

According to the modeling method in references [11], the exponential fitting relation between P_m and y can be obtained as shown in formula (6).

$$y = -1.2939 + 0.0064 \times P_m - 2.5855 \times 10^{-7} \times P_m^2 \tag{6}$$

In order to verify the generalization ability of constructed neural network, two test samples are selected to test the training accuracy. The relative error between the test sample values and the corresponding fitting values of neural networks, polynomial fitting values and exponential fitting values are given in table 4.

Table 4. The computing values of different methods

	Peak pressure 331.51(MPa)		Peak pressure 612.73(MPa)	
	deformation of copper cylinders /mm	Relative error/%	deformation of copper cylinders /mm	Relative error/%
Test samples	0.767	—	2.530	—
Neural network model	0.764	0.39	2.518	0.47
Second order polynomial fitting	0.785	2.35	2.504	1.03
Exponential fitting	0.492	35.85	2.903	14.74

From the comparison results in Table 4, the maximum relative error predicted by the neural network model of copper cylinders based on quasi-dynamic calibration is no more than 0.5%. The fitting value of RBF neural network is closest to the experimental value, and the precision of the second order polynomial fitting model is second, the precision of the exponential fitting model is the worst.

5. Conclusion

1) Based on the characteristic of RBF neural network that can approximate any nonlinear function with arbitrary precision, a neural network fitting model is established, which takes the peak value of pressure pulse as the input and the deformation of the plastic sensitive element as the output.

2) According to the neural network model of a batch of copper cylinder based on quasi-dynamic calibration, the prediction value of the neural network model is closest to the test sample value, and the accuracy of the prediction model is better than 0.5%. The predicted value of the second-order polynomial fitting model is close to the experimental sample value, and prediction accuracy of the second-order polynomial fitting model is better than 3%. Hence the exponential fitting model has the worst prediction accuracy. Therefore, applying the neural network in the quasi-dynamic calibration experiment of plastic sensitive elements can obtain a quasi-dynamic calibration model with higher accuracy, which lays a foundation for making the high accuracy pressure conversion table.

6. References

[1] Longdon. L. W, Textbook of ballistics and gunnery, Vol. 2, Part III, Her Majesty's StationeryOffice, London 1984, p 17.
[2] Li-xia Y 2013 a study on the measurement and development of plastic pressure Measurement of gun barrel in China. China. J. *Metrology & Measurement Technology*. 33(suppl) 1-4.
[3] De-ren K, Chang-an D, and Qi-sheng F 2006 *Plastic pressure measurement technique* (Beijing: The Publishing House of Ordnance Industry)
[4] De-ren K, Ming-wu Z and Shu-ning Z 2001 Simple discussion of the actuality & the development of copper cylinders. China. J. *Metrology & Measurement Technology*. 28(5) 15-16

[5] Gui-yi T, Pei-hua L 2004 the dynamic error analyse about the copper columniation measuring pressure equipment in interior ballistic. China. J. *Journal of Test and Measurement Technology.* 18(Suppl) 121-124.
[6] Wen-bin Y, Tie-hua M, Yong-hong D, et al. 2015 Research on the deviation mechanism in artillery chamber pressure test with copper cylinder pressure gage. China. J. *Chinese Journal of Scientific Instrument.* 36(10) 2290-96.
[7] De-ren K, 2011 Analysis of the difference between static and dynamic performance of copper Cylinder pressure measurement and the correcting method. China. J. Advanced Materials Research, 346 772-779.
[8] De-ren K and Ming-wu Z 2004 Mechanism of Quasi-dynamic Calibration for Copper Cylinder and Copper Ball. *Journal of Nangjing University of Science and technology.* 28(4) 375-379.
[9] Su-ning Z Experimental study on quasi dynamic calibration technology [D]. Nanjing: Nanjing University of Science & Technology, 2003.
[10] Ming-wu Z, Chang-ming W, Guang-liao L et al. 1993 the quasi dynamic calibration practice of Copper cylinder. China. J. *Acta Armamentarii.* (3) 80-83.
[11] De-ren K, Ming-wu Z, Shu-ning Z et al. 2001 Discussion on the application scope of quasi dynamic calibration table of copper cylinders. China. J. *Journal of Ballistics.* 13(3) 92-96.
[12] KONG D R, LI Y X and ZHU M W 2001 the experiential model of quasi-dynamic calibration on copper balls based on drop hammer dynamic calibration system. China. J. *Journal of Nanjing University of Science and Technology* 25(5) 475-479.
[13] Li W and Xun-tao L 2006 Calibration of pressure sensor based on neural network. China. J.*Coal Technology.* 25(7) 109-110.
[14] WANG R, JIANG J CH and PAN Y 2012 Prediction of Impact Sensitivity of Nonheterocyclic Nitroenergetic Compounds Using Genetic Algorithm and Artificial Neural Network. J. *Journal of Energetic Materials.* 30(30) 135-155.
[15] Ahmed S and Ferri B 2014 Prediction-based asynchronous Cpu-budget allocation for soft-real-time applications. USA. J. *IEEE Transaction on Computers.* 63(9) 2343-55.
[16] Elliott C, Vijayakumar V, Zink W, et al. 2007 National Instruments LabVIEW: A Programming Environment for Laboratory Automation and Measurement. *Journal of the Association for Laboratory Automation.* 12(1):17-24.
[17] Ming C 2014 *MATLAB neural network theory and example* (Beijing: Tsinghua University Press)

A image watermark resisting geometrical attacks based on quaternion moments

Ge Wu

Department of Electronic Engineering, Changchun University of Science and Technology, Changchun, China, 130022
E-mail: 651837271@qq.com

Abstract. This paper, proposes a method that by generating a quaternion moments, which will work as the watermark by multiply the quaternion orthogonal moments. Firstly, quaternion orthogonal moments matrix is gotten from source image and watermarking image separately.secondly extract N dimension matrix of two matrixes, multiply two matrixes to get a new matrix as the copyright information.Finally the matrix is sent to the copyright center. the watermark can be detected effectively, and it is robust against RST attacking such as rotating attack, cropping and scaling attack.

1. Introduction

Classically two types of watermarks are presently under consideration, namely robust watermarking and fragile watermarking. Researches are mostly focused on the development of robust watermarking, designed for the copyright protection of multimedia content.

Now more watermarking algorithms are applied, quaternion moments algorithms have been applied popularly in recent years. He Bing[1] proposed a zero watermarking algorithm based on Radon transform and real matrix of quternion, in the algorithm three components are extracted from color image and each component is decomposed, then the invariant moments of the real matrix is calculated and used to design zero watermarking. Zhong Minjie [2] proposed a RST invariant color image watermarking based on quaternion zernike moments, the algorithm can diffuse the error caused by watermark embedding to three component images.

This paper describes an image watermarking algorithm based on quaternion orthogonal moments. The algorithm is related in various degrees to much earlier work on scalable image compression.

2. Basic Methods

2.1. Quaternion Orthogonal Moments

The image orthogonal moments are projective coefficients that are projected onto a set of orthogonal polynomials , and image noise and other related operations has higher robustness.

Real moment of RGB image is ususlly described as:

$$I = \begin{bmatrix} 0 & R & G & B \\ -R & 0 & B & -G \\ -G & -B & 0 & R \\ -B & G & -R & 0 \end{bmatrix} \tag{1}$$

Left quaternion orthogonal moments [3] in polar coordinate is shown as formula 1,

$$Q^{l}_{n,m} = \int_{0}^{+\infty}\int_{0}^{2\pi} r^{n+m} e^{-\lambda(n-m)\theta} f(r,\theta) r dr d\theta \quad (2)$$

$f(x,y)$ denotes the form in rectangular coordinate, $f(r,\theta)$ denotes the form in polar coordinate, λ is Scaling factor. Suppose the image is rotated by the angle α ,it becomes

$$Q^{l}_{n,m}[f^{'}(r,\theta)] = k^{-n-m-2} \cdot e^{-\lambda(n-m)\alpha} \cdot Q^{l}_{n,m}[f(r,\theta)] \quad (3)$$

Left quaternion Fourier-Mellin moment [4]:

$$F^{l}_{n,m} = \int_{0}^{\infty}\int_{0}^{2\pi} r^{n-1} e^{-\lambda m\theta} f(r,\theta) dr d\theta \quad (4)$$

Suppose the image is rotated by the angle α , it becomes

$$F^{l}_{n,m}[f^{'}(r,\theta)] = \lambda^{-n} \cdot e^{\lambda m\alpha} \cdot F^{l}_{n,m}[f(r,\theta)] \quad (5)$$

According to the conjugate property of the product of quaternion[5],the phase is gotten,

$$\omega_{n,m} = \frac{1}{2\pi}\sum_{m=1}^{M}\sum_{n=1}^{N}(\psi_{n,m} - (m-1)\alpha) \quad (6)$$

Because the phase information contained in quaternion orthogonal moments is more important than the corresponding amplitudes information, the quaternion orthogonal moments could be used as the important factor to identify the image.

Right quaternion orthogonal moments is as follow,

$$OF^{'}_{n,m} = \frac{p+1}{\pi}\int_{0}^{2\pi}\int_{0}^{1} f(r,\theta) Q_n(r) e^{-\mu m\theta} r dr d\theta \quad (7)$$

$$Q_n(r) = \sum_{s=0}^{n}(-1)^{n+s}\frac{(n+s+1)!}{(n-s)!s!(s+1)!}$$

With the increasing orders of quaternion orthogonal moment, the NIRE values of the reconstructed color images showed a decreasing trend.

3. Proposed Algorithm

3.1. Watermarking Embedding Algorithm

Fig.1 depicts the watermark embedding algorithm.

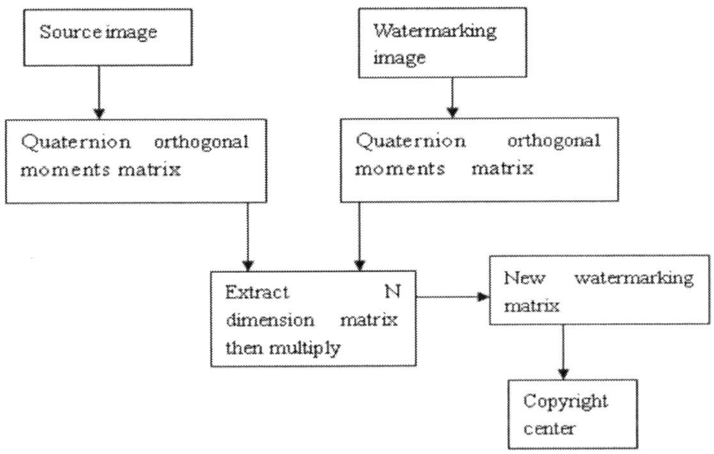

Figure 1. Watermarking Embedding Algorithm

Firstly, Quaternion orthogonal moments matrix is gotten from source image and watermarking image separately.secondly extract N dimension matrix of two matrixes, multiply the two matrixes to get a new matrix as the copyright information.Finally the matrix is sent to the copyright center.

For $N \times N$ image, its N+M rank Tchebichef moment is defined as following:

$$\mathrm{T}_{n,m} = \frac{1}{\tilde{\rho}(n,N)\tilde{\rho}(m,N)} \sum_{x=0}^{N-1}\sum_{y=0}^{N-1} \tilde{t}_n(x)\tilde{t}_m(y) f(x,y) \tag{8}$$

In which $\tilde{t}_n(x) = \frac{t_n(x)}{N^n}$, $t_n(x)$ is the polynominal of n rank Tchebichef, it is defined as following:

$$t_n(x) = n!\sum_{k=0}^{n}(-1)^{n-k}\binom{N-1-k}{n-k}\binom{n+k}{n}\binom{x}{k} \tag{9}$$

In fact, discrete image could be gotten by next formula,

$$f(x,y) = \sum_{n=0}^{N-1}\sum_{m=0}^{N-1} T_{nm}\tilde{t}_n(x)\tilde{t}_m(y) \tag{10}$$

The watermark image is calculated according to the same way to get $W(x,y)$,Then watermarking matrix $W'(x,y)$ is gotten from

$$W'(x,y) = W(x,y) \times f(x,y) \tag{11}$$

$W'(x,y)$ is sent to the copyright center to save.

3.2. Watermarking Detection Algorithm

Given a watermarked image,because the stability of quaternion orthogonal moments, the detector still needs two other inputs for the watermarking and source image, the process is the same as the first two steps in generating watermarking.Then we can get the result of the copyright of image.

Step 1: Calculating the quaternion orthogonal moments matrix of the source image and extract the first $N \times N$ dimension of it,that is $W(x,y)$.

Step 2: Getting watermarking matrix from copyright center,that is $W^{'}(x,y)$.and execute dividing algorithm of the two matrix to get a new matrix.

Step 3: Recover the source watermarking image from the new quaternion orthogonal moments matrix according to formula 11.

4. Experimental Results

Firstly, we extract 256×256 orthgonal matrix from the source image and watermarking image, then multiply two matrixs to get the copyright watermarking matrix. From Table 1 and Fig.2 ,it is clear that when the watermarked image undergo the rotation which angle is less than 2 degree, the algorithm can detect the watermark correctly ,which is better than EZW algorithm.

Table 1. Comparison of small rotation attack and the detection result

Nomalized correlation coefficient			Nomalized correlation coefficient		
Rotation angle	algorithm	Ezw	Rotation angle	algorithm	Ezw
0.5	0.42	0.37	-0.5	0.46	0.32
1.0	0.34	0.31	-1.0	0.36	0.30
1.5	0.28	0.25	-1.5	0.31	0.25
2.0	0.26	0.21	-2.0	0.27	0.20

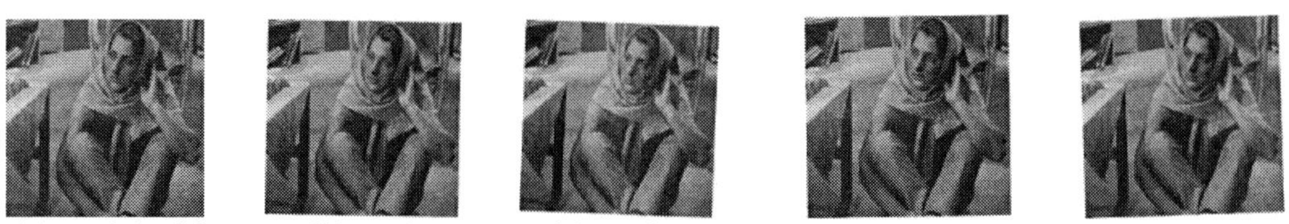

(a) Original image (b) angle 1.0 (c) angle 2.0 (d) angle -1.0 (e) angle -2.0

Figure 2. Watermarked image after rotating attacking

From fig.3 and Table2,the algorithm is robust to the cropping and scaling attack,when the remaining area after cropping is only 49% of the original image ,the watermarking could be detected correctly.

Table 2. Detection result under cropping and scaling attack

Nomalized correlation coefficient			Nomalized correlation coefficient		
Rotation angle	THE algorithm	EZW	Rotation angle	THE algorithm	EZW
0.5	0.42	0.37	-0.5	0.46	0.32
1.0	0.34	0.31	-1.0	0.36	0.30
1.5	0.28	0.25	-1.5	0.31	0.25
2.0	0.26	0.21	-2.0	0.27	0.20

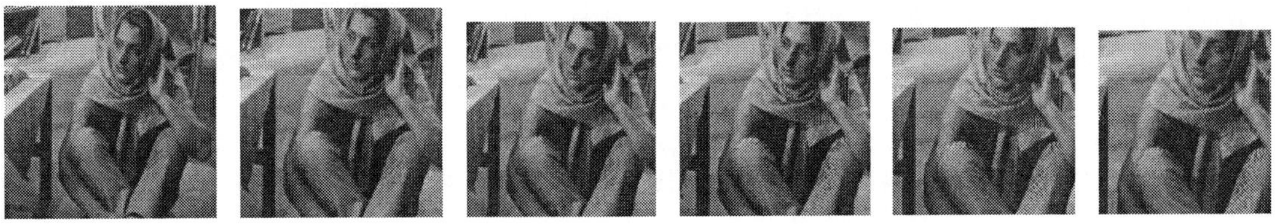

(a) Original image (b) cropping 10% (c) cropping 20% (d) cropping 34% (e) cropping 51% f.cropping 60%

Figure 3. Watermarked image after cropping and scaling attacking

5. Conclusion

In this paper ,we have presented a digital watermarking scheme for both watermarking embedding and detection, our scheme is effective. Our experiments confirmed that the watermarked image is nearly rate-distortion optimized, the decoder can recover the original image, the watermark can be detected effectively, and it is robust against geometric attacking such as rotating attack ,cropping and scaling attacking.

6. References

[1] He Bing,"Zero color image watermarking based on Radon transform and quaternion real matrix,"Optical Technique pp. 166-171,March ,2016.

[2] Zhong Minjie, " RST Invariant Color Image Watermarking Algorithm Based on Quaternion Zernike Moments,"Ji Suan Ji and Xian Dai Hua. pp. 243-250, June 2015.

[3] Guo L Q,Dai M,Zhu M.Quaternion moment and its invariants for color object classification[J].Inform Sci,2014,273:132-143

[4] Ping Z L,Ren H P,Zou J,et al.Generic orthogonal moments:Jacobi-Fourier moments for invariant image description[J].Pattern Recong,2007,40:1254.

[5] Liu Q,Zhu H Q,Li Q.Object recongnition by combined invariants of rothogonal Fourier-Mellin moments[C].In:Proc 8th Int Conf Information,Communications and Signal Processing,2011:1-5

[6] Chen B J,Shu H Z,Coatrieux G,et al.Color image analysis by quaternion-type moments[J].Math Imaging Vis,2014,DOI:10.1007/s10851-014-0511-6.

IOP Conf. Series: Journal of Physics: Conf. Series **887** (2017) 012096 doi:10.1088/1742-6596/887/1/012096

Identification of maturity of cuiguan pear based on RobustICA and PSO-ELM

Yu Zhang[1], Yongmei Huang[1,*], Min Lin[1], Huijun Liu[1] and Xinqi Zhou[2]

[1]College of Metrology and Measurement Engineering, China Jiliang University Hangzhou, China
[2]Focused Photonics (Hangzhou),Inc.Hangzhou 310052,China
* Corresponding author. Email:ymhuang@cjlu.edu.cn

Abstract. In order to improve the stability of extreme learning machine (ELM), it was proposed to use the combination of robust independent component analysis (RobustICA) and particle swarm optimization ELM algorithm. It was applied to identify the maturity of cuiguan pear at different growing periods. First the two order derivative and the discrete wavelet transform were adopted to remove the background noise and compression spectra. Then RobustICA was used to decompose the spectra to extract independent components and mixed coefficient matrices. Finally, by using particle swarm optimization, the best input layer weight and hidden layer deviation were obtained and adjusted the number of hidden layer nodes to get the best ELM model to differentiate maturity. The method was used to identify the maturity of cuiguan pear at four different stages, and has strong robustness. The results showed that RobustICA-PSO-ELM had better discrimination ability compared with RobustICA - ELM and ELM, and the accuracy increases from 75% to 94.4%.

1. Introduction

Cuiguan pear is one of the most widely distributed fruit trees, which is rich in nutrition and taste well. After harvesting, the photosynthesis of the cuiguan pear basically stops. The nutrients consumed in the process of storage and transportation which lead to the quality drop and lose the commodity value, resulting in the loss rate as high as 30%. Therefore, strengthening the identification of the fruit maturity of the cuiguan pear is meaningful to reduce the loss and increase the economic benefits. Near infrared spectroscopy is a fast and non-destructive method. Compared with the traditional method, it is more convenient which widely used in many aspects of the analysis.

Traditional methods such as PCA and SVD can only decompose uncorrelated components, and they can't guarantee that these components are independent of each other. Therefore it reduces the typicality of the extracted features. Independent component analysis (ICA) is a kind of blind source separation (BBS), which can separate signals into independent components. Under the premise of the independence of spectral components, ICA can be used to decompose spectra. Fang Limin established a prediction model of moisture, protein and starch content in corn by using ICA method and neural network. [1] Sun Tong use uninformative variable elimination method to eliminate irrelevant variables. Then he used ICA method to establish the soluble solid model of NanFeng Orange. [2] Shao Yongni, who applied the ICA method to obtain different vintages of rice, establish the identification model of rice year by using the characteristic bands as input to artificial neural network and got sensitive bands corresponding to major components of paddy rice.[3]

Extreme learning machine (ELM) is a new neural network algorithm. It can get the unique optimal solution in the training process when the input layer weight and the hidden layer bias are given randomly. However, the hidden layer parameters generated by the ELM will lead to poor generalization performance. In order to improve the prediction accuracy, the number of hidden layer nodes needs to be increased which will increase the complexity of the network and lead to overfit. Aiming at the shortage of ELM, a particle swarm optimization algorithm is proposed to optimize the input layer weight and the hidden layer bias, and an optimal model is obtained.

Through the analysis of near infrared spectroscopy of cuiguan pear, RobustICA was used to extract the independent components and corresponding mixing. Particle swarm optimization was used to optimize the input layer weight and hidden layer deviation of ELM and established the cuiguan pear maturity model which can identificate of maturity successfully.

2. Basic Theory

2.1. Robust Independent Component Analysis

The article uses the RobustICA arithmetic which is based on the kurtosis. The RobustICA algorithm takes kurtosis as the control function, then optimizes the kurtosis by searching the best step length, then get the independent component (basic component, spectral matrix) and mixed coefficient matrix with kurtosis that the kurtosis is not 0. [4]It extracts the main components of the spectrum and discards the unwanted information. The advantage is that the stronger robust and the faster converges.

2.2. ELM Optimization Based on PSO

This paper presents the PSO-ELM algorithm. The concrete steps are as follows:

(a) First, it needs to initial population and set particle number m to generate population. Then it initializes PSO particle velocity v and maximum/ minimum inertia w_{max} / w_{min}.Last learning factor c1 or c2 and maximum iteration time k are set up.

(b) It selects the inertia weight according to $w = w_{max} - \frac{w_{max}-w_{min}}{k} \times a$ (a is the current number of iterations). The fitness value of the particles is calculated according to the fitness function. The fitness values of the particles and the global fitness values are constantly compared until the end condition is satisfied.

(c) It needs to update the position and speed of particles constantly.

(d) When the optimization satisfies the maximum iteration number k or the evaluation value which is less than the specified precision b (b>0), the search is finished and the optimal input weight and the hidden layer bias are obtained.

(e) The best ELM model is obtained by substituting the optimal solution into ELM, and the model is trained by test set.

3. Experiment

3.1. Instrument and Equipment

The near infrared spectra of cuiguan pear samples were collected by using MCS600 array optical fiber spectrometer of Cai Si (Zeiss, Germany) in Germany. The wavelength range was 800-1700nm, and the spectral collection software was Aspect Plus. Matlab2015a was used to analysis Spectral data.

3.2. Sample Preparation and Spectra Acquisition

The 144 samples of cuiguan pear used in this experiment were all from a fruit industry park in Hangzhou, Binjiang. Samples were collected at two weeks before the samples normal maturation, and 36 samples were collected every seven days. The samples were gathered four times, and were divided into I, II, III and IV. After simple cleaning, the spectra were got. The samples were collected several times, and the average spectra were used in the experiment. All samples were divided into calibration set and prediction set by using kennard stone algorithm. The calibration set contained 108 samples for modeling, and the prediction set contained 36 samples for detecting model accuracy.

From Figure 1.can been seen that the spectra of different maturity of cuiguan pear were different and some spectral regions overlapped. Because of the change of sugar and acid of cuiguan pear, it appeared the difference in near infrared spectral absorption. This is the basis for identification of maturity and showed that the identification of cuiguan pear maturity is meaningful.

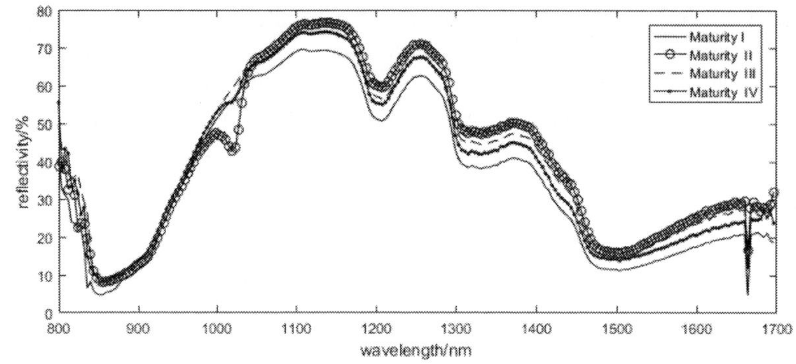

Figure 1. Original spectra of four maturity of cuiguan pear

3.3. Spectral Data Processing and Modeling

Because of the size and surface roughness of cuiguan pear samples were different, and at the same time a large number of useless information in the spectra will reduce the accuracy of the model and spend more time and energy. So this paper uses the two order derivative and discrete wavelet transform to preprocess the original spectra in order to eliminate interference from external factors interference and improve the identification ability of mathematical model. Wavelet base function selected db2 and decomposed to 4 layers. The compressed data is about 7% of the original amount of data.

Mixing coefficient matrix and independent components corresponding are extracted by RobustICA. PSO is used to optimize the input layer weight and the hidden layer deviation to get the optimal ELM model. On the other hand, RobustICA combined with the ELM model, and only through the identification model by pretreatment, the direct use of ELM to establish the comparison. The classification accuracy (CR) was used to evaluate the model. The higher the classification accuracy (CR), the model is better.

$$CR = \frac{\text{Number of samples which correctly classified}}{\text{total sample}} \times 100\% \tag{1}$$

3.4. Experimental Process

From figure 2 we can see the experimental process as follows:

(a) The samples of cuiguan pear were collected several times, and the average value was taken as the original spectra. The original spectra were processed by two order derivative and discrete wavelet to eliminate the interference of spectral baseline drift and background light and compressed data to improve modeling efficiency. Wavelets used dbn and n is vanishing moments.

(b) The processed spectra are decomposed by RobustICA and the best independent component number (ICs) is selected to obtain the best independent component and its mixing coefficient matrix.

(c) The input layer weight and hidden layer deviation are as the particle in the particle swarm search space. The relevant parameters were initialized and particle iteration optimize by algorithm until a termination condition is satisfied: when search the maximum number of iterations is reached k or less than a specified value evaluation accuracy of a (a>0), the search is over and get the best input layer and hidden layer deviation.

(d) The optimal input layer weight and hidden layer bias are substituted into the ELM model, and the optimal ELM model is obtained. The mixed coefficient matrix is used as the input, and the actual maturity is taken as the output to get the identification model of cuiguan pear.

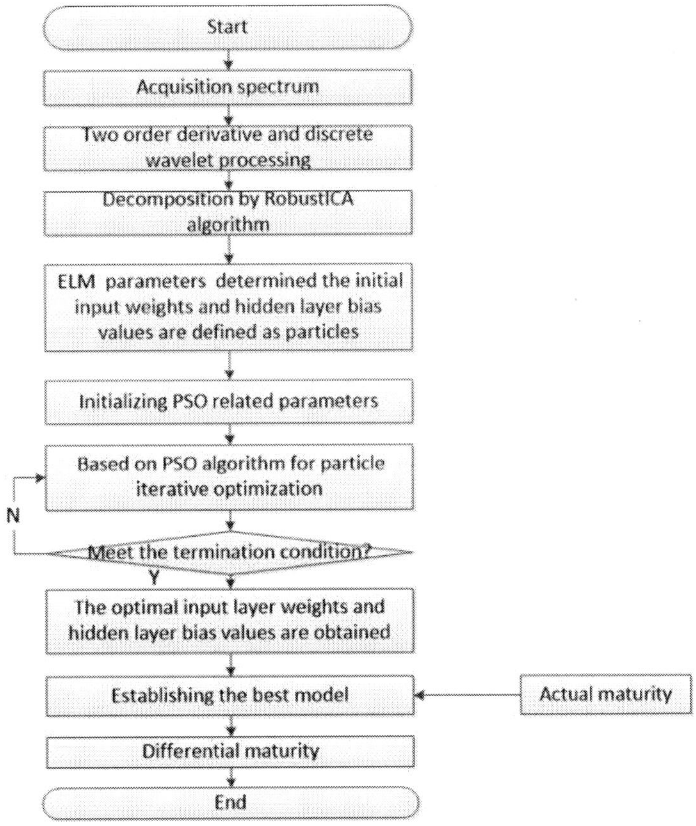

Figure 2. Experimental process

4. Results of Experiment and Discussions

4.1. Independent Component Selection of RobustICA

After two order derivative and discrete wavelet pretreatment, the spectrum had basically removed the useless information and the interference of the background light. When RobustICA was used to extract independent components and its corresponding mixed coefficient matrices, the influence of the number of independent components on the model results were investigated. In this paper, the initial independent component number was 5, and 1 is added as step length until get to the 12. The other parameters of the model remain unchanged, which is used to select the best independent components of RobustICA-PSO-ELM and RobustICA-ELM. As shown in figure.3, the RobustICA-PSO-ELM model selects the best independent component number of 7, and the RobustICA-ELM model has the best independent component number of 10.

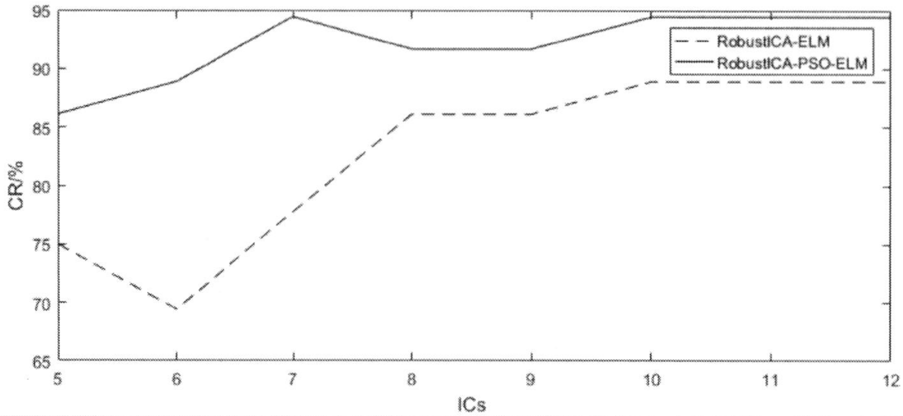

Figure 3. The accuracy of different choice of ICs in different models

4.2. PSO-ELM Model

The ELM hidden layer excitation can choose 'Sigmoidal' 'Gauss' etc. by taking the mixed coefficient matrix which is obtained by RobustICA as the input of the limit learning machine model and using the maturity as output. The article chooses the function named 'Sigmoidal'. Initialize the parameters of the particle swarm algorithm, set the particle number m=30, set all the learning factors as 1.5, the maximum number of iterations k=30, set the maximum inertia weight to 0.9 and the minimum inertia weight to 0.4, set the minimum accuracy b to 0.01.

The selection of the hidden layer nodes in ELM is critical. Set the initial number of the hidden layer nodes to 6 then increase to 24 by using the step size of 2, and train the model 15 times. The picture below shows the accuracy of near infrared model maturity in RobustICA-PSO-ELM, RobustICA-ELM and ELM under different hidden layer nodes.It shows that the accuracy decreases dramatically when just using the ELM to build the model with the number of hidden layers increases because of the model has overfitted. After the particle swarm optimization, the situation improves obviously, and the model accuracy also improved.

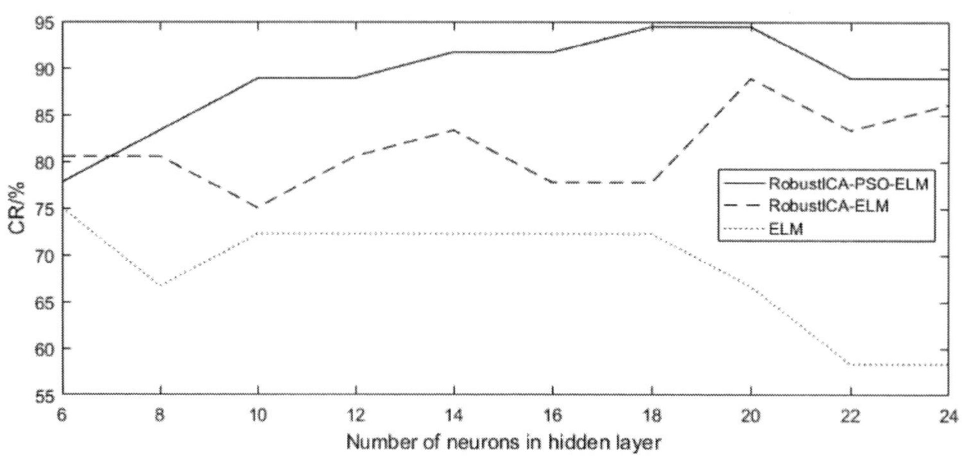

Figure 4. The accuracy of different number of neurons in different models

From figure 4.the results showed that the RobustICA-PSO-ELM results were optimal when the number of nodes in the hidden layer is 18. The identification maturity model of cuiguan pear which established by RobustICA-PSO-ELM was better than other model methods. Particle swarm optimization was used to find the input layer weight and hidden layer deviation of ELM, and the effect was better, faster. The results of model classification were more accurate. Using the RobustICA algorithm, the output was non-Gauss which could extract more useful information and increase the accuracy of the model. Particle swarm optimization was used to find the best input layer weight and hidden layer deviation which avoid random selection and obtaining the best ELM model.

4.3. Comparison of Modeling and Prediction Results

Table 1. The accuracy of different models

Model name	Accuracy
RobustICA-PSO-ELM(ICs=7,nods=18)	94.44%
RobustICA-ELM(ICs=10,nods=20)	88.89%
ELM(nods=6)	75%

As shown in the table 1. The accuracy was only 75% when only used the ELM method. After using RobustICA method, the model had some improvement .When RobustICA-PSO-ELM method was employed, the accuracy was best. When the number of independent components is 7 and the number of nodes is 18, the performance of the ELM model was improved and the accuracy rate rose to 94.4%.So it can be seen that the model is reliable and significant effect, improve the accuracy which can be effectively applied to the business and many other things.

5. Conclusion

The independent component was extracted by RobustICA and the mixed coefficient matrix was obtained. The input layer weight and hidden layer deviation of ELM were optimized by PSO, and the maturity of cuiguan pear was identified. The results showed that the accuracy of RobustICA-PSO-ELM is the highest compared with the RobustICA-ELM model and the ELM model alone. When the number of independent components was 7 and the number of nodes in ELM hidden layer was 18, the accuracy of identification was 94.44%. Therefore, RobustICA-PSO-ELM has excellent modeling and identification ability, and it can be used for the accurate identification of cuiguan pear. At the same time, RobustICA-PSO-ELM modeling method can also provide reference for other modeling research methods.

6. Acknowledgment

This work is supported by National major scientific instruments and equipment development projects (2014YQ470377).The authors awfully thanks for anonymous reviewers who made constructive comments.

7. References

[1] Fang Limin, Lin Min. A Method of Near Infrared Multi-component Analysis Based on Independent Component Analysis and Neural Networks Model [J]. *Spectroscopy and Spectral Analysis*. 2008, 36(**6**).

[2] Sun Tong, Xu Wenli, Hu Tian, Liu Muhua. Determination of Soluble Solids Content in Nanfeng Mandarin by Vis/NIR Spectroscopy and UVE-ICA-LS-SVM [J]. *Spectroscopy and Spectral Analysis*. 2013, **12:3235**-3239.

[3] Shao Yongni, Cao Fang, He Yong. Discrimination Years of Rough Rice by Using Visible/near Infrared Spectroscopy Based on Independent Component Analysis and BP Neural Network [J]. *Journal of infrared and millimeter waves*. 2007(**6**).

[4] Zhao Taoyu, Shao Penghua. Robust Independent Component Algorithm Based on Kurtosis [J]. *Computer Development & Applications*.2012, 25(**8**).

Institute of Physics
Dirac House, Temple Back
Bristol BS1 6BE UK

ISSN: 1742-6588
ISBN 978-1-5108-4791-0

CURRAN ASSOCIATES INC.
proceedings.com